Gauge Theory of Weak Decays

This is the first advanced, systematic, and comprehensive look at weak decays in the framework of gauge theories. Included is a large spectrum of topics, both theoretical and experimental. In addition to explicit advanced calculations of Feynman diagrams and the study of renormalization group strong interaction effects in weak decays, the book is devoted to the Standard Model Effective Field Theory, dominating present phenomenology in this field, and to new physics models with the goal of searching for new particles and interactions through quantum fluctuations. This book will benefit theorists, experimental researchers, and PhD students working on flavor physics and weak decays as well as physicists interested in physics beyond the Standard Model. In its concern for the search for new phenomena at short-distance scales through the interplay between theory and experiment, this book constitutes a travel guide to physics far beyond the scales explored by the Large Hadron Collider at CERN.

Andrzej J. Buras is one of the most cited particle theorists in Europe and the most cited theoretical flavor physicist worldwide. He has written extensively on weak decays through lecture notes and review articles. He has been awarded the Smoluchowski-Warburg Medal of German and Polish Physics Societies, a Senior Carl von Linde Fellowship at TUM-IAS, an Advanced ERC Grant, and the 2020 Max Planck Medal. He is an ordinary member of the Bavarian Academy of Sciences and foreign member of two Academies in Poland.

Gauge Theory of Weak Decays

The Standard Model and the Expedition to New Physics Summits

ANDRZEJ J. BURAS

Technical University Munich

CAMBRIDGE
UNIVERSITY PRESS

CAMBRIDGE
UNIVERSITY PRESS

University Printing House, Cambridge CB2 8BS, United Kingdom

One Liberty Plaza, 20th Floor, New York, NY 10006, USA

477 Williamstown Road, Port Melbourne, VIC 3207, Australia

314–321, 3rd Floor, Plot 3, Splendor Forum, Jasola District Centre, New Delhi – 110025, India

79 Anson Road, #06–04/06, Singapore 079906

Cambridge University Press is part of the University of Cambridge.

It furthers the University's mission by disseminating knowledge in the pursuit of education, learning, and research at the highest international levels of excellence.

www.cambridge.org
Information on this title: www.cambridge.org/9781107034037
DOI: 10.1017/9781139524100

© Andrzej J. Buras 2020

First published 2020

Printed in the United Kingdom by TJ International Ltd. Padstow Cornwall

A catalogue record for this publication is available from the British Library.

Library of Congress Cataloging-in-Publication Data
Names: Buras, Andrzej J. (Andrzej Jerzy), 1946– author.
Title: Gauge theory of weak decays : the standard model and the expedition to new physics summits / Andrzej J. Buras.
Description: [New York, New York] : [Cambridge University Press], [2020] | Includes bibliographical references and index.
Identifiers: LCCN 2019040866 (print) | LCCN 2019040867 (ebook) | ISBN 9781107034037 (hardback) | ISBN 9781139524100 (epub)
Subjects: LCSH: Weak interactions (Nuclear physics) | Gauge fields (Physics) | Standard model (Nuclear physics) | Particles (Nuclear physics)–Flavor. | Mathematical physics.
Classification: LCC QC794.8.W4 B84 2020 (print) | LCC QC794.8.W4 (ebook) | DDC 539.7/544–dc23
LC record available at https://lccn.loc.gov/2019040866
LC ebook record available at https://lccn.loc.gov/2019040867

ISBN 978-1-107-03403-7 Hardback

**To Gurli,
Robert, Karin and Allan,
Franziska and Ute
Freya, Falk, Elisabeth, Inga and Janosch**

Contents

Preface

The present theory of elementary particles and of their interactions known under the name of the Standard Model (SM) [1–3] is a relativistic quantum field theory with a specific local gauge symmetry dictated by nature. It incorporates the theory of strong interactions (quantum chromodynamics, QCD) and the unified theory of electroweak interactions (quantum flavor dynamics, QFD). It is a theoretically adequate description of leptons, quarks and their fundamental interactions in accordance with the principles of unitarity, causality, Lorentz invariance, quantum mechanics, and gauge invariance. As such it combines the main achievements of the physics of the twentieth century and includes the Maxwell theory of electromagnetism formulated already in 1865.

The dream of physicists for many centuries was to find a "world formula" by help of which all the phenomena surrounding us could be described and explained. In particle physics the first step toward such a formula is the Lagrangian of the SM, which summarizes the particle content of the present theory and describes the basic structure of particle interactions as we know them at present. As the SM is unable to explain all phenomena around us, other world formulas in the form of Lagrangians appear in the literature. They generally contain all particles and interactions of the SM, but in addition new particles and new interactions are present in these models. As of 2020, we do not know which of these models will turn out to be the next step toward our dream, but it is exciting that in the coming years we might know it.

To a person, not familiar with the subject, these world formulas look like hieroglyphics that appear to us when we visit Egyptian pyramids. Yet, to trained particle physicists, similar to trained archaeologists, these formulas reveal a vast amount of information about the physics of elementary particles. In these formulas the basic dynamics of particle interactions in a given model are compactly encoded, and it is in principle only a matter of will and time to translate this information into the more common language and to derive testable predictions.

In practice the derivation of physical predictions from the Lagrangian of the SM and of its extensions is often not an easy task. It requires generally certain mathematical skills and the knowledge of sophisticated field theoretical methods. Moreover on many occasions the tools that we presently have at our disposal are not yet powerful enough to allow for accurate predictions. Fortunately there exist also many quantities for which accurate calculations have been already made and can be already compared with experimental results. Such a comparison is in fact most exciting as it allows testing a given theory. But this is only possible if experimental results for a given quantity are also available, and this is not always the case. In fact, some phenomena take place very rarely, and this requires large and often expensive experiments to measure the probability for seeing them.

The first main goal of this book is to describe weak decays of mesons, the bound states of quarks and antiquarks and of leptons within the SM. Starting from the Lagrangian of this model we will derive, as far as it is possible, predictions for decay widths, branching ratios, and various observables that can be used to confront this theory with experimental data. Therefore, in addition to more technical, field theoretical aspects of the SM, we will also devote a large part of this book to phenomenological applications of the derived formulas and to the comparison of the SM predictions with experimental data.

The physics of weak decays is a fascinating subject and constitutes an important part of the SM and particle physics in general. There are several reasons for this:

- This sector probes in addition to weak interactions also electromagnetic and strong interactions at short- and long-distance scales. As such it involves the dominant part of the dynamics of the SM.
- It contains most of the free parameters of the SM and consequently plays an important role in their determination.
- The occurrence of a large class of processes that take place only as "loop effects" automatically tests the quantum structure of the theory.
- The renormalization group effects known also from statistical physics play here an important role in view of the vast difference between the weak interaction $O(100 \text{ GeV})$ and strong interaction $O(1 \text{ GeV})$ scales.
- The nature of the violation of various symmetries such as charge conjugation (C), parity (P), time reversal (T), and CP can be studied very efficiently.
- It is an ideal laboratory for nonperturbative techniques.
- Very importantly, it is a "window" through virtual effects to very short distances, which may shed light on some outstanding questions in particle physics such as the origin of masses and the number of fermion generations as well as hierarchies in the strength of their interactions. In particular weak decays could help us to identify the origin of the matter-antimatter asymmetry observed in the universe, which is necessary for our existence and which is not understood within the SM. They could also shed some light on dark matter.

The last point brings us to the second main goal of this book: the presentation of weak decays within the most popular extensions of the SM. In particular we will develop efficient methods that will allow us to distinguish the predictions of these extensions for weak decays from the SM ones. This may help us to identify new physics (NP) at very short-distance scales well beyond the reach of the Large Hadron Collider (LHC). But even if the LHC would discover NP in the coming years, a detailed study of the properties of discovered new particles and interactions cannot be made only through high-energy collisions but requires very strong involvement of low-energy processes like weak decays of mesons where these new phenomena manifest themselves through quantum effects.

Similar to the SM, its extensions have also been the subject of intensive research in the last forty years. This implies that it is a challenge to describe adequately only the most important advances in this field in one book. Therefore as in any serious expedition, a strategy for reaching our goals is unavoidable. Before presenting this strategy and the related map of our expedition, it is necessary to present two Grand Views: the first one dealing with the SM and the second one dealing with some aspects of NP. These grand views will be brief and superficial. They are only meant to help nonexperts to understand at least roughly the map of this book, which will be presented subsequently and most importantly to motivate any reader to join the author in this expedition.

In the latter context let me remark that the failure of the LHC to discover any new particles until now resulted in some frustration in the particle physics community. Often at conferences the speakers show pictures of the Sahara, meaning that until a new very high energy collider is built, no new particles will be discovered. My view, as a flavor physicist, is much more optimistic. It is represented well by the photo on the book's front cover. We are standing on the mainland representing the SM. There is an energy gap represented by the water, which we have to cross in order to reach eventually the NP summits in the far distance. They represent different possibilities for the grander theory. But in order to find out which summits will answer all our questions, we have to cross the glacier and all of its crevasses, which represent various sophisticated technologies in flavor physics necessary to conquer these summits. But the main reason for showing this photo is its stunning beauty similar to the beauty of flavor physics.

Acknowledgments

This book was written between the summer 2013 and the summer 2019. Initially, it was planned to be complete in 2015, but fortunately it did not happen as in the last four years many exciting things have happened in flavor physics, which allowed me to include them in this book.

First of all there are four people whom I want to thank most for helping me reaching my goals.

Jennifer Girrbach-Noe: We started writing this book together. Jennifer being much stronger in LATEX than me set up all the files, which I used until I completed this book. She also wrote the grand view of the Standard Model and also the grand view of new physics. Part II has been written fully by her, although at view places I added new material and made a few modifications in her writing. But this part of the book should be credited to her. She also produced several figures and read roughly 25 percent of the book, which was the status at the end of 2014. Unfortunately, in January 2015, Jennifer decided not to continue reseach in particle physics, so the rest of the book was written by me alone. Yet, these first steps made with her were very important, and foremost I want to thank her. As Jennifer was one of the most efficient collaborators among about 120 collaborators I had in my research and we wrote twenty papers within three and a half years together, I am sure the book would be better than it is now if she had continued this expedition with me.

Robert Buras-Schnell: After Jennifer left, Robert was responsible for almost all figures in the book. Moreover, he was of great help in any LATEX issues. Without him the progress in writing this book would be very slow. Many thanks to Robert.

Jason Aebischer: Jason was the only one who read almost the full book before it was sent to Cambridge University Press for printing. In particular he demonstrated Swiss precision in reading Chapter 14, where one can get easily lost in the indices present in the renormalization group equations of the SMEFT. He checked many equations in the full book, found misprints, and made suggestions that in my view improved the clarity of the text. Many thanks to Jason.

Christoph Bobeth: The sections on leptoquarks and vectorlike quarks resulted from our intensive study of these models in the context of the SMEFT. Despite two joint publications on these models a significant fraction of these two sections was not published and is presented here for the first time. Moreover, Christoph and Jason helped in doing some numerics relevant for the book.

But there are still several of my colleagues who helped me with advices on the literature, updates of our previous joined papers, and also updates of their own papers, which I asked them to do in connection with my book. These are in alphabetic order: Monika Blanke, Gerhard Buchalla, Marcin Chrzaszcz, Vincenzo Cirigliano, Andreas Crivellin, Sébastien Descotes-Genon, Svjetlana Fajfer, Fulvia de Fazio, Jean-Marc Gérard, Martin Jung, Alexander Lenz, Emilie Passemar, Janusz Rosiek, Luca Silvestrini, David Straub, and Robert Szafron. In particular, I benefited enormously from Robert's expertise in $(g-2)_{\mu,e}$.

This book was written entirely at the TUM Institute for Advanced Study. The fantastic atmosphere at this Institute, created by the directors Gerhard Abstreiter and Ernst Rank and by the wonderful, in many respects, IAS-Team, was very helpful in writing this book. Particular thanks go also to the Clusters of Excellence: *Universe* and *Origins*, for financial support and Stephan Paul for being such a wonderful boss after my retirement in 2012. Many thanks to my secretary Elke Hutsteiner for help in preparing PowerPoint presentations in the last 20 years that were indirectly important for writing this book.

Particular thanks go to Simon Capelin, Roisin Munnelly, Henry Cockburn, Sarah Lambert, and Dinesh S. Negi from Cambridge University Press for many advices and help during the writing of my book. Very special thanks go to Karen Slaght, an impressive copyeditor, who improved my English. Many thanks to Sapphire Duveau and in particular to Neena S. Maheen for the great help during the final preparations of the book for printing and to the LATEX team for an impressive job while introducing corrections.

Finally, I would like to thank my family, in particular my wife Gurli and my daughter Karin, for the great encouragement for completing this book during all these years and to Robert as already mentioned earlier. Special thanks go to my second son, Allan Buras, for providing the stunning photo on the front cover, which he took during one of his expeditions to the far north.

Abbreviations

Our conventions, notations, and useful general formulas are collected at the end of the book. Here we just list those abbreviations that are used frequently in our book.

2HDM = Two Higgs-Doublet Model
ADM = Anomalous Dimension Matrix
BSM = Beyond the Standard Model
ChPT = Chiral Perturbation Theory
CKM = Cabibbo-Kobayashi-Maskawa
CLFV = Charged Lepton Flavor Violation
CMFV = Constrained Minimal Flavor Violation
CP = CP-invariance
CPV = CP-invariance Violation
DHP = Double Higgs-penguins
DR = Dimensional Regularization
DRED = Dimensional Reduction
DQCD = Dual QCD
EDM = Electric Dipole Moment
EFT = Effective Field Theory
EWP = Electroweak Penguin
EWSB = Electroweak Spontaneous Symmetry Breakdown
FBP = Flavor Blind Phases
FC = Flavor Changing
FCNC = Flavor Changing Neutral Currents
FLAG = Flavor Lattice Averaging Group
FR = Feynman rules
FSI = Final State Interactions
GIM = Glashow-Iliopoulos-Maiani
GUT = Grand Unified Theory
HFLAV = Heavy Flavor Averages
HP = Higgs-penguins
HQE = Heavy Quark Expansion
HV = 't Hooft-Veltman
I.B. = Isospin Breaking
KG = Klein-Gordon
LCSR = Light-Cone Sum Rules

LD = Long Distance
LEFT = Low Energy Effective Field Theory
LFU = Lepton Flavor Universality
LFUV = Lepton Flavor Universality Violation
LFV = Lepton Flavor Violation
LH = Left-handed
LHC = Large Hadron Collider
LHS = Left-handed Scenario
LHT = Littlest Higgs Model with T-Parity
LO = Leading Order
LQCD = Lattice QCD
LR = Left-Right
LQ = Leptoquark
MFV = Minimal Flavor Violation
MS = Minimal Scheme
$\overline{\textbf{MS}}$ = Modified Minimal Scheme
MSSM = Minimal Supersymmetric Standard Model
NDR = Naive Dimensional Regularization
NL = Non-Leptonic
NLO = Next to Leading Order
NNLO = Next to Next Leading Order
NP = New physics
OPE = Operator Product Expansion
PBE = Penguin-Box Expansion
PDG = Particle Data Group
PMNS = Pontecorvo-Maki-Nakagawa-Sakata
QCD = Quantumchromodynamics
QCDF = QCD Factorization
QCDP = QCD Penguins
QED = Quantumelectrodynamics
QFT = Quantum Field Theory
RG = Renormalization Group
RH = Right-handed
RHS = Right-handed Scenario
RS = Renormalization Scheme
SCET = Soft-collinear effective theory
SD = Short Distance
SL = Semi-leptonic
SM = Standard Model
SM4 = Standard Model with 4 Generations
SMEFT = Standard Model Effective Field Theory
SSB = Spontaneous Symmetry Breakdown
SUSY = Supersymmetry

UV = Ultraviolet
UT = Unitarity Triangle
UUT = Universal Unitarity Triangle
VLL = Vectorlike Lepton
VLQ = Vectorlike Quark
WC = Wilson Coefficient

Introduction

Grand View of the Standard Model

What are the fundamental building blocks of matter? This question is easy to understand, but it took centuries and required tremendous theoretical and experimental efforts to approach an answer. The desire to know what everything is made of led to big scientific achievements, new technologies, and a better understanding of our universe. Since the 1860s we have known that atoms are basic building blocks of matter. But opposed to what was originally thought, they are neither indestructible nor fundamental as they are composed of protons, neutrons, and electrons. The finding that the whole periodic table is made of only three basic ingredients was already a big step forward toward answering the question of what everything is made of. However protons and neutrons are again not fundamental particles but rather complicated objects. Simplified one can say that they consist of three quarks: two up quarks and one down quark for the proton and vice versa for the neutron. It is amazing to know at first sight that everyday objects are made of only up and down quarks and electrons. Nevertheless this is not the end of the story as these quarks have to be bound in protons and neutrons, and this happens with the help of strong interactions that bring in new particles: gluons. Moreover pairs of quarks and antiquarks, not only of up and down quarks but also heavier quarks can also be present deep in protons and neutrons. This shows that the true picture of protons and neutrons is not as simple as often presented to pedestrians. What is amazing is the fact that this picture has been developed only in the last fifty years. This is now where the Standard Model of particle physics (SM) enters the stage.

The SM describes the properties of all elementary particles and three out of the four fundamental forces in nature: the weak force, electromagnetic force, and strong force. From a quantum field theoretical point of view the latter two are called, respectively, quantum electrodynamics (QED) and quantum chromodynamics (QCD). These forces are mediated by the exchange of gauge bosons. The weak force is mediated by the exchange of $W^{\pm}$ and Z^0 boson, the photon γ is the mediator of electromagnetic interactions, and gluons G^a are mediators of strong interactions. Gravity is not included in the SM. It is by far the weakest of the fundamental forces and not relevant for particle physics at the energy scales presently explored in experiments.

The SM sorts the elementary particles by their properties: there are the matter particles with spin-$\frac{1}{2}$ (fermions), the gauge bosons with spin-1, and one scalar particle with spin-0, the so-called Higgs boson. The fermions can be divided into quarks and leptons. The six

Table I.1 Particle content of the SM.			
Fermions: three generations			Gauge Bosons
ν_e	ν_μ	ν_τ	γ
e	μ	τ	G^a
u	c	t	$W^\pm$
d	s	b	Z^0
	Higgs		

quarks – up u, down d, charm c, strange s, top t, and bottom b – feel all three interactions, as they have a weak charge (called weak isospin), an electric charge, and a color charge related to strong interactions. Leptons do not participate in the strong interactions as they are color neutral. The three charged leptons – electron e, muon μ, and tau τ – participate in the electromagnetic and weak interactions, whereas the neutral leptons – the three neutrinos ν_e, ν_μ, and ν_τ – only feel the weak force.

Similar to the periodic table of elements where you have both a horizontal and a vertical order, the SM collects leptons and quarks into three generations as seen in Table I.1. The first generation consists of electron neutrino ν_e, electron e, up u, and down quark d. It is the first generation of particles that all the matter around us is made of. Nothing else at first sight is needed. However there exist also a second (ν_μ, μ, c, s) and a third generation (ν_τ, τ, t, b), which are in principle exact copies of the first generation (same charges and properties as seen in Table 2.1) except that they are heavier and thus unstable. At the end they decay into particles of the first generation. That is why the muon is sometimes called the heavy brother of the electron: They have the same quantum numbers, but differ in mass. Flavor physics studies such transitions between different generations that are mediated by the weak interaction. We cannot answer yet why there are exactly three generations, but so far we could already learn a lot about their properties in studying weak decays.

The SM is a theoretical model based on a (gauge) symmetry principle and very successful in describing and predicting experimental results. From this symmetry principle it follows that all fermions and gauge bosons must be massless, which is obviously not the case. Only photons and gluons are massless. This is now where the Higgs boson (or more precisely the Higgs field together with spontaneous symmetry breaking) enters the scene. It is responsible within the SM for the masses of quarks, charged leptons, and in particular gauge bosons of weak interactions. Even if the discovered nonvanishing neutrino masses around the turn of the millennium were not predicted within this model, its success in describing most of the data made the majority of particle physicists believe in the existence of the Higgs boson, and a lot of effort was put into finding this last missing piece of the SM. Its discovery in 2012 was a great scientific achievement – a true milestone.

The development of the SM was a big step forward to answer our initial question. Nevertheless, we still do not think that the story ends here. Despite its great success the SM does not answer several important questions about the nature surrounding us, and most physicists think that it is only an effective theory of a more fundamental theory. We will now address this important topic. Before continuing we would like to stress that this Grand View and the following one has been written by Jennifer Girrbach-Noe.

Grand View of New Physics

The ultimate question of elementary particle physics is: What is the fundamental Lagrangian of nature surrounding us? The Lagrangian of the SM is very successful in describing nature at the currently available energy range. The discovery of the Higgs boson completed the particle spectrum of the SM and it is another proof of how well the SM, works. Nevertheless, the SM cannot be the end of the story, and it is for sure not the fundamental Lagrangian of nature. The Lagrangian of the SM loses its validity at the latest at the Planck scale where gravitational effects become noticeable. Most physicists think of the SM as an *effective theory* that has to be replaced by a more fundamental theory above the TeV scale. What the word *effective* really means will hopefully be clear at later stages of our book. For the time being we will list some problems and open questions of the SM.

What Is Dark Matter Made Of?

We know that dark matter-energy makes up around 27 percent of the universe. Ordinary matter that is built of particles of the SM account for only 5 percent of the mass-energy of the universe. The remaining part is dark energy about which we know even less than about dark matter. Consequently we do not understand 95 percent of the universe's energy budget. Dark matter is called dark because it does not interact with photons and thus is electrically neutral and cannot be seen directly with telescopes. However, dark matter has gravitational effects that can be observed. This additional matter is needed to explain the rotational speeds of galaxies, gravitational lensing, and the anisotropies in the cosmic microwave background. Consequently the existence of dark matter is generally accepted, but we have no idea what dark matter is really made of. No particle of the SM can explain it. Some models beyond the SM have appropriate particle candidates for dark matter, e.g., the lightest neutralino in Supersymmetry with R-parity, and the search for this particle is one of major efforts in particle physics.

Why Is There More Matter Than Antimatter in the Universe?

Everything around us is made of matter, from the smallest life forms on Earth to the largest stellar objects. Antimatter, produced, for example, in particle collisions in experiments or in the atmosphere, can only exist for a very short time because matter and antimatter annihilate each other, and only pure energy in the form of photons is left over. According to Einstein's famous equation $E = mc^2$, mass and energy can be transformed into each other. In the big bang the same amount of matter and antimatter must have been created out of energy in the early universe. Matter and antimatter particles are produced in pairs out of energy, and if they come in contact, they annihilate each other. To create the observed matter abundance and not a universe made of pure energy, a mechanism that creates a tiny asymmetry between matter and antimatter is needed: Only about one particle per billion must have managed to survive; all others particles annihilated with their antiparticle. In the SM there actually is a mechanism that treats matter and antimatter differently, called CP violation. However, its strength is much too small to explain the observed matter-antimatter asymmetry, which is essential for our existence. Combined with the puzzling aspects of dark matter and dark energy, in principle we do not understand the 5 percent of ordinary matter either!

Why Are Atoms Electrically Neutral?

The simplest atom is a hydrogen atom: It consists of a proton and an electron, and both have exactly the same charge but with opposite sign such that hydrogen is electrically neutral. Although this might sound plausible, it is highly nontrivial. A proton consists of three quarks: two up quarks and one down quark with charges $+\frac{2}{3}e$ and $-\frac{1}{3}e$, respectively. The electron has charge $-e$. Now in the SM there is at least at first sight no connection or relation between quarks and electrons although the self-consistency of the theory at the quantum level requires some relations between quarks and leptons in order to remove the so-called *gauge anomalies* to be discussed briefly at later stages of our book. In any case, the fact that the charges of the proton and electron sum up to zero exactly is remarkable. Is it just an accident or is a deeper reason behind this? In principle the electric charge of a particle could be anything.[1] So why are the electric charges quantized? This problem can be solved if the SM is embedded in a grand unified theory.[2]

Why Are the Forces of Nature of Such Different Strengths?

The four forces of nature are of very different strengths. Gravity is by far the weakest of all forces. Is it possible that all forces have the same origin and thus the same strength at a certain energy scale. Did they develop from the same force present in the early universe? The idea to unify different forces or interactions goes back very far in the history of physics. Already Isaac Newton (1642–1727) linked the planetary motions with the falling of the stone on earth. James C. Maxwell (1831–1879) accomplished unifying electricity and magnetism to electromagnetism. Nearly 100 years later Sheldon Glashow, Abdus Salam, and Steven Weinberg achieved unifying QED with the weak interactions to the electroweak interactions. Together with QCD they build the SM of particle physics. Is it now also possible to unify strong and electroweak force at a higher energy? The energy dependence of the strength of the forces in a quantum field theory suggests that this might be possible, although an exact unification cannot be achieved without modifying the theory along all the energy range. Embedding the SM into a Grand Unified Theory (GUT) would also solve the problem of why atoms are neutral as discussed earlier. Such GUT models further predict correlations between quantum numbers and masses of quarks and leptons not present in the SM. Theoretically many GUT models were studied and their predictions compared with experimental measurements. But so far no clear hints for the existence of such a unified interaction are found.

---

[1] This is due to the fact that the electromagnetic force is based on an abelian U(1) symmetry in which the normalization of the coupling and charges are arbitrary. However, in the case of nonabelian symmetries like SU(5) or SO(10) the electric charges are fixed.

[2] As soon as the U(1) force is embedded in a larger nonabelian symmetry group also the U(1) charges are automatically quantized.

Why Are Neutrino Masses Nonzero and So Small?

Neutrino masses are special in two ways. First of all, it is because neutrinos are exactly massless in the SM. Consequently we already found physics beyond the SM! It is possible to incorporate neutrino masses in the Lagrangian of the SM but there are different ways to do this. Either new particles as right-handed neutrinos or Higgs-triplets have to be added to the particle spectrum, and an additional heavy mass scale for Majorana masses is needed or a symmetry that forbids this mass term. Either way, neutrino masses require a nontrivial extension of the SM. The second thing is that neutrino masses are much smaller than the masses of other fermions. There is already a large mass hierarchy between the electron and the top quark of roughly six orders of magnitude, and neutrino masses extend this hierarchy to another six orders. This is a part of the so-called *flavor problem*.

Why Are There Three Generations of Particles?

The objects around us are all made up of particles of the first generation, namely electron e, up u, and down d. Out of up and down quarks we can build protons p and neutrons n and that is all we need for the atoms of the periodic table of elements. The electron neutrino ν_e is emitted in radioactive decays of those atoms. If that's all we need, why are there more matter particles? The muon μ, the heavy brother of the electron was first found in cosmic rays and belongs to the second particle generation. It has exactly the same properties as the electron but it is roughly 200 times heavier and consequently unstable. It decays and finally ends up in particles of the first generations. The same applies to the other particles of the second and third generation. The strange quark s and the bottom quark b are heavier versions of the down quark d. All these particles were not discovered at once. When "half" of a generation was found, it was expected that the other half must also exist due to symmetry principles. For example the charm quark was predicted at times when the strange quark was already found and similarly for the top quark when the bottom quark was already known. We do not know why there are exactly three generations and not two, four, five . . . or only one. But it is very satisfying that one could exclude additional generations through the study of Higgs decays. On the other hand, in a SM with only two or one generation there would be no CP violation,[3] thus no matter-antimatter asymmetry. This is only possible with more than two generations. However as already mentioned earlier, the CP violation present in a SM with three generations is still way too small to explain the matter abundance of the universe. On the other hand, it is sufficiently large to accommodate other phenomena observed in nature, and we will discuss them in detail in this book.

How Can We Stabilize the Electroweak Scale?

This is also called the hierarchy problem and is often seen as one of the biggest (theoretical) problems of the SM. If the SM is considered as an effective theory we get an additional higher energy scale Λ, which characterizes the valid energy range of the SM. Λ corresponds to the mass of new heavy particles. Under the hierarchy problem one understands the

---

[3] We put here aside, the so-called strong CP violation to be briefly discussed in Section 17.3.

difficulty to stabilize – under the presence of this scale Λ – the electroweak scale in every order of perturbation theory and to explain the large gap between the electroweak scale $v = 246$ GeV and the GUT scale $M_{\mathrm{GUT}} = 10^{16}$ GeV or Planck scale $M_{\mathrm{Pl}} = 10^{19}$ GeV. Fermion masses m_f are proportional to the only scale in the SM, $m_f \propto v$. They are protected by the chiral symmetry, which is only broken by the mass parameter such that radiative corrections stay small: $\delta m_f \propto m_f \log \frac{\Lambda}{m_f}$. These corrections vanish for $m_f \rightarrow 0$, that is in the limit of exact chiral symmetry. The mass of the photon on the other hand is also protected, namely by the gauge symmetry, and stays zero in all orders of perturbation theory. This relation between symmetry and a small parameter is summarized in the naturalness principle by 't Hooft [4]: A small parameter of a theory is natural if and only if the symmetry of the system enlarges when this parameter is set to zero. Fine-tuning between different large contributions to a given quantity by hand with the goal to obtain a much smaller total contribution than the individual ones is regarded as unnatural. The problem here is of course how to quantify this. How much fine-tuning is still natural? In the SM this problem arises in connection with the Higgs mass that is not protected by any symmetry. The Higgs mass gets quadratic corrections proportional to the highest present scale, e.g., $\delta m_h^2 / m_h^2 \propto \Lambda^2$ and thus destabilizes the electroweak scale.

Naturalness is of course not a necessary condition for a theory to be internally consistent, but it is also not only an aesthetic requirement. It assumes the lack of certain conspiracies between phenomena occurring at very different energy scales. Naturalness might be – and often was in the past – a helpful guiding principle to infer the energy scale at which an effective theory breaks down. The separation of scales is inherent in effective theories, and applying the naturalness principle could help to find the energy scale where new physics (NP) enters, e.g., the problem of the electromagnetic energy of a classical electron is cured by the positron.

There are different possibilities to end up with a Higgs mass at the electroweak scale and not at a much higher scale:

- Fine-tuning: It could be that the parameters of the Lagrangian are fine-tuned such that the physical Higgs mass is at 125 GeV. However, this accidental cancellations up to 32 significant digits ($M_{\mathrm{Pl}}^2 / M_W^2 \approx 10^{32}$) must happen in every order of perturbation theory (and not only once).
- Symmetry: There might be an additional symmetry that protects the Higgs mass from large radiative corrections. Supersymmetry, which connects bosons and fermions, is the most prominent example for this.
- The Higgs is not a fundamental scalar particle, but composed of further particles.
- A softly broken scale invariance: In the SM only the mass parameter μ of the Higgs potential breaks scale invariance. If this is the only symmetry breaking parameter all corrections would be proportional to μ^2.

These are some of the reasons why one can think that the SM should be extended to a more fundamental theory. However, it is not clear in which way NP will manifest itself. In principle only the hierarchy problem requires new phenomena not too far away from the electroweak scale and thus reachable by the LHC. Also dark matter studies suggest that the new dark matter particles should be weakly interacting and around the electroweak scale (so called WIMP, weakly interacting massive particles). It is also possible that some of the other problems might be solved at a much higher energy scale not reachable within the near future with the help of a high energy collider.

After this list of problems and open questions of the SM, we shortly want to indicate how to look for new particles and new interactions beyond the SM. In order to discover NP there are in principle two strategies:

- **Direct Detection**

    In collider experiments new particles could be produced and directly detected. The limiting factor here is the center of mass energy of the experiment. According to $E = mc^2$, if this energy is too small or the new particle too heavy, it cannot be produced as a real particle but only exists as a virtual particle. The advantage of a direct detection is, of course, that one really sees the new particle. The direct detection of the Higgs boson was the final proof that it really exists. Prior to its detection some indirect hints and limits where known, and from the theory side it was predicted to exist somewhere around the electroweak scale. With the LHC it is possible to reach distance scales of $5 \cdot 10^{-20}$ m or equivalent energy scales in the ballpark of 5–7 TeV if sufficient statistics are available.

- **Indirect Detection**

    The existence of new particles can become noticeable indirectly through quantum fluctuations, which results in deviation (often a very small one) between SM prediction and measurement. To trace such small effects a very high precision both from theory and experiment is required. This way is followed by weak decay experiments LHCb, SuperKEKB and NA62 and as we will demonstrate in this book these decays are in principle sensitive to distance scales as small as 10^{-21} m and even smaller scales. The influence of new particles as a sign of physics beyond the SM, is for example, possible in some weak decays that are suppressed in the SM such that NP effects can compete with the SM contributions. A historic example not related to particle physics is the discovery of the Neptune by Le Verrier in 1846. He studied the deviations of the orbit of the Uranus from the predicted one and concluded that it must be due to gravitational effects of a further planet whose position he predicted with an accuracy of $1°$. Similarly one can try to track new particles, and it was often done in the past.

    Indeed, already in 1987 a heavy top quark was predicted at the B factory DORIS at DESY with the ARGUS experiment, where for the first time $B_d - \overline{B}_d$ oscillations were discovered. Similarly, the existence of charm quark was predicted by measuring and calculating the rate for $K_L \rightarrow \mu^+ \mu^-$. Without the charm quark the theoretical prediction for the rate of the decay $K_L \rightarrow \mu^+ \mu^-$ was predicted to be much larger than experimentally measured. Including the charm quark, which together with the strange quark builds a doublet under the electroweak gauge symmetry, the branching ratio for $K_L \rightarrow \mu^+ \mu^-$ could be suppressed and made consistent with data. Even more, Gaillard and Lee [5], calculating the $K_L - K_S$ mass difference ΔM_K and comparing it with its measured value, could predict in 1974 the mass of the charm quark to be in the ballpark of 2 GeV before its discovery. But even in 2019 we do not know whether at some level NP enters both $B_d - \overline{B}_d$ oscillations and ΔM_K because of theoretical and parametric uncertainties.

In this context of particular interest are rare decays of leptons, like $\mu \rightarrow e\gamma$ or $\tau^- \rightarrow \mu^- e^+ e^-$ for which the predicted branching ratios are so small in the SM that they probably will never be measured if the SM is the whole story. Here NP contributions could be much larger than the SM ones making their measurement possible. Observation of such decays would be then a clear signal of NP. In fact, the next years could bring discoveries in this sector of particle physics. Similar comments apply to electric dipole moments of leptons, the neutron, the proton, and atoms and anomalous magnetic moments of leptons and quarks. Also here new discoveries could take place in the coming years.

After these two grand views we are almost ready to start our expedition. It will first dominantly take place within the SM reaching distance scales of the order of 10^{-18} m (*the Attouniverse*) corresponding roughly to the mass of the heaviest SM particle, the top quark. However, already in this part we will develop a powerful technology that will allow us to go in the later parts of our book to much shorter distances scale, reaching eventually the (*Zeptouniverse*) and possibly even shorter distance scales. What we still need is an efficient strategy and corresponding outline of our expedition. The construction of such a strategy and the outline in question is our next task.

The Grand View of the Expedition and the Strategy

Writing a long book reminds me of expeditions to Himalayas, which I know only from several books. Reading books about great composers I see some similarities to composing symphonies or operas. I believe that despite differences in these three fields, it is crucial to have a plan for these expeditions and the strategy for reaching the goals. The construction of the plan for this book was rather challenging in view of many topics present in the field of weak decays. While the number of topics discussed in this book is large, some topics will be only mentioned or discussed very briefly. The principles that guided me in choosing the topics were as follows:

- I wanted to present in detail first of all the material that I know and that I developed over many years with many collaborators, in particular with many of my PhD students and postdoctoral fellows. This is quark flavor physics in the SM and beyond.
- This includes in particular calculations of QCD corrections to weak decay processes at the leading and next-to-leading order, which dominated the topics in my research group at TUM in the 1990s.
- However, I included also topics on which I wrote only few papers until now but which in my view are very important. Hopefully not only the topics but also my papers. These are in particular lepton flavor violating decays, electric dipole moments of atoms, molecules, leptons, and nucleons, and of course anomalous magnetic moments like $g-2$ of the muon. Here I benefited from many reviews that I will list in the relevant chapters.
- I put significant effort into explaining how the derivations of predictions for most important observables both within the SM and its extensions are made. This requires

often tedious Feynman diagram calculations, and the related quantum field technology had to be developed to reach this goal. However, only a subset of results could be derived here as otherwise the book would be much longer.

- But what is equally important is the development of skills in connecting the results of such calculations to observables measured by experimentalists. In this context I will stress the correlations between many observables that turn out often crucial not only for the tests of the SM but also to distinguish between its various possible extensions.

- Concerning the latter, we will first devote one chapter to the technology of the so-called *SM gauge invariant effective field theory* (SMEFT), which became very popular in recent years. This tool allows in a model independent manner to look beyond the SM, which is useful but in my view contains too many free parameters so that additional model assumptions have to be made in order to reach some definite conclusions.

- In the latter context we will discuss first the so-called *simplified models*, which contain only few parameters. Models with the so-called constrained *Minimal Flavor Violation* are the best examples here. But also models with a new heavy neutral gauge boson Z', new heavy scalars, and those with flavor changing Z couplings are sufficiently simple that we can discuss them in some detail. Already these simple models will give us some idea on what experimentalists are telling us and hopefully some hints for the construction of new theories and the identification of NP at very short distance scales beyond the reach of the LHC. Yet, we will warn the reader that very simple models can often miss the physics hidden in the complicated quantum effects described by SMEFT, and we will try to expose this problem in a few examples.

- Few chapters will be devoted to concrete nonsupersymmetric models, like models with vectorlike fermions, leptoquark models, and explicit models with new heavy-gauge bosons and scalars. In these cases no derivations will be present. Supersymmetric models will not be discussed in this book because this would require too many pages for its introduction. Instead we will give a collection of references to useful papers.

- We will also give some outlook for the future of this field.

With these goals in mind the book consists, similar to a symphony, of four parts:

PART I: Basics of Gauge Theories
PART II: The Standard Model
PART III: Weak Decays in the Standard Model
PART IV: Weak Decays beyond the Standard Model

The last part includes also observables like electric dipole moments and anomalous magnetic moments. These four parts consist of twenty chapters, which are the steps in our expedition. We are now ready to list these steps and indicate how they are related to each other. This will hopefully make our book more transparent. In this context it is advisable to read the description of a given step, presented next, before making that step. Therefore here comes an important message:

PLEASE READ THE OUTLINE OF EACH STEP BEFORE MAKING IT!

Step 1: Chapter 1

We will begin by presenting the general structure of gauge theories, stressing their symmetries and their breakdown. The structure of the Lagrangians in this step will be fundamental for the full book. We will encounter Lagrangians involving spin-0, spin-$\frac{1}{2}$ and spin-1 fields. We will also collect literature in which further details can be found. But the information provided in this step should be sufficient for following the next steps.

Step 2: Chapter 2

We will next move to discuss the SM of electroweak and strong interactions. We will present the basic SM Lagrangian, and we will discuss various properties of this simplest theory but we will not do any phenomenology at this stage. There are many excellent books on the dynamics of the SM, and we will in most cases skip derivations of well-known formulas to save space and in particular energy for the later steps that are not covered in detail in textbooks. The most important result of this step will be collection of the Feynman rules for those interactions present in the SM that we will need for explicit calculations as we proceed. For readers' convenience we will collect these rules in Appendix B.

Step 3: Chapter 3

Having the Feynman rules at hand, we will present the simplest calculations, the so-called *tree-level* calculations that do not yet require a deep knowledge of quantum field theory like renormalization and renormalization group. This step will give us leading formulas for the determination of various parameters of the SM like the Fermi constant G_F and the parameters of the CKM matrix, which will be introduced and discussed in detail. These formulas will also give us a transparent picture of the leading decays of mesons and leptons and will allow approximate determination of the parameters in question. Yet, for a serious phenomenology we will need various corrections to these leading expressions. They will be calculated or simply listed in later steps when we turn to the phenomenology at the frontiers of weak decays. In this step we will also find out that in the case of meson decays some entries in the decay amplitudes cannot be calculated in perturbation theory. These are hadronic matrix elements of quark currents and in particular hadronic matrix elements of four-quark operators. This task will be left for other steps. But seeing how these matrix elements enter the decay amplitudes will turn out to be useful when our climb will become technically more difficult, and we will have to concentrate on other matters.

Step 4: Chapter 4

This step will show us the technology of quantum field theory at the one-loop level, including general comments and results of two and higher loop calculations that we will need later on. But the first goal of this step is the detailed presentation of the dimensional regularization and the development of the technology for one-loop calculations. The next goals in this chapter are the presentation of the renormalization and the renormalization group methods. This step is rather technical but crucial. While our presentation is rather general, we will show explicit results only in QCD and QED. The case of the renormalization of electroweak interactions is much more involved because of spontaneous symmetry breakdown. It deserves a separate book, and definitely we do not want to get too exhausted before entering the most interesting parts of this book related to flavor physics. Yet, we will have a brief compact presentation of the most important issues involved here. We will need this information later on. One of the important results of this step will be the collection of integrals necessary not only to perform the renormalization of QED and QCD, but also those that will be crucial for one-loop calculations at all stages of our expedition. We collected these integrals in Appendix C.

Step 5: Chapter 5

This is one of the most important steps in our expedition. Here we will first discuss the concept of the operator product expansion (OPE) and the renormalization group equations for the Wilson coefficient functions of the operators present in the SM. In this part we will first concentrate our discussion on the simplest operators: current-current operators. We will calculate their coefficients including not only leading QCD effects (LO) but also non-leading ones (NLO). In more complicated cases we will indicate how such calculations should be done and present only the results for them.

Step 6: Chapter 6

Step 5 will be very tough, and after completing it we will first relax a bit discussing *flavor changing neutral current* (FCNC) processes. In the SM such processes can only occur first at the one-loop level due to Glashow–Iliopoulos–Maiani (GIM) mechanism [6], a very important mechanism in the field of weak decays. We will first present the general structure of these processes, which will involve a set of *basic master one-loop functions*. We will calculate most of these functions explicitly without including first QCD corrections. Subsequently we will show how these functions enter the operator product expansion. Using then technology of Step 5, we will be able to include QCD and electroweak corrections to a number of FCNC processes. This will result in a number of effective Hamiltonians for most important processes that involve new operators not encountered in Step 5: the so-called penguin operators. In this step we will study several properties of these Hamiltonians, but we will postpone detailed phenomenology of these processes to later steps.

Step 7: Chapter 7

We will see in Step 6 that the phenomenological application of the collected formulas will require input from lattice QCD and other nonperturbative approches. In particular we will need the values of hadronic matrix elements of quark currents and those of four-quark operators. While we cannot go into details here, we will collect all results for hadronic matrix elements, weak decay constants, and other nonperturbative objects in one place so that they can be looked up quickly whenever necessary. But before presenting the results obtained by purely numerical lattice QCD calculations, we will discuss the application of the so-called *dual QCD approach* to weak decays that was developed by Bardeen, Gérard, and myself over many years. While not as accurate as the lattice QCD method, it offers in my view a better insight in the dominant QCD dynamics at long-distance scales than lattice QCD (LQCD) itself. But it can only be applied efficiently to K meson physics, while LQCD can be applied in principle to all low-energy processes. Yet, the application of LQCD to nonleptonic decays of B mesons is very difficult, and therefore in this part we will also devote a section to the QCD factorization approach, stressing its successes and limitations.

Step 8: Chapter 8

In this step we will present a detailed discussion of the particle-antiparticle mixing in the SM and perform the classification of various types of CP violation. We will discuss the determination of the CKM parameters in tree-level decays and also the determination of the so-called Unitarity triangle from particle-antiparticle mixing both from $B_{s,d}$ and K meson systems and tree-level decays in question. In this context we will present various methods for the determination of the angles α, β, and γ in this triangle from certain leading decays. The considered decays are subject to only small contributions from NP so that the CKM parameters can be extracted without knowledge of what happens beyond the SM. This is useful as then the results of tree-level determinations of CKM parameters can be used to find SM predictions for particle-antiparticle mixing observables considered already in this step and for rare processes considered in the next two steps. As both particle-antiparticle mixing and rare processes are subject to potential NP contributions, this could help us to identify NP through the pattern of deviations of the data for them from SM predictions.

Step 9: Chapter 9

In this step we will discuss most important rare K and $B_{s,d}$ decays and related observables. We will base our discussion on the information gained in previous steps. In particular we will collect many formulas that are necessary to perform the phenomenology of weak decays of mesons within the SM. This step will be rather long but a very important one for the subsequent steps. The general expressions for various observables will remain often unchanged in many NP models, and only the master functions of Step 6 will be modified. But in order to be prepared for the study of the physics beyond the SM we will already here present rather general formulas that contain new operators that are absent in the SM or are very strongly suppressed.

Step 10: Chapter 10

This step deals exclusively with $K \rightarrow \pi\pi$ decays and in particular with their two pillars: the so-called $\Delta I = 1/2$ rule and the ratio ε'/ε. We devote a separate chapter to it because the QCD analysis of both perturbative and nonperturbative contributions to these decays is very challenging. Moreover, it appears that the SM has significant problems in explaining the data for ε'/ε, and presently it is not yet clear whether SM dynamics are fully responsible for the $\Delta I = 1/2$ rule. We present several analytical formulas for ε'/ε and its numerical value as a function of most important hadronic parameters that should allow us to monitor future progress in the evaluation of this ratio.

Step 11: Chapter 11

This step deals with basic aspects of flavor physics in the charm sector. I have written only two papers on charm physics in the last thirty years, and consequently this step will not be as detailed as other ones. Yet, it will contain some fundamentals related to CP violation in the charm sector.

Step 12: Chapter 12

After the long investigations presented in Steps 8–11 we will be able to collect various lessons and thereby gain a grand picture of how the SM faces the data in 2019. In particular it will be important to collect possible deviations from its predictions that these days carry the name of *anomalies*. This will show us that indeed the SM has significant difficulties in describing all the flavor data simultaneously. Not all of these anomalies are still fully convincing because of theoretical and experimental uncertainties, but they give us strong motivations for making an important step toward the identification of NP that could be responsible for them.

Step 13: Chapter 13

We will next go beyond the SM. It will be strategically useful to present first the general structure of effective Hamiltonians beyond the SM and identify new operators that are absent in the SM. This step will deal with physics at the electroweak scale and below it as we already encountered in previous chapters, but now particular emphasis will be put on new operators generated by some NP at much higher scale than at the electroweak scale and below it will contribute to various processes. What will be discussed here is the low-energy effective field theory that carries the name LEFT in order to distinguish it from the effective theory discussed in the next chapter. Here the basic symmetry for finding possible operators is simply the symmetry $SU(3)_C \times U(1)_Q$, that are the symmetries of QCD and QED. Simply the color and electric charges have to be conserved.

Step 14: Chapter 14

To properly describe the physics above the electroweak scale, we will discuss next the so-called gauge invariant *Standard Model Effective Field Theory* (SMEFT). This theory became very popular in view of the fact that the LHC until now did not discover directly any new particles. It offers a model independent formulation of NP beyond the SM in terms of operators that are invariant under the full SM gauge group, which implies, among other things, that not only color and charge have to be conserved but also hypercharge.

This eliminates certain operators allowed in the previous chapter. Moreover, the $SU(2)_L$ invariance implies certain relations between observables.

While more constrained than LEFT, SMEFT is in my view too general to give us a definite view about NP beyond the LHC scales without additional dynamic assumptions. But this is a very important framework, and many techniques developed in this decade in the context of SMEFT are crucial for serious phenomenological applications to obtain model independent results. Moreover, SMEFT technology plays an important role in the derivation of predictions in concrete models as well. We will see this in this chapter and subsequent chapters of the book. But already in this step we will discuss implications of SMEFT in a number of general NP scenarios in which only a subset of operators is relevant. Moreover, we will present a few general formulas, in particular the one for the ratio ε'/ε.

Step 15: Chapter 15

Chapter 14 is rather general, and it will be strategically useful to look at specific simple examples of NP. Having already all relevant SM formulas for flavor observables at hand it will be of interest to see how they generalize beyond the SM. In this step we will first discuss the concept of minimal flavor violation (MFV), based on the flavor symmetry $U(3)^3$, that is the most modest modification of the flavor violating effects found in the SM. In fact, there is the full class of MFV models, which can be characterized by a number of correlations between flavor observables, which depend often on only few new parameters and occasionaly are parameter free. We will see that the method of master one-loop functions introduced in Step 6 will offer a very transparent and simple description of MFV models, but we will also find that the anomalies found in Steps 8–10 cannot be explained in this class of models.

Therefore already in this step we will go beyond MFV models beginning first with models based on reduced flavor symmetry $U(2)^3$. Subsequently we will present the so-called *simplified models* in which NP contributions to flavor violating processes are dominated by tree-level exchanges of neutral bosons, in particular the SM Z-boson, a heavy neutral gauge boson Z', and a heavy scalar S or pseudoscalar A. In these models, then, the GIM mechanism is not working, and FCNC processes appear already at tree-level modifying in a striking manner the pattern of flavor violation found in the SM and in MFV models. Yet, also in this case the method of master functions will turn out to be very efficient in exhibiting these differences. We will see that some of these simplified models will give us some insight into the origin of anomalies found in previous steps. However, in contrast to the SM, MFV models, and $U(2)^3$ models, the simplified models discussed in this step will offer correlations between various flavor observables only within a given meson system. While already quite restrictive, they cannot be the whole story and one should expect that a fundamental theory of flavor would imply correlations between observables of different meson systems. While such a fundamental theory has still to be constructed, several models in the literature, that go beyond MFV, imply correlations between observables of different meson systems.

Step 16: Chapter 16

To illustrate what happens in specific models, we will discuss three classes of models: the so-called 331 models, models with heavy vector like quarks, and models with leptoquarks, either of spin-0 or spin-1. Leptoquarks are known from grand unified theories in which the strong and electroweak forces are unified and in which leptoquarks have typical masses of $O(10^{16})$ GeV. But here we will concentrate on models with leptoquarks having sufficiently low masses so that they could be in principle discovered by the LHC. But even if the masses of new particles in the models considered will turn out to be beyond the reach of the LHC, their presence can still be identified in the rare processes discussed by us.

These three classes of models are sufficiently simple so that we can present them here in some details. The new aspect of these models will be larger involvement of leptons than in previous chapters. Yet, in this chapter we will confine the phenomenology to quark flavor observables, postponing the discussion of lepton flavour violation to the next step. There are several more complicated models that we will only briefly mention here. These are the Littlest Higgs model with T-parity (LHT), Randall–Sundrum models with custodial protection (RSc), left-right symmetric models, models with gauged flavour symmetries and supersymmetric models. As the structure of these models and their phenomenology are rather involved we will mainly refer to papers investigating weak decays in these models.

Step 17: Chapter 17

Until now we have discussed dominantly flavor physics of quarks. But our book must also contain flavour physics of leptons, in particular charged lepton decays mediated by flavor changing neutral currents. While very strongly suppressed within the SM, lepton flavor violation (LFV) can be by many orders of magnitude larger in NP models. Therefore finding such transitions would be a clear signal of NP. The nice feature of most of these decays is the absence of QCD corrections making the theoretical analysis much simpler than was the case of meson decays. Another important topic is electric dipole moments (EDMs) of various atoms, molecules, nuclei, and nucleons that all are very strongly suppressed in the SM. They test CP violation without flavor violation. Similar to LFV, a measurement of any nonvanishing EDM would be a clear signal of NP at work. But in contrast to LFV there are large nonperturbative uncertainties in this case. In this part we will also discuss anomalous magnetic moments of charged leptons.

These topics are somehow related to each other, and we will begin this part with a general look on them. Subsequently we will present a compendium of formulas for the transitions like $\mu \rightarrow e\gamma$ and processes involving only leptons. These formulas are quite general so that one can use them for various models. We will illustrate them on a few examples. Again, one of the goals of this step will be the identification of correlations between various observables. It will be a good exercise in using the collected formulas in simple setting.

We will then move to discuss EDMs and anomalous magnetic moments, collecting formulas that are useful for phenomenology and summarizing the present status of these topics. Finally, we will only mention some aspects of neutrino oscillations and stress the importance of neutrinoless double β decay.

Step 18: Chapter 18

We will next attempt a grand summary of NP models discussed by us comparing the different patterns of flavor violation present in these models. While they have been discussed already previously in some detail, it will be instructive to compare how different models face various anomalies listed in Step 12. This chapter is really a general view on the literature related to various anomalies and after being completed will hopefully motivate the readers to study numerous papers containing many ideas with the goal to construct their own models.

Step 19: Chapter 19

With all this knowledge, we will show how a "Flavor Expedition to the Zeptouniverse" could in principle take place. Here we will address the question which of the processes considered in our book could give us best information about NP beyond the SM. In this context we will collect low-energy observables that could be useful for the search for NP, in particular beyond the reach of the LHC. Subsequently, with the goal to distinguish various NP models, we will discuss DNA charts proposed already in a number of papers.

Step 20: Chapter 20

Finally, we will be able to present a grand summary of weak decays, list open questions, and present a shopping list for the coming years.

A careful reader certainly noted that in listing the steps we did not mention detailed expositions on P, C, CP, and T transformations. They are presented in detail in textbooks on relativistic quantum mechanics and on advanced quantum field theory. In particular the book [7] contains an impressive amount of information on this topic and should be looked up whenever necessary. But in order to facilitate the reading of our book, we have made a collection of our phase conventions in Appendix A.2. A number of appendices collects useful formulas that we will need as we proceed. These are in particular Fierz identities and integrals for one-loop calculations among others. Two older books on weak decays [8, 9] are also very useful. Other books will be recommended in the course of our expedition.

There are still the following points about this book that should be emphasized. While the first chapters are devoted to the introduction of the field in simple terms, they assume that readers have some elementary knowledge of the field theory and in particular of the SM. Similarly in subsequent chapters we will refer from time to time to some concepts that cannot be explained in any detail. Otherwise this book would be much longer. Most importantly, we will present in detail only topics that the author thinks to have fully under control and is familiar with through his publications and lectures. But as we will see, this material will already cover a significant part of the field of weak decays. The presentation of remaining topics will be rather brief, but should provide sufficient input for the phenomenological analyses performed in this book. Moreover, we will refer to the rich literature where more details on these topics can be found. This applies in particular to the papers that appeared in the last five years, up to the fall of 2019.

Introducing Main Players

The title of our book makes it clear who the main players are in our long expedition. These are weak decays of various particles as described by gauge theories. In addition to global quantities like branching ratios for the weak decays in question, there is a multitude of more local quantities that are related to them, and in addition there are transitions, like particle-antiparticle oscillations, known under the name of particle-antiparticle mixing. Also a number of measurable quantities (observables) are related to them.

While, these objects will enter the scene in various chapters one by one, it is useful to see them already now together, even if for some readers, not familiar with the subject, the next pages will be difficult to follow. They can then skip them and return later to test how much they have learned. But a better attitude is to go through these pages in order to become acquainted at least with the names of main players and realize their importance. Reference to literature related to statements made soon will be given in the course of our expedition.

In my view the processes listed next will turn out to be the superstars and stars of 2020–2030 in quark flavor and lepton flavor physics and belong to the most promising observables in the search for NP. There are many measurements one can do and many observables one can calculate, but in my view the following ones will lead the tests of the SM and searches for NP in the coming ten years within the *quark flavor physics*:

- Particle-Antiparticle mixing or in short $\Delta F = 2$ observables with F meaning any quark flavor:

$$\Delta M_s, \qquad \Delta M_d, \qquad S_{\psi K_S} \qquad S_{\psi \phi}, \qquad \varepsilon_K, \qquad \Delta M_K ,$$

with ΔM_s and ΔM_d denoting mass differences between mass eigenstates in $B^0_{s,d} - \bar{B}^0_{s,d}$ mixings and $S_{\psi K_S}$ and $S_{\psi \phi}$ being mixing induced CP-asymmetries that are measured in $B^0_d \to \psi K_S$ and $B^0_s \to \psi \phi$ decays. ε_K is a measure of CP violation in $K_L \to \pi\pi$ decays induced by $K^0 - \bar{K}^0$ mixing and ΔM_K is the mass difference in $K^0 - \bar{K}^0$ system, analogous to $\Delta M_{s,d}$. Despite the nonleptonic nature of all these observables, we expect that during the coming years all of them will be known very precisely within the SM due to advances in LQCD. ΔM_s, ΔM_d, ε_K, and ΔM_K are already very well measured. $S_{\psi K_S}$ and $S_{\psi \phi}$ should be precisely measured in the coming years.

- Angular observables and branching ratios in the decays ($\ell = e, \mu, \tau$)

$$\bar{B} \to \bar{K}^* \ell^+ \ell^-, \qquad \bar{B} \to \bar{K} \ell^+ \ell^- , \qquad B_s \to \phi \ell^+ \ell^- ,$$

with $B = B_d, B^\pm$ and correspondingly for K^* and K. The ones with $\ell = e, \mu$ are presently among the stars of flavor physics with the data signaling intriguing violation of $\mu - e$ universality, that is, the difference between rates for decays into muons and electrons.

- The branching ratios

$$\mathcal{B}(B_s \to \mu^+\mu^-), \qquad \mathcal{B}(B_d \to \mu^+\mu^-),$$

that should be known very precisely in the SM once the accuracy on the CKM elements $|V_{ts}|$ and $|V_{td}|$ will be improved at Belle II at the beginning of this decade. Experimentally $\mathcal{B}(B_s \to \mu^+\mu^-)$ should be known rather precisely by 2025, while it may take longer time for $\mathcal{B}(B_d \to \mu^+\mu^-)$. These decays, in particular taken together, offer very powerful tests of the SM and of its extensions. We do not list yet $\mathcal{B}(B_d \to \tau^+\tau^-)$, as the present upper bound is larger by several orders than its SM expectation. Moreover, its measurement is very challenging. We will encounter few other observables in these decays related to the time evolution in $B_s \to \mu^+\mu^-$.

- The branching ratios

$$\mathcal{B}(\bar{B} \to \bar{K}\nu\bar{\nu}), \qquad \mathcal{B}(\bar{B} \to \bar{K}^*\nu\bar{\nu}), \qquad \mathcal{B}(B \to X_s\nu\bar{\nu}),$$

which offer powerful means to study the effects of right-handed currents.

- The branching ratios for $K^+ \to \pi^+\nu\bar{\nu}$, $K_L \to \pi^0\nu\bar{\nu}$, $K_{L,S} \to \mu^+\mu^-$ and the ratio ε'/ε with ε' being another measure of CP violation in $K_L \to \pi\pi$ decays in addition to ε_K:

$$\mathcal{B}(K^+ \to \pi^+\nu\bar{\nu}), \qquad \mathcal{B}(K_L \to \pi^0\nu\bar{\nu}), \qquad \mathcal{B}(K_{L,S} \to \mu^+\mu^-), \qquad \varepsilon'/\varepsilon.$$

The first two branching ratios are basically free from hadronic uncertainties as opposed to the remaining three observables, which suffer from these uncertainties, but they should be reduced in the coming years. These five quantities taken together and in particular in correlation with all observables listed earlier offer very powerful means to test the SM and in identifying NP beyond the LHC scales. Moreover, present theoretical calculations of ε'/ε signal significant departures from the SM.

- The branching ratios

$$\mathcal{B}(B^+ \to \tau^+\nu_\tau), \qquad \mathcal{B}(\bar{B} \to D\tau\nu_\tau), \qquad \mathcal{B}(\bar{B} \to D^*\tau\nu_\tau)$$

play important roles in searching for NP effects mediated by charged scalars, charged gauge bosons and leptoquarks. The present data on the last two signal the largest departures from the SM among decays that occur at tree-level in this model. Not only branching ratios but in particular various differential decay rates will be crucial for the identification of NP in these decays.

- Finally the branching ratios

$$\mathcal{B}(B \to X_s\gamma), \qquad \mathcal{B}(B \to K^*(\varrho)\gamma)$$

should not be forgotten with the first one providing already for years a very stringent test of the SM and bounding NP contributions.

There is no question that *lepton flavor physics* will also play a very important role in identifying NP beyond the SM. In particular, the following decays should provide a deep insight into the dynamics at short-distance scales:

- First of all the branching ratios

$$\mathcal{B}(\mu \to e\gamma), \qquad \mathcal{B}(\tau \to e\gamma), \qquad \mathcal{B}(\tau \to \mu\gamma),$$

that similar to $B \to X_s\gamma$ are governed by dipole operators. They are tiny within the SM, and any observation of these decays would be a clear signal of NP. They already now put significant constraints on the parameters of various extensions of the SM. Improved upper bounds and even their measurements are expected from PSI, Belle II, and the LHCb.

- Next come

$$\mathcal{B}(\mu^- \to e^- e^+ e^-), \qquad \mathcal{B}(\tau^- \to \mu^- \mu^+ \mu^-), \qquad \mathcal{B}(\tau^- \to e^- e^+ e^-).$$

These decays are very interesting as they are strongly correlated with $\mu \to e\gamma$, $\tau \to e\gamma$, and $\tau \to \mu\gamma$, and these correlations are different for different NP models.

- Also the four branching ratios

$$\mathcal{B}(\tau^- \to e^- \mu^+ e^-), \qquad \mathcal{B}(\tau^- \to \mu^- e^+ \mu^-),$$

$$\mathcal{B}(\tau^- \to \mu^- e^+ e^-), \qquad \mathcal{B}(\tau^- \to e^- \mu^+ \mu^-)$$

will enrich the search for NP.

- $\mu - e$ conversion in nuclei, even if subject to hadronic uncertainties, could become the star of lepton flavor physics in the coming decade. The dedicated J-PARC experiment PRISM/PRIME should reach a sensitivity of $O(10^{-18})$. Also, semileptonic τ decays like $\tau \to \pi\mu e$ should not be forgotten.

Of special interest are decays that proceed through both quark flavor and lepton flavor violating transitions. These are in particular

$$K_{L,S} \to \mu e, \qquad K_{L,S} \to \pi^0 \mu e,$$

$$B_{d,s} \to \mu e, \qquad B_{d,s} \to \tau e, \qquad B_{d,s} \to \tau \mu,$$

$$B_d \to K^{(*)} \tau^\pm \mu^\mp, \qquad B_d \to K^{(*)} \mu^\pm e^\mp.$$

A natural mechanism responsible for such transitions are tree-level exchanges of leptoquarks or tree-level Z' exchanges. But they can also be generated in certain models at one-loop level.

Another set of observables is

$$\text{Electric Dipole Moments(EDMs)}, \qquad (g-2)_{e,\mu}.$$

Even if these observables are flavor conserving, they put strong bounds on extensions of the SM. Both are governed by dipole operators. EDMs are CP-violating, $(g-2)_{e,\mu}$ conserve CP. In certain models there are correlations between them. The $(g-2)_\mu$ anomaly (departure from SM prediction) found at Brookhaven should be clarified by Fermilab in 2020. As EDMs are very strongly suppressed in the SM, any measurement of a nonvanishing EDM would signal NP.

In addition to all these processes, charm physics will play a significant role in constraining NP models, but I did not list any processes here to emphasize that from my point of view the observables listed earlier will be more powerful in searching for NP because of smaller hadronic uncertainties. An exception here is CP violation in D decays as it is predicted to be tiny in the SM.

This list contains very many processes and including, in addition to branching ratios, various local observables, that is, various distributions and asymmetries related to them, gives us great opportunities for making progress in the search for fundamental laws of nature. Measuring all these observables precisely one day will surely allow us to get a deep insight into the NP dynamics beyond the LHC scales provided the departures from the SM predictions in at least some of them will be significant. But to reach this goal we have to develop strategies for a transparent use of these future data. This will also be one of the important goals of this book.

How to Use This Book Optimally

Writing this book was a real challenge for two reasons:

- There are many important weak decays that have to be considered.
- There are many NP models describing these decays often in a rather different manner.

As weak decays of mesons and leptons, EDMs and anomalous magnetic moments will play the crucial role in the search for new dynamics beyond LHC scales before a new collider will be built, it was mandatory to extensively discuss NP models independent of whether presently observed departures from SM predictions (anomalies) will remain or not. The point is that there is no doubt that some presently unknown dynamics exist at very short-distance scales, and I am confident that with improved precision of experiments and theory we will one day find out with the help of decays discussed by us what this dynamics is.

Of course, we had to present first of all weak decays within the SM, and it is very important to read this part first. It is not crucial to do all the loop calculations presented in this part, even if this should be useful for theorists. More important is to get an idea about the general structure of effective Hamiltonians governing weak decays and about the pattern of flavor violation in the SM that is governed by

- Left-handed nature of charged current interactions
- GIM mechanism at tree-level and its breakdown at one-loop that is enhanced in K and $B_{s,d}$ decays and mixings through large top Yukawa coupling
- CKM matrix that contains the weak phase necessary for the description of CP violation observed in several processes

In the beyond SM (BSM) part, it is probably advisable for beginners to first have a look at the simplest extensions of the SM that could be considered as *simplified models*. Several of them are presented in Chapter 15. This allows one to get an idea what could happen beyond the SM in simple settings. For those readers who know already a lot about flavor physics, the natural starting point is Chapter 13, in which the reader will meet a number of new operators that are absent in the SM. However, even more important is Chapter 14, in which the SMEFT is discussed in some detail. Seeing so many operators in these two chapters at once for the first time is a bit of a shock, but SMEFT is presently a popular language to discuss BSM physics, and it is advisable to learn this language as soon as possible. To make this chapter hopefully transparent and to facilitate reaching this important goal, I presented a number of examples in the form of various NP scenarios in which only a subset of operators plays the dominant role and the implications are simpler. With the knowledge gained in Chapters 13–15, it should not be too difficult to follow the material presented in the rest of the book.

In particular in Chapter 19 we will develop specific strategies for learning about NP beyond the LHC scales on the basis of processes presented in this book. These strategies, involving the so-called *DNA-charts*, will certainly turn out to be more useful in this decade than in the previous one, after hopefully many deviations from SM predictions for observables considered by us will be found in the coming years.

The BSM part contains many references to papers published in the last five years. They are only briefly described in our book but hopefully will give readers a general view on the efforts made these days by many theorists and experimentalists to find out which summits on the front cover of our book play the crucial role in nature. As the titles of all references have been exhibited, it is possibly a good idea to get a general view of the material presented in our book by first reading the titles of all references. This could appear to some readers as a crazy idea, but it is an original idea, and I have no doubts some readers will enjoy performing this experiment.

In the spirit of the last sentence, most importantly, I hope reading this book will be fun. Therefore our book does not contain any exercises so that the readers will not feel obliged to work if they just want to get general ideas and simply read this book. But if they are sufficiently motivated, they can check several calculations either performed in detail or sketched by the author. The collection of useful integrals and the list of papers in which Feynman rules for several concrete models can be found should make this task doable. Moreover, I tried to avoid long and complicated derivations of formulas that can be found in the literature to which we will refer frequently. More important are final results of these derivations and their phenomenological implications.

But there are additional reasons for not having exercises in our book. First of all, they take a lot of space, in particular if also solutions to them are provided. Having them would not allow me to present a number of important topics. Finally, in the last forty years I have read many books on field theory and particle physics that contained many exercises. I ignored them altogether. The reason is simple: I did not want anybody, even if much more experienced than me, to tell me what I should do. The great joy in research is to find problems and exercises by ourselves and have fun in solving them. In the spirit of the last statements, the readers could ignore all my advice on how to use optimally this book, even if I hope for some beginners this advice may turn out to be useful. This is in particular the case of recommendations for a number of papers and reviews, made rather frequently in this book, that I hope will help the readers to get their own ideas for the exploration of flavor physics. It will also motivate the readers to begin the day with the opening of **http://arxiv.org/** in order to find new papers and of course **http://www.slac.stanford.edu/ spires/hep/** in order to find out the status of their own citations. The latter database allows also to find any paper listed at the end of this book by simply inserting there its identification number instead of the author's name. However, one can reach this goal even faster by inserting this number into the Google machine.

While for some readers the absence of exercises will be considered as a deficiency, it is compensated by several author's ideas for research projects, which could not be realized before the submission of the text for printing. I hope that at least some of these projects will be completed in the year 2019 and will appear in the hep-arxiv by the time this book has been printed. Combining them with the present version will extend the book beyond the assigned page limit.

We will encounter many equations on our route. I have put the most important ones in shaded boxes or white frames. There is no strict logic between these two choices. Roughly speaking, a given group of formulas has been collected in a shaded box, while the separate results have been put in white frames. Shaded boxes and white frames appeared already earlier. I hope this procedure makes the presentation more transparent.

We have mentioned two websites earlier. A very useful collection of other relevant websites can be found at the end of Langacker's book [10]. For our expedition, the following three are most important, and we should have them in our smartphones. These are

HFLAV = Heavy Flavour Averages: **http://hflav.web.cern.ch/**
PDG= Particle Data Group: **http://pdg.lbl.gov/**
FLAG = Flavour Lattice Averaging Group: **http://flag.unibe.ch/**

WE ARE NOW READY TO START OUR EXPEDITION!

PART I

BASICS OF GAUGE THEORIES

1 Fundamentals

1.1 Preliminaries

We have mentioned already at the beginning of the book that the fundamental role in elementary particle physics, that is, the SM and its extensions, is played by Lagrangians. They encode the information about the particle content of a given theory and of fundamental interactions between these particles that are characteristic for this theory. Therefore, it is essential to start our presentation by discussing the general structure of various Lagrangians that we will encounter in this book.

The theories we will discuss are relativistic quantum field theories, and it would appear at first sight that this first step of our expedition is extremely difficult. Yet, the seminal observation of Feynman that a given classical theory can be quantized by means of the path integral method simplifies things significantly. We can formulate the quantum field theory with the help of a Lagrangian of a classical field theory without introducing operators as done in canonical quantization. Having it, a simple set of steps allows us to derive the so-called Feynman rules and use them to calculate the implications of a given theory for various observables that can be compared with experiment.

A very important role in particle physics is played by symmetries. They increase significantly the predictive power of a given theory, in particular by reducing the number of free parameters. In this context a very good example is quantum chromodynamics (QCD), the theory of strong interactions. With eight gluons and three colors for quarks, there is a multitude of interactions that, without the SU(3) symmetry governing them, could be rather arbitrary. But the SU(3) symmetry implies certain conservation laws, and at the end there is only a single parameter in QCD: the value of the strong coupling evaluated at some energy scale that can be determined in experiment. Once this is done, all effects of strong interactions can be uniquely predicted, even if this requires often very difficult calculations.

The case of QCD is however special as it is based on an exact nonabelian symmetry. We will be more specific about this terminology later. In quantum electrodynamics (QED), which is based on an exact abelian symmetry U(1), in addition to the value of the electromagnetic coupling also the electric charges of quarks and leptons and generally fermions, scalars, and vector particles in a given theory are free. They have to be determined in experiment. In QCD all *color charges* are fixed by the SU(3) symmetry.

Yet QED, similar to QCD, is a very predictive theory because it is based on an exact symmetry. This is generally not the case in nature, and on many occasions the symmetries that we encounter in particle physics are only approximate, and the manner in which they are broken has an impact on physical implications. Moreover, models with broken symmetries often contain many new free parameters beyond the couplings, significantly lowering the predictive power of the theory. A very prominent example is supersymmetry.

Our goal for the next pages is to write down Lagrangians, in fact Lagrangian densities,[1] for simplest theories involving spin-0 particles (scalars), spin-$\frac{1}{2}$ particles (fermions), and spin-1 particles (vectors or vector bosons). In this context we will discuss various symmetries that we will encounter later at various places in our book.

1.2 Lagrangians for Scalar Fields

1.2.1 Real Scalar Field

Let us consider the real scalar field $\varphi \equiv \varphi(x)$ for which the Lagrangian, neglecting interactions, reads

$$\mathscr{L}(\varphi, \partial_\mu \varphi) = \frac{1}{2} \left(\partial_\mu \varphi \right) \left(\partial^\mu \varphi \right) - \frac{1}{2} m^2 \varphi^2. \tag{1.1}$$

Inserting this Lagrangian into the Euler–Lagrange equation,

$$\frac{\delta \mathscr{L}}{\delta \varphi} = \partial_\mu \frac{\delta \mathscr{L}}{\delta \left(\partial_\mu \varphi \right)}, \tag{1.2}$$

where

$$\delta \mathscr{L}(\varphi) = \mathscr{L}(\varphi + \delta \varphi) - \mathscr{L}(\varphi), \tag{1.3}$$

we find the Klein–Gordon (KG) equation

$$\left(\Box + m^2 \right) \varphi = 0 \tag{1.4}$$

so that the only parameter in (1.1) m can be interpreted as the mass of the spin-0 particle corresponding to the field φ.

1.2.2 Complex Scalar Field

We next promote φ in (1.1) to a complex scalar field. The Lagrangian takes now the following form:

$$\mathscr{L}(\varphi, \partial_\mu \varphi, \varphi^*, \partial_\mu \varphi^*) = \frac{1}{2} \left(\partial_\mu \varphi^* \right) \left(\partial^\mu \varphi \right) - \frac{1}{2} m^2 \varphi^* \varphi. \tag{1.5}$$

---

[1] For simplicity, we will call these Lagrangian densities just Lagrangians.

If φ and φ^* are written as

$$\varphi = \frac{1}{\sqrt{2}} \left(\varphi_1 + i\varphi_2 \right), \qquad \varphi^* = \frac{1}{\sqrt{2}} \left(\varphi_1 - i\varphi_2 \right), \tag{1.6}$$

where $\varphi_{1,2}$ are real, then these real fields satisfy separately the KG equation in (1.4). Equivalently we have

$$\left(\Box + m^2 \right) \varphi = 0, \qquad \left(\Box + m^2 \right) \varphi^* = 0. \tag{1.7}$$

Before continuing, we have to discuss an important topic.

1.3 First Encounter with Symmetries

A symmetry is a transformation on the fields and x, which leaves the Lagrangian invariant. Thus if there is a transformation R:

$$\varphi(x) \xrightarrow{R} \varphi'(x'), \tag{1.8}$$

so that

$$\mathcal{L}(\varphi(x)) \xrightarrow{R} \mathcal{L}(\varphi'(x')) = \mathcal{L}(\varphi(x)), \tag{1.9}$$

then R is a symmetry of the Lagrangian.

The symmetry transformations that we will encounter in this book can be grouped in three classes:

- Continuous transformations in space-time

$$x \to x' + \delta x, \qquad \varphi(x) \to \varphi'(x'). \tag{1.10}$$

These are the Lorentz transformations. We will not discuss them in this book as they are the topic of introductory lectures on field theory. However, we will make sure that our Lagrangians and their implications are consistent with Lorentz invariance. In the Lagrangians in (1.1) and (1.5) this is achieved by contracting the indices μ.

- Continuous internal symmetries

$$\varphi(x) \to \varphi(x) + \delta\varphi(x) = \varphi'(x). \tag{1.11}$$

These symmetries will play a crucial role in our book. One distinguishes between *global* and *local* internal symmetries, and each of them can be either *abelian* or *nonabelian*. What this really means will be explained in detail as we proceed.

- Discrete symmetries

A typical example of a discrete transformation is the flip of the sign of the field φ:

$$\varphi \to -\varphi. \tag{1.12}$$

However, the best-known discrete transformations are

Parity (P): $\qquad\qquad\qquad\qquad$ $t \rightarrow t' = t, \quad \vec{x} \rightarrow \vec{x}' = -\vec{x},$ $\qquad$ (1.13)

Charge conjugation (C): $\qquad\qquad$ $Q \rightarrow -Q,$ $\qquad\qquad\qquad$ (1.14)

Time reversal (T): $\qquad\qquad\quad$ $t \rightarrow t' = -t, \quad \vec{x} \rightarrow \vec{x}' = \vec{x}.$ $\qquad$ (1.15)

As we will see, QED and QCD interactions are invariant under these three transformations, while this is not the case of weak interactions.

1.4 Checking the Symmetries of Scalar Lagrangians

Armed with this elementary knowledge of symmetries, we can now investigate which internal symmetries are present in the Lagrangians in (1.1) and (1.5). Specifically, let us ask whether these Lagrangians are invariant under the following transformations:

$i)$ $\qquad$ $\varphi(x) \rightarrow \varphi'(x) = -\varphi(x)$ $\qquad$ (discrete), $\qquad\qquad$ (1.16)

$ii)$ $\qquad$ $\varphi(x) \rightarrow \varphi'(x) = e^{i\theta r}\varphi(x)$ $\qquad$ (continuous global), $\qquad$ (1.17)

$iii)$ $\qquad$ $\varphi(x) \rightarrow \varphi'(x) = e^{i\theta(x)r}\varphi(x)$ $\qquad$ (continuous local), $\qquad$ (1.18)

where θ is a phase that is either independent of x (global transformation) or dependent on x (local transformation). The parameter r is introduced to characterize a property of the field under this transformation, not the transformation itself, and can be interpreted as the conserved "charge," as we will see later on.

The Lagrangian in (1.1) is clearly invariant under the discrete transformation in (1.16) but fails completely with respect to the transformations in (1.17) and (1.18) as

$$\varphi^2 \rightarrow e^{2i\theta r}\varphi^2. \qquad\qquad (1.19)$$

This is not surprising. The transformations in (1.17) and (1.18) promoted the real field to a complex field and thus changed the nature of the field. Because the Lagrangian for a complex field in (1.5) looks different than the Lagrangian in (1.1), it is not surprising that (1.1) is not invariant under (1.17) and (1.18).

Yet there is a solution to this problem. We just set $r = 0$ so that φ does not transform at all. One says it is a *singlet* under transformation (1.17) and (1.18), and its charge r vanishes.

We next investigate the Lagrangian (1.5) to find that indeed it is invariant under the global transformation (1.17) as

$$\varphi \rightarrow e^{i\theta r}\varphi, \qquad \varphi^* \rightarrow \varphi^* e^{-i\theta r} \qquad\qquad (1.20)$$

leaves (1.5) unchanged provided θ is independent of x. In a more group theoretical language the phase transformation in (1.20) is the simplest unitary transformation related to the group U(1). The r can then be interpreted as a conserved charge. Indeed, the sign in front of r in case of φ^* is opposite to the one in the transformation of φ: the charges of φ

and φ^* differ by sign and as known from elementary relativistic field theory φ^* represents the antiparticle to φ.

What about the last transformation (1.18) on our list? The last term in (1.5) is clearly invariant under (1.18), but the first one is not! Indeed

$$\partial_\mu \varphi' = \partial_\mu \left(e^{i\theta(x)r} \varphi(x) \right) = e^{i\theta(x)r} \partial_\mu \varphi(x) + ir \partial_\mu \theta(x) \left(e^{i\theta(x)r} \varphi(x) \right), \qquad (1.21)$$

$$\partial_\mu \varphi'^* = \partial_\mu \left(\varphi(x)^* e^{-i\theta(x)r} \right) = \partial_\mu \varphi(x)^* e^{-i\theta(x)r} - ir \partial_\mu \theta(x) \left(\varphi(x)^* e^{-i\theta(x)r} \right). \qquad (1.22)$$

Inserting these expressions into (1.5) we readily verify that the terms proportional to $\partial_\mu \theta(x)$ break the symmetry. Thus (1.5) is not invariant under the transformation (1.18), the local U(1) transformation.

There is no way out. We have to modify (1.5) to make it invariant under (1.18), or in other words we have to promote the global U(1) symmetry present already in (1.5) to a local U(1) symmetry.

There is a well-known procedure for how to perform this task with a very remarkable result. The requirement of a local U(1) symmetry implies automatically the existence of a new particle with spin-1 and a specific structure of the interaction of this new particle with the original complex field φ that we introduced from the start. Let us present this procedure that consists of four steps. We refrain from profound geometrical interpretations of transformations encountered here, as they can be found in many textbooks on field theory.

1.5 Promotion of a Global U(1) Symmetry to a Local U(1) Symmetry

- **Step 1**

    We introduce a vector particle A_μ to be called *gauge boson* in what follows. It is not surprising that we have to introduce a vector field $A_\mu(x)$. After all, we have to cancel terms involving $\partial_\mu \theta(x)$.

- **Step 2**

    We replace the derivative ∂_μ by a *covariant derivative* D_μ, which transforms under U(1) as the fields φ and φ^*:

$$D_\mu \varphi \to e^{i\theta(x)r} D_\mu \varphi, \qquad \left(D_\mu \varphi \right)^* \to \left(D_\mu \varphi \right)^* e^{-i\theta(x)r}. \qquad (1.23)$$

It is given by

$$D_\mu \varphi = \left(\partial_\mu - irg A_\mu \right) \varphi, \qquad \left(D_\mu \varphi \right)^* = \left(\partial_\mu + irg A_\mu \right) \varphi^*, \qquad (1.24)$$

where g is a real parameter, the gauge coupling characterizing the strength of the interaction, and r can be again interpreted as the charge of φ. It is not necessarily an electric charge but a charge related to a given local U(1) symmetry.

- **Step 3**

    In order to satisfy (1.23) also A_μ has to transform under U(1) in a special manner:

$$A_\mu \to A_\mu + \frac{1}{g} \partial_\mu \theta(x). \qquad (1.25)$$

- **Step 4**

For A_μ to be interpreted as a physical particle, a kinetic term describing the motion of this new particle has to be added to the original Lagrangian. This term must be invariant under the transformation (1.25). This additional term is familiar from QED:

$$\Delta\mathscr{L} = -\frac{1}{4}F_{\mu\nu}F^{\mu\nu}, \qquad F_{\mu\nu} = \partial_\mu A_\nu - \partial_\nu A_\mu. \tag{1.26}$$

The resulting Lagrangian takes the form

$$\mathscr{L}_{\text{gauged}} = \left(D_\mu\varphi\right)^*\left(D^\mu\varphi\right) - m^2\varphi^*\varphi - \frac{1}{4}F_{\mu\nu}F^{\mu\nu}. \tag{1.27}$$

One can easily verify that it is invariant under the following set of transformations:

$$\varphi(x) \to e^{i\theta(x)r}\varphi(x), \qquad \varphi^*(x) \to \varphi^*(x)e^{-i\theta(x)r} \tag{1.28}$$

for scalar fields and (1.25) for A_μ.

The requirement that the Lagrangian is invariant under such a local symmetry is called gauge principle. It is very restrictive for the Lagrangian, and some consequences are discussed in the next section. Nowadays gauge invariance is not just a property of the Lagrangian but rather the fundamental principle that determines the structure of the Lagrangian.

1.6　A Closer Look at the U(1) Gauge Theory

The Lagrangian in (1.27) has certain properties that we would like to emphasize here.

While the scalar particle has a mass term consistent with the U(1) gauge symmetry, the corresponding mass term for the gauge boson,

$$(\Delta\mathscr{L})_{\text{mass}}^{\text{gauge}} = \frac{M^2}{2}A_\mu A^\mu, \tag{1.29}$$

is clearly not invariant under (1.25) and is absent in (1.27). Consequently, A_μ is a massless gauge boson. The most prominent example of such a gauge boson is the photon.

The imposition of (local) gauge symmetry implies particular structure of the interactions between φ and A_μ. This structure can be found by decomposing (1.27) into a free and interacting Lagrangian:

$$\mathscr{L}_{\text{gauged}} = \mathscr{L}_{\text{free}} + \mathscr{L}_{\text{int}}, \tag{1.30}$$

where

$$\mathscr{L}_{\text{free}} = \partial_\mu\varphi^*\partial^\mu\varphi - m^2\varphi^*\varphi - \frac{1}{4}F_{\mu\nu}F^{\mu\nu}. \tag{1.31}$$

$$\mathscr{L}_{\text{int}} = -irgA_\mu\varphi\partial^\mu\varphi^* + irgA_\mu\varphi^*\partial^\mu\varphi + g^2r^2A_\mu A^\mu\varphi^*\varphi. \tag{1.32}$$

$\mathscr{L}_{\text{free}}$ describes the propagation of the fields φ and A_μ. These propagations can be represented by simple lines (propagators) for which mathematical expressions can be found. In the case at hand these propagators are given in momentum space as follows

$$\begin{array}{c} \varphi \\ \text{------}\blacktriangleright\text{------} \\ k \end{array} \quad = \frac{i}{k^2 - m^2 + i\varepsilon}$$

$$\begin{array}{c} A_\mu \\ \sim\sim\sim\sim\sim \\ k \end{array} \quad = -\frac{ig_{\mu\nu}}{k^2 + i\varepsilon}$$

and what we have written down are the simplest Feynman rules with k_μ the four-momentum of the propagating particle. We do not derive these rules, as explicit derivations can be found in textbooks on quantum field theory, see, for instance [11].

In $\mathscr{L}_{\text{int}}$ all terms involve A_μ, φ, and φ^* at one point x, and these terms describe simply the local interactions between the particles in question. Then g is the gauge coupling describing the strength of the interaction and r the charge of φ.

With the help of path integral methods one can derive Feynman rules for the interactions in (1.32). In fact, these rules can be read of from (1.32) by simply multiplying $\mathscr{L}_{\text{int}}$ by i and replacing $i\partial_\mu$ by k_μ. We find then for the vertices representing the first and the last term in (1.32):

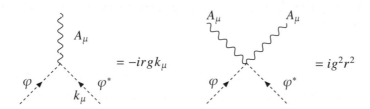

$$= -irgk_\mu \qquad\qquad = ig^2 r^2$$

and $irgk_\mu$ for the second term. From the propagator and vertices, Feynman diagrams can be constructed. Two examples are given in Figure 1.1. The first one represents the scattering of φ and φ^* with exchange of a photon, the second one the annihilation of these two fields into the photon followed by their regeneration.

Finally, we note that there are no vertices involving A_μ only. This is typical for a U(1) symmetry, which is an abelian symmetry: The gauge boson related to this symmetry carries no charge, and consequently there are no interactions between A_μ. In fact, in the absence of the field φ a theory based on a local U(1) symmetry is a free theory.

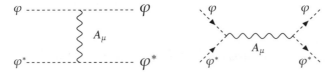

Figure 1.1   Feynman diagrams for a gauged U(1) theory for scalars where φ is a scalar particle and A_μ the gauge boson.

On the other hand, pure interactions between the fields φ and φ^* consistent with U(1) symmetry can be introduced by adding the following term to $\mathscr{L}_{\text{int}}$:

$$\Delta\mathscr{L}_{\text{int}} = -\frac{1}{4}\lambda\,(\varphi^*\varphi)^2 \tag{1.33}$$

with λ describing the strength of this interaction. We will encounter this term elsewhere in this book.

1.7 Nonabelian Global Symmetries

1.7.1 General Considerations

So far we have considered a very simple theory that contained the scalar particle φ, its anti-particle φ^* and one gauge boson A_μ. In general, theories contain several fields $\varphi_1,\ldots\varphi_N$ and also several gauge bosons A_μ^a, where the index a distinguishes between different gauge bosons. If there are symmetries that involve these particles simultaneously, then the corresponding transformations of the fields are more complicated, and also the structure of the Lagrangian is modified relative to the U(1) theory considered until now.

To be able to efficiently discuss such theories, we have to recall certain elements of group theory. The most common groups encountered in particle physics are U(N), SU(N), and SO(N) groups:

- U(N): group of unitary $N \times N$ matrices,
- SU(N): as U(N) but with detU = 1,
- SO(N): group of orthogonal $N \times N$ matrices with determinant 1.

Here we will discuss only U(N) and SU(N) symmetries and postpone the discussion of orthogonal transformations until later.

While the U(1) transformation on a given field had the simple form of just multiplication by a phase factor

$$U(\theta) = e^{i\theta}, \tag{1.34}$$

in the case of U(N) and SU(N) transformations we have

$$U(\theta_1,\ldots\theta_m) = e^{i\sum_{a=1}^m \theta^a T^a}, \tag{1.35}$$

where θ^a are the parameters of the group and T^a the corresponding generators. We have $m = N^2$ and $m = N^2 - 1$ for U(N) and SU(N), respectively. The transformations $U(\theta_1,\ldots\theta_m)$ with θ_i being real satisfy the relations

$$U(\theta_1,\ldots\theta_m) \cdot U(\theta_1',\ldots\theta_m') = U(\theta_1'',\ldots\theta_m''), \tag{1.36}$$

where θ_i'' is analytic in θ_j and θ_k'. The T^a are the generators of SU(N) and satisfy commutation relations

$$\left[T^a, T^b\right] = if^{abc}T^c, \tag{1.37}$$

with f^{abc} being the group structure constants.[2] They are simply numbers that for a given group can be found in any book on group theory but I can also recommend the review in [12]. In practice summations over indices a, b, c like the one over c in (1.37) are performed, and we do not have to remember the values of f^{abc}, except that they vanish for two indices being equal. Useful formulas involving f^{abc} will be presented in subsequent chapters, when we will start doing explicit calculations. The fact that generally $f^{abc} \neq 0$ expresses the nonabelian nature of the transformation: T^a and T^b do not commute, and consequently the order of two unitary transformations matters. This fact has profound physical implications, in particular when the global nonabelian symmetry is promoted to the local one. Every set of $N^2 - 1$ $M \times M$ matrices T_r^a that fulfill (1.37) generates a representation r of the Lie algebra. The index r in T_r^a labels the representation. For example, for the fundamental representation of SU(N) one writes $r = \mathbf{N}$ and for the adjoint $r = G$. By an unitary transformation of T_r^a, every representation of SU(N) can be such that all T_r^a are block diagonal:

$$
T_r^a = \begin{pmatrix} T_r^{a(1)} & 0 & \cdots & 0 \\ 0 & T_r^{a(2)} & \ddots & \vdots \\ \vdots & \ddots & \ddots & 0 \\ 0 & \cdots & 0 & T_r^{a(n)} \end{pmatrix}.
\tag{1.38}
$$

If there is only one block ($n = 1$), then the representation is called irreducible.

In contrast to U(1) transformations, which acted separately on each field φ, nonabelian transformations act simultaneously on a set of fields that from theory group point of view are the basis vectors of a given irreducible representation. These basis vectors denote a set of quantum mechanical states and are said to constitute a multiplet: doublet, triplet, octet, decouplet, etc. The transformations of the group transform a given field φ_i into linear combinations of the fields belonging to a given multiplet.

The important group theoretical property of a nonabelian symmetry is the existence of multiplets of a size characteristic for a given symmetry. For instance, while a doublet is the smallest multiplet that transforms nontrivially in the case of SU(2) symmetry, in the case of SU(3), it is a triplet. A singlet is of course always smaller. Moreover, while in the case of SU(3), triplets, sextets, octets, and decouplets and specific larger multiplets are present, a quartet or fiveplet is not possible in the case of SU(3).

This discussion shows that if already discovered particles do not fill out the full multiplet of a given nonabelian symmetry, this symmetry predicts the existence of new particles necessary to complete the multiplet in question. This was the case of Ω that was the missing member of the baryon decouplet of the global flavor SU(3) proposed by Gell–Mann or the case of the charm quark in the SU(2) doublet involving also the strange quark.

---

[2] One can also say that the T^a are the elements of a Lie algebra that generate the Lie algebra (1.37). The number of linear independent generators T^a is the dimension of the Lie algebra or Lie group. A Lie group is called nonabelian if at least one f^{abc} is nonvanishing.

If we restrict our attention first to the simplest multiplets of the SU(2) and SU(3) groups, we can represent the doublets and triplets by column vectors

$$
\vec{\varphi} = \begin{pmatrix} \varphi_1 \\ \vdots \\ \varphi_N \end{pmatrix}, \qquad \vec{\varphi}^\dagger = \left(\varphi_1^\dagger, \dots \varphi_N^\dagger \right), \tag{1.39}
$$

with $N = 2$ and $N = 3$ for SU(2) and SU(3), respectively. The SU(N) transformations of $\vec{\varphi}$ and $\vec{\varphi}^\dagger$ read

$$
\vec{\varphi}' = \exp\left(i\theta^a T^a\right) \vec{\varphi}, \tag{1.40}
$$

$$
\vec{\varphi}^{\dagger\prime} = \vec{\varphi}^\dagger \exp\left(-i\theta^a T^a\right), \tag{1.41}
$$

where we have used $T^{a\dagger} = T^a$ and summation over $a = 1, \dots N^2 - 1$ is understood.

The generators T^a can be represented by Hermitian $N \times N$ matrices, which we recall here for completeness for $N = 2$ and $N = 3$:

$$
T^a = \begin{cases} \sigma^a/2 & a = 1, 2, 3 & \text{SU(2)}, \\ \lambda^a/2 & a = 1, \dots 8 & \text{SU(3)} \end{cases}. \tag{1.42}
$$

Here σ^a are the Pauli matrices and λ^a satisfying

$$
\text{Tr}(\lambda^a \lambda^b) = 2\delta^{ab} \tag{1.43}
$$

are Gell–Mann matrices. Their explicit expressions are given in Appendix A.2. These matrices are Hermitian and traceless, which follows from the unitarity of the transformations U and the requirement that in the case of SU(N), detU = 1:

$$
\det\left(e^{iT}\right) = e^{i\,\text{Tr}(T)} = 1 \quad \Rightarrow \quad \text{Tr}(T) = 0. \tag{1.44}
$$

As

$$
U(N) = SU(N) \otimes U(1), \tag{1.45}
$$

a U(N) symmetry has an additional generator that is represented by a unit matrix.

We have used here the words *representation* and *multiplet*. As a given irreducible representation of a given group implies automatically the size of a multiplet, both names are used often in particle physics to denote a multiplet. We will follow this terminology here as well.

The so-called adjoint representation is important for particle physics because the gauge bosons A_μ^a of a SU(N) gauge theory belong to it. In the case of SU(3)$_C$, with the subscript C standing for color, we have eight gluons and in the case of SU(2)$_L$ three weak gauge bosons. The generators of the adjoint representation are simply given by the structure constants of the group. One can check that with $(T_r^b)_{ac} := i f^{abc}$ the commutation relation in (1.37) is fulfilled. The complex conjugated representation $\bar{r}$ to a given representation r is generated by

$$
T_{\bar{r}}^a := -T_r^{a*}. \tag{1.46}
$$

In the case of the fundamental representation of SU(N), which we denote by **N**, the complex conjugated representation is denoted by **N̄**. For example quarks transform as triplets **3** under $SU(3)_C$ while their antiparticles transform as **3̄**. If two representations r and $\bar{r}$ are equivalent, i.e., there exists a unitary matrix U such that $T_{\bar{r}}^a = U T_r^a U^\dagger$, then we call r real. For example, all representations of SU(2), especially the fundamental representation **2**, are real, whereas **3**, the fundamental representation of SU(3), is not. A consequence is that something like an "anti(iso)spin," related to SU(2), does not exist, whereas antiquarks have anticolors. Particles that are their own antiparticles transform under real representations of the symmetry group.

1.7.2 Lagrangian and First Implications

The Lagrangian invariant under the transformations (1.40) and (1.41) is given as follows

$$\mathscr{L} = \left(\partial_\mu \vec{\varphi}^\dagger \right) \left(\partial^\mu \vec{\varphi} \right) - \vec{\varphi}^\dagger \mathsf{M}^2 \vec{\varphi}, \tag{1.47}$$

where M^2 is the mass matrix (squared) of the field $\vec{\varphi}$. The fields in (1.47) belong to a single multiplet. If there are several multiplets, for each of them (1.47) applies except that M^2 could be different for different multiplets. There is an immediate consequence of a nonabelian symmetry. The masses of particles belonging to a given multiplet must be degenerate. This means that M^2 must be a unit matrix multiplied by m^2. In order to see this, we perform the transformations (1.40) and (1.41) on the last term in (1.47):

$$\vec{\varphi}^\dagger \mathsf{M}^2 \vec{\varphi} \rightarrow \vec{\varphi}^\dagger e^{-iT^a \theta^a} \mathsf{M}^2 e^{iT^a \theta^a} \vec{\varphi} \tag{1.48}$$

and find that this term is invariant under these transformations if and only if M^2 commutes with all generators T^a (Schur's lemma):

$$\left[T^a, \mathsf{M}^2 \right] = 0. \tag{1.49}$$

This is only possible if M^2 is proportional to the unit matrix.

1.7.3 Explicit Breakdown of a Global Nonabelian Symmetry

In reality the masses of physical particles belonging to a given multiplet are not exactly equal to each other even if the mass splittings could be very small. This means that whereas in this case the first term in (1.47) is still invariant under global nonabelian transformation, the mass term in this equation is not. This type of symmetry breaking is called explicit because it takes place at the level of the Lagrangian. In contrast, in Section 1.10.4 we will discuss spontaneous symmetry breaking where the Lagrangian is still symmetric but only the ground state breaks the symmetry. It is instructive to consider two examples of explicit symmetry breaking: one of SU(2) symmetry, the other of SU(3). In the SU(2) case let us assume that

$$\mathsf{M}^2 = \begin{pmatrix} m_1^2 & 0 \\ 0 & m_2^2 \end{pmatrix}, \qquad m_1 \neq m_2. \tag{1.50}$$

Inserting this matrix into (1.48) with $T^a = \sigma^a/2$ we find

$$\left[\sigma^{1,2}, M^2\right] \neq 0, \qquad \left[\sigma^3, M^2\right] = 0. \tag{1.51}$$

Evidently only one generator commutes with M^2, and the symmetry has been reduced from SU(2) down to U(1):

$$SU(2) \rightarrow U(1), \tag{1.52}$$

with $T^3 = \sigma^3/2$ playing the role of the generator of the leftover U(1) symmetry.

One easily finds that the nonvanishing commutator in (1.51) is proportional to $m_1^2 - m_2^2$ and if this difference is small the breaking of symmetry can be regarded as a small perturbation. Now the disparity of masses in the mass matrix led to the breakdown of SU(2) to a U(1) symmetry and so at first sight the situation looks like in an abelian U(1) theory: the fields φ_1 and φ_2 are unrelated to each other. However not everything has been lost and the presence of the initial SU(2) symmetry left footprints which we would like to identify. Indeed the charges of the fields φ_1 and φ_2 with respect to the leftover U(1) group are not independent of each other, which would not be the case if φ_1 and φ_2 did not sit in an SU(2) doublet together. In order to see this, we note that the leftover U(1) symmetry is summarized by

$$\vec{\varphi} \rightarrow \exp\left(i\frac{\sigma_3}{2}\theta^{(3)}\right)\vec{\varphi}, \qquad \vec{\varphi} = \begin{pmatrix} \varphi_1 \\ \varphi_2 \end{pmatrix}, \tag{1.53}$$

or equivalently

$$\varphi_1 \rightarrow \exp\left(i\frac{1}{2}\theta^{(3)}\right)\varphi_1, \qquad \varphi_2 \rightarrow \exp\left(-i\frac{1}{2}\theta^{(3)}\right)\varphi_2. \tag{1.54}$$

Comparing with the U(1) transformation in (1.20) we note that the charges r are fixed and related to each other. Denoting this quantum number by T_3 we have

$$T_3(\varphi_1) = \frac{1}{2}, \qquad T_3(\varphi_2) = -\frac{1}{2}. \tag{1.55}$$

This result is not surprising. T_3 is the third component of the isospin.

This consideration can be extended to SU(3) by choosing for the mass matrix

$$M^2 = \begin{pmatrix} m^2 & 0 & 0 \\ 0 & m^2 & 0 \\ 0 & 0 & m_3^2 \end{pmatrix}. \tag{1.56}$$

Inserting this matrix into (1.48) with $\vec{\varphi}$ being this time three-dimensional vector, we find that λ^1, λ^2, λ^3, and λ^8 commute still with M^2, while this is not the case of λ^4, λ^5, λ^6, and λ^7. Thus the symmetry has been reduced as follows

$$SU(3) \rightarrow SU(2) \otimes U(1), \tag{1.57}$$

with $(\lambda_1, \lambda_2, \lambda_3)$ being the generators of the leftover SU(2) and λ_8 of U(1). Writing then

$$\vec{\varphi} \to \exp\left(i\frac{\lambda_8}{2}\theta^{(8)}\right)\vec{\varphi}, \qquad \vec{\varphi} = \begin{pmatrix} \varphi_1 \\ \varphi_2 \\ \varphi_3 \end{pmatrix}, \tag{1.58}$$

with λ_8 given in (A.21), we find the charges of $\varphi_{1,2,3}$ with respect to the leftover U(1) symmetry:

$$T_8(\varphi_1) = T_8(\varphi_2) = \frac{1}{2\sqrt{3}}, \qquad T_8(\varphi_3) = -\frac{1}{\sqrt{3}}. \tag{1.59}$$

Note that the T_8 charges of φ_1 and φ_2 are equal to each other as these two fields form a doublet under the leftover SU(2) symmetry.

1.7.4 More Complicated Global Symmetries

In general there can be several multiplets. For instance, three doublets

$$\vec{\varphi}_A = \begin{pmatrix} \varphi_1 \\ \varphi_2 \end{pmatrix}, \qquad \vec{\varphi}_B = \begin{pmatrix} \varphi_3 \\ \varphi_4 \end{pmatrix}, \qquad \vec{\varphi}_C = \begin{pmatrix} \varphi_5 \\ \varphi_6 \end{pmatrix}. \tag{1.60}$$

The SU(2) symmetric Lagrangian is given then by

$$\mathscr{L} = \sum_{s=A,B,C} \left(\partial_\mu \vec{\varphi}_s^\dagger\right)\left(\partial^\mu \vec{\varphi}_s\right) - \sum_{s=A,B,C} m_s^2 \vec{\varphi}_s^\dagger \vec{\varphi}_s, \tag{1.61}$$

with the mass m_s different for different doublets.

These examples should be sufficient to illustrate the basic features of global nonabelian symmetries involving spinless particles. Some new features appear in the case of fermions, as the left-handed fermions can generally transform differently than the right-handed ones. We will discuss this issue soon.

1.8 Promotion of a Global Nonabelian Symmetry to a Local One

In Section 1.5 we have made this promotion for U(1) finding that this required the introduction of a gauge boson that was massless and neutral with respect to the conserved charge connected with the U(1) symmetry. We now want to see what happens when a nonabelian global symmetry becomes a local symmetry, or in short it is gauged.

In order to have a transparent discussion let us just consider SU(N) symmetry. The transformations in (1.40) and (1.41) become local transformations

$$\vec{\varphi}'(x) = \exp\left(i\theta^a(x)T^a\right)\vec{\varphi}(x), \tag{1.62}$$

$$\vec{\varphi}^{\dagger\prime}(x) = \vec{\varphi}^\dagger(x)\exp\left(-i\theta^a(x)T^a\right), \tag{1.63}$$

and we find that the Lagrangian (1.47) is not invariant under these transformations due to the appearance of the terms $\partial_\mu\theta^a(x)$. These terms have to be cancelled, and this requires the introduction of new particles, gauge bosons, one for every $\partial_\mu \theta^a(x)$ or equivalently one for every generator of SU(N). The four steps of Section 1.5 can also be made here. Only formulas are more complicated and, as we will see the resulting dynamics, differs profoundly from the one of an abelian gauge theory. The four steps in the nonabelian case are then as follows:

- **Step 1**

  We introduce a vector particle A_μ^a for every generator T^a. Thus in the case of the SU(N) group the local gauge invariance requires the existence of $N^2 - 1$ gauge bosons.
- **Step 2**

  We replace the derivatives $\partial_\mu \vec{\varphi}$ and $\partial_\mu \vec{\varphi}^\dagger$ by covariant derivatives

$$D_\mu \vec{\varphi} = \left(\partial_\mu - ig A_\mu^a T^a\right) \vec{\varphi}, \qquad \left(D_\mu \vec{\varphi}\right)^\dagger = \vec{\varphi}^\dagger \left(\partial_\mu + ig A_\mu^a T^a\right), \tag{1.64}$$

  with g denoting the gauge coupling corresponding to the SU(N) group.
- **Step 3**

  The transformation for the gauge fields is given by

$$A_\mu^a \to A_\mu^{a\prime} = A_\mu^a + \frac{1}{g}\partial_\mu \theta^a(x) - f^{abc}\theta^b(x)A_\mu^c \tag{1.65}$$

  with f^{abc} being the structure constants introduced in (1.37).
- **Step 4**

  The gauge invariant strength tensor corresponding to A_μ^a is given as follows

$$F_{\mu\nu}^a = \partial_\mu A_\nu^a - \partial_\nu A_\mu^a + g f^{abc} A_\mu^b A_\nu^c. \tag{1.66}$$

The resulting Lagrangian that is invariant under the transformations (1.62), (1.63), and (1.65) is finally given as follows:

$$\mathscr{L}_{\text{gauged}} = \left(D_\mu \vec{\varphi}\right)^\dagger (D^\mu \vec{\varphi}) - m^2 \vec{\varphi}^\dagger \vec{\varphi} - \frac{1}{4}F_{\mu\nu}^a F^{\mu\nu,a}. \tag{1.67}$$

At this point we want to mention that there exist at least two different conventions in the literature, which results in some sign flips. More details on conventions can be found in Appendix B. Here we only discuss two of them. If the transformation in (1.62) has opposite sign in the exponential, then also the sign in front of g in the covariant derivative has to be changed. We collect here the differences between these two conventions

- Convention 1 (used by us)

$$\varphi(x) \to \varphi(x)' = \exp\left(i\theta^a(x)T^a\right)\varphi(x) \tag{1.68}$$

$$D_\mu = \partial_\mu - ig A_\mu^a T^a \tag{1.69}$$

$$A_\mu^a(x) \to A_\mu^{a\prime}(x) = A_\mu^a(x) + \frac{1}{g}\partial_\mu \theta^a(x) - f^{abc}\theta^b(x)A_\mu^c \tag{1.70}$$

$$F_{\mu\nu}^a = \partial_\mu A_\nu^a - \partial_\nu A_\mu^a + g f^{abc} A_\mu^b A_\nu^c \tag{1.71}$$

- Convention 2

$$\varphi(x) \rightarrow \varphi(x)' = \exp\left(-i\theta^a(x)T^a\right)\varphi(x) \tag{1.72}$$

$$D_\mu = \partial_\mu + ig A_\mu^a T^a \tag{1.73}$$

$$A_\mu^a(x) \rightarrow A_\mu^{a\prime}(x) = A_\mu^a(x) + \frac{1}{g}\partial_\mu\theta^a(x) + f^{abc}\theta^b(x)A_\mu^c \tag{1.74}$$

$$F_{\mu\nu}^a = \partial_\mu A_\nu^a - \partial_\nu A_\mu^a - g f^{abc} A_\mu^b A_\nu^c \tag{1.75}$$

Conventions 1 and 2 can be transformed into each other by flipping simultaneously the signs of θ^a and g.

Let us then list the properties of the Lagrangian in (1.67) paying in particular attention to those properties that distinguish it from the one in (1.27):

- The appearance of several gauge bosons, one for each generator of the symmetry group.
- All these gauge bosons are massless as the mass term

$$\Delta\mathscr{L}_{\text{mass}} = \frac{M^2}{2}\sum_a A_\mu^a A^{\mu,a} \tag{1.76}$$

is clearly not invariant under (1.65), even if all these gauge bosons were degenerate in mass. Thus the exact nonabelian gauge symmetry SU(3) of strong interactions implies that all eight gluons are massless. On the other hand, this is a problem for heavy $W^{\mu\pm}$ and Z^μ bosons. Indeed, the gauge bosons were introduced to be able to change the phase of the wave function of all particles independently at each space-time point without any observable consequence. Thus the gauge boson has to reconcile such phase changes over arbitrary large distances. A force with an infinite range is associated with a massless gauge boson. However, the weak force is short-ranged due to the masses of $W^{\mu\pm}$ and Z^μ bosons. We will address this problem in Sections 1.10 and 1.11.

- In order to see the structure of the interactions present in (1.67), we separate the free Lagrangian

$$\mathscr{L}_{\text{free}} = \partial_\mu\vec{\varphi}^\dagger\partial^\mu\vec{\varphi} - m^2\vec{\varphi}^\dagger\vec{\varphi} - \frac{1}{4}\left(\partial_\mu A_\nu^a - \partial_\nu A_\mu^a\right)\left(\partial^\mu A^{\nu,a} - \partial^\nu A^{\mu,a}\right). \tag{1.77}$$

The interactions of $\vec{\varphi}$ with gauge bosons are then described by

$$\mathscr{L}_{\text{int}}^{(1)} = -ig\partial_\mu\vec{\varphi}^\dagger T^a\vec{\varphi}A^{\mu,a} + ig\vec{\varphi}^\dagger T^a\partial_\mu\vec{\varphi}A^{\mu,a} + g^2\vec{\varphi}^\dagger T^a T^b\vec{\varphi}A_\mu^a A^{\mu,b}, \tag{1.78}$$

where summation over repeated indices is understood. It is convenient to rewrite this Lagrangian in the component form ($i, j = 1, \ldots N$)

$$\mathscr{L}_{\text{int}}^{(1)} = -ig\partial_\mu\varphi_i^\dagger T_{ij}^a\varphi_j A^{\mu,a} + ig\varphi_i^\dagger T_{ij}^a\partial_\mu\varphi_j A^{\mu,a} + g^2\varphi_i^\dagger\left(T^a T^b\right)_{ij}\varphi_j A_\mu^a A^{\mu,b}. \tag{1.79}$$

Using the matrix representations for the generators T^a, like the ones in terms of σ^a and λ^a in (1.42), it is an easy matter to derive the Feynman rules for the relevant vertices. One just uses $k_\mu = i\partial_\mu$, multiplies $\mathscr{L}_{\text{int}}^{(1)}$ by i, and simply reads of the coefficients in front of the products of fields. While at first sight, in view of $i, j = 1, \ldots N$, these interactions look rather complicated, they have a very simple structure due to the SU(N) symmetry. We will exhibit this in a moment. Moreover the indices i, j, the colors in the case of

SU(3)$_C$ for the strong interactions, are seldom seen in practical calculations as one can use "color algebra," which we will develop in Section 2.7.

The remarkable facts about these interactions are

- there is only a single coupling g describing them,
- all "nonabelian" charges entering these interactions are fixed by the symmetry. They are simply given in terms of the elements of the matrices T^a.

- Finally, we discuss the last term in (1.67), which involves the gauge bosons only. We find two types of interactions

$$\mathscr{L}^{(2)}_{\text{int}} = -\frac{1}{4} g f^{abc} \left(\partial_\mu A^a_\nu \right) A^{\mu,b} A^{\nu,c} + \cdots \tag{1.80}$$

$$\mathscr{L}^{(3)}_{\text{int}} = -\frac{1}{4} g^2 f^{abd} f^{ade} A^b_\mu A^c_\nu A^{\mu,d} A^{\nu,e}. \tag{1.81}$$

Evidently $\mathscr{L}^{(2)}_{\text{int}}$ and $\mathscr{L}^{(3)}_{\text{int}}$ describe triple and quartic gauge boson vertices, which again are fully fixed by the symmetry up to the coupling g.

The presence of such couplings is a very important difference from the case of abelian gauge theories, which are free theories in the absence of matter fields φ_i. The nonabelian theories are interacting even when the matter fields are absent. This is because the gauge bosons A^a_μ carry "color charges." The presence of interactions between gauge bosons in nonabelian theories has very profound dynamical implications to which we will return at several places in this book.

1.9 Lagrangians for Fermions

1.9.1 Preliminaries

Until now we have considered only scalars as matter fields. Even if the Higgs particle has been discovered recently, most of the known elementary particles are fermions: quarks and leptons. In order to incorporate them into the framework presented until now, we have to construct Lagrangians involving spin-$\frac{1}{2}$ particles that possess global and local, abelian and nonabelian symmetries. To this end it will be useful to recall some of the important properties of fermions that distinguish them from scalars. In this context we follow the conventions of Bjorken and Drell [13]. See also [14]. They differ from Weinberg's well-known book [15].

A fermion is described by a four-component spinor and its adjoint:

$$\psi, \qquad \bar{\psi} \equiv \psi^\dagger \gamma^0, \tag{1.82}$$

where γ^0 is one of the Dirac matrices. There are different representations of these matrices, which can be found in many textbooks. In particular we have

$$\gamma^0 = \begin{pmatrix} \mathbb{1} & 0 \\ 0 & -\mathbb{1} \end{pmatrix}, \qquad [\gamma^i] = \vec{\gamma} = \begin{pmatrix} 0 & \vec{\sigma} \\ -\vec{\sigma} & 0 \end{pmatrix}. \tag{1.83}$$

Here $\vec{\sigma} = \left[\sigma^1, \sigma^2, \sigma^3 \right]$ are 2×2 Pauli matrices given in Appendix A.2, and $\mathbb{1}$ stands for the 2×2 unit matrix. We will use

$$\gamma^\mu \quad (\mu = 0, 1, 2, 3), \qquad \gamma_5 = \begin{pmatrix} 0 & \mathbb{1} \\ \mathbb{1} & 0 \end{pmatrix} = i\gamma^0 \gamma^1 \gamma^2 \gamma^3 \tag{1.84}$$

with

$$\{\gamma^\mu, \gamma^\nu\} = 2g^{\mu\nu}, \qquad \{\gamma_5, \gamma^\mu\} = 0. \tag{1.85}$$

Note that

$$\gamma_5^\dagger = \gamma_5, \qquad \gamma_5^2 = \mathbb{1}. \tag{1.86}$$

Some other properties of Dirac matrices are collected in Appendix A, and the full set can be found in any textbook for quantum field theory, in particular in [13, 14].

We next introduce left-handed (LH) and right-handed (RH) fermion fields

$$\psi_L = \frac{1}{2} (1 - \gamma_5) \psi \equiv P_L \psi, \qquad \psi_R = \frac{1}{2} (1 + \gamma_5) \psi \equiv P_R \psi. \tag{1.87}$$

Their adjoints are given by

$$\bar{\psi}_L = \bar{\psi} \frac{1}{2} (1 + \gamma_5) = \bar{\psi} P_R, \qquad \bar{\psi}_R = \bar{\psi} \frac{1}{2} (1 - \gamma_5) = \bar{\psi} P_L. \tag{1.88}$$

Consequently, using

$$\psi = \psi_L + \psi_R, \qquad \bar{\psi} = \bar{\psi}_L + \bar{\psi}_R, \tag{1.89}$$

we find very important properties for Dirac structures that will appear at many places in this book:

$$\bar{\psi}\gamma_\mu\psi = \bar{\psi}_L\gamma_\mu\psi_L + \bar{\psi}_R\gamma_\mu\psi_R \qquad \text{(vector)}, \qquad (1.90)$$

$$\bar{\psi}\psi = \bar{\psi}_L\psi_R + \bar{\psi}_R\psi_L \qquad \text{(scalar)}, \qquad (1.91)$$

$$\bar{\psi}\gamma_\mu\gamma_5\psi = -\bar{\psi}_L\gamma_\mu\psi_L + \bar{\psi}_R\gamma_\mu\psi_R \qquad \text{(axial vector)}, \qquad (1.92)$$

$$\bar{\psi}\gamma_5\psi = \bar{\psi}_L\psi_R - \bar{\psi}_R\psi_L \qquad \text{(pseudo scalar)}. \qquad (1.93)$$

We note that vector-type terms with γ_μ and $\gamma_\mu\gamma_5$ connect only fields of the same helicity, whereas the scalar-type terms connect only fields of opposite helicity. The latter case is often called *helicity flip*. With this information at hand we can present Lagrangians involving fermions that possess global and local, abelian and nonabelian symmetries.

1.9.2 Abelian Global Symmetry

Our starting point is the free Lagrangian for a fermion with mass m. It can be found by demanding that through Euler–Lagrangian equations the Dirac equation follows:

$$\left(i\gamma_\mu\partial^\mu - m\right)\psi = 0. \qquad (1.94)$$

This turns out to be

$$\mathscr{L}_{\text{free}} = \bar{\psi}\left(i\gamma_\mu\partial^\mu\right)\psi - m\bar{\psi}\psi. \qquad (1.95)$$

We note the appearance of a vector and a scalar structure discussed earlier. However, for the time being we will not decompose ψ into ψ_L and ψ_R and present first Lagrangians and their symmetries for the full ψ. That is, we assume first that ψ_L and ψ_R transform identically under symmetries considered by us. We will see that this works for QCD and QED but fails for weak interactions that break within the SM parity P maximally. There are also important global (flavor) symmetries under which ψ_L and ψ_R transform differently. These will also be discussed in our book.

Evidently the Lagrangian in (1.95) is invariant under a global U(1) transformation

$$\psi \to e^{i\theta r}\psi, \qquad \bar{\psi} \to \bar{\psi}e^{-i\theta r}. \qquad (1.96)$$

1.9.3 Nonabelian Global Symmetry

If several fermionic fields are present, so that

$$\psi = \begin{pmatrix} \psi_1 \\ \vdots \\ \psi_N \end{pmatrix}, \qquad \bar{\psi} = \left(\bar{\psi}_1, \ldots, \bar{\psi}_N\right), \qquad (1.97)$$

the Lagrangian in (1.95) is invariant under SU(N) or U(N) global symmetry:

$$\psi \to \exp\left(i\theta^a T^a\right)\psi, \qquad \bar{\psi} \to \bar{\psi}\exp\left(-i\theta^a T^a\right), \tag{1.98}$$

where again summation over repeated indices is understood. One of the implications of nonabelian symmetries is the equality of masses of $\psi_1, \ldots \psi_N$ belonging to a multiplet.

1.9.4 Abelian Local Symmetry

The Lagrangian having a local U(1) symmetry can be constructed following the steps of Section 1.5 with the result

$$\mathscr{L}_{\text{gauged}} = \bar{\psi}\left(i\gamma^\mu D_\mu\right)\psi - m\bar{\psi}\psi - \frac{1}{4}F_{\mu\nu}F^{\mu\nu}, \tag{1.99}$$

where

$$D_\mu = \partial_\mu - irg A_\mu \tag{1.100}$$

is the covariant derivative, g the gauge coupling, and r the charge of ψ under the U(1) symmetry. $F_{\mu\nu}$ is given in (1.26). The local U(1) transformation is given by (1.98) with θ^a replaced by $\theta^a(x)$. The transformation on A_μ is as in (1.25), and also the covariant derivative in (1.100) is identical to the scalar case in (1.24). However as we deal now with the interaction of a vector particle A_μ with a fermion and not a scalar, the structure of the first two terms in (1.99), the so-called kinetic terms, differs from the one in (1.27). In particular the fermionic Lagrangian is linear in the covariant derivative, while the scalar Lagrangian involves a product of D_μ and D^μ necessary to obtain a Lorentz invariant Lagrangian. This, of course, has direct implications on the structure of interactions.

Similar to the scalar case, we can derive the Feynman rules for this theory. For the fermion propagator we just have

$$\frac{i}{\not{k} - m + i\varepsilon} = i\frac{\not{k} + m}{k^2 - m^2 + i\varepsilon},$$

where $\not{k} = \gamma_\mu k^\mu$. The interaction vertex is then

$$= igr\gamma_\mu$$

where as commonly done we do not show explicitly external fields, $A_\mu, \bar{\psi}, \psi$ on the right-hand side of this rule.

1.9.5 Nonabelian Local Symmetry

Proceeding as in Section 1.8 we find the Lagrangian

$$\mathscr{L} = \bar{\psi} \left(i\gamma_\mu D^\mu \psi \right) - m\bar{\psi}\psi - \frac{1}{4} F^a_{\mu\nu} F^{\mu\nu,a}, \tag{1.101}$$

where ψ and $\bar{\psi}$ are given in (1.97) and

$$D_\mu \psi = \left(\partial_\mu - ig A^a_\mu T^a \right) \psi, \tag{1.102}$$

with g being the gauge coupling and A^a_μ the gauge bosons corresponding to the generators T^a. This Lagrangian is invariant under the transformation (1.98) for fermions with θ^a replaced by $\theta^a(x)$ and (1.65) for gauge bosons. The pure gauge sector, in particular interactions among different gauge bosons, did not change relative to the scalar case.

The interaction of a given gauge boson with fermions belonging to a given multiplet takes the form

$$= ig\gamma_\mu \left(T^a \right)_{ij}.$$

We again note that all these interactions are given entirely in terms of a single-gauge coupling g and various "charges" $(T^a)_{ij}$, which are fully fixed by the symmetry.

1.9.6 Important Properties

Let us end the first discussion of fermionic Lagrangians by stressing several properties of both the interaction and the mass term. To this end it is sufficient to rewrite the interaction and mass terms in (1.99) in terms of ψ_L and ψ_R:

$$\mathscr{L}_{\text{int}} = +gr \left(\bar{\psi}_L \gamma_\mu A^\mu \psi_L + \bar{\psi}_R \gamma_\mu A^\mu \psi_R \right) \tag{1.103}$$

$$\mathscr{L}_{\text{mass}} = -m \left(\bar{\psi}_R \psi_L + \bar{\psi}_L \psi_R \right). \tag{1.104}$$

We observe

- The gauge interactions connect only fields of the same helicity. There is no helicity flip.
- A mass term connects fields of opposite helicity. Thus, in order to dynamically generate a mass term, a helicity flip is needed. An exception is the so-called Majorana mass term in the case of neutrinos or generally neutral fermions.
- The two properties imply that if the interactions in a given theory are just gauge interactions and there is no mass term, there is no way to generate masses of fermions radiatively through interactions.

- The strength of gauge interactions of ψ_L and ψ_R is the same if ψ_L and ψ_R transform identically under the gauge group. This is the case of vectorial gauge theories (only γ_μ matters). Prominent examples are QED and QCD.
- However, quite often ψ_L and ψ_R transform differently under global and gauge symmetry groups. Such symmetries are called chiral. The $SU(3)_L \times SU(3)_R$ chiral global symmetry of QCD and the $SU(2)_L$ gauge symmetry of the SM are well-known examples.
- If ψ_L and ψ_R transform differently under a given symmetry group, local or global, fermion mass terms are forbidden by the symmetry. This is also the case for gauge bosons in general, as we have already seen on previous pages.

As we will see in the next chapter, both the masses of weak gauge bosons $W^\pm$ and Z^0 as well as of all quarks and leptons are forbidden by $SU(2)_L$. This is clearly a disaster, and one could ask whether one could simply introduce some explicit terms in the Lagrangian that would break $SU(2)_L$ so that masses of weak gauge bosons and of quarks and leptons would become nonzero. Fortunately it turns out that this is not possible without destroying other properties of the SM related in particular to its renormalizability, a topic of Section 4.2. It is fortunate because otherwise the theory would not be as predictive as it is. Yet, the problem of the generation of gauge boson and fermion masses in the presence of exact symmetry in a given Lagrangian must be solved somehow. This is what we will do next.

1.10 Spontaneous Symmetry Breakdown (SSB)

1.10.1 Preliminaries

If the Lagrangian possesses a certain symmetry S but the ground state (vacuum) is not invariant under such symmetry transformation, then we call S a spontaneously broken symmetry. Both discrete and continuous symmetries can be spontaneously broken. In Section 1.10.2 we will first discuss SSB of a discrete symmetry and in Section 1.10.3 of a continuous symmetry. But first we want to outline here the connection between vacuum expectation values (vev) and SSB.

Let us consider an infinitesimal transformation of a continuous symmetry as in (1.40) by expanding the exponential and keeping only the term linear in θ^a

$$\varphi_i \to \varphi_i' = \varphi_i + i\theta^a T_{ij}^a \varphi_j. \tag{1.105}$$

For an unbroken symmetry we have

$$\langle 0|\varphi_i|0\rangle \overset{!}{=} \langle 0|\varphi_i'|0\rangle = \langle 0|\varphi_i|0\rangle + i\theta^a T_{ij}^a \langle 0|\varphi_j|0\rangle. \tag{1.106}$$

For an irreducible representation T^a this implies

$$\langle 0|\varphi_j|0\rangle = 0. \tag{1.107}$$

According to the Noether theorem there is a conserved charge Q^a that corresponds to the generator T^a of the symmetry. As shown in many books on quantum field theory, these charges also generate the symmetry transformation in the following manner

$$\varphi_i' = e^{i\theta^a Q^a} \varphi_i e^{-i\theta^a Q^a}. \tag{1.108}$$

Considering only infinitesimal transformations and comparing with (1.106) this means

$$i\theta^a T_{ij}^a \varphi_j = i\theta^a [Q^a, \varphi_i]. \tag{1.109}$$

Now for an unbroken symmetry the vacuum is neutral, i.e., $Q^a|0\rangle = 0$. Together with (1.109) we conclude that for an uncharged vacuum the vacuum expectation value has to vanish:

$$Q^a|0\rangle = 0 \qquad \Rightarrow \qquad \langle 0|\varphi_j|0\rangle = 0. \tag{1.110}$$

If, on the other hand, at least one component φ_j has a nonvanishing vacuum expectation value, we have

$$v_j := \langle 0|\varphi_j|0\rangle \neq 0, \qquad Q^a|0\rangle \neq 0. \tag{1.111}$$

Consequently the vacuum is charged under the symmetry, and all symmetries under which φ_j transforms nontrivially are spontaneously broken while the symmetries under which φ_j transforms as a singlet are unbroken. Because we do not want to break Lorentz invariance φ_j can only be a scalar field and, for example, not a vector field.

1.10.2 Spontaneous Breakdown of a Discrete Symmetry

In order to introduce the concept of spontaneous symmetry breakdown of a symmetry we consider the Lagrangian for a real scalar field

$$\mathscr{L} = \frac{1}{2} \left(\partial_\mu \varphi \right) \left(\partial^\mu \varphi \right) - V(\varphi), \tag{1.112}$$

where the potential is given by

$$V(\varphi) = \frac{1}{2}\mu^2 \varphi^2 + \frac{1}{4}\lambda \varphi^4, \qquad \lambda > 0. \tag{1.113}$$

The parameter λ describes the strength of the scalar interactions with itself. The condition $\lambda > 0$ ensures that the potential is bounded from below. The parameter μ will play a crucial role in a moment. Evidently the Lagrangian (1.112) is invariant under the discrete symmetry

$$\varphi \rightarrow -\varphi. \tag{1.114}$$

We next look at the term quadratic in φ and consider two cases.

- $\mu^2 > 0$

    Comparing (1.113) with (1.1) we conclude that $\mu = m$ is just the mass of φ. In Fig. 1.2 we show $V(\varphi)$ for this case. We observe that $V(\varphi)$ has a unique minimum at $\varphi = 0$. This is the ground state of the theory, which as usually done will be called vacuum. Thus

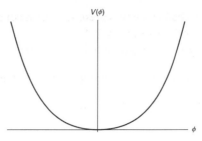

Figure 1.2 Potential for a real scalar field with $\mu^2 > 0$.

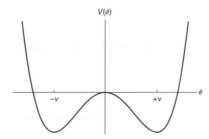

Figure 1.3 Potential for a real scalar field with $\mu^2 < 0$.

$$\varphi_{\text{vac}} = 0, \qquad \mu^2 > 0. \tag{1.115}$$

Moreover the vacuum is symmetric with respect to $\varphi \to -\varphi$. Nothing exciting so far.

- $\mu^2 < 0$

   This case is more interesting. Indeed, from

$$\frac{\partial V}{\partial \varphi} = \varphi \left(\mu^2 + \lambda \varphi^2\right) = 0, \tag{1.116}$$

we learn as seen in Fig. 1.3 that there are two minima so that

$$\varphi_{\text{vac}} = \pm v, \qquad v = \sqrt{-\frac{\mu^2}{\lambda}}. \tag{1.117}$$

The potential looks still symmetric under $\varphi \to -\varphi$ but in order to calculate predictions of the theory we have to choose the ground state. This breaks the symmetry. Indeed sitting in one of the two vacuua the world does not look symmetric anymore. The best proof of this is that flipping the sign of φ we move to a different world with φ_{vac} having opposite sign.

Let us investigate the consequences of SSB. To this end we expand $\varphi(x)$ around the vacuum state $\varphi = v$

$$\varphi(x) = v + \eta(x), \qquad \eta_{\text{vac}} = 0, \tag{1.118}$$

with η describing fluctuations around this vacuum.

The Lagrangian (1.112) expressed in terms of η is given as follows

$$\mathscr{L} = \frac{1}{2}(\partial_\mu \eta)(\partial^\mu \eta) - \lambda v^2 \eta^2 - \lambda v \eta^3 - \frac{1}{4}\lambda \eta^4 + \text{const.}, \qquad (1.119)$$

where the constant terms do not involve η and are physically irrelevant. From the second term we find the mass of η:

$$m_\eta = \sqrt{2\lambda v} = \sqrt{-2\mu^2}. \qquad (1.120)$$

The same result can also be obtained from

$$\frac{\partial^2 V}{\partial \varphi^2} = \mu^2 + 3\lambda \varphi^2 \Big|_{\varphi=v} = -2\mu^2 = m_\eta^2. \qquad (1.121)$$

We collect a few lessons from this simple exercise:

- Useful formula for scalar masses

$$m_\eta^2 = \frac{\partial^2 V}{\partial \varphi^2}\Big|_{\varphi=v}. \qquad (1.122)$$

- There are two degenerate vacua connected by the original symmetry $\varphi \to -\varphi$.
- η has a nonvanishing mass given in (1.120).
- The Lagrangian in (1.119) is clearly not invariant under $\eta \to -\eta$ because of the η^3 term.

While these results are still not terribly exciting, we will see that they will turn out to be useful in the context of the spontaneous breakdown of continuous symmetries.

1.10.3 Spontaneous Breakdown of a Continuous Abelian Global Symmetry

We next consider the generalization of (1.112) to a complex scalar field φ:

$$\mathscr{L} = \frac{1}{2}\left(\partial_\mu \varphi^*\right)\left(\partial^\mu \varphi\right) - V(\varphi^*, \varphi), \qquad (1.123)$$

where the potential is given by

$$V(\varphi^*, \varphi) = \mu^2 \varphi^* \varphi + \frac{1}{4}\lambda(\varphi^* \varphi)^2, \qquad \lambda > 0. \qquad (1.124)$$

This Lagrangian is invariant under the U(1) symmetry

$$\varphi \to e^{i\theta}\varphi, \qquad \varphi^* \to \varphi^* e^{-i\theta}. \qquad (1.125)$$

This symmetry is spontaneously broken for $\mu^2 < 0$. Indeed, we find now

$$|\varphi_{\text{vac}}|^2 = -\frac{2\mu^2}{\lambda} \equiv \frac{v^2}{2}, \qquad v = \sqrt{-4\frac{\mu^2}{\lambda}}. \qquad (1.126)$$

The resulting potential is shown in Fig. 1.4. It looks like a Mexican hat.

The profound difference from the case of the discrete symmetry is the full circle of degenerate minima connected by the original symmetry as the minimum condition (1.126) does not fix the phase of φ_{vac}. We choose now one of this vacua as our ground state, and

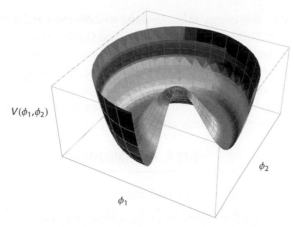

$V(\phi_1,\phi_2)$

ϕ_2

ϕ_1

Figure 1.4   Potential for a complex scalar field with $\mu^2 < 0$.

this breaks the U(1) symmetry spontaneously. Let us find the mass spectrum after SSB. To this end we write

$$\varphi(x) = \frac{1}{\sqrt{2}}\left(\varphi_1(x) + i\varphi_2(x)\right), \tag{1.127}$$

with φ_1 and φ_2 being real. The vacuum condition now reads

$$\varphi_1^2 + \varphi_2^2\big|_{\text{vac}} = v^2. \tag{1.128}$$

We next choose the vacuum to be

$$(\varphi_1, \varphi_2)_{\text{vac}} = (v, 0), \tag{1.129}$$

and having two degrees of freedom we introduce two fields η and ξ, which describe fluctuations around the vacuum (1.129)

$$\varphi(x) = \frac{1}{\sqrt{2}}\left(v + \eta(x) + i\xi(x)\right). \tag{1.130}$$

Inserting this expression into (1.123) we find after some algebra

$$\mathcal{L} = \left[\frac{1}{2}(\partial_\mu\eta)(\partial^\mu\eta) - \frac{1}{2}m_\eta^2\eta^2\right] + \left[\frac{1}{2}(\partial_\mu\xi)(\partial^\mu\xi)\right] + \text{interactions}, \tag{1.131}$$

with

$$m_\eta^2 = \frac{\partial^2 V}{\partial\varphi_1^2}\bigg|_{(v,0)} = -2\mu^2, \tag{1.132}$$

$$m_\xi^2 = \frac{\partial^2 V}{\partial\varphi_2^2}\bigg|_{(v,0)} = 0. \tag{1.133}$$

The striking difference from the breakdown of a discrete symmetry is the appearance of a massless particle in addition to a massive one. The appearance of a massless particle can easily be understood by noting that the potential V is flat in the φ_2 direction. It does not cost any energy to move along this direction and this is only possible for a massless particle. This massless particle is called Goldstone boson. In the φ_1 direction the potential is not flat, and it costs some energy to move along it: the particle η has a mass.

This important result of the appearance of a massless particle as the consequence of a spontaneous breakdown of a continuous global symmetry is a special case of the Goldstone theorem, which states that for each broken symmetry there is one Goldstone boson. As the U(1) symmetry has only one generator, we have only one massless boson in our example.

1.10.4 Generalization to the Nonabelian Case

In order to better understand the implications of the Goldstone theorem, we will now generalize our considerations to a nonabelian global symmetry. In principle we could consider the breakdown of an SU(N) group, but it is easier to consider in this case first the breakdown of an orthogonal group, simply because in this case the fields are real. This was also the strategy in the book of Bailin and Love [16] from which we benefited a lot when presenting the following material and where further details can be found.[3] SU(2) and SU(3) groups will be worked out in detail in the context of the SM and its extensions later in our book.

We consider then n real fields, which form an n-dimensional representation described by a column vector

$$\vec{\varphi}(x) = \begin{pmatrix} \varphi_1(x) \\ \vdots \\ \varphi_n(x) \end{pmatrix}. \tag{1.134}$$

The relevant Lagrangian

$$\mathcal{L} = \frac{1}{2} \left(\partial_\mu \vec{\varphi}^\top \right) \left(\partial^\mu \vec{\varphi} \right) - V(\vec{\varphi}^\top \vec{\varphi}) \tag{1.135}$$

is invariant under infinitesimal global transformation

$$\vec{\varphi} \rightarrow \vec{\varphi} + \delta\vec{\varphi}, \qquad \delta\vec{\varphi} = i\theta^a T^a \vec{\varphi}, \tag{1.136}$$

where $a = 1, \ldots N$. In component form we have

$$\delta\varphi_i = i \left(\theta^a T^a \right)_{ij} \varphi_j. \tag{1.137}$$

T^a are $n \times n$ Hermitian matrices, but iT^a must be real to keep the real character of the fields φ_i, and consequently T^a must be antisymmetric, precisely what the generators of an orthogonal group are. We have seen in the previous example that the invariance of $\mathcal{L}$ under a given symmetry still played an important role after SSB. Let us then investigate the implications of the invariance in this more complicated case.

From the invariance of V we have

$$\delta V = \frac{\partial V}{\partial \varphi_i} \delta\varphi_i = i \frac{\partial V}{\partial \varphi_i} \left(\theta^a T^a \right)_{ij} \varphi_j = 0. \tag{1.138}$$

---

[3] See chapter 13 in [16].

But the θ^a are arbitrary, and consequently for every generator T^a we have

$$\frac{\partial V}{\partial \varphi_i} (T^a)_{ij} \varphi_j = 0 \qquad a = 1, \ldots N. \tag{1.139}$$

We next consider SSB, that is, in the vacuum

$$\langle 0 | \vec{\varphi} | 0 \rangle = \vec{v} \qquad \text{or} \qquad \langle 0 | \varphi_i | 0 \rangle = v_i, \tag{1.140}$$

and

$$\left. \frac{\partial V}{\partial \vec{\varphi}} \right|_{\vec{\varphi} = \vec{v}} = \vec{0} \qquad \text{or} \qquad \left. \frac{\partial V}{\partial \varphi_i} \right|_{\varphi_i = v_i} = 0. \tag{1.141}$$

Differentiating (1.139) with φ_k we find

$$\frac{\partial^2 V}{\partial \varphi_k \partial \varphi_i} (T^a)_{ij} \varphi_j + \frac{\partial V}{\partial \varphi_i} (T^a)_{ik} = 0, \tag{1.142}$$

and evaluating it at the minimum (1.141), we find

$$\left(\frac{\partial^2 V}{\partial \varphi_k \partial \varphi_i} \right)_{\vec{\varphi} = \vec{v}} (T^a)_{ij} v_j = 0, \tag{1.143}$$

which summarizes the implications of a global invariance around the ground state.

We next expand around the vacuum

$$\vec{\varphi} = \vec{v} + \tilde{\vec{\varphi}} \qquad \text{or} \qquad \varphi_i = v_i + \tilde{\varphi}_i, \tag{1.144}$$

with

$$\langle 0 | \tilde{\varphi}_i | 0 \rangle = 0. \tag{1.145}$$

Expanding in small fluctuations $\tilde{\varphi}_i$ and using (1.141), we find

$$\mathscr{L} = \frac{1}{2} \left[(\partial_\mu \tilde{\varphi}_i)(\partial^\mu \tilde{\varphi}_i) - \tilde{\varphi}_i \tilde{\varphi}_j \left(\frac{\partial^2 V}{\partial \varphi_i \partial \varphi_j} \right)_{\vec{\varphi} = \vec{v}} \right] - V(\vec{v}) + O(\tilde{\varphi}^3). \tag{1.146}$$

Consequently, the mass spectrum after SSB is described on the basis of $(\tilde{\varphi}_1, \ldots, \tilde{\varphi}_n)$ by $n \times n$ mass matrix squared

$$\left(\mathsf{M}^2 \right)_{ij} = \left(\frac{\partial^2 V}{\partial \varphi_i \partial \varphi_j} \right)_{\vec{\varphi} = \vec{v}}. \tag{1.147}$$

But according to (1.143), which followed from global invariance, this matrix satisfies the equations

$$\left(\mathsf{M}^2 \right)_{ki} (T^a)_{ij} v_j = 0, \qquad a = 1, \ldots N \tag{1.148}$$

or more compactly

$$\mathsf{M}^2 T^a \vec{v} = 0, \qquad a = 1, \ldots N. \tag{1.149}$$

We now denote the symmetry group by G and assume that it is broken spontaneously to its subgroup $H \subset G$. The generators T^a can now be divided into X^a and Y^a

$$Y^a \subset H; \qquad Y^a = T^a \qquad a = 1, \ldots M \tag{1.150}$$

$$X^a \subset G/H; \qquad X^a = T^a \qquad a = M + 1, \ldots N, \tag{1.151}$$

with Y^a building the subgroup H and the broken generators X^a belonging to the so-called coset space. They are just the remaining generators, but by itself they do not build a subgroup of G.

Now in the case of the subgroup H the vacuum is invariant, and this means

$$Y^a \vec{v} = 0 \qquad \text{or} \qquad (Y^a)_{ij} v_j = 0. \tag{1.152}$$

This allows to satisfy (1.149) trivially and gives no constraint on M^2. On the other hand, for broken generators

$$X^a \vec{v} \neq 0 \qquad \text{or} \qquad (X^a)_{ij} v_j \neq 0, \tag{1.153}$$

and the conditions in (1.149) imply zero eigenvalues in the mass matrix M^2. In other words,

$$\vec{U}^a \equiv X^a \vec{v} \qquad \text{or} \qquad (U^a)_i = (X^a)_{ij} v_j \tag{1.154}$$

are the eigenvectors corresponding to zero masses.

In summary, for every broken generator X^a there is a massless Goldstone boson that is a linear combination of the fields $\tilde{\varphi}_i$ and given by

$$(\tilde{\varphi}_1, \ldots \tilde{\varphi}_n) \, X^a \begin{pmatrix} v_1 \\ \vdots \\ v_n \end{pmatrix} = \left(\vec{\tilde{\varphi}}\right)^\top X^a \vec{v}. \tag{1.155}$$

1.10.5 Summary

We conclude that a spontaneous breakdown of a global symmetry generated another problem: new massless particles for each broken generator. Such massless particles, if they would exist in nature, would have been discovered already a long time ago. On the other hand, the lightest mesons like pions and kaons could be regarded as nearly Goldstone bosons of a broken global $\mathrm{SU}(3)_L \times \mathrm{SU}(3)_R$ symmetry, which would be exact at the level of Lagrangian if pions and kaons were massless. But as pions and kaons have masses, $\mathrm{SU}(3)_L \times \mathrm{SU}(3)_R$ has to be broken explicitly so that these mesons at the end obtain small masses. Consequently they are not true Goldstone bosons but the so-called pseudo-Goldstone bosons. A very nice article on such bosons in general terms is the one by Steven Weinberg [17].

An explicit breakdown of a global symmetry has no theoretical problems and combined with spontaneous symmetry turns out to be useful for the description of the physics of lightest mesons as we just mentioned.

On the other hand, explicit breakdown of a gauge symmetry is not allowed as it spoils the renormalization of such theories. Only spontaneous breakdown is admitted. From the Goldstone theorem it follows that if a global symmetry is spontaneously broken physical massless, spin-0 bosons, the Goldstone bosons, emerge. What happens now if one breaks a local symmetry spontaneously? We will discover soon that the "flat" directions $(T^a)_{ij}v_j$ of a local symmetry correspond to unphysical gauge degrees of freedom, i.e., the gauge symmetry is broken but we do not get physical Goldstone bosons. Yet at first sight the appearance of these Goldstone bosons (one for each broken generator) seems to be a new problem, but it turns out to be a way to generate the masses of gauge bosons without breaking explicitly the gauge symmetry of the Lagrangian. Indeed, a massless gauge boson, like the photon, has only two degrees of freedom corresponding to its two transverse polarizations. On the other hand, a massive gauge boson like $W^\pm$ and Z^0 has three degrees of freedom, the third one corresponding to its longitudinal polarization. It is the Goldstone boson of a spontaneously broken gauge symmetry that provides the third degree of freedom to every gauge boson corresponding to a broken generator. Thus at the end $W^\pm$ and Z^0 are massive, and Goldstone bosons do not appear in the particle spectrum. That's why the Goldstone bosons of a local symmetry are unphysical. One can say that they have been eaten by the gauge bosons. In the next section we will discuss this mechanism for generation of masses of gauge bosons in explicit terms.

1.11 Higgs Mechanism

1.11.1 U(1) Symmetry

Let us then gauge the Lagrangian in (1.123) so that

$$\mathcal{L}_{\text{gauged}} = \left(D_\mu \varphi\right)^* (D^\mu \varphi) - V(\varphi^*, \varphi) - \frac{1}{4}F_{\mu\nu}F^{\mu\nu}. \tag{1.156}$$

This Lagrangian is invariant under simultaneous U(1) transformations of φ and A_μ:

$$\varphi \to \varphi e^{i\theta(x)}, \qquad A_\mu \to A_\mu + \frac{1}{g}\partial_\mu \theta(x). \tag{1.157}$$

This invariance will be crucial for the removal of Goldstone bosons from the physical mass spectrum. We again write after SSB

$$\varphi(x) = \frac{1}{\sqrt{2}}\left(v + \eta(x) + i\xi(x)\right) \approx \frac{1}{\sqrt{2}}\exp\left(i\frac{\xi(x)}{v}\right)\left(v + \eta(x)\right), \tag{1.158}$$

but this time it will be useful to also have the last expression in (1.158).

For the discussion of the mass generation for gauge bosons only the covariant derivative $D_\mu \varphi$ is of interest as only there the scalars and gauge bosons interact with each other. We have then

$$D_\mu \varphi = \frac{1}{\sqrt{2}} \left[\partial_\mu \eta - i \left(vg A_\mu - \partial_\mu \xi \right) - ig A_\mu \left(\eta + i\xi \right) \right]. \tag{1.159}$$

Of particular interest is the second term on the r.h.s. of this equation. We can rewrite it as follows

$$vg A_\mu - \partial_\mu \xi = gv \left(A_\mu - \frac{1}{gv} \partial_\mu \xi \right) \equiv gv A'_\mu. \tag{1.160}$$

Note that the original field A_μ and A'_μ are related by gauge transformation in (1.157) with

$$\theta(x) = -\frac{\xi(x)}{v}. \tag{1.161}$$

This transformation performed on the field φ in (1.158) results in

$$\varphi \to \varphi' = \frac{1}{\sqrt{2}} \left(v + \eta \right). \tag{1.162}$$

But we know that such a gauge transformation leaves the Lagrangian invariant, and consequently this Lagrangian rewritten in terms of A'_μ and φ' describes the same physics as the original Lagrangian. In this new Lagrangian the Goldstone boson is not seen; it has been gauged away.

Dropping now the prime in (1.162) and denoting $\eta = H$ with H standing for a Higgs particle, this discussion shows that due to the gauge invariance of $\mathscr{L}$ we are allowed to write

$$\varphi = \frac{1}{\sqrt{2}} \left(v + H \right), \tag{1.163}$$

$$D_\mu \varphi = \frac{1}{\sqrt{2}} \left(\partial_\mu H - ivg A_\mu - ig A_\mu H \right). \tag{1.164}$$

Consequently we obtain

$$\mathscr{L}_{\text{gauged}} = \frac{1}{2} \partial_\mu H \partial^\mu H - \frac{1}{4} F_{\mu\nu} F^{\mu\nu} - \frac{1}{2} m_H^2 H^2 + \frac{1}{2} M_A^2 A_\mu A^\mu + \cdots, \tag{1.165}$$

which implies

$$M_A^2 = v^2 g^2, \qquad m_H^2 = -2\mu^2. \tag{1.166}$$

Indeed, the gauge boson is now massive. Its mass depends on the gauge coupling g and v. In a given theory M_A can be predicted if g and v have been determined somewhere else. On the other hand, m_H is rather arbitrary as μ is a parameter in the potential V.

It should be remarked that the explicit disappearance of the Goldstone boson from the theory is only possible in the unitary gauge in which φ takes the form (1.162). This gauge is very useful for exhibiting the physical spectrum, but it is less convenient for Feynman diagram calculations. Therefore quite generally the latter calculations are done in other gauges, the so-called covariant gauges, in which Goldstone bosons appear in loop diagrams and one has to take these contributions into account in order to obtain a physical result that is gauge independent. The Feynman rules involving Goldstone bosons in the case of the SM can be found in Appendix B.

1.11.2 Nonabelian Gauge Symmetry

We next gauge the Lagrangian (1.135) to obtain

$$\mathscr{L} = \frac{1}{2}\left(D_\mu\vec{\varphi}^\top\right)(D^\mu\vec{\varphi}) - V(\vec{\varphi}^\top\vec{\varphi}) - \frac{1}{4}F_{\mu\nu}^a F^{\mu\nu,a}. \tag{1.167}$$

All these symbols have been defined earlier. We now consider as in Section 1.10.4 spontaneous symmetry breakdown $G \to H$ with unbroken generators denoted by Y^a and the broken ones by X^a. We have

$$Y^a\vec{v} = 0, \qquad a = 1,\dots M \tag{1.168}$$

$$X^a\vec{v} \neq 0, \qquad a = M+1,\dots N, \tag{1.169}$$

and the Goldstone bosons are given in (1.155). Expanding around the vacuum as in (1.144)–(1.146) we find

$$\begin{aligned}
\left(D_\mu\vec{\varphi}\right)^\top(D^\mu\vec{\varphi}) &= \left(\partial_\mu\vec{\tilde{\varphi}}\right)^\top\left(\partial^\mu\vec{\tilde{\varphi}}\right) + g^2 A_\mu^a A^{\mu,b}\vec{v}^\top T^a T^b\vec{v} \\
&\quad - ig\left(\partial_\mu\vec{\tilde{\varphi}}\right)^\top T^b\vec{v}A^{\mu,b} + ig\vec{v}^\top T^a\partial_\mu\vec{\tilde{\varphi}}A^{\mu,a} + \cdots
\end{aligned} \tag{1.170}$$

The last two terms vanish for $T^a = Y^a$ but are nonvanishing for $T^a = X^a$. From (1.155) we know that these terms for $a = M+1,\dots N$ represent mixed terms involving Goldstone bosons and the gauge bosons corresponding to broken generators X^a. As in the case of abelian symmetry, they can be gauged away by going to the unitary gauge.

The mass spectrum of this theory after SSB is as follows

- The masses of gauge bosons can be read of from the second term on the r.h.s. of (1.170). The gauge boson mass matrix is simply given as follows

$$\frac{1}{2}\left(M_A^2\right)^{ab} = g^2\vec{v}^\top T^a T^b\vec{v}. \tag{1.171}$$

As T^a includes both Y^a and X^a satisfying (1.168) and (1.169), this matrix has both vanishing and nonvanishing entries so that after diagonalization we will find massless gauge bosons corresponding to the generators of the unbroken subgroup H and the massive gauge bosons corresponding to the remaining generators of G.

- The masses of physical Higgs particles are the nonvanishing eigenvalues of the matrix in (1.147):

$$\left(M^2\right)_{ij} = \left(\frac{\partial^2 V}{\partial\varphi_i\partial\varphi_j}\right)_{\vec{\varphi}=\vec{v}}, \qquad i,j = 1,\dots n. \tag{1.172}$$

The vanishing entries correspond to the Goldstone bosons. The number of physical Higgs particles, N_H, is just given by

$$N_H = n - N_{\mathrm{GB}} = n - (N - M) \tag{1.173}$$

with n denoting the number of φ_i fields and $N - M$ the number of broken generators. As N_H is a positive number, it is evident that in order to break a given group G down to H, sufficient food in the form of φ_i has to be provided such that the gauge bosons corresponding to the broken generators become massive.

1.11.3 Summary

In this chapter we have presented the most important aspects of gauge theories that are necessary in order to follow the next chapters. In this context we have discussed also global symmetries, which, similar to local symmetries, play an important role in particle physics. In fact the first symmetries discussed in particle physics like SU(2), SU(2)$_L$ × SU(2)$_R$, SU(3)$_L$ × SU(3)$_R$, and SU(3) were all global symmetries. Even if in the 1970s gauge symmetries took over the leadership, due to significant development of flavor physics in the late 1980s and the following three decades, global symmetries play these days again a very important role, and we will discuss other important consequences of them in later chapters of this book.

With all this information at hand we are ready to move to Part II of our book in order to describe the SM of electroweak and strong interactions. This will allow us to begin to discuss the main topic of our book, namely weak decays of mesons and later leptons as well as other interesting rare processes.

WE ARE READY TO CLIMB TO THE BASE CAMP: THE STANDARD MODEL!

PART II

THE STANDARD MODEL

The Standard Model of Electroweak and Strong Interactions

2.1 Particle Content and Gauge Group of the SM

The SM of particle physics is based on the gauge group[1]

$$SU(3)_C \times SU(2)_L \times U(1)_Y. \tag{2.1}$$

It is a very successful model for the description of three out of the four fundamental interactions in nature. The electromagnetic, weak, and strong interactions between the known elementary particles at the currently available energy scales are predicted and described by the SM in terms of a number of parameters that can be determined in experiment or calculated from a more fundamental theory.

The gauge symmetry $SU(3)_C$ is the symmetry of QCD, the theory of strong interactions. $SU(2)_L$ is the gauge group of weak interactions, and the $U(1)_Y$ force is felt by all particles with a nonzero hypercharge quantum number Y. Whereas $SU(3)_C$ remains exact, $SU(2)_L \times U(1)_Y$ is spontaneously broken down to $U(1)_{em}$, the symmetry group of QED (see Section 2.3 for details). The quantum numbers of $SU(3)_C$, $SU(2)_L$, and $U(1)_Y$ are called color, weak isospin, and hypercharge, respectively. While hypercharge can in principle be any real number, there are three colors (red, green, blue) and two weak isospin states ($\pm\frac{1}{2}$). The transformation properties of all SM fermion and gauge fields under the symmetry group $SU(3)_C \times SU(2)_L \times U(1)_Y$ are given in Table 2.1, where we show the representations in the case of $SU(3)_C$ and $SU(2)_L$ and the hypercharge Y in the case of $U(1)_Y$, which is normalized as follows. Note that in the literature Y is sometimes replaced by Y/2.

$$Q = T_3 + Y. \tag{2.2}$$

This is the well-known Gell-Mann–Nishima relation. Here Q is the electric charge and T_3 the third component of the weak isospin belonging to $SU(2)_L$. The fermions of the SM are grouped in several ways: On the one hand we distinguish between quarks and leptons, where only quarks feel QCD and thus transform nontrivially under $SU(3)_C$. Quarks are triplets **3** under $SU(3)_C$ with quantum numbers red, green, and blue; antiquarks are antitriplets **3̄** with quantum numbers antired, antigreen, and antiblue, whereas leptons are $SU(3)_C$ singlets and thus color neutral. On the other hand, there are $SU(2)_L$ doublets (left-handed fields) and $SU(2)_L$ singlets (all right-handed fields). Therefore only the doublets

---

[1] This chapter has been written by Jennifer Girrbach-Noe.

Table 2.1 Particle content of the SM and transformation properties.

Fermions: three generations			$SU(3)_C$	$SU(2)_L$	$U(1)_Y$
e_R	μ_R	τ_R	1	1	-1
$L_1 = (\nu_e, e_L)^\top$	$L_2 = (\nu_\mu, \mu_L)^\top$	$L_3 = (\nu_\tau, \tau_L)^\top$	1	2	$-\frac{1}{2}$
u_R	c_R	t_R	3	1	$\frac{2}{3}$
d_R	s_R	b_R	3	1	$-\frac{1}{3}$
$Q_1 = (u_L, d_L)^\top$	$Q_2 = (c_L, s_L)^\top$	$Q_3 = (t_L, b_L)^\top$	3	2	$\frac{1}{6}$
Gauge bosons					
	G_μ^a	$a = 1\text{--}8$	8	1	0
	W_μ^a	$a = 1, 2, 3$	1	3	0
	B_μ		1	1	0
Higgs					
	$\Phi = \left(\phi^+, \phi^0\right)^\top$		1	2	$\frac{1}{2}$

participate in the charged weak interactions mediated by $W^\pm$ bosons. The first (second) component of an $SU(2)_L$ doublet has weak isospin $T_3 = \frac{1}{2}$ ($T_3 = -\frac{1}{2}$). Because all representations of $SU(2)$ are real (see Section 1.7.1), antiparticles also have isospin and not anti-isospin. Furthermore the fermions exist in three copies, called generations. The particles of the second and third generations have exactly the same quantum numbers as the corresponding particles of the first generation (as shown in Table 2.1) but they are different in mass. The requirement of local gauge invariance predicts the existence of massless gauge bosons for each gauge group that belong to adjoint representations of these groups. We have then eight gluons G_μ^a for $SU(3)_C$, three W_μ^a for $SU(2)_L$, and one B_μ for $U(1)_Y$. After electroweak symmetry breaking W_μ^a and B_μ mix into $W_\mu^\pm$, Z^0 and γ where only γ stays massless and belongs to the unbroken $U(1)_{\text{em}}$ symmetry, the symmetry of QED.

It is now time to encode all this information in an appropriate Lagrangian.

2.2　Short Overview: Lagrangian of the SM

The SM Lagrangian can be split into four parts given in (2.3), (2.8), (2.12), and (2.13). The kinetic terms of the SM fermions together with their couplings to gauge bosons follow from

$$\mathscr{L}_{\text{fermion}}^{\text{SM}} = \sum_{j=1,2,3} \bar{L}_j i \slashed{D} L_j + \bar{e}_{Rj} i \slashed{D} e_{Rj} + \bar{Q}_j i \slashed{D} Q_j + \bar{u}_{Rj} i \slashed{D} u_{Rj} + \bar{d}_{Rj} i \slashed{D} d_{Rj}, \qquad (2.3)$$

where we used the shorthand notation $\slashed{D} = D_\mu \gamma^\mu$. The covariant derivatives are given as

$$D_\mu = \partial_\mu - ig_1 Y B_\mu - ig_2 \frac{\sigma^a}{2} W_\mu^a \qquad\qquad \text{for LH lepton doublets } L_j, \qquad (2.4)$$

$$D_\mu = \partial_\mu - ig_1 Y B_\mu \qquad\qquad\qquad\qquad \text{for RH lepton singlets } e_{Rj}, \qquad (2.5)$$

$$D_\mu = \partial_\mu - ig_1 Y B_\mu - ig_2 \frac{\sigma^a}{2} W_\mu^a - ig_s \frac{\lambda^a}{2} G_\mu^a \quad \text{for LH quark doublets } Q_j, \qquad (2.6)$$

$$D_\mu = \partial_\mu - ig_1 Y B_\mu - ig_s \frac{\lambda^a}{2} G_\mu^a \qquad\quad \text{for RH quark singlets } u_{Rj}, d_{Rj}, \qquad (2.7)$$

and are the same for all three generations $j = 1, 2, 3$. Here Y denotes the hypercharge quantum number, σ^a are the Pauli matrices defined in (A.17), λ^a are the Gell-Mann matrices defined in (A.19)–(A.21), and G_μ^a are the gluon fields. The couplings for $SU(3)_C$, $SU(2)_L$, and $U(1)_Y$ are denoted by g_s, g_2, and g_1, respectively. In the more general considerations of Chapter 1 we used always g as a coupling constant.

The kinetic term of the gauge bosons is given by

$$\mathscr{L}_{\text{gauge}}^{\text{SM}} = -\frac{1}{4} G_{\mu\nu}^a G^{\mu\nu,a} - \frac{1}{4} W_{\mu\nu}^a W^{\mu\nu,a} - \frac{1}{4} B_{\mu\nu} B^{\mu\nu}, \qquad (2.8)$$

where

$$G_{\mu\nu}^a = \partial_\mu G_\nu^a - \partial_\nu G_\mu^a + g_s f^{abc} G_\mu^b G_\nu^c, \qquad (2.9)$$

$$W_{\mu\nu}^a = \partial_\mu W_\nu^a - \partial_\nu W_\mu^a + g_2 \varepsilon^{abc} W_\mu^b W_\nu^c, \qquad (2.10)$$

$$B_{\mu\nu} = \partial_\mu B_\nu - \partial_\nu B_\mu. \qquad (2.11)$$

From the first two terms in the Lagrangian (2.8) three gauge boson and four gauge boson self-couplings follow. The couplings of the gauge bosons to the fermions come from the covariant derivatives D_μ.

Explicit mass terms for both fermions and gauge bosons are not allowed in the SM because $-m\bar{\psi}\psi = -m(\bar{\psi}_L \psi_R + \bar{\psi}_R \psi_L)$ couples left- and right-handed fields that transform differently under $SU(2)_L$ and also $M^2 A_\mu^a A^{\mu,a}$ for gauge bosons breaks the gauge symmetry (see discussion in Sections 1.6 and 1.8). Thus, from (2.3) and (2.8) alone we do not get any mass terms. The Higgs mechanism, discussed in general terms in the previous chapter, provides a solution to this problem, and we will discuss it in some detail in Section 2.3. For completeness we give here only the relevant contribution of the Higgs doublet Φ to the Lagrangian of the SM:

$$\mathscr{L}_\Phi = \left(D_\mu \Phi\right)^\dagger (D^\mu \Phi) - V(\Phi). \qquad (2.12)$$

It contains the kinetic term for the Higgs with its couplings to gauge bosons and the Higgs potential. The covariant derivative for the Higgs is the same as for left-handed lepton doublets in (2.4) except for the value of the hypercharge that differs by sign as seen in Table 2.1.

The Yukawa sector of the SM contains the couplings of the Higgs to the fermions. It is very crucial for this book because it determines the flavor structure. The quantum numbers of the Higgs are such that it couples in a gauge invariant manner to an $SU(2)_L$ doublet and singlet so that the Yukawa part of the Lagrangian reads:

$$\mathcal{L}_Y = -\bar{Q}\Phi Y^D d_R - \bar{Q}\Phi^c\, Y^U u_R - \bar{L}\Phi Y^E\, e_R + h.c. \tag{2.13}$$

Here the fermion fields are three-component vectors such that all three generations are included, $Y^{D,U,E}$ are 3×3 Yukawa-matrices, and Φ^c is the charged conjugated Higgs field. More details will be given in Section 2.5. There is no Yukawa interactions with neutrinos because of the absence of right-handed neutrinos in the SM.

2.3 Spontaneous Symmetry Breakdown in the SM

In the SM the Higgs field $\Phi = \left(\phi^+, \phi^0\right)^\top$ is responsible for the breakdown of the electroweak to the electromagnetic gauge symmetry

$$SU(2)_L \times U(1)_Y \rightarrow U(1)_{em}, \tag{2.14}$$

and for generating $W^\pm$, Z^0, and fermion masses. The Higgs field is a color neutral $SU(2)_L$ doublet with hypercharge $Y = \frac{1}{2}$ whose neutral component develops a vacuum expectation value (vev)

$$\langle 0|\Phi|0\rangle = \frac{1}{\sqrt{2}}\begin{pmatrix} 0 \\ v \end{pmatrix}, \qquad v = \frac{\mu}{\sqrt{\lambda}}, \tag{2.15}$$

which minimizes the Higgs potential of the SM[2]

$$V(\Phi) = -\mu^2 \Phi^\dagger\Phi + \lambda\left(\Phi^\dagger\Phi\right)^2, \quad \mu, \lambda > 0. \tag{2.16}$$

The Higgs potential introduces two new parameters: the "negative" mass term $-\mu^2$ and the Higgs self coupling λ. The charge operator,

$$\hat{Q} = \hat{T}_3 + \hat{Y} = \frac{1}{2}\sigma_3 + \frac{1}{2}\mathbb{1}_{2\times 2} = \begin{pmatrix} 1 & 0 \\ 0 & 0 \end{pmatrix}, \tag{2.17}$$

leaves $(0, v)^\top$ invariant, and thus $U(1)_{em}$ stays unbroken. As we will see in Section 3.1 the mass of the $W^\pm$ boson M_W can be related to the Fermi constant G_F, which is

---

[2] In Sections 1.10.2 and 1.10.3 the sign in front of μ in the Higgs potential is different. There we demonstrated that only for $\mu^2 < 0$ spontaneous symmetry breaking was possible. Here we make the replacement $\mu^2 \rightarrow -\mu^2$ such that now SSB arises for $\mu^2 > 0$. Then μ^2 is directly related to the squared Higgs mass m_h^2.

known very precisely from the muon decay (see (3.3)). This important relation is given as follows

$$\frac{G_F}{\sqrt{2}} = \frac{g_2^2}{8M_W^2}.$$ (2.18)

But as we will soon derive, the vev v is related to M_W in such a manner that eventually the value of G_F gives uniquely the numerical value of v. In this chapter we do not include any higher-order electroweak corrections so that the following expressions, although rather accurate, represent the SM at the so-called tree-level. We will improve on it in later chapters in cases in which it is necessary.

To derive the formulas for the masses M_W and M_Z we will now look closer at the Higgs part of the SM Lagrangian in (2.12). The covariant derivative is given explicitly as follows

$$D_\mu = \partial_\mu - ig_1 Y B_\mu - ig_2 \frac{\sigma^a}{2} W_\mu^a = \partial_\mu - i \begin{pmatrix} g_1 Y B_\mu + \frac{g_2}{2} W_\mu^3 & \frac{g_2}{\sqrt{2}} W_\mu^+ \\ \frac{g_2}{\sqrt{2}} W_\mu^- & g_1 Y B_\mu - \frac{g_2}{2} W_\mu^3 \end{pmatrix},$$ (2.19)

where we replaced the real $W_\mu^{1,2}$ by $W_\mu^\pm$

$$W_\mu^- = \frac{1}{\sqrt{2}} \left(W_\mu^1 + iW_\mu^2 \right), \quad W_\mu^+ = \left[W_\mu^- \right]^\dagger = \frac{1}{\sqrt{2}} \left(W_\mu^1 - iW_\mu^2 \right),$$ (2.20)

which are eigenvectors to T_3 and also to Q due to $Y = 0$. As mentioned in Section 1.7.1 gauge bosons transform under the adjoint representations of the gauge group, which is given by the structure constants. Thus in SU(2) we have $(T_3)_{ab} = i\varepsilon_{a3b}$:

$$T_3 \begin{pmatrix} W_\mu^1 \\ W_\mu^2 \\ W_\mu^3 \end{pmatrix} = i \begin{pmatrix} 0 & -1 & 0 \\ 1 & 0 & 0 \\ 0 & 0 & 0 \end{pmatrix} \begin{pmatrix} W_\mu^1 \\ W_\mu^2 \\ W_\mu^3 \end{pmatrix} = \begin{pmatrix} -iW_\mu^2 \\ iW_\mu^1 \\ 0 \end{pmatrix}.$$ (2.21)

Consequently

$$T_3 W_\mu^\pm = Q W_\mu^\pm = \pm W_\mu^\pm,$$ (2.22)

i.e., $W_\mu^\pm$ have electric charges ± 1, whereas W_μ^3 and B_μ are electrically neutral.

We next define

$$\phi^0(x) = \frac{1}{\sqrt{2}} \left(v + h(x) + iG^0(x) \right),$$ (2.23)

$$\phi^+(x) = G^+(x), \quad G^-(x) = [G^+(x)]^\dagger,$$ (2.24)

where $G^\pm$ and G^0 are the Goldstone bosons and h will become the physical Higgs boson field. Inserting Φ into $V(\Phi)$ and keeping only terms proportional to h^2, we find as the Higgs mass

$$m_h = \sqrt{2}\mu = \sqrt{2\lambda}v.$$ (2.25)

One complication occurs, namely mixing between B_μ and W_μ^3, so that they cannot be mass eigenstates. Because U(1)$_{\rm em}$ is unbroken one of the mass eigenstates must be massless and identified with the photon A_μ. The second one is massive. It is Z_μ. As we have to keep the normalization of the kinetic terms of the gauge fields in (2.8) only an orthogonal transformation from (B_μ, W_μ^3) to (A_μ, Z_μ) characterized by the Weinberg mixing angle ϑ_W is allowed:[3]

$$
\begin{pmatrix} A_\mu \\ Z_\mu \end{pmatrix} = \begin{pmatrix} \cos \vartheta_W & \sin \vartheta_W \\ -\sin \vartheta_W & \cos \vartheta_W \end{pmatrix} \begin{pmatrix} B_\mu \\ W_\mu^3 \end{pmatrix},
\tag{2.26}
$$

or its inverse

$$
\begin{pmatrix} B_\mu \\ W_\mu^3 \end{pmatrix} = \begin{pmatrix} \cos \vartheta_W & -\sin \vartheta_W \\ \sin \vartheta_W & \cos \vartheta_W \end{pmatrix} \begin{pmatrix} A_\mu \\ Z_\mu \end{pmatrix}.
\tag{2.27}
$$

Inserting this back into (2.12) and (2.19) and requiring that A_μ couples only to electrically charged particles and not to ϕ^0, we get the following relations for the Weinberg angle and the couplings e, g_1 and g_2:

$$
\tan \vartheta_W = \frac{g_1}{g_2}, \quad e = g_2 \sin \vartheta_W = g_1 \cos \vartheta_W.
\tag{2.28}
$$

Gauge invariant mass terms for the gauge bosons arise from "mixed" terms in the covariant derivative where one picks up a vev v from Φ and, for example, a W^+. Thus the gauge boson masses can be calculated from

$$
\left[D_\mu \begin{pmatrix} 0 \\ v \end{pmatrix} \right]^\dagger D^\mu \begin{pmatrix} 0 \\ v \end{pmatrix} \overset{!}{=} M_W^2 W_\mu^- W^{\mu+} + \frac{1}{2} M_Z^2 Z_\mu Z^\mu + \frac{1}{2} m_\gamma A_\mu A^\mu,
\tag{2.29}
$$

$$
\Rightarrow M_W = \frac{g_2 v}{2}, \quad M_Z = \frac{g_2 v}{2 \cos \vartheta_W}, \quad m_\gamma = 0.
\tag{2.30}
$$

In summary, then,

$$
M_W = \frac{g_2 v}{2}, \quad M_Z = \frac{g_Z v}{2}, \quad g_Z = \frac{g_2}{\cos \vartheta_W} = \sqrt{g_1^2 + g_2^2},
\tag{2.31}
$$

and consequently

$$
\frac{M_W}{M_Z} = \cos \vartheta_W.
\tag{2.32}
$$

---

[3] Note that the sign in front of $\sin \vartheta_W$ is not unique, and different conventions appear in the literature. Appendix B should be useful in this respect.

Inserting the expression for M_W in (2.30) into (2.18), we find with our normalization in (2.15)

$$v = 2^{-\frac{1}{4}} G_F^{-\frac{1}{2}} = 246 \text{ GeV.} \tag{2.33}$$

The numerical value follows from (3.10) to which we will return in the next chapter. However, we should warn the reader that instead of (2.15) also $\langle \Phi \rangle = (0, v)^\top$ is common in the literature, such that in this case one has $v = 174$ GeV.

We observe that SSB implies relations between M_W, M_Z, e, $g_{1,2}$, which can be tested experimentally. Another important test is the manner in which the fermion masses are generated through the Higgs mechanism. While this topic will be discussed at later stages in this book, let us note that replacing Φ by its vev in (2.13) we get terms like $y_e \bar{L} \phi e_R \rightarrow \frac{1}{\sqrt{2}} y_e v e_L e_R$, where we set $y_e = Y_{11}^E$ such that we can identify the electron mass as $m_e = \frac{1}{\sqrt{2}} y_e v$. In the case of quarks the complication arises due to mixing between different quarks carrying the same electric charge. This implies that Yukawa matrices for up-quarks and down-quarks cannot be diagonalized simultaneously with very profound phenomenological consequences. This will be the subject of Section 2.5.

2.4 Gauge Boson Self-Interactions

The gauge bosons of the SM are the eight gluons G_μ^a of SU(3)$_C$, $W_\mu^{1,2,3}$ of SU(2)$_L$, and B_μ of U(1)$_Y$. As seen in Section 2.3 after electroweak symmetry breaking, W_μ^3 and B_μ mix into mass eigenstates Z^0 and A_μ. In this section we have a closer look at the kinetic terms of the gauge bosons as written in the Lagrangian in (2.8). We start with the first term for gluons because here the problem of mixing does not occur. It can be rewritten as (see also general discussion in Section 1.8):

$$\mathcal{L}_{\text{gauge}}^g = -\frac{1}{2} \left(\partial_\mu G_\nu^a \right) \left(\partial^\mu G^{\nu,a} \right) + \frac{1}{2} \left(\partial_\mu G_\nu^a \right) \left(\partial^\nu G^{\mu,a} \right) + \mathcal{L}_{\text{fix}} \tag{2.34}$$

$$- g_s f^{abc} \left(\partial_\mu G^{\nu,a} \right) G^{\mu,b} G^{\nu,c} \tag{2.35}$$

$$- \frac{1}{4} g_s^2 f^{abc} f^{ade} G_\mu^b G_\nu^c G^{\mu,d} G^{\nu,e}. \tag{2.36}$$

The first line determines the propagator, the second line describes triple gluon couplings, and the third line describes four-gluon couplings.

As discussed in any book on gauge field theories, to derive a propagator for a massless gauge boson we need an additional term $\mathcal{L}_{\text{fix}}$ that fixes the gauge. A_μ and G_μ^a, being massless, have two unphysical degrees of freedom. The spin-0 contribution is set to zero through the Lorenz condition $\partial_\mu A^\mu = 0$. However, we still have one degree of freedom left. The generalization of the Lorenz gauge are the so-called R_ξ gauges. Instead of fixing the gauge we add a Lagrange multiplier $1/(2\xi)$ to the Lagrangian

$$\mathcal{L}_{\text{fix}} = \frac{1}{2\xi} \left(\partial_\mu G^{\mu,a} \right)^2. \tag{2.37}$$

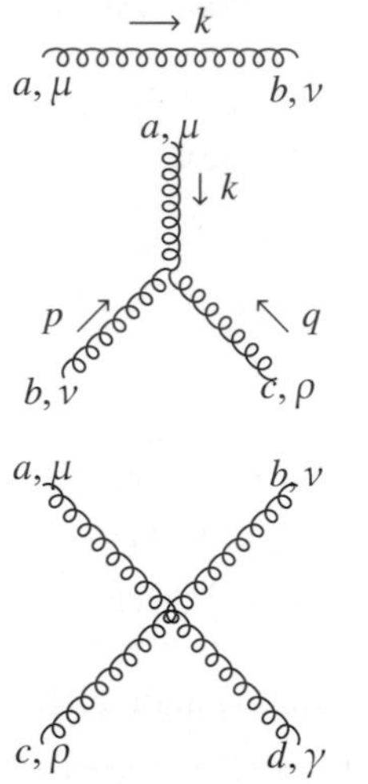

$$\frac{-i\delta_{ab}}{k^2+i\delta}\left(g^{\mu\nu}-(1-\xi)\frac{k^\mu k^\nu}{k^2}\right)$$

$$g_s f^{abc}\left[g^{\mu\nu}(k-p)^\rho + g^{\nu\rho}(p-q)^\mu + g^{\rho\mu}(q-k)^\nu\right]$$

$$-ig_s^2\Big[f^{abe}f^{cde}\left(g^{\mu\rho}g^{\nu\sigma}-g^{\mu\sigma}g^{\nu\rho}\right)+ f^{ace}f^{bde}\left(g^{\mu\nu}g^{\rho\sigma}-g^{\mu\sigma}g^{\nu\rho}\right)+ f^{ade}f^{bce}\left(g^{\mu\nu}g^{\rho\sigma}-g^{\mu\rho}g^{\nu\sigma}\right)\Big]$$

Figure 2.1 Propagator, triple vertex, and quartic vertex of the Gluon.

The Laundau gauge corresponds to $\xi \to 0$, the Feynman–'t Hooft gauge to $\xi = 1$. We get the Feynman rules as shown in Figure 2.1.

For $SU(2)_L$ in the basis of $W_\mu^{1,2,3}$ the Lagrangian and the triple and quartic coupling look exactly the same with the replacement $f^{abd} \to \varepsilon^{abd}$ and $g_s \to g_2$. Furthermore one simplification occurs because $f^{abe}f^{cde} = \varepsilon^{abe}\varepsilon^{cde} = \delta_{ac}\delta_{bd} - \delta_{ad}\delta_{bc}$ such that only the $W_\mu^1 W_\mu^2 W_\mu^3$ coupling is left over. However mixing makes it a little more complicated. For example, the triple W^+W^-A vertex has the coupling e in front and the $W^+W^-Z^0$ vertex $e\cot\vartheta_W$. Because it is an unitary rotation the kinetic terms for the propagator do not contain mixed terms:

$$\begin{aligned}
&\frac{1}{4}\left(\partial_\mu W_\nu^a - \partial_\nu W_\mu^a\right)\left(\partial^\mu W^{\nu,a} - \partial^\nu W^{\mu,a}\right) + \frac{1}{4}\left(\partial_\mu B_\nu - \partial_\nu B_\mu\right)\left(\partial^\mu B^\nu - \partial^\nu B^\mu\right)\\
&\quad = \frac{1}{2}\left(\partial_\mu W_\nu^+ - \partial_\nu W_\mu^+\right)\left(\partial^\mu W^{\nu-} - \partial^\nu W^{\mu-}\right)\\
&\quad + \frac{1}{4}\left(\partial_\mu Z_\nu - \partial_\nu Z_\mu\right)\left(\partial^\mu Z^\nu - \partial^\nu Z^\mu\right) + \frac{1}{4}\left(\partial_\mu A_\nu - \partial_\nu A_\mu\right)\left(\partial^\mu A^\nu - \partial^\nu A^\mu\right).
\end{aligned} \tag{2.38}$$

Because the electromagnetic force is based on a U(1) symmetry, there are no triple and quartic photon couplings, but only W^+W^-A, W^+W^-AZ, and W^+W^-AA vertices. We collect the Feynman rules in the SM in Appendix B.

2.5 Flavor Structure of the SM

2.5.1 Rotation from Flavor to Mass Eigenstates

The Yukawa sector of the SM contains most of its free parameters: six quark masses, three charged lepton masses, three mixing angles, and one phase. Neutrinos are massless in the

SM, although we know that they have masses. We will improve on it soon but this will imply new parameters.

The interaction Lagrangian of (2.3) is invariant under a $[U(3)]^5$ flavor symmetry

$$Q_L \to V_L^u Q, \quad u_R \to V_R^u u_R, \quad d_R \to V_R^d d_R, \tag{2.39}$$

$$L \to V_L^e L, \quad e_R \to V_R^e e_R, \tag{2.40}$$

where $V_L^u, V_R^u, V_R^d, V_L^e$, and V_R^e are unitary 3×3 matrices. However, the Yukawa sector (2.13) breaks $[U(3)]^5$ down to $[U(1)]^4$, which results in the conservation of baryon number B and of the three individual lepton flavor numbers $L_{e,\mu,\tau}$. These are, however, accidental symmetries of the SM, i.e., they are not imposed from the beginning and do not have to be an exact symmetry in models beyond the SM. We will discuss these symmetries in more detail in Section 15.1.2.

As a side remark, let us recall that the gauge group of the SM in (2.1) is a subgroup of the $[U(3)]^5$ flavor symmetry and with the quantum numbers of all SM particles in Table 2.1 is free from the so-called triangle anomalies as required by the renormalizability of the SM. It is interesting that among the leftover U(1) global symmetries only $U(1)_{B-L}$ is anomalyfree and as such could be gauged. Even $B + L$ is broken by anomalies. However, Majorana masses for neutrinos discussed later break L and consequently also $B - L$. The issue of triangle anomalies and their cancellations is a very important constraint on NP models, and we will return to it in Chapter 15 illustrating it on a few examples. General discussion can be found in many textbooks on field theory. See in particular [18].

The Yukawa matrices in (2.13) are a priori arbitrary 3×3 matrices. But we can use the $[U(3)]^5$ flavor symmetry of the interaction Lagrangian in (2.3) to simplify them. The question then arises whether we could diagonalize all Yukawa matrices simultaneously by just using the rotations in (2.39).

It is easy to convince oneself that this is impossible. In fact, quite generally Yukawa matrices can be diagonalized through biunitary transformations:

$$(V_L^d)^\dagger Y^D V_R^d = \hat{Y}^D, \quad (V_L^u)^\dagger Y^U V_R^u = \hat{Y}^U, \quad (V_L^e)^\dagger Y^E V_R^e = \hat{Y}^E. \tag{2.41}$$

Because left-handed fields are embedded into doublets Q and L, we have to rotate the members of the doublets with the same matrix, i.e., u_L and d_L are rotated in the same way and also ν_L and e_L. Usually one rotates Q with V_L^u and L with V_L^e as written in (2.39) and (2.40). Once this is done these rotations are fixed.

Now from (2.41) we can see that four matrices in the quark sector are needed to diagonalize both $Y^{U,D}$, but we have only three of them: V_L^u, V_R^u, and V_R^d. The matrix V_L^d is missing in (2.39). Equivalently we could perform the first rotation in (2.39) with the help of V_L^d but then V_L^u would be missing. We thus conclude that the matrices Y^D and Y^U cannot be simultaneously diagonalized by rotations, which leaves the interaction Lagrangian invariant. On the other hand, without right-handed neutrinos that are not present in the original version of the SM, the lepton Yukawa matrix Y^E can be diagonalized.

In summary after these rotations we get

$$\mathscr{L}_Y = -\bar{Q}\Phi(V_L^u)^\dagger V_L^d \hat{Y}^D d_R - \bar{Q}\Phi^c \hat{Y}^U u_R - \bar{L}\Phi\hat{Y}^E e_R + h.c. \tag{2.42}$$

with the first term being nondiagonal. To diagonalize this term and consequently get the mass eigenstate basis we must perform an additional rotation of the down-quarks

$$d_L \to d'_L = (V^u_L)^\dagger V^d_L \, d_L, \tag{2.43}$$

where d'_L are the original flavor eigenstates from where we started and d_L the mass eigenstates. But this additional rotation modifies the interaction Lagrangian in (2.3), which written in terms of d_L involves the matrix $(V^u_L)^\dagger V^d_L$.

This mismatch between mass and flavor eigenstates is responsible for flavor transitions in the SM, and the matrix $(V^u_L)^\dagger V^d_L$ is just the CKM matrix, named after Cabibbo, Kobayashi, and Maskawa [19, 20]:

$$V_{\text{CKM}} = (V^u_L)^\dagger V^d_L, \qquad d'_L = V_{\text{CKM}} \, d_L. \tag{2.44}$$

Using the common notation for the entries of the CKM matrix we have then a very important relation between flavor and mass eigenstates

$$\begin{pmatrix} d'_L \\ s'_L \\ b'_L \end{pmatrix} = \begin{pmatrix} V_{ud} & V_{us} & V_{ub} \\ V_{cd} & V_{cs} & V_{cb} \\ V_{td} & V_{ts} & V_{tb} \end{pmatrix} \begin{pmatrix} d_L \\ s_L \\ b_L \end{pmatrix}. \tag{2.45}$$

Note that we now resolved d_L into its three components $(d, s, b)_L$, and we will do the same with $u_L = (u, c, t)_L$ and corresponding right-handed fields. The elements of the CKM matrix will play a very profound role in this book. As we will see later V_{ud} determines the strength of $d \to u$ flavor transitions and similar for other elements. The basis where all Yukawa couplings are simultaneously diagonal is called mass eigenstate basis. If we now insert the vev of Φ into (2.42) we get diagonal 3×3 mass matrices:

$$\hat{M}^U = \text{diag} \, (m_u, m_c, m_t) = \frac{v}{\sqrt{2}} \hat{Y}^U = \frac{v}{\sqrt{2}} \cdot \text{diag} \, (y_u, y_c, y_t), \tag{2.46}$$

$$\hat{M}^D = \text{diag} \, (m_d, m_s, m_b) = \frac{v}{\sqrt{2}} \hat{Y}^D = \frac{v}{\sqrt{2}} \cdot \text{diag} \, (y_d, y_s, y_b), \tag{2.47}$$

$$\hat{M}^E = \text{diag} \, (m_e, m_\mu, m_\tau) = \frac{v}{\sqrt{2}} \hat{Y}^E = \frac{v}{\sqrt{2}} \cdot \text{diag} \, (y_e, y_\mu, y_\tau). \tag{2.48}$$

2.5.2 Including Right-Handed Neutrinos

From the experimental observation of neutrino oscillations we know that neutrinos must have a small mass. However, from the construction of the SM, as we have seen earlier, neutrinos stay exactly massless. This fact tells us already that the SM is not the whole story and that there must be some kind of new physics beyond the SM that allows to generate neutrino masses. The literature on this topic is very rich. Here we describe first the simplest solutions to this problem. We will briefly discuss the role of neutrinos in weak decays in Section 17.5.

There are various possibilities to generate massive neutrinos. While it is possible to generate them through a coupling of an additional Higgs triplet to the usual left-handed neutrinos, let us have a look first at how neutrino masses can be generated with the help of the SM Higgs doublet. This requires the existence of right-handed neutrinos. Right-handed neutrinos, if they exist, are singlets under the SM gauge group and transform as

$$\nu_R \sim (\mathbf{1}, \mathbf{1}, 0). \tag{2.49}$$

Thus, in principle they do not feel the SM gauge interactions at all. However they do have an effect on the $W^{\pm}$ couplings as soon as neutrinos are massive. In analogy to the quark sector we can add an additional neutrino Yukawa term in (2.13), which will generate a Dirac neutrino mass term after electroweak symmetry breaking in the same manner as in the quark sector. Thus we add a term

$$-\bar{L} \Phi^c \, Y^{\nu} \nu_R \tag{2.50}$$

to (2.13). The neutrino Yukawa matrix can also be diagonalized by a biunitary transformation

$$(V_L^{\nu})^{\dagger} Y^{\nu} V_R^{\nu} = \hat{Y}^{\nu}, \tag{2.51}$$

and in addition to (2.40) we have an additional rotation for the right-handed neutrinos $\nu_R \to V_R^{\nu} \nu_R$. With this setup this works exactly as for quarks, and in (2.42) we have to add $-\bar{L} \Phi^c (V_L^e)^{\dagger} V_L^{\nu} \hat{Y}^{\nu} \nu_R$. As at this stage neutrinos are still in the flavor basis, additional rotation on neutrino fields has to be performed bringing us to the mass eigenstates. Evidently this rotation is made by the mixing matrix in the lepton sector

$$U_{\text{PMNS}} = (V_L^e)^{\dagger} V_L^{\nu}, \tag{2.52}$$

the Pontecorvo–Maki–Nakagawa–Sakata (PMNS) matrix [21, 22]. Rotation to the mass eigenstates in the lepton sector is then done through

$$\begin{pmatrix} \nu_e \\ \nu_{\mu} \\ \nu_{\tau} \end{pmatrix} = \begin{pmatrix} U_{e1} & U_{e2} & U_{e3} \\ U_{\mu 1} & U_{\mu 2} & U_{\mu 3} \\ U_{\tau 1} & U_{\tau 2} & U_{\tau 3} \end{pmatrix} \begin{pmatrix} \nu_1 \\ \nu_2 \\ \nu_3 \end{pmatrix}, \tag{2.53}$$

where ν_{ℓ} with $\ell = e, \mu, \tau$ are flavor eigenstates and ν_i, with $i = 1, 2, 3$ mass eigenstates. Comparing with the CKM matrix in (2.45), we note that whereas the PMNS matrix relates neutrinos in the mass and interaction (flavor) bases, the CKM matrix is doing it for down-quarks. This is only the convention used in the literature.

This generation of neutrino masses assumes that neutrinos are Dirac particles and is a straightforward generalization of the one for quark masses. But the tiny masses of neutrinos, when compared even with electron mass, suggest that a different mechanism for mass generation involving right-handed neutrinos could be at work here. Indeed as neutrinos carry no electric charge they could also be Majorana particles. As right-handed neutrinos are SM singlets in this case, it is also allowed to add a Majorana mass term to (2.13):

$$-\frac{1}{2}\nu_R^\top C \, \mathsf{M}_N \nu_R. \tag{2.54}$$

Here C is the charge conjugation operator and M_N is an arbitrary Majorana mass matrix. Because this mass term is not protected by any symmetry (the Dirac mass terms are protected by the chiral symmetry) M_N can be much larger than the electroweak scale. This Majorana mass term together with (2.50) leads to the seesaw mechanism that can explain the light neutrino masses observed in nature with a high Majorana mass scale. For the PMNS matrix one just has to exchange the matrix V_L^ν in (2.52) by another matrix U_χ that diagonalizes the effective neutrino mass matrix that one gets through the seesaw mechanism. We refer to an excellent book of Bilenky [23], in which this sector of particle physics is described.

While the latter mechanism for mass generation has no direct analog in the quark sector because quarks carry electric charges, there exist other means of generating quark masses then just coupling quarks to the Higgs. This brings us to one of the most important puzzles in the SM.

2.5.3 The Flavor Puzzle in the SM

We have seen earlier how quarks and leptons get their masses in the SM and from where the CKM and PMNS matrices arise. However, this does not explain the values of the quark and lepton masses and of the observed structure of the unitary CKM and PMNS matrices. With the discovery of the Higgs boson and the measurement of Yukawa couplings we know that this mechanism for mass generation seems to be correct. Yet the Yukawa couplings and the elements of CKM and PMNS matrices are still free parameters, and in particular we would like to understand the observed mass spectrum. Indeed, the masses of the SM particles span a very wide range: from a heavy top quark of roughly 170 GeV to a tiny electron mass in the MeV region and even much lighter neutrino masses in the eV region (see Table D.1).

Qualitatively the CKM and PMNS matrices look as follows[4]

$$V_{\text{CKM}} = \begin{pmatrix} \bullet & \cdot & \cdot \\ \cdot & \bullet & \cdot \\ \cdot & \cdot & \bullet \end{pmatrix}, \qquad U_{\text{PMNS}} = \begin{pmatrix} \bullet & \bullet & \cdot \\ \cdot & \bullet & \bullet \\ \cdot & \bullet & \bullet \end{pmatrix}. \tag{2.55}$$

The hierarchical structure of the CKM matrix is clearly visible, whereas the magnitudes of the entries of the PMNS matrix are of the same order. The explanation of the origin of these patterns and in particular why they differ from each other remains an important goal of the theorists.

---

[4] The radius of the circles illustrate the absolute values of the entries, i.e., $r \simeq \sqrt{|V_{ij}|}$.

The flavor puzzle of the SM can be summarized as follows: Why do the Yukawa couplings cover a large range of $O(10^{-6})$ to $O(1)$? Why are the mixing angles of the CKM matrix so small and those of the PMNS matrix rather large? Why are there exactly three particle generations?

There is a very rich literature on the generation of fermion masses and related CKM and PMNS matrices by some new dynamics, but we cannot describe these efforts here. Useful references can be found in the first paragraphs of [24]. Moreover, one should keep in mind that quark and lepton masses and the parameters of CKM and PMNS matrices could remain as given parameters of nature forever, and there is no dynamical explanation of their values. More important for the time being is the derivation of the Feynman rules for electroweak interactions of quarks and leptons. We begin with neutral currents coupled to the heavy gauge boson Z and the photon A. Subsequently we will discuss charged currents coupled to $W^{\pm}$.

2.5.4 Neutral Currents

To derive the couplings of quarks and leptons to the photon A_{μ} and to Z_{μ} we first insert the mixing in (2.26) back into the covariant derivatives (2.4)–(2.7) using the relations in (2.28). For doublets we can also use directly (2.19) with appropriate values of the hypercharge Y. For *lepton doublets* L_j we get, for example,

$$D_{\mu} = \partial_{\mu} - ie \begin{pmatrix} K_{11} & K_{12} \\ K_{21} & K_{22} \end{pmatrix}, \tag{2.56}$$

where

$$K_{11} = \left(Y + \frac{1}{2}\right) A_{\mu} - \frac{2Y \tan^2 \vartheta_W - 1}{2 \tan \vartheta_W} Z_{\mu} = \frac{1}{2 \sin \vartheta_W \cos \vartheta_W} Z_{\mu}, \tag{2.57}$$

$$K_{12} = \frac{1}{\sqrt{2} \sin \vartheta_W} W_{\mu}^{+}, \tag{2.58}$$

$$K_{21} = \frac{1}{\sqrt{2} \sin \vartheta_W} W_{\mu}^{-}, \tag{2.59}$$

$$K_{22} = \left(Y - \frac{1}{2}\right) A_{\mu} - \frac{2Y \tan^2 \vartheta_W + 1}{2 \tan \vartheta_W} Z_{\mu} = -A_{\mu} - \frac{-\sin^2 \vartheta_W + \frac{1}{2}}{\sin \vartheta_W \cos \vartheta_W} Z_{\mu}, \tag{2.60}$$

and in evaluating K_{11} and K_{22}, we set $Y = -1/2$ in accordance with Table 2.1.
For *lepton singlets* e_{Rj} we find

$$D_{\mu} = \partial_{\mu} - i \frac{e}{\cos \vartheta_W} Y \left(\cos \vartheta_W A_{\mu} - \sin \vartheta_W Z_{\mu}\right)\Big|_{Y=-1} = \partial_{\mu} + ie A_{\mu} - ie \tan \vartheta_W Z_{\mu}. \tag{2.61}$$

Inserting this back into (2.3) and only considering the couplings to neutral gauge bosons Z_μ and A_μ we get

$$\mathscr{L}_I^{\mathrm{SM}} \supset \sum_{j=1,2,3} \bar{L}_j i \slashed{D} L_j + \bar{e}_{Rj} i \slashed{D} e_{Rj} \supset$$

$$\sum_{j=1,2,3} \left[\frac{e}{2 \sin \vartheta_W \cos \vartheta_W} \bar{\nu}_{Lj} \gamma^\mu Z_\mu \nu_{Lj} - e\bar{e}_{Lj} \gamma^\mu A_\mu e_{Lj} - e\bar{e}_{Rj} \gamma^\mu A_\mu e_{Rj} \right. \tag{2.62}$$

$$\left. + \frac{e}{\sin \vartheta_W \cos \vartheta_W} \left(-\frac{1}{2} + \sin^2 \vartheta_W \right) \bar{e}_{Lj} \gamma^\mu Z_\mu e_{Lj} + \frac{e}{\sin \vartheta_W \cos \vartheta_W} \sin^2 \vartheta_W \bar{e}_{Rj} \gamma^\mu Z_\mu e_{Rj} \right].$$

As it should be, the photon couples in the same manner to left- and right-handed particles. The same procedure can be used for quarks, and the result can be simplified and summarized as follows:

$$\mathscr{L}_I^{\mathrm{SM}} \supset \frac{e}{\sin \vartheta_W \cos \vartheta_W} \left(T_3 - \sin^2 \vartheta_W Q_f \right) \bar{f} \gamma^\mu Z_\mu f + e Q_f \bar{f} \gamma^\mu A_\mu f, \tag{2.63}$$

where f is RH or LH fermion (quark or lepton) with weak isospin T_3 ($T_3 = 0$ for RH fermions and $T_3 = \pm \frac{1}{2}$ for LH fermions) and electric charge Q_f in units of e, e.g., $Q_e = -1$ for the electron. If one multiplies (2.63) by i one gets directly the Feynman rules for the interaction of Z_μ and A_μ with SM fermions. We list these rules in Appendix B.

The expression in (2.63) is valid for all three generations. The CKM matrix that occurred in the Yukawa sector as explained in Section 2.5 does not occur in interactions with the Z boson and the photon. This is due to the unitarity of the CKM matrix. For $V_\mu = Z_\mu$, $V_\mu = G_\mu$, or $V_\mu = A_\mu$ neutral currents involving down-type quarks do not change after the rotation to mass eigenstates in (2.43):

$$\sum_{j=1,2,3} \bar{d}_{Lj} \gamma_\mu d_{Lj} V^\mu \rightarrow \sum_{j,k=1,2,3} \bar{d}_{Lj} \underbrace{\left(V_{\mathrm{CKM}}^\dagger V_{\mathrm{CKM}} \right)_{jk}}_{= \delta_{jk}} \gamma_\mu d_{Lk} V^\mu. \tag{2.64}$$

This is a very important result and is called tree-level GIM mechanism:

There are no flavor-changing neutral currents (FCNCs) at tree level in the SM. This consequence of the unitarity of the CKM matrix is called GIM mechanism, named after Glashow, Iliopoulos, and Maiani [6].

However the GIM mechanism is also active in many loop processes. FCNCs occur at loop level but not at tree level within the SM. In the so-called penguin diagrams, transitions like $b \rightarrow s$ and $s \rightarrow d$ are possible (see Figure 2.2). In the loop all three up-type quarks appear. The structure of such a penguin diagram is

$$P = \lambda_u f(m_u, M_W) + \lambda_c f(m_c, M_W) + \lambda_t f(m_t, M_W) \tag{2.65}$$

$$= \lambda_c \left(f(m_c, M_W) - f(m_u, M_W) \right) + \lambda_t \left(f(m_t, M_W) - f(m_u, M_W) \right), \qquad \lambda_i = V_{id} V_{is}^*,$$

with a loop function f that depends on M_W and the quark mass entering the loop. In the second line we used the unitarity of the CKM matrix

$$\lambda_u + \lambda_c + \lambda_t = 0. \tag{2.66}$$

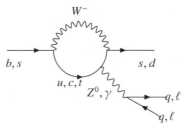

Figure 2.2 FCNC at loop level: penguin diagram for a $b \to s$ or $s \to d$ transition.

The important virtue of the GIM mechanism is that it does not only forbid tree-level FCNCs but also suppresses FCNCs at loop level for observables in which only light-quarks matter. It would also work if the masses of quarks with a given charge would be the same. Indeed for degenerate masses $m_u = m_c = m_t$ the penguin diagram in (2.65) would vanish, and this is a general property of FCNCs in the SM as we will see in other chapters of this book. For different masses a GIM suppression factor occurs when only the c and u quark contributions are relevant. Indeed, in this case

$$(f(m_c, M_W) - f(m_u, M_W)) \propto \frac{m_c^2 - m_u^2}{M_W^2} \approx 2 \cdot 10^{-4}, \tag{2.67}$$

which allows to suppress naturally certain observables in the K meson system to a desired level. However, due to $m_t > M_W \gg m_{u,c}$ there is no GIM suppression when the contributions of the top quark in the loops dominate a given observable. What is interesting is that the pattern of the violation of GIM mechanism through the top quark contributions appears until now to be confirmed within experimental and theoretical uncertainties by present data. We will witness this in the phenomenological parts of this book. But as we will stress there, beyond SM this structure could be violated, and FCNCs could take place already at tree level.

2.5.5 Charged Currents

The couplings of charged gauge bosons $W_\mu^\pm$ to fermions result from the covariant derivative in (2.3) and can be directly derived from (2.58) and (2.59). When rotating to the mass eigenstates the CKM and PMNS matrices occur:

$$\mathcal{L}^{CC} = \frac{g_2}{\sqrt{2}} \left(\bar{u}_j \gamma^\mu P_L d_j W_\mu^+ + \bar{\nu}_j \gamma^\mu P_L \ell_j W_\mu^+ \right) + h.c.$$

$$\to \frac{g_2}{\sqrt{2}} \left(\bar{u}_j \gamma^\mu V_{jk} P_L d_k W_\mu^+ + \bar{\nu}_j \gamma^\mu U_{kj}^* P_L \ell_k W_\mu^+ \right) + h.c. \tag{2.68}$$

$$= \frac{g_2}{\sqrt{2}} (\bar{u} \gamma^\mu V_{\text{CKM}} P_L d\, W_\mu^+ + \bar{\nu} \gamma^\mu U_{\text{PMNS}}^\dagger P_L \ell\, W_\mu^+) + h.c.$$

The complex conjugation in the PMNS matrix should be noticed, but this should be clear from the definition of this matrix in (2.52).

In the original formulation of the SM without neutrino oscillations one has to set $U_{\text{PMNS}} = 1$. The corresponding Feynman rules for the mass eigenstates are shown in

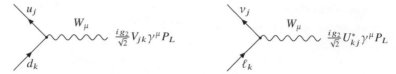

Figure 2.3 Feynman rules for charged W-fermion couplings. Here we already take into account neutrino mixing.

Figure 2.3. As one can see flavor violation or flavor transitions occur now in gauge interactions. However, one should emphasize that the gauge sector is in principle flavor blind: (2.3) doesn't distinguish between the different generations, and it is invariant under a rotation in flavor space as in (2.39) and (2.40). But the Yukawa sector is no longer flavor blind. So the origin of flavor violation comes from the Yukawa interactions. Only after diagonalizing all Yukawa matrices, i.e., in the mass eigenstate basis, flavor violation is transferred to the charged currents of the gauge sector.

2.5.6 CKM Matrix, CP Violation, and the Unitarity Triangle

The CKM matrix was already defined in (2.44). As a unitary 3×3 matrix, it can be parametrized in general by three angles and six phases. But not all of them are physical. The diagonal Yukawa couplings $\hat{Y}^U$, $\hat{Y}^D$ possess a $[U(1)]^6$ symmetry, i.e.,

$$u_j^{L,R} \to e^{i\varphi_j^u} u_j^{L,R}, \quad d_j^{L,R} \to e^{i\varphi_j^d} d_j^{L,R}, \tag{2.69}$$

leave $\bar{u}_L \hat{Y}^U u_R$ and $\bar{d}_L \hat{Y}^D d_R$ invariant but not the W couplings. With this we can rotate away five out of the six phases via

$$V_{ij}^{\text{CKM}} \to e^{i(\varphi_j^d - \varphi_i^u)} V_{ij}^{\text{CKM}}. \tag{2.70}$$

The only leftover phase in the SM is called Kobayashi–Maskawa phase [20]. It is the only origin of CP violation in the SM[5] and always connected with flavor changing transitions. Because there is only one single phase in the CKM matrix, CP violating processes are correlated with each other in the SM. The standard parametrization [25] of the CKM matrix with three angles and one phase is given as

$$V_{\text{CKM}} = \begin{pmatrix} 1 & 0 & 0 \\ 0 & c_{23} & s_{23} \\ 0 & -s_{23} & c_{23} \end{pmatrix} \begin{pmatrix} c_{13} & 0 & s_{13}e^{-i\delta} \\ 0 & 1 & 0 \\ -s_{13}e^{i\delta} & 0 & c_{13} \end{pmatrix} \begin{pmatrix} c_{12} & s_{12} & 0 \\ -s_{12} & c_{12} & 0 \\ 0 & 0 & 1 \end{pmatrix} \tag{2.71}$$

and consequently

$$V_{\text{CKM}} = \begin{pmatrix} c_{12}c_{13} & s_{12}c_{13} & s_{13}e^{-i\delta} \\ -s_{12}c_{23} - c_{12}s_{23}s_{13}e^{i\delta} & c_{12}c_{23} - s_{12}s_{23}s_{13}e^{i\delta} & s_{23}c_{13} \\ s_{12}s_{23} - c_{12}c_{23}s_{13}e^{i\delta} & -c_{12}s_{23} - s_{12}c_{23}s_{13}e^{i\delta} & c_{23}c_{13} \end{pmatrix}. \tag{2.72}$$

[5] We do not consider here the θ term of QCD, which is related to the strong CP violation that we will briefly discuss in Section 17.3.

The abbreviations s_{ij} and c_{ij} stand for $\sin\theta_{ij}$ and $\cos\theta_{ij}$. If there were only two generations, no physical phase would be left over but only one mixing angle, the Cabibbo angle θ_C.

From experiment we know that the CKM matrix is a hierarchical matrix. It is close to the unit matrix with small off-diagonal entries. In particular to an excellent approximation:

$$c_{12} = 1 - \frac{\lambda^2}{2}, \qquad c_{13} = 1, \qquad c_{23} = 1, \tag{2.73}$$

with λ being one of the Wolfenstein parameters, discussed later, that is equal to the $\sin\theta_C$ or equivalently to s_{12}. The virtue of the parametrization (2.72) supplemented by (2.73) is that by measuring $|V_{us}|$, $|V_{ub}|$, and $|V_{cb}|$ in tree-level decays one can determine s_{12}, s_{13}, and s_{23} simply through:

$$s_{12} = |V_{us}|, \qquad s_{13} = |V_{ub}|, \qquad s_{23} = |V_{cb}|. \tag{2.74}$$

This relation and (2.73) express the known fact that θ_{ij} can be chosen without losing the generality to be in the first quadrant. While then the phase δ could have in principle any value in the range $[0, 2\pi]$, we know already for some time that nature chooses it to be in the first quadrant as well. The values of all these parameters will play an important role in this book, and we will discuss their determination in the next chapters.

The hierarchical structure of the CKM matrix is nicely represented by its Wolfenstein parametrization [26]. It is an expansion in the parameter $\lambda = |V_{us}| = 0.225$. The other three parameters A, ρ, and η of this parametrization lie between zero and one. Thus

$$V_{\mathrm{CKM}} = \begin{pmatrix} 1 - \frac{\lambda^2}{2} & \lambda & A\lambda^3(\rho - i\eta) \\ -\lambda & 1 - \frac{\lambda^2}{2} & A\lambda^2 \\ A\lambda^3(1 - \rho - i\eta) & -A\lambda^2 & 1 \end{pmatrix} + O(\lambda^4). \tag{2.75}$$

In the flavor precision era one should at least include $O(\lambda^4)$ corrections and possibly higher-order terms. An efficient and systematic way of finding higher-order terms in λ has been proposed in [27]. One goes back to the standard parametrization (2.72) and *defines* the parameters $(\lambda, A, \varrho, \eta)$ through [27]

$$s_{12} = \lambda, \qquad s_{23} = A\lambda^2, \qquad s_{13}e^{-i\delta} = A\lambda^3(\varrho - i\eta), \tag{2.76}$$

to *all orders* in λ. It follows that

$$\varrho = \frac{s_{13}}{s_{12}s_{23}}\cos\delta, \qquad \eta = \frac{s_{13}}{s_{12}s_{23}}\sin\delta. \tag{2.77}$$

Equations (2.76) and (2.77) represent simply the change of variables from the standard ones to the ones of Wolfenstein. Making this change of variables in the standard parametrization (2.72) we find the CKM matrix as a function of $(\lambda, A, \varrho, \eta)$, which satisfies unitarity exactly! Expanding next each element in powers of λ, we recover the matrix in (2.75) and in addition find explicit corrections of $O(\lambda^4)$ and higher-order terms.[6]

---

[6] A similar parametrization has been proposed by Branco and Lavoura [28]. See [7] for a general discussion.

The result of this exercise can be found in [29]. One finds that by definition V_{ub} remains unchanged, and the corrections to V_{us} and V_{cb} appear only at $O(\lambda^7)$ and $O(\lambda^8)$, respectively. Consequently to a very good accuracy we have:

$$V_{us} = \lambda, \qquad V_{cb} = A\lambda^2, \tag{2.78}$$

$$V_{ub} = A\lambda^3(\varrho - i\eta), \qquad V_{td} = A\lambda^3(1 - \bar{\varrho} - i\bar{\eta}) \tag{2.79}$$

with

$$\boxed{\bar{\varrho} = \varrho\left(1 - \frac{\lambda^2}{2}\right), \qquad \bar{\eta} = \eta\left(1 - \frac{\lambda^2}{2}\right).} \tag{2.80}$$

Useful analytic expressions for $\lambda_i = V_{id}V_{is}^*$ with $i = c, t$ are the following ones

$$\mathrm{Im}\lambda_t = -\mathrm{Im}\lambda_c = \eta A^2 \lambda^5 = |V_{ub}||V_{cb}|\sin\delta, \tag{2.81}$$

$$\mathrm{Re}\lambda_c = -\lambda\left(1 - \frac{\lambda^2}{2}\right), \tag{2.82}$$

$$\mathrm{Re}\lambda_t = -\left(1 - \frac{\lambda^2}{2}\right)A^2\lambda^5(1 - \bar{\varrho}), \tag{2.83}$$

and a more accurate formula for $\mathrm{Re}\lambda_t$ can be derived ($\gamma = \delta$)

$$\mathrm{Re}\lambda_t \simeq |V_{ub}||V_{cb}|\cos\gamma(1 - 2\lambda^2) + (|V_{ub}|^2 - |V_{cb}|^2)\lambda\left(1 - \frac{\lambda^2}{2}\right). \tag{2.84}$$

While these expressions are useful for the derivation of analytic formulas for various observables, in numerical calculations these days it is better to use exact standard parametrization as given in (2.72). To this end, as suggested to my knowledge by CKMfitters, $\bar{\varrho}$ and $\bar{\eta}$ can be defined in a phase convention-independent manner through

$$\boxed{\bar{\varrho} + i\bar{\eta} = -\frac{V_{ud}V_{ub}^*}{V_{cd}V_{cb}^*}.} \tag{2.85}$$

One can check that in the standard parametrization this definition agrees to an excellent accuracy with the original one in (2.80).

Before continuing our discussion of the CKM matrix, it should be emphasized that in contrast to strong and electromagnetic interactions, which are both invariant under parity P, charge conjugation C, and time reversal T, the weak interaction violates P and C maximally, and the combination CP is violated by the phase δ of the CKM matrix. In the standard phase convention we have under CP

$$\bar{u}_L\gamma_\mu d_L \xrightarrow{CP} -\bar{d}_L\gamma^\mu u_L, \qquad W_\mu^\pm \xrightarrow{CP} -W^{\mp\mu}. \tag{2.86}$$

Thus, this means for charged currents:

$$\mathcal{L}_q^{CC} = -\frac{g_2}{\sqrt{2}}\left(\bar{u}_j\gamma^\mu V_{jk}P_L d_k W_\mu^+ + \bar{d}_k\gamma^\mu V_{jk}^* P_L u_j W_\mu^-\right)$$
$$\xrightarrow{CP} -\frac{g_2}{\sqrt{2}}\left(\bar{d}_k\gamma_\mu V_{jk}P_L u_j W^{\mu-} + \bar{u}_j\gamma_\mu V_{jk}^* P_L d_k W^{\mu+}\right). \tag{2.87}$$

This Lagrangian is only invariant under CP if $V_{jk} = V_{jk}^*$ for all $j, k = 1, 2, 3$. Furthermore, to get CP violation all three generations must be involved in the process because with only two generations there exist no physical phases.

The unitarity of the CKM matrix leads first to the following set of equations:

$$|V_{ud}|^2 + |V_{cd}|^2 + |V_{td}|^2 = 1, \tag{2.88}$$

$$|V_{us}|^2 + |V_{cs}|^2 + |V_{ts}|^2 = 1, \tag{2.89}$$

$$|V_{ub}|^2 + |V_{cb}|^2 + |V_{tb}|^2 = 1, \tag{2.90}$$

$$|V_{ud}|^2 + |V_{us}|^2 + |V_{ub}|^2 = 1, \tag{2.91}$$

$$|V_{cd}|^2 + |V_{cs}|^2 + |V_{cb}|^2 = 1, \tag{2.92}$$

$$|V_{td}|^2 + |V_{ts}|^2 + |V_{tb}|^2 = 1, \tag{2.93}$$

that describe the normalization of the columns and rows of the CKM matrix, respectively. They are sometimes used to test the unitarity of this matrix.

The remaining six relations originate from the orthogonality of different columns and rows and read

$$V_{ik} V_{jk}^* = \delta_{ij}, \qquad V_{ik} V_{ij}^* = \delta_{kj}, \qquad i \neq j. \tag{2.94}$$

They can be represented as six "unitarity" triangles in the complex plane [30, 31]. The area of all triangles is the same and a measure for CP violation in the SM. It is described by the Jarlskog invariant J [32, 33]

$$J = \mathrm{Im} \left[V_{td}^* V_{tb} V_{ub}^* V_{ud} \right] = c_{12} c_{23} c_{13}^2 s_{12} s_{23} s_{13} \sin \delta \simeq A^2 \lambda^6 \eta. \tag{2.95}$$

The best-known *unitarity triangle* (UT) is the one following from the relation

$$\boxed{V_{ud} V_{ub}^* + V_{cd} V_{cb}^* + V_{td} V_{tb}^* = 0.} \tag{2.96}$$

Phenomenologically this relation played an important role in the last twenty-five years because it involves simultaneously the elements V_{ub}, V_{cb}, and V_{td}, which were and still are under extensive discussion.

The sides of *this* unitarity triangle are all of the order $O(\lambda^3)$, and it is constructed from the relation (2.96) divided by $V_{cd} V_{cb}^*$. The resulting triangle is shown in Figure 2.4.

It will be useful to have a collection of the most important properties of this triangle:

- Using simple trigonometry one can express $\sin(2\phi_i)$, $\phi_i = \alpha, \beta, \gamma$, in terms of $(\bar{\varrho}, \bar{\eta})$ as follows:

$$\sin(2\alpha) = \frac{2\bar{\eta}(\bar{\eta}^2 + \bar{\varrho}^2 - \bar{\varrho})}{(\bar{\varrho}^2 + \bar{\eta}^2)((1 - \bar{\varrho})^2 + \bar{\eta}^2)}, \tag{2.97}$$

$$\sin(2\beta) = \frac{2\bar{\eta}(1 - \bar{\varrho})}{(1 - \bar{\varrho})^2 + \bar{\eta}^2}, \tag{2.98}$$

$$\sin(2\gamma) = \frac{2\bar{\varrho}\bar{\eta}}{\bar{\varrho}^2 + \bar{\eta}^2} = \frac{2\varrho\eta}{\varrho^2 + \eta^2}. \tag{2.99}$$

- The lengths CA and BA in the rescaled triangle to be denoted by R_b and R_t, respectively, are given by

$$R_b \equiv \frac{|V_{ud}V_{ub}^*|}{|V_{cd}V_{cb}^*|} = \sqrt{\bar{\varrho}^2 + \bar{\eta}^2} = \left(1 - \frac{\lambda^2}{2}\right)\frac{1}{\lambda}\left|\frac{V_{ub}}{V_{cb}}\right|, \tag{2.100}$$

$$R_t \equiv \frac{|V_{td}V_{tb}^*|}{|V_{cd}V_{cb}^*|} = \sqrt{(1 - \bar{\varrho})^2 + \bar{\eta}^2} = \frac{1}{\lambda}\left|\frac{V_{td}}{V_{cb}}\right|. \tag{2.101}$$

- R_b and R_t can also be determined by measuring two of the angles of the UT:

$$R_b = \frac{\sin(\beta)}{\sin(\alpha)} = \frac{\sin(\alpha + \gamma)}{\sin(\alpha)} = \frac{\sin(\beta)}{\sin(\gamma + \beta)}, \tag{2.102}$$

$$R_t = \frac{\sin(\gamma)}{\sin(\alpha)} = \frac{\sin(\alpha + \beta)}{\sin(\alpha)} = \frac{\sin(\gamma)}{\sin(\gamma + \beta)}. \tag{2.103}$$

- The angles β and γ of the unitarity triangle are related directly to the complex phases of the CKM elements V_{td} and V_{ub}, respectively, through

$$V_{td} = |V_{td}|e^{-i\beta}, \quad V_{ub} = |V_{ub}|e^{-i\gamma}. \tag{2.104}$$

- The angle α can be obtained through the relation

$$\alpha + \beta + \gamma = 180° \tag{2.105}$$

expressing the unitarity of the CKM matrix.

Now note that the unitarity relation (2.96) can also be rewritten as

$$R_b e^{i\gamma} + R_t e^{-i\beta} = 1. \tag{2.106}$$

This formula shows transparently that the knowledge of (R_t, β) allows to determine (R_b, γ) through

$$R_b = \sqrt{1 + R_t^2 - 2R_t \cos \beta}, \qquad \cot \gamma = \frac{1 - R_t \cos \beta}{R_t \sin \beta}. \tag{2.107}$$

Similarly, (R_t, β) can be expressed through (R_b, γ):

$$R_t = \sqrt{1 + R_b^2 - 2R_b \cos \gamma}, \qquad \cot \beta = \frac{1 - R_b \cos \gamma}{R_b \sin \gamma}. \tag{2.108}$$

These relations are remarkable. They imply that the knowledge of the coupling V_{td} between t and d quarks allows to deduce the strength of the corresponding coupling V_{ub} between u and b quark and vice versa.

There are other useful trygonometrical relations that can be used to construct the rescaled unitarity triangle by measuring e.g., the angle γ and β as given in (2.102) and (2.103). The complete list of such relations can be found in [34], and we will encounter some of them later in this book.

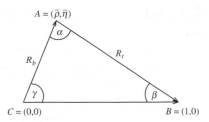

Figure 2.4   Unitarity Triangle. In Japan $\beta = \phi_1, \alpha = \phi_2$, and $\gamma = \phi_3$.

The triangle depicted in Figure 2.4 together with $|V_{us}|$ and $|V_{cb}|$ gives a full description of the CKM matrix. Looking at the expressions for R_b and R_t, we observe that within the SM the measurements of four CP *conserving* decays sensitive to $|V_{us}|$, $|V_{ub}|$, $|V_{cb}|$, and $|V_{td}|$ can tell us whether CP violation ($\eta \neq 0$) is predicted in the SM. This is a very remarkable property of the Kobayashi–Maskawa picture of CP violation: Quark mixing and CP violation are closely related to each other. Overconstraining the sides and angles of the unitarity triangle with experimental data is an important test of the CKM mechanism of the SM. Possible deviations are hints of contributions from NP. The processes used presently in the construction of this triangle will be discussed in Section 8.4 and its present status in Chapter 12. Yet already now we list the values of four CKM parameters obtained from present SM global fits. The CKMfitter collaboration finds

$$A = 0.8403^{+0.0056}_{-0.0201}, \quad \lambda = 0.224747^{+000254}_{-000059}, \quad \bar{\varrho} = 0.1577^{0.0096}_{-0.0074}, \quad \bar{\eta} = 0.3493^{0.0095}_{-0.0071},$$
(2.109)

and the UTfitter collaboration

$$A = 0.826 \pm 0.012, \quad \lambda = 0.2255 \pm 0.0005, \quad \bar{\varrho} = 0.148 \pm 0.013, \quad \bar{\eta} = 0.348 \pm 0.010,$$
(2.110)

in good agreement with the CKMfitter determination. More details are given in Table 8.1.

2.6 Lepton Flavor Violation

The PMNS matrix defined in (2.52) and (2.53) that occurs in charged currents in (2.68) describes flavor transitions in the lepton sector in the same manner as the CKM matrix describes quark flavor transitions. The standard parametrization of the PMNS matrix is the same as for the CKM matrix and given in (2.72). But due to the possibility of Majorana neutrinos one has less phase rotation available such that two additional Majorana phases are left over. In this case one has to multiply (2.72) with the diagonal phase matrix

$$P = \mathrm{diag}\left(e^{\frac{i\alpha_1}{2}}, e^{\frac{i\alpha_2}{2}}, 1\right).$$
(2.111)

Experiments observed a deficit in muon antineutrinos that are produced in cosmic rays in the atmosphere, which was known as *atmospheric neutrino problem*. Similarly less electron neutrinos as produced in the sun were finally detected on earth (*solar neutrino problem*). Both problems can be explained by neutrino oscillations, which necessarily implies nonvanishing mass differences of the neutrinos $\Delta m_{ij}^2 = m_{\nu_i}^2 - m_{\nu_j}^2$ and a violation of the individual lepton number – both in contradiction to the SM. However the absolute mass scale for neutrino masses is still unknown, and only upper bounds are currently available. Furthermore we do not know if ν_3 is heavier (normal hierarchy) or lighter (inverted hierarchy) than ν_1 and ν_2 (ν_1 and ν_2 are defined to be those neutrinos that describe the solar mixing). Compared with the mixing angles of the CKM matrix, the PMNS matrix has two large mixing angles, the atmospheric $\theta_{23} = \theta_{atm}$ and solar mixing angle $\theta_{12} = \theta_{sol}$ and is thus not hierarchical at all. The third mixing angle, the reactor angle θ_{13}, was long thought to be around zero, but recent experiments measured $\theta_{13} \approx 8°$, which is even larger than the corresponding mixing angle in the CKM matrix. The CP phase δ of the PMNS matrix hasn't been measured so far, but some information on it can be obtained from global fits [1335]. The latter paper contains also a very useful material for mixing angles and mass differences. The current experimental values are [35]

$$\sin^2 \theta_{12} = 0.310 \pm 0.013, \qquad \Delta m_{21}^2 = (7.53 \pm 0.18) \cdot 10^{-5} \text{ eV}^2,$$

$$\sin^2 \theta_{23} = 0.582 \pm 0.019, \qquad \Delta m_{32}^2 = \begin{cases} -(2.56 \pm 0.04) \cdot 10^{-3} \text{ eV}^2 & \text{inverted,} \\ (2.51 \pm 0.05) \cdot 10^{-3} \text{ eV}^2 & \text{normal,} \end{cases}$$

$$\sin^2 \theta_{13} = 0.0220 \pm 0.0007. \tag{2.112}$$

So far only lepton flavor violation in neutrino physics has been measured, namely in neutrino oscillations. However, no charged lepton-flavor violating decays have been observed to date. In the SM enriched by light neutrino masses lepton-flavor violating decays $\ell_j \to \ell_i \gamma$ occur at unobservable small rates because the transition amplitudes are suppressed by a factor of $(m_{\nu_j}^2 - m_{\nu_i}^2)/M_W^2$.

In Section 17.2 we will discuss charged lepton-flavor violating decays in some detail, demonstrating explicitly that their branching ratios in the SM are tiny and that any observation of such decays would be a clear signal of physics beyond the SM. On the other hand, we will basically not discuss neutrino oscillations that are already elaborated in many textbooks. An excellent account of this subject is given in [23]. However, we will list in Section 17.5 processes related to neutrinos that are important in the search for NP. Here neutrinoless double β decay plays a prominent role.

2.7 Quantumchromodynamics (QCD)

2.7.1 QCD Lagrangian

At the beginning of this chapter we looked briefly at QCD. Even if weak interactions are the driving force in weak decays, strong interactions play a very important role in this book simply because the decaying objects in the case of quark physics are mesons, bound states of quarks and antiquarks. Therefore we have to discuss in more detail this theory. Here we only collect a few basic facts about QCD that will be discussed in more detail in later chapters. We begin with the QCD Lagrangian that has the structure of a nonabelian gauge theory based on the SU(3) symmetry group usually denoted by $SU(3)_C$. In the process of quantization ghost fields have to be introduced. We will not expose them here. Also the gauge has to be fixed. We will dominantly work in the Feynman gauge and consequently set the gauge parameter to $\xi = 1$. The QCD Lagrangian then reads

$$\mathscr{L}_{QCD} = -\frac{1}{4}(\partial_\mu A^a_\nu - \partial_\nu A^a_\mu)(\partial^\mu A^{a\nu} - \partial^\nu A^{a\mu}) - \frac{1}{2}(\partial^\mu A^a_\mu)^2$$

$$+ \bar{q}_\alpha(i\,\slashed{\partial} - m_q)q_\alpha + g_s\bar{q}_\alpha T^a_{\alpha\beta}\gamma^\mu q_\beta A^a_\mu$$

$$- \frac{g_s}{2}f^{abc}(\partial_\mu A^a_\nu - \partial_\nu A^a_\mu)A^{b\mu}A^{c\nu} - \frac{g_s^2}{4}f^{abe}f^{cde}A^a_\mu A^b_\nu A^{c\mu}A^{d\nu}. \qquad (2.113)$$

Here A^a_μ are the gluon fields with $a, b, c = 1, \dots 8$ and $q = (q_1, q_2, q_3)$ is the color triplet of quark flavor q, $q = u, d, s, c, b, t$. We will denote the colors of quarks by Greek letters α, β, γ. The parameter g_s is the QCD coupling so that

$$\alpha_s = \frac{g_s^2}{4\pi}. \qquad (2.114)$$

Finally, T^a and f^{abc} are the generators and structure constants of $SU(3)_C$, respectively. From this Lagrangian one can derive the Feynman rules for QCD. They are given in Appendix B.

2.7.2 Asymptotic Freedom

Probably the most important property of QCD is asymptotic freedom [36–39], which means that at large-energy scales or equivalently short-distance scales the QCD coupling α_s is small. This implies that calculations of strong interaction effects can be performed at short-distance scales within perturbation theory or more precisely in renormalization group improved perturbation theory. For the time being this statement is only a warning that we will have to go beyond usual perturbation theory to obtain useful results. The full technology will be exposed in Section 4.3, and subsequent sections related to renormalization group methods. On the other hand at large distances QCD is supposed to confine quarks and gluons in hadrons, and α_s must be large at low-energy scales. Thus α_s must be a scale-dependent quantity, and in fact QCD predicts this dependence.

We have already stated before that in an unbroken nonabelian gauge theory there is only one free parameter, the gauge coupling constant, which has to be determined from experiment. Now, the gauge coupling in a four-dimensional gauge theory is a dimensionless quantity, and its dependence on the energy scale Q implies that there must exist still another scale Λ_{QCD} so that α_s depends only on Q/Λ_{QCD} and remains dimensionless despite the dependence on the energy scale. Therefore instead of the gauge coupling also Λ_{QCD} can be regarded as the sole free parameter of QCD and as the fundamental scale of QCD. The fact that the description of QCD can be made either in terms a fundamental scale Λ_{QCD}, which obviously is dimensionful, or a dimensionless quantity α_s is sometimes called *dimensional transmutation*. It is the property of renormalizable theories, which have an intrinsic scale, the renormalization scale.

In the late 1970s and in 1980s it was Λ_{QCD} or more precisely the so-called $\Lambda_{\overline{MS}}$, to be introduced later, that played the role in the phenomenology as it could be extracted from violation of Bjorken scaling in deep-inelastic electron-proton collisions. However, in the 1990s the strength of strong interactions could be determined more precisely at LEP in e^+e^- annihilation into hadrons and from lattice QCD calculations. It turned out then to be more useful to quote the value of α_s at the scale $\mu = M_Z$ instead of the value of $\Lambda_{\overline{MS}}$. But once this is done, the theory predicts uniquely the value of α_s at any other scale at which perturbative calculations make sense. This typically means $\mu \geq 1\,\text{GeV}$. Of course, this information can also be translated into the corresponding value of $\Lambda_{\overline{MS}}$.

The scale dependence of α_s is governed by a renormalization group equation, which we will derive and discuss in Section 4.3.

Quark masses enter the SM Lagrangian for electroweak interactions and also the QCD Lagrangian (2.113). We did not count them as parameters of QCD because they depend on flavor, and strong interactions are flavor blind. The SM cannot explain their values, but there are models trying to explain them. We will not discuss such models in this book as their number is very large. They can all be found with the help of INSPIRE or even Google.

Similar to α_s, quark masses in the Lagrangian depend on the energy scale and are renormalization scheme dependent. We will discuss this in Section 4.3.

2.7.3 Hadrons: Mesons and Baryons

The strong force can bound quarks into hadrons. Hadrons are strongly interacting particles and can be sorted into mesons and baryons. Mesons consist of a quark and an antiquark $q_i \bar{q}_j$, have integer spin, and are thus bosons. Two groups of lightest mesons containing (u, d, s) quarks are pions and kaons, which will play an important role in this book. However, since the middle of the 1970s also heavier mesons containing heavier quarks, $D^0 (D_s^\pm)$ and $B_{d,s}^0, B^\pm$ mesons became even more important in testing the SM. Baryons consist of three quarks $q_i q_j q_k$ (or three antiquarks $\bar{q}_i \bar{q}_j \bar{q}_k$), have half-integer spin, and are thus fermions. The two lightest baryons are the proton and the neutron. Quarkonia are bound states as $c\bar{c}$ (charmonium) and $b\bar{b}$ (bottomium) and can be treated similar to the hydrogen atom (pe^-) or positronium (e^+e^-). Hadrons including top quarks do not exist because the top quark decays too fast before it could form a bound state. In this book we will dominantly deal with the decays of mesons. Neutron and proton will play a role in the context of electric dipole moments in Section 17.3. We will not discuss quarkonia in this book and refer to [40–42]. Looking at citations of these reviews one can easily find out more recent reviews.

We will mainly discuss mesons here. The decays of mesons, the bound states of quarks, and antiquarks, will play a very prominent role in this book, and it is important to recall their valence structure.

π **mesons:**

$$\pi^+ = (u\bar{d}), \quad \pi^0 = \frac{u\bar{u} - d\bar{d}}{\sqrt{2}}, \quad \pi^- = (\bar{u}d). \tag{2.115}$$

K **mesons:**

$$K^+ = (u\bar{s}), \quad K^- = (\bar{u}s), \quad K^0 = (d\bar{s}), \quad \bar{K}^0 = (\bar{d}s). \tag{2.116}$$

B mesons:

$$B^+ = (u\bar{b}), \quad B^- = (\bar{u}b), \quad B^0_d = (d\bar{b}), \quad \bar{B}^0_d = (\bar{d}b) \qquad (2.117)$$

$$B^0_s = (s\bar{b}), \quad \bar{B}^0_s = (\bar{s}b), \quad B^+_c = (c\bar{b}), \quad B^-_c = (\bar{c}b). \qquad (2.118)$$

D mesons:

$$D^0 = (c\bar{u}), \quad \bar{D}^0 = (\bar{c}u), \quad D^+ = (c\bar{d}), \quad D^- = (\bar{c}d), \qquad (2.119)$$

$$D^+_s = (c\bar{s}), \quad D^-_s = (\bar{c}s). \qquad (2.120)$$

Note that K and B mesons contain a $\bar{s}$ and $\bar{b}$, respectively, and not an s and b quark. This has historical reasons. However, the D meson contains a c and not $\bar{c}$.

Mesons can be characterized by their J^{PC} quantum number, where J is the spin of the meson

$$\vec{J} = \vec{L} + \vec{S}, \qquad \vec{S} = \vec{s}_q + \vec{s}_{\bar{q}}, \qquad (2.121)$$

composed of the angular momentum $\vec{L}$ and the total spin of the two quarks $\vec{S}$. For the lighter meson systems the quark and antiquark do not have a relative angular momentum and are in a state with $L = 0$. We will mostly consider such mesons. Parity and the charge conjugation quantum number are given, respectively, as

$$P = (-1)^{L+1}, \qquad C = (-1)^{L+S}. \qquad (2.122)$$

The mesons built out of the three lightest quarks u, d, s can be grouped into multiplets of an approximate flavor SU(3) symmetry. If this symmetry would be exact, then there would be a singlet and an octet according to

$$3_q \otimes \bar{3}_{\bar{q}} = 1 \oplus 8. \qquad (2.123)$$

But because the flavor symmetry is broken, the singlet and the octet mix. For $L = 0$ we have pseudoscalar mesons, where the spins of the two quarks sum to zero: $J^{PC} = 0^{-+}$. Vector mesons have $S = 1$ such that $J^{PC} = 1^{--}$. In Figure 2.5 we arrange the pseudoscalar mesons and vector mesons in the (S, T^s_3) plane where S is the strangeness quantum number and T^s_3 the third component of the *strong* isospin with

$$T^s_3(u) = -T^s_3(\bar{u}) = \frac{1}{2}, \quad T^s_3(d) = -T^s_3(\bar{d}) = -\frac{1}{2}, \quad T^s_3(s) = T^s_3(\bar{s}) = 0. \qquad (2.124)$$

The ρ mesons have the same quark content as the pions just as the spins of the two quarks point in the same direction. Similar for the K^* mesons. The pseudoscalars η and η' are given in the SU(3) symmetry limit ($m_u = m_d = m_s$) by:

$$\eta \simeq \eta_8 = \frac{1}{\sqrt{6}} \left(u\bar{u} + d\bar{d} - 2s\bar{s} \right) \qquad \text{member of SU(3) octet} \qquad (2.125)$$

$$\eta' \simeq \eta_0 = \frac{1}{\sqrt{3}} \left(u\bar{u} + d\bar{d} + s\bar{s} \right), \qquad \text{SU(3) singlet.} \qquad (2.126)$$

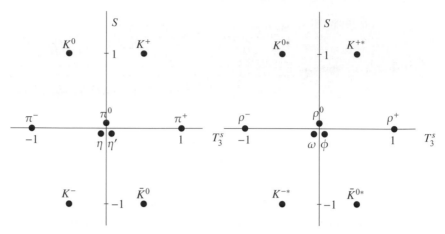

Figure 2.5   Meson multiplets: pseudoscalar mesons (left) and vector mesons (right).

Once the SU(3) is broken, a mixing angle of order $-20°$ appears between η_8 and η_0 such that physical mass eigenstates η and η' are mixtures of η_8 and η_0. However, for the vector mesons we have a larger mixing between the singlet and the octet and the flavor content of vector mesons differs from the one of η and η'.

$$\omega = \frac{1}{\sqrt{2}} \left(u\bar{u} + d\bar{d} \right), \qquad \phi = s\bar{s}. \tag{2.127}$$

Neutral mesons as K^0, B_d^0, B_s^0, and D^0 have an interesting property: Via the weak, interactions they can oscillate into their antiparticles. This will be discussed in detail in Chapters 6 and 8.

Our discussion of baryons will be very limited with the exception of the electric dipole moments (EDMs) in Section 17.3. The reason is that hadronic uncertainties in baryon decays are significantly larger than in the case of mesons, which is related to their more complicated structure. Consequently the tests of the SM in this sector are much more difficult, although the decays of Λ can also provide useful tests of the SM and its extensions. Rather exceptional are EDMs as here SM contributions are so small that any measurement of a nonvanishing EDM would be a clear signal of new physics.

2.8 Final Remarks

The information on the SM given in this chapter is sufficient for making the next steps and will be extended in the rest of the book. Yet, not all aspects of the SM can be discussed by us. There are numerous books on the dynamics of the SM where information on other topics, including baryons, can be found. See in particular [8]. Among recent ones, I can strongly recommend Langacker book [10]. Nice short introductions to flavor physics including most recent developments can be found in [43, 44]. More detailed aspects will be found in later chapters. Some of them can be found in a number of reviews and lectures, in particular in [7, 9, 29, 45] and very recently in [46]. Other lectures and reviews will be quoted as we proceed.

PART III

WEAK DECAYS IN THE STANDARD MODEL

Weak Decays at Tree Level

3.1 Muon Decay

3.1.1 The Decay Amplitude

We begin our discussion of weak decays with the simplest but a very important decay:

$$\mu^- \to \nu_\mu e^- \bar{\nu}_e, \tag{3.1}$$

for which the Feynman diagram is given in Figure 3.1 We just use the Feynman rules for the vertices and the W propagator collected in Appendix B. Multiplying the result by i to obtain the decay amplitude, we find

$$\mathcal{A}(\mu^- \to \nu_\mu e^- \bar{\nu}_e) = -\frac{1}{8} \frac{g_2^2}{k^2 - M_W^2} [\bar{\nu}_\mu \gamma_\mu (1 - \gamma_5) \mu][\bar{e} \gamma^\mu (1 - \gamma_5) \nu_e], \tag{3.2}$$

where the Lorentz index $\mu = 0, 1, 2, 3$ in γ_μ should not be confused with μ in the names of the involved particles. The four-momentum squared, k^2, in the W-propagator is due to the smallness of the muon mass negligible when compared with M_W^2 and can be set to zero to an excellent approximation. Comparing then the result with the one from the Fermi theory

$$\mathcal{A}(\mu^- \to \nu_\mu e^- \bar{\nu}_e) = \frac{G_F}{\sqrt{2}} [\bar{\nu}_\mu \gamma_\mu (1 - \gamma_5) \mu][\bar{e} \gamma^\mu (1 - \gamma_5) \nu_e], \tag{3.3}$$

we obtain an important relation between the Fermi constant G_F and the parameters of the SM

$$\boxed{\frac{G_F}{\sqrt{2}} = \frac{g_2^2}{8M_W^2}.} \tag{3.4}$$

This amplitude has a very simple structure. It is a product of $G_F/\sqrt{2}$, which as seen in (3.4) contains the information about the strength of the interaction and the four-fermion operator composed of charged leptons and corresponding neutrino fields. We will see that such products will also be fundamental for other decays.

There is one important point to be made. As seen in (3.1) there is $\bar{\nu}_e$ in the final state, while both in (3.3) and in the diagram in Figure 3.1 we see ν_e. This is not a misprint. In popular talks in which no arrows are indicated, and even if they are indicated, one would see $\bar{\nu}_e$ in Figure 3.1. But in this book, in which the diagrams will be evaluated, we will

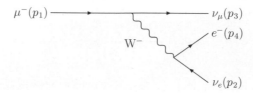

Figure 3.1   Muon decay

proceed as done in this figure: Neutrino moving back in time represents an antineutrino and similar in other diagrams and other particles in our book.

At this stage a few additional remarks should be made:

- Muon decay is strongly suppressed not because of the smallness of g_2, which in fact is $O(1)$, but because of the large $W^\pm$-mass. This is the main reason why G_F is so small.
- The four-fermion operator is a product of two left-handed currents, signaling the maximal parity violation in decays mediated by $W^\pm$. This is one of the crucial properties of electroweak interactions in the SM.
- This operator has energy dimension six.
- As leptons do not carry color, the result in (3.3) is not affected by strong interactions. While the diagrams with additional exchanges of photons $W^\pm$ and Z^0 have to be taken into account for precision test of the SM, they are small and can be calculated in perturbation theory.
- Very importantly, the branching ratio for this decay is more than 0.99, so that the muon lifetime is almost entirely determined by this decay.

3.1.2 The Decay Width, Muon Lifetime, and G_F

Having the decay amplitude in (3.3) at hand, we can calculate the decay width. This calculation can be found in many textbooks, and to save few pages we will only present those details that can be useful for other decays considered by us. A more detailed derivation can be found, e.g., in Langacker's book [10]. In any case, any reader who was exposed to elementary relativistic quantum field theory should be able to perform this calculation without any help from Langacker or the present author. I have chosen Langacker's book because after the routine calculation of the total decay rate in the leading order, he also discusses higher-order QED and electroweak corrections to this decay, which are beyond the scope of our book but relevant for electroweak precision tests. Useful calculations can also be found in other textbooks, in particular in the one by Schwartz [47].

If we are interested only in the total decay rate, we have to average over the decaying muon polarization states and sum over electron polarization states. Denoting the momenta of involved particles as in Figure 3.1 and neglecting the electron mass for the moment gives

$$\overline{|\mathcal{A}(\mu^- \to \nu_\mu e^- \bar{\nu}_e)|^2} = \frac{1}{2}\frac{G_F^2}{2}\mathrm{Tr}[\gamma_\mu(1-\gamma_5)(\not{p}_1 + m_\mu)\gamma_\nu(1-\gamma_5)\,\not{p}_3]$$
$$\times \mathrm{Tr}[\gamma^\mu(1-\gamma_5)\,\not{p}_2\gamma^\nu(1-\gamma_5)\,\not{p}_4]. \tag{3.5}$$

With the standard technology for the evaluation of traces, it is straightforward to find

$$\overline{|\mathcal{A}(\mu^- \to \nu_\mu e^- \bar{\nu}_e)|^2} = 64 G_F^2 (p_1 \cdot p_2)(p_3 \cdot p_4). \tag{3.6}$$

Detailed steps are given on page 251 in [10]. The integration over momenta, performed in detail on page 252 in [10], is a bit tedious but the final result is very simple and depends only on G_F and m_μ:

$$\Gamma_\mu \equiv \Gamma(\mu^- \to \nu_\mu e^- \bar{\nu}_e) = \frac{G_F^2 m_\mu^5}{192\pi^3}. \tag{3.7}$$

General formulas for phase space integration can be found in any edition of PDG, but the explicit examples in [10] are certainly very useful.

Of course, to increase precision electron mass effects, radiative QED corrections and W propagator effects have to be included with the result summarized in (6.51) of [10]. Here we only show the leading contributions of these three corrections:

$$\Gamma_\mu = \frac{G_F^2 m_\mu^5}{192\pi^3} \left(1 - 8\frac{m_e^2}{m_\mu^2}\right) \left[1 + \frac{\alpha}{2\pi} \left(\frac{25}{4} - \pi^2\right)\right] \left(1 + \frac{3}{5}\frac{m_\mu^2}{M_W^2}\right). \tag{3.8}$$

The radiative QED correction is obtained by adding photon exchange between initial muon and the electron in the final state and other QED corrections at this order in α. Note that this is an 0.2% correction to the leading contribution from the diagram in Figure 3.1. But it has to be included along with higher QED corrections to match the precision of the experimental value of τ_μ given in Appendix D, which is used to determine G_F. The other inputs m_e, m_μ, α, and M_W, also given in Appendix D, are very precise as well.

Now the muon lifetime is given by

$$\tau_\mu = \frac{1}{\Gamma_\mu}. \tag{3.9}$$

Consequently (3.8) amended by additional nonleading corrections not explicitly shown here allows us together with (3.9) to determine G_F with high precision:

$$G_F = 1.166367(5) \times 10^{-5} \text{ GeV}^{-2}. \tag{3.10}$$

G_F is one of the fundamental input parameters in the SM and will appear frequently in this book.

In principle we could next discuss the decays of the τ lepton, but because of its high mass it can also decay to mesons, and it is strategically useful to postpone this discussion and proceed with the leptonic decays of charged mesons.

3.2 Leptonic Decays of Charged Mesons

3.2.1 Leptonic Decays of π^+ and K^+

Let us next extend our tree-level calculations to the leptonic decays

$$\pi^+ \to \mu^+ \nu_\mu, \qquad K^+ \to \mu^+ \nu_\mu. \tag{3.11}$$

The diagrams for these decays are given in Figure 3.2. The results for the corresponding amplitudes expressed first in terms of quark and lepton fields have a similar structure to (3.3). However, the quarks are confined in π^+ and K^+, and the final amplitudes for the decays in question read

$$\mathcal{A}(\pi^+ \to \mu^+ \nu_\mu) = \frac{G_F}{\sqrt{2}} V^*_{ud} (\bar{\nu}_\mu \mu^-)_{V-A} \langle 0|(\bar{d}u)_{V-A}|\pi^+\rangle, \tag{3.12}$$

$$\mathcal{A}(K^+ \to \mu^+ \nu_\mu) = \frac{G_F}{\sqrt{2}} V^*_{us} (\bar{\nu}_\mu \mu^-)_{V-A} \langle 0|(\bar{s}u)_{V-A}|K^+\rangle, \tag{3.13}$$

where we have introduced the shorthand notation

$$\bar{d}_\alpha \gamma^\mu (1 - \gamma_5) u_\alpha \equiv (\bar{d}_\alpha u_\alpha)_{V-A} \equiv (\bar{d}u)_{V-A} \tag{3.14}$$

and similar for other quark and lepton currents. In what follows, summation over repeated color indices (α in this case) will be understood.

We note the appearance of the CKM factors V_{ud} and V_{us} and the Feynman rules for the quark-W-boson vertices in Figure 3.2 require that the complex conjugates of these factors enter this amplitude. If we considered the corresponding decays of π^- and K^-, no complex conjugation would be required. As V_{ud} and V_{us} are real in our conventions, this difference is immaterial in the decays considered but would be crucial if in particular V_{ub} or V_{td} would be present instead.

In comparison with the muon decay, a complication arises. The last objects on the r.h.s. in (3.12) and (3.13) are matrix elements between the initial mesons and the vacuum. The flavors in the quark currents have been chosen appropriately so that indeed in each case the meson is destroyed by the current, leaving no meson in the final state. But as the initial meson is a bound state of quarks, these matrix elements cannot be evaluated in perturbation theory, and nonperturbative methods, in particular lattice QCD, are required to evaluate them.

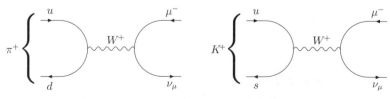

Figure 3.2 $\pi^+ \to \mu^+ \nu_\mu$ and $K^+ \to \mu^+ \nu_\mu$.

Yet, to make a step forward, we can parametrize these matrix elements as follows

$$\langle 0|(\bar{d}u)_{V-A}|\pi^+\rangle = iF_\pi p_\pi^\mu, \qquad \langle 0|(\bar{s}u)_{V-A}|K^+\rangle = iF_K p_K^\mu, \tag{3.15}$$

where p_π^μ and p_K^μ are the four momenta of the decaying mesons. Recall that a γ^μ is hidden in the matrix element of the current. As this matrix element carries the dimension of the energy squared and we have only the four-momenta p_π^μ and p_K^μ to our disposal, this is a unique parametrization consistent with Lorentz invariance up to an overall numerical factor. In some papers a factor of $\sqrt{2}$ on the r.h.s. of these equations can be found, and this implies a reduction of the numerical values of F_π and F_K by $\sqrt{2}$. When using results found in the literature it is crucial to check which definitions of F_π and F_K are used in a given paper.

F_π and F_K are the so-called weak decay constants for pion and kaon, respectively. In the past these constants were usually extracted from the data by combining the leptonic decays in question with the semileptonic ones that we will discuss soon. But the advances in lattice calculations allow us presently to calculate F_π and F_K directly from QCD [48]

$$F_\pi = 129.8(15)\,\mathrm{MeV}, \qquad F_K = 156.1(11)\,\mathrm{MeV}. \tag{3.16}$$

The evaluation of the decay widths for these two decays is easy as the final states contain only two particles and not three as in the muon decay. Let us do it for the pion decay without being confused that the indices on the four-momenta p_μ and p_ν denote *muon* and *neutrino* and are not Lorentz indices.

Inserting (3.15) into (3.12), we find first

$$\mathcal{A}(\pi^+ \to \mu^+\nu_\mu) = i\frac{G_F}{\sqrt{2}}V_{ud}^*F_\pi[\bar{v}_\mu\,\slashed{p}_\pi(1-\gamma_5)\mu^-]. \tag{3.17}$$

But using Dirac equation, we have

$$\bar{v}_\mu\,\slashed{p}_\nu = m_\nu\bar{v}_\mu, \qquad \slashed{p}_\nu\mu^- = m_\mu\mu^-. \tag{3.18}$$

Setting neutrino mass to zero, we can use $\slashed{p}_\nu = 0$ to obtain

$$\slashed{p}_\pi = \slashed{p}_\nu + \slashed{p}_\mu = \slashed{p}_\mu, \qquad \slashed{p}_\mu(1-\gamma_5)\mu^- = m_\mu(1+\gamma_5)\mu^-. \tag{3.19}$$

Consequently

$$\mathcal{A}(\pi^+ \to \mu^+\nu_\mu) = i\frac{G_F}{\sqrt{2}}V_{ud}^*F_\pi m_\mu[\bar{v}_\mu(1+\gamma_5)\mu^-]. \tag{3.20}$$

Thus denoting this amplitude simply by $\mathcal{A}$, we have

$$|\mathcal{A}|^2 = \frac{G_F^2}{2}|V_{ud}|^2F_\pi^2m_\mu^2|\bar{v}_\mu(1+\gamma_5)\mu^-|^2. \tag{3.21}$$

Summing over polarizations of final particles, we find using standard methods (see [10])

$$\overline{|\mathcal{A}|^2} = \frac{G_F^2}{2}|V_{ud}|^2F_\pi^2m_\mu^2[8p_\nu p_\mu] \tag{3.22}$$

with the last factor in square brackets resulting from

$$\mathrm{Tr}[\slashed{p}_\nu(1+\gamma_5)(\slashed{p}_\mu+m_\mu)(1-\gamma_5)] = 2\mathrm{Tr}[\slashed{p}_\nu\,\slashed{p}_\mu(1-\gamma_5)] = 8p_\nu p_\mu. \tag{3.23}$$

Denoting by E_ν the neutrino energy in the pion rest frame, we have $p_\nu p_\mu = E_\nu m_\pi$ and

$$\overline{|\mathcal{A}|^2} = 4\, G_F^2 |V_{ud}|^2 F_\pi^2 m_\mu^2 m_\pi E_\nu. \tag{3.24}$$

The decay width is given by

$$\Gamma(\pi^+ \to \mu^+ \nu_\mu) = \frac{\overline{|\mathcal{A}|^2}}{2m_\pi} \left[\frac{E_\nu}{4\pi m_\pi} \right], \tag{3.25}$$

with the last factor in square brackets coming from phase-space volume.
  But simple kinematics tells us that

$$E_\nu = \frac{m_\pi}{2} \left(1 - \frac{m_\mu^2}{m_\pi^2} \right). \tag{3.26}$$

Therefore, combining the last three formulas, we reproduce the known result

$$\Gamma(\pi^+ \to \mu^+ \nu_\mu) = \frac{G_F^2}{8\pi} m_\pi m_\mu^2 \left(1 - \frac{m_\mu^2}{m_\pi^2} \right)^2 F_\pi^2 |V_{ud}|^2. \tag{3.27}$$

The result for K^+ decay is then obtained from (3.27) in no time by just replacing properly various indices and changing the CKM factor from $|V_{ud}|$ to $|V_{us}|$. We find

$$\Gamma(K^+ \to \mu^+ \nu_\mu) = \frac{G_F^2}{8\pi} m_K m_\mu^2 \left(1 - \frac{m_\mu^2}{m_K^2} \right)^2 F_K^2 |V_{us}|^2. \tag{3.28}$$

These two formulas are very important as they can be used for the leptonic decays of heavy mesons with proper replacement of masses, weak decay constants, and CKM factors.
  The important property of these formulas is the helicity suppression factor m_μ^2 that is related to the $V - A$ structure of the quark current. In the rest frame of a meson that has spin-0, the momenta of the muon and of the neutrino must have opposite directions, and their spins must add to zero as shown in Figure 3.3. This means that the projection of the spin onto the momentum $\vec{\sigma}\vec{p}/|\sigma p|$ is the same for both particles. However, because neutrinos are always left-handed (assuming $m_{\nu_\mu} = 0$), only the left-handed part of μ^+ can contribute. If both neutrino and muon were exactly massless, this decay would not be possible because then the W boson only couples to left-handed particles (and right-handed antiparticles). But due to the nonvanishing $\mu^\pm$ mass ($m_\mu \neq 0$) the μ^+ couples to W through its left-handed component. Thus the amplitude must be proportional to m_μ.
  With the values of G_F, F_π, and F_K in (3.10) and (3.16), the only unknowns in the decays widths are the CKM elements $|V_{ud}|$ and $|V_{us}|$. They can be determined from semileptonic decays of π^+ and K^+, respectively. Before doing this let us turn our attention to leptonic decays of heavy mesons.

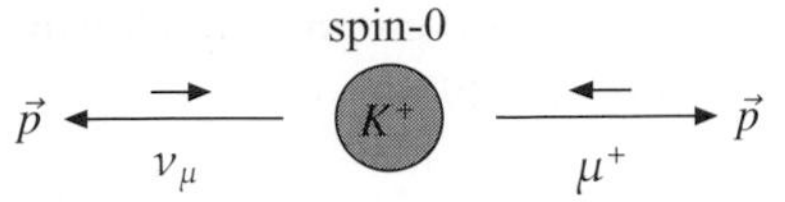

Figure 3.3 Helicity suppression of $K^+ \to \mu^+ \nu_\mu$ in the SM.

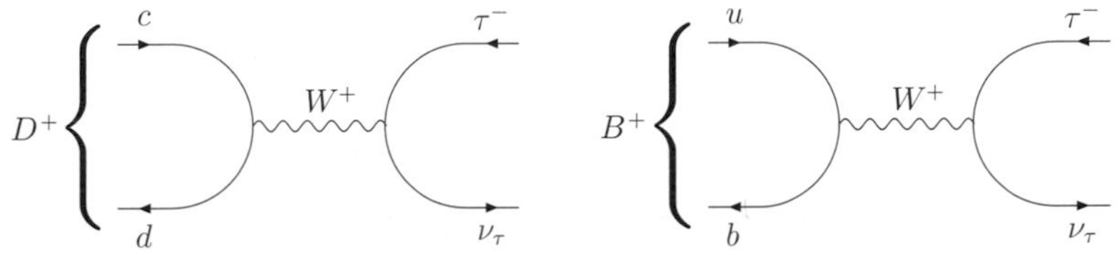

Figure 3.4 $D^+ \to \tau^+ \nu_\tau$ and $B^+ \to \tau^+ \nu_\tau$.

3.2.2 Leptonic Decays of D^+ and B^+

We will next consider the decays

$$D^+ \to \tau^+ \nu_\tau, \qquad B^+ \to \tau^+ \nu_\tau \qquad (3.29)$$

with $D^+ = (c\bar{d})$ and $B^+ = (u\bar{b})$. Analogous calculations can be done for decays into $\mu^+ \nu_\mu$ and $e^+ \nu_e$. Only the branching ratios in these cases are much smaller because of the helicity suppression. The Feynman diagrams for these two decays are given in Figure 3.4. There is no need to repeat the calculation of the decay widths as they can be directly obtained from previous formulas by just changing flavor indices. As both decays do not represent the dominant decay channels of D^+ and B^+ it is customary to consider the branching ratios instead of the width for them. They are obtained by dividing the width for these particular decays by the total width or equivalently by multiplying the width for a given decay with the lifetime of the decaying meson. We find then

$$\mathcal{B}(D^+ \to \tau^+ \nu_\tau)_{\text{SM}} = \frac{G_F^2 m_{D^+} m_\tau^2}{8\pi} \left(1 - \frac{m_\tau^2}{m_{D^+}^2} \right)^2 F_{D^+}^2 |V_{cd}|^2 \tau_{D^+}, \qquad (3.30)$$

$$\mathcal{B}(B^+ \to \tau^+ \nu_\tau)_{\text{SM}} = \frac{G_F^2 m_{B^+} m_\tau^2}{8\pi} \left(1 - \frac{m_\tau^2}{m_{B^+}^2} \right)^2 F_{B^+}^2 |V_{ub}|^2 \tau_{B^+}. \qquad (3.31)$$

The subscript SM reminds us that these formulas apply only to the SM. As discussed in later chapters of this book, these branching ratios, being strongly suppressed, can be modified in the extensions of this model. While this is in principle also possible for π^+ and K^+ leptonic decays, one expects that new physics effects there are very small because the corresponding branching ratios are already measured very precisely, and the SM describes them very well.

The relevant weak decay constants are already accurately known from lattice calculations and given as follows [49]

$$F_{D^+} = 212.7(5)\,\text{MeV}, \qquad F_{B^+} = 189.4(13)\,\text{MeV}. \tag{3.32}$$

To predict these branching ratios, we need the values of the CKM parameters. Here semileptonic decays of mesons play an important role, and we will discuss them next.

3.3 Semileptonic Decays of Charged Mesons

3.3.1 Semileptonic Decays of K^+

We next consider the semileptonic decay

$$K^+ \to \pi^0 e^+ \nu_e, \tag{3.33}$$

for which the Feynman diagram is given in Figure 3.5. The part of this diagram that involves the exchange of the $W^\pm$ boson is familiar from previous decays, but in addition there is a quark, this time the u-quark, that belongs to K^+ but does not couple to $W^\pm$. Such a quark is called spectator quark as its role is passive in this decay. In the final state this quark belongs to π^0. With the experience gained so far, we can immediately write down the relevant decay amplitude

$$\mathcal{A}(K^+ \to \pi^0 e^+ \nu_e) = \frac{G_F}{\sqrt{2}} V_{us}^* (\bar{\nu}_e e^-)_{V-A} \langle \pi^0 | (\bar{s}u)_{V-A} | K^+ \rangle. \tag{3.34}$$

This decay can be used to determine $|V_{us}|$ provided the last factor can be determined by nonperturbative methods.

Similar to what we did in (3.15), we parametrize this factor through

$$\langle \pi^0 | (\bar{s}u)_{V-A} | K^+ \rangle = q_+^\mu f_+(q_-^2) + q_-^\mu f(q_-^2), \tag{3.35}$$

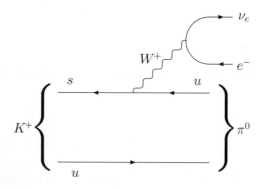

Figure 3.5   Decay $K^+ \to \pi^0 e^+ \nu_e$.

where

$$q_+^\mu = p_K^\mu + p_\pi^\mu, \qquad q_-^\mu = p_K^\mu - p_\pi^\mu = p_{\nu_e}^\mu + p_e^\mu. \tag{3.36}$$

We recall that the Lorentz index μ is hidden in $(\bar{s}u)_{V-A}$. But using Dirac equation, we find that

$$(\bar{\nu}_e e^-)_{V-A} q_-^\mu \propto m_e, \tag{3.37}$$

and consequently the second term in (3.35) can be neglected to first approximation so that only the knowledge of the formfactor $f_+(q_-^2)$ is relevant for the determination of $|V_{us}|$. The evaluation of the decay rate requires the knowledge of the q^2 dependence of $f_+(q_-^2)$ and is discussed in detail in section IV of [50], where also QED corrections and hadronic effects affecting the determination of $|V_{us}|$ are discussed in detail.

This presentation provides just an idea what is required to be able to extract the value of $|V_{us}|$ from the experimental branching ratio for this decay. Usually what one finds in the literature is the value of the product $f_+(0)|V_{us}|$ extracted from the data. Calculating then $f_+(0)$ by means of lattice QCD and using some information from chiral perturbation theory, it is possible to find the value of $|V_{us}|$.

The second decay used for the determination of $|V_{us}|$ is $K_L \rightarrow \pi^+ e^- \bar{\nu}_e$. The relevant amplitude is easily obtained from (3.34), and a different hadronic matrix element of the current $(\bar{s}u)_{V-A}$ must be known to extract $|V_{us}|$: this time between K_L and π^+. We again refer to [50] and the references given there.

The determination of $|V_{us}|$ is very important for any considerations in flavor physics as it determines the parameter λ in the Wolfenstein parametrization. By now there have been many sophisticated analyses that include corrections to the tree-level formulas given here and use other decays. In the CKM workshops that take place every second year, a separate working group is devoted to this determination and to the determination of $|V_{ud}|$. We will not go into details of this topic as it is adequately covered in [50] and in any proceedings of the CKM workshops. For us what is most important is the value of $|V_{us}|$, which we will use in the rest of this book, and the fact that $|V_{ud}|$, determined independently agrees well with the value obtained from the unitarity of the CKM matrix neglecting the contribution from $|V_{ub}|$. As of 2019 the CKM element $|V_{us}|$ is known very precisely, and we will use

$$\boxed{|V_{us}| = \lambda = 0.2252(9), \qquad |V_{ud}| = 1 - \frac{\lambda^2}{2} = 0.9746} \tag{3.38}$$

in analytic expressions, although in final numerical computations it is common these days to use the standard parametrization of the CKM as given in (2.72), which imposes the exact unitarity.

We can now insert these values of CKM parameters into (3.27) and (3.28) to see whether the relevant branching ratios agree with experiment. Using the input from Table D.1 and the conversion from sec to GeV in (D.1), we find

$$\mathcal{B}(\pi^+ \rightarrow \mu^+ \nu_\mu) = \Gamma(\pi^+ \rightarrow \mu^+ \nu_\mu)\, \tau(\pi^+) = 1.031|V_{ud}|^2 = 0.979, \tag{3.39}$$

$$\mathcal{B}(K^+ \rightarrow \mu^+ \nu_\mu) = \Gamma(K^+ \rightarrow \mu^+ \nu_\mu)\, \tau(K^+) = 12.45|V_{ud}|^2 = 0.631, \tag{3.40}$$

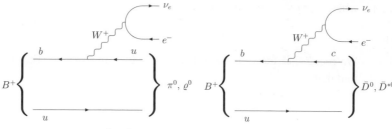

Figure 3.6 $B^+ \to \pi^0(\rho^0)e^+\nu_e$ and $B^+ \to \bar{D}^0(\bar{D}^{*0})e^+\nu_e$.

both in remarkable agreement with the data in Table D.2. Small differences result from QED corrections.

3.3.2 Semileptonic Decays of B^+

We can next consider the decays

$$B^+ \to \pi^0 e^+\nu_e, \qquad B^+ \to \rho^0 e^+\nu_e, \tag{3.41}$$

for which the Feynman diagrams are given in Figure 3.6, and the amplitude in the first case reads

$$\mathcal{A}(B^+ \to \pi^0 e^+\nu_e) = \frac{G_F}{\sqrt{2}} V_{ub}^* (\bar{\nu}_e e^-)_{V-A} \langle \pi^0 | (\bar{b}u)_{V-A} | B^+ \rangle. \tag{3.42}$$

In the second case π^0 in the last factor should be replaced by ρ^0. Evidently these decays allow the determination of $|V_{ub}|$ provided the relevant hadronic matrix elements of the current $(\bar{b}u)_{V-A}$ can be evaluated by lattice QCD.

Similarly for the decays

$$B^+ \to \bar{D}^0 e^+\nu_e, \qquad B^+ \to \bar{D}^{*0} e^+\nu_e, \tag{3.43}$$

shown also in Figure 3.6, we find

$$\mathcal{A}(B^+ \to \bar{D}^0 e^+\nu_e) = \frac{G_F}{\sqrt{2}} V_{cb}^* (\bar{\nu}_e e^-)_{V-A} \langle \bar{D}^0 | (\bar{b}c)_{V-A} | B^+ \rangle \tag{3.44}$$

with $\bar{D}^0$ replaced by $\bar{D}^{*0}$ in the second decay in (3.43). These decays allow the determination of $|V_{cb}|$ provided the hadronic matrix elements of the current $(\bar{b}c)_{V-A}$ can be evaluated by lattice QCD.

3.4 The Determination of $|V_{cb}|$ and $|V_{ub}|$

The extraction of $|V_{ub}|$ and $|V_{cb}|$ from the decays listed here, the so-called exclusive decays, in which the final state is specified, developed over the years to a big industry. In the case of $|V_{cb}|$ it is based on the distributions as given, e.g., in [51]

$$\frac{d\Gamma(B \to D^*\ell\nu)}{d\omega} = \frac{G_F^2 m_{D^*}^3}{48\pi^3}(m_B - m_{D^*})^2 \sqrt{\omega^2 - 1}\chi(\omega)\mathcal{F}^2(\omega)|V_{cb}|^2, \tag{3.45}$$

$$\frac{d\Gamma(B \to D\ell\nu)}{d\omega} = \frac{G_F^2 m_D^3}{48\pi^3}(m_B + m_D)^2(\omega^2 - 1)^{3/2}\mathcal{G}^2(\omega)|V_{cb}|^2, \tag{3.46}$$

with $\mathcal{F}(\omega)$ and $\mathcal{G}(\omega)$ being formfactors related to the matrix element in (3.44) for D^* and D, respectively. Their expressions are given in (19) of [51]. Moreover,

$$\chi(\omega) = 1 + \frac{4\omega}{\omega + 1}\frac{1 - 2\omega r_{D^*} + r_{D^*}^2}{(1 - r_{D^*})^2}, \tag{3.47}$$

with

$$r_{D^*} = \frac{m_{D^*}}{m_B}, \qquad \omega = \frac{p_B \cdot p_{D^{(*)}}}{m_B m_{D^{(*)}}} = \frac{m_B^2 + m_{D^{(*)}}^2 - q^2}{2m_B m_{D^{(*)}}}. \tag{3.48}$$

But there is another route to $|V_{ub}|$ and $|V_{cb}|$. It uses the so-called inclusive decays in which one sums over all final states that are reached through the $b \to u$ or $b \to c$ transitions driven by $|V_{ub}|$ and $|V_{cb}|$, respectively. Here very much developed methods of *heavy quark expansions* are essential for the determination of $|V_{ub}|$ and $|V_{cb}|$. We will describe them briefly in Section 7.5.

Although the extraction of these two CKM matrix elements using these decays is in principle the optimal strategy, being basically independent of new physics contributions, it has been weakened by disagreements between the exclusive and inclusive determinations of $|V_{ub}|$ and to a lesser extent in the case of $|V_{cb}|$. This is clearly seen in a number of reviews and discussions, which make efforts to understand this disagreement. A subset of relevant papers can be found in [52–60].

As an example for this disagreement we quote the *exclusive* determinations from lattice QCD form factors [61–64]

$$|V_{ub}|_{\text{excl}} = (3.72 \pm 0.14) \times 10^{-3}, \qquad |V_{cb}|_{\text{excl}} = (39.36 \pm 0.75) \times 10^{-3}, \tag{3.49}$$

and the *inclusive* ones given by [64, 65]

$$|V_{ub}|_{\text{incl}} = (4.40 \pm 0.25) \times 10^{-3}, \qquad |V_{cb}|_{\text{incl}} = (42.21 \pm 0.78) \times 10^{-3}, \tag{3.50}$$

typical values that one can encountered in the literature.

The problem with $|V_{ub}|$ will hopefully be solved in the BELLE II era. On the other hand, a breakthrough in the case of $|V_{cb}|$ has been made in [55, 57, 58] by realizing that the so-called CLN parametrization [66] of $\mathcal{F}(\omega)$ and $\mathcal{G}(\omega)$ formfactors on which the exclusive determinations of $|V_{cb}|$ were based, being model-dependent, is not optimal, and the model-independent BGL parametrization [67] should be favored instead. As an example, one finds [55]

$$|V_{cb}|_{\text{excl}}^{\text{CLN}} = (37.4 \pm 1.3) \times 10^{-3}, \qquad |V_{cb}|_{\text{excl}}^{\text{BGL}} = (41.9 \pm 1.9) \times 10^{-3}, \tag{3.51}$$

with the second value very close to the inclusive determination in (3.50).

The story of both $|V_{ub}|$ and $|V_{cb}|$ determinations is not over. In particular, there is still a significant difference between the exclusive and inclusive determination of $|V_{ub}|$.

Interested readers should have a look at the talks presented at the CKM 2018 workshop, in particular those of Gambino and Schwanda, from various lattice QCD analyses, the summary of the 2018 CKM working group II [68] and [69].

For this book we will take as nominal values

$$\boxed{|V_{ub}|^{\text{nom}} = 3.7 \times 10^{-3}, \qquad |V_{cb}|^{\text{nom}} = 42.0 \times 10^{-3},} \qquad (3.52)$$

with the one for $|V_{ub}|$ close to exclusive determinations and the one for $|V_{cb}|$ closer to the inclusive ones. However, on various places, we will stress which SM predictions are particularly sensitive to the values of $|V_{ub}|$ and $|V_{cb}|$ so that their precise determinations are very important. The choice of exclusive value of $|V_{ub}|$ is the author's prejudice that with improved lattice QCD calculations its exclusive value in (3.49) will be favored.

3.5 Leptonic and Semileptonic Decays of Neutral Mesons

A quick look at Table D.2 reveals the following facts about these decays:

- Purely leptonic decays are absent in this table because they are highly suppressed. Best-known examples are $K_{L,S} \to \mu^+ \mu^-$ and $B^0_{s,d} \to \mu^+ \mu^-$, which will appear in this book but they can only proceed through loop diagrams in the SM.
- Semileptonic decays of K_L dominate the decays of this meson with a significant branching ratio coming from K_L decaying into three pions. This pattern originates in the fact that the K_L meson, as we will discuss in Chapter 8, is dominantly a CP= − eigenstate and can easily decay into an odd number of pions. The decay into two pions can only proceed when CP symmetry is violated.
- As K_S is dominantly a CP= + eigenstate, semileptonic decays in this case are very strongly suppressed, and weak decays of K_S to two pions dominate.
- We list there two branching ratios for B^0_d and B^+ decays. Other can be found in PDG.

We do not show the leading diagrams for semileptonic decays, as they can be obtained from the ones for charged meson decays by changing the *spectator* quark that does not couple to $W^\pm$ from the up-quark u to down-quark d. We only warn the reader that when doing calculations with K_L and K_S mesons factors of $\sqrt{2}$ appear in their definitions and have to be taken properly into account in the evaluation of decay rates.

3.6 Nonleptonic Decays of Mesons

3.6.1 Preliminaries

Finally let us consider nonleptonic decays of mesons like

$$K^+ \to \pi^+ \pi^0, \qquad D^+ \to \pi^0 K^+, \qquad B^+ \to D^0 K^+. \qquad (3.53)$$

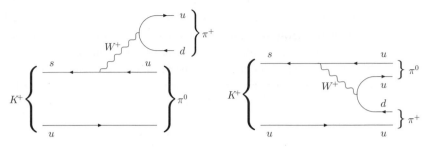

Figure 3.7   Decay $K^+ \to \pi^+ \pi^0$.

The flavor structure of the first decay is represented by the left diagram in Figure 3.7. Proceeding like in previous cases, we can immediately write down the corresponding decay amplitude

$$\mathcal{A}(K^+ \to \pi^+ \pi^0) = \frac{G_F}{\sqrt{2}} V_{us}^* V_{ud} \langle \pi^+ \pi^0 | (\bar{u}d)_{V-A} (\bar{s}u)_{V-A} | K^+ \rangle. \tag{3.54}$$

The new feature relative to the decays considered so far is the appearance of the product of two quark currents, the so-called four-quark operator

$$Q_2 = (\bar{u}d)_{V-A}(\bar{s}u)_{V-A} = (\bar{u}_\alpha d_\alpha)_{V-A}(\bar{s}_\beta u_\beta)_{V-A}. \tag{3.55}$$

The index "2" and not "1" is kept to conform with the literature. Moreover, we exposed this time color indices α and β as they will be relevant soon.

We know already from previous decays that the hadronic matrix element of Q_2 cannot be calculated in perturbation theory, and we will need lattice QCD or some other nonperturbative method to obtain its value. But it is tempting to factorize this matrix element as follows:

$$\langle \pi^+ \pi^0 | (\bar{u}d)_{V-A}(\bar{s}u)_{V-A} | K^+ \rangle = \langle \pi^+ | (\bar{u}d)_{V-A} | 0 \rangle \langle \pi^0 | (\bar{s}u)_{V-A} | K^+ \rangle \tag{3.56}$$

so that each factor could be extracted from previous decays. Yet as seen in Figure 3.7, this can only be at best an approximation as the gluon exchanges between the quarks on the two ends of the $W^\pm$ propagator spoil this factorization. Moreover, the appearance of new types of diagrams like the second one in Figure 3.7, not encountered in leptonic and semileptonic decays, clearly shows that (3.56) cannot be true. We conclude therefore that the study of nonleptonic decays is much harder than of leptonic and semileptonic decays because of the breakdown of factorization of a matrix element of a four-quark operator into the product of matrix elements of two quark currents.

For the remaining two decays in (3.53) we find

$$\mathcal{A}(D^+ \to \pi^0 K^+) = \frac{G_F}{\sqrt{2}} V_{cd}^* V_{us} \langle \pi^0 K^+ | (\bar{u}s)_{V-A}(\bar{d}c)_{V-A} | D^+ \rangle, \tag{3.57}$$

$$\mathcal{A}(B^+ \to D^0 K^+) = \frac{G_F}{\sqrt{2}} V_{cb}^* V_{us} \langle D^0 K^+ | (\bar{u}s)_{V-A}(\bar{b}c)_{V-A} | B^+ \rangle. \tag{3.58}$$

Also in these cases factorization of the matrix element of a four-quark operator into a product of matrix elements of currents is broken. The additional complication arises because in this case lattice QCD alone, as of 2020, is of little help, and other methods like the ones based on flavor symmetries and those using the fact that now heavy mesons are involved have to be used. One of these methods is the so-called QCD factorization, which goes beyond the *naive* factorization in (3.56). We will describe this method briefly in Section 7.4. But let us first look a bit closer at QCD effects in nonleptonic meson decays to be able to make progress in this book.

3.6.2 QCD Effects

As we have seen, in leptonic and semileptonic decays of mesons the decay amplitude involves the product of the matrix element of a quark current between meson states and of the leptonic current. Even if such matrix elements cannot be calculated in perturbation theory, they compactly summarize all QCD effects in these decays and are sufficiently simple to be calculated these days by lattice methods with respectable precision.

However, this is no longer possible in the case of nonleptonic decays, and one has to face a nonperturbative evaluation of hadronic matrix element of the four quark operator involving three external meson states. But this is not the end of the story. Once the diagrams with both $W^\pm$ and gluon exchanges are considered, we find that the operator Q_2 in (3.55) is not the only operator responsible for the decay $K^+ \to \pi^+\pi^0$. Generally there are several additional operators that are generated by QCD effects. One of them is

$$Q_1 = (\bar{u}_\alpha d_\beta)_{V-A}(\bar{s}_\beta u_\alpha)_{V-A}, \tag{3.59}$$

where again summation over color indices is understood. Note that the flavor structure of Q_1 is the same as the one of Q_2, but the color structure differs. We will demonstrate the existence of Q_1 in Section 5.1.

If the four-quark flavors in Q_1 and Q_2 were all different from each other, then indeed these operators would be the only four-quark operators of dimension six we would have to consider in nonleptonic K decays and analogous operators in B and D nonleptonic decays. In fact, there are meson decays, to be discussed later, which involve only such *current-current* operators. But Q_1 and Q_2 contain two up-quarks and in such a case new operators, the so-called penguin operators, are generated through gluon and photon exchanges, and they also have to be included in the analysis. We demonstrate their existence in Chapter 6.

Thus in general several four-quark operators have to be taken into account. Denoting these operators by Q_i we find that the amplitudes for $K \to \pi\pi$ decays with K and π being charged or neutral have the general structure

$$\mathcal{A}(K \to \pi\pi) = \frac{G_F}{\sqrt{2}} \sum_i V^i_{\text{CKM}} C_i \langle \pi\pi | Q_i | K \rangle, \tag{3.60}$$

with V^i_{CKM} being appropriate CKM factors that can be different for different operators. Analogous expressions for other nonleptonic decays exist and will be discussed later.

The coefficients C_i, the so-called Wilson Coefficients, describe the strength with which a given operator contributes to the decay amplitude, although this strength depends also on the size of the matrix elements $\langle \pi\pi | Q_i | K \rangle$. The coefficients C_i could at first sight be evaluated in the ordinary perturbation theory in α_s but due to the fact that the mass of $W^\pm$ is vastly different from the low-energy scales involved in the decays of mesons, large logarithms of the ratio of M_W and these low-energy scales multiply α_s. Therefore, even in the presence of asymptotic freedom in QCD, in nonleptonic decays not only factorization of matrix elements into products of matrix elements of currents but also the ordinary perturbation theory breaks down.

Yet, in the last forty years powerful methods like effective theories, operator product expansion, and renormalization group among others have been developed, and to continue our presentation of the gauge theory of weak decays, we have to learn what all these methods are. We must also develop efficient technology to apply these methods not only to make predictions for weak decays within the SM but also for its extensions. This is the prime goal of the next three chapters.

3.7 Summary and Motivation

We have seen that tree-level decays play an important role in the determination of the elements of the CKM matrix. In this chapter we have briefly described how

$$|V_{us}|, \qquad |V_{ub}|, \qquad |V_{cb}|, \tag{3.61}$$

can be determined from leptonic and semileptonic decays of mesons. This determination is important as it is believed to be not polluted by contributions from unknown dynamics at very short distance scales, which still has to be discovered. We have also discussed the muon decay, which allows to extract the Fermi constant, the crucial parameter in the phenomenology of weak decays in the SM.

We have also stressed that the study of nonleptonic decays is much harder. Yet, such a study is indispensable for many reasons. Here are few of them

- We have not determined the KM phase in the CKM matrix or equivalently the angle γ in the unitarity triangle and also other angles in this triangle. As we will see in Section 8.5, nonleptonic decays of B mesons play a very important role in these determinations.
- The ratio of the real parts of the isospin amplitudes A_0 and A_2 in $K \to \pi\pi$ decays is measured to be 22.4, while a simple application of factorization of hadronic matrix elements gives $\sqrt{2}$ for this ratio. As we will see in Sections 7.2 and 7.3, nonfactorizable QCD affects both at short-distance scales and in particular long distances are responsible dominantly for this rule.
- CP-violation was discovered in 1964 in $K_L \to \pi\pi$ decays, and its size summarized by the value of the parameter ε_K is an important constraint for the SM and NP beyond it. The same applies to the ratio ε'/ε measured also in these decays. It describes the size of another type of CP violation, to be defined in Section 8.1.

- Nonleptonic transitions between particles and antiparticles known under the name of mixing play crucial role in the tests of the SM and of its extensions. These are $K^0 - \bar{K}^0$, $B_s^0 - \bar{B}_s^0$, $B_d^0 - \bar{B}_d^0$, and $D^0 - \bar{D}^0$ mixings.

But before we can begin to study all these phenomena we have to collect various tools that will allow us to go beyond the tree diagrams and also properly include strong interaction effects.

> **We are starting collecting tools for our expedition!**

Technology beyond Trees

4.1 Loop Calculations

4.1.1 General View

Until now the calculations of Feynman diagrams were easy. Knowing the Feynman rules for vertices and propagators a simple multiplication of the corresponding expressions for them allowed to calculate the contribution of a given tree diagram to the decay amplitude we were interested in. However, generally more complicated diagrams, like the ones in Figure 6.9 have to be considered. If the interactions are sufficiently weak, we know that these *loop diagrams* involving higher powers of couplings will result in contributions that are smaller than the tree-level ones. Yet, we have to calculate them at least for three reasons:

- We have to check the accuracy of the tree-level predictions and whether the perturbation theory works at all.
- The true tests of a given quantum field theory start at one-loop level.
- Certain decays are absent at tree-level, and the leading contributions come from diagrams involving one loop.

We will soon see that there are other reasons for performing these, often tedious, calculations. But already these three reasons are sufficient motivations to learn how to efficiently calculate these diagrams.

The first obstacle one encounters in evaluating the diagrams in Figure 6.9 is not only the integration over the internal momentum in the loop but also the fact that the result for the left diagram is meaningless; it is divergent. This is bad news. The good news is that instead of trying to improve on this by studying the full diagram in Figure 6.9 or other diagrams with different loops, it is sufficient to look first at those subdiagrams (parts) of these diagrams, which are responsible for the divergences in question. The simplest of them are shown in Figure 4.1

Our goals for the first sections of this chapter are as follows.

- To learn the technology for loop calculations.
- To train Dirac algebra in $D \neq 4$ dimensions. This technology is crucial in addressing the divergences in loop diagrams.
- To train color algebra relevant for QCD calculations.
- To exhibit the calculations of the first two diagrams in Figure 4.1 in explicit terms and to indicate the structure of the calculation of the remaining diagrams.

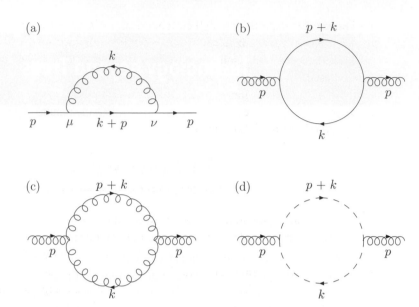

Figure 4.1 Self-energy diagrams.

- To exhibit differences between QED and QCD already at this stage by considering the exchanged gauge boson to be a photon and a gluon, respectively. These differences are important in the case of the first two diagrams but not as profound as the ones when the third and fourth diagram are considered: They are absent in QED, and in QCD they have important implications for the short-distance structure of the theory.

Once this is done we can address in subsequent sections the important topics of renormalization and renormalization group methods. While we will present all these topics in some detail, a very useful book in this context is the one by Muta [70].

At this stage the following comment should be made. Some of the younger readers may consider all these details unnecessary as these days several computer programs can perform complicated analytic calculations without entering the details presented next. Yet, such an approach prohibits a deep insight in what is going on and in my view takes away the pleasure of doing these calculations by hand. In fact, all one-loop calculations in the SM can be done easily by hand and the ones in QED and QCD also at the two-loop level. A different story are two-loop calculations of electroweak contributions and higher loop diagrams for which computer calculations are unavoidable.

4.1.2 Dimensional Regularization

Preliminaries

To deal with divergences that appear in loop diagrams we have to regularize the theory to have an explicit parametrization of the singularities. In this book we will employ *dimensional regularization* (DR), which is used basically by everybody these days except for lattice QCD experts, who in their numerical nonperturbative calculations use different regularization by discretizing the space. We will not discuss how these two regularizations are related, as the final results for physical quantities cannot depend on the regularization method used.

In DR Feynman diagrams are evaluated in $D = 4 - 2\varepsilon$ space-time dimensions, and singularities are extracted as poles for $\varepsilon \to 0$. Thus the results of one-loop and two-loop calculations have the following general structure:

$$\text{One-loop result} = \frac{a_1}{\varepsilon} + b_1 , \tag{4.1}$$

$$\text{Two-loop result} = \frac{a_2}{\varepsilon^2} + \frac{b_2}{\varepsilon} + c_2 , \tag{4.2}$$

where a_i, b_i, and c_2 are finite.

Several useful formulas for the evaluation of Feynman diagrams in $D = 4 - 2\varepsilon$ dimensions will be given soon. Here we only stress the following important point. Let us consider the second term in the second line in the QCD Lagrangian in (2.113). It describes the interaction of the gluon field A_μ^a with quarks q_i. The energy dimensions of q_i, A_μ^a and of $\mathcal{L}$ itself in D dimensions are $(D-1)/2$, $(D-2)/2$, and D, respectively. Consequently, the dimension of g_s in $D = 4 - 2\varepsilon$ dimensions is simply equal to ε. It is more useful to work with a dimensionless coupling constant in arbitrary D dimensions. To this end we make the replacement in (2.113):

$$\boxed{g_s \quad \to \quad g_s \, \mu^\varepsilon, \qquad \varepsilon = \frac{4 - D}{2}, } \tag{4.3}$$

where μ is an arbitrary parameter with the dimension of energy and g_s on the r.h.s. is dimensionless. The appearance of the scale μ has profound impact on this and subsequent sections.

D-Dimensional Integrals

In the calculations of one-loop Feynman diagrams we will encounter integrals over the internal four-momentum k_μ, and it is useful to have the results for them at hand. We will not derive these expressions as the derivations can be found in many field theory books. In particular we refer to pages 112–113 of Muta's book [70].

The three integrals we will need first are as follows

$$\boxed{\int \frac{d^D k}{(2\pi)^D} \frac{1}{(k^2 - V + i\epsilon)^N} = i \frac{(-1)^N}{(4\pi)^{D/2}} \frac{\Gamma(N - D/2)}{\Gamma(N)} \frac{1}{(V - i\epsilon)^{N-D/2}}, } \tag{4.4}$$

$$\boxed{\int \frac{d^D k}{(2\pi)^D} \frac{k^2}{(k^2 - V + i\epsilon)^N} = i \frac{(-1)^{N-1}}{(4\pi)^{D/2}} \frac{\Gamma(N - 1 - D/2)}{\Gamma(N)} \frac{D}{2} \frac{1}{(V - i\epsilon)^{N-1-D/2}}, } \tag{4.5}$$

$$\boxed{\int \frac{d^D k}{(2\pi)^D} \frac{k_\mu k_\nu}{(k^2 - V + i\epsilon)^N} = i g_{\mu\nu} \frac{(-1)^{N-1}}{(4\pi)^{D/2}} \frac{\Gamma(N - 1 - D/2)}{\Gamma(N)} \frac{1}{2} \frac{1}{(V - i\epsilon)^{N-1-D/2}}. } \tag{4.6}$$

Here N can be a complex number, and V is a function of masses of exchanged particles and momenta of external ones. $\Gamma(\alpha)$ is one of Euler functions for which useful expressions will be given soon. Also the following two relations are useful:

$$\int \frac{d^D k}{(2\pi)^D} k_\mu k_\nu f(k^2) = \frac{1}{D} g_{\mu\nu} \int \frac{d^D k}{(2\pi)^D} k^2 f(k^2), \qquad (4.7)$$

$$g_{\mu\nu} g^{\mu\nu} = D, \qquad \int \frac{d^D k}{(2\pi)^D} k_\mu f(k^2) = 0. \qquad (4.8)$$

Euler-Beta and Euler-Gamma Functions

While doing simple one-loop calculations, we will encounter integrals that can be expressed in terms of Euler-beta function $B(\alpha, \beta)$ and Euler-gamma function $\Gamma(\alpha)$, where α and β are complex numbers. We have

$$\int_0^1 dx\, x^{\alpha-1}(1-x)^{\beta-1} = B(\alpha, \beta) = \frac{\Gamma(\alpha)\Gamma(\beta)}{\Gamma(\alpha+\beta)}, \qquad (4.9)$$

with

$$\Gamma(1) = \Gamma(2) = 1, \qquad \Gamma(1+\alpha) = \alpha\Gamma(\alpha). \qquad (4.10)$$

Of interest for us are various expansions of both functions in powers of the small parameter ε. In particular we have

$$\Gamma(\varepsilon) = \frac{1}{\varepsilon} - \gamma_E + O(\varepsilon), \qquad \gamma_E = 0.57721\ldots, \qquad (4.11)$$

$$\Gamma(1-\varepsilon) = 1 + O(\varepsilon^2), \qquad \Gamma(2-\varepsilon) = 1 - \varepsilon, \qquad (4.12)$$

$$B(2-\varepsilon, 1-\varepsilon) = \frac{1}{2}[1 + 2\varepsilon] + O(\varepsilon^2), \qquad (4.13)$$

$$B(2-\varepsilon, 2-\varepsilon) = \frac{1}{6}\left[1 + \frac{5}{3}\varepsilon\right] + O(\varepsilon^2), \qquad (4.14)$$

$$B(3-\varepsilon, 1-\varepsilon) = \left(\frac{2-\varepsilon}{1-\varepsilon}\right) B(2-\varepsilon, 2-\varepsilon) + O(\varepsilon^2), \qquad (4.15)$$

where γ_E is the Euler constant. Finally an important quantity is

$$P_{div} \equiv \frac{\Gamma(\varepsilon)}{(4\pi)^{2-\varepsilon}} \left(\frac{\mu^2}{-p^2}\right)^\varepsilon = \frac{1}{16\pi^2}[\frac{1}{\varepsilon} + \ln 4\pi - \gamma_E + \ln \frac{\mu^2}{-p^2} + O(\varepsilon)], \qquad (4.16)$$

where p (with $p^2 < 0$) is an external momentum, and we have performed the expansion around $\varepsilon = 0$. We will encounter P_{div} in the following one-loop calculations. The $1/\varepsilon$ is the divergence we mentioned earlier.

Feynman Integrals

When writing the mathematical expression corresponding to a given diagram, we encounter the products of various propagators. To bring these products to the structure of the momentum integrals listed earlier we can use the following relations found by Feynman himself:

$$\frac{1}{AB} = \int_0^1 dx \frac{1}{[xA + (1-x)B]^2}, \qquad \frac{1}{A^\alpha B^\beta} = \int_0^1 dx \frac{x^{\alpha-1}(1-x)^{\beta-1}}{[xA + (1-x)B]^{\alpha+\beta}}, \quad (4.17)$$

$$\frac{1}{ABC} = \int_0^1 dx \int_0^1 dy \frac{y}{[xyA + y(1-x)B + (1-y)C]^3}. \qquad (4.18)$$

Here A, B, and C represent the denominators in the propagators in loop diagrams. More general expressions can be found in Muta's book.

Dirac Algebra in D Dimensions

Here we give the most important relations between Dirac matrices γ_μ in D dimensions. The three basic relations are

$$g_{\mu\nu}g^{\mu\nu} = D, \qquad \{\gamma_\mu, \gamma_\nu\} = 2g_{\mu\nu}, \qquad \gamma_\mu\gamma^\mu = D. \qquad (4.19)$$

Next

$$\gamma_\nu\gamma_\mu\gamma^\nu = (2-D)\gamma_\mu = 2(\varepsilon-1)\gamma_\mu, \qquad (4.20)$$

$$\gamma_\lambda\gamma_\mu\gamma_\nu\gamma^\lambda = 4g_{\mu\nu} - 2\varepsilon\gamma_\mu\gamma_\nu, \qquad (4.21)$$

$$\gamma_\lambda\gamma_\mu\gamma_\nu\gamma_\rho\gamma^\lambda = -2\gamma_\rho\gamma_\nu\gamma_\mu + 2\varepsilon\gamma_\mu\gamma_\nu\gamma_\rho, \qquad (4.22)$$

and finally

$$\mathrm{Tr}[\hat{1}] = 4, \qquad \mathrm{Tr}[\gamma_\mu\gamma_\nu] = 4g_{\mu\nu}, \qquad (4.23)$$

$$\mathrm{Tr}[\gamma_\mu\gamma_\nu\gamma_\lambda\gamma_\rho] = 4[g_{\mu\nu}g_{\lambda\rho} + g_{\mu\rho}g_{\nu\lambda} - g_{\mu\lambda}g_{\nu\rho}]. \qquad (4.24)$$

Setting $\varepsilon = 0$ one recovers the usual relations for $D = 4$ known from any book on relativistic quantum mechanics or relativistic quantum field theory.

Color Algebra

Here we will collect some useful formulas that we will need in our QCD calculations. The review in [12] is very useful in this context. In view of the large N approach to QCD, which we will discuss in Section 7.2, we will present these formulas for the group SU(N) and begin our presentation with SU(N) Gell-Mann matrices λ_a for which explicit expressions in the case $N = 3$ are given in (A.19)–(A.21) of Appendix A.2. But we will not need these expressions here as only the general relations listed by us will be relevant for our calculations.

There are $N^2 - 1$ matrices λ^a for the SU(N) group; they are Hermitian and traceless and normalized as

$$\text{Tr}(\lambda_a \lambda_b) = 2\delta_{ab}, \qquad \delta_{ab}\delta_{ab} = N^2 - 1, \tag{4.25}$$

with δ_{ab} being Kronecker delta. The tensors f_{abc} and d_{abc} are defined by

$$\lambda_a \lambda_b = \frac{2}{N}\delta_{ab} + (if_{abc} + d_{abc})\lambda_c \tag{4.26}$$

and have the following properties

$$d_{aab} = 0, \qquad f_{aab} = 0, \qquad d_{acd}f_{bcd} = 0, \tag{4.27}$$

$$f_{acd}f_{bcd} = N\delta_{ab}, \qquad d_{acd}d_{bcd} = \left(N - \frac{4}{N}\right)\delta_{ab}, \tag{4.28}$$

$$(\lambda_a)_{\alpha\beta}(\lambda_a)_{\gamma\delta} = 2\left(\delta_{\alpha\delta}\delta_{\gamma\beta} - \frac{1}{N}\delta_{\alpha\beta}\delta_{\gamma\delta}\right), \tag{4.29}$$

$$\lambda_a \lambda_a = 4C_F \hat{1}, \qquad C_F = \frac{N^2 - 1}{2N}, \tag{4.30}$$

where $\hat{1}$ is the unit matrix in the color space. In all these formulas summation over repeated color indices is understood.

The color charges in the quark-gluon vertices are summarized by

$$t_a = \frac{\lambda_a}{2}, \qquad [t_a, t_b] = if_{abc}t_c. \tag{4.31}$$

Frequently used relations are

$$t_a t_a = C_F \hat{1}, \qquad t_a t_b t_a = -\frac{1}{2N}t_b, \qquad t_a t_b t_a t_b = -\frac{C_F}{2N}\hat{1}, \tag{4.32}$$

$$t_a t_b t_c t_a t_b = \frac{1}{4}\left(1 + \frac{1}{N^2}\right)t_c, \qquad t_a t_b t_c t_a = \frac{1}{4}\delta_{bc}\hat{1} - \frac{1}{2N}t_b t_c, \tag{4.33}$$

$$t_a t_b t_c t_b t_a = \frac{1}{4N^2}t_c. \tag{4.34}$$

Fermion Self-Energy Diagram in QED

With this technology at hand we will now calculate explicitly the one-loop self-energy diagram (a) of Figure 4.1 in QED so that the exchanged gauge boson is photon and not gluon as shown there. Setting the fermion mass to zero, denoting the external fermion momentum by p (with $p^2 < 0$), and using QED Feynman rules, given in Appendix B, we find for this diagram

$$I_1 = N_1 \int \frac{d^D k}{(2\pi)^D} \frac{\gamma_\nu (\not{k} + \not{p}) \gamma_\mu}{(k + p)^2 k^2} g^{\mu\nu} = 2N_1 (\varepsilon - 1) \int \frac{d^D k}{(2\pi)^D} \frac{(\not{k} + \not{p})}{(k + p)^2 k^2}, \quad (4.35)$$

where

$$N_1 = -e^2 Q_f^2 \mu^{2\varepsilon}, \qquad \not{p} \equiv p_\mu \gamma^\mu, \quad (4.36)$$

Q_f is the electric charge of the fermion, and e the QED coupling constant. It should be noted that we deal here with the so-called amputated Green function (see any book on field theory and Section 4.2) without external propagators.

We use next the first formula in (4.17) to obtain

$$\frac{1}{(k + p)^2 k^2} = \int_0^1 dx \frac{1}{[x(k + p)^2 + (1 - x)k^2]^2} = \int_0^1 dx \frac{1}{[l^2 + p^2 x(1 - x)]^2}, \quad (4.37)$$

where we made a shift in the internal momentum to bring the integral into the desired form so that we can calculate it in no time:

$$l = k + px, \qquad k + p = l + p(1 - x). \quad (4.38)$$

Indeed, using (4.4) and (4.8), we obtain right away

$$I_1 = 2N_1 (\varepsilon - 1) \not{p} \int_0^1 dx \int \frac{d^D l}{(2\pi)^D} \frac{(1 - x)}{[l^2 + p^2 x(1 - x)]^2}, \quad (4.39)$$

then

$$I_1 = 2N_1 (\varepsilon - 1) \not{p} \int_0^1 dx(1 - x) \left[i \frac{(-1)^2}{(4\pi)^{D/2}} \frac{\Gamma(2 - D/2)}{\Gamma(2)} \frac{1}{[-p^2 x(1 - x)]^{2-D/2}} \right], \quad (4.40)$$

and subsequently

$$I_1 = i \not{p} Q_f^2 e^2 [2(1 - \varepsilon)] P_{div} B(2 - \varepsilon, 1 - \varepsilon), \quad (4.41)$$

with P_{div} given in (4.16) and $B(2 - \varepsilon, 1 - \varepsilon)$ in (4.13). Expanding in ε we finally find

$$\boxed{I_a^{em} = i \not{p} Q_f^2 \frac{\alpha_{em}}{4\pi} [\frac{1}{\varepsilon} + \ln 4\pi - \gamma_E + \ln \frac{\mu^2}{-p^2} + 1],} \quad (4.42)$$

where $O(\varepsilon)$ terms have been set to zero, and we added a superscript "em" to indicate that the calculation was done in QED. We have thus extracted the singularity as a $1/\varepsilon$ pole and have obtained a well-defined finite part. The appearance of the first four terms in the square bracket in (4.42), originating from P_{div} in (4.16), is characteristic for all divergent one-loop calculations. We will use this result in the process of renormalization of QED in the next section, but after this useful practice in one-loop calculations let us practice them a bit more.

Quark Self-Energy Diagram in QCD

This is the diagram (a) of Figure 4.1 in which this time a gluon is exchanged. This calculation is identical to the one just performed except that α_{em} should be replaced by α_s and Q_f^2 by the relevant color factor. This factor is calculated as follows

$$t_{\alpha\delta}^a t_{\delta\beta}^b \delta^{ab} = (t^a t^a)_{\alpha\beta} = C_F \delta_{\alpha\beta}, \qquad C_F = \frac{4}{3}, \tag{4.43}$$

where t^a come from quark-gluon vertices and δ^{ab} from gluon propagator. Making these changes in (4.42) we arrive at

$$\boxed{(I_a^{QCD})_{\alpha\beta} = i \not{p} \, C_F \, \delta_{\alpha\beta} \frac{\alpha_s}{4\pi} [\frac{1}{\varepsilon} + \ln 4\pi - \gamma_E + \ln \frac{\mu^2}{-p^2} + 1].} \tag{4.44}$$

Again we will use this result in the next section where we will renormalize QCD but to be able to do this we have to work harder and to calculate first few additional diagrams.

Vacuum Polarization in QED

We calculate next the second diagram (b) in Figure 4.1 in QED. We find first

$$(I_b^{em})_{\mu\nu} = N_2 \int \frac{d^D k}{(2\pi)^D} \frac{\text{Tr}(\gamma_\nu(\not{k}+\not{p})\gamma_\mu \not{k})}{(k+p)^2 k^2} = N_2 \int \frac{d^D k}{(2\pi)^D} \frac{(k+p)_{\sigma_1} k_{\sigma_2}}{(k+p)^2 k^2} T_{\nu\mu}^{\sigma_1\sigma_2}, \tag{4.45}$$

where

$$N_2 = -e^2 Q_f^2 \mu^{2\varepsilon}, \qquad T_{\nu\mu}^{\sigma_1\sigma_2} = \text{Tr}(\gamma_\nu \gamma^{\sigma_1} \gamma_\mu \gamma^{\sigma_2}). \tag{4.46}$$

The additional minus sign comes from the closed fermion loop. The structure of the denominator is the same as in the self-energy diagram calculations, and consequently the same shift in momenta as given in (4.38) should be made to bring the four-momentum integral in the desired form. We find then

$$(I_b^{em})_{\mu\nu} = N_2 T_{\mu\nu}^{\sigma_1\sigma_2} \left[G_1^{\sigma_1\sigma_2} + G_2^{\sigma_1\sigma_2} \right], \tag{4.47}$$

where

$$G_1^{\sigma_1\sigma_2} = \int_0^1 dx \int \frac{d^D l}{(2\pi)^D} \frac{l_{\sigma_1} l_{\sigma_2}}{[l^2 + p^2 x(1-x)]^2}, \tag{4.48}$$

$$G_2^{\sigma_1\sigma_2} = -p_{\sigma_1} p_{\sigma_2} \int_0^1 dx \int \frac{d^D l}{(2\pi)^D} \frac{x(1-x)}{[l^2 + p^2 x(1-x)]^2}. \tag{4.49}$$

With the list of integrals given a few pages earlier, these integrals can be straightforwardly evaluated to find

$$\mu^{2\varepsilon} G_1^{\sigma_1\sigma_2} = \frac{i}{2} g_{\sigma_1\sigma_2} \frac{(-p^2)}{(1-\varepsilon)} P_{div} B(2-\varepsilon, 2-\varepsilon), \tag{4.50}$$

$$\mu^{2\varepsilon} G_2^{\sigma_1\sigma_2} = -i p_{\sigma_1} p_{\sigma_2} P_{div} B(2-\varepsilon, 2-\varepsilon). \tag{4.51}$$

Next the products of the tensor $T^{\sigma_1\sigma_2}_{\nu\mu}$ and $g_{\sigma_1\sigma_2}$ and $p_{\sigma_1}p_{\sigma_2}$ have to be evaluated. We find

$$T^{\sigma_1\sigma_2}_{\nu\mu} g_{\sigma_1\sigma_2} = \text{Tr}(\gamma_\nu \gamma^{\sigma_1} \gamma_\mu \gamma_{\sigma_1}) = 2(\varepsilon - 1)\text{Tr}(\gamma_\nu \gamma_\mu) = 8(\varepsilon - 1)g_{\mu\nu}, \tag{4.52}$$

$$T^{\sigma_1\sigma_2}_{\nu\mu} p_{\sigma_1}p_{\sigma_2} = \text{Tr}(\gamma_\nu \not{p}\gamma_\mu \not{p}) = 4(2p_\mu p_\nu - p^2 g_{\mu\nu}) = 8p_\mu p_\nu - 4p^2 g_{\mu\nu}. \tag{4.53}$$

What remains to be done is to insert the results of all these intermediate calculations into (4.47) and expand in powers of ε. We obtain first

$$(I^{\text{em}}_b)_{\mu\nu} = -ie^2 Q_f^2 8[p^2 g_{\mu\nu} - p_\mu p_\nu]P_{div}B(2 - \varepsilon, 2 - \varepsilon), \tag{4.54}$$

and expanding in powers of ε and dropping $O(\varepsilon)$ terms, we find

$$\boxed{(I^{\text{em}}_b)_{\mu\nu} = -iQ_f^2 \frac{\alpha_{\text{em}}}{4\pi} \frac{4}{3}[p^2 g_{\mu\nu} - p_\mu p_\nu]\left[\frac{1}{\varepsilon} + \ln 4\pi - \gamma_E + \ln \frac{\mu^2}{-p^2} + \frac{5}{3}\right].} \tag{4.55}$$

From gauge invariance we know that the following relation has to be satisfied

$$p^\mu (I^{\text{em}}_b)_{\mu\nu} = 0. \tag{4.56}$$

Fortunately our result satisfies this relation.

Vacuum Polarization in QCD

Again the calculation is identical to the one just performed except that α_{em} should be replaced by α_s and Q_f^2 by the relevant color factor. This factor is this time

$$t^a_{\beta\alpha} t^b_{\alpha\beta} = \text{Tr}(t^a t^b) = \frac{1}{2}\delta^{ab}, \tag{4.57}$$

and using (4.54) we can first write

$$(I^{\text{QCD}}_b)^{ab}_{\mu\nu} = -ig_s^2 \delta^{ab} n_F 4[p^2 g_{\mu\nu} - p_\mu p_\nu]P_{div}B(2 - \varepsilon, 2 - \varepsilon), \tag{4.58}$$

where we included the contribution of n_F massless quarks. Finally we find

$$\boxed{(I^{\text{QCD}}_b)^{ab}_{\mu\nu} = -in_F \delta^{ab} \frac{\alpha_s}{4\pi} \frac{2}{3}[p^2 g_{\mu\nu} - p_\mu p_\nu]\left[\frac{1}{\varepsilon} + \ln 4\pi - \gamma_E + \ln \frac{\mu^2}{-p^2} + \frac{5}{3}\right].} \tag{4.59}$$

Yet, whereas the result in (4.55) was the final result in QED, the one just obtained is only one contribution to vacuum polarization in QCD. We have to include additional diagrams (c) and (d) shown in Figure 4.1, which are typical for a nonabelian gauge theory. The one involving the so-called tadpole, in which two fermion lines and two gauge boson lines meet at one point, vanishes in dimensional regularization as demonstrated on page 172 in Muta's book [70]. But the other two have to be calculated. In particular the calculation of the one involving triple gluon vertices is rather tough, and we will not be able to present it in all details here. But let us show at least the most important steps.

The integral to be evaluated is this time

$$(I^{\text{QCD}}_c)^{ab}_{\mu\nu} = N_3 \int \frac{d^D k}{(2\pi)^D} \frac{1}{(k + p)^2 k^2} V_{\mu\rho\sigma}(p, k, -k - p)V^{\sigma\rho}_\nu(-p, p + k, -k), \tag{4.60}$$

where

$$N_3 = (-i)^2 g_s^2 f_{bcd} f_{adc} \frac{1}{2} \mu^{2\varepsilon} = g_s^2 \frac{N}{2} \mu^{2\varepsilon} \delta_{ab} \tag{4.61}$$

with $1/2$ being a symmetry factor explained on page 91 of Muta's book. Now the general formula for the triple gluon vertex involves, in addition to $g_s f_{abc}$, the function

$$V_{\mu\nu\lambda}(k_1, k_2, k_3) = (k_1 - k_2)_\lambda g_{\mu\nu} + (k_2 - k_3)_\mu g_{\nu\lambda} + (k_3 - k_1)_\nu g_{\mu\lambda}, \tag{4.62}$$

where all momenta are incoming. The crucial thing in the next steps is not to get lost in indices. While a computer program like *Mathematica* can help us here, it is useful once in your life to do such calculation by hand. We find then

$$V_{\mu\rho\sigma}(p, k, -k - p) = (p - k)_\sigma g_{\mu\rho} + (2k + p)_\mu g_{\rho\sigma} + (-2p - k)_\rho g_{\mu\sigma}, \tag{4.63}$$

$$V_{\nu\sigma\rho}(-p, p + k, -k) = (-2p - k)_\rho g_{\nu\sigma} + (2k + p)_\nu g_{\rho\sigma} + (p - k)_\sigma g_{\nu\rho}. \tag{4.64}$$

Subsequently the product of these two expressions and contraction over repeated ρ and σ indices gives

$$V_{\mu\rho\sigma} V_\nu^{\sigma\rho} = (2k^2 + 2pk + 5p^2) g_{\mu\nu} + (4D - 6) k_\mu k_\nu + (D - 6) p_\mu p_\nu + (2D - 3)(k_\mu p_\nu + k_\nu p_\mu). \tag{4.65}$$

The suppressed arguments in the vertex functions are as in the previous two expressions. Inserting this result into (4.60), we are faced with a number of integrals that can be easily performed using the technology developed by us before. We quote here the final result

$$(I_c^{\text{QCD}})_{\mu\nu}^{ab} = i\delta^{ab} g_s^2 N [g_{\mu\nu} p^2 (19 - 12\varepsilon) - 2(11 - 7\varepsilon) p_\mu p_\nu] P_{div} \frac{B(2 - \varepsilon, 2 - \varepsilon)}{2(1 - \varepsilon)}. \tag{4.66}$$

The striking feature of this result is that it does not satisfy the condition (4.56).

Yet, the diagram (d) in Figure 4.1, in which a ghost is exchanged, removes this problem. Its contribution is

$$(I_d^{\text{QCD}})_{\mu\nu}^{ab} = N_4 \int \frac{d^D k}{(2\pi)^D} \frac{(k + p)_\mu k_\nu}{(k + p)^2 k^2}, \tag{4.67}$$

where

$$N_4 = (-1)(-i)^2 g_s^2 \mu^{2\varepsilon} f_{bdc} f_{acd} = -g_s^2 \mu^{2\varepsilon} N \delta_{ab}. \tag{4.68}$$

The first minus sign comes from the ghost loop. The next steps are similar to the ones we already made in the process of the calculation of the fermion loop (integral I_2), and we will not repeat them here. We find then

$$(I_d^{\text{QCD}})_{\mu\nu}^{ab} = i\delta^{ab} g_s^2 N [g_{\mu\nu} p^2 + 2(1 - \varepsilon) p_\mu p_\nu] P_{div} \frac{B(2 - \varepsilon, 2 - \varepsilon)}{2(1 - \varepsilon)}. \tag{4.69}$$

This result does not satisfy the gauge invariance condition in (4.56) either. But the gluon and ghost loops together satisfy it:

$$(I_c^{QCD} + I_d^{QCD})_{\mu\nu}^{ab} = i\delta^{ab}g_s^2 N[g_{\mu\nu}p^2 - p_\mu p_\nu]P_{div}\frac{B(2-\varepsilon,2-\varepsilon)}{(1-\varepsilon)}(10-6\varepsilon). \qquad (4.70)$$

Adding the quark loop contribution in (4.59), we complete the calculation of the vacuum polarization in QCD ($K_{\mu\nu} = [g_{\mu\nu}p^2 - p_\mu p_\nu]$)

$$(I_b^{QCD} + I_c^{QCD} + I_d^{QCD})_{\mu\nu}^{ab} = i\delta^{ab}g_s^2 K_{\mu\nu}P_{div}\frac{B(2-\varepsilon,2-\varepsilon)}{(1-\varepsilon)}[N(10-6\varepsilon)-4n_F(1-\varepsilon)].$$

$$(4.71)$$

This result will play an important role in the next sections. But let us first address an important issue present in the dimensional regularization.

4.1.3 The Issue of γ_5 in D Dimensions

Preliminaries

The dimensional regularization is the favorite regularization in gauge theories as it preserves all symmetries of the theory. Possible problems are connected with the treatment of γ_5 in $D \neq 4$ dimensions. This issue did not play any role in QED and QCD, in which the fermion-gauge boson couplings are vectorlike (γ_μ), but it is important in the study of weak interactions, where γ_5 plays a prominent role. Let us then discuss this issue now. Even if the material in this section is rather technical and could be skipped, motivated readers should at least look briefly at the following pages. We follow here [71].

Let us describe the three distinct sets of computational rules, for the manipulation of covariants and Dirac matrices, most commonly used in perturbative calculations in the SM. These schemes all employ the method of dimensional regularization of the Feynman integrals used already earlier. We will not discuss other regularization schemes such as BPHZ and lattice. These work directly in four dimensions and hence don't have algebraic consistency problems with respect to γ_5, but their use introduces other subtleties, and already two-loop calculations therewith are extremely tedious.

Naive Dimensional Regularization

The most commonly used set of rules is one we shall call "naive dimensional regularization" (NDR). This is the one we used already earlier, but now our presentation will be more formal and will be generalized to include γ_5. Only the D-dimensional metric tensor g is introduced satisfying,

$$g_{\mu\nu} = g_{\nu\mu}, \qquad g_{\mu\rho}g_\nu^\rho = g_{\mu\nu}, \qquad g_\mu^\mu = D, \qquad (4.72)$$

and the Dirac matrices γ_μ and γ_5 obey

$$\{\gamma_\mu, \gamma_\nu\} = 2g_{\mu\nu}, \qquad \{\gamma_\mu, \gamma_5\} = 0. \qquad (4.73)$$

It is standard (but inessential) to set the trace of the unit matrix to equal 4; we shall adopt this convention in this and other schemes discussed by us. When γ_5 appears in the Feynman vertices the manipulation rule adopted in this scheme is, as given here, that it anticommutes with the Dirac matrices.

It has repeatedly been emphasized in the literature [72–76] that the latter rule leads to obvious algebraic inconsistencies. Nevertheless this scheme has been most widely employed for most calculations because of its ease to incorporate standard software in computer programs. It is known to lead to incorrect results in certain cases, e.g., the axial anomaly is not reproduced. On the other hand, in many cases it does reproduce the correct results. A necessary condition for this seems to be that the calculated amplitude does not involve the evaluation of a closed odd parity fermion loop. Indeed, with the NDR rules one does not know how to unambiguously handle the expression $\mathrm{Tr}(\gamma_5\gamma_\mu\gamma_\nu\gamma_\rho\gamma_\lambda)$.

Fortunately beginning with the work in [71] it has been demonstrated in many explicit calculations that the NDR scheme gives correct results, consistent with the schemes without the γ_5 problems, provided one can avoid the calculations of traces like the one just mentioned. In fact, all the higher-order QCD calculations for weak decays performed in the NDR scheme in more than twenty-five years and reviewed in [77] could avoid the direct calculation of such traces.

Dimensional Reduction

A second set of manipulation rules initially introduced in [78] for the renormalization of supersymmetric theories goes under the name of dimensional reduction (DRED). Here the Dirac matrices $\tilde\gamma$ are taken to be in four dimensions, thus

$$\{\tilde\gamma_\mu, \tilde\gamma_\nu\} = 2\tilde g_{\mu\nu}, \tag{4.74}$$

where $\tilde g$ is the four dimensional metric tensor,

$$\tilde g_{\mu\nu} = \tilde g_{\nu\mu}, \qquad \tilde g_{\mu\rho}\tilde g_\nu^\rho = \tilde g_{\mu\nu}, \qquad \tilde g_\mu^\mu = 4. \tag{4.75}$$

When evaluating the Feynman integrals the D-dimensional, $g_{\mu\nu}$ inevitably makes its appearance, and it is necessary to supplement the rules with one that stipulates the result of contraction of the four- and D-dimensional metric tensors. To preserve gauge invariance and in apparent concord with the reduction to $D < 4$ dimensions the rule employed is

$$\tilde g_{\mu\rho}g_\nu^\rho = g_{\mu\nu}. \tag{4.76}$$

The advantage of this scheme is that the four-dimensional Dirac algebra can be used to reduce the algebraic complexity of the amplitudes. However, there is a price to be paid, which involves a number of field theoretical subtleties, some of which are already present in the pure QCD part of the dynamics. These are discussed in [79]. Again, this scheme has been criticized [72–76] because it leads to similar difficulties as the NDR described earlier. In particular, it implies that identities homogeneous in the metric tensor in four dimensions are also satisfied in generic D dimensions, which is manifestly algebraically inconsistent. Although the axial anomaly can be reproduced [80, 81], and although there is to our knowledge as yet no known explicit calculation using DRED that gives the wrong result, it has not yet been established as a consistent scheme and thus maintains at present merely the status of a prescription.

In the field of weak decays the DRED scheme has been used for the first time in [79] for the calculation of higher-order QCD corrections to weak decays. This result has been confirmed in [71] and shown to be compatible with the NDR scheme as well as with the 't Hooft–Veltman (HV) scheme discussed next. Similarly the initial problem of calculating higher-order QCD corrections to the $B \to X_s \gamma$ in the DRED scheme [82] has been resolved by Misiak [83]. These days the DRED scheme is less popular, and most calculations of QCD corrections are performed in the NDR scheme and sometimes in the 't Hooft–Veltman scheme, to which we turn our attention now.

The 't Hooft–Veltman Rules

The third set of rules is the one originally proposed by 't Hooft and Veltman [84] and by Akyeampong and Delbourgo [85–87] and systematized by Breitenlohner and Maison [72–74]. The latter authors showed that this is a consistent formulation of dimensional regularization even when γ_5 couplings are present.

To write down the rules it is convenient to introduce, in addition to the D- and four-dimensional metric tensors g and $\tilde{g}$ satisfying (4.72) and (4.75), respectively, the -2ε- dimensional tensor $\hat{g}$ satisfying,

$$\hat{g}_{\mu\nu} = \hat{g}_{\nu\mu}, \qquad \hat{g}_{\mu\rho}\hat{g}_\nu^\rho = \hat{g}_{\mu\nu}, \qquad \hat{g}_\mu^\mu = -2\varepsilon. \tag{4.77}$$

The important difference with respect to DRED is that instead of the rule (4.76) for contracting the different metric tensors one imposes

$$\tilde{g}_{\mu\rho}g_\nu^\rho = \tilde{g}_{\mu\nu}, \tag{4.78}$$

which does not lead to manifest algebraic inconsistencies. In addition to (4.78), one has

$$\hat{g}_{\mu\rho}g_\nu^\rho = \hat{g}_{\mu\nu}, \qquad \hat{g}_{\mu\rho}\tilde{g}_\nu^\rho = 0. \tag{4.79}$$

The D-dimensional Dirac matrix is now split into a 4- and -2ε-dimensional part,

$$\gamma_\mu = \tilde{\gamma}_\mu + \hat{\gamma}_\mu, \tag{4.80}$$

with γ and $\tilde{\gamma}$ obeying the anticommutation relations (4.73) and (4.74), respectively. $\hat{\gamma}$ on the other hand satisfies

$$\{\hat{\gamma}_\mu, \hat{\gamma}_\nu\} = 2\hat{g}_{\mu\nu}, \tag{4.81}$$

and it anticommutes with $\tilde{\gamma}$

$$\{\hat{\gamma}_\mu, \tilde{\gamma}_\nu\} = 0. \tag{4.82}$$

Note also by virtue of (4.79) that it follows

$$\hat{\gamma}_\mu\tilde{\gamma}^\mu = 0, \qquad \hat{g}_\mu^\nu\tilde{\gamma}_\nu = 0, \qquad \tilde{g}_\mu^\nu\hat{\gamma}_\nu = 0. \tag{4.83}$$

In [72–74] it is shown that a γ_5 can be introduced, which anticommutes with $\tilde{\gamma}$ but commutes with $\hat{\gamma}$,

$$\gamma_5^2 = 1, \qquad \{\gamma_5, \tilde{\gamma}_\nu\} = 0, \qquad [\gamma_5, \hat{\gamma}_\nu] = 0. \tag{4.84}$$

Because γ_5 does not have simple commutation properties with γ_μ it is important to consistently define the coupling to chiral fields in a model such as the SM; e.g., for couplings to left-handed fields the symmetrically defined vertex

$$\frac{1}{2}(1 + \gamma_5)\gamma_\mu(1 - \gamma_5) = \tilde{\gamma}_\mu(1 - \gamma_5), \tag{4.85}$$

should be used [88].

This scheme has admittedly some rather unattractive features. In particular, it is more inconvenient to implement in algebraic computer programs than the NDR scheme. Nevertheless it must be stressed again that it is to date the only known scheme (within the framework of dimensional regularization) that has been demonstrated to be consistent [72–76], and thus its inconvenience must be tolerated. For this reason a computer package for Dirac algebra manipulation in the HV and NDR schemes called TRACER has been developed by Jamin and Lautenbacher [89]. Using this program one can appreciate the simplicity of the NDR scheme compared with the HV scheme for which the computer calculations turned out in the 1990s to be really time consuming. With much improved computers this is not the case now. But as we mentioned earlier, most researchers use the NDR scheme making sure that no mathematical inconsistencies appear. In particular, even two-loop calculations are doable by hand in the NDR scheme, whereas in the HV scheme already one-loop calculations are rather involved.

In this book we will exclusively work in the NDR scheme, but we will occasionally quote results obtained in the HV scheme. Of course physical results cannot depend on the scheme used, but one should always remember that in field theory there are many intermediate objects that are scheme and also gauge dependent, and only after these objects are combined into a physical quantity that can be measured in experiments these nonphysical dependences drop out. Therefore it is often useful to perform calculations in different schemes and different gauges to verify the correctness of obtained results.

4.2 Renormalization

4.2.1 General Remarks

We have seen in the previous section that the simplest one-loop diagrams are divergent, and these divergences are then also present in Green functions[1] and finally in decay amplitudes. The process of renormalization allows to remove these divergences and to obtain finite decay amplitudes and consequently finite predictions for the observables as functions of

---

[1] The Green functions are just propagators, vertices and more complicated diagrams and are discussed formally in various books. We will define them later.

the parameters of a given theory. There are many books discussing renormalization of gauge theories, among them [14, 16, 18, 90], but for this book I found most useful the one of Muta [70], where further details can be found.

The idea of renormalization is actually rather simple. A given theory depends on

- Fields, as φ_0, ψ_0, A_0^μ etc.
- Parameters, as couplings g_0, masses m_0 etc.

The index "0" indicates that these are the so-called bare fields and parameters. Working with the bare fields and parameters, as we did until now, suppressing the index "0," results in divergent Green functions. The renormalization program is just a set of rules for replacing the bare fields and bare parameters by the so-called renormalized fields and renormalized parameters so that the Green functions and decay amplitudes written in terms of them are finite and can be compared with experiment.

The renormalized fields φ, ψ, and A^μ and renormalized parameters g, m, etc. are related to the bare ones as follows

$$A_{0\mu}^a = Z_3^{1/2} A_\mu^a, \qquad q_0 = Z_q^{1/2} q,$$
$$g_{0,s} = Z_g g_s \, \mu^\varepsilon, \qquad m_0 = Z_m \, m, \tag{4.86}$$

where we have just shown the most important examples. In particular g_s is the renormalized QCD coupling and m the renormalized quark mass.

The factors Z_i are the renormalization constants. They are divergent quantities, chosen in such a manner that the divergences disappear once the Greens functions have been expressed in terms of renormalized quantities only and a particular rescaling of them, exhibited soon, has been made. We have gauge field renormalization (Z_3), quark field renormalization (Z_q), coupling renormalization (Z_g), and mass renormalization (Z_m). There will be other quantities that we will have to renormalize later.

Within DR the renormalization constants have the following structure

$$Z_i = 1 + \frac{\alpha_s}{4\pi} \left(\frac{a_i}{\varepsilon} + b_i \right) + O(\alpha_s^2) \tag{4.87}$$

with the following properties:

- a_i are characteristic for a given theory, here QCD, and are independent of the renormalization scheme used.
- b_i are not characteristic for a given theory but depend on the renormalization scheme.

It should be stressed that the unrenormalized parameters $g_{0,s}$ and m_0 are independent of the scale μ. This implies, in particular, that g_s must be μ-dependent. Because Z_i have a perturbative expansion in g_s they must also depend on μ. Consequently, also the renormalized mass m is μ-dependent.

Before entering the details let us stress that not all theories can be renormalized in this manner. Generally one encounters:

- *Super renormalizable theories*: Only a finite number of Feynman diagrams diverge in all orders of perturbation theory.

- *Renormalizable theories*: Only a finite number of Feynman subdiagrams diverge, but they occur at each order of perturbation theory.
- *Nonrenormalizable theories*: All amplitudes are divergent at a sufficiently high order of perturbation theory.

The SM is a renormalizable theory, but as we will see at various places in this book, in particular in Chapter 14, also effective theories that look nonrenormalizable can provide useful results if properly interpreted [91].

4.2.2 MS and $\overline{\text{MS}}$ Renormalization Schemes

The simplest renormalization scheme is the *Minimal Subtraction scheme* MS [92] in which only divergences are subtracted. In this scheme, the renormalization constants are given by

$$Z_i = 1 + \frac{\alpha_s}{4\pi} \frac{a_{1i}}{\varepsilon} + \left(\frac{\alpha_s}{4\pi}\right)^2 \left(\frac{a_{2i}}{\varepsilon^2} + \frac{b_{2i}}{\varepsilon}\right) + O(\alpha_s^3), \tag{4.88}$$

where a_{ji} and b_{ji} are μ-independent constants. The fact that in this scheme the renormalization constants do not have any explicit μ-dependence and depend on μ only through g_s is an important virtue of this scheme. This property is discussed in the context of renormalization group equations in the next section. Similarly the renormalization constants Z_i do not depend on masses. Therefore the MS-scheme and the schemes discussed next belong to the class of mass independent renormalization schemes [93].

Now, starting with the MS scheme, one can construct a whole class of subtraction schemes that differ from MS by a different continuation of the renormalized coupling constant to D dimensions. For these MS-like schemes we have

$$g_{0,s} = Z_g^k g_s^k \mu_k^\varepsilon, \qquad \mu_k = \mu f_k, \tag{4.89}$$

where f_k is an arbitrary number that defines the particular scheme "k." Because different schemes in this class differ from the MS scheme only by a shift in μ, the renormalization constants for these schemes can be obtained from (4.88) by replacing α_s by α_s^k characteristic for a given scheme. The constants a_{ji} and b_{ji}, being μ-independent, remain unchanged.

Of particular interest is the so-called $\overline{\text{MS}}$ scheme [94] in which

$$\mu_{\overline{MS}} = \mu e^{\gamma_E/2}(4\pi)^{-1/2} \tag{4.90}$$

and P_{div} in (4.16) is replaced by

$$\bar{P}_{div} \equiv \frac{\Gamma(\varepsilon)}{(4\pi)^{2-\varepsilon}} \left(\frac{\mu_{\overline{MS}}^2}{-p^2}\right)^\varepsilon = \frac{1}{16\pi^2} \left[\frac{1}{\varepsilon} + \ln\frac{\mu^2}{-p^2} + O(\varepsilon)\right]. \tag{4.91}$$

We observe that in this scheme the terms $\ln 4\pi - \gamma_E$, the artifacts of the dimensional regularization, are absent! In summary, then:

$$\{\text{MS} \to \overline{\text{MS}}\} \equiv \{\mu \to \mu_{\overline{MS}}\} \tag{4.92}$$

$$\{Z_i^{MS} \to Z_i^{\overline{MS}}\} \equiv \{\alpha_s^{MS} \to \alpha_s^{\overline{MS}}\}. \tag{4.93}$$

In this book, we will exclusively work with the $\overline{\text{MS}}$ scheme. To simplify the notation we will denote $\mu_{\overline{\text{MS}}}$ simply by μ and simultaneously drop the $\ln 4\pi - \gamma_E$ terms in any finite contribution. Similarly α_s in this book will always stand for $\alpha_s^{\overline{\text{MS}}}$.

In later subsections of this section we will use the results of the previous section to calculate explicitly Z_q, Z_m, Z_3, and Z_{g_s} in the $\overline{\text{MS}}$ scheme, but it is useful to see the result already now. We have in the case of QCD ·

$$Z_q = 1 - \frac{\alpha_s}{4\pi} C_F \frac{1}{\varepsilon} + O(\alpha_s^2), \tag{4.94}$$

$$Z_m = 1 - \frac{\alpha_s}{4\pi} 3 C_F \frac{1}{\varepsilon} + O(\alpha_s^2), \tag{4.95}$$

$$Z_3 = 1 - \frac{\alpha_s}{4\pi} \left[\frac{2}{3} f - \frac{5}{3} N \right] \frac{1}{\varepsilon} + O(\alpha_s^2), \tag{4.96}$$

$$Z_{g_s} = 1 - \frac{\alpha_s}{4\pi} \left[\frac{11}{6} N - \frac{2}{6} f \right] \frac{1}{\varepsilon} + O(\alpha_s^2), \tag{4.97}$$

where N denotes the number of colors ($N = 3$ in QCD), C_F is given in (4.30), and f stands for the number of quark flavors. Z_q and Z_3 are gauge dependent and are given here in the Feynman gauge ($\xi = 1$). But this gauge dependence is cancelled by other contributions to physical amplitudes.

There are two routes to find the Z_i. One uses the relation between renormalized and unrenormalized Green functions, and the second one is the so-called counter method. We will first introduce these methods and subsequently use them to calculate Z_i in QCD and QED.

4.2.3 Renormalization of Green Functions

To discuss all this in explicit terms let us denote by

$$G^{(n_F, n_G)}(p_j, g_s, m, \mu, \varepsilon) \equiv \langle 0 | T(q_1, \ldots q_{n_F}, A_1^\mu, \ldots A_{n_G}^\mu) | 0 \rangle \tag{4.98}$$

a *connected renormalized* Green function with n_F quark and n_G gluon external legs carrying momenta p_j. Here m indicates general dependence on masses. The corresponding *amputated renormalized one-particle irreducible* Green function is given by

$$\Gamma^{(n_F, n_G)} = \frac{G^{(n_F, n_G)}}{\prod^{n_F} G^{(2,0)} \prod^{n_G} G^{(0,2)}} . \tag{4.99}$$

The division by the products of propagators of external fields amputates the Green function. Similar expressions exist for the unrenormalized Green functions $G_0^{(n_F, n_G)}$ and $\Gamma_0^{(n_F, n_G)}$ with all renormalized parameters and fields replaced by the corresponding bare quantities. With (4.86), $\Gamma^{(n_F, n_G)}$ and $\Gamma_0^{(n_F, n_G)}$ are related to each other by

$$\Gamma^{(n_F, n_G)}(p_j, g_s, m, \mu, \varepsilon) = Z_q^{n_F/2} Z_3^{n_G/2} \Gamma_0^{(n_F, n_G)}(p_j, g_{0,s}, m_0, \varepsilon) . \tag{4.100}$$

The factors in front of $\Gamma_0^{(n_F,n_G)}$ result from the rescaling of the fields in (4.86). The renormalization then means that when $g_{0,s}$ and m_0 on the r.h.s. of (4.100) are expressed through g and m according to (4.86), $\Gamma^{(n_F,n_G)}$ are finite and the limit

$$\lim_{\varepsilon \to 0} \Gamma^{(n_F,n_G)}(p_j, g_s, m, \mu, \varepsilon) = \Gamma^{(n_F,n_G)}(p_j, g_s, m, \mu) \qquad (4.101)$$

exists.

4.2.4 The Counter-Term Method

A straightforward way to implement renormalization is provided by the counter-term method. Thereby parameters and fields in the original Lagrangian, considered as unrenormalized (bare) quantities, are reexpressed through renormalized ones by means of (4.86). Thus

$$\mathcal{L}_{QCD}^0 = \mathcal{L}_{QCD} + \mathcal{L}_C, \qquad (4.102)$$

where $\mathcal{L}_{QCD}$ is given in (2.113). $\mathcal{L}_{QCD}^0$ is also given by (2.113) but with q replaced by q_0 and similarly for A_μ^a, g_s, and m. $\mathcal{L}_C$ is the *counter-term* Lagrangian. It is simply defined by (4.102). For instance:

$$\mathcal{L}_q = \bar{q}_0 i \not{\partial} q_0 - m_0 \bar{q}_0 q_0 \equiv \bar{q} i \not{\partial} q - m\bar{q}q + (Z_q - 1)\bar{q} i \not{\partial} q - (Z_q Z_m - 1)m\bar{q}q. \quad (4.103)$$

$\mathcal{L}_{QCD}$ given entirely in terms of renormalized quantities leads to the usual Feynman rules of Appendix B. The counter-terms ($\sim (Z - 1)$) can be formally treated as new interaction terms that contribute to Green functions calculated in perturbation theory. For these new interactions also Feynman rules can be derived. For instance, the Feynman rule for the counter-terms in (4.103) reads (p is the quark momentum)

$$i\delta_{\alpha\beta}[(Z_q - 1)\not{p} - (Z_q Z_m - 1)m] . \qquad (4.104)$$

The constants Z_i are determined such that the contributions from these new interactions cancel the divergences in the Green functions resulting from the calculations based on $\mathcal{L}_{QCD}$ in (4.102) only.

There is some arbitrariness how this can be done because a given renormalization prescription can in general subtract not only the divergences but also finite parts. The subtractions of finite parts is, however, not uniquely defined, which results in the *renormalization scheme dependence* of Z_i and of the renormalized fields and parameters as we already announced on previous pages.

4.2.5 Explicit Calculation of Renormalization Constants (QCD)

Let us first calculate Z_q and Z_m. To this end we repeat the calculation of the self-energy diagram of Figure 4.1, this time keeping the quark mass m. Dropping finite terms, which are of no concern for finding Z_i in the $\overline{\text{MS}}$ scheme, we find

$$(i\Sigma_{\alpha\beta})_{div} = iC_F \delta_{\alpha\beta} \frac{\alpha_s}{4\pi}(\not{p} - 4m)\frac{1}{\varepsilon} + O(\alpha_s^2). \qquad (4.105)$$

The l.h.s. is just fancy notation for $(I_1^{QCD})_{\alpha\beta}$ in (4.44). Note that for $m = 0$ (4.105) gives the divergent part in (4.44). Adding to this result the counter-term (4.104) and requiring the final result to be zero, we readily find Z_q and Z_m in (4.94) and (4.95), respectively.

As an exercise we can check whether Z_q in (4.94) indeed renormalizes the two-point Green function, which for $m = 0$ is simply given by $(I_1^{QCD})_{\alpha\beta}$ in (4.44). In the notation of (4.100) its divergent part added to the "tree level" propagator is given by

$$\Gamma_0^{(2,0)} = iC_F\delta_{\alpha\beta}\not{p}\left(1 + \frac{\alpha_s}{4\pi}\frac{1}{\varepsilon}\right).\tag{4.106}$$

The corresponding renormalized two-point function is given by

$$\Gamma^{(2,0)} = Z_q\Gamma_0^{(2,0)},\tag{4.107}$$

which with (4.94) is indeed finite. In this case at $O(\alpha_s)$ only quark field renormalization is needed to obtain a finite result. Coupling renormalization is necessary first at $O(\alpha_s^2)$.

Let us next calculate Z_3 in QCD. The result in (4.71) is just the $O(\alpha_s)$ contribution to the unrenormalized Green function $\Gamma_0^{(0,2)}$. Adding the tree-level part and expanding the expression in (4.71) in powers of ε and keeping only the divergent part we find

$$\Gamma_0^{(0,2)} = i\delta^{ab}[p_\mu p_\nu - g_{\mu\nu}p^2]\left(1 + \frac{\alpha_s}{4\pi}\frac{1}{\varepsilon}r\right), \qquad r = \frac{2}{3}f - \frac{5}{3}N.\tag{4.108}$$

But the corresponding renormalized Green function $\Gamma^{0,2}$ is given by

$$\Gamma^{(0,2)} = Z_3\Gamma_0^{(0,2)}.\tag{4.109}$$

Requiring it to be finite gives Z_3 in (4.96).

The calculation of Z_g is much harder as it involves either the diagrams in Figure 4.2 or the diagrams in Figure 4.3. There is no space for doing these calculations in detail here,

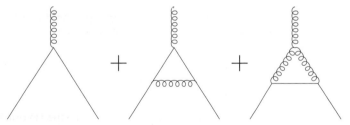

Figure 4.2   The diagrams contributing to Z_g.

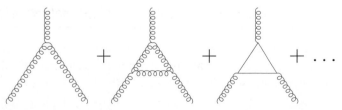

Figure 4.3   Different route to Z_g.

and it will suffice to quote only the results for them. Keeping only the divergent parts one finds first

$$\Gamma_0^{(2,1)} = ig_s Z_g \gamma_\mu t^a (Z_1^F)^{-1}, \qquad Z_1^F = 1 - \frac{\alpha_s}{4\pi} \frac{(N + C_F)}{\varepsilon}, \tag{4.110}$$

$$\Gamma_0^{(0,3)} \propto ig_s Z_g (Z_1^G)^{-1}, \qquad Z_1^G = 1 - \frac{\alpha_s}{4\pi} \frac{2}{3\varepsilon}(f - N). \tag{4.111}$$

In writing these expressions we replaced the bare coupling constant in the tree-level terms by the renormalized coupling. In the correction terms this replacement is immaterial at this order in perturbation theory. Moreover, we introduced Z_1 familiar from QED. Due to the complicated Lorenz structure of the triple gluon vertex, we suppressed it as it does not play any role in finding Z_g.

But the corresponding renormalized Green functions are given by

$$\Gamma^{(2,1)} = Z_2 Z_3^{1/2} \Gamma_0^{(2,1)}, \qquad \Gamma^{(0,3)} = Z_3^{3/2} \Gamma_0^{(0,3)}. \tag{4.112}$$

Using (4.110) and (4.111) and requiring these Green functions to be finite implies, respectively,

$$Z_g = Z_q^{-1} Z_3^{-1/2} Z_1^F, \qquad Z_g = Z_3^{-3/2} Z_1^G. \tag{4.113}$$

Inserting the already known expressions for Z_q, Z_3, Z_1^F, and Z_1^G and keeping only the $O(\alpha_s)$ terms we find from both expressions the same Z_g given in (4.97).

The fact that we have obtained the same Z_g from two different Green functions follows from gauge invariance of QCD as expressed by the Slavnov–Taylor identity:

$$\frac{Z_1^F}{Z_2} = \frac{Z_1^G}{Z_3}. \tag{4.114}$$

4.2.6 Renormalization Constants in QED

After this practice of calculating the renormalization constants in QCD, it is straightforward to find corresponding results in QED. We leave it as an exercise and just list the results that are analogous to the ones in (4.94)–(4.97),

$$Z_f = 1 - \frac{\alpha_{em}}{4\pi} Q_f^2 \frac{1}{\varepsilon} + O(\alpha_{em}^2), \tag{4.115}$$

$$Z_m = 1 - \frac{\alpha_{em}}{4\pi} 3Q_f^2 \frac{1}{\varepsilon} + O(\alpha_{em}^2), \tag{4.116}$$

$$Z_3 = 1 - \frac{\alpha_{em}}{4\pi} \left[\frac{4}{3}\sum_f Q_f^2\right] \frac{1}{\varepsilon} + O(\alpha_{em}^2), \tag{4.117}$$

$$Z_e = 1 - \frac{\alpha_{em}}{4\pi} \left[\frac{2}{3}\sum_f Q_f^2\right] \frac{1}{\varepsilon} + O(\alpha_{em}^2). \tag{4.118}$$

We note that for fermion and mass renormalization, in going from QCD to QED one just replaces C_F by Q_f^2. Also in QED one has relations resulting from gauge invariance. These are the so-called Ward identities:

$$Z_1^F = Z_f, \qquad Z_e = Z_3^{-1/2}. \tag{4.119}$$

On the other hand, the change in coupling renormalization when going from QCD to QED is very profound. To appreciate it we will in the next section look at the dependence of QCD and QED couplings on the scale μ, which is governed by renormalization group equations.

4.3 Renormalization Group Equations

4.3.1 The Basic Equations

In the process of renormalization we have introduced an arbitrary mass parameter μ. The μ-dependence of the renormalized coupling constant g_s and of the renormalized quark mass m is governed by the renormalization group equations. These equations are derived from the definitions (4.86) using the fact that bare quantities are μ-independent. One finds ($g \equiv g_s$):

$$\frac{dg(\mu)}{d\ln\mu} = \beta(g(\mu), \varepsilon), \tag{4.120}$$

$$\frac{dm(\mu)}{d\ln\mu} = -\gamma_m(g(\mu))m(\mu), \tag{4.121}$$

where

$$\beta(g, \varepsilon) = -\varepsilon g + \beta(g), \tag{4.122}$$

$$\beta(g) = -g\frac{1}{Z_g}\frac{dZ_g}{d\ln\mu}, \qquad \gamma_m(g) = \frac{1}{Z_m}\frac{dZ_m}{d\ln\mu}. \tag{4.123}$$

Equation (4.122) is valid in arbitrary dimensions. In four dimensions $\beta(g, \varepsilon)$ reduces to $\beta(g)$. Let us prove (4.122). Using (4.86) we have

$$\beta(g, \varepsilon) = g_0\mu\frac{d}{d\mu}[\mu^{-\varepsilon}Z_g^{-1}] = g_0\left[-\varepsilon\mu^{-\varepsilon}Z_g^{-1} + \mu^{-\varepsilon+1}\frac{dZ_g^{-1}}{d\mu}\right]$$

$$= -\varepsilon g - g_0\mu^{-\varepsilon+1}\frac{1}{Z_g^2}\frac{dZ_g}{d\mu} = -\varepsilon g - g\mu\frac{dZ_g}{d\mu}\frac{1}{Z_g}. \tag{4.124}$$

Similarly, one can derive the expression for γ_m in (4.123) by inserting $m = m_0/Z_m$ into (4.121).

$\beta(g)$ and $\gamma(g)$ are called *renormalization group functions*. $\beta(g)$ governs the μ-dependence of $g(\mu)$. γ_m, the *anomalous dimension* of the mass operator, governs the μ-dependence of $m(\mu)$. In the MS ($\overline{\text{MS}}$)-scheme they depend only on g. In particular they carry no explicit μ-dependence and are independent of masses. Writing

$$Z_i = 1 + \sum_{k=1}^{\infty} \frac{1}{\varepsilon^k} Z_{i,k}(g), \tag{4.125}$$

and using (4.122) and (4.123) one finds

$$\beta(g) = 2g^3 \frac{dZ_{g,1}(g)}{dg^2}, \tag{4.126}$$

$$\gamma_m(g) = -2g^2 \frac{dZ_{m,1}(g)}{dg^2}. \tag{4.127}$$

Thus $\beta(g)$ and $\gamma_m(g)$ can be directly obtained from the $1/\varepsilon$-pole parts of the renormalization constants Z_g and Z_m, respectively. This is a very useful property of the MS-like schemes. Let us demonstrate that (4.126) is indeed true. We follow Muta [70] and write

$$\beta(g, \varepsilon) = -\varepsilon g - g f(g), \qquad f(g) = \frac{\mu}{Z_g} \frac{dZ_g}{d\mu}. \tag{4.128}$$

Specializing the expansion (4.125) to Z_g and inserting it into the formula for $f(g)$ gives

$$f(g) \left(1 + \frac{Z_{g,1}}{\varepsilon} + \frac{Z_{g,2}}{\varepsilon^2} + \cdots \right) = \frac{1}{\varepsilon} \beta(g, \varepsilon) \left(\frac{dZ_{g,1}}{dg} + \frac{1}{\varepsilon} \frac{dZ_{g,2}}{dg} + \cdots \right). \tag{4.129}$$

Now finiteness of $\beta(g)$ implies finiteness of $f(g)$. Consequently, the equality (4.129) should hold for each coefficient of the power $1/\varepsilon$. In particular the nonsingular terms give

$$f(g) = -g \frac{dZ_{g,1}}{dg}, \tag{4.130}$$

which with $\beta(g) = -g f(g)$ gives (4.126). The proof of (4.127) can be done in an analogous manner using the finiteness of γ_m. It is left as a homework problem.

With Z_g and Z_m in (4.97) and (4.95), respectively, the formulas (4.126) and (4.127) give immediately the leading terms for $\beta(g)$ and $\gamma_m(g)$:

$$\beta(g) = -\frac{g^3}{16\pi^2} \left[\frac{11}{3} N - \frac{2}{3} f \right], \tag{4.131}$$

$$\gamma_m(g) = \frac{g^2}{16\pi^2} 6 C_F. \tag{4.132}$$

With this technique it is also easy to show that the anomalous dimensions of the quark field (γ_q) and the qluon field (γ_G) defined by

$$\gamma_q(g) = \frac{1}{2} \frac{1}{Z_q} \frac{dZ_q}{d \ln \mu}, \qquad \gamma_G(g) = \frac{1}{2} \frac{1}{Z_3} \frac{dZ_3}{d \ln \mu}, \tag{4.133}$$

are given by

$$\gamma_i(g) = -g^2 \frac{dZ_{i,1}(g)}{dg^2}, \qquad (i = q, G). \tag{4.134}$$

4.3.2 Compendium of Useful Results

It will be useful to have a collection of results for $\beta(g)$, $\gamma(\alpha_s)$, and $Z_{q,1}(\alpha_s)$, including also two-loop contributions. They are

$$\beta(g) = -\beta_0 \frac{g^3}{16\pi^2} - \beta_1 \frac{g^5}{(16\pi^2)^2}, \tag{4.135}$$

$$\gamma_m(\alpha_s) = \gamma_m^{(0)} \frac{\alpha_s}{4\pi} + \gamma_m^{(1)} \left(\frac{\alpha_s}{4\pi}\right)^2, \tag{4.136}$$

$$Z_{q,1}(\alpha_s) = a_1 \frac{\alpha_s}{4\pi} + a_2 \left(\frac{\alpha_s}{4\pi}\right)^2, \tag{4.137}$$

where we dropped higher-order terms in $g = g_s$ and

$$\beta_0 = \frac{11N - 2f}{3}, \qquad \beta_1 = \frac{34}{3}N^2 - \frac{10}{3}Nf - 2C_F f, \tag{4.138}$$

$$\gamma_m^{(0)} = 6C_F, \qquad \gamma_m^{(1)} = C_F \left(3C_F + \frac{97}{3}N - \frac{10}{3}f\right), \tag{4.139}$$

$$a_1 = -C_F, \qquad a_2 = C_F \left(\frac{3}{4}C_F - \frac{17}{4}N + \frac{1}{2}f\right), \tag{4.140}$$

$$C_F = \frac{N^2 - 1}{2N}. \tag{4.141}$$

These results are valid in the MS ($\overline{MS}$) scheme. N is the number of colors and f the number of quark flavors. Whereas β_0, β_1, $\gamma_m^{(0)}$, and $\gamma_m^{(1)}$ are gauge independent, a_1 and a_2 given here are valid only in the $\xi = 1$ gauge. The two-loop β_1 has been calculated in [95], the two-loop $\gamma_m^{(1)}$ in [96] and the three-loop $\gamma_m^{(2)}$ in [1336].

4.3.3 Running Coupling Constant

With the expansion (4.135), the renormalization group equation (4.120) for $g(\mu)$ can be written as follows:

$$\frac{d\alpha_s}{d\ln\mu} = -2\beta_0 \frac{\alpha_s^2}{4\pi} - 2\beta_1 \frac{\alpha_s^3}{(4\pi)^2}. \tag{4.142}$$

Solving it, one finds [94]:

$$\frac{\alpha_s(\mu)}{4\pi} = \frac{1}{\beta_0 \ln(\mu^2/\Lambda_{\overline{MS}}^2)} - \frac{\beta_1}{\beta_0^3} \frac{\ln[\ln(\mu^2/\Lambda_{\overline{MS}}^2)]}{\ln^2(\mu^2/\Lambda_{\overline{MS}}^2)}. \tag{4.143}$$

Let us make a few comments:

- $\Lambda_{\overline{MS}}$ is a QCD scale characteristic for the $\overline{MS}$ scheme. It can be determined by measuring $\alpha_s(\mu)$ at a single value of μ. To this end the quantity used to determine $\alpha_s(\mu)$ has to be calculated in the $\overline{MS}$ scheme. Strictly speaking $\alpha_s(\mu)$ should really read $\alpha_{s,\overline{MS}}$, but we will work exclusively in the $\overline{MS}$ scheme, and this complication of the notation is unnecessary. Yet it is useful to quote the relation to the MS scheme. Inserting (see (4.90))

$$\mu \equiv \mu_{\overline{MS}} = \mu_{MS} e^{\gamma_E/2} (4\pi)^{-1/2} \tag{4.144}$$

into (4.143), one finds the relation between α_s in the MS and $\overline{MS}$ schemes:

$$\alpha_{s,MS} = \alpha_{s,\overline{MS}} \left(1 + \beta_0 (\gamma_E - \ln 4\pi) \frac{\alpha_{s,\overline{MS}}}{4\pi} \right) \tag{4.145}$$

or

$$\Lambda_{\overline{MS}}^2 = 4\pi e^{-\gamma_E} \Lambda_{MS}^2. \tag{4.146}$$

- $\Lambda_{\overline{MS}}$ and $\alpha_s(\mu)$ depend on f, the number of "effective" flavors present in β_0 and β_1. These effective flavors are defined by

$$\begin{cases} f = 6 & \mu \geq m_t, \\ f = 5 & m_b \leq \mu \leq m_t, \\ f = 4 & m_c \leq \mu \leq m_b, \\ f = 3 & \mu \leq m_c. \end{cases} \tag{4.147}$$

Denoting by $\alpha_s^{(f)}$ the effective coupling constant for a theory with f effective flavors and by $\Lambda_{\overline{MS}}^{(f)}$ the corresponding QCD scale parameter, we have the following boundary conditions that follow from the continuity of α_s:

$$\alpha_s^{(6)}(m_t) = \alpha_s^{(5)}(m_t), \qquad \alpha_s^{(5)}(m_b) = \alpha_s^{(4)}(m_b), \qquad \alpha_s^{(4)}(m_c) = \alpha_s^{(3)}(m_c). \tag{4.148}$$

The preceding continuity conditions allow to find values of $\Lambda_{\overline{MS}}^{(f)}$ for different f once one particular $\Lambda_{\overline{MS}}^{(f)}$ is known. In Table 4.1 we show different $\alpha_s^{(f)}(\mu)$ and $\Lambda_{\overline{MS}}^{(f)}$ corresponding to a much larger range than the present world average

$$\alpha_s^{(5)}(M_Z) = 0.1181 \pm 0.0006, \tag{4.149}$$

extracted from different processes, partly to allow the reader to reproduce this table and to underline the big progress in the determination of $\alpha_s^{(5)}(M_Z)$ in the last twenty years. But our central value is in fact very close to the present world average.

To this end we have set $m_c = 1.3$ GeV, $m_b = 4.4$ GeV, and $m_t = 170$ GeV. We observe that for $\mu \geq m_c$ the values of $\alpha_s(\mu)$ are sufficiently small that the effects of strong interactions can be treated in perturbation theory. When one moves to low-energy scales, α_s increases and at $\mu \approx O(1 \text{ GeV})$ and high values of $\Lambda_{\overline{MS}}^{(3)}$ one finds $\alpha_s^{(3)}(\mu) > 0.5$. This signals breakdown of perturbation theory for scales lower than 1 GeV. Yet it is gratifying that strong interaction contributions to weak decays coming from scales higher than 1 GeV can be treated by perturbative methods.

Table 4.1 Values of $\alpha_s^{(f)}(\mu)$ and $\Lambda_{\overline{MS}}^{(f)}$ corresponding to given values of $\alpha_s^{(5)}(M_Z)$.

$\alpha_s^{(6)}(m_t)$	0.1037	0.1054	0.1079	0.1104	0.1120
$\Lambda_{\overline{MS}}^{(6)}[\text{MeV}]$	66	76	92	110	123
$\alpha_s^{(5)}(M_Z)$	0.113	0.115	0.118	0.121	0.123
$\Lambda_{\overline{MS}}^{(5)}[\text{MeV}]$	169	190	226	267	296
$\alpha_s^{(5)}(m_b)$	0.204	0.211	0.222	0.233	0.241
$\Lambda_{\overline{MS}}^{(4)}[\text{MeV}]$	251	278	325	376	413
$\alpha_s^{(4)}(m_c)$	0.336	0.357	0.396	0.443	0.482
$\Lambda_{\overline{MS}}^{(3)}[\text{MeV}]$	297	325	372	421	457
$\alpha_s^{(3)}(1\,\text{GeV})$	0.409	0.444	0.514	0.605	0.690

Finally, we would like to give an equivalent expression for α_s, which allows to calculate $\alpha_s(\mu)$ directly from the experimental value given in (4.149):

$$\alpha_s(\mu) = \frac{\alpha_s(M_Z)}{v(\mu)}\left[1 - \frac{\beta_1}{\beta_0}\frac{\alpha_s(M_Z)}{4\pi}\frac{\ln v(\mu)}{v(\mu)}\right], \qquad v(\mu) = 1 - \beta_0\frac{\alpha_s(M_Z)}{2\pi}\ln\left(\frac{M_Z}{\mu}\right).$$

$$\text{(4.150)}$$

Strictly speaking, (4.150) is valid for the $f = 5$ theory. To find $\alpha_s(\mu)$ for $f \neq 5$ one has to proceed as in (4.147) and (4.148).

We have discussed here only two-loop contributions to the QCD β function. By now also three-loop [97, 98], four-loop [99, 100], and five-loop [101, 102] contributions are known.

4.3.4 Running Quark Mass

Let us next find the μ-dependence of $m(\mu)$. With $dg/d\ln\mu = \beta(g)$ the solution of

$$\frac{dm(\mu)}{d\ln\mu} = -\gamma_m(g)m(\mu) \tag{4.151}$$

is obviously

$$m(\mu) = m(\mu_0)\exp\left[-\int_{g(\mu_0)}^{g(\mu)} dg'\,\frac{\gamma_m(g')}{\beta(g')}\right]. \tag{4.152}$$

Here $m(\mu_0)$ is the value of the running mass at the scale μ_0. For instance: $m_s(2\text{ GeV})$. Inserting the expansions for $\gamma_m(g)$ and $\beta(g)$ into (4.152) and expanding in α_s gives:

$$m(\mu) = m(\mu_0)\left[\frac{\alpha_s(\mu)}{\alpha_s(\mu_0)}\right]^{\frac{\gamma_m^{(0)}}{2\beta_0}}\left[1 + \left(\frac{\gamma_m^{(1)}}{2\beta_0} - \frac{\beta_1\gamma_m^{(0)}}{2\beta_0^2}\right)\frac{\alpha_s(\mu) - \alpha_s(\mu_0)}{4\pi}\right]. \tag{4.153}$$

In the literature the running quark mass is often denoted by $\overline{m}(\mu)$, but in this book we will use simply $m(\mu)$.

Because formulas similar to (4.151)–(4.153) will often appear in subsequent sections, it is useful to derive at least the leading term in (4.153). Keeping the leading terms in $\gamma_m(g)$ and $\beta(g)$, we have

$$-\int_{g(\mu_0)}^{g(\mu)} dg' \frac{\gamma_m(g')}{\beta(g')} = \int_{g(\mu_0)}^{g(\mu)} dg' \frac{\gamma_m^{(0)}}{\beta_0} \frac{1}{g'} = \frac{1}{2} \frac{\gamma_m^{(0)}}{\beta_0} \ln \frac{g^2(\mu)}{g^2(\mu_0)}, \qquad (4.154)$$

which inserted into (4.152) gives the leading term in (4.153). Keeping also the NLO terms in $\gamma_m(g)$ and $\beta(g)$ and proceeding in a similar manner, one readily finds the NLO term in (4.153).

Because the power $\gamma_m^{(0)}/2\beta_0$ is positive, $m(\mu)$ similar to α_s decreases with increasing μ. For low μ this dependence is sizable so that between $1\,\text{GeV}$ and $2\,\text{GeV}$ the light quark masses decrease typically by 25%. On the other hand the μ dependence of the top quark mass $m_t(\mu_t)$ is much weaker. In the range $100\,\text{GeV} \leq \mu_t \leq 300\,\text{GeV}$ the top quark mass decreases only by 8%.

By now also three-loop [103], four-loop [104, 105], and five loop [106] contributions to $\gamma_m(\alpha_s)$ are known.

The most difficult part in this analysis is the determination of the mass $m(\mu_0)$ for a given quark from the data. For the top quark, which is not confined, this can be done through high-energy processes, but in the case of the remaining five quarks, which are confined in hadrons, nonperturbative methods are required to find $m(\mu_0)$. The most accurate values are presently obtained from lattice simulations. One finds then in FLAG and PDG reports

$$m_u(2\,\text{GeV}) = (2.16 \pm 0.11)\,\text{MeV}, \qquad m_d(2\,\text{GeV}) = (4.68 \pm 0.15)\,\text{MeV}, \qquad (4.155)$$

$$m_c(m_c) = (1.279 \pm 0.013)\,\text{GeV}, \qquad m_s(2\,\text{GeV}) = (93.8 \pm 2.4)\,\text{MeV}, \qquad (4.156)$$

$$m_b(m_b) = 4.19^{+0.18}_{-0.06}\,\text{GeV}, \qquad m_t(m_t) = 163(1)\,\text{GeV}, \qquad (4.157)$$

where we also list the value of $m_t(m_t)$, which is found from collider experiments. In Table 4.2 we show the values of m_i for different scales μ using as input the values in (4.155)–(4.157).[2]

4.3.5   RG Improved Perturbation Theory

The structure of (4.150) and (4.153) makes it clear that the RG approach goes beyond the usual perturbation theory. To see what is going on, let us consider the leading term in (4.150):

$$\alpha_s(\mu) = \frac{\alpha_s(M_Z)}{1 - \beta_0 \frac{\alpha_s(M_Z)}{2\pi} \ln\left(\frac{M_Z}{\mu}\right)}. \qquad (4.158)$$

---

[2] The author thanks Jason Aebischer and Christoph Bobeth for providing this table.

Table 4.2 Central values of quark masses at different scales.[a]							
μ [TeV]	α_s	m_u [MeV]	m_d [MeV]	m_s [MeV]	m_c [GeV]	m_b [GeV]	m_t [GeV]
1	0.0886	1.05	2.27	45.5	0.535	2.41	143.8
5	0.0762	0.96	2.07	41.5	0.488	2.20	131.2
10	0.0719	0.92	2.00	40.1	0.471	2.13	126.6
20	0.0680	0.89	1.94	38.8	0.456	2.06	122.5
50	0.0635	0.86	1.86	37.2	0.437	1.97	117.5
100	0.0605	0.83	1.80	36.2	0.425	1.92	114.1
200	0.0578	0.81	1.75	35.2	0.413	1.86	111.0

[a] The input values are given in (4.155)–(4.157). The second column gives the corresponding values of α_s.

Expanding it in $\alpha_s(M_Z)$, we find:

$$\alpha_s(\mu) = \alpha_s(M_Z) \left[1 + \sum_{n=1}^{\infty} \left(\beta_0 \frac{\alpha_s(M_Z)}{2\pi} \ln\left(\frac{M_Z}{\mu}\right) \right)^n \right]. \qquad (4.159)$$

We conclude that the solution of the renormalization group equations sums automatically large logarithms $\log(M_Z/\mu)$, which appear for $\mu \ll M_Z$. More generally

$$\text{LO}: \quad \text{Summation of } \left[\alpha_s(M_Z) \ln\left(\frac{M_Z}{\mu}\right) \right]^n, \qquad (4.160)$$

$$\text{NLO}: \quad \text{Summation of } \alpha_s(M_Z)^n \left[\ln\left(\frac{M_Z}{\mu}\right) \right]^{n-1}. \qquad (4.161)$$

In particular we note that the expansion (4.153) in terms of $\alpha_s(\mu)$ does not involve large logarithms, and a few terms suffice to obtain reliable results. Equation (4.153) is an example of a *renormalization group improved perturbative expansion*. We will encounter similar expansions for other quantities in the later chapters.

In this chapter we have collected certain information about QCD and tools like RG methods that allow us to sum large logarithms. We have also discussed the μ dependences of the running QCD coupling and the running quark masses. Yet all these nice and powerful tools are still insufficient to discuss properly weak decays. Yes, what we still need is *the operator product expansion* and more generally the technology of *effective field theories*.

THE REAL FUN BEGINS ONLY NOW!

Short-Distance Structure of Weak Decays

5.1 Operator Product Expansion in Weak Decays

5.1.1 Basic Idea

Weak decays of mesons are driven by weak interactions of quarks, whose strong interactions, binding the quarks into mesons, are characterized by a typical hadronic energy scale $\mu \leq O(1\,\text{GeV})$, much lower than the scale of weak interactions: $O(M_W)$. Evidently the problem of calculating the amplitudes for weak decays of mesons involves within the SM a number of energy scales to be denoted by μ:

- Scales $\mu = O(M_W)$ at which the fundamental weak transition between quarks responsible for weak decays of mesons takes place and at which the QCD coupling $\alpha_s(\mu)$ is rather small.
- Scales $O(1\,\text{GeV}) \leq \mu \leq M_W$ in which significant variation of $\alpha_s(\mu)$ with μ has to be taken into account, and large logarithms multiplying $\alpha_s(\mu)$ must be summed to all orders in α_s.
- Scales $\mu \leq O(1\,\text{GeV})$ where confinement-effects binding quarks into mesons have to be taken into account.

Beyond the SM higher scales have to be considered, but the technology developed in this chapter can also be used there. For the time being we stay within the SM, which has the advantage that the model is well defined and moreover describes the data rather well.

Our goal is then to derive an effective low-energy theory that would allow us to calculate efficiently the amplitudes for weak decays, taking the dynamical effects at all scales listed earlier into account. The formal framework to achieve this is the Operator Product Expansion (OPE) [107–110]. I follow here to a large extent my Les Houches lectures [29] and the review in [111] making updates after twenty years. A very nice introduction to this field with many explicit calculations and references can also be found in [46]. Many aspects and technology of effective field theories are also presented in [112].

Let us first present the basic idea of the OPE. Consider the quark level transition $c \rightarrow su\bar{d}$. Disregarding QCD effects for the moment, the corresponding tree-level W-exchange amplitude (Figure 5.1a multiplied by "i") is given by

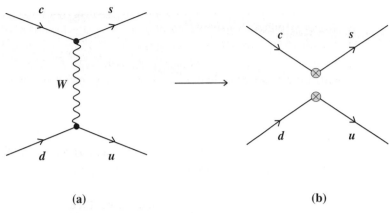

(a)   (b)

Figure 5.1 $c \to s\bar{u}d$ at tree level. From [29].

$$A = -\frac{G_F}{\sqrt{2}} V_{cs}^* V_{ud} \frac{M_W^2}{k^2 - M_W^2} (\bar{s}c)_{V-A}(\bar{u}d)_{V-A}$$

$$= \frac{G_F}{\sqrt{2}} V_{cs}^* V_{ud} (\bar{s}c)_{V-A}(\bar{u}d)_{V-A} + O\left(\frac{k^2}{M_W^2}\right), \tag{5.1}$$

where k, the momentum transfer through the W propagator, is very small as compared to M_W. Consequently, terms of order $O(k^2/M_W^2)$ can safely be neglected, and the full amplitude A can be approximated by the first term on the r.h.s. of (5.1).

As in Section 3.1, we expressed the result in terms of the Fermi constant G_F and the product of quark currents $(\bar{s}c)_{V-A}$ and $(\bar{u}d)_{V-A}$ using the shorthand notations as

$$(\bar{s}c)_{V-A} \equiv \bar{s}\gamma_\mu(1 - \gamma_5)c \equiv \bar{s}_\alpha\gamma_\mu(1 - \gamma_5)c_\alpha \equiv (\bar{s}_\alpha c_\alpha)_{V-A}, \tag{5.2}$$

where $\alpha = 1, 2, 3$ is the color index. As the W boson is color blind, the summation over repeated color indices is straightforward, and we can just work with the first expression for the quark current. The index $V - A$ reminds us of the Dirac structure of this current.

Now the result in (5.1) may also be obtained from

$$\mathcal{H}_{\text{eff}} = \frac{G_F}{\sqrt{2}} V_{cs}^* V_{ud} (\bar{s}c)_{V-A}(\bar{u}d)_{V-A} + \text{High D Operators}, \tag{5.3}$$

where the higher-dimension operators, typically involving derivative terms, correspond to the terms $O(k^2/M_W^2)$ in (5.1). Neglecting the latter terms corresponds to the neglect of higher-dimensional operators. In what follows we will always neglect the higher-dimensional operators keeping only the operators with dimension six or having lower dimensions. Then we can just write

$$\mathcal{H}_{\text{eff}} = \frac{G_F}{\sqrt{2}} V_{cs}^* V_{ud} C Q, \qquad Q = (\bar{s}c)_{V-A}(\bar{u}d)_{V-A}. \tag{5.4}$$

This simple example illustrates the basic idea of OPE: The product of two charged current operators is expanded into a series of local operators (Q), whose contributions are weighted by effective coupling constants C, the Wilson coefficients. In this particular example we have only one operator, represented by the diagram (b) in Figure 5.1, and its Wilson coefficient is simply equal unity: $C = 1$. This value will be changed by QCD corrections. Moreover, QCD corrections to the diagrams in Figure 5.1 will generate another operator.

The example presented here illustrates also the so-called Appelquist–Carazzone decoupling theorem [113]:

> Heavy fields of mass M decouple at low-energy, generating operators suppressed by powers of $1/M$, except for their contributions to renormalization effects.

In our case $M = M_W$ and the suppression in question is hidden in G_F. The renormalization, or rather renormalization group effects, will be encountered soon. In the latter context the review by Wilson [114] can be strongly recommended.

5.1.2 Formal Approach

Let us be a bit more formal for a moment and investigate whether the same result can be obtained using the path integral formalism. We will see that this is indeed the case. Readers not familiar with this approach can skip this presentation and just read the last paragraph of this section, which will certainly delight them.

This discussion will on the one hand provide a formal basis for the simple procedure given earlier and on the other hand will give us more insight into the virtues of the OPE. Simultaneously we will discover that there is no need to be very formal for the rest of this book, and we can proceed by simply generalizing our simple procedure of just looking at diagrams to more complicated situations in which also QCD effects and more complicated diagrams are present.

Our formal discussion follows [111] and consists of four steps.

Step 1

Consider the generating functional for Green functions in the path integral formalism. The relevant part for our discussion is

$$Z_W \sim \int [dW^+][dW^-] \exp(i \int d^4 x \mathscr{L}_W), \tag{5.5}$$

where

$$\mathscr{L}_W = -\frac{1}{2}(\partial_\mu W_\nu^+ - \partial_\nu W_\mu^+)(\partial^\mu W^{-\nu} - \partial^\nu W^{-\mu}) + M_W^2 W_\mu^+ W^{-\mu} + \frac{g_2}{2\sqrt{2}}(J_\mu^+ W^{+\mu} + J_\mu^- W^{-\mu}), \tag{5.6}$$

$$J_\mu^+ = V_{pn}\bar{p}\gamma_\mu(1 - \gamma_5)n, \qquad p = (u, c, t), \qquad n = (d, s, b), \qquad J_\mu^- = (J_\mu^+)^\dagger. \tag{5.7}$$

Step 2

We use the unitary gauge for the W field. Introducing the operator:

$$K_{\mu\nu}(x, y) = \delta^{(4)}(x - y)\left[g_{\mu\nu}(\partial^2 + M_W^2) - \partial_\mu \partial_\nu\right], \tag{5.8}$$

we have, after discarding a total derivative in the W kinetic term,

$$Z_W \sim \int [dW^+][dW^-] \exp\left[i \int d^4x\,d^4y\, W_\mu^+(x) K^{\mu\nu}(x,y) W_\nu^-(y) \right.$$
$$\left. + i\frac{g_2}{2\sqrt{2}} \int d^4x (J_\mu^+ W^{+\mu} + J_\mu^- W^{-\mu}) \right]. \tag{5.9}$$

The inverse of $K_{\mu\nu}$, denoted by $\Delta_{\mu\nu}$, and defined through

$$\int d^4y\, K_{\mu\nu}(x,y) \Delta^{\nu\lambda}(y,z) = g_\mu^{\ \lambda} \delta^{(4)}(x-z), \tag{5.10}$$

is given by

$$\Delta_{\mu\nu}(x,y) = \int \frac{d^4k}{(2\pi)^4} \Delta_{\mu\nu}(k) e^{-ik(x-y)}, \tag{5.11}$$

with

$$i\Delta_{\mu\nu}(k) = \frac{-i}{k^2 - M_W^2}\left(g_{\mu\nu} - \frac{k_\mu k_\nu}{M_W^2}\right). \tag{5.12}$$

It is the W propagator in the unitary gauge

Step 3

Performing the Gaussian functional integration over $W^\pm(x)$ in (5.9) explicitly, we arrive at

$$Z_W \sim \exp\left[-i\int \frac{g_2^2}{8} J_\mu^-(x) \Delta^{\mu\nu}(x,y) J_\nu^+(y) d^4x\,d^4y \right]. \tag{5.13}$$

This result implies a nonlocal action functional for the quarks:

$$S_{nl} = \int d^4x\, \mathcal{L}_{\text{kin}} - \frac{g_2^2}{8} \int d^4x\,d^4y\, J_\mu^-(x) \Delta^{\mu\nu}(x,y) J_\nu^+(y), \tag{5.14}$$

where the second term represents charged current interactions of quarks.

Step 4

Finally, we expand this second, nonlocal term in powers of $1/M_W^2$ to obtain a series of local interaction operators of dimensions that increase with the order in $1/M_W^2$. To lowest order

$$\Delta^{\mu\nu}(x,y) \approx \frac{g^{\mu\nu}}{M_W^2} \delta^{(4)}(x-y), \tag{5.15}$$

and the second term in (5.14) becomes

$$-\frac{g_2^2}{8M_W^2} \int d^4x\, J_\mu^-(x) J^{+\mu}(x), \tag{5.16}$$

corresponding to the usual effective charged current interaction Lagrangian

$$\mathcal{L}_{\text{int,eff}} = -\frac{G_F}{\sqrt{2}} J_\mu^-(x) J^{+\mu}(x) = -\frac{G_F}{\sqrt{2}} V_{pn}^* V_{p'n'} (\bar{n}p)_{V-A} (\bar{p}'n')_{V-A}, \tag{5.17}$$

which contains, among other terms, the leading contribution to (5.3). The minus sign follows simply from $\mathscr{L} = -\mathscr{H}$.

Let us note several basic aspects of this approach:

- Formally, the procedure to approximate the interaction term in (5.14) by (5.16) is an example of short-distance OPE. The product of the local operators $J_\mu^-(x)$ and $J_\nu^+(y)$, to be taken at short distances due to the convolution with the massive, short-range W propagator $\Delta^{\mu\nu}(x, y)$, is expanded into a series of composite local operators. The leading term is shown in (5.16).

- The dominant contributions in the short-distance expansion come from the operators of lowest dimension (six in the present example). The operators of higher dimensions can usually be neglected in weak decays, although in the literature some studies of them can be found. Not in this book.

- OPE series is equivalent to the original theory, when considered to all orders in $1/M_W^2$. The truncation of the operator series yields a systematic approximation scheme for low-energy processes, neglecting contributions suppressed by powers of k^2/M_W^2.

- In going from the full to the effective theory the W boson is removed as an explicit, dynamical degree of freedom: It is "integrated out" in step 3 of our procedure. Alternatively in the canonical operator formalism the W field gets "contracted out" through the application of Wick's theorem. From the point of view of low-energy dynamics, the effects of a short-range exchange force mediated by a heavy boson approximately corresponds to a point interaction familiar from the Fermi theory.

- Similarly one can "integrate out" or "contract out" heavy quarks. This gives *effective f-quark theories* where f denotes the "light" quarks that have not been integrated out. We now understand what the effective number of flavors introduced in connection with the formula (4.147) really means. By going from higher to lower μ scales one integrates out systematically flavors with masses higher than the actual value of μ. However, as we will stress later, in connection with renormalization group ideas, there is some freedom at which μ a given flavor is integrated out. For instance, one can extend the five-flavor theory down to $\mu = m_b/2$. Such unphysical dependences on the choice of μ are eliminated by considering sufficient order in perturbation theory, but it is too early to discuss this important topic already here.

All this was a bit formal but fortunately we make still another observation. The approach of evaluating the relevant Green functions (or amplitudes) directly to construct the OPE, as in (5.1), gives the same result as the more formal technique employing path integrals. Consequently we can return, putting aside path integrals, to our Feynman diagram calculations. Our first task is to investigate how (5.1) or (5.3) changes when QCD effects are included.

5.1.3 OPE and Short-Distance QCD Effects

Preliminaries

Due to the asymptotic freedom of QCD, the short-distance QCD corrections to weak decays, that is the contribution of hard gluons at energies of the order $O(M_W)$ down to hadronic scales $O(1\,\text{GeV})$, can be treated in the renormalization group (RG) improved perturbation theory. We will illustrate this on our simple example of the $c \rightarrow su\bar{d}$ transition beginning with the ordinary perturbation theory, subsequently summing leading logarithms by the RG method and finally in the next section generalizing the result to include next-to-leading logarithms. We will do this in some detail, emphasizing certain characteristic features of this approach. In particular we will discuss at length the *scale and renormalization scheme dependences* of various elements of the calculations and their cancellations in physical quantities to be termed observables. Once all these features are well understood it will be straightforward to proceed to other transitions and to generalize this approach to more exciting situations involving penguin and box diagrams.

For the $c \rightarrow su\bar{d}$ transition we found without QCD effects the effective Hamiltonian (5.4), which we now rewrite as follows

$$\mathcal{H}_{\text{eff}}^{(0)} = \frac{G_F}{\sqrt{2}} V_{cs}^* V_{ud} (\bar{s}_\alpha c_\alpha)_{V-A} (\bar{u}_\beta d_\beta)_{V-A}. \tag{5.18}$$

Here the summation over repeated color indices is understood, but we exposed now these indices as gluons will soon enter the scene, and they are not color blind. The superscript (0) reminds us that QCD effects have not been included yet.

We will soon discover that when QCD effects are taken into account $\mathcal{H}_{\text{eff}}^{(0)}$ is generalized to

$$\mathcal{H}_{\text{eff}} = \frac{G_F}{\sqrt{2}} V_{cs}^* V_{ud} (C_1(\mu) Q_1 + C_2(\mu) Q_2), \tag{5.19}$$

where

$$Q_1 = (\bar{s}_\alpha c_\beta)_{V-A} (\bar{u}_\beta d_\alpha)_{V-A}, \tag{5.20}$$

$$Q_2 = (\bar{s}_\alpha c_\alpha)_{V-A} (\bar{u}_\beta d_\beta)_{V-A}. \tag{5.21}$$

The essential features of this Hamiltonian are

- In addition to the original operator Q_2 (with index 2 for historical reasons) a new operator Q_1 with the *same flavor structure* but *different color structure* is generated. That a new operator has to be introduced is evident if we inspect the color structure of the diagrams (b) and (c) in Figure 5.2. They contain the product of the color charges $t_{\alpha\beta}^a$ and $t_{\gamma\delta}^a$, which, using the color algebra developed in Section 4, can be rewritten as follows

$$t_{\alpha\beta}^a t_{\gamma\delta}^a = -\frac{1}{2N} \delta_{\alpha\beta} \delta_{\gamma\delta} + \frac{1}{2} \delta_{\alpha\delta} \delta_{\gamma\beta}. \tag{5.22}$$

The first term on the r.h.s. gives a correction to the coefficient of the operator Q_2, and the second term gives life to the new operator Q_1. Indeed δ and β have been interchanged in this term relative to the l.h.s. as is the case of Q_1 compared with Q_2. The diagram (a) in Figure 5.2 gives again Q_2.
- The Wilson coefficients C_1 and C_2, the coupling constants for the interaction terms Q_1 and Q_2, become calculable nontrivial functions of α_s, M_W, and the renormalization scale μ.
- If QCD is neglected, $C_1 = 0$, $C_2 = 1$, and (5.19) reduces to (5.18).

Calculation of Wilson Coefficients

Our first task is the calculation of the coefficients $C_{1,2}$ in the ordinary perturbation theory. $C_{1,2}$ can be determined by the requirement that the amplitude A_{full} in the full theory be reproduced by the corresponding amplitude in the effective theory (5.19):

$$A_{\text{full}} = A_{\text{eff}} = \frac{G_F}{\sqrt{2}} V_{cs}^* V_{ud} (C_1 \langle Q_1 \rangle + C_2 \langle Q_2 \rangle). \tag{5.23}$$

This procedure is called "the *matching* of the full theory onto the effective theory." We recall that the full theory is the one in which all particles appear as dynamical degrees of freedom. In the case at hand the effective theory is constructed by integrating out the W field only. The matching procedure, which gives the values of C_1 and C_2, proceeds in three steps and has been first outlined not in the context of weak decays but in the context of QCD effects in deep-inelastic electron-proton scattering [94]. The explicit three steps presented here are sufficient for the subsequent summation of the leading logarithms or equivalently for the leading order (LO) term of the RG improved perturbation theory. We will generalize these steps in the next section to be able to include also the next-to-leading order (NLO) term in this expansion. Here we go:

Step 1: Calculation of A_{full}

The current-current diagrams of Figure 5.2 (a)–(c) and their symmetric counterparts give for the full amplitude A_{full} to $O(\alpha_s)$ ($m_i = 0$, $p^2 < 0$):

$$A_{\text{full}} = \frac{G_F}{\sqrt{2}} V_{cs}^* V_{ud} \left[\left(1 + 2 C_F \frac{\alpha_s}{4\pi} \left(\frac{1}{\varepsilon} + \ln\left(\frac{\mu^2}{-p^2} \right) \right) \right) S_2 + \frac{3}{N} \frac{\alpha_s}{4\pi} \ln\left(\frac{M_W^2}{-p^2} \right) S_2 \right.$$

$$\left. - 3 \frac{\alpha_s}{4\pi} \ln\left(\frac{M_W^2}{-p^2} \right) S_1 \right]. \tag{5.24}$$

Here:

$$S_1 \equiv \langle Q_1 \rangle_{tree} = (\bar{s}_\alpha c_\beta)_{V-A} (\bar{u}_\beta d_\alpha)_{V-A}, \tag{5.25}$$

$$S_2 \equiv \langle Q_2 \rangle_{tree} = (\bar{s}_\alpha c_\alpha)_{V-A} (\bar{u}_\beta d_\beta)_{V-A}, \tag{5.26}$$

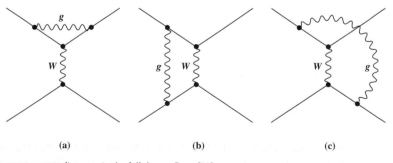

(a)   **(b)**   **(c)**

Figure 5.2   One-loop current-current diagrams in the full theory. From [29].

are just the tree-level matrix elements of Q_1 and Q_2. We recommend this calculation and the ones of Step 2 to motivated readers. With the technology developed in the previous chapter this is a rather easy task.

A few comments should be made.

- We use the term "amplitude" in the meaning of an "amputated Green function" (multiplied by "i"). Correspondingly operator matrix elements are amputated Green functions with operator insertion. Thus gluonic self-energy corrections on external legs are not included.
- For simplicity we have chosen all external momenta p to be equal and set all quark masses to zero. As we will see soon this choice has no impact on the coefficients C_i. We could, for instance, set all external quark momenta to zero and give all quarks a universal mass. We could also set all quark masses and momenta to zero but give the gluon a mass. All these tricks represent particular regularization of infrared divergences that have nothing to do with Wilson coefficients representing short-distance behavior of the theory. We stress this point because in the past in this part several mistakes have been done in the literature.
- We have kept only logarithmic corrections $\sim \alpha_s \cdot \log$ and discarded constant contributions of order $O(\alpha_s)$, which corresponds to the leading log approximation (LO). In the next section, which deals with NLO corrections, we will have to keep these terms.
- The singularity $1/\varepsilon$ can be removed by quark field renormalization. The interested reader can check it by using the procedure of renormalizing the amputated Green functions that we outlined in Section 4.2. This is, however, not necessary for finding C_i as we will see soon.

Step 2: Calculation of matrix elements $\langle Q_i \rangle$

The unrenormalized current-current matrix elements of Q_1 and Q_2 are found at $O(\alpha_s)$ by calculating the diagrams in Figure 5.3 (a)–(c) and their symmetric counterparts. Adding the contributions without QCD corrections (S_1 and S_2, respectively) and using the same assumptions about the external legs as in step 1 (in general the same infrared regularization), we find

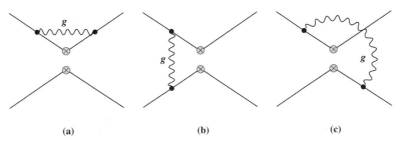

(a)   (b)   (c)

Figure 5.3 One-loop current-current diagrams in the effective theory. The four-vertex "$\otimes \ \otimes$" denotes the insertion of a four-fermion operator Q_i. From [29].

$$\langle Q_1 \rangle^{(0)} = \left(1 + 2C_F \frac{\alpha_s}{4\pi}\left(\frac{1}{\varepsilon} + \ln\left(\frac{\mu^2}{-p^2}\right)\right)\right) S_1 + \frac{3}{N}\frac{\alpha_s}{4\pi}\left(\frac{1}{\varepsilon} + \ln\left(\frac{\mu^2}{-p^2}\right)\right) S_1$$
$$- 3\frac{\alpha_s}{4\pi}\left(\frac{1}{\varepsilon} + \ln\left(\frac{\mu^2}{-p^2}\right)\right) S_2, \tag{5.27}$$

$$\langle Q_2 \rangle^{(0)} = \left(1 + 2C_F \frac{\alpha_s}{4\pi}\left(\frac{1}{\varepsilon} + \ln\left(\frac{\mu^2}{-p^2}\right)\right)\right) S_2 + \frac{3}{N}\frac{\alpha_s}{4\pi}\left(\frac{1}{\varepsilon} + \ln\left(\frac{\mu^2}{-p^2}\right)\right) S_2$$
$$- 3\frac{\alpha_s}{4\pi}\left(\frac{1}{\varepsilon} + \ln\left(\frac{\mu^2}{-p^2}\right)\right) S_1. \tag{5.28}$$

The divergences in the first terms can again be eliminated through the quark field renormalization. However, in contrast to the full amplitude in (5.24), the resulting expressions are still divergent after this renormalization. To remove these additional divergences *operator renormalization* is necessary:

$$Q_i^{(0)} = Z_{ij} Q_j \ . \tag{5.29}$$

We observe that the renormalization constant is in this case a 2×2 matrix $\hat{Z}$. Using (4.100) with $(n_F, n_G) = (4, 0)$, we find the relation between the unrenormalized ($\langle Q_i \rangle^{(0)}$) and the renormalized amputated Green functions ($\langle Q_i \rangle$):

$$\langle Q_i \rangle^{(0)} = Z_q^{-2} Z_{ij} \langle Q_j \rangle \ . \tag{5.30}$$

Z_q^{-2} removes the $1/\varepsilon$ divergences in the first terms in (5.27) and (5.28). Z_{ij} remove the remaining divergences. From (5.27) and (5.28) we read off ($\overline{\text{MS}}$-scheme)

$$\hat{Z} = 1 + \frac{\alpha_s}{4\pi}\frac{1}{\varepsilon}\left(\begin{array}{cc} 3/N & -3 \\ -3 & 3/N \end{array}\right). \tag{5.31}$$

Thus the renormalized matrix elements $\langle Q_i \rangle$ are given by

$$\langle Q_1 \rangle = \left(1 + 2C_F \frac{\alpha_s}{4\pi}\ln\left(\frac{\mu^2}{-p^2}\right)\right) S_1 + \frac{3}{N}\frac{\alpha_s}{4\pi}\ln\left(\frac{\mu^2}{-p^2}\right) S_1 - 3\frac{\alpha_s}{4\pi}\ln\left(\frac{\mu^2}{-p^2}\right) S_2 , \tag{5.32}$$

$$\langle Q_2 \rangle = \left(1 + 2C_F \frac{\alpha_s}{4\pi}\ln\left(\frac{\mu^2}{-p^2}\right)\right) S_2 + \frac{3}{N}\frac{\alpha_s}{4\pi}\ln\left(\frac{\mu^2}{-p^2}\right) S_2 - 3\frac{\alpha_s}{4\pi}\ln\left(\frac{\mu^2}{-p^2}\right) S_1 . \tag{5.33}$$

Step 3: Extraction of C_i

Inserting $\langle Q_i \rangle$ into (5.23), and comparing with (5.24), we can now extract the coefficients C_1 and C_2. Yet, we have to be a bit careful. In the full theory we did not perform any quark field renormalization, whereas we did this renormalization in the effective theory as signaled by the first factor on the r.h.s. of (5.30). This is clearly inconsistent, and this inconsistency is signaled by the divergent Wilson coefficient, which is clearly wrong. To proceed correctly we have to either remove the divergence in (5.24) by performing quark field renormalization as in (5.30) or to leave (5.24) as it is and remove the quark field renormalization from (5.30). In both cases the matching (5.23) gives the same result

$$C_1(\mu) = -3\frac{\alpha_s}{4\pi}\ln\left(\frac{M_W^2}{\mu^2}\right), \qquad C_2(\mu) = 1 + \frac{3}{N}\frac{\alpha_s}{4\pi}\ln\left(\frac{M_W^2}{\mu^2}\right). \tag{5.34}$$

In particular the infrared regulator $-p^2$ disappeared. It canceled out in the process of matching.

This simple example shows that it is essential in the process of matching to treat the external states in the full and the effective theory in the same manner to obtain the correct result for the Wilson coefficients. In this example we were lucky. The inconsistency, which we made for pedagogical reasons, was signaled by a leftover divergence. In the case of NLO calculations where also finite nonlogarithmic corrections have to be kept, a possible inconsistency in matching is much harder to see, and it is crucial that at all stages of matching the treatment of the external legs on both sides of (5.23) is the same. For this reason we are free to decide whether we perform external field renormalization or not. In the latter case the leftover divergences in the full and the effective theory will simply cancel each other in the process of matching. We discussed here the issue of the cancelation of the *ultraviolet* divergences related to external fields. The same comments, made already after (5.34), apply to the *infrared* divergences. For strategic reasons we will now discuss something else and will return to the issue of infrared divergences in the context of matching soon after.

A Different Look at Wilson Coefficients

The renormalization of the interaction terms $C_i Q_i$ in the effective theory can also be achieved in a different, but equivalent, way by using the standard counter-term method. Here C_i are treated as coupling constants, which have to be renormalized. We follow here [111].

To this end let us consider $\mathscr{H}_{\text{eff}}$ as the starting point with fields and "coupling constants" C_i regarded as bare quantities. They are renormalized according to $(q = s, c, u, d)$

$$q^{(0)} = Z_q^{1/2} q, \qquad C_i^{(0)} = Z_{ij}^c C_j, \tag{5.35}$$

where $\hat{Z}^c$ denotes the renormalization matrix for the couplings C_j. It is evident that $\hat{Z}^c$ must be somehow related to the renormalization matrix $\hat{Z}$ in (5.31). Let us find this relation.

Omitting the factor $(G_F/\sqrt{2}) V_{cs}^* V_{ud}$ we have

$$\mathscr{H}_{\text{eff}} = C_i^{(0)} Q_i(q^{(0)}) \equiv Z_q^2 Z_{ij}^c C_j Q_i \equiv C_i Q_i + (Z_q^2 Z_{ij}^c - \delta_{ij}) C_j Q_i, \tag{5.36}$$

where the first term on the r.h.s. is written in terms of renormalized couplings and fields $(C_i Q_i)$ and the second term is a counter–term. The argument $q^{(0)}$ on the l.h.s. of (5.36) indicates that the interaction vertices Q_i are composed of bare fields. Using (5.36) we get the finite renormalized result

$$A_{\text{eff}} = Z_q^2 Z_{ij}^c C_j \langle Q_i \rangle^{(0)}. \tag{5.37}$$

On the other hand, using (5.30), we have

$$A_{\text{eff}} = C_j \langle Q_j \rangle = C_j Z_{ji}^{-1} Z_q^2 \langle Q_i \rangle^{(0)}. \tag{5.38}$$

Hence comparing the last two equations we finally find the relation

$$Z_{ij}^c = Z_{ji}^{-1}. \tag{5.39}$$

This relation will turn out to be very useful in deriving the renormalization group equations for the couplings C_i.

Operator Mixing and Diagonalization

We have just seen that gluonic corrections to the matrix element of the original operator Q_2 are not just proportional to Q_2 itself, but involve the additional structure Q_1. Therefore, besides a Q_2-counter-term, a counter-term involving Q_1 is needed to renormalize this matrix element. Similarly the renormalization of Q_1 requires both Q_1 and Q_2 counter-terms. We say that the operators Q_1 and Q_2 mix under renormalization.

For the study of the renormalization group properties of the system (Q_1, Q_2), it is useful to diagonalize it by going to a different operator basis defined by

$$Q_\pm = \frac{Q_2 \pm Q_1}{2}, \qquad C_\pm = C_2 \pm C_1 . \tag{5.40}$$

The new operators Q_+ and Q_- are renormalized independent of each other:

$$Q_\pm^{(0)} = Z_\pm Q_\pm, \tag{5.41}$$

where

$$Z_\pm = 1 + \frac{\alpha_s}{4\pi} \frac{1}{\varepsilon} \left(\mp 3 \frac{N \mp 1}{N} \right) . \tag{5.42}$$

In this new basis the OPE reads

$$A \equiv A_+ + A_- = \frac{G_F}{\sqrt{2}} V_{cs}^* V_{ud} (C_+(\mu)\langle Q_+(\mu)\rangle + C_-(\mu)\langle Q_-(\mu)\rangle) , \tag{5.43}$$

where $(S_\pm = (S_2 \pm S_1)/2)$,

$$A_\pm = \frac{G_F}{\sqrt{2}} V_{cs}^* V_{ud} \left[\left(1 + 2C_F \frac{\alpha_s}{4\pi} \ln\left(\frac{\mu^2}{-p^2}\right) \right) S_\pm + \left(\frac{3}{N} \mp 3 \right) \frac{\alpha_s}{4\pi} \ln\left(\frac{M_W^2}{-p^2}\right) S_\pm \right] \tag{5.44}$$

and

$$\langle Q_\pm(\mu)\rangle = \left(1 + 2C_F \frac{\alpha_s}{4\pi} \ln\left(\frac{\mu^2}{-p^2}\right) \right) S_\pm + \left(\frac{3}{N} \mp 3 \right) \frac{\alpha_s}{4\pi} \ln\left(\frac{\mu^2}{-p^2}\right) S_\pm , \tag{5.45}$$

$$C_\pm(\mu) = 1 + \left(\frac{3}{N} \mp 3 \right) \frac{\alpha_s}{4\pi} \ln\left(\frac{M_W^2}{\mu^2}\right) . \tag{5.46}$$

Factorization of SD and LD

We have just witnessed in explicit terms the most important feature of the OPE: factorization of short-distance (coefficients) and long-distance (operator matrix elements) contributions. Schematically, this factorization has the following structure at $O(\alpha_s)$ (r is a coefficient):

$$\left(1 + \alpha_s r \ln\left(\frac{M_W^2}{-p^2}\right) \right) \doteq \left(1 + \alpha_s r \ln\left(\frac{M_W^2}{\mu^2}\right) \right) \cdot \left(1 + \alpha_s r \ln\left(\frac{\mu^2}{-p^2}\right) \right) , \tag{5.47}$$

which is achieved by the following splitting of the logarithm

$$\ln \frac{M_W^2}{-p^2} = \ln \left(\frac{M_W^2}{\mu^2} \right) + \ln \left(\frac{\mu^2}{-p^2} \right), \tag{5.48}$$

or from the point of view of the integration over some virtual momenta through the splitting

$$\int_{-p^2}^{M_W^2} \frac{dk^2}{k^2} = \int_{\mu^2}^{M_W^2} \frac{dk^2}{k^2} + \int_{-p^2}^{\mu^2} \frac{dk^2}{k^2}. \tag{5.49}$$

In particular the last formula makes it clear that the Wilson coefficients contain the contributions from large virtual momenta of the loop correction from scales $\mu = O(1\,\text{GeV})$ to M_W, whereas the low-energy contributions are separated into the matrix elements. The renormalization scale μ acts as the scale at which the full contribution to the amplitude is separated into a low-energy and a high-energy part.

Independence of C_i from External States

Let us next return to the issue of the infrared divergences in the process of matching. In the matching discussed explicitly earlier, they are regulated by taking $p^2 \neq 0$. They appear both in A_{full} and A_{eff}. Yet as we have shown earlier, the dependence of A_{full} on p^2, representing the long-distance structure of A, is, from the point of view of the effective theory, fully contained in $\langle Q_i \rangle$, and the Wilson coefficients C_i are free from this dependence.

Because the coefficient functions do not depend on the external states, any external state can be used for their extraction, the only requirement being that the infrared (and mass) singularities are properly regularized. In our example an off-shell momentum p for massless external quarks has been used, but such a choice is clearly one of several possibilities. In general one could work with any other arbitrary momentum configuration, on-shell or off-shell, with or without external quark mass, with infrared divergences regulated by off-shell momenta, quark masses, a fictitious gluon mass, or by dimensional regularization. All these methods would give the same results for C_i.

In particular the dimensional regularization of infrared divergences is very convenient as many integrals simplify considerably. Older discussions of dimensional infrared regularization can be found in Muta's book [70] and also in a paper by Marciano [115]. This method has been used, for instance, in calculating NLO corrections to $K \to \pi \nu \bar{\nu}$ [116] and also for the matching conditions in $B \to X_s \gamma$ [117, 118]. In particular as we stressed in [118], the distinction of $1/\varepsilon$ ultraviolet divergences from the infrared ones is not necessary because after proper renormalization of ultraviolet singularities, the leftover divergences are of infrared origin only. These singularities cancel then automatically in the process of matching. To this end, however, it is essential to perform the matching at all stages in $D = 4 - 2\varepsilon$ dimensions. This implies that already at the NLO level, $O(\varepsilon)$ terms in Wilson coefficients have to be kept at the intermediate stages of the calculation. More details on this efficient technique can be found in the papers just quoted.

5.1.4 OPE and the Renormalization Group

Preliminaries

So far we have computed

$$C_\pm(\mu) = 1 + \left(\frac{3}{N} \mp 3\right) \frac{\alpha_s}{4\pi} \ln\left(\frac{M_W^2}{\mu^2}\right), \tag{5.50}$$

in ordinary perturbation theory. Unfortunately for $\mu = 1\,\text{GeV}$ the first-order correction term amounts to $65 - 130\%$, although $\alpha_s/4\pi \approx 4\%$. This finding illustrates explicitly the breakdown of the naive perturbative expansion caused by the appearance of large logarithms originating in the presence of largely disparate scales M_W and $\mu = 1\,\text{GeV}$.

Clearly, the result in (5.50) can only be used for $\mu = O(M_W)$. For $\mu \ll M_W$ we have to sum the large logarithms to all orders of perturbation theory before we can trust our result for $C_\pm$. Fortunately we have developed in Section 4.3 a very powerful technique to sum such logarithms, and we know exactly what we have to do. Yes, to sum these large logs we have to find renormalization group equations for $C_\pm$ and solve them.

Renormalization Group Equations for $C_\pm$

The renormalization group equations for $C_\pm$ follow from the fact, that the unrenormalized Wilson coefficients $C_\pm^{(0)}$ do not depend on μ. Using the relations (5.35) and (5.39), properly adapted to the diagonal basis, we have first

$$C_\pm = Z_\pm C_\pm^{(0)}, \qquad Q_\pm^{(0)} = Z_\pm Q_\pm, \tag{5.51}$$

and subsequently

$$\boxed{\frac{dC_\pm(\mu)}{d\ln\mu} = \gamma_\pm(g)C_\pm(\mu).} \tag{5.52}$$

Here $\gamma_\pm$ is the anomalous dimension of the operator $Q_\pm$ and given by

$$\boxed{\gamma_\pm(g) = \frac{1}{Z_\pm}\frac{dZ_\pm}{d\ln\mu}.} \tag{5.53}$$

Comparing (5.52) and (5.53) with (4.121) and (4.123), respectively, we see great similarities with the case of the running quark mass. The only modification is the opposite sign in (5.52). Consequently, many relevant formulas of Section 4.3 can be immediately employed. Here we go:

- In the MS ($\overline{\text{MS}}$)-scheme

$$Z_\pm = 1 + \sum_{k=1}^{\infty} \frac{1}{\varepsilon^k} Z_{\pm,k}(g), \tag{5.54}$$

and consequently

$$\gamma_\pm(g) = -2g^2 \frac{\partial Z_{\pm,1}(g)}{\partial g^2}.$$

(5.55)

- Using then

$$Z_\pm = 1 + \frac{\alpha_s}{4\pi} \frac{1}{\varepsilon} \left(\mp 3 \frac{N \mp 1}{N} \right),$$

(5.56)

as obtained in (5.42) gives the one-loop anomalous dimensions of $Q_\pm$:

$$\gamma_\pm(\alpha_s) = \frac{\alpha_s}{4\pi} \gamma_\pm^{(0)}, \qquad \gamma_\pm^{(0)} = \pm 6 \frac{N \mp 1}{N}.$$

(5.57)

- The solution of (5.52) is given as for $m(\mu)$ in (4.152):

$$C_\pm(\mu) = U_\pm(\mu, \mu_W) C_\pm(\mu_W),$$

(5.58)

where $\mu_W = O(M_W)$ and $U_\pm(\mu, \mu_W)$ is the evolution function:

$$U_\pm(\mu, \mu_W) = \exp \left[\int_{g(\mu_W)}^{g(\mu)} dg' \frac{\gamma_\pm(g')}{\beta(g')} \right].$$

(5.59)

- Using (5.57) and $\beta(g) = -\beta_0 g^3/16\pi^2$ we can now find $C_\pm(\mu)$ by using the leading term in the formula (4.153) for $m(\mu)$. Setting $\mu_0 = M_W$ and taking into account the relative sign between (4.151) and (5.52), we have

$$C_\pm(\mu) = \left[\frac{\alpha_s(M_W)}{\alpha_s(\mu)} \right]^{\frac{\gamma_\pm^{(0)}}{2\beta_0}} C_\pm(M_W).$$

(5.60)

- To complete the calculation, we use the fact that at $\mu = M_W$ no large logarithms are present, and $C_\pm(M_W)$ can be calculated in ordinary perturbation theory. From (5.46) we have in LO

$$C_\pm(M_W) = 1,$$

(5.61)

and consequently for $\mu = \mu_b = O(m_b)$

$$C_\pm(\mu_b) = \left[\frac{\alpha_s(M_W)}{\alpha_s(\mu_b)} \right]^{\frac{\gamma_\pm^{(0)}}{2\beta_0}}.$$

(5.62)

We have now summed all leading logarithms, and the important formula (5.62) gives the coefficients $C_\pm$ in the leading log approximation or in other words the leading term of the RG improved perturbation theory. For instance, specializing to the case of $f = 5$ and $\mu_b = O(m_b)$ we obtain

$$C_+(\mu_b) = \left[\frac{\alpha_s(M_W)}{\alpha_s(\mu_b)}\right]^{\frac{6}{23}}, \qquad C_-(\mu_b) = \left[\frac{\alpha_s(M_W)}{\alpha_s(\mu_b)}\right]^{\frac{-12}{23}}, \qquad (5.63)$$

with α_s given by the leading expression (4.158). For $\mu_b = 5.0$ GeV and $\Lambda_{\overline{MS}}^{(5)} = 225$ MeV, one finds $C_+(\mu_b) = 0.847$ and $C_-(\mu_b) = 1.395$, i.e., suppression of C_+ and an enhancement of C_- relative to $C_- = C_+ = 1$ without QCD corrections. The corresponding enhancements and suppressions for scales $O(1$ GeV$)$ reflect to some extent the dominance of the $\Delta I = 1/2$ transitions over $\Delta I = 3/2$ transitions in $K \to \pi\pi$ decays (the $\Delta I = 1/2$ rule) first analyzed in QCD in [119, 120]. These short-distance effects are insufficient, however, to explain the dominance of $\Delta I = 1/2$ transitions observed experimentally. We will return to this issue in Section 7.2.3.

Choice of the Matching Scale

In calculating (5.60) we have set the high-energy matching scale to M_W. The choice of the high-energy matching scale, to be denoted by μ_W, is, of course, not unique. The only requirement is that $\mu_W = O(M_W)$ to avoid large logarithms $\ln(M_W/\mu_W)$. However, we know from (5.50) that in the LO approximation, in which $O(\alpha_s)$ terms are dropped in $C_\pm(\mu_W)$, we have using (5.61)

$$C_\pm(\mu_W) = C_\pm(M_W) + O(\alpha_s) = 1. \qquad (5.64)$$

Consequently in this approximation we also have

$$C_\pm(\mu_b) = \left[\frac{\alpha_s(\mu_W)}{\alpha_s(\mu_b)}\right]^{\frac{\gamma_\pm^{(0)}}{2\beta_0}} = \left[\frac{\alpha_s(M_W)}{\alpha_s(\mu_b)}\right]^{\frac{\gamma_\pm^{(0)}}{2\beta_0}} (1 + O(\alpha_s)), \qquad (5.65)$$

which differs from (5.62) by $O(\alpha_s)$ corrections.

We observe that a change of μ_W around the value of M_W causes an ambiguity of $O(\alpha_s)$ in the Wilson coefficient. This ambiguity represents a theoretical uncertainty in the determination of $C_\pm(\mu_b)$. To reduce it, it is necessary to go beyond the leading order. We will do this in the next section. Similar ambiguity exists in the choice of the low-energy scale $\mu_b = O(m_b)$, as we will discuss at various places in this book. In fact, the uncertainty here is larger as $\alpha_s(\mu_b)$ is larger than $\alpha_s(\mu_W)$.

Threshold Effects in LO

The evolution function U depends on f through $\alpha_s^{(f)}$, and β_0 in the exponent. One can generalize the renormalization group evolution from M_W down to say $\mu_c = O(m_c)$ to include the threshold effect of the b-quark as follows

$$C_\pm(\mu_c) = U_\pm^{(f=4)}(\mu_c, \mu_b) U_\pm^{(f=5)}(\mu_b, M_W) C_\pm(M_W), \qquad (5.66)$$

which is valid in LO. Thus (5.63) generalizes to

$$
C_+(\mu_c) = \left[\frac{\alpha_s^{(4)}(\mu_b)}{\alpha_s^{(4)}(\mu_c)}\right]^{\frac{6}{25}} \left[\frac{\alpha_s^{(5)}(M_W)}{\alpha_s^{(5)}(\mu_b)}\right]^{\frac{6}{23}} , \quad C_-(\mu_c) = \left[\frac{\alpha_s^{(4)}(\mu_b)}{\alpha_s^{(4)}(\mu_c)}\right]^{\frac{-12}{25}} \left[\frac{\alpha_s^{(5)}(M_W)}{\alpha_s^{(5)}(\mu_b)}\right]^{\frac{-12}{23}} .
$$

$$(5.67)$$

Again also here there is an ambiguity in μ_c, which can only be reduced by going to NLO.

RGE for C_i: The Case of Operator Mixing

The coefficients $C_i(\mu)$ can be now calculated by inverting (5.40) with the result

$$
C_1(\mu) = \frac{C_+(\mu) - C_-(\mu)}{2} , \quad C_2(\mu) = \frac{C_+(\mu) + C_-(\mu)}{2} ,
$$

$$(5.68)$$

where $C_\pm(\mu)$ are given in (5.62) or (5.67).

Yet, it is instructive to derive (5.68) by using a procedure that one can also apply to more complicated situations in which several operators mix under renormalization. To this end we write

$$
\vec{C}^T = (C_1, C_2) , \qquad \vec{Q}^T = (Q_1, Q_2).
$$

$$(5.69)$$

Then

$$
\vec{C}^{(0)} = \hat{Z}_c \vec{C}, \qquad \vec{Q}^{(0)} = \hat{Z} \vec{Q},
$$

$$(5.70)$$

with $\hat{Z}_c^T = \hat{Z}^{-1}$. Defining next the anomalous dimension matrix $\hat{\gamma}$ by

$$
\hat{\gamma} = \hat{Z}^{-1} \frac{d\hat{Z}}{d \ln \mu},
$$

$$(5.71)$$

the μ-independence of $\vec{C}^{(0)}$ implies

$$
\frac{d\vec{C}(\mu)}{d \ln \mu} = \hat{\gamma}^T(\alpha_s)\vec{C}(\mu).
$$

$$(5.72)$$

The solution of this equation is

$$
\vec{C}(\mu) = \hat{U}(\mu, M_W)\vec{C}(M_W),
$$

$$(5.73)$$

where

$$
\hat{U}(\mu, M_W) = \exp\left[\int_{g(M_W)}^{g(\mu)} dg' \frac{\hat{\gamma}^T(g')}{\beta(g')}\right],
$$

$$(5.74)$$

is the RG-evolution matrix.

In the MS ($\overline{\text{MS}}$)-scheme we have

$$\hat{Z} = \hat{1} + \sum_{k=1}^{\infty} \frac{1}{\varepsilon^k} \hat{Z}_k(g),$$

(5.75)

and

$$\hat{\gamma}(g) = -2g^2 \frac{\partial \hat{Z}_1(g)}{\partial g^2}.$$

(5.76)

Consequently, using

$$\hat{Z} = 1 + \frac{\alpha_s}{4\pi} \frac{1}{\varepsilon} \begin{pmatrix} 3/N & -3 \\ -3 & 3/N \end{pmatrix}$$

(5.77)

we have to first order in α_s [119, 120]

$$\hat{\gamma}(\alpha_s) = \frac{\alpha_s}{4\pi} \hat{\gamma}^{(0)} = \frac{\alpha_s}{4\pi} \begin{pmatrix} -6/N & 6 \\ 6 & -6/N \end{pmatrix}.$$

(5.78)

To find $C_i(\mu)$ let us write the LO evolution matrix as

$$\hat{U}^{(0)}(\mu, M_{\text{W}}) = \hat{V} \left(\left[\frac{\alpha_s(M_{\text{W}})}{\alpha_s(\mu)} \right]^{\frac{\vec{\gamma}^{(0)}}{2\beta_0}} \right)_D \hat{V}^{-1},$$

(5.79)

where $\hat{V}$ diagonalizes $\hat{\gamma}^{(0)T}$

$$\hat{\gamma}_D^{(0)} = \hat{V}^{-1} \hat{\gamma}^{(0)T} \hat{V},$$

(5.80)

and $\vec{\gamma}^{(0)}$ is the vector containing the diagonal elements of the diagonal matrix:

$$\hat{\gamma}_D^{(0)} = \begin{pmatrix} \gamma_+^{(0)} & 0 \\ 0 & \gamma_-^{(0)} \end{pmatrix},$$

(5.81)

with $\gamma_\pm^{(0)}$ given in (5.57). Using

$$\hat{V} = \hat{V}^{-1} = \frac{1}{\sqrt{2}} \begin{pmatrix} 1 & 1 \\ 1 & -1 \end{pmatrix},$$

(5.82)

and

$$\vec{C}^T(M_{\text{W}}) = (C_1(M_{\text{W}}), C_2(M_{\text{W}})) = (0, 1),$$

(5.83)

we reproduce (5.68) with $C_\pm(\mu)$ given by (5.63).

The threshold effects can be incorporated as in (5.66)

$$\vec{C}(\mu_c) = \hat{U}^{(f=4)}(\mu_c, \mu_b) \hat{U}^{(f=5)}(\mu_b, \mu_W) \vec{C}(\mu_W).$$

(5.84)

It is evident that this procedure is valid for arbitrary number of operators mixing under renormalization. However, for more complicated situations one has to use computer programs like *Mathematica* to obtain analytic formulas like (5.68). We will give some examples later.

5.1.5 Summary of the Basic Formalism

It is a good moment to make a break and to summarize what we have achieved in our expedition so far. This will also allow us to make a strategy for the next steps, which as we will see are technically more advanced.

Ultimately, our goal is the evaluation of weak decay amplitudes involving hadrons in the framework of a low-energy effective theory, of the form

$$\langle \mathcal{H}_{\text{eff}} \rangle = \frac{G_F}{\sqrt{2}} V_{CKM} \langle \vec{Q}^T(\mu) \rangle \vec{C}(\mu), \tag{5.85}$$

where μ denotes a scale of the order of the mass of the decaying hadron. The procedure for this calculation can be divided into the following three steps.

Step 1: Matching in perturbation theory
Calculate the Wilson coefficients $\vec{C}(\mu_W)$ at $\mu_W = O(M_W)$ to the desired order in α_s. Because logarithms of the form $\ln(\mu_W/M_W)$ are not large, this can be performed in ordinary perturbation theory. In the case of the operators $Q_{1,2}$ and in the LO approximation, we simply have $C_1(\mu_W) = 0$, $C_2(\mu_W) = 1$, or $C_{\pm}(\mu_W) = 1$.

This step amounts to matching the full theory onto a five-quark effective theory. In this process $W^{\pm}$, Z^0, the top-quark, and generally all heavy particles with masses higher than M_W are integrated out unless they have been already integrated out at a much higher scale. In the case of $Q_{1,2}$ analyzed so far, the effect of integrating out the top-quark is signaled by the use of $\alpha^{(5)}(\mu)$ instead of $\alpha^{(6)}(\mu)$ for $\mu \leq M_W$. Later when we move to other decays, the effect of integrating out the top quark will be more profound.

The matching in question is achieved using the following procedure:

- Calculation of the amplitude in the full theory,
- Calculation of the operator matrix elements,
- Extraction of $C_i(\mu_W)$ from $A_{\text{full}} = A_{\text{eff}}$.

The resulting $C_i(\mu_W)$ depend generally on the masses of the heavy particles that have been integrated out. Again in the special case of $Q_{1,2}$, this dependence is absent in C_i, although it is present in G_F as G_F depends on M_W.

Step 2: RG improved perturbation theory
- Calculation of the anomalous dimensions of the operators,
- Solution of the renormalization group equation for $\vec{C}(\mu)$,
- Evolution of the coefficients from μ_W down to the appropriate low-energy scale μ

$$\vec{C}(\mu) = \hat{U}(\mu, \mu_W) \vec{C}(\mu_W) . \tag{5.86}$$

Step 3: Nonperturbative regime

Calculation of the hadronic matrix elements $\langle \vec{Q}(\mu) \rangle$, normalized at the appropriate low-energy scale μ, by means of some nonperturbative method. These days lattice gauge theories dominate these calculations although other methods can still provide useful results in cases where lattice methods still have some difficulties. We will briefly discuss these methods in Chapter 7.

Important issues in this procedure are

- **Factorization** of short- and long-distance contributions:

  - $\vec{C}(\mu)$: contributions from scales *higher* than μ
  - $\langle \vec{Q}(\mu) \rangle$: contributions from scales *lower* than μ
  - Cancellation of the μ-dependence between $C_i(\mu)$ and $\langle Q_i(\mu) \rangle$.

- **Summation of large logs** by means of the RG method. More specifically, in the nth order of RG improved perturbation theory the terms

$$\alpha_s^n(\mu) \left(\alpha_s(\mu) \ln \frac{M_W}{\mu} \right)^k \tag{5.87}$$

are summed to all orders in k ($k = 0, 1, 2,\ldots$). This approach is justified as long as $\alpha_s(\mu)$ is small enough. The leading order corresponds in most cases to $n = 0$, the NLO to $n = 1$. In certain processes these canonical values of n may change.

5.1.6 Future Generalizations

Until now, our application of the basic formalism just summarized, concentrated on the current-current operators Q_1 and Q_2 in the LO approximation. In the following sections we will generalize this discussion in several aspects:

- We will generalize the calculation of the couplings $C_i(\mu)$ beyond the LO approximation.
- We will include new operators originating in penguin diagrams of various sort (gluon-penguins, photon-penguins, Z^0-penguins, chromo-magnetic penguins). These operators are generally called *penguin operators*. The new feature will be the m_t-dependence of $C_i(\mu_W)$ of these new operators.
- We will also include new operators originating in *box diagrams*. Also in this case the coefficients $C_i(\mu_W)$ will be m_t-dependent.
- In the process of including new operators it will turn out to be necessary in certain cases to consider also renormalization group equations involving simultaneously C_i of order $O(1)$, $O(\alpha_s)$, and $O(\alpha)$ with $\alpha = \alpha_{\text{QED}}$ and even $O(\alpha \alpha_s)$.
- Finally, we will develop efficient methods for the calculation of the anomalous dimensions of the operators Q_i.

In the next section we will begin these generalizations by including NLO QCD corrections to $C_{\pm}(\mu)$. We will do this in such a manner that the generalization of the formulas, listed in the next section to more complicated processes involving penguin operators will be straightforward.

5.1.7 Motivations for NLO Calculations

Going beyond the LO approximation is certainly an important but a nontrivial step. For this reason we need some motivations to perform this step. Here are the main reasons for going beyond LO:

- The NLO is first of all necessary to test the validity of the renormalization group improved perturbation theory.
- Without going to NLO the QCD scale $\Lambda_{\overline{MS}}$ or equivalently $\alpha_s^{\overline{MS}}$ extracted from various high-energy processes cannot be used meaningfully in weak decays.
- Due to renormalization group invariance the physical amplitudes do not depend on the scales μ present in α_s or in the running quark masses, in particular $m_t(\mu)$, $m_b(\mu)$ and $m_c(\mu)$. However, in perturbation theory this property is broken through the truncation of the perturbative series. Consequently one finds sizable scale ambiguities in the leading order, which can be reduced considerably by going to NLO. An example of such an ambiguity is the choice of the high-energy matching scale μ_W discussed earlier. Larger-scale ambiguities appear at the b-quark and charm-quark thresholds due to larger values of $\alpha_s(\mu)$ at these scales.
- The Wilson Coefficients are renormalization scheme dependent quantities. This scheme dependence appears first at NLO. For a proper matching of the short-distance contributions to the long-distance matrix elements obtained from lattice calculations, it is essential to calculate NLO corrections. The same is true for inclusive heavy quark decays in which the hadron decay can be modeled by a decay of a heavy quark and the matrix elements of Q_i can be effectively calculated in an expansion in $1/m_b$.
- In several cases the central issue of the top quark mass dependence is strictly an NLO effect.

The first NLO calculations for weak decays were performed in the 1980s, and an important progress in calculating them for most interesting decays has been done in 1990s. First NNLO calculations were performed around the year 2000. A history of the NLO and NNLO efforts has been summarized in [77] with the last update in 2014.

It should be remarked that in the 1990s such involved calculations were really not required because of large experimental errors and theoretical and parametric uncertainties. Therefore the early NLO calculations could not be fully appreciated at the time of their performance. But in the present days they are mandatory in view of flavor precision era that we entered few years ago and in view of the impressive progress made by lattice calculations. It is therefore mandatory to discuss this important topic next.

5.2 Current-Current Operators beyond Leading Order

5.2.1 Preliminaries

We will now generalize the formulas of the previous section beyond the LO approximation concentrating on the Wilson coefficients $C_\pm$ and $C_{1,2}$. We will begin with the case without operator mixing. Subsequently we will generalize our discussion to the case of operator mixing. Next we will develop methods for the calculation of anomalous dimensions both at the one-loop level and the two-loop level. While we do not have space to present an explicit two-loop calculation of anomalous dimensions, we will derive explicitly the one-loop anomalous dimension matrix (5.78). A detailed discussion of renormalization scheme and renormalization scale dependences and of their cancellations in physical amplitudes is an important part of this section. We will also discuss the issue of the so-called evanescent operators, which have to be taken into account in a proper calculation of the anomalous dimensions at the two-loop level. This section ends with a compendium of NLO formulas, which allow the calculation of Wilson coefficients of current-current four-quark operators at the NLO level.

5.2.2 The Case without Operator Mixing

Let us consider the coefficients $C_\pm(\mu)$ for which we have the general expression:

$$C_\pm(\mu) = U_\pm(\mu, M_W)C_\pm(M_W),$$

(5.88)

where

$$U_\pm(\mu, M_W) = \exp\left[\int_{g(M_W)}^{g(\mu)} dg' \frac{\gamma_\pm(g')}{\beta(g')}\right],$$

(5.89)

and we have set $\mu_W = M_W$ to simplify the formulas listed here. This restriction will be relaxed whenever it will turn out to be important.

At NLO we have to include nonlogarithmic $O(\alpha_s)$ correction to $C_\pm(M_W)$, which we have neglected in our one-loop calculations in the previous section. Therefore we write

$$C_\pm(M_W) = 1 + \frac{\alpha_s(M_W)}{4\pi}B_\pm,$$

(5.90)

with the coefficients $B_\pm$ still to be determined.

Moreover, we have to include two-loop contributions to $\gamma_\pm(\alpha_s)$ and $\beta(g)$:

$$\gamma_\pm(\alpha_s) = \gamma_\pm^{(0)}\frac{\alpha_s}{4\pi} + \gamma_\pm^{(1)}\left(\frac{\alpha_s}{4\pi}\right)^2,$$

(5.91)

$$\beta(g) = -\beta_0\frac{g^3}{16\pi^2} - \beta_1\frac{g^5}{(16\pi^2)^2},$$

(5.92)

which are represented by the second terms on the r.h.s. of these two equations. The fact that for the NLO analysis we need simultaneously one-loop contributions to $C_\pm(M_W)$ but two-loop contributions for $\gamma_\pm(\alpha_s)$ and $\beta(g)$ shows that we are going beyond the usual perturbation theory. But this should not be surprising. At LO we used $\gamma_\pm(\alpha_s)$ and $\beta(g)$ at one-loop level and could set $B_\pm = 0$. The one-loop calculations we performed in the previous section served only to determine $\gamma_\pm(\alpha_s)$. A more efficient method for calculating $\gamma_\pm(\alpha_s)$ will be developed in this section.

Inserting the last two formulas into (5.89) and expanding in α_s we find

$$U_\pm(\mu, M_W) = \left[1 + \frac{\alpha_s(\mu)}{4\pi} J_\pm\right] \left[\frac{\alpha_s(M_W)}{\alpha_s(\mu)}\right]^{d_\pm} \left[1 - \frac{\alpha_s(M_W)}{4\pi} J_\pm\right] \qquad (5.93)$$

with

$$J_\pm = \frac{d_\pm}{\beta_0} \beta_1 - \frac{\gamma_\pm^{(1)}}{2\beta_0}, \qquad d_\pm = \frac{\gamma_\pm^{(0)}}{2\beta_0}. \qquad (5.94)$$

This is similar to the μ-dependence of the quark mass discussed previously and given in (4.153) except that we have written the evolution function in a particular way: The couplings $\alpha_s(\mu)$ increase by going from right to left. This is clearly not necessary in the absence of operator mixing, but it will turn out to be useful for the future generalization to the case of operator mixing. One deals then with matrices $\hat{\gamma}$ and $\hat{J}$, and the order of matrices in a product matters.

Inserting (5.93) and (5.90) into (5.88), we find an important formula for $C_\pm(\mu)$ in the NLO approximation:

$$C_\pm(\mu) = \left[1 + \frac{\alpha_s(\mu)}{4\pi} J_\pm\right] \left[\frac{\alpha_s(M_W)}{\alpha_s(\mu)}\right]^{d_\pm} \left[1 + \frac{\alpha_s(M_W)}{4\pi} (B_\pm - J_\pm)\right]. \qquad (5.95)$$

Let us next outline the procedure for finding $B_\pm$. Because the operators Q_+ and Q_- do not mix under renormalization, B_+ and B_- can be found separately. The procedure for finding $B_\pm$ amounts to the generalization of the matching procedure in LO to include in addition to logarithms also constant $O(\alpha_s)$ terms. The diagrams to be calculated are again the ones in Figures 5.2 and 5.3.

Step 1

$$A_{full}^\pm = \frac{G_F}{\sqrt{2}} \left(1 + \frac{\alpha_s(\mu_W)}{4\pi} \left[-\frac{\gamma_\pm^{(0)}}{2} \ln \frac{M_W^2}{-p^2} + \tilde{A}_\pm^{(1)}\right]\right) S_\pm, \qquad (5.96)$$

where $S_\pm$ are the tree matrix elements and $\tilde{A}_\pm^{(1)}$ the terms that we dropped in the LO calculation.

Step 2

$$A_{eff}^\pm = \frac{G_F}{\sqrt{2}} C_\pm(\mu_W)\langle Q_\pm(\mu_W)\rangle \qquad (5.97)$$

$$= \frac{G_F}{\sqrt{2}} C_\pm(\mu_W) \left(1 + \frac{\alpha_s(\mu_W)}{4\pi} \left[\frac{\gamma_\pm^{(0)}}{2} \ln \frac{-p^2}{\mu_W^2} + \tilde{r}_\pm\right]\right) S_\pm,$$

where $\tilde{r}_\pm$ are nonlogarithmic $O(\alpha_s)$ terms in $\langle Q_\pm(\mu_W)\rangle$, which we have to keep now.

Step 3

Comparison of (5.96) and (5.97) yields

$$C_\pm(\mu_W) = 1 + \frac{\alpha_s(\mu_W)}{4\pi} \left[-\frac{\gamma_\pm^{(0)}}{2} \ln \frac{M_W^2}{\mu_W^2} + B_\pm \right], \tag{5.98}$$

where

$$B_\pm = \tilde{A}_\pm^{(1)} - \tilde{r}_\pm . \tag{5.99}$$

Setting $\mu_W = M_W$, we reproduce (5.90). Any infrared dependence like $\ln(-p^2)$ or any special properties of the external quark states present in $\tilde{A}_\pm^{(1)}$ and $\tilde{r}_\pm$ cancel in the difference (5.99) so that $B_\pm$ are just numerical constants independent of external states. We will give the numerical values of $B_\pm$ later.

Needless to say these formulas apply to any operator that does not mix with other operators under renormalization. The operator dependence enters only through the values of $\gamma_\pm^{(0)}$ and $B_\pm$.

5.2.3 The Case of Operator Mixing

Preliminaries

Let us generalize the preceding discussion to the case of operator mixing and any number of operators that mix under renormalization. Now

$$\mathscr{H}_{\text{eff}} = \frac{G_F}{\sqrt{2}} \sum_i C_i(\mu) Q_i(\mu) \equiv \frac{G_F}{\sqrt{2}} \vec{Q}^T(\mu) \vec{C}(\mu) , \tag{5.100}$$

where the index i runs over all contributing operators, in our case Q_1 and Q_2.

The Wilson coefficient functions are given then by

$$\vec{C}(\mu) = \hat{U}(\mu, \mu_W) \vec{C}(\mu_W) . \tag{5.101}$$

Our goal is to find $\vec{C}(\mu_W)$ and the evolution matrix $\hat{U}(\mu, \mu_W)$ keeping NLO corrections.

Determination of $\vec{C}(\mu_W)$

The procedure for finding $\vec{C}(\mu_W)$ proceeds again in three steps:

Step 1

The amplitude in the full theory after field renormalization is given by:

$$A_{\text{full}} = \frac{G_F}{\sqrt{2}} \vec{S}^T \left(\vec{A}^{(0)} + \frac{\alpha_s(\mu_W)}{4\pi} \vec{A}^{(1)} \right). \tag{5.102}$$

Here $\vec{S}$ denotes the tree-level matrix elements of the operators $\vec{Q}$. To simplify the presentation we have absorbed the logarithms in the $O(\alpha_s)$ term $\vec{A}^{(1)}$, which also contains nonlogarithmic terms as in (5.96). Later we will discuss a specific example that will exhibit the detailed structure of (5.102) more transparently.

Step 2

In the effective theory, after quark field renormalization and the renormalization of the operators through

$$\vec{Q}^{(0)} = \hat{Z}\vec{Q},$$ (5.103)

the renormalized matrix elements of the operators are

$$\langle \vec{Q}(\mu_W) \rangle = \left(\hat{1} + \frac{\alpha_s(\mu_W)}{4\pi} \hat{r} \right) \vec{S},$$ (5.104)

and consequently

$$A_{\text{eff}} = \frac{G_F}{\sqrt{2}} \vec{S}^T \left(\hat{1} + \frac{\alpha_s(\mu_W)}{4\pi} \hat{r}^T \right) \vec{C}(\mu_W).$$ (5.105)

Again $\hat{r}$ contains the relevant logarithms together with the nonlogarithmic terms as in (5.97).

Step 3

Equating (5.102) and (5.105), we obtain

$$\boxed{\vec{C}(\mu_W) = \vec{A}^{(0)} + \frac{\alpha_s(\mu_W)}{4\pi} \left(\vec{A}^{(1)} - \hat{r}^T \vec{A}^{(0)} \right).}$$ (5.106)

Renormalization Group Evolution

The next goal is the calculation of the evolution matrix $\hat{U}(\mu, \mu_W)$, including NLO corrections. We will do it for a general anomalous dimension matrix, so that our formulas can also be used in later sections in which new operators, in particular QCD penguin operators, will enter the scene.

The renormalization group equation for $\vec{C}$,

$$\boxed{\frac{d\vec{C}(\mu)}{d \ln \mu} = \hat{\gamma}^T(g)\vec{C}(\mu),}$$ (5.107)

has to be solved now with the boundary condition (5.106).

The general solution can be written down iteratively

$$\hat{U}(\mu, \mu_W) = 1 + \int_{g(\mu_W)}^{g(\mu)} dg_1 \frac{\hat{\gamma}^T(g_1)}{\beta(g_1)} + \int_{g(\mu_W)}^{g(\mu)} dg_1 \int_{g(\mu_W)}^{g_1} dg_2 \frac{\hat{\gamma}^T(g_1)}{\beta(g_1)} \frac{\hat{\gamma}^T(g_2)}{\beta(g_2)} + \cdots$$ (5.108)

which using $dg/d \ln \mu = \beta(g)$ solves

$$\frac{d}{d \ln \mu} \hat{U}(\mu, \mu_W) = \hat{\gamma}^T(g)\hat{U}(\mu, \mu_W).$$ (5.109)

The series in (5.108) can be written more compactly:

$$\hat{U}(\mu, \mu_W) = T_g \exp\left[\int_{g(\mu_W)}^{g(\mu)} dg' \frac{\hat{\gamma}^T(g')}{\beta(g')}\right],$$ (5.110)

where in the case $g(\mu) > g(\mu_W)$ the g-ordering operator T_g is defined through

$$T_g f(g_1) \ldots f(g_n) = \sum_{perm} \Theta(g_{i_1} - g_{i_2}) \ldots \Theta(g_{i_{n-1}} - g_{i_n}) f(g_{i_1}) \ldots f(g_{i_n}).$$ (5.111)

It brings ordering of the functions $f(g_i)$ such that the coupling constants increase from right to left. The sum in (5.111) runs over all permutations $\{i_1, \ldots, i_n\}$ of $\{1, 2, \ldots, n\}$. The T_g ordering is necessary because at NLO $[\hat{\gamma}(g_1), \hat{\gamma}(g_2)] \neq 0$. Indeed the matrices $\hat{\gamma}^{(0)}$ and $\hat{\gamma}^{(1)}$ in the perturbative expansion of the anomalous dimension matrix

$$\hat{\gamma}(\alpha_s) = \hat{\gamma}^{(0)} \frac{\alpha_s}{4\pi} + \hat{\gamma}^{(1)} \left(\frac{\alpha_s}{4\pi}\right)^2,$$ (5.112)

do not commute with each other.

Inserting (5.112) and the expansion (5.92) for $\beta(g)$ into (5.110), we can write the evolution matrix in analogy to (5.93) as

$$\hat{U}(\mu, \mu_W) = \left[\hat{1} + \frac{\alpha_s(\mu)}{4\pi} \hat{J}\right] \hat{U}^{(0)}(\mu, \mu_W) \left[\hat{1} - \frac{\alpha_s(\mu_W)}{4\pi} \hat{J}\right].$$ (5.113)

Now it is clear why we have written (5.93) in a special manner. It can be nicely generalized to the mixing case where the ordering of matrices matters. $\hat{U}^{(0)}$ in (5.113) is the leading evolution matrix, which we already encountered in (5.79):

$$\hat{U}^{(0)}(\mu, \mu_W) = \hat{V} \left(\left[\frac{\alpha_s(\mu_W)}{\alpha_s(\mu)}\right]^{\frac{\vec{\gamma}^{(0)}}{2\beta_0}}\right)_D \hat{V}^{-1},$$ (5.114)

where $\hat{V}$ diagonalizes $\hat{\gamma}^{(0)T}$

$$\hat{\gamma}_D^{(0)} = \hat{V}^{-1} \gamma^{(0)T} \hat{V},$$ (5.115)

and $\vec{\gamma}^{(0)}$ is the vector containing the diagonal elements of the diagonal matrix $\hat{\gamma}_D^{(0)}$.

The derivation of the analytic expression for the matrix $\hat{J}$ follows [121] and is also given in the appendix of [122]. As it is rather involved, we give here only the final result. To write down the expression for the matrix $\hat{J}$, we define the matrix

$$\hat{G} = \hat{V}^{-1} \hat{\gamma}^{(1)T} \hat{V},$$ (5.116)

and a matrix $\hat{H}$ whose elements are

$$H_{ij} = \delta_{ij} \gamma_i^{(0)} \frac{\beta_1}{2\beta_0^2} - \frac{G_{ij}}{2\beta_0 + \gamma_i^{(0)} - \gamma_j^{(0)}}.$$ (5.117)

Then $\hat{J}$ can be written compactly as follows

$$\boxed{\hat{J} = \hat{V}\hat{H}\hat{V}^{-1}.}$$

(5.118)

The formula for H_{ij} is not valid for $2\beta_0 + \gamma_i^{(0)} - \gamma_j^{(0)} = 0$, which happens for certain values of (i, j) for $f = 3$. For this case the correct formula can be found in [123].

Final Result for $\vec{C}(\mu)$

Putting all things together, we obtain the final result

$$\boxed{\vec{C}(\mu) = \left(1 + \frac{\alpha_s(\mu)}{4\pi}\hat{J}\right)\hat{U}^{(0)}(\mu, \mu_W)\left(\vec{A}^{(0)} + \frac{\alpha_s(\mu_W)}{4\pi}\left[\vec{A}^{(1)} - (\hat{r}^T + \hat{J})\vec{A}^{(0)}\right]\right).}$$

(5.119)

We will discuss this important result in more detail later stressing in particular its μ and renormalization scheme dependences and their cancellations by the ones of the hadronic matrix elements.

In the case of (Q_1, Q_2) the inclusion of the flavor thresholds in (5.101) is very similar to the LO case:

$$\vec{C}(\mu) = \hat{U}_3(\mu, \mu_c)\hat{U}_4(\mu_c, \mu_b)\hat{U}_5(\mu_b, \mu_W)\vec{C}(\mu_W),$$

(5.120)

where $\hat{U}_f$ is the evolution matrix for f effective flavors given in (5.113). This formula has to be slightly modified if the penguin operators are present. We will return to this point in Section 6.4.

The expression (5.120) applies to $\mu \leq m_c$ and could be used for K meson decays. When the D meson system is considered, the first factor on the r.h.s. of this equation is absent. For B meson physics only $\hat{U}_5(\mu_b, \mu_W)$ is relevant.

5.2.4 Calculation of Anomalous Dimensions

Master Formulas

In the previous section we have calculated the anomalous dimension matrix in the process of the matching of the full and effective theories. In fact, comparing (5.31) with the last two terms in (5.32) and (5.33) one can see that the anomalous dimension matrix for $Q_{1,2}$ can be read off from the coefficients of the logarithms in the matrix elements $\langle Q_{1,2}\rangle$ in (5.32) and (5.33). In more complicated situations, where many operators are present and two-loop and higher-loop diagrams are involved, such a method is not very useful, and it is important to develop an efficient method for the calculation of anomalous dimensions. Here it comes:

- The evaluation of the amputated Green functions with insertion of the operators $\vec{Q}$ as in Figure 5.3 gives the relation

$$\langle \vec{Q}\rangle^{(0)} = Z_q^{-2}\hat{Z}\langle \vec{Q}\rangle \equiv \hat{Z}_{GF}\langle \vec{Q}\rangle,$$

(5.121)

where $\hat{Z}$ is the renormalization constant matrix for the operators $\vec{Q}$, Z_q for the external quark fields, and $\hat{Z}_{GF}$ is just defined earlier. As the matrix elements $\langle \vec{Q} \rangle$ are finite, $\hat{Z}_{GF}$ collects all the divergences present in $\langle \vec{Q} \rangle^{(0)}$.

- Next, the anomalous dimension matrix is given by

$$\hat{\gamma}(g) = \hat{Z}^{-1} \frac{d\hat{Z}}{d\ln\mu}. \tag{5.122}$$

- In the MS (or $\overline{\text{MS}}$) scheme we have

$$\hat{Z} = \hat{1} + \sum_{k=1}^{\infty} \frac{1}{\varepsilon^k} \hat{Z}_k(g), \tag{5.123}$$

and consequently as derived in the previous section

$$\hat{\gamma}(g) = -2g^2 \frac{\partial \hat{Z}_1(g)}{\partial g^2} = -2\alpha_s \frac{\partial \hat{Z}_1(\alpha_s)}{\partial \alpha_s}. \tag{5.124}$$

- For Z_q and $\hat{Z}_{GF}$ we have

$$Z_q = 1 + \sum_{k=1}^{\infty} \frac{1}{\varepsilon^k} Z_{q,k}(g), \qquad \hat{Z}_{GF} = \hat{1} + \sum_{k=1}^{\infty} \frac{1}{\varepsilon^k} \hat{Z}_{GF,k}(g). \tag{5.125}$$

As the matrix elements $\langle \vec{Q} \rangle$ are finite, the singularities in $\hat{Z}_{GF}$ are found directly from the calculation of the unrenormalized Green functions (5.121).

- From (5.121), (5.123), and (5.125) we find

$$\hat{Z}_1 = 2Z_{q,1}\hat{1} + \hat{Z}_{GF,1}. \tag{5.126}$$

- With

$$Z_{q,1} = a_1 \frac{\alpha_s}{4\pi} + a_2 \left(\frac{\alpha_s}{4\pi}\right)^2, \qquad \hat{Z}_{GF,1} = \hat{b}_1 \frac{\alpha_s}{4\pi} + \hat{b}_2 \left(\frac{\alpha_s}{4\pi}\right)^2, \tag{5.127}$$

we obtain by means of (5.124) *master formulas* for the elements of the one- and two-loop anomalous dimension matrices:

$$\boxed{\gamma_{ij}^{(0)} = -2[2a_1\delta_{ij} + (b_1)_{ij}], \qquad \gamma_{ij}^{(1)} = -4[2a_2\delta_{ij} + (b_2)_{ij}].} \tag{5.128}$$

- In the case without mixing between operators, these expressions reduce to:

$$\boxed{\gamma^{(0)} = -2[2a_1 + b_1], \qquad \gamma^{(1)} = -4[2a_2 + b_2].} \tag{5.129}$$

How to Use One-Loop Master Formulas

Let us illustrate how the first formula in (5.128) can be used to obtain the one-loop anomalous dimension matrix (5.78). From (5.27) we extract the coefficients of the $1/\varepsilon$ singularities to be

$$(b_1)_{11} = 2C_F + \frac{3}{N}, \qquad (b_1)_{12} = -3. \tag{5.130}$$

Similarly from (5.28) we find

$$(b_1)_{21} = -3, \qquad (b_1)_{22} = 2C_F + \frac{3}{N} . \tag{5.131}$$

Now $a_1 = -C_F$. Consequently, the term $2a_1$ in (5.128) cancels precisely the $2C_F$ term present in $(b_1)_{11}$ and $(b_1)_{22}$. The leftover entries give the one-loop anomalous dimension matrix (5.78).

The fact that the renormalization of the external quark fields cancels the terms $2C_F$ in $(b_1)_{11}$ and $(b_1)_{22}$ is by no means accidental. It is a consequence of the vanishing of the anomalous dimension of the conserved weak current. Indeed, the master formula for the anomalous dimension of a current can be obtained by considering a two-point Green function instead of four-point functions considered in deriving the master formulas (5.129). One finds this time

$$\gamma_c^{(0)} = -2[a_1 + b_1^c], \qquad \gamma_c^{(1)} = -4[a_2 + b_2^c], \tag{5.132}$$

where b_1^c and b_2^c are obtained by calculating the relevant one-loop and two-loop diagrams, respectively. b_1^c is simply obtained by calculating the one-loop upper vertex of the diagram (a) in Figure 5.3. b_2^c is found by calculating the corresponding two-loop generalization of this vertex. One finds $b_1^c = C_F$. The factor of two in $(b_1)_{11}$ in front of C_F in (5.130) comes from a symmetric diagram to the diagram (a) in Figure 5.3 with the gluon exchanged between the lower quark legs. With $a_1 = -C_F$ we find $\gamma_c^{(0)} = 0$ as it should be. At two loops $\gamma_c^{(1)}$ is renormalization scheme dependent and in certain schemes can be nonzero. As we will see later, it vanishes in the NDR scheme but is nonzero in the HV scheme.

We get the following useful message from this discussion that applies to schemes in which the anomalous dimension of a quark current vanishes at any order of perturbation theory. The only diagrams responsible for the nonvanishing anomalous dimensions of current-current operators Q_1 and Q_2 are the diagrams in which the gluons connect the quark legs belonging to different weak currents. At the one-loop level these are the diagrams (b) and (c) in Figure 5.3 and the corresponding symmetric diagrams. One should stress that this simple rule is not valid for the insertion of penguin operators into current-current diagrams. We will see this explicitly in Section 6.4.

How to Use the Two-Loop Master Formulas

The calculation of $(b_2)_{ij}$ in (5.128) is a bit trickier and technically more difficult. To this end one has to calculate first two-loop diagrams with Q_i insertions. Examples are given in Figure 5.4. Next the corresponding two-loop counter-diagrams have to be *subtracted*. In the MS-like schemes the latter are obtained by retaining only the $1/\varepsilon$ parts in the subdiagrams. The counter-diagrams corresponding to the two-loop diagrams in Figure 5.4 are shown in Figure 5.5, where the small boxes stand for the singular parts of the corresponding subdiagrams in Figure 5.4. For instance, the Feynman rule for the small box in diagram (b) of Figure 5.5 is, in accordance with (4.42), given by

$$iC_F \delta_{\alpha\beta} \frac{\alpha_s}{4\pi} \not{p} \frac{1}{\varepsilon}, \tag{5.133}$$

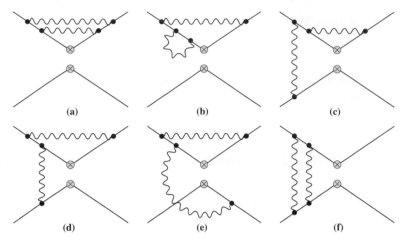

Figure 5.4   Examples of two-loop current-current diagrams contributing to the NLO anomalous dimensions of the operators Q_1 and Q_2. From [29].

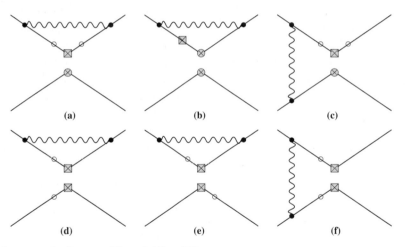

Figure 5.5   Counter-diagrams to the diagrams of Figure 5.4. From [29].

where p is the four-momentum of the quark. Because MS-like schemes are the so-called mass-independent renormalization schemes, the quark masses can be set to zero in evaluating anomalous dimensions.

Dropping color factors and Dirac tensors the result for each diagram including the counter-diagram has the structure

$$I - I_C = \left(\frac{\alpha}{4\pi}\right)^2 \left[\frac{\mu^2}{-p^2}\right]^{2\varepsilon} \left[\frac{F}{\varepsilon^2} + \frac{G}{\varepsilon} + \cdots\right] - \left(\frac{\alpha}{4\pi}\right)^2 \left[\frac{\mu^2}{-p^2}\right]^{\varepsilon} \left[\frac{F_C}{\varepsilon^2} + \frac{G_C}{\varepsilon} + \cdots\right],$$

$$(5.134)$$

where the second term represents the counter-diagram. Note that the power of μ^2 in the counter-term is ε and not 2ε as in I. It turns out that for diagrams with nonvanishing F

one has diagram by diagram the relation $F_C = 2F$, and consequently the pole part does not depend on μ as it should be:

$$I - I_C = \left(\frac{\alpha}{4\pi}\right)^2 \left[-\frac{F}{\varepsilon^2} + \frac{G - G_C}{\varepsilon} + \cdots\right] \quad \to \quad b_2 = G - G_C. \qquad (5.135)$$

The coefficient $G - G_C$ can then be identified with the contribution of a given diagram (after the inclusion of color factors) to the coefficient $(b_2)_{ij}$ entering the master formula (5.128).

Again, as in the one-loop case, the diagrams with gluons exchanged only between quark legs belonging to the same weak current can be omitted in the evaluation of two-loop anomalous dimensions of the operators Q_1 and Q_2, provided the anomalous dimension of the weak current vanishes at two-loop level. In this case their contributions to $(b_2)_{ij}$ are canceled by the a_2 term in the master formula (5.128). It should be stressed that this feature might not be preserved by some regularization schemes. In particular, it depends on the treatment of γ_5 in $D \neq 4$ dimensions. In the NDR scheme, in which γ_5 anticommutes with γ_μ in $D \neq 4$ dimensions, one has indeed $\gamma_c^{(1)} = 0$. But this is not true in the HV scheme, where γ_5 has more complicated properties. Indeed one finds [71]

$$[\gamma_c^{(1)}]_{HV} = 4C_F \beta_0. \qquad (5.136)$$

We will return to this issue in a moment. For the time being we give the two-loop generalizations of the one-loop anomalous dimension matrix (5.78) in the NDR scheme [71]:

$$\hat{\gamma}_{NDR}^{(1)} = \begin{pmatrix} -\frac{22}{3} - \frac{57}{2N^2} - \frac{2f}{3N} & \frac{39}{N} - \frac{19N}{6} + \frac{2f}{3} \\ \frac{39}{N} - \frac{19N}{6} + \frac{2f}{3} & -\frac{22}{3} - \frac{57}{2N^2} - \frac{2f}{3N} \end{pmatrix}, \qquad (5.137)$$

and in the HV scheme [71]:

$$\hat{\gamma}_{HV}^{(1)} = \begin{pmatrix} -\frac{110}{3} - \frac{57}{2N^2} + \frac{44N^2}{3} + \left(\frac{14}{3N} - \frac{8N}{3}\right)f & \frac{39}{N} + \frac{23N}{2} - 2f \\ \frac{39}{N} + \frac{23N}{2} - 2f & -\frac{110}{3} - \frac{57}{2N^2} + \frac{44N^2}{3} + \left(\frac{14}{3N} - \frac{8N}{3}\right)f \end{pmatrix}. \qquad (5.138)$$

The corresponding result in the DRED scheme has been first calculated in [79] and confirmed in [71]. We observe substantial renormalization scheme dependence in $\hat{\gamma}^{(1)}$. In particular the diagonal elements in (5.138) contain terms $O(N^2)$, whereas such terms are absent in (5.137). The origin of these terms can be traced back to the nonvanishing of $[\gamma_c^{(1)}]_{HV}$. Indeed these terms cancel in the difference

$$\hat{\gamma}_{HV}^{(1)} - 2[\gamma_c^{(1)}]_{HV}\hat{1} \equiv [\hat{\gamma}_{HV}^{(1)}]_{\text{eff}}. \qquad (5.139)$$

As we will discuss later, such a shift of two-loop anomalous dimensions is always possible provided also the matching conditions for the Wilson coefficients at μ_W are appropriately changed. Yet eventually this shift modifies the Wilson coefficients, and this modification will be compensated by the corresponding change in the matrix elements of the operators so that physical quantities are independent of these manipulations. All this should be clearer after we discussed cancellation of these scheme dependences in physical quantities in explicit terms.

In this context we should warn the reader that the numerical values of the Wilson coefficients in the HV scheme presented here and in [111, 123] correspond to the choice $[\hat{\gamma}_{HV}^{(1)}]_{\text{eff}}$. This differs from the treatment in [124, 125], who use $\hat{\gamma}_{HV}^{(1)}$ of (5.138) instead. For this reason the NLO corrections to Wilson coefficients in the HV scheme presented here are generally smaller than the ones found by the Rome group. The final physical results for decay amplitudes, and observables are, however, the same.

5.2.5 Explicit Calculation of 2 × 2 Anomalous Dimension Matrix

Current-Current Insertions: Generalities

Until now we have discussed the structure of the calculation of the anomalous dimensions of current-current operators, but it is time to enter some details. The set of six diagrams contributing to one-loop anomalous dimension matrix through operator insertions into current-current topologies is given by the diagrams in Figure 5.3 and their symmetric counterparts. We begin by developing the technology for the calculation of insertions of any operator with arbitrary color and Dirac structure into the diagrams of Figure 5.3. This will allow us to calculate later also insertions of penguin operators into the current-current topologies of Figure 5.3.

Let us then denote the color and Dirac structure of any operator by

$$\hat{V}_1 \otimes \hat{V}_2, \qquad \Gamma_1 \otimes \Gamma_2 , \tag{5.140}$$

respectively, so that an operator can be generally written as follows:

$$O = (\bar{s}_\alpha \Gamma_1 \hat{V}_1^{\alpha\beta} c_\beta) \otimes (\bar{u}_\gamma \Gamma_2 \hat{V}_2^{\gamma\delta} d_\delta). \tag{5.141}$$

Here we have made specific choice of quark flavors adapted to the operators discussed in the previous section, but it is trivial to generalize the following discussion to any other choice of flavors.

Let us consider a few examples of the color structures:

$$\hat{V}_1^{\alpha\beta} \otimes \hat{V}_2^{\gamma\delta} = \delta_{\alpha\beta} \otimes \delta_{\gamma\delta} \equiv \mathbf{1}_{\alpha\beta,\gamma\delta} , \tag{5.142}$$

$$\hat{V}_1^{\alpha\beta} \otimes \hat{V}_2^{\gamma\delta} = \delta_{\alpha\delta} \otimes \delta_{\gamma\beta} \equiv \tilde{\mathbf{1}}_{\alpha\beta,\gamma\delta} , \tag{5.143}$$

$$\hat{V}_1^{\alpha\beta} \otimes \hat{V}_2^{\gamma\delta} = (t^a)_{\alpha\beta} \otimes (t^a)_{\gamma\delta} \equiv \Pi_{\alpha\beta,\gamma\delta} . \tag{5.144}$$

Then the color identity (5.22) is simply given by

$$\Pi = \frac{1}{2}\left(\tilde{\mathbf{1}} - \frac{1}{N}\mathbf{1}\right) . \tag{5.145}$$

In this notation the operators Q_1 and Q_2 have both the Dirac structure $\Gamma_1 = \Gamma_2 = \gamma_\mu(1-\gamma_5)$. With the ordering of flavors as in (5.141), the color structure of Q_1 is $\tilde{\mathbf{1}}$. The one of Q_2 is $\mathbf{1}$.

To gain some insight into the calculation of the diagrams in Figure 5.3, let us consider diagram (a) with the insertion of the operator Q_2. The flavor labels are as in Figure 5.1. For this diagram we have then

$$\mathcal{D}_a^{(1)} = -ig^2\mu^{2\varepsilon}C_F \int \frac{d^D k}{(2\pi)^D} \frac{[\bar{s}T_\nu c] \otimes [\bar{u}\Gamma^\nu d]}{k^2[(k+p)^2]^2}, \qquad (5.146)$$

where

$$T_\nu = \gamma_\mu(\slashed{k}+\slashed{p})\Gamma_\nu(\slashed{k}+\slashed{p})\gamma^\mu = \gamma_\mu\gamma_\rho\Gamma_\nu\gamma_\sigma\gamma^\mu(k+p)^\rho(k+p)^\sigma, \qquad (5.147)$$

and $\Gamma_\nu = \gamma_\nu(1-\gamma_5)$. We have used $t^a t^a = C_F$.

Keeping only the divergent part in the relevant D-dimensional integral, we find

$$\int \frac{d^D k}{(2\pi)^D} \frac{(k+p)^\rho(k+p)^\sigma}{k^2[(k+p)^2]^2} = ig^{\rho\sigma}\frac{1}{16\pi^2}\frac{1}{4}\frac{1}{\varepsilon} + \text{finite}. \qquad (5.148)$$

Thus the divergent part of $\mathcal{D}_a^{(1)}$ is given by

$$\mathcal{D}_a^{(1)} = C_F\frac{\alpha_s}{4\pi}\left[\frac{1}{4}\frac{1}{\varepsilon}\right][\bar{s}\gamma_\mu\gamma_\rho\Gamma_\nu\gamma^\rho\gamma^\mu c] \otimes [\bar{u}\Gamma^\nu d]. \qquad (5.149)$$

It is straightforward to extend this calculation to other diagrams in Figure 5.3 and to the arbitrary operator given in (5.141). To this end we fix the ordering of the four flavors as in (5.141) and drop the external spinors. We find then

$$\mathcal{D}_a = \frac{\alpha_s}{4\pi}\left[\frac{1}{4}\frac{1}{\varepsilon}\right]\left(C_a^{(1)}\gamma_\mu\gamma_\rho\Gamma_1\gamma^\rho\gamma^\mu \otimes \Gamma_2 + C_a^{(2)}\Gamma_1 \otimes \gamma_\mu\gamma_\rho\Gamma_2\gamma^\rho\gamma^\mu\right), \qquad (5.150)$$

$$\mathcal{D}_b = -\frac{\alpha_s}{4\pi}\left[\frac{1}{4}\frac{1}{\varepsilon}\right]\left(C_b^{(1)}\Gamma_1\gamma_\rho\gamma_\mu \otimes \Gamma_2\gamma^\rho\gamma^\mu + C_b^{(2)}\gamma_\mu\gamma_\rho\Gamma_1 \otimes \gamma^\mu\gamma^\rho\Gamma_2\right), \qquad (5.151)$$

$$\mathcal{D}_c = \frac{\alpha_s}{4\pi}\left[\frac{1}{4}\frac{1}{\varepsilon}\right]\left(C_c^{(1)}\Gamma_1\gamma_\rho\gamma_\mu \otimes \gamma^\mu\gamma^\rho\Gamma_2 + C_c^{(2)}\gamma_\mu\gamma_\rho\Gamma_1 \otimes \Gamma_2\gamma^\rho\gamma^\mu\right), \qquad (5.152)$$

where the index (1) stands for the diagrams shown in Figure 5.3 and the index (2) for their symmetric counterparts. The color factors are given by

$$C_a^{(1)} = t^a\hat{V}_1 t^a \otimes \hat{V}_2, \qquad C_a^{(2)} = \hat{V}_1 \otimes t^a\hat{V}_2 t^a, \qquad (5.153)$$

$$C_b^{(1)} = \hat{V}_1 t^a \otimes \hat{V}_2 t^a, \qquad C_b^{(2)} = t^a\hat{V}_1 \otimes t^a\hat{V}_2, \qquad (5.154)$$

$$C_c^{(1)} = \hat{V}_1 t^a \otimes t^a\hat{V}_2, \qquad C_c^{(2)} = t^a\hat{V}_1 \otimes \hat{V}_2 t^a. \qquad (5.155)$$

Anomalous Dimensions of Q_1 and Q_2

Let us apply these general formulas to the case of the operator Q_2 for which we have

$$\hat{V}_1 \otimes \hat{V}_2 = \mathbf{1}, \qquad \Gamma_1 = \Gamma_2 = \gamma_\nu (1 - \gamma_5) \equiv \Gamma. \tag{5.156}$$

Because we are interested only in the $1/\varepsilon$, singularity we can use in one-loop formulas the rules for γ-algebra valid in four dimensions. Then

$$\gamma_\mu \gamma_\rho \Gamma \gamma^\rho \gamma^\mu \otimes \Gamma = \Gamma \otimes \gamma_\mu \gamma_\rho \Gamma \gamma^\rho \gamma^\mu = 4\Gamma \otimes \Gamma, \tag{5.157}$$

$$\Gamma \gamma_\rho \gamma_\mu \otimes \Gamma \gamma^\rho \gamma^\mu = \gamma_\mu \gamma_\rho \Gamma \otimes \gamma^\mu \gamma^\rho \Gamma = 16\Gamma \otimes \Gamma, \tag{5.158}$$

$$\Gamma \gamma_\rho \gamma_\mu \otimes \gamma^\mu \gamma^\rho \Gamma = \gamma_\mu \gamma_\rho \Gamma \otimes \Gamma \gamma^\rho \gamma^\mu = 4\Gamma \otimes \Gamma. \tag{5.159}$$

These results can be most efficiently reproduced by using a trick that we will call the *Greek Method* [126] from now on. Let us illustrate this method by deriving (5.158). Following [126] let us write

$$\gamma_\mu \gamma_\rho \Gamma \otimes \gamma^\mu \gamma^\rho \Gamma = A \, \Gamma \otimes \Gamma, \tag{5.160}$$

where A is the coefficient we are looking for. To find it we replace $\otimes$ in (5.160) by a matrix γ_τ to obtain

$$\gamma_\mu \gamma_\rho \Gamma \gamma_\tau \gamma^\mu \gamma^\rho \Gamma = A \, \Gamma \gamma_\tau \Gamma. \tag{5.161}$$

Inserting Γ of (5.156) into this equality and contracting indices, we determine A to be 16. The reader can check that the same result is obtained by using the Clifford algebra, which allows the reduction of three γ_μ matrices to a single one.

Using (5.157)–(5.159) in our master formulas (5.150)–(5.152) and summing all diagrams, we find

$$\sum_i \mathcal{D}_i = \frac{\alpha_s}{4\pi} \frac{1}{\varepsilon} \Gamma \otimes \Gamma \left[C_a^{(1)} + C_a^{(2)} - 4(C_b^{(1)} + C_b^{(2)}) + C_c^{(1)} + C_c^{(2)} \right]. \tag{5.162}$$

For the color structure in (5.156), we then find

$$C_a^{(1)} = C_a^{(2)} = C_F \mathbf{1}, \tag{5.163}$$

$$C_b^{(1)} = C_b^{(2)} = C_c^{(1)} = C_c^{(2)} = \frac{1}{2} \left(\tilde{\mathbf{1}} - \frac{1}{N} \mathbf{1} \right). \tag{5.164}$$

Now $\tilde{\mathbf{1}}$ stands for the operator Q_1. Consequently, inserting (5.163) and (5.164) into (5.162) and comparing the coefficient of $1/\varepsilon$ with $\hat{Z}_{GF,1}$ in (5.127), we extract

$$(b_1)_{21} = -3, \qquad (b_1)_{22} = 2C_F + \frac{3}{N}. \tag{5.165}$$

The insertion of the operator Q_1 represented by $\tilde{\mathbf{1}}$ into diagrams of Figure 5.3 can be evaluated in an analogous manner by using the master formulas (5.150)–(5.152). Because the color structure is more complicated, the calculation is now a bit more involved. To avoid this complication it is useful to make a Fierz reordering in Q_1 and Q_2 so that

$$Q_1 = (\bar{s}_\alpha d_\alpha)_{V-A}(\bar{u}_\beta c_\beta)_{V-A} \qquad Q_2 = (\bar{s}_\alpha d_\beta)_{V-A}(\bar{u}_\beta c_\alpha)_{V-A} . \qquad (5.166)$$

Now the roles of Q_1 and Q_2 are interchanged: Q_1 is $\mathbf{1}$, and Q_2 is $\tilde{\mathbf{1}}$. Because gluons are flavor blind, we find immediately

$$(b_1)_{11} = 2C_F + \frac{3}{N}, \qquad (b_1)_{12} = -3 . \qquad (5.167)$$

Equations (5.165) and (5.167) are precisely the values given in (5.131) and (5.130), respectively. Upon inserting them into the one-loop master formula (5.128) and using $a_1 = -C_F$, we finally reproduce the anomalous dimension matrix (5.78). We will extend this calculation to penguin operators in Section 6.4.

5.2.6 Mixing of Operators with Different Dimensions

It is useful to know the following properties of mixing of operators with different canonical dimensions

- The operators of a given dimension mix only into operators of the same or lower dimension. In a more formal terminology: To renormalize an operator of a given dimension one needs only operators as counter-terms of the same or lower dimension.
- This means in particular that the operators of dimension six, as Q_1 and Q_2, can mix into other six-dimensional operators and five dimensional magnetic penguin operators, which we will encounter in the next section. On the other hand, the magnetic penguin operators cannot mix into dimension six operators.
- Consequently, whereas the Q_1 and Q_2 operators influence the Wilson coefficients of the magnetic penguin operators, the latter operators have no impact on C_1 and C_2.

The proof of these properties is based essentially on dimensional analysis. It can be found on page 149 of the book by Collins [90].

Here comes another useful remark. As we will discuss in Section 6.5, the mixing between the operators (Q_1, Q_2) and the magnetic penguin operators appears first at the two-loop level. That is, the leading anomalous dimension is obtained by calculating two-loop diagrams and not one-loop diagrams as discussed so far. The next-to-leading anomalous dimensions are then obtained from three-loop calculations. In this particular case our master formulas in (5.128) change to ($i \neq j$)

$$\gamma_{ij}^{(0)} = -4[(b_2)_{ij}], \qquad \gamma_{ij}^{(1)} = -6[(b_3)_{ij}], \qquad (5.168)$$

with $(b_2)_{ij}$ and $(b_3)_{ij}$ obtained from $1/\varepsilon$ singularities in two-loop and three-loop diagrams, respectively. Motivated readers should be able to derive these formulas without any efforts.

5.2.7 Renormalization Scheme Dependence

At NLO various quantities like the Wilson coefficients and the anomalous dimensions depend on the renormalization scheme (RS) for operators. This dependence arises because the renormalization prescription involves an arbitrariness in the finite parts to be subtracted along with the ultraviolet singularities. Two different schemes are then related by a finite renormalization.

A particular example of the RS dependence is the dependence on the treatment of γ_5 in D-dimensions. We have seen in (5.137) and (5.138) that the two-loop anomalous dimension matrix for the operators (Q_1, Q_2) in the NDR scheme differs from the one in the HV scheme.

Returning back to our discussion of the NLO corrections to the Wilson coefficients of Section 5.2.3, we find that

$$\beta_0, \quad \beta_1, \quad \gamma^{(0)}, \quad \vec{A}^{(0)}, \quad \vec{A}^{(1)}, \quad \hat{r}^T + \hat{J}, \quad \langle \vec{Q} \rangle^T \vec{C} \tag{5.169}$$

are *scheme independent*, whereas

$$\hat{r}, \quad \gamma^{(1)}, \quad \hat{J}, \quad \vec{C}, \quad \langle \vec{Q} \rangle \tag{5.170}$$

are *scheme dependent*. Let us demonstrate this.

First of all, it is clear that the product

$$\langle \vec{Q}(\mu) \rangle^T \vec{C}(\mu), \tag{5.171}$$

representing the full amplitude, is independent of the RS. The factorization of the amplitude into $\vec{C}$ and $\langle \vec{Q} \rangle$ makes them, however, scheme dependent. Explicitly, for two different schemes (primed and unprimed), we have

$$\langle \vec{Q} \rangle' = \left(\hat{1} + \frac{\alpha_s}{4\pi} \hat{s} \right) \langle \vec{Q} \rangle, \qquad \vec{C}' = \left(\hat{1} - \frac{\alpha_s}{4\pi} \hat{s}^T \right) \vec{C}, \tag{5.172}$$

where $\hat{s}$ is a constant matrix representing a finite renormalization of $\vec{C}$ and $\langle \vec{Q} \rangle$.

Having the relations (5.172) at hand it is straightforward to find relations between various quantities in the primed and unprimed schemes. From

$$\langle \vec{Q}(\mu_W) \rangle = \left(\hat{1} + \frac{\alpha_s(\mu_W)}{4\pi} \hat{r} \right) \vec{S}, \tag{5.173}$$

where $\vec{S}$ is a vector of tree-level matrix elements, we immediately obtain

$$\hat{r}' = \hat{r} + \hat{s}. \tag{5.174}$$

Next, from

$$\langle \vec{Q}(\mu) \rangle^T \vec{C}(\mu) \equiv \langle \vec{Q}(\mu) \rangle^T \hat{U}(\mu, M_W) \vec{C}(M_W) \tag{5.175}$$

and (5.172), we have

$$\hat{U}'(\mu, M_W) = \left(\hat{1} - \frac{\alpha_s(\mu)}{4\pi} \hat{s}^T \right) \hat{U}(\mu, M_W) \left(\hat{1} + \frac{\alpha_s(M_W)}{4\pi} \hat{s}^T \right). \tag{5.176}$$

A comparison with

$$\hat{U}(\mu, \mu_W) = \left(\hat{1} + \frac{\alpha_s(\mu)}{4\pi} \hat{J} \right) \hat{U}^{(0)}(\mu, \mu_W) \left(\hat{1} - \frac{\alpha_s(\mu_W)}{4\pi} \hat{J} \right) \tag{5.177}$$

yields then

$$\hat{J}' = \hat{J} - \hat{s}^T. \tag{5.178}$$

Next from (5.30) and (5.172) we clearly have

$$\hat{Z}' = \hat{Z}\left(\hat{1} - \frac{\alpha_s}{4\pi}\hat{s}\right). \tag{5.179}$$

Using next the definition of the anomalous dimension matrix (5.122) and the expansion (5.112), we find

$$\hat{\gamma}^{(0)\prime} = \hat{\gamma}^{(0)}, \qquad \hat{\gamma}^{(1)\prime} = \hat{\gamma}^{(1)} + [\hat{s}, \hat{\gamma}^{(0)}] + 2\beta_0\hat{s}. \tag{5.180}$$

Let us make a few observations:

- From (5.174) and (5.178) follows the scheme independence of $\hat{r}^T + \hat{J}$. Next, $\vec{A}^{(0)}$ and $\vec{A}^{(1)}$, obtained from the calculation in the full theory, are clearly independent of the renormalization of operators. Consequently, the factor on the right-hand side of $\hat{U}^{(0)}$ in $\vec{C}(\mu)$ in (5.119), related to the "upper end" of the evolution, is independent of the RS.
- The same is true for $\hat{U}^{(0)}$, as $\hat{\gamma}^{(0)}$ and β_0 are scheme independent.
- $\vec{C}$ depends on the RS through $\hat{J}$ to the left of $\hat{U}^{(0)}$. This dependence is compensated by the corresponding scheme dependence of $\langle\vec{Q}\rangle$ in (5.172).

In the absence of operator mixing the relations between various quantities in two different schemes simplify. Going back to $C_\pm(\mu)$ in (5.95), we have

$$\gamma_+^{(1)\prime} = \gamma_\pm^{(1)} + 2\beta_0 s_\pm, \qquad B'_\pm = B_\pm - s_\pm, \qquad J'_\pm = J_\pm - s_\pm, \tag{5.181}$$

where $s_\pm$ are constant numbers analogous to $\hat{s}$ in (5.172).

Recalling

$$J_\pm = \frac{1}{2\beta_0}\left(\frac{\beta_1}{\beta_0}\gamma_\pm^{(0)} - \gamma_\pm^{(1)}\right), \tag{5.182}$$

we verify the scheme independence of $B_\pm - J_\pm$ in (5.95). Again the scheme dependence of $C_\pm(\mu)$ originates in the scheme dependence of $J_\pm$ present in the first factor in (5.95).

We should emphasize that the renormalization scheme dependence discussed here refers to the renormalization of operators and should be distinguished from the renormalization scheme dependence of α_s. The issue of the latter scheme dependence in the context of OPE is discussed in detail at the end of section III in [111] and will not be repeated here.

5.2.8 Renormalization Scale Dependence

A physical amplitude cannot depend on the arbitrary renormalization scale μ. The μ-dependence of the Wilson coefficients has to be canceled by the μ-dependence of the matrix elements $\langle Q_i(\mu)\rangle$. Due to the mixing under renormalization this cancellation may involve simultaneously several operators. Now, whereas $C_i(\mu)$ can be calculated in perturbation theory, this is not the case for the matrix elements $\langle Q_i(\mu)\rangle$. Yet, assuming that we know these matrix elements at a given scale, their μ–dependence is again governed by renormalization group equations with the solution given by

$$\langle \vec{Q}(M_W) \rangle^T = \langle \vec{Q}(\mu) \rangle^T \hat{U}(\mu, M_W). \tag{5.183}$$

Here $\hat{U}(\mu, M_W)$ is the same evolution matrix used for the evolution of WCs. But it should be noted that whereas the evolution of WCs, as seen in (5.107), is governed by $\hat{\gamma}^T$, the one of operator matrix elements by $\hat{\gamma}$.

Moreover, in inclusive decays of heavy mesons like $B \rightarrow X_s\gamma$ and $B \rightarrow X_s e^+ e^-$, one can analyze the cancellation of the μ-dependence using perturbative calculations of the relevant matrix elements $\langle Q_i(\mu) \rangle$. We refer to [127, 128] for the full exposition of this issue in $B \rightarrow X_s\gamma$ and $B \rightarrow X_s e^+ e^-$. Here it suffices to illustrate the cancellation of the μ dependence by considering a toy model in which only a single operator Q is present, and its matrix element $\langle Q(\mu) \rangle$ is calculated in perturbation theory. In fact, the μ-dependence of this matrix element is correctly described within perturbation theory and can be obtained from the more general formula (5.183).

Let us consider then the amplitude

$$A = \langle \mathcal{H}_{\text{eff}} \rangle = \frac{G_F}{\sqrt{2}} \langle Q(\mu_b) \rangle C(\mu_b) \tag{5.184}$$

with

$$C(\mu_b) = U(\mu_b, \mu_W) C(\mu_W), \tag{5.185}$$

where $\mu_b = O(m_b)$ and $\mu_W = O(M_W)$. We want to discuss the cancellation of the μ_b and μ_W dependences in (5.184) and (5.185) in explicit terms.

Beginning with the leading logarithmic approximation, we have

$$U(\mu_b, \mu_W) = U^{(0)}(\mu_b, \mu_W) = \left[\frac{\alpha_s(\mu_W)}{\alpha_s(\mu_b)}\right]^{\frac{\gamma^{(0)}}{2\beta_0}}, \qquad C(\mu_W) = 1. \tag{5.186}$$

Moreover, $\langle Q(\mu_b) \rangle = \langle Q \rangle_{\text{tree}}$ carries no μ_b dependence. Consequently, in LO the amplitude depends on μ_b and μ_W. Because $\alpha_s(\mu_b) \gg \alpha_s(\mu_W)$, the μ_b-dependence is stronger than the μ_W-dependence. If $\gamma^{(0)}/(2\beta_0)$ is $O(1)$, the μ_b-dependence of the resulting amplitudes and branching ratios may be rather disturbing. A known example of such a situation is the strong μ_b-dependence of the branching ratio $\mathcal{B}(B \rightarrow X_s\gamma)$ at LO. We will discuss this in detail in Section 6.5.

Let us next include NLO corrections. Now the various entries in (5.186) are generalized as follows:

$$U(\mu_b, \mu_W) = \left(1 + \frac{\alpha_s(\mu_b)}{4\pi} J\right) U^{(0)}(\mu_b, \mu_W) \left(1 - \frac{\alpha_s(\mu_W)}{4\pi} J\right), \tag{5.187}$$

$$C(\mu_W) = 1 + \frac{\alpha_s(\mu_W)}{4\pi} \left(\frac{\gamma^{(0)}}{2} \ln\left(\frac{\mu_W^2}{M_W^2}\right) + B\right), \tag{5.188}$$

$$\langle Q(\mu_b) \rangle = \langle Q \rangle_{\text{tree}} \left[1 + \frac{\alpha_s(\mu_b)}{4\pi} \left(\frac{\gamma^{(0)}}{2} \ln\left(\frac{m_b^2}{\mu_b^2}\right) + \tilde{r}\right)\right]. \tag{5.189}$$

We will now show that the amplitude A in (5.184) is independent of μ_b and μ_W at $O(\alpha_s)$. Using the following useful formula,

$$\frac{\alpha_s(m_1)}{\alpha_s(m_2)} = 1 + \frac{\alpha_s}{4\pi}\beta_0 \ln\left(\frac{m_2^2}{m_1^2}\right), \qquad (5.190)$$

where only $O(\alpha_s)$ terms have been retained, and keeping only logarithmic terms we can rewrite (5.187) as

$$U(\mu_b, \mu_W) = \left(1 + \frac{\alpha_s(\mu_b)}{4\pi}\frac{\gamma^{(0)}}{2}\ln\left(\frac{\mu_b^2}{m_b^2}\right)\right)\left[\frac{\alpha_s(M_W)}{\alpha_s(m_b)}\right]^{\frac{\gamma^{(0)}}{2\beta_0}}\left(1 + \frac{\alpha_s(\mu_W)}{4\pi}\frac{\gamma^{(0)}}{2}\ln\left(\frac{M_W^2}{\mu_W^2}\right)\right).$$
$$(5.191)$$

Inserting (5.188), (5.189), and (5.191) into (5.184), we find that μ_b and μ_W dependences cancel at $O(\alpha_s)$.

This simple example illustrates very clearly the virtue of NLO corrections. They reduce considerably various μ-dependences present in the LO approximation. On the other hand, we recover the well-known fact that at fixed order of perturbation theory there remain unphysical μ-dependences, which are of the order of the neglected higher-order contributions. In our simple example the leftover μ_b and μ_W dependences can be investigated numerically by inserting expressions (5.187)–(5.189) into (5.184) and varying μ_b and μ_W say in the ranges $m_b/2 \leq \mu_b \leq 2m_b$ and $M_W/2 \leq \mu_W \leq 2M_W$, respectively. By comparing the result of this exercise with an analogous exercise in LO, one can on the one hand appreciate the importance of NLO calculations. On the other hand, the leftover μ_W and μ_b dependences at NLO give a rough estimate of the theoretical uncertainty due to the truncation of the perturbative series. Such estimates can be found in any NLO analysis presented in the literature and will not be presented here except for a few examples in later chapters.

The μ-dependences discussed here are related to the renormalization group evolution of the Wilson coefficients from high- to low-energy scales. This evolution originates in the nonvanishing of the anomalous dimensions of the corresponding operators Q_i. On the other hand, as we have seen in Section 4.3, the nonvanishing of the anomalous dimension γ_m of the mass operator implies the μ-dependence of the quark masses, in particular $m_t(\mu_t)$, $m_b(\mu_b)$, and $m_c(\mu_c)$. These μ-dependences and their cancellation in decay amplitudes will be discussed in later chapters of this book.

The following section is rather technical and can be skipped by the readers who do not plan to do two-loop calculations.

5.2.9 Evanescent Operators

Origin of Evanescent Operators

In evaluating the anomalous dimensions of Q_1 and Q_2, we have used the Greek Method to reduce the complicated Dirac structures given in (5.157)–(5.159) to $\Gamma \otimes \Gamma$. Because we were only interested in the $1/\varepsilon$ singularity in a one-loop diagram, this reduction has been performed in $D = 4$ dimensions. In the case of two-loop calculations, in which the diagrams of Figure 5.3 are subdiagrams of the diagrams in Figure 5.4, this reduction has to be performed in arbitrary D-dimensions. Indeed, in a two-loop diagram, the leading singularity is $1/\varepsilon^2$. The $O(\varepsilon)$ terms arising from reductions like (5.157)–(5.159) in arbitrary D-dimensions, when multiplied by $1/\varepsilon^2$, will contribute to the $1/\varepsilon$ singularity relevant for the calculation of the two-loop anomalous dimensions.

The question then arises of how to find $O(\varepsilon)$ corrections to (5.157)–(5.159). We will follow here a simple method for finding these terms that has been proposed in [71]. Although more general methods have been developed subsequently, this method appears to be most useful for practical purposes. Yet other methods [129, 130], in particular the one of Herrlich and Nierste [130], give a deeper insight into these matters, and we will briefly discuss them at the end of this section.

The simplest method to find the $O(\varepsilon)$ terms in question would be to apply the Greek Method in D-dimensions. That is, evaluate (5.161) in D-dimensions to determine the coefficient A. For instance in the case of (5.158), we would find $16 - 4\varepsilon$ instead of 16 when using the NDR scheme for γ_5. This is what has been done in [126]. Yet as pointed out in [71], the mere replacement of 16 in (5.158) by $16 - 4\varepsilon$ with analogous replacements in (5.157) and (5.159) would eventually give incorrect two-loop anomalous dimensions. As demonstrated in [71] the correct procedure is to supplement the Greek Method in D-dimensions by the addition of other operators to the r.h.s. of (5.157)–(5.159), which vanish in $D = 4$ dimensions. Such operators are called *evanescent* operators.

We will soon explain the role of evanescent operators in the calculation of two-loop anomalous dimensions. While these matters are rather technical, we think it is useful to present them here to illustrate how careful one has to be already at two-loop level to obtain correct results. First, however, let us generalize (5.157)–(5.159) to $D \neq 4$ dimensions. In the case of the NDR scheme for γ_5, one finds [71]

$$\gamma_\mu \gamma_\rho \Gamma \gamma^\rho \gamma^\mu \otimes \Gamma = 4(1 - 2\varepsilon)\Gamma \otimes \Gamma, \tag{5.192}$$

$$\Gamma \gamma_\rho \gamma_\mu \otimes \Gamma \gamma^\rho \gamma^\mu = 4(4 - \varepsilon)\Gamma \otimes \Gamma + E^{\text{NDR}}, \tag{5.193}$$

$$\Gamma \gamma_\rho \gamma_\mu \otimes \gamma^\mu \gamma^\rho \Gamma = 4(1 - 2\varepsilon)\Gamma \otimes \Gamma - E^{\text{NDR}}, \tag{5.194}$$

where $O(\varepsilon^2)$ terms have been dropped. Identical results are found for the structures in the second column of the set (5.157)–(5.159). E^{NDR} stands for the evanescent operator given explicitly by

$$E^{\text{NDR}} = \frac{1}{2}[\gamma_\mu \gamma_\rho \Gamma \gamma^\rho \gamma^\mu \otimes \Gamma + \Gamma \otimes \gamma_\mu \gamma_\rho \Gamma \gamma^\rho \gamma^\mu - \Gamma \gamma_\rho \gamma_\mu \otimes \gamma^\mu \gamma^\rho \Gamma - \gamma_\mu \gamma_\rho \Gamma \otimes \Gamma \gamma^\rho \gamma^\mu]. \tag{5.195}$$

As one can verify using the Greek Method, E^{NDR} vanishes in $D = 4$. In the case of HV and DRED schemes, the formulas (5.192)–(5.194) are modified and the evanescent operators have more complicated structures. They can be found in [71]. It should be noted that there is no contribution from evanescent operators to (5.192) in the NDR scheme.

From calculational point of view the insertions of evanescent operators into the relevant diagrams are most efficiently evaluated by defining E^{NDR} simply as the difference between the structures on the l.h.s. of (5.193) and (5.194) and the respective terms on the r.h.s. involving $\Gamma \otimes \Gamma$. We will demonstrate this explicitly later.

Including Evanescent Operators in the Master Formulas

In Section 5.2.4 we have derived the master formulas (5.128) and (5.129) for the computation of two loop anomalous dimensions. This derivation did not take into account the presence of evanescent operators.[1] Therefore in cases in which the contributions of these operators matter, our formulas are strictly speaking incomplete. It is the purpose of the next few pages to correct for it and to derive a procedure for the calculation of two-loop anomalous dimensions, which takes into account the evanescent operators. We follow here again [71].

Let us go back to (5.134), which involves the coefficients (F, G) and (F_C, G_C) in the singularities of the two-loop diagrams and the corresponding counter-diagrams, respectively. The evaluation of F and G is still straightforward. Having the final result for a two-loop diagram with complicated Dirac structure, one can simply project on the space of physical operators, denoted generically by $\Gamma \otimes \Gamma$, by using the Greek method. In this way one can easily deduce the coefficients of the terms proportional to $\Gamma \otimes \Gamma$. As an example, let us consider the Dirac structure resulting from the diagram (f) in Figure 5.4. Then the projection by means of the Greek Method gives:

$$\Gamma \gamma_\mu \gamma_\nu \gamma_\rho \gamma_\tau \otimes \Gamma \gamma^\mu \gamma^\nu \gamma^\rho \gamma^\tau = 16 \left(16 - 14\varepsilon\right) \Gamma \otimes 1 . \tag{5.196}$$

The treatment of counter-diagrams needs more care. After the evaluation of the subdiagrams, the $1/\varepsilon$ contributions are multiplied by the structures in (5.192)–(5.194), i.e., they include evanescent (E) operators. Making the projection onto $\Gamma \otimes \Gamma$ already at this stage would be incorrect. Indeed, inserting E into counter-diagrams of Figure 5.5, generates back the original operator $\Gamma \otimes \Gamma$ and introduces a correction to G_C and consequently a correction to the two-loop anomalous dimension of the original operator. It is precisely this correction that we have neglected in our master formulas. We will now find how our method has to be modified to include the effects of E-operators. The next two pages are very technical, and some readers may prefer to skip them and go directly to the three-step procedure presented subsequently.

For a two-loop computation, it is sufficient to consider the effects of mixing with the evanescent operators specified in the previous section. However, higher-loop computations would require, in the NDR and HV schemes, consideration of an ever-increasing number of independent operators. Thus generally the renormalized operators in generic D- dimensions are given by

$$O_i = (\hat{Z}^{-1})_{ij} O_j^{(0)} . \tag{5.197}$$

---

[1] The results in (5.137) and (5.138) include these contributions.

Let us consider the case where the set $O_i^{(0)}$ includes the initial bare operators Q_+ and Q_- introduced in the previous section, which are expected not to mix under renormalization. In what follows we will denote them here by $O_j^{(0)}$ with $j = 1, 2$, respectively. All operators O_j with $j > 2$ correspond to evanescent operators. It is convenient to choose the basis such that the operators $O_j^{(0)}$ with $j = 3, 4$, respectively are the evanescent operators E^+, E^-. As mentioned earlier, it is, for our purposes, not necessary to specify the basis further nor to give explicit formulas of $E^{\pm}$.

The renormalization matrix $\hat{Z}$ has a perturbative expansion of the form,

$$\hat{Z} = \hat{1} + \frac{\alpha_s}{4\pi}\hat{Z}^{(1)} + \frac{\alpha_s^2}{(4\pi)^2}\hat{Z}^{(2)} + \cdots \tag{5.198}$$

Only the first four columns and first four rows of these (a priori infinite dimensional) matrices are of interest here. The understanding of the form of the matrices $\hat{Z}^{(1)}, \hat{Z}^{(2)}$ is crucial. First we have

$$Z^{(1)} = \begin{pmatrix} * & 0 & * & 0 \\ 0 & * & 0 & * \\ * & * & - & - \\ * & * & - & - \end{pmatrix}, \tag{5.199}$$

where a $*$ denotes nonzero entries and the elements "$-$" are of no interest to us. In particular we have $Z_{12}^{(1)} = Z_{21}^{(1)} = 0$. This situation need not, however, continue at higher loops because in generic D-dimensions the bare operators $Q_{\pm}^{(0)}$ do not have definite Fierz transformation properties in the NDR and HV schemes. Hence it can, and in fact does in the NDR and HV schemes, happen that,

$$Z_{12}^{(2)} \neq 0, \qquad Z_{21}^{(2)} \neq 0. \tag{5.200}$$

At the same time, at the one-loop level not only do we have $Z_{31}^{(1)} \neq 0$, $Z_{42}^{(1)} \neq 0$ but to define renormalized evanescent operators, which can really be neglected in precisely four dimensions one must, in general, take into account the mixing with operators of differing "naive" Fierz symmetry, i.e., it can happen that

$$\hat{Z}_{32}^{(1)} \neq 0, \qquad \hat{Z}_{41}^{(1)} \neq 0. \tag{5.201}$$

We will soon see that this is necessary so that the renormalized operators when restricted to precisely four dimensions have the correct Fierz symmetry.

Consider now the renormalization group equations for Green functions containing one renormalized operator O_j insertion, in the regularized D-dimensional theory. They take the standard form but due to the mixing with the evanescent bare operators, an anomalous dimension matrix occurs

$$\hat{\gamma} = \hat{Z}^{-1}\mu\frac{\partial}{\partial\mu}\hat{Z} = \hat{Z}^{-1}(-\varepsilon g + \beta(g))\frac{\partial}{\partial g}\hat{Z}. \tag{5.202}$$

Expanding $\beta(g)$, $\hat{\gamma}$ and $\hat{Z}$ in powers of the renormalized coupling g as in the previous sections we obtain from (5.202)

$$\hat{\gamma}^{(0)} = -2\varepsilon\hat{Z}^{(1)}, \tag{5.203}$$

and at two loops,

$$\hat{\gamma}^{(1)} = -4\varepsilon\hat{Z}^{(2)} - 2\beta_0\hat{Z}^{(1)} + 2\varepsilon\hat{Z}^{(1)}\hat{Z}^{(1)}. \tag{5.204}$$

Expanding the $\hat{Z}^{(r)}$ in inverse powers of ε,

$$\hat{Z}^{(1)} = \hat{Z}_0^{(1)} + \frac{1}{\varepsilon}\hat{Z}_1^{(1)}, \qquad \hat{Z}^{(2)} = \hat{Z}_0^{(2)} + \frac{1}{\varepsilon}\hat{Z}_1^{(2)} + \frac{1}{\varepsilon^2}\hat{Z}_2^{(2)} \tag{5.205}$$

and using the fact that the anomalous dimension matrix has a finite limit for $\varepsilon \to 0$, we must have the relation,

$$4\hat{Z}_2^{(2)} + 2\beta_0\hat{Z}_1^{(1)} - 2\hat{Z}_1^{(1)}\hat{Z}_1^{(1)} = 0, \tag{5.206}$$

which has been explicitly checked for the physical $\pm$ submatrix in [71]. We also get

$$\gamma^{(1)} = -4\hat{Z}_1^{(2)} - 2\beta_0\hat{Z}_0^{(1)} + 2(\hat{Z}_1^{(1)}\hat{Z}_0^{(1)} + \hat{Z}_0^{(1)}\hat{Z}_1^{(1)}). \tag{5.207}$$

Note that we have introduced in (5.205) the nonsingular terms $\hat{Z}_0^{(1)}$ and $\hat{Z}_0^{(2)}$ to be able to incorporate the effects of evanescent operators. In particular the presence of the finite renormalization $\hat{Z}_0^{(1)}$ allows in the approach of [71] to remove the finite contributions from evanescent operators to the matrix elements of physical operators. On the other hand, as we will see in a moment, this finite renormalization has an impact on the two-loop anomalous dimensions of the physical operators and consequently on their Wilson coefficients. In this context we note that

$$(\hat{Z}_0^{(1)})_{ij} = 0 \quad (i, j = 1, 2), \qquad (\hat{Z}_1^{(1)})_{31} = (\hat{Z}_1^{(1)})_{42} = 0. \tag{5.208}$$

The latter property ensures that $1/\varepsilon^2$ terms are not affected by the evanescent operators at the two-loop level. Finally using the properties (5.199) and (5.208) in (5.207), we find

$$\gamma_+^{(1)} = \gamma_{11}^{(1)} = -4(\hat{Z}_1^{(2)})_{11} + 2(\hat{Z}_1^{(1)})_{13}(\hat{Z}_0^{(1)})_{31}, \tag{5.209}$$

$$\gamma_-^{(1)} = \gamma_{22}^{(1)} = -4(\hat{Z}_1^{(2)})_{22} + 2(\hat{Z}_1^{(1)})_{24}(\hat{Z}_0^{(1)})_{42}, \tag{5.210}$$

$$\gamma_{+-}^{(1)} = \gamma_{12}^{(1)} = -4(\hat{Z}_1^{(2)})_{12} + 2(\hat{Z}_1^{(1)})_{13}(\hat{Z}_0^{(1)})_{32}, \tag{5.211}$$

$$\gamma_{-+}^{(1)} = \gamma_{21}^{(1)} = -4(\hat{Z}_1^{(2)})_{21} + 2(\hat{Z}_1^{(1)})_{24}(\hat{Z}_0^{(1)})_{41}. \tag{5.212}$$

The first term in (5.209) and (5.210) represents (after addition of wave function renormalization) our master formula (5.128), which was obtained neglecting the mixing with the E-operators. The remaining terms reflecting the mixing in question are the corrections we were looking for.

Without these corrections and corresponding corrections in (5.211) and (5.212), the renormalized operators $Q_\pm$ would not have the correct Fierz symmetry, and they would mix under renormalization at the two-loop level, i.e., $\gamma_{+-}^{(1)}$ and $\gamma_{-+}^{(1)}$ would be nonzero. The inclusion of E-operators restores the Fierz symmetry and removes this mixing, i.e., $\gamma_{+-}^{(1)} = \gamma_{-+}^{(1)} = 0$. This is explicitly demonstrated in [71].

Looking at the "extra contribution" from the evanescent operators, one realizes that it is proportional to the contribution that the counter-terms involving an evanescent operator insertion yield to the computation of $\hat{Z}^{(2)}$. This is precisely what we stated at the beginning of this section. Note, however, that the correction terms in (5.209) and (5.210) are by a factor of 2 smaller than the corresponding counter-terms (involving $Q_\pm$ operators) present in the main terms. In the language of diagrams the result just means that in calculating $\gamma^{(1)}$ the contributions to counter-term diagrams involving an evanescent operator should be multiplied by a factor of $1/2$.

How to Use the Improved Master Formulas

We can now summarize the improved procedure for the calculation of the two-loop anomalous dimensions.

Step 1

Calculate the full two-loop diagrams and project the Dirac structures onto the physical operators by means of the Greek Method. This gives in particular the coefficient G in (5.134).

Step 2

Calculate the usual contribution to the counter-term by taking the relevant subdiagram of a given two-loop diagram, projecting it onto the physical operators $\Gamma \otimes \Gamma$ by means of the Greek Method, inserting the result of this projection into the remaining subdiagram of a given two-loop diagram, and projecting the resulting expression again onto the physical operators by means of the Greek Method. This step gives the first part of G_C in (5.134). We will denote it by G_C^a.

Step 3

Calculate the contribution of the evanescent operator to the counter-diagram by simply inserting the difference of the two structures in (5.193) or (5.194), which define E, into the remaining one-loop subdiagram of a given two-loop diagram and project the result onto the physical operators by means of the Greek Method. This step gives the correction to the counter-term. We will denote the coefficient of $1/\varepsilon$ from this part by G_C^b.

Then the two-loop anomalous dimension matrix is found by calculating

$$\gamma_{ij}^{(1)} = -4[2a_2\delta_{ij} + (b_2)_{ij}], \qquad (b_2)_{ij} = \left(G - G_C^a - \frac{1}{2}G_C^b\right)_{ij}. \qquad (5.213)$$

Note the factor $1/2$ in the evanescent contribution. Formula (5.213) generalizes the master formula (5.128) with b_{ij} in (5.135) to include the contributions of evanescent operators.

Let us illustrate this procedure by calculating the contribution of the diagram (f) in Figure 5.4 and of its counter-diagram (f) in Figure 5.5 to the two-loop anomalous dimension of the operator with the Dirac structure $\gamma_\mu(1 - \gamma_5) \otimes \gamma^\mu(1 - \gamma_5)$. We drop the color factors in what follows.

Step 1

Calculating the diagram (f) in Figure 5.4, using the projection (5.196) and multiplying by two (inclusion of the symmetric counterpart) one finds [71]

$$F = 16 , \qquad G = 66 \tag{5.214}$$

with (F, G) defined in (5.134).

Step 2

We first calculate the diagram b in Figure 5.3, as in (5.151). We find

$$\mathcal{D}_b^{(1)} = -\frac{\alpha_s}{4\pi} \left[\frac{1}{4} \frac{1}{\varepsilon} \right] (1 + 2\varepsilon) \left[\Gamma\gamma_\rho\gamma_\mu \otimes \Gamma\gamma^\rho\gamma^\mu \right] , \tag{5.215}$$

where $(1 + 2\varepsilon)$ is an additional correction to the integral (5.148), which has to be kept now.

To find G_C^a we first project $\mathcal{D}_b^{(1)}$ on $\Gamma \otimes \Gamma$ by using the Greek Method and keeping only the divergent part:

$$[\mathcal{D}_b^{(1)}]_{div} = -\frac{\alpha_s}{4\pi} \left[\frac{1}{4} \frac{1}{\varepsilon} \right] 16 \, \Gamma \otimes \Gamma . \tag{5.216}$$

Inserting this into diagram f in Figure 5.5 gives by means of (5.215)

$$I_C^{(a)} = 2 \left(\frac{\alpha_s}{4\pi} \right)^2 \left[\frac{1}{4} \frac{1}{\varepsilon} \right]^2 (1 + 2\varepsilon) \left[\Gamma\gamma_\rho\gamma_\mu \otimes \Gamma\gamma^\rho\gamma^\mu \right] , \tag{5.217}$$

where the overall factor 2 takes into account the symmetric counterpart of this diagram. Using next the projection

$$\Gamma\gamma_\rho\gamma_\mu \otimes \Gamma\gamma^\rho\gamma^\mu = 4(4 - \varepsilon)\Gamma \otimes \Gamma \tag{5.218}$$

gives

$$I_C^{(a)} = \left(\frac{\alpha_s}{4\pi} \right)^2 \left[\frac{32}{\varepsilon^2} + \frac{56}{\varepsilon} \right] \Gamma \otimes \Gamma, \tag{5.219}$$

where finite terms have been dropped. Consequently,

$$F_C^a = 32 , \qquad G_C^a = 56. \tag{5.220}$$

Step 3

To calculate G_C^b we take first the evanescent part of $\mathcal{D}_b^{(1)}$. Dropping the $O(\varepsilon)$ from the integral (it contributes only to finite parts of $I_C^{(b)}$), we have

$$[\mathcal{D}_b^{(1)}]_{ev} = -\frac{\alpha_s}{4\pi} \left[\frac{1}{4} \frac{1}{\varepsilon} \right] \left[\Gamma\gamma_\rho\gamma_\mu \otimes \Gamma\gamma^\rho\gamma^\mu - 4(4 - \varepsilon)\Gamma \otimes \Gamma \right] , \tag{5.221}$$

where we have used (5.193) to express E^{NDR} in terms of the difference of Dirac structures between the square brackets in (5.221). Inserting $[\mathcal{D}_b^{(1)}]_{ev}$ into the diagram f of Figure 5.5 and multiplying by two for the symmetric counterpart, we get

$$I_C^{(b)} = 2\left(\frac{\alpha_s}{4\pi}\right)^2 \left[\frac{1}{4}\frac{1}{\varepsilon}\right]^2 (1+2\varepsilon)\left[\Gamma\gamma_\rho\gamma_\mu\gamma_\nu\gamma_\tau \otimes \Gamma\gamma^\rho\gamma^\mu\gamma^\nu\gamma^\tau - 4(4-\varepsilon)\Gamma\gamma_\nu\gamma_\tau \otimes \Gamma\gamma^\nu\gamma^\tau\right].$$
(5.222)

Projecting the two Dirac structures on $\Gamma \otimes \Gamma$ by means of (5.196) and (5.193), respectively, we obtain

$$I_C^{(b)} = \left(\frac{\alpha_s}{4\pi}\right)^2 \left[-\frac{12}{\varepsilon}\right]\Gamma \otimes \Gamma$$
(5.223)

and

$$F_C^b = 0, \qquad G_C^b = -12.$$
(5.224)

We observe that the $1/\varepsilon^2$ singularity is unaffected by the evanescent contribution. We can now calculate the relevant combination in (5.213) to be

$$G - G_C^a - \frac{1}{2}G_C^b = 16,$$
(5.225)

which is precisely the $1/\varepsilon$ singularity given in table 3 (diagram 5) of [71]. Also the $1/\varepsilon^2$ singularity $F - F_C^a = -16$ agrees with [71] and $F_C = 2F$ as promised after (5.134). Great! Everything works! We hope the reader is now motivated to calculate the remaining twenty-six two-loop diagrams and corresponding counter-diagrams necessary to reproduce the matrix (5.137). Actually the calculation of counter-diagrams is rather straightforward. The difficult part is the calculation of the two-loop diagrams, like the ones in Figure 5.4.

Evanescent Scheme Dependences

The definition of evanescent operators is not unique as stressed by Dugan and Grinstein [129] and in particular by Herrlich and Nierste [130]. As an example consider the Dirac structure on the l.h.s. of (5.193). Following [130] we can generalize this formula to

$$\Gamma\gamma_\rho\gamma_\mu \otimes \Gamma\gamma^\rho\gamma^\mu = (16 + a\varepsilon)\Gamma \otimes \Gamma + E^{\mathrm{NDR}}(a),$$
(5.226)

where "a" is an arbitrary parameter, which defines the evanescent operator $E^{\mathrm{NDR}}(a)$. For $a = -4$ the definition in (5.193) is chosen.

Now as the preceding discussion has shown, the presence of evanescent operators influences the two-loop anomalous dimensions of physical operators $\Gamma \otimes \Gamma$. Consequently, as emphasized in [130], the arbitrariness in the definition of the evanescent operators translates into an additional scheme dependence of two-loop anomalous dimensions, which can be effectively parametrized by "a" in (5.226). Therefore, when giving the results for two-loop anomalous dimensions, it is not sufficient to state simply that they correspond to NDR, HV or any other renormalization scheme. One has to specify in addition the definition of evanescent operators. This is essential as this scheme dependence of two-loop anomalous dimensions can only be canceled in physical amplitudes by the corresponding scheme dependences present in the matching conditions (for instance, $B_\pm$) at scales $O(M_W)$ and by the one present in the finite matrix elements of operators at scales $O(\mu)$.

This means that the treatment of evanescent operators in the process of matching and in the calculation of matrix elements of operators at scales $O(\mu)$ must be consistent with the one used in the calculation of two-loop anomalous dimensions. This issue is elaborated at length in [130] and in appendix B of [131].

There are two virtues of the definition of evanescent operators proposed in [71] and discussed in detail by us:

- The evanescent operators defined in this manner influence only two-loop anomalous dimensions. By definition they do not contribute to the matching and to the finite corrections to matrix elements at scales $O(\mu)$. They are simply subtracted away in the process of renormalization.
- As a consequence of this, the Fierz symmetry is preserved separately in two-loop anomalous dimensions, matching conditions, and matrix elements at scales $O(\mu)$.

The second property ensures that the operators Q_+ and Q_- do not mix under renormalization separately in two-loop anomalous dimensions, in the matching conditions, and in the matrix elements so that objects like B_{+-}, B_{-+}, $\gamma_{+-}^{(1)}$, $\gamma_{-+}^{(1)}$ are ensured to vanish in this scheme. In other schemes (see [129]) this is not the case, and the Fierz symmetry is only recovered after the two-loop anomalous dimensions are combined with the matching conditions, which makes the calculations unnecessarily rather involved.

Now comes the most important message of this section: Most of the existing NLO calculations adopt the definition of evanescent operators in [71], and all the two-loop anomalous dimensions and matching conditions given in this book, in my Les Houches lectures [29], and in the review [111] correspond to this definition.

5.2.10 The Effective $\Delta F = 1$ Hamiltonian: Current-Current Operators

Basic Formalism

Let us summarize the results for the coefficients $C_{1,2}(\mu)$ of the current-current operators $Q_{1,2}$ discussed extensively in this section, and let us evaluate them for the cases of $\Delta B = 1$, $\Delta C = 1$ and $\Delta S = 1$ decays.

To be specific, let us consider

$$Q_1 = (\bar{b}_\alpha c_\beta)_{V-A}(\bar{u}_\beta d_\alpha)_{V-A}, \qquad Q_2 = (\bar{b}_\alpha c_\alpha)_{V-A}(\bar{u}_\beta d_\beta)_{V-A}, \qquad (5.227)$$

$$Q_1 = (\bar{s}_\alpha c_\beta)_{V-A}(\bar{u}_\beta d_\alpha)_{V-A}, \qquad Q_2 = (\bar{s}_\alpha c_\alpha)_{V-A}(\bar{u}_\beta d_\beta)_{V-A}, \qquad (5.228)$$

$$Q_1 = (\bar{s}_\alpha u_\beta)_{V-A}(\bar{u}_\beta d_\alpha)_{V-A}, \qquad Q_2 = (\bar{s}_\alpha u_\alpha)_{V-A}(\bar{u}_\beta d_\beta)_{V-A}, \qquad (5.229)$$

for $\Delta B = 1$, $\Delta C = 1$, and $\Delta S = 1$ decays, respectively. Note, that these operators are responsible for the decays of mesons with b quark, c quark, and s quark, respectively. Looking at the flavor structure of Q_2 operators, one can easily reconstruct Feynman diagrams for various decays analogous to the ones in Chapter 3. We leave it as a useful exercise.

We set $f = 5$ in the preceding formulas and use the two-loop $\alpha_s(\mu)$ of (4.143) with $\Lambda_{\overline{MS}}^{(5)} = 225\,\text{MeV}$ corresponding to $\alpha_s(M_Z) = 0.118$. Choosing $\mu_b = 4.40\,\text{MeV}$, we find

$$C_1(\mu_b) = -0.184, \qquad C_2(\mu_b) = 1.078, \qquad \text{(NDR)}, \qquad (5.245)$$

$$C_1(\mu_b) = -0.226, \qquad C_2(\mu_b) = 1.100, \qquad \text{(HV)}. \qquad (5.246)$$

In the case of D decays and K decays, the relevant scales are $\mu = O(m_c)$ and $\mu = O(1\,\text{GeV})$, respectively, and consequently a more complicated formula with thresholds should in principle be used. Yet it is possible to use the following trick, which avoids these complications. We can simply use our master formulas with $\Lambda_{\overline{MS}}^{(5)}$ replaced by $\Lambda_{\overline{MS}}^{(4)}$ and an "effective" number of active flavors $f = 4.15$. The latter effective value for f allows to obtain an agreement with the exact results to better than 1.5%. We find then with $\Lambda_{\overline{MS}}^{(4)} = 325\,\text{MeV}$, $\mu_K = 1.0\,\text{GeV}$, and $\mu_c = 1.50\,\text{GeV}$

$$C_1(\mu_K) = -0.510, \qquad C_2(\mu_K) = 1.275, \qquad \text{(NDR)}, \qquad (5.247)$$

$$C_1(\mu_K) = -0.631, \qquad C_2(\mu_K) = 1.358, \qquad \text{(HV)}, \qquad (5.248)$$

$$C_1(\mu_c) = -0.378, \qquad C_2(\mu_c) = 1.188, \qquad \text{(NDR)}, \qquad (5.249)$$

$$C_1(\mu_c) = -0.457, \qquad C_2(\mu_c) = 1.237, \qquad \text{(HV)}. \qquad (5.250)$$

6 Effective Hamiltonians for FCNC Processes

6.1 Overture: General View of FCNC Processes

6.1.1 General Remarks

The pattern of flavor violation in the SM is governed by the $V - A$ structure of $W^\pm$ interactions with quarks and leptons and equally importantly by the natural suppression of flavor-changing neutral current processes with the help of the GIM mechanism that we introduced briefly in Chapter 2. The flavor diagonal structure of the basic vertices involving γ, Z, and G in the SM forbids the appearance of FCNC processes at the tree level. With the help of the flavor-changing $W^\pm$-vertex one can, however, construct one-loop and higher-order diagrams that mediate FCNC processes. Our next goal is then to describe first the general structure of these processes within the SM and subsequently to cast the gained knowledge into effective Hamiltonians given in the form of various OPEs. This will allow us to include QCD effects by means of the technology developed in the last chapters in a straightforward manner. We will soon see that the structure of these processes is much richer than the one of tree-level decays considered so far, which involved only the current-current operators Q_1 and Q_2.

The fact that FCNCs take place in the SM only as loop effects makes them particularly useful for testing the quantum structure of the theory and the search for NP beyond the SM. There they can enter in principle already at treelevel being suppressed by propagators of very heavy new particles. But in the SM at the one-loop level they can be described by a set of basic triple and quartic effective vertices. In the literature they appear under the names of penguin and box diagrams, respectively. These vertices depend on electroweak parameters like G_F, α and $\sin\vartheta_W$, the CKM parameters, and very importantly on the masses of quarks that appear virtually in these diagrams together with $W^\pm$. Therefore, once these parameters are determined in other processes, the rate for FCNC decays in the SM can be predicted.

The description of this class of processes in terms of effective vertices is useful as it allows to exhibit very strong correlations between various decays within the SM, which provide a number of excellent tests of this model. Let us then discuss the structure of FCNCs in the SM in explicit terms. In the next section we will calculate some of these vertices in detail and then move to include QCD corrections to FCNCs. These considerations will be generalized to extensions of the SM in the last part of this book.

6.1.2 Effective Vertices

Penguin Vertices

These vertices involve only quarks and can be depicted as in Figure 6.1, where i and j have the same charge but different flavor, and k denotes the internal quark whose charge is different from that of i and j. These effective vertices can be calculated by using the Feynman rules for elementary vertices and propagators in the SM. Important examples are given in Figure 6.2. The diagrams with Goldstone boson exchanges in place of $W^{\pm}$ have not been shown. They have to be included to obtain a gauge independent result, where gauge refers to the $W^{\pm}$ propagator. Only in the unitary gauge the diagrams with Goldstone bosons are absent but then the loop calculations are more difficult because the $W^{\pm}$ propagator in this gauge is more complicated than in the Feynman gauge used for most calculations in this book. Strictly speaking, also self-energy corrections on external lines have to be included to make the effective vertices finite. These technicalities are discussed explicitly in Section 6.2.

Figure 6.1   Penguin vertices. From [29].

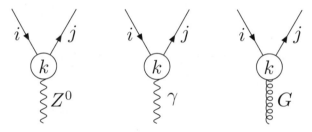

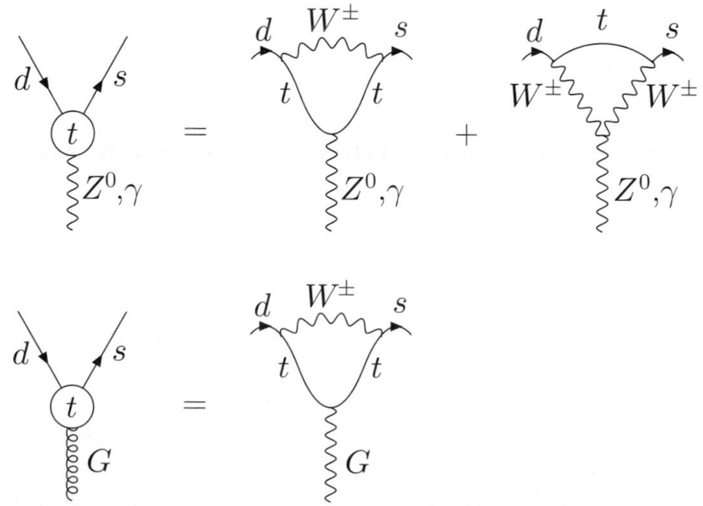

Figure 6.2   Penguin vertices resolved in terms of basic vertices. From [29].

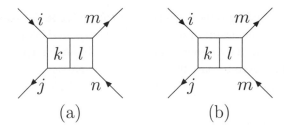

$$(a) \qquad\qquad (b)$$

Figure 6.3   Box vertices. From [29].

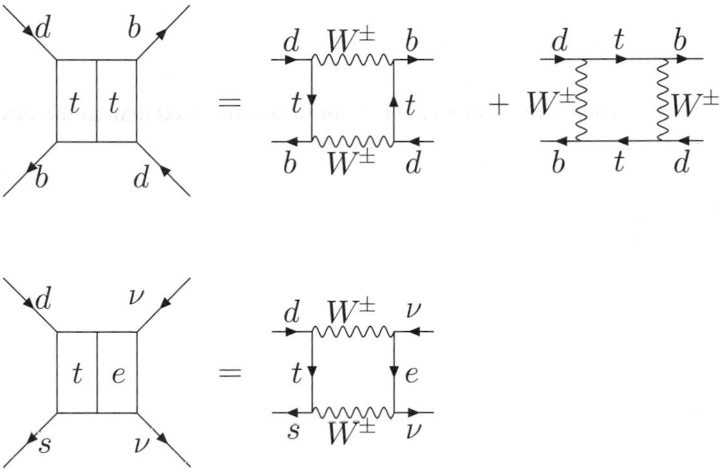

$$
\begin{array}{c}
\text{(box diagram)} = \text{(diagram 1)} + \text{(diagram 2)}
\end{array}
$$

Figure 6.4   Box vertices resolved in terms of elementary vertices. From [29].

Box Vertices

These vertices involve in general both quarks and leptons and can be depicted as in Figure 6.3, where again i, j, m, and n stand for external quarks or leptons and k and l denote the internal quarks and leptons. In the vertex (a) the flavor violation takes place on both sides (left and right) of the box, whereas in (b) the right-hand side is flavor conserving. These effective quartic vertices can also be calculated using the Feynman rules for elementary vertices and propagators in the SM. We have, for instance, the vertices in Figure 6.4, which contribute to $B_d^0 - \bar{B}_d^0$ mixing and $K^+ \to \pi^+ \nu \bar{\nu}$, respectively. Other interesting examples will be discussed later. Again, the explicit calculation of these vertices in the SM is performed in Section 6.2.

Effective Feynman Rules

Using the Feynman rules for elementary vertices and propagators in the SM, one can derive "Feynman rules" for the effective vertices in question. This derivation is obtained by simply calculating the diagrams on the r.h.s. of the equations in Figures 6.2 and 6.4 and similar for other vertices. In Figure 6.2 the Z^0, γ, and *gluon* are off-shell. In the case of inclusive decays $B \to X\gamma$ and $B \to XG$, with G standing for gluons, we need also corresponding vertices with on-shell photons and gluons. For these two cases it is essential to keep the masses of the external quarks, as otherwise the corresponding vertices would vanish. Moreover, momentum flowing through gluons and the photon propagator cannot be set to zero because these particles are massless. This means that also external momenta of quarks have to be kept, which makes the calculations more difficult than in the case of Z-penguin diagrams and box diagrams. We will return to these technicalities in Section 6.2. Here we just present the results so that readers not interested in spending days and nights in going through the calculations in Section 6.2 can also have some fun.

The rules for effective vertices are in the 't Hooft–Feynman gauge for the $W^\pm$ propagator as follows:

$$\text{Box}(\Delta S = 2) = \lambda_i^2 \frac{G_F^2}{16\pi^2} M_W^2 S_0(x_i)(\bar{s}d)_{V-A}(\bar{s}d)_{V-A}, \tag{6.1}$$

$$\text{Box}(T_3 = 1/2) = \lambda_i \frac{G_F}{\sqrt{2}} \frac{\alpha}{2\pi \sin^2 \vartheta_W}[-4B_0(x_i)](\bar{s}d)_{V-A}(\bar{\nu}\nu)_{V-A}, \tag{6.2}$$

$$\text{Box}(T_3 = -1/2) = \lambda_i \frac{G_F}{\sqrt{2}} \frac{\alpha}{2\pi \sin^2 \vartheta_W} B_0(x_i)(\bar{s}d)_{V-A}(\bar{\mu}\mu)_{V-A}, \tag{6.3}$$

$$\bar{s}Zd = i\lambda_i \frac{G_F}{\sqrt{2}} \frac{e}{2\pi^2} M_Z^2 \frac{\cos \vartheta_W}{\sin \vartheta_W} C_0(x_i)\bar{s}\gamma_\mu(1-\gamma_5)d, \tag{6.4}$$

$$\bar{s}\gamma d = -i\lambda_i \frac{G_F}{\sqrt{2}} \frac{e}{8\pi^2} D_0(x_i)\bar{s}(q^2\gamma_\mu - q_\mu \not{q})(1-\gamma_5)d, \tag{6.5}$$

$$\bar{s}G^a d = -i\lambda_i \frac{G_F}{\sqrt{2}} \frac{g_s}{8\pi^2} E_0(x_i)\bar{s}_\alpha(q^2\gamma_\mu - q_\mu \not{q})(1-\gamma_5)t_{\alpha\beta}^a d_\beta, \tag{6.6}$$

$$\bar{s}\gamma' b = i\bar{\lambda}_i^* \frac{G_F}{\sqrt{2}} \frac{e}{8\pi^2} D_0'(x_i)\bar{s}[i\sigma_{\mu\lambda}q^\lambda[m_b(1+\gamma_5)]]b, \tag{6.7}$$

$$\bar{s}G'^a b = i\bar{\lambda}_i^* \frac{G_F}{\sqrt{2}} \frac{g_s}{8\pi^2} E_0'(x_i)\bar{s}_\alpha[i\sigma_{\mu\lambda}q^\lambda[m_b(1+\gamma_5)]]t_{\alpha\beta}^a b_\beta, \tag{6.8}$$

where we simplified the notation for CKM factors

$$\lambda_i \equiv \lambda_i^{(K)} = V_{is}^* V_{id}, \qquad \bar{\lambda}_i \equiv \lambda_i^{(s)} = V_{ib}^* V_{is}. \tag{6.9}$$

The first six rules are relevant for K meson physics. For B meson physics only external flavors in the operators should be changed. The functions $F(x_i)$ are the same. The last two rules are relevant for $b \to s$ transitions. The generalization to $b \to d$ transitions should be obvious. These two rules are less relevant in K physics because these vertices are suppressed by $m_{s,d}$ in this case.

In these rules q_μ is the *outgoing* gluon or photon momentum, and T_3 indicates whether $\nu\bar{\nu}$ or $\ell^+\ell^-$ leaves the box diagram. The last two rules involve on-shell photon and gluon. We have set $m_s = 0$ in these rules.

These rules for effective vertices together with the rules for the propagators of the gauge boson involved allow the calculation of the effective Hamiltonians for FCNC processes, albeit without the inclusion of QCD corrections. The way these rules should be used requires some care:

- The penguin vertices should be used in the same manner as the elementary Feynman rules, which follow from $i\mathcal{L}$. Once a mathematical expression corresponding to a given diagram has been found, the contribution of this diagram to the relevant effective Hamiltonian is obtained by multiplying this mathematical expression by "i."
- Our conventions for the box vertices are such that they directly give the contributions to the effective Hamiltonians.

We will give an example later by calculating the internal top-quark contributions to $K^+ \to \pi^+ \nu \bar{\nu}$. First, however, let us make a few general remarks emphasizing the new features of these effective vertices as compared to the elementary ones.

- They are higher order in the gauge couplings and consequently suppressed relative to the elementary transitions. This is consistent with experimental findings, which show very strong suppression of FCNC transitions relative to tree-level processes.
- Because of the internal $W^\pm$ exchanges, all penguin vertices in Figure 6.2 are purely $V - A$, i.e., the effective vertices involving γ and G are parity violating as opposed to their elementary interactions in Figure 6.1! Also, the Dirac structure of the flavor-violating Z coupling generated at one-loop level differs from the elementary one because now only $V - A$ couplings are involved. The box vertices are of the $(V - A) \otimes (V - A)$ type.
- The effective vertices depend on the masses of internal quarks or leptons and consequently are calculable functions of

$$x_i = \frac{m_i^2}{M_W^2}, \qquad i = u, c, t. \tag{6.10}$$

The masses of leptons except for the τ in internal charm contributions are set to zero, as this is an excellent approximation. A set of basic universal functions can be found. These functions govern the physics of all FCNC processes in the SM. They are given subsequently.

- The effective vertices depend on the elements of the CKM matrix, and this dependence can be found directly from the diagrams of Figures 6.2 and 6.4.
- The dependences of a given vertex on the CKM factors and the masses of internal fermions govern the strength of the vertex in question.
- Another new feature of the effective vertices as compared to the elementary ones is their dependence on the gauge used for the $W^\pm$ propagator. We will return to this point soon.

Basic Functions

The basic functions present in (6.1)–(6.8) were calculated by various authors, in particular by Inami and Lim [132]. We will present some of these calculations using our methods in Section 6.2. They are given explicitly as follows:

$$B_0(x_t) = \frac{1}{4}\left[\frac{x_t}{1-x_t} + \frac{x_t \ln x_t}{(x_t-1)^2}\right], \tag{6.11}$$

$$C_0(x_t) = \frac{x_t}{8}\left[\frac{x_t-6}{x_t-1} + \frac{3x_t+2}{(x_t-1)^2}\ln x_t\right], \tag{6.12}$$

$$D_0(x_t) = -\frac{4}{9}\ln x_t + \frac{-19x_t^3 + 25x_t^2}{36(x_t-1)^3} + \frac{x_t^2(5x_t^2 - 2x_t - 6)}{18(x_t-1)^4}\ln x_t, \tag{6.13}$$

$$E_0(x_t) = -\frac{2}{3}\ln x_t + \frac{x_t^2(15 - 16x_t + 4x_t^2)}{6(1-x_t)^4}\ln x_t + \frac{x_t(18 - 11x_t - x_t^2)}{12(1-x_t)^3}, \tag{6.14}$$

$$D_0'(x_t) = -\frac{(8x_t^3 + 5x_t^2 - 7x_t)}{12(1-x_t)^3} + \frac{x_t^2(2 - 3x_t)}{2(1-x_t)^4}\ln x_t, \tag{6.15}$$

$$E_0'(x_t) = -\frac{x_t(x_t^2 - 5x_t - 2)}{4(1-x_t)^3} + \frac{3}{2}\frac{x_t^2}{(1-x_t)^4}\ln x_t, \tag{6.16}$$

$$S_0(x_t) = \frac{4x_t - 11x_t^2 + x_t^3}{4(1-x_t)^2} - \frac{3x_t^3 \ln x_t}{2(1-x_t)^3}, \tag{6.17}$$

$$S_0(x_c) = x_c, \tag{6.18}$$

$$S_0(x_c, x_t) = x_c\left[\ln\frac{x_t}{x_c} - \frac{3x_t}{4(1-x_t)} - \frac{3x_t^2 \ln x_t}{4(1-x_t)^2}\right]. \tag{6.19}$$

We would like to make a few comments:

- In the last two expressions we have kept only linear terms in $x_c \ll 1$, but of course all orders in x_t. The last function generalizes $S_0(x_t)$ in (6.17) to include box diagrams with simultaneous top-quark and charm-quark exchanges.
- The subscript "0" indicates that these functions do not include QCD corrections to the relevant penguin and box diagrams. These corrections will be discussed in detail in subsequent sections.
- In writing the expressions in (6.11)–(6.19), we have omitted x_t-independent terms, which do not contribute to decays due to the GIM mechanism. We will discuss this issue in more detail later. Moreover,

$$S_0(x_t) \equiv F(x_t, x_t) + F(x_u, x_u) - 2F(x_t, x_u) \tag{6.20}$$

and

$$S_0(x_i, x_j) = F(x_i, x_j) + F(x_u, x_u) - F(x_i, x_u) - F(x_j, x_u), \tag{6.21}$$

where $F(x_i, x_j)$ is the true function corresponding to a given box diagram with i and j quark exchanges. These particular combinations can be found by drawing all possible box diagrams (also those with u-quark exchanges), setting $m_u = 0$, and using unitarity of the CKM-matrix, which implies in particular the relations:

$$\lambda_u^{(i)} + \lambda_c^{(i)} + \lambda_t^{(i)} = 0, \qquad i = K, s, d \tag{6.22}$$

for K, B_s^0 and B_d^0 systems with explicit expressions for CKM factors given in (6.9) and (6.41). In this way the effective Hamiltonians for FCNC transitions can be directly obtained by summing only over t and c quarks.

- The expressions given for $B_0(x_t)$, $C_0(x_t)$, and $D_0(x_t)$ correspond to the 't Hooft–Feynman gauge ($\xi = 1$). In an arbitrary R_ξ gauge they look different. In phenomenological applications it is therefore useful to work instead with the following gauge independent combinations [133]:

$$C_0(x_t, \xi) - 4B_0(x_t, \xi, 1/2) = C_0(x_t) - 4B_0(x_t) = X_0(x_t), \tag{6.23}$$

$$C_0(x_t, \xi) - B_0(x_t, \xi, -1/2) = C_0(x_t) - B_0(x_t) = Y_0(x_t), \tag{6.24}$$

$$C_0(x_t, \xi) + \frac{1}{4}D_0(x_t, \xi) = C_0(x_t) + \frac{1}{4}D_0(x_t) = Z_0(x_t). \tag{6.25}$$

- $X_0(x_t)$ and $Y_0(x_t)$ are linear combinations of the $V - A$ components of Z^0-penguin and box diagrams with final quarks or leptons having weak isospin T_3 equal to $1/2$ and $-1/2$, respectively.
- $Z_0(x_t)$ is a linear combination of the vector component of the Z^0-penguin and the γ-penguin.
- These new functions are given explicitly as follows:

$$X_0(x_t) = \frac{x_t}{8} \left[\frac{x_t + 2}{x_t - 1} + \frac{3x_t - 6}{(x_t - 1)^2} \ln x_t \right], \tag{6.26}$$

$$Y_0(x_t) = \frac{x_t}{8} \left[\frac{x_t - 4}{x_t - 1} + \frac{3x_t}{(x_t - 1)^2} \ln x_t \right], \tag{6.27}$$

$$Z_0(x_t) = -\frac{1}{9} \ln x_t + \frac{18x_t^4 - 163x_t^3 + 259x_t^2 - 108x_t}{144(x_t - 1)^3}$$

$$+ \frac{32x_t^4 - 38x_t^3 - 15x_t^2 + 18x_t}{72(x_t - 1)^4} \ln x_t. \tag{6.28}$$

Thus the set of gauge independent basic functions that govern the FCNC processes in the SM is given by:

$$S_0(x_t), \quad X_0(x_t), \quad Y_0(x_t), \quad Z_0(x_t), \quad E_0(x_t), \quad D_0'(x_t), \quad E_0'(x_t). \tag{6.29}$$

For numerical calculations one should, of course, use the exact expressions just given but to get an idea on the size of these functions and for derivation of analytical formulas one can use the following approximate but simple expressions

$$S_0(x_t) = 0.784\, x_t^{0.76}, \quad X_0(x_t) = 0.660\, x_t^{0.575}, \tag{6.30}$$

$$Y_0(x_t) = 0.315\, x_t^{0.78}, \quad Z_0(x_t) = 0.175\, x_t^{0.93}, \quad E_0(x_t) = 0.564\, x_t^{-0.51}, \tag{6.31}$$

$$D_0'(x_t) = 0.244\, x_t^{0.30}, \quad E_0'(x_t) = 0.145\, x_t^{0.19}. \tag{6.32}$$

In the range $160\,\text{GeV} \leq m_t(m_t) \leq 165\,\text{GeV}$, these approximations reproduce the exact expressions to an accuracy better than 0.2%, which is sufficient for deriving analytic formulas. We have then

$$S_0(x_t) = 2.31 \left(\frac{m_t(m_t)}{163\,\text{GeV}} \right)^{1.52}, \tag{6.33}$$

$$X_0(x_t) = 1.49 \left(\frac{m_t(m_t)}{163\,\text{GeV}} \right)^{1.15}, \qquad Y_0(x_t) = 0.95 \left(\frac{m_t(m_t)}{163\,\text{GeV}} \right)^{1.56}, \tag{6.34}$$

$$Z_0(x_t) = 0.66 \left(\frac{m_t(m_t)}{163\,\text{GeV}} \right)^{1.86}, \qquad E_0(x_t) = 0.27 \left(\frac{m_t(m_t)}{163\,\text{GeV}} \right)^{-1.02}, \tag{6.35}$$

$$D_0'(x_t) = 0.37 \left(\frac{m_t(m_t)}{163\,\text{GeV}} \right)^{0.60}, \qquad E_0'(x_t) = 0.19 \left(\frac{m_t(m_t)}{163\,\text{GeV}} \right)^{0.38}. \tag{6.36}$$

These formulas will allow us to exhibit elegantly the m_t dependence of various branching ratios in the phenomenological sections of this book. But as m_t is in 2020 rather precisely known, such representation of the m_t dependence of various observables does not play as important role as it played in 1990, that is, before the top-quark discovery, when the basic gauge independent functions were introduced.

An Example

With the help of the elementary Feynman rules for gauge boson propagators and the effective rules in (6.2) and (6.4), it is an easy matter to construct the effective Hamiltonian for the decay $K^+ \to \pi^+ \bar{\nu}_e \nu_e$ to which the diagrams in Figure 6.5 contribute.

We show this for the top-quark contribution. Neglecting the momentum in the Z^0 propagator in the first diagram, that is, replacing it by $i g_{\mu\nu}/M_Z^2$, multiplying it by the vertex in (6.4) and by "i," we find Z^0-penguin contribution. Subsequently adding the box diagram in (6.2), we find the well-known result for the top contribution to this decay:

$$\mathcal{H}_{\text{eff}}(K^+ \to \pi^+ \nu_e \bar{\nu}_e) = \frac{G_F}{\sqrt{2}} \frac{\alpha}{2\pi \sin^2 \vartheta_W} V_{ts}^* V_{td} \, X_0(x_t) \, (\bar{s}d)_{V-A} (\bar{\nu}_e \nu_e)_{V-A}. \tag{6.37}$$

Here we have expressed the combination $C_0(x_t) - 4B_0(x_t)$ through the function $X_0(x_t)$.

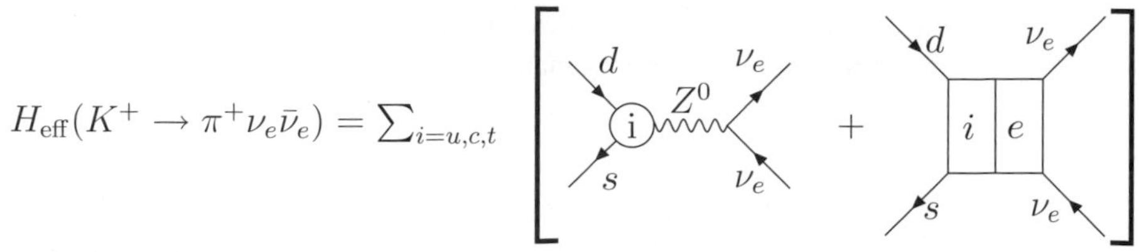

$$H_{\text{eff}}(K^+ \to \pi^+ \nu_e \bar{\nu}_e) = \sum_{i=u,c,t} \left[\quad \right.$$

Figure 6.5   Calculation of $\mathcal{H}_{\text{eff}}(K^+ \to \pi^+ \nu_e \bar{\nu}_e)$. From [29].

Penguin-Box Expansion

One can generalize this calculation to other processes in which other basic effective vertices are present. For decays involving photonic and/or gluonic penguin vertices, the $1/q^2$ in the propagator cancels the q^2 in the vertex, and the resulting effective Hamiltonian can again be written in terms of local four-fermion operators. Thus generally an effective Hamiltonian for any decay considered can be written in the absence of QCD corrections as

$$\mathcal{H}_{\text{eff}}^{\text{FCNC}} = \sum_k C_k O_k, \qquad (6.38)$$

where O_k denote local operators such as $(\bar{s}d)_{V-A}(\bar{s}d)_{V-A}$, $(\bar{s}d)_{V-A}(\bar{u}u)_{V-A}$, etc. The coefficients C_k of these operators are simply linear combinations of the functions in (6.29) times the corresponding CKM factors, which can be read from our rules. Consequently, it is possible to write down the SM amplitudes for all FCNC decays and transitions as linear combinations of the basic, process-independent m_t-dependent functions $F_r(x_t)$ in (6.29) with corresponding coefficients P_r characteristic for the decay under consideration. This "Penguin-Box-Expansion" (PBE) [133] takes the following general form:

$$\boxed{A(\text{decay}) = P_0(\text{decay}) + \sum_r P_r(\text{decay})\, F_r(x_t),} \qquad (6.39)$$

where the sum runs over all possible functions contributing to a given amplitude. P_0 summarizes contributions stemming from internal quarks other than the top, in particular the charm quark. It will be evident from our presentation that the general expansion in (6.39) can be derived from the OPE and is valid also in the presence of QCD corrections with the dominant ones present in coefficients P_0 and P_r. At higher orders in gauge couplings new basic functions can enter, but we will leave them aside for now.

We will encounter many examples of the expansion (6.39) in this book, and generally several basic functions contribute to a given decay. In particular, in the SM we have:

$K^0 - \bar{K}^0$-mixing	$S_0(x_t),\ S_0(x_c, x_t)$
$B^0 - \bar{B}^0$-mixing	$S_0(x_t)$
$K \to \pi\nu\bar{\nu},\ B \to K(K^*)\nu\bar{\nu}$	$X_0(x_t)$
$K_L \to \mu\bar{\mu},\ B_{s,d} \to \ell^+\bar{\ell}^-$	$Y_0(x_t)$
$K_L \to \pi^0 e^+ e^-$	$Y_0(x_t),\ Z_0(x_t),\ E_0(x_t)$
ε'	$X_0(x_t),\ Y_0(x_t),\ Z_0(x_t),\ E_0(x_t)$
$B \to X_{s,d}\,\gamma$	$D_0'(x_t),\ E_0'(x_t)$
$B \to K(K^*)\,\ell^+\bar{\ell}^-$	$Y_0(x_t),\ Z_0(x_t),\ E_0(x_t),\ D_0'(x_t),\ E_0'(x_t)$

In the last part of this book we will see that beyond the SM (BSM) other functions and other Dirac structures (operators) contribute to FCNCs. Moreover, the functions listed here will become in some BSM models complex quantities and the flavor independence of these functions will no longer be valid.

6.1.3 More about the GIM Mechanism

At this stage it is useful to return to the GIM mechanism [6], which did not allow tree-level FCNC transitions. This mechanism is also felt in the Hamiltonian of (6.38), and in fact it is fully effective when the masses of the internal quarks of a given charge in loop diagrams are set to be equal, e.g., $m_u = m_c = m_t$. Indeed, the CKM factors in any FCNC process enter in the combinations

$$C_k \propto \sum_{i=u,c,t} \lambda_i\, F(x_i) \quad \text{or} \quad \sum_{i,j=u,c,t} \lambda_i \lambda_j\, \tilde{F}(x_i, x_j), \tag{6.40}$$

where $F, \tilde{F}$ denote any of the functions of (6.29), and the λ_i are given in the case of K and $B_{s,d}$ meson decays and particle-antiparticle mixing as follows ($i = u, c, t$):

$$\lambda_i^{(K)} = V_{is}^* V_{id}, \qquad \lambda_i^{(d)} = V_{ib}^* V_{id}, \qquad \lambda_i^{(s)} = V_{ib}^* V_{is}. \tag{6.41}$$

They satisfy the unitarity relation (6.22), which implies vanishing coefficients C_k in (6.40) if $x_u = x_c = x_t$. For this reason the mass-independent terms in the calculation of the basic functions in (6.29) can always be omitted. In this limit, FCNC decays and transitions are absent. Thus beyond tree level the conditions for a complete GIM cancellation of FCNC processes are

- Unitarity of the CKM matrix.
- Horizontal flavor symmetry ensuring the equality of quark masses of a given charge.

Now in nature such a horizontal symmetry, even if it could exist at very short distance scales, is certainly broken at low energies by the disparity of masses of quarks of a given charge. This in fact is the origin of the breakdown of the GIM mechanism at the one-loop level and the appearance of FCNC transitions. The size of this breakdown, and consequently the size of FCNC transitions, depends on the disparity of masses, on the behavior of the basic functions in (6.29), and can be affected by QCD corrections as we will see in later chapters. Let us make two observations:

- For small $x_i \ll 1$, relevant for $i \neq t$, the functions (6.11)–(6.19) behave as follows:

$$S_0(x_i) \propto x_i, \quad B_0(x_i) \propto x_i \ln x_i, \quad C_0(x_i) \propto x_i \ln x_i, \tag{6.42}$$

$$D_0(x_i) \propto \ln x_i, \quad E_0(x_i) \propto \ln x_i, \quad D_0'(x_i) \propto x_i, \quad E_0'(x_i) \propto x_i. \tag{6.43}$$

  This implies "hard" (quadratic) GIM suppression of FCNC processes governed by the functions $S_0, B_0, C_0, D_0', E_0'$ provided the top-quark contributions due to small CKM factors can be neglected. In the case of $D_0(x_i)$ and $E_0(x_i)$, only "soft" (logarithmic) GIM suppression is present.
- For large x_t we have

$$S_0(x_t) \propto x_t, \quad B_0(x_t) \propto \text{const}, \quad C_0(x_t) \propto x_t \tag{6.44}$$

$$D_0(x_t) \propto \ln x_t, \quad E_0(x_t) \propto \text{const}, \quad D_0'(x_t) \propto \text{const}, \quad E_0'(x_t) \propto \text{const}. \tag{6.45}$$

Thus for FCNC processes governed by top-quark contributions, the GIM suppression is not effective at the one-loop level and in fact in the case of decays and transitions receiving contributions from $S_0(x_t)$ and $C_0(x_t)$ important enhancements of FCNCs are possible.

The latter property emphasizes the special role of K and $B_{s,d}$ decays with regard to FCNC transitions. In these decays the appearance of the top quark in the loops with $m_t > M_W \gg m_c, m_u$ removes the GIM suppression, making K and $B_{s,d}$ decays a particularly useful place to test FCNC transitions and to study the physics of the top quark. Of course, the hierarchy of various FCNC transitions is also determined by the hierarchy of the elements of the CKM matrix allowing in this manner to perform sensitive tests of this sector of the SM. It is remarkable that this structure is presently compatible within experimental and theoretical uncertainties with the data, although here and there some deviations from theory are observed.

The FCNC decays of D-mesons are much stronger suppressed in the SM because only d, s, and b quarks with $m_d, m_s, m_b \ll M_W$ enter the loops, and the GIM mechanism is much more effective. Also, the known structure of the CKM matrix is less favorable than in K and $B_{s,d}$ decays. For these reasons this book will be dominated by the latter, although some aspects of D decays will be discussed in Chapter 11 as well. In particular in the extensions of the SM, FCNC transitions are possible at the tree level, and the hierarchies discussed here may not apply. Moreover, the fact that FCNCs in D decays are so strongly suppressed in the SM allows in principle to see NP effects easier than in the other meson systems. Unfortunately, hadronic uncertainties in these decays are much larger than in K and $B_{s,d}$ decays so that the latter decays presently dominate flavor physics. Advances in lattice QCD could improve the status of the D meson system, but it will still take some years.

6.1.4 Final Comments

The discussion of FCNCs presented in this section left out completely QCD effects that we discussed in previous chapters in such a great detail in the case of tree-level decays. Still, we hope that this overture to FCNCs has shown the richness of the field of these processes. In the next section explicit calculations of the basic master one-loop functions will be presented. This particular section should not be skipped by students as it teaches them how to do loop calculations by hand. But it is not important for reading the subsequent sections in which using the technology of OPE and RG, developed in previous chapters, we will include QCD effects and electroweak effects to most interesting FCNC transitions and subsequently derive the relevant effective Hamiltonians that will be fundamental for more phenomenological parts of this book.

6.2 Calculations of Basic One-Loop Functions

6.2.1 Explicit Calculation of $\Delta F = 2$ Box Diagrams

The box diagrams contributing to $\Delta F = 2$ processes in the SM are given in Figure 6.6. We have specified the flavors so that these diagrams contribute to $K^0 - \bar{K}^0$ mixing, but analogous diagrams contribute to $B_{s,d}^0 - \bar{B}_{s,d}^0$ mixings. There are also diagrams in which

(a)

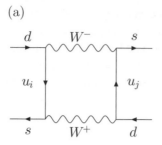

(b)

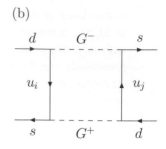

(c)

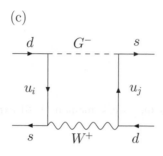

(d)

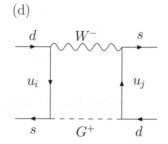

Figure 6.6 $\Delta F = 2$ Box Diagrams. From [29].

the quark lines are on horizontal propagators and W and Goldstone bosons on the vertical ones as the last diagram in the first row in Figure 6.4. But they give the same result, and it is sufficient to consider only the ones shown in Figure 6.6. In the unitary gauge only the first diagram is present, but the integrals, involving $W^\pm$ propagator in this gauge, although doable with the help of the list given in Appendix C.4, are not easy, and we will show here the explicit calculation in the Feynman–'t Hooft gauge $\xi = 1$. Then all diagrams contribute, but the calculation directly uses the first list of integrals in Appendix C and is straightforward. This also stresses the point that in the R_ξ gauges also diagrams with Goldstone bosons have to be included. We set the masses and momenta of external quarks to zero.

WW-Diagram

Using the rules in Appendix B, we find for a given pair of up-quarks ($i, j = u, c, t$)

$$\mathcal{D}_a = \left(\frac{g_2}{2\sqrt{2}} \right)^4 \lambda_i^{(K)} \lambda_j^{(K)} T_{\sigma\tau}^{(1)} R_1^{\sigma\tau}, \qquad \lambda_i^{(K)} = V_{is}^* V_{id}, \tag{6.46}$$

where

$$R_1^{\sigma\tau} = \int \frac{d^4 k}{(2\pi)^4} \frac{k^\sigma k^\tau}{[k^2 - m_i^2][k^2 - m_j^2][k^2 - M_W^2]^2} \tag{6.47}$$

and

$$T_{\sigma\tau}^{(1)} = 4\bar{s}\gamma_\mu\gamma_\sigma\gamma_\nu(1 - \gamma_5)d \otimes \bar{s}\gamma^\nu\gamma_\tau\gamma^\mu(1 - \gamma_5)d. \tag{6.48}$$

Here the factor 4 comes after shifting $(1 - \gamma_5)$ on both sides of $\otimes$ to the right end of the chain and using

$$(1 - \gamma_5)^2 = 2(1 - \gamma_5). \tag{6.49}$$

But $R_1^{\sigma\tau}$ is equal to the integral in (C.26)

$$R_1^{\sigma\tau} = -\frac{ig^{\sigma\tau}}{64\pi^2} \frac{1}{M_W^2} \left[\frac{x_i^2 \log x_i}{(1 - x_i)^2(x_i - x_j)} + \frac{x_j^2 \log x_j}{(1 - x_j)^2(x_j - x_i)} + \frac{1}{(1 - x_i)(1 - x_j)} \right], \tag{6.50}$$

where

$$x_i = \frac{m_i^2}{M_W^2}. \tag{6.51}$$

Contracting $g^{\sigma\tau}$ with $T_{\sigma\tau}$ and using the Greek method [126] explained in (5.160) and (5.161), we find

$$g^{\sigma\tau} T_{\sigma\tau}^{(1)} = 16 \, (\bar{s}d)_{V-A}(\bar{s}d)_{V-A}. \tag{6.52}$$

We are now done with the calculation of $\mathcal{D}_a$ but to find its contribution to the effective Hamiltonian for $\Delta S = 2$ transitions, we have to multiply the result for $\mathcal{D}_a$ by i and divide it by a combinatoric factor 2, which takes into account both the diagram with vertical quark lines replaced by the horizontal lines and the fact that we deal with a product of two identical quark currents. The most straightforward way to find this factor is to look at a tree-level exchange of a heavy neutral particle that mediates a $\Delta F = 2$ transition and integrate it out. We will illustrate it in Section 15.4.3 on the example of a heavy Z' gauge boson.

Using

$$\left(\frac{g_2}{2\sqrt{2}} \right)^4 = \frac{1}{2} G_F^2 M_W^4, \tag{6.53}$$

we finally obtain

$$\mathcal{H}_{WW} = \frac{G_F^2}{16\pi^2} M_W^2 \lambda_i^{(K)} \lambda_j^{(K)} T(x_i, x_j)(\bar{s}d)_{V-A}(\bar{s}d)_{V-A}, \tag{6.54}$$

$$T(x_i, x_j) = \left[\frac{x_i^2 \log x_i}{(1 - x_i)^2(x_i - x_j)} + \frac{x_j^2 \log x_j}{(1 - x_j)^2(x_j - x_i)} + \frac{1}{(1 - x_i)(1 - x_j)} \right], \tag{6.55}$$

$$T(x_i, x_i) = \left[\frac{2x_i \log x_i}{(1 - x_i)^3} + \frac{1 + x_i}{(1 - x_i)^2} \right]. \tag{6.56}$$

GG-Diagram

We next calculate the diagram (b) in Figure 6.6, which contains two Goldstone bosons. We have first

$$\mathcal{D}_b = \frac{G_F^2}{2} \lambda_i^{(K)} \lambda_j^{(K)} T_{\sigma\tau}^{(2)} R_1^{\sigma\tau}, \tag{6.57}$$

with $R_1^{\sigma\tau}$ given in (6.50) and

$$T_{\sigma\tau}^{(2)} = 4m_i^2 m_j^2 \bar{s} \gamma_\sigma (1 - \gamma_5) d \otimes \bar{s} \gamma_\tau (1 - \gamma_5) d. \tag{6.58}$$

Multiplying the result by $i/2$, as explained earlier, we obtain

$$\mathcal{H}_{GG} = \frac{G_F^2}{16\pi^2} M_W^2 \lambda_i^{(K)} \lambda_j^{(K)} \left[\frac{x_i x_j}{4} \right] T(x_i, x_j)(\bar{s}d)_{V-A}(\bar{s}d)_{V-A}, \tag{6.59}$$

with $T(x_i, x_j)$ given in (6.55).

WG-Diagrams

There are two diagrams in Figure 6.6 with one W and one Goldstone boson, but they give the same result. Incorporating this factor of 2 we first have

$$\mathcal{D}_{c+d} = G_F^2 M_W^2 \lambda_i^{(K)} \lambda_j^{(K)} T^{(3)} R_2, \tag{6.60}$$

with

$$R_2 = \int \frac{d^4 k}{(2\pi)^4} \frac{1}{[k^2 - m_i^2][k^2 - m_j^2][k^2 - M_W^2]^2}, \tag{6.61}$$

and

$$T^{(3)} = -4m_i^2 m_j^2 (\bar{s}d)_{V-A}(\bar{s}d)_{V-A}. \tag{6.62}$$

But R_2 is equal to the integral in (C.20)

$$R_2 = -\frac{i}{16\pi^2} \frac{1}{M_W^4} \left[\frac{x_i \log x_i}{(1 - x_i)^2(x_i - x_j)} + \frac{x_j \log x_j}{(1 - x_j)^2(x_j - x_i)} + \frac{1}{(1 - x_i)(1 - x_j)} \right]. \tag{6.63}$$

Multiplying the result for $\mathcal{D}_{c+d}$ by $i/2$ we find

$$\mathcal{H}_{WG} = -\frac{G_F^2}{8\pi^2} M_W^2 \lambda_i^{(K)} \lambda_j^{(K)} x_i x_j \tilde{T}(x_i, x_j)(\bar{s}d)_{V-A}(\bar{s}d)_{V-A}, \tag{6.64}$$

$$\tilde{T}(x_i, x_j) = \left[\frac{x_i \log x_i}{(1 - x_i)^2(x_i - x_j)} + \frac{x_j \log x_j}{(1 - x_j)^2(x_j - x_i)} + \frac{1}{(1 - x_i)(1 - x_j)} \right], \tag{6.65}$$

$$\tilde{T}(x_i, x_i) = \left[\frac{(1 + x_i) \log x_i}{(1 - x_i)^3} + \frac{2}{(1 - x_i)^2} \right]. \tag{6.66}$$

Total Effective Hamiltonian

We add now the contributions in (6.54), (6.59), and (6.64) to obtain the total effective Hamiltonian for $\Delta S = 2$ transitions. We obtain first for fixed (i, j)

$$\mathcal{H}_{\text{tot}}^{ij} = \frac{G_F^2}{16\pi^2} M_W^2 \lambda_i^{(K)} \lambda_j^{(K)} \left[\left(1 + \frac{x_i x_j}{4} \right) T(x_i, x_j) - 2 x_i x_j \tilde{T}(x_i, x_j) \right] (\bar{s}d)_{V-A} (\bar{s}d)_{V-A}.$$

(6.67)

Summing over the internal up-quarks and using

$$\lambda_u^{(K)} = -\lambda_c^{(K)} - \lambda_t^{(K)},$$

(6.68)

that follows from the unitarity of the CKM matrix, we find after some algebra

$$\mathcal{H}_{\text{tot}} = \frac{G_F^2}{16\pi^2} M_W^2 \left[\lambda_c^2 S_0(x_c) + 2 \lambda_c \lambda_t S_0(x_c, x_t) + \lambda_t^2 S(x_t) \right] (\bar{s}d)_{V-A} (\bar{s}d)_{V-A},$$

(6.69)

where we suppressed the index (K) for artistic reasons.

The functions $S_0(x_i)$ and $S_0(x_i, x_j)$ are given in (6.17)–(6.19). They are just linear combinations in (6.20) and (6.21) with

$$F(x_i, x_j) = \left(1 + \frac{x_i x_j}{4} \right) T(x_i, x_j) - 2 x_i x_j \tilde{T}(x_i, x_j).$$

(6.70)

The index 0 indicates that QCD corrections have not been included.

The result in (6.67) can be extended to other meson systems by changing the quark flavors. Moreover, each term should be multiplied by QCD factors η_{cc}, η_{ct}, and η_{tt} to be given later on. We will also see that in $B_{s,d}^0$ systems only the last term in (6.69) is relevant. We will show this in the next section.

6.2.2 Explicit Calculation of $\Delta F = 1$ Box Diagrams

Let us next calculate box diagrams contributing to the transitions

$$d \to s \mu \bar{\mu}, \qquad d \to s \nu \bar{\nu}.$$

(6.71)

The relevant diagram for $\nu \bar{\nu}$ case is shown in Figure 6.4. We see the absence of diagrams with Goldstone bosons. They are present but can be neglected because the approximation of setting the internal lepton masses to zero is very good. An exception is the τ lepton, and we will take this into account later. Setting $m_\nu = m_\mu = 0$ the contributions of Goldstone boson exchanges $\phi^\pm$ are also set to zero, and we are left only with the $W^\pm$ exchanges.

We could proceed as in the case of $\Delta F = 2$ transitions and calculate these diagrams from scratch. But a smarter way to proceed is to get the result for these diagrams from the result in (6.54) directly. This indeed saves time.

$d \to s\mu\bar{\mu}$

Beginning with the first transition we note the following changes relative to the WW-case considered earlier:

- While the integral is the same, we can set $x_j = 0$.
- Next

$$T_{\sigma\tau} = 4\bar{s}\gamma_\mu\gamma_\sigma\gamma_\nu(1 - \gamma_5)d \otimes \bar{\mu}\gamma^\nu\gamma_\tau\gamma^\mu(1 - \gamma_5)\mu \tag{6.72}$$

and consequently

$$g^{\sigma\tau}T_{\sigma\tau} = 16\,(\bar{s}d)_{V-A}(\bar{\mu}\mu)_{V-A}, \tag{6.73}$$

which is exactly the result in (6.52) except that the operator is different.

- CKM factor is $\lambda_i^{(K)}$, and there is no combinatoric factor 1/2 because the second diagram with a horizontal fermion line does not exist, and the fermions in the initial and final state are all different from each other.

Taking these changes into account, we can immediately write down the effective Hamiltonian for the box contribution to the $d \to s\mu\bar{\mu}$ transition, by simply multiplying the effective Hamiltonian in (6.54) by 2, setting $x_j = 0$, removing the CKM factor $\lambda_j^{(K)}$, and changing the operator with the result:

$$\mathcal{H}_{\text{Box}}^{\mu\bar{\mu}} = \frac{G_F^2}{8\pi^2}M_W^2 \sum_{i=u,c,t} \lambda_i^{(K)} T(x_i, 0)(\bar{s}d)_{V-A}(\bar{\mu}\mu)_{V-A}, \tag{6.74}$$

where we took the summation over the internal up-quarks into account. We note next that

$$T(x_i, 0) = 4B_0(x_i) + 1, \tag{6.75}$$

where $B_0(x_i)$ is one of the basic functions given in (6.11). Using the unitarity relation (6.68) and noting that $B_0(x_u) = 0$ to an excellent approximation, we find finally

$$\mathcal{H}_{\text{Box}}^{\mu\bar{\mu}} = \frac{G_F^2}{2\pi^2}M_W^2\left[\lambda_c^{(K)}B_0(x_c) + \lambda_t^{(K)}B_0(x_t)\right](\bar{s}d)_{V-A}(\bar{\mu}\mu)_{V-A}. \tag{6.76}$$

Note that the mass independent term "1" in (6.75) cancelled in the sum as a result of the GIM mechanism. We will see in later sections that in those BSM models in which the GIM mechanism is broken and $W^\pm$ is replaced by a very heavy particles with masses $O(\text{TeV})$ only this term matters, as it is not suppressed by very small x_i in this case.

The overall factor in (6.76) can be rewritten by using the usual tree-level relations between SM parameters:

$$\frac{G_F^2}{2\pi^2}M_W^2 = \frac{G_F}{\sqrt{2}}\frac{\alpha}{2\pi\sin^2\vartheta_W} \equiv \frac{g_{\text{SM}}^2}{4}, \tag{6.77}$$

which is useful when comparing with formulas given in the literature. g_{SM}^2 will appear frequently in the BSM part of this book.

$$d \to s \nu \bar{\nu}$$

Having the result in (6.76), we can easily obtain box diagram contribution to the effective Hamiltonian for $d \to s \nu \bar{\nu}$ transitions. We note that the momentum on the internal lepton propagator has opposite direction to the momentum k_μ. As in the WW box, the masses in the numerator of the fermion propagators do not contribute; this brings an additional overall minus sign. In addition this change implies a different order of gamma matrices so that including this sign we have

$$\tilde{T}_{\sigma\tau} = -4\bar{s}\gamma_\mu\gamma_\sigma\gamma_\nu(1-\gamma_5)d \otimes \bar{\nu}\gamma^\mu\gamma_\tau\gamma^\nu(1-\gamma_5)\nu \tag{6.78}$$

and consequently

$$g^{\sigma\tau}\tilde{T}_{\sigma\tau} = -64\,(\bar{s}d)_{V-A}(\bar{\nu}\nu)_{V-A}. \tag{6.79}$$

Concentrating first on internal e and μ leptons and setting their masses to zero, we obtain then

$$\mathcal{H}_{\text{Box}}^{\nu\bar{\nu}} = [-4]\frac{G_F^2}{2\pi^2}M_W^2\left[\lambda_c^{(K)}B_0(x_c) + \lambda_t^{(K)}B_0(x_t)\right](\bar{s}d)_{V-A}(\bar{\nu}\nu)_{V-A}, \quad (\nu_e, \nu_\mu).$$

$$\tag{6.80}$$

In the case of τ we have to include its mass. We find then

$$\mathcal{H}_{\text{Box}}^{\nu\bar{\nu}} = [-4]\frac{G_F^2}{2\pi^2}M_W^2\left[\lambda_c^{(K)}\tilde{B}_0(x_c) + \lambda_t^{(K)}\tilde{B}_0(x_t)\right](\bar{s}d)_{V-A}(\bar{\nu}\nu)_{V-A}, \quad (\nu_\tau), \tag{6.81}$$

where

$$\tilde{B}_0(x_i) = T(x_i, x_\tau) - T(0, x_\tau), \qquad x_\tau = \frac{m_\tau^2}{M_W^2}, \tag{6.82}$$

with $T(x_i, x_\tau)$ given in (6.55). As the function T is symmetric in the arguments, we have using (6.75)

$$T(0, x_\tau) = 4B_0(x_\tau) + 1, \tag{6.83}$$

and again 1 cancels out in the differences in (6.81). One can check that this modification is numerically relevant only in the term involving x_c so that in the rest of this book we can set

$$\tilde{B}_0(x_t) = B_0(x_t) \tag{6.84}$$

and include the effect of m_τ only in the charm contribution.

6.2.3 Explicit Calculation of Z-Penguin Diagrams

Z-Vertex

Our first goal is the calculation of the function $C_0(x_i)$, which enters the induced flavor-violating vertex $\bar{s}Z^\mu d$

$$\Gamma_Z^\mu \equiv i\frac{g_2^3}{16\pi^2}\frac{1}{\cos\vartheta_W}V_{id}V_{is}^*C_0(x_i)\bar{s}\gamma^\mu(1-\gamma_5)d, \qquad (6.85)$$

with $i = u, c, t$. For B_s and B_d meson systems, the flavors have to be changed, but $C_0(x_i)$ remains unchanged.

The full set of the diagrams contributing to Γ_Z^μ in the $\xi = 1$ gauge is given in Figures 6.7 and 6.8. We will see that several of these diagrams are divergent, but their sum will be finite. To regulate the divergences in the individual diagrams, we will work in $D = 4 - 2\varepsilon$ dimensions. All the integrals necessary to complete this task are collected in Appendix C so that the calculation is easy despite several contributing diagrams.

It will be useful to work with the following form of Z coupling

$$i\bar{f}Z_\mu f = i\frac{g_2}{2\cos\vartheta_W}\gamma_\mu\left[a_f(1+\gamma_5) + b_f(1-\gamma_5)\right], \qquad (6.86)$$

where

$$a_f = -Q_f\sin^2\vartheta_W, \qquad b_f = T_3(f) - Q_f\sin^2\vartheta_W, \qquad (6.87)$$

with f being the internal quark. In the following formulas $f = i$ represents the internal u, c, t. Moreover in the intermediate formulas we will suppress external spinors and the CKM factors, which will be exposed in the final formula for every diagram.

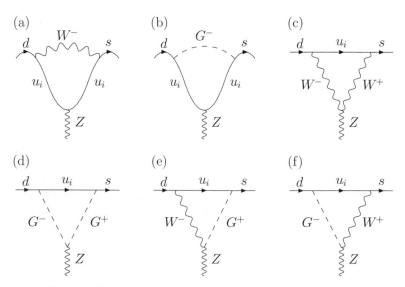

Figure 6.7   Vertex diagrams contributing to Z-penguin.

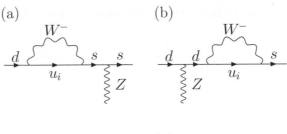

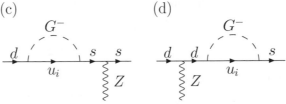

Figure 6.8   Self-energy diagrams contributing to Z-penguin.

The contribution of the diagram (a) is given by

$$\Delta_a \Gamma^\mu(Z) = \frac{g_2^3}{16 \cos \vartheta_W} \int \frac{d^D k}{(2\pi)^D} \frac{k_\rho k_\lambda S_1^{\mu\rho\lambda} + m_i^2 S_2^\mu \cdot}{[k^2 - m_i^2]^2 [k^2 - M_W^2]}, \qquad (6.88)$$

where after shifting $(1 - \gamma_5)$ to the right of all γ_μ matrices we find

$$S_1^{\mu\rho\lambda} = 4 b_i \gamma_\nu \gamma^\rho \gamma^\mu \gamma^\lambda \gamma^\nu (1 - \gamma_5) \qquad (6.89)$$

and

$$S_2^\mu = 4 a_i \gamma_\nu \gamma^\mu \gamma^\nu (1 - \gamma_5) = -8 a_i (1 - \varepsilon) \gamma^\mu (1 - \gamma_5). \qquad (6.90)$$

Note that we have dropped the terms linear in momentum k_μ as they would vanish anyway after performing the integration.

The two integrals can be found in Appendix C:

$$\int \frac{d^D k}{2\pi)^D} \frac{k_\rho k_\lambda}{[k^2 - m_i^2]^2 [k^2 - M_W^2]} = I_4(m_i, m_i, M_W) = \frac{i g_{\rho\lambda}}{32\pi^2} \left[\frac{1}{\bar{\varepsilon}} + \frac{3}{4} + F_1(x_i) \right], \quad (6.91)$$

and

$$\int \frac{d^D k}{2\pi)^D} \frac{1}{[k^2 - m_i^2]^2 [k^2 - M_W^2]} = I_3(m_i, m_i, M_W) = \frac{i}{16\pi^2} \frac{1}{M_W^2} R_1(x_i), \qquad (6.92)$$

where

$$R_1(x_i) = \frac{\log x_i}{(1 - x_i)^2} + \frac{1}{(1 - x_i)}. \qquad (6.93)$$

Now

$$g_{\rho\lambda} S_1^{\mu\rho\lambda} = 16 b_i (1 - 2\varepsilon) \gamma^\mu (1 - \gamma_5), \qquad (6.94)$$

where we kept $O(\varepsilon)$ terms, as they will contribute after the multiplication by $1/\bar{\varepsilon}$ with $\bar{\varepsilon}$ defined in (C.5).

Inserting all these expressions into (6.88), we find

$$\Delta_a \Gamma^\mu(Z) = i\frac{g_2^3}{32\pi^2}\frac{1}{\cos\vartheta_W}\left[b_i\left(\frac{1}{\bar{\varepsilon}} - \frac{1}{4} + F_1(x_i)\right) - a_i x_i R_1(x_i)\right]V_{id}V_{is}^*\bar{s}\gamma^\mu(1-\gamma_5)d,$$

(6.95)

where we dropped terms $O(\varepsilon)$, and the functions F_1 and R_1 are defined in (C.3) and (6.93). The contribution of the diagram (b) is given by

$$\Delta_b \Gamma^\mu(Z) = -x_i\frac{g_2^3}{16\cos\vartheta_W}\int\frac{d^D k}{(2\pi)^D}\frac{k_\rho k_\lambda \tilde{S}_1^{\mu\rho\lambda} + m_i^2\tilde{S}_2^\mu}{[k^2 - m_i^2]^2[k^2 - M_W^2]},$$

(6.96)

where

$$\tilde{S}_1^{\mu\rho\lambda} = 4a_i\gamma^\rho\gamma^\mu\gamma^\lambda(1-\gamma_5), \qquad \tilde{S}_2^\mu = 4b_i\gamma^\mu(1-\gamma_5).$$

(6.97)

The integrals to be evaluated are the same as in the first diagram and

$$g_{\rho\lambda}\tilde{S}_1^{\mu\rho\lambda} = -8a_i(1-\varepsilon)\gamma^\mu(1-\gamma_5).$$

(6.98)

Inserting all these expressions into (6.96), we find

$$\Delta_b \Gamma^\mu(Z) = ix_i\frac{g_2^3}{64\pi^2}\frac{1}{\cos\vartheta_W}\left[a_i\left(\frac{1}{\bar{\varepsilon}} + \frac{1}{4} + F_1(x_i)\right) - b_i x_i R_1(x_i)\right]V_{id}V_{is}^*\bar{s}\gamma^\mu(1-\gamma_5)d.$$

(6.99)

The contribution of the diagram (c) is given by

$$\Delta_c \Gamma^\mu(Z) = -\frac{g_2^3}{8}\cos\vartheta_W\int\frac{d^D k}{(2\pi)^D}\frac{U^\mu}{[k^2 - m_i^2][k^2 - M_W^2]^2},$$

(6.100)

where

$$U_\mu = 2k_\rho\gamma_\nu\gamma^\rho\gamma_\beta V_{\mu\beta\nu}(1-\gamma_5)$$

(6.101)

with the momentum dependent part of the triple gauge boson vertex listed in Appendix B and given by

$$V_{\mu\beta\nu}(0,-k,k) = k_\nu g_{\mu\beta} - 2k_\mu g_{\nu\beta} + k_\beta g_{\mu\nu}.$$

(6.102)

We find then

$$U_\mu = 4k^2\gamma_\mu(1-\gamma_5) + 8(1-\varepsilon)k_\mu k_\beta\gamma^\beta(1-\gamma_5).$$

(6.103)

The two integrals can be found in Appendix C:

$$\int\frac{d^D k}{(2\pi)^D}\frac{k_\mu k_\beta}{[k^2 - m_i^2][k^2 - M_W^2]^2} = I_4(m_i, M_W, M_W) = \frac{ig_{\mu\beta}}{32\pi^2}\left[\frac{1}{\bar{\varepsilon}} + \frac{3}{4} + F_2(x_i)\right],$$

(6.104)

and

$$\int \frac{d^D k}{(2\pi)^D} \frac{k^2}{[k^2 - m_i^2][k^2 - M_W^2]^2} = \frac{i}{8\pi^2} \left[\frac{1}{\bar{\varepsilon}} + \frac{1}{2} + F_2(x_i) \right], \qquad (6.105)$$

with F_2 given in (C.4).

Inserting all these expressions into (6.100), we find

$$\boxed{\Delta_c \Gamma^\mu(Z) = -i \frac{g_2^3}{32\pi^2} \cos \vartheta_W \left[\frac{3}{\bar{\varepsilon}} + \frac{5}{4} + 3F_2(x_i) \right] V_{id} V_{is}^* \bar{s} \gamma^\mu (1 - \gamma_5) d.} \qquad (6.106)$$

The calculation of the vertex diagram (d) is rather easy

$$\Delta_d \Gamma^\mu(Z) = \frac{g_2^3}{4} \frac{\left[\sin^2 \vartheta_W - \cos^2 \vartheta_W \right]}{\cos \vartheta_W} x_i \int \frac{d^D k}{(2\pi)^D} \frac{k^\mu k^\nu}{[k^2 - m_i^2][k^2 - M_W^2]^2} \gamma_\nu (1 - \gamma_5).$$

$$\qquad (6.107)$$

The relevant integral is just the one in (6.104) so that we readily find

$$\boxed{\Delta_d \Gamma^\mu(Z) = i \frac{g_2^3}{128\pi^2} \frac{\left[\sin^2 \vartheta_W - \cos^2 \vartheta_W \right]}{\cos \vartheta_W} x_i \left[\frac{1}{\bar{\varepsilon}} + \frac{3}{4} + F_2(x_i) \right] V_{id} V_{is}^* \bar{s} \gamma^\mu (1 - \gamma_5) d V_{id} V_{is}^*.}$$

$$\qquad (6.108)$$

The final two diagrams in Figure 6.7 are very simple. They give an identical result and are finite. Including a factor of two the contribution of these two diagrams is given by

$$\Delta_{e+f} \Gamma^\mu(Z) = -\frac{g_2^3}{2} \frac{\sin^2 \vartheta_W}{\cos \vartheta_W} \int \frac{d^D k}{(2\pi)^D} \frac{m_i^2}{[k^2 - m_i^2][k^2 - M_W^2]^2} \gamma^\mu (1 - \gamma_5). \qquad (6.109)$$

The integral can be found in (C.10), but in view of other contributions, it is useful to express the result in terms of the functions F_1 and F_2 so that addition of diagrams will be easier if the calculation is done by hand. Otherwise it is irrelevant. We find

$$\boxed{\Delta_{e+f} \Gamma^\mu(Z) = -i \frac{g_2^3}{32\pi^2} \frac{\sin^2 \vartheta_W}{\cos \vartheta_W} x_i \left[F_2(x_i) - F_1(x_i) - \frac{1}{2} \right] V_{id} V_{is}^* \bar{s} \gamma^\mu (1 - \gamma_5) d.}$$

$$\qquad (6.110)$$

This completes the calculation of the vertex. It should be noted that the singularities in (6.95) and (6.106) are cancelled separately due to the unitarity of the CKM matrix when summation over internal quarks is performed. Also, the contribution in (6.110) is finite. But in (6.99) and (6.108) singularities depend on the masses of exchanged quarks, and unitarity of the CKM matrix would only remove them for $m_u = m_c = m_t$, which is clearly not the case. It is also evident that these singularities do not cancel each other so that the sum of five contributions is divergent. We will now demonstrate that these singularities disappear when contributions from the diagrams in Figure 6.8 are taken into account.

Self-Energy Diagrams

We first calculate only the self-energy diagram present in diagram (a) with external momentum p and denote it by

$$K_a = \frac{g_2^2}{2}(1-\varepsilon)\int \frac{d^D k}{(2\pi)^D} \frac{\not{p}+\not{k}}{[(k+p)^2 - m_i^2][k^2 - M_W^2]}(1-\gamma_5). \tag{6.111}$$

The integral is precisely the one in (C.2), and therefore we can right away write down the result remembering that the factor $(1-\varepsilon)$ will modify the constant term in (C.2). Including CKM factor and external spinors, we find

$$K_a = i\frac{g_2^2}{32\pi^2}\left[\frac{1}{\varepsilon} + \frac{1}{4} + F_2(x_i)\right] V_{id}V_{is}^* \,\bar{s}\,\not{p}(1-\gamma_5)d. \tag{6.112}$$

This contribution is finite after summation over internal quarks, and this is also the case of self-energy diagram (b) that gives the same result $K_b = K_a$. But the rescue comes from the diagrams (c) and (d). For (c) we have

$$K_c = \frac{g_2^2}{4}x_i\int \frac{d^D k}{(2\pi)^D} \frac{\not{p}+\not{k}}{[(k+p)^2 - m_i^2][k^2 - M_W^2]}(1-\gamma_5). \tag{6.113}$$

The integral is as in the first self-energy diagram, but the overall factor is different and most important depends on quark masses. We find then

$$K_c = i\frac{g_2^2}{64\pi^2}x_i\left[\frac{1}{\varepsilon} + \frac{3}{4} + F_2(x_i)\right] V_{id}V_{is}^* \,\bar{s}\,\not{p}(1-\gamma_5)d. \tag{6.114}$$

K_c is divergent, and this is also the case of $K_d = K_c$. We will see soon that the divergences in these diagrams remove the ones resulting from vertex diagrams. We will show this in two ways beginning with the heuristic one. We calculate the diagrams in Figure 6.8 by just inserting the results for K_a and K_c into diagrams (a) and (c), respectively. It is sufficient to calculate only these diagrams, as the result is independent of which external line is considered, and the diagram (b) gives the same result as diagram (a) and similar for (c) and (d). As we are doing some kind of renormalization like $Z_2^{1/2}$ there is a factor $1/2$ that cancels the one coming from considering two diagrams.

In calculating the diagrams, the internal s and d propagators are massless, but we have to remember that the Z coupling to down-quarks differs from the one to up-quarks present in vertex diagrams.

We thus find the following additional contributions to Γ_Z^μ

$$\Delta_{a+b}\Gamma^\mu(Z) = -i\frac{g_2^3}{32\pi^2}\frac{b_d}{\cos\vartheta_w}\left[\frac{1}{\varepsilon} + \frac{1}{4} + F_2(x_i)\right] V_{id}V_{is}^* \,\bar{s}\gamma^\mu(1-\gamma_5)d, \tag{6.115}$$

and

$$\Delta_{c+d}\Gamma^\mu(Z) = -i\frac{g_2^3}{64\pi^2}\frac{b_d}{\cos\vartheta_w}x_i\left[\frac{1}{\varepsilon} + \frac{3}{4} + F_2(x_i)\right] V_{id}V_{is}^* \,\bar{s}\gamma^\mu(1-\gamma_5)d. \tag{6.116}$$

Both contributions modify the results from the vertex by finite terms, but the second one cancels in addition the divergence from the diagrams (6.99) and (6.108) so that the resulting Γ_Z^μ is finite.

Final Result for Penguin Diagrams

Adding all contributions, we find Γ_Z^μ in (6.85) with $C_0(x_i)$ given in (6.12). Having this vertex, it is now an easy matter to calculate the full Z-penguin diagram by calculating tree-level diagram with the upper vertex given simply by (6.85) and the lower by (6.86). This is a tree-level calculation that everybody can do by now. Multiplying the result by i, we find the contribution of the Z-penguin to the effective Hamiltonian:

$$
\mathcal{H}_{ff}^Z = \frac{G_F^2}{2\pi^2} M_W^2 \left[\lambda_c^{(K)} C_0(x_c) + \lambda_t^{(K)} C_0(x_t) \right] (\bar{s}d)_{V-A}
$$

$$
\times \left[2T_3(f)(\bar{f}f)_{V-A} - 4Q_f \sin^2 \vartheta_W (\bar{f}f)_V \right], \tag{6.117}
$$

with f being this time quark or lepton with the weak isospin $T_3(f)$ and electric charge Q_f.

It should be remarked that the overall factor in this formula found in the literature is not the preceding one but the one on the r.h.s. of (6.77). At this level both expressions give the same result, but when higher order electroweak interactions are included and one enters the issue of definitions of $\sin^2 \vartheta_W$, similar to definition of α_s, these two versions of $\mathcal{H}_{ff}^Z$ differ by a few percent with the difference cancelled by higher-order electroweak corrections. As G_F and not $\sin^2 \vartheta_W$ is commonly used as input, we prefer to use in this book the overall factor as given in (6.117). Higher-order electroweak corrections have been calculated for a large m_t in [134] and for an arbitrary m_t in [135]. They will be taken into account in Section 9.5. But now let us look closer at (6.117).

In addition to the operators encountered already in (6.76) and (6.80), there is an additional operator with the vector coupling $(\bar{f}f)_V$ that is proportional to Q_f. Its Wilson coeffecient vanishes only for neutrinos but is nonzero for charged leptons and quarks. Such a vector coupling would result if the Z boson would be replaced by the photon, but then also the function C_0 would be replaced by D_0, and some factors would be changed. We will return to the calculation of the photon-penguin diagrams and of D_0 at the end of this section.

Here it is sufficient to state that the functions B_0, C_0, and D_0 are gauge dependent, where gauge dependence is related to the W propagator and only their linear combinations represented by the functions X_0, Y_0, and Z_0:

$$
X_0 = C_0 - 4B_0, \qquad Y_0 = C_0 - B_0, \qquad Z_0 = C_0 + \frac{1}{4}D_0 \tag{6.118}
$$

given in (6.26), (6.27), and (6.28) are gauge independent. The term involving vector coupling just discussed is represented by C_0 in Z_0. The $V - A$ part contributes to X_0 and Y_0. We will not demonstrate it here as this issue has been discussed in detail in [133], where the functions X_0, Y_0, and Z_0 were introduced. These functions are sometimes called *Inami–Lim functions*, but Inami and Lim really worked with B_0, C_0, and D_0, not with X_0, Y_0, and Z_0 that have been introduced in the context of the *Penguin-Box Expansion* in [133] and are used commonly in the literature these days.

A Different Look at the Divergence in Z-Penguins

We will now consider a different approach to the issue of the removal of the singularities in the penguin diagrams. It considers only one-particle irreducible diagrams so that the diagrams in Figure 6.8 are not involved and only our results for self-energies in (6.112) and (6.114) will enter. We follow [116].

The main philosophy of this different, more sophisticated route, is as follows.

We have seen that the self-energy diagrams induce flavor nondiagonal propagation in the (s, d) sector. This propagation must be absent for the renormalized fields if we work in the flavor eigenbasis. This requirement uniquely determines electroweak counterterm that should be added to the Lagrangian resulting from the vertex calculation. Let us show this procedure in explicit terms and demonstrate that also this procedure gives the same function $C_0(x_i)$.

We begin with the kinetic term for d and s quarks in the unrenormalized Lagrangian

$$\mathcal{L}_{\text{kin}} = \bar{d}_L^0 i\partial_\mu \gamma^\mu d_L^0 + \bar{s}_L^0 i\partial_\mu \gamma^\mu s_L^0 \tag{6.119}$$

with s^0 and d^0 being unrenormalized fields.

We perform next field renormalization, allowing for nondiagonal pieces in the (d, s) sector

$$d_L^0 = Z_{11} d_L + Z_{12} s_L, \qquad s_L^0 = Z_{21} d_L + Z_{22} s_L. \tag{6.120}$$

Rewritten in terms of renormalized fields, the Lagrangian in (6.119) reads

$$\mathcal{L}_{\text{kin}} = (Z_{11}\bar{d}_L + Z_{12}\bar{s}_L)i\partial_\mu\gamma^\mu(Z_{11}d_L + Z_{12}s_L) + (Z_{21}\bar{d}_L + Z_{22}\bar{s}_L)i\partial_\mu\gamma^\mu(Z_{21}d_L + Z_{22}s_L). \tag{6.121}$$

Keeping only nondiagonal $\bar{s}d$ terms, we find

$$\mathcal{L}_{\text{kin}}^{sd} = (Z_{11}Z_{12} + Z_{22}Z_{21})\bar{s}_L i\partial_\mu\gamma^\mu d_L \equiv \kappa \bar{s} i\partial_\mu\gamma^\mu(1 - \gamma_5)d, \tag{6.122}$$

where

$$\kappa = \frac{1}{2}(Z_{11}Z_{12} + Z_{22}Z_{21}). \tag{6.123}$$

Next consider the unrenormalized neutral current Lagrangian

$$\mathcal{L}_{\text{NC}} = -\frac{g_2}{4\cos\vartheta_W}\left[1 - \frac{2}{3}\sin^2\vartheta_W\right]\left[\bar{d}^0\gamma_\mu(1 - \gamma_5)d^0 + \bar{s}^0\gamma_\mu(1 - \gamma_5)s^0\right]Z^\mu. \tag{6.124}$$

Rewriting it in terms of renormalized fields and using (6.123), we find for the $\bar{s}d$ term:

$$\Delta\mathcal{L}_{\text{NC}} = -2\kappa\frac{g_2}{4\cos\vartheta_W}\left[1 - \frac{2}{3}\sin^2\vartheta_W\right]\bar{s}\gamma_\mu(1 - \gamma_5)d\,Z^\mu. \tag{6.125}$$

We have omitted right-handed fields, which do not contribute at one-loop level to Z-vertex. This shift in the Lagrangian has to be added to the result coming from the calculation of vertex diagrams and if everything works should be equal to the contribution coming from self-energy diagrams calculated by us earlier. This is in fact the case.

Indeed the expression in (6.122) is really a counter-term added to the initial Lagrangian

$$\mathcal{L}^{sd}_{\text{counter}} = \kappa \bar{s} i \partial_\mu \gamma^\mu (1 - \gamma_5) d, \tag{6.126}$$

which implies the Feynman rule

$$K_{\text{counter}} = i\kappa \bar{s} \, \not{p}(1 - \gamma_5)d. \tag{6.127}$$

The parameter κ can then be determined by requiring

$$K_1 + K_2 + i\kappa \bar{s} \, \not{p}(1 - \gamma_5)d = 0 \tag{6.128}$$

with K_1 and K_2 given in (6.112) and (6.114), respectively. The reader may check that with κ determined in this manner, the shift in (6.125) added to the result from the vertex diagrams implies the finite expression for $C_0(x_i)$ equal to the one calculated previously.

6.2.4 Effective Hamiltonians with Boxes and Z-Penguins

Postponing the discussion of the second operator in (6.117) to the next section, we can already combine the effective Hamiltonians in (6.80) and (6.117) specified to neutrinos to obtain the effective Hamiltonian for the decays $K^+ \to \pi^+ \nu \bar{\nu}$ and $K_L \to \pi^0 \nu \bar{\nu}$, which with the change of flavors can also be used for $B \to K(K^*)\nu\bar{\nu}$. We have

$$\mathcal{H}^{\nu\bar{\nu}}_{\text{eff}} = \frac{G_F^2}{2\pi^2} M_W^2 \left[\lambda_c^{(K)} X_0(x_c) + \lambda_t^{(K)} X_0(x_t) \right] (\bar{s}d)_{V-A}(\bar{\nu}\nu)_{V-A}, \tag{6.129}$$

with X_0 given in (6.26). For $B \to K(K^*)\nu\bar{\nu}$ decays, the first term is strongly suppressed relative to the second one because $X(x_t) \gg X(x_c)$, while the corresponding CKM factors are comparable.

Similarly combining the effective Hamiltonian in (6.76) and the one in (6.117) specified to muons we obtain the effective Hamiltonian for the decay $K_L \to \mu^+\mu^-$, which with the change of flavors can also be used for $B_{s,d} \to \mu^+\mu^-$ decays. We have

$$\mathcal{H}^{\mu\mu}_{\text{eff}} = -\frac{G_F^2}{2\pi^2} M_W^2 \left[\lambda_c^{(K)} Y_0(x_c) + \lambda_t^{(K)} Y_0(x_t) \right] (\bar{s}d)_{V-A}(\bar{\mu}\mu)_{V-A}, \tag{6.130}$$

with Y_0 given in (6.27). For $B_{s,d} \to \mu^+\mu^-$ decays, the first term is strongly suppressed relative to the second one because of $Y(x_t) \gg Y(x_c)$.

Our calculation has been performed within the SM. Formulas for Z-penguin contributions in generic extensions of the SM that involve arbitrary charged fermions, scalars, and gauge bosons but similar to the SM do not generate FCNC mediated by Z at tree level, have been presented in [136]. Because of the gauge invariance also the relevant box contributions are given there. This paper is particularly important for models with RH charged currents as in this case one-loop calculations require even more care than in the presence of only LH currents as illustrated earlier. It is a very useful paper. Other useful papers related to one-loop calculations will be listed in Section 14.6.2.

The inclusion of QCD and electroweak corrections to these Hamiltonians and the related phenomenology will be discussed in Chapter 9.

6.2.5   QED and QCD Penguins

General Comments

The calculation of the QED and QCD penguin diagrams and the derivation of the rules (6.5)–(6.8) is more difficult and longer because in the case of (6.5) and (6.6) one has to keep external momenta and in the case of (6.7) and (6.8) also external masses. Therefore we will not present explicit derivation of these rules but will only indicate a method that we hope will be helpful for readers that decide to do this calculation.

The result for the induced flavor-violating vertex involving photon or gluon can be written in the case of $b \to s$ transitions generally as follows

$$\Gamma_P^\mu = F_1 \bar{s}(q^2 \gamma^\mu - q^\mu \not{q}) P_L b + m_b F_2 \bar{s}(i\sigma^{\mu\nu} q_\nu P_R)b, \tag{6.131}$$

with

$$P_{L,R} = \frac{1}{2}(1 \mp \gamma_5), \qquad \sigma^{\mu\nu} = \frac{i}{2}[\gamma^\mu, \gamma^\nu], \qquad q^\mu = p_b^\mu - p_s^\mu. \tag{6.132}$$

p_b^μ and p_s^μ are four-momenta of the b quark and s quark, respectively. We set $m_s = 0$. We suppress t^a in the case of gluons as it does not play any role in the arguments given next.

To calculate the functions in the rules (6.5)–(6.8), it is useful to set $q^2 = 0$. Using next the identities

$$\not{p}_b \gamma^\mu = -\gamma^\mu \not{p}_b + 2p_b^\mu, \qquad \gamma^\mu \not{p}_s = -\not{p}_s \gamma^\mu + 2p_s^\mu \tag{6.133}$$

and the Dirac equation

$$\bar{s} \not{p}_s = \bar{s}m_s = 0, \qquad \not{p}_b b = m_b b, \tag{6.134}$$

the vertex Γ_P^μ can be expressed in terms of

$$O_1^\mu = 2m_b^2 \gamma^\mu P_L, \qquad O_2^\mu = 2p_b^\mu m_b P_R, \qquad O_3^\mu = 2p_s^\mu m_b P_R. \tag{6.135}$$

We obtain

$$\Gamma_P^\mu = \frac{1}{2}F_1 \bar{s}(O_3^\mu - O_2^\mu)b + \frac{1}{2}F_2 \bar{s}(O_2^\mu + O_3^\mu - O_1^\mu)b. \tag{6.136}$$

Calculating the relevant diagrams in Figure 6.13, we find next

$$\Gamma_P^\mu = i\frac{G_F}{\sqrt{2}} \frac{g}{8\pi^2} \bar{s}(B_1 O_1^\mu + B_2 O_2^\mu + B_3 O_3^\mu), \tag{6.137}$$

with B_i being functions of x_t.

But it turns out that B_i are related through

$$B_1 + B_3 = -(B_1 + B_2), \tag{6.138}$$

which constitutes a useful test of the calculation. Using this relation, we obtain

$$\Gamma_P^\mu = i\frac{G_F}{\sqrt{2}}\frac{g}{8\pi^2}\bar{s}[B_1(O_1^\mu - O_2^\mu - O_3^\mu) + (B_1 + B_2)(O_2^\mu - O_3^\mu)]b, \tag{6.139}$$

where the coupling g is either g_s for QCD or e for QED.

Comparing with (6.136), we find

$$F_1 = -\frac{G_F}{\sqrt{2}}\frac{g}{4\pi^2}(B_1 + B_2), \qquad F_2 = -\frac{G_F}{\sqrt{2}}\frac{g}{4\pi^2}B_1. \tag{6.140}$$

Calculating B_1 and B_2, we find F_1 and F_2 and

$$F_1 = -i\frac{G_F}{\sqrt{2}}\frac{g_s}{4\pi^2} E_0(x_t), \qquad F_2 = i\frac{G_F}{\sqrt{2}}\frac{g_s}{4\pi^2} E_0'(x_t), \tag{6.141}$$

in the case of the gluon penguin and

$$F_1 = -i\frac{G_F}{\sqrt{2}}\frac{e}{4\pi^2} D_0(x_t), \qquad F_2 = i\frac{G_F}{\sqrt{2}}\frac{e}{4\pi^2} D_0'(x_t), \tag{6.142}$$

in the case of the photon penguin.

Inserting these results in (6.131), we reproduce the rules in (6.5)–(6.8).

Useful Relations

The QED and QCD contributions up to color and electric charge factors differ mainly through the absence of nonabelian vertices involving $W^\pm$ and external gauge boson in the QCD case. As the color factors do not enter the function $E_0'(x_t)$, explicit calculation allows to relate the function $D_0'(x_t)$ to $E_0'(x_t)$ and the electric charges of the internal and external quark, Q_u and Q_d, respectively.

First, it is useful to decompose $D_0'(x_t)$ into contribution from abelian and nonabelian diagrams:

$$D_0'(x_t) = [D_0'(x_t)]_{\text{abelian}} + [D_0'(x_t)]_{\text{nonabelian}}. \tag{6.143}$$

We find then

$$[D_0'(x_t)]_{\text{abelian}} = (2Q_d - Q_u) E_0'(x_t), \tag{6.144}$$

and consequently

$$[D_0'(x_t)]_{\text{nonabelian}} = D_0'(x_t) - (2Q_d - Q_u) E_0'(x_t). \tag{6.145}$$

The terms involving Q_d come from the self-energy diagrams. These relations simplify the calculation of the rate for $\mu \to e\gamma$ decay, as we will see in Section 17.2.

6.3 $\Delta F = 2$ Transitions

6.3.1 Preliminaries

We will next investigate the structure of effective Hamiltonians for particle-antiparticle mixing like $K^0 - \bar{K}^0$ mixing and $B_q^0 - \bar{B}_q^0$ mixings with $q = d, s$. These transitions already played a crucial role in the tests of the SM and its extensions for decades. Being all FCNC transitions, they are absent at tree level in the SM and at one-loop level proceed within the SM to an excellent approximation only through box diagrams with internal $W^{\pm}$ and charge $+2/3$ quark exchanges. We have calculated these diagrams in the previous section, and we have to establish now how our results enter mixing observables.

We will begin with the presentation of $B_q^0 - \bar{B}_q^0$ mixings as the effective Hamiltonians for them have simpler structure than the one for $K^0 - \bar{K}^0$ mixing. The point is that $B_q^0 - \bar{B}_q^0$ mixings are to an excellent approximation described by box diagrams with internal top-quark exchanges. The contributions of the internal u and c quarks are only needed to remove with the help of the GIM mechanism mass-independent contributions. We have seen this explicitly in the previous section. Otherwise they can be set to zero due to the smallness of m_u and m_c relative to m_t because the relevant one-loop function increases strongly with the mass of exchanged quarks. In the case of $K^0 - \bar{K}^0$ mixing this increase is compensated partly by the size of the CKM elements and box diagrams with two charm quarks, and diagrams with one charm and one top quark have to be taken into account. We will elaborate on the size of these different contributions after a detailed presentation of $B_q^0 - \bar{B}_q^0$ mixings.

In what follows we will discuss explicitly the effective Hamiltonian for $B_d^0 - \bar{B}_d^0$ mixing. The one for $B_s^0 - \bar{B}_s^0$ mixing can then be obtained by just replacing everywhere the flavor d by s. The effective Hamiltonian for $B_d^0 - \bar{B}_d^0$ mixing, relevant for scales $\mu_b = O(m_b)$, is then given by

$$\mathcal{H}_{\text{eff}}^{\Delta B = 2} = \frac{G_F^2}{16\pi^2} M_W^2 \left(V_{tb}^* V_{td} \right)^2 C_Q(\mu_b) Q(\Delta B = 2) + h.c., \tag{6.146}$$

where

$$Q(\Delta B = 2) = (\bar{b}_\alpha d_\alpha)_{V-A} (\bar{b}_\beta d_\beta)_{V-A} . \tag{6.147}$$

In the absence of QCD corrections (6.146) can be easily derived by calculating the relevant box diagrams as done in Section 6.2.1. Thus the Wilson coefficient C_Q is given at $\mu_W = O(M_W)$ as follows

$$C_Q(\mu_W) = S_0(x_t), \qquad S_0(x_t) = \frac{4x_t - 11x_t^2 + x_t^3}{4(1 - x_t)^2} - \frac{3x_t^3 \ln x_t}{2(1 - x_t)^3}. \tag{6.148}$$

But the matrix element of the operator $Q(\Delta B = 2)$ between $\bar{B}_d^0$ and B_d^0 states is evaluated by lattice methods at scales $\mu_b = O(m_b)$, and we really need $C_Q(\mu_b)$ and not $C_Q(\mu_W)$ that serves only as the initial condition for C_Q at $\mu_W = O(M_W)$ for the RG evolution down to low-energy scale μ_b. Therefore we have to include QCD corrections: first at LO and afterwards at least at NLO to reduce various scale uncertainties and have proper matching to hadronic matrix elements obtained from lattice QCD.

6.3.2 LO Analysis: $B^0_{d,s} - \bar{B}^0_{d,s}$ Mixing

From our previous studies we already know that at LO we only have to calculate the anomalous dimension of the operator $Q(\Delta B = 2)$. In fact we can calculate this anomalous dimension in no time by noting that it is equal to the anomalous dimension of the operator Q_+ considered in previous chapter. Indeed, in the case at hand the Q_1 and Q_2 operators are given by

$$Q_1(\Delta B = 2) = (\bar{b}_\alpha d_\beta)_{V-A}(\bar{b}_\beta d_\alpha)_{V-A}, \qquad Q_2(\Delta B = 2) = Q(\Delta B = 2) . \tag{6.149}$$

and using one of Fierz identities in Appendix A.3 we also have

$$Q_1(\Delta B = 2) = Q(\Delta B = 2). \tag{6.150}$$

Consequently

$$Q_+ = \frac{Q_2 + Q_1}{2} = Q(\Delta B = 2) , \qquad Q_- = \frac{Q_2 - Q_1}{2} = 0 \tag{6.151}$$

and $\gamma_Q = \gamma_+$. In particular, in LO

$$\gamma_Q = \gamma_Q^{(0)} \frac{\alpha_s}{4\pi} , \qquad \gamma_Q^{(0)} = \gamma_+^{(0)} = 4 . \tag{6.152}$$

This in turn implies

$$C_Q(\mu_b) = U^{(0)}(\mu_b, \mu_W)C_Q(\mu_W), \qquad U^{(0)}(\mu_b, \mu_W) = \left[\frac{\alpha_s(\mu_W)}{\alpha_s(\mu_b)} \right]^{\frac{\gamma_Q^{(0)}}{2\beta_0}} , \tag{6.153}$$

where $\beta_0 = 23/3$ for $f = 5$. Thus in LO, with (6.148), the Wilson coefficient $C_Q(\mu_b)$ is given by

$$C_Q(\mu_b) = \left[\frac{\alpha_s(\mu_W)}{\alpha_s(\mu_b)} \right]^{6/23} S_0(x_t). \tag{6.154}$$

Before going to the NLO case, let us calculate the matrix element

$$\langle \bar{B}^0_d | \mathcal{H}^{\Delta B=2}_{\text{eff}} | B^0_d \rangle = \frac{G^2_F}{16\pi^2} M^2_W \left(V^*_{tb} V_{td} \right)^2 C_Q(\mu_b) \langle \bar{B}^0_d | Q(\Delta B = 2)(\mu_b) | B^0_d \rangle, \tag{6.155}$$

where

$$\langle \bar{B}^0_d | Q(\Delta B = 2)(\mu_b) | B^0_d \rangle \equiv \frac{4}{3} B_{B_d}(\mu_b) F^2_{B_d} m_{B_d} \tag{6.156}$$

and F_{B_d} is the B_d-meson decay constant. Both F_{B_d} and $B_{B_d}(\mu_b)$ have been calculated by now with good precision by means of lattice QCD. We quote their values in Table D.3.

The μ_b-dependent parameter $B_B(\mu_b)$ parametrizes the nonperturbative effects in the hadronic matrix element of the operator $Q(\Delta B)$. Its value is $O(1)$. In phenomenological applications it is useful to define two μ_b-independent quantities:

$$\eta_B^{(0)} = \left[\alpha_s(\mu_W)\right]^{6/23}, \qquad \hat{B}_{B_d} = B_{B_d}(\mu_b)\left[\alpha_s(\mu_b)\right]^{-6/23}. \tag{6.157}$$

Then:

$$\langle \bar{B}_d^0|\mathcal{H}_{\text{eff}}^{\Delta B=2}|B_d^0\rangle = \frac{G_F^2}{12\pi^2} M_W^2 \left(V_{tb}^* V_{td}\right)^2 \hat{B}_{B_d} F_{B_d}^2 m_{B_d} \eta_B^{(0)} S_0(x_t). \tag{6.158}$$

For B_s^0 the index d should be replaced by s, but η_B remains unchanged.

We note that there is a leftover μ_W-dependence in η_B and μ_t dependence in $S_0(x_t(\mu_t))$. The latter dependence has been suppressed until now, but we know that quark masses depend on scales and $\mu_t = O(m_t)$. To reduce these scale dependences we have to include NLO corrections.

6.3.3 NLO Analysis: $B_{d,s}^0 - \bar{B}_{d,s}^0$ Mixing

From previous chapters we know that the inclusion of NLO QCD corrections amounts in the case at hand to the calculation of $O(\alpha_s)$ corrections to the box diagrams, matching the result to the effective theory without the top quark and W boson, and finally calculating the two-loop anomalous dimension of the operator $Q(\Delta B = 2)$. Fortunately, we do not have to do the latter calculation because to all orders of perturbation theory the anomalous dimension of $Q(\Delta B = 2)$ equals the one of Q_+. Consequently, only the calculation of $O(\alpha_s)$ corrections to $C_Q(\mu_W)$ is new.

Applying the standard procedure of matching one finds [137]

$$C_Q(\mu_W) = S_0(x_t) + \frac{\alpha_s(\mu_W)}{4\pi}\left[S_1(x_t) + F(\mu_W, \mu_t) + B_t S_0(x_t)\right], \tag{6.159}$$

where

$$F(\mu_W, \mu_t) = \gamma_m^{(0)} x_t \frac{\partial S_0(x_t)}{\partial x_t} \ln\frac{\mu_t^2}{M_W^2} + \frac{\gamma_Q'^{(0)}}{2} \ln\frac{\mu_W^2}{M_W^2} S_0(x_t), \tag{6.160}$$

$$B_t = 5\frac{N-1}{2N} + 3\frac{N^2-1}{2N} \qquad \text{(NDR)}, \tag{6.161}$$

$$S_1(x_t) = \text{Complicated Function}. \tag{6.162}$$

The function $S_1(x_t)$ given in (XII.12) of [111] is a result of a two-loop calculation [137] involving gluon corrections to the box diagrams. Typical diagrams are shown in Figure 6.9. The interested reader should consult the detailed analysis in [137], where a spectacular cancellation of infrared divergences and gauge dependences present in the diagrams of the full theory is achieved by the corresponding diagrams in the effective theory.

Note that in addition to the function $S_1(x_t)$, the QCD corrections in (6.159) involve one term proportional to the derivative of the leading term $S_0(x_t)$ with respect to x_t and two terms proportional to $S_0(x_t)$. This structure allows us to expect that these terms remove at

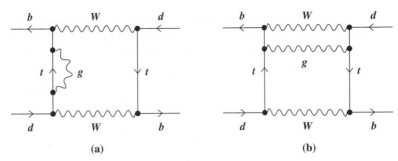

Figure 6.9 Examples of two-loop diagrams contributing to $B_d^0 - \bar{B}_d^0$ mixing. From [29].

this order some nonphysical dependences present in the LO term. Indeed the first log in $F(\mu_W, \mu_t)$ cancels the μ_t dependence in $S_0(x_t)$. The second logarithm in (6.160) cancels the μ_W dependence present in $U^{(0)}(\mu_b, \mu_W)$. For $\mu_W = \mu_t$ the formulas just given reduce to the ones given in [137] and [111]. But as discussed already there, μ_W and μ_t can differ from each other, and for pedagogical reasons we do not put them equal here. The role of the last term in (6.159) will be exhibited in a moment.

The NLO evolution function is given simply by

$$U(\mu_b, \mu_W) = \left[1 + \frac{\alpha_s(\mu_b)}{4\pi} J_5\right] U^{(0)}(\mu_b, \mu_W) \left[1 - \frac{\alpha_s(\mu_W)}{4\pi} J_5\right], \tag{6.163}$$

with

$$J_5 = J_+ = 1.627 \qquad (\text{NDR}, \ f = 5). \tag{6.164}$$

J_5 is renormalization scheme dependent, but this dependence is canceled by the one of B_t.

We can now define μ_b and μ_W independent quantities at the NLO level, which moreover are renormalization scheme independent:

$$\eta_B = \left[\alpha_s(\mu_W)\right]^{6/23} \left[1 + \frac{\alpha_s(\mu_W)}{4\pi}\left(\frac{S_1(x_t)}{S_0(x_t)} + F(\mu_W, \mu_t) + B_t - J_5\right)\right], \tag{6.165}$$

$$\boxed{\hat{B}_{B_d} = B_{B_d}(\mu_b) \left[\alpha_s^{(5)}(\mu_b)\right]^{-6/23} \left[1 + \frac{\alpha_s^{(5)}(\mu_b)}{4\pi} J_5\right].} \tag{6.166}$$

Then: [137]

$$\boxed{\mathcal{H}_{\text{eff}}^{\Delta B = 2} = \frac{G_F^2}{16\pi^2} M_W^2 \left(V_{tb}^* V_{td}\right)^2 \eta_B S_0(x_t) Q(\Delta B = 2) + h.c.,} \tag{6.167}$$

and

$$\boxed{\langle \bar{B}_d^0 | \mathcal{H}_{\text{eff}}^{\Delta B = 2} | B_d^0 \rangle = \frac{G_F^2}{12\pi^2} M_W^2 \left(V_{tb}^* V_{td}\right)^2 \hat{B}_{B_d} F_{B_d}^2 m_{B_d} \eta_B S_0(x_t).} \tag{6.168}$$

It should be noted that both η_B and $S_0(x_t)$ depend on μ_t but the product $\eta_B \cdot S_0(x_t)$ is μ_t-independent in $O(\alpha_s)$ as the first logarithm in (6.160) cancels the μ_t dependence in $S_0(x_t(\mu_t))$. If one varies μ_t in the range 100 GeV $\leq \mu_t \leq 300$ GeV, the μ_t dependence of $\langle \bar{B}^0 | \mathcal{H}_{\text{eff}}^{\Delta B=2} | B^0 \rangle$ amounts in LO to $\pm 9\%$ and is reduced to $\pm 1\%$ in NLO.

It is customary to evaluate η_B at $\mu_t = \mu_W = m_t$, then practically η_B is independent of m_t and the full m_t dependence of $B_d^0 - \bar{B}_d^0$ mixing resides in $S_0(x_t)$ with $m_t(m_t)$. Then using these formulas one finds [137, 138]

$$\eta_B = 0.5510 \pm 0.0022, \tag{6.169}$$

where the error includes also the leftover scale uncertainties, which can only be reduced by calculating $O(\alpha_s^2)$ corrections.

6.3.4 LO and NLO Analysis: $K^0 - \bar{K}^0$ Mixing

The case of $K^0 - \bar{K}^0$ mixing is more involved. Investigating the CKM factors in the box diagrams in Figure 6.10, we observe that we cannot neglect this time the charm contributions even after the GIM mechanism has been used. Simply, the CKM factors in diagrams involving u and c quarks compensate significantly the mass suppression of these contributions relative to the top contribution that, although still enhanced by the large top-quark mass, is suppressed by small CKM factors. Indeed, after the unitarity of the CKM matrix has been used, the relevant CKM factors are

$$\lambda_c = V_{cs}^* V_{cd} = O(\lambda), \qquad V_{ts}^* V_{td} = O(\lambda^5), \tag{6.170}$$

and consequently the three terms λ_c^2, $\lambda_c \lambda_t$, and λ_t^2 have to be kept. We will see that their importance will depend on whether we will consider CP-conserving or CP-violating observables.

The second difficulty relative to $B_{d,s}^0 - \bar{B}_{d,s}^0$ mixings is the calculation of QCD corrections to the parts involving charm. The point is that the charm quark cannot be integrated out together with W or top quark at scales $O(\mu_W)$. Consequently, after W and top quark have been integrated out, the operators relevant for scales $m_c \leq \mu \leq M_W$, the so-called bilocal operators, have a more complicated structure than the operator in (6.147). Therefore, the renormalization group evolution for these scales for charm-charm and charm-top contributions is much more complicated than the one for $B_{d,s}^0 - \bar{B}_{d,s}^0$ mixings, and we

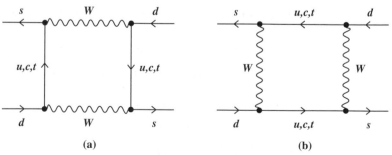

(a)   (b)

Figure 6.10   Box diagrams contributing to $K^0 - \bar{K}^0$ mixing in the SM. From [29].

cannot present it here. But, for scales $\mu < \mu_c = O(m_c)$, at which the charm quark is absent in the effective theory, we again find a single operator

$$Q(\Delta S = 2) = (\bar{s}d)_{V-A}(\bar{s}d)_{V-A}. \qquad (6.171)$$

On the other hand, the renormalization group evolution in the pure top-quark part has the same structure as in $B^0_{d,s} - \bar{B}^0_{d,s}$ mixings. Only the evolution does not stop at μ_b but is continued down to $\mu < \mu_c$, implying a different value of the Wilson coefficient of $Q(\Delta S = 2)$ that in addition receives contributions from charm.

The final effective Hamiltonian for $\Delta S = 2$ transitions for $\mu < \mu_c = O(m_c)$ is given then as follows [137]

$$\mathcal{H}_{\text{eff}}^{\Delta S=2} = \frac{G_F^2}{16\pi^2} M_W^2 \left[\lambda_c^2 \eta_1 S_0(x_c) + \lambda_t^2 \eta_2 S_0(x_t) + 2\lambda_c \lambda_t \eta_3 S_0(x_c, x_t) \right]$$

$$\times \left[\alpha_s^{(3)}(\mu) \right]^{-2/9} \left[1 + \frac{\alpha_s^{(3)}(\mu)}{4\pi} J_3 \right] Q(\Delta S = 2) + h.c. \qquad (6.172)$$

It contains a charm-, a top-, and a mixed charm-top contribution. This form is obtained upon eliminating λ_u by means of the unitarity of the CKM matrix and setting $x_u = 0$. The function $S_0(x_i)$ has already been given in (6.17), and $S_0(x_c, x_t)$ is given in (6.19).

Short-distance QCD effects are described through the correction factors η_1, η_2, η_3, and the explicitly α_s-dependent terms in (6.172). η_2 is the analogue of η_B discussed already. All three η_i have been evaluated at the NLO level a long time ago in [137, 139–142]. For η_1 and η_3 also NNLO corrections have been calculated in [143, 144].

η_{1-3} are defined in analogy to (6.165). This means that in $O(\alpha_s)$ they are independent of the renormalization scales and the renormalization scheme for the operator $Q(\Delta S)$. The present best values of η_i are given as follows (see however [1317])

$$\eta_1 = 1.87(76), \qquad \eta_2 = 0.5765(65), \qquad \eta_3 = 0.496(47). \qquad (6.173)$$

The quoted errors reflect the remaining theoretical uncertainties due to leftover μ-dependences at $O(\alpha_s^2)$ in η_2 and $O(\alpha_s^3)$ in $\eta_{1,2}$. The larger uncertainties in $\eta_{1,3}$ than η_2 reflect the fact that due to charm contribution lower scales in the process of matching are involved than in η_2.

We next define in analogy to (6.166), the renormalization group invariant parameter $\hat{B}_K$ by

$$\hat{B}_K = B_K(\mu) \left[\alpha_s^{(3)}(\mu) \right]^{-2/9} \left[1 + \frac{\alpha_s^{(3)}(\mu)}{4\pi} J_3 \right], \qquad (6.174)$$

with

$$J_3 = J_+ = 1.895 \qquad (\text{NDR}, \ f = 3). \qquad (6.175)$$

We will summarize the status of $\hat{B}_K$ in Chapter 7.

Then

$$\langle \bar{K}^0|(\bar{s}d)_{V-A}(\bar{s}d)_{V-A}|K^0\rangle \equiv \frac{4}{3}B_K(\mu)F_K^2 m_K, \qquad (6.176)$$

where F_K is the K-meson decay constant and m_K the K-meson mass.

Finally, using (6.172), one finds

$$\langle \bar{K}^0|\mathcal{H}_{\text{eff}}^{\Delta S=2}|K^0\rangle = \frac{G_F^2}{12\pi^2}F_K^2 \hat{B}_K m_K M_W^2\left[\lambda_c^2 \eta_1 S_0(x_c) + \lambda_t^2 \eta_2 S_0(x_t) + 2\lambda_c\lambda_t\eta_3 S_0(x_c,x_t)\right].$$
$$(6.177)$$

6.3.5 Summary

The most important formulas in this section are the final formulas for the matrix elements of effective Hamiltonians in (6.168) and (6.177). They will play a fundamental role in tests of the SM by means of $B_{d,s}^0 - \bar{B}_{d,s}^0$ and $K^0 - \bar{K}^0$ mixings in later steps of our expedition. But also the corresponding formulas for effective Hamiltonians in (6.167) and (6.172) will be important when we will enter the discussion of BSM physics.

6.4 The World of Penguins

6.4.1 Preliminaries

Until now we have visited in detail only the world of four-quark current-current operators Q_1 and Q_2 and just before the world of four-quark operators responsible for $B_{s,d}^0 - \bar{B}_{s,d}^0$ and $K^0 - \bar{K}^0$ mixings. We have presented effective Hamiltonians for various meson systems that involved these operators. Both LO and NLO effects in renormalization group improved perturbation theory have been taken into account, and various scale and renormalization scheme dependences including their cancelations have been discussed in detail. But as we have seen in Section 6.2.2, these operators are not the whole story, and we have to generalize the effective Hamiltonians to include many more operators already in the SM. In BSM models still more operators will be present but for the time being let us confine our discussion to the SM.

In principle the presence of new operators could have an impact on the Wilson coefficients of Q_1 and Q_2, but in fact this does not happen, and the values of $C_1(\mu)$ and $C_2(\mu)$, which we calculated in previous sections, will not be modified by the presence of new operators. This is clearly good news. But more good news is that many of the general properties related to scale and renormalization scheme dependence can be taken over to the more complicated situations where in addition to Q_1 and Q_2 other operators are present. In this and the next three sections we will discuss

- QCD penguin and electroweak penguin operators involving four quarks (this section), which enter the analysis not only of nonleptonic decays of mesons but through QCD effects and electroweak effects also those of semileptonic decays of mesons.
- Dipole operators relevant for $B \to X_s \gamma$ decay (Section 6.5).
- Semileptonic four-fermion operators involving a quark current and a leptonic current like $(\ell^+ \ell^-)_V$ and $(\ell^+ \ell^-)_A$, which enter the analyses of $B \to K(K^*)\ell^+\ell^-$, $B_{s,d} \to \ell^+\ell^-$ and $K_{L,S} \to \ell^+\ell^-$ decays (Section 6.6).
- Semileptonic operators involving a quark current and $(\nu\bar{\nu})_{V-A}$, which enter the analyses of $K \to \pi\nu\bar{\nu}$ and $B \to K(K^*)\nu\bar{\nu}$ decays (Section 6.7).

Except for an explicit calculation of one-loop anomalous dimensions of four-quark current-current and QCD penguin operators in this section, we will mainly discuss new features skipping often derivations. Indeed, these four sections should be considered as a guide to the weak effective Hamiltonians that we will use for the SM phenomenology in later chapters of this book, where further details on the physics behind them will be given. We should stress again that this and the following three sections deal exclusively with effective Hamiltonians in the SM. In the generalizations of the SM often new operators have to be included, and the effective Hamiltonians look more complicated. We will devote to them additional sections at later stages of this book stressing new features of these new operators, in particular those containing right-handed currents. Also scalar and tensor operators will sometimes play a significant role.

Finally, charged lepton flavor-violating decays involving in most cases only leptons is still another story. The relevant formulas, mostly given in the extensions of the SM, will be presented in Section 17.2. In the SM such processes are suppressed to a level that they cannot be measured even in the next decades, and an observation of any of these decays would automatically signal NP beyond the SM.

6.4.2 QCD Penguins

Operators

We will now have a close look at QCD penguin operators that contribute to nonleptonic $\Delta F = 1$ transitions. They originate in the gluon penguin diagram (a) of Figure 6.11. Evaluating this diagram one can clearly see that there are two color structures as in the case of Q_1 and Q_2. They follow simply from the decomposition

$$t^a_{\alpha\beta} t^a_{\gamma\delta} = -\frac{1}{2N}\delta_{\alpha\beta}\delta_{\gamma\delta} + \frac{1}{2}\delta_{\alpha\delta}\delta_{\gamma\beta}, \tag{6.178}$$

where t^a enters the two quark-gluon vertices. This formula is just (5.22) of Section 5.1.3, which we used to prove the existence of the operator Q_1. But as we will see soon, due to the presence of gluon and later photon exchanges and different topology of penguin diagrams, the numbers of penguin operators at the end will amount to eight: four QCD penguin operators and four electroweak penguin operators. The reason is as follows.

The upper effective FCNC vertex in the gluon penguin diagram has $V - A$ structure as can be verified by using the Feynman rules without fully calculating the diagram. This

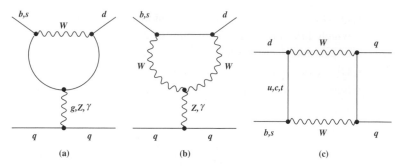

Figure 6.11 One-loop penguin and box diagrams in the full theory. From [29].

fact is related to $V - A$ structure of charged current interactions in the SM mediated by the W. The lower vertex is vectorial (V), and with the two color structures in (6.178) we have two $(V - A) \otimes V$ operators. It turns out, however, that the renormalization of these two operators requires the introduction of two new operators with the Dirac structure $(V - A) \otimes A$ and the color structures as in (6.178). Indeed, when the $(V - A) \otimes V$ operators are inserted into the one-loop diagrams of Figure 6.12 used to calculate anomalous dimensions, two new operators in question are generated. The full set of operators that closes under renormalization consists then of two current-current operators (Q_1, Q_2) and the four QCD penguin operators. It is customary to work with the $(V - A) \otimes (V - A)$ and $(V - A) \otimes (V + A)$ penguin operators rather than with the $(V - A) \otimes V$ and $(V - A) \otimes A$ structures. Then the basis of the operators necessary for the description of $\Delta B = 1$ decays with $\Delta S = 1$ is given (in the limit $\alpha \equiv \alpha_{\text{QED}} = 0$) as follows:

Current-current:

$$Q_1^u = (\bar{u}_\alpha b_\beta)_{V-A} \, (\bar{s}_\beta u_\alpha)_{V-A}, \quad Q_2^u = (\bar{u}b)_{V-A} \, (\bar{s}u)_{V-A}, \tag{6.179}$$

$$Q_1^c = (\bar{c}_\alpha b_\beta)_{V-A} \, (\bar{s}_\beta c_\alpha)_{V-A}, \quad Q_2^c = (\bar{c}b)_{V-A} \, (\bar{s}c)_{V-A}. \tag{6.180}$$

QCD penguins:

$$Q_3 = (\bar{s}b)_{V-A} \sum_{q=u,d,s,c,b} (\bar{q}q)_{V-A}, \quad Q_4 = (\bar{s}_\alpha b_\beta)_{V-A} \sum_{q=u,d,s,c,b} (\bar{q}_\beta q_\alpha)_{V-A}, \tag{6.181}$$

$$Q_5 = (\bar{s}b)_{V-A} \sum_{q=u,d,s,c,b} (\bar{q}q)_{V+A}, \quad Q_6 = (\bar{s}_\alpha b_\beta)_{V-A} \sum_{q=u,d,s,c,b} (\bar{q}_\beta q_\alpha)_{V+A}. \tag{6.182}$$

The corresponding operators for other B decays and the D and K decays can be obtained from this basis by an appropriate change of flavors. We will meet these operators later on, as well as the electroweak penguin operators, and they will have to wait for us.

Effective Hamiltonian

The effective Hamiltonian for $\Delta B = 1$ decays with $\Delta S = 1$ is given then by

$$
\mathcal{H}_{\text{eff}}(\Delta B = 1) = \frac{G_F}{\sqrt{2}} \left[\lambda_u (C_1(\mu_b) Q_1^u + C_2(\mu_b) Q_2^u) + \lambda_c (C_1(\mu_b) Q_1^c + C_2(\mu_b) Q_2^c) \right.
$$
$$
\left. - \lambda_t \sum_{i=3}^{6} C_i(\mu_b) Q_i \right], \qquad \lambda_q = V_{qs}^* V_{qb}, \tag{6.183}
$$

where the CKM factors λ_q are just complex conjugates of the CKM factors used in $\Delta B = 2$ processes. This is merely the result of the definition of operators without any physical implications but should be kept in mind. To simplify notation we will not express λ_q in terms of the CKM factors used for $\Delta B = 2$ processes.

The reason why the operators $Q_{1,2}^q$ with $q = u, c$ have to be included is simple. The internal fermion line in the gluon penguin diagram is a quark with charge $+2/3$ and consequently u, c, and t quarks have to be included. But at scale $\mu < m_t$ the top quark has been integrated out and does not appear as a dynamical degree of freedom so that $Q_{1,2}^t$ are absent in the effective Hamiltonian. The presence of the top quark is signaled by the last term in (6.183), which is simply obtained by calculating, as discussed in Section 6.2, the gluon penguin diagram (a) of Figure 6.11 with internal top quark. Indeed λ_t is just the CKM factor in this diagram.

Wilson Coefficients

The calculation of the Wilson coefficients $C_i(\mu_b)$ of the QCD penguin operators proceeds as outlined in previous sections. The matching at $\mu_W = M_W$ gives, in the presence of the penguin diagrams, the values of $\vec{C}(M_W)$. In the NDR scheme they are given by [123]:

$$
C_1(M_W) = \frac{11}{2} \frac{\alpha_s(M_W)}{4\pi}, \tag{6.184}
$$

$$
C_2(M_W) = 1 - \frac{11}{6} \frac{\alpha_s(M_W)}{4\pi}, \tag{6.185}
$$

$$
C_3(M_W) = -\frac{\alpha_s(M_W)}{24\pi} \widetilde{E}_0(x_t), \tag{6.186}
$$

$$
C_4(M_W) = \frac{\alpha_s(M_W)}{8\pi} \widetilde{E}_0(x_t), \tag{6.187}
$$

$$
C_5(M_W) = -\frac{\alpha_s(M_W)}{24\pi} \widetilde{E}_0(x_t), \tag{6.188}
$$

$$
C_6(M_W) = \frac{\alpha_s(M_W)}{8\pi} \widetilde{E}_0(x_t), \tag{6.189}
$$

where

$$\widetilde{E}_0(x_t) = E_0(x_t) - \frac{2}{3} \tag{6.190}$$

$$E_0(x_t) = -\frac{2}{3}\ln x_t + \frac{x_t(18 - 11x_t - x_t^2)}{12(1 - x_t)^3} + \frac{x_t^2(15 - 16x_t + 4x_t^2)}{6(1 - x_t)^4}\ln x_t, \tag{6.191}$$

with

$$x_t = \frac{m_t^2}{M_W^2}. \tag{6.192}$$

$C_{1,2}(M_W)$ are simply obtained using (5.90) and (5.236). We will derive the QCD penguin coefficients $C_i(M_W)$ ($i = 4 - 6$) soon. The constant $-2/3$ in (6.190) is characteristic for the NDR scheme. It is absent in the HV scheme. In LO $C_2(M_W) = 1$ with all remaining coefficients set to zero. We observe that the m_t-dependence in the case at hand enters first at the NLO level. It results from the gluon penguin diagram with internal top-quark propagators considered in Section 6.2.

The anomalous dimension matrix is 6×6:

$$\hat{\gamma}_s(\alpha_s) = \hat{\gamma}_s^{(0)}\frac{\alpha_s}{4\pi} + \hat{\gamma}_s^{(1)}\left(\frac{\alpha_s}{4\pi}\right)^2. \tag{6.193}$$

The one-loop coefficient $\hat{\gamma}_s^{(0)}$ is given for $N = 3$ by [145]

$$\hat{\gamma}_s^{(0)} = \begin{pmatrix} -2 & 6 & 0 & 0 & 0 & 0 \\ 6 & -2 & \frac{-2}{9} & \frac{2}{3} & \frac{-2}{9} & \frac{2}{3} \\ 0 & 0 & \frac{-22}{9} & \frac{22}{3} & \frac{-4}{9} & \frac{4}{3} \\ 0 & 0 & 6 - \frac{2f}{9} & -2 + \frac{2f}{3} & \frac{-2f}{9} & \frac{2f}{3} \\ 0 & 0 & 0 & 0 & 2 & -6 \\ 0 & 0 & \frac{-2f}{9} & \frac{2f}{3} & \frac{-2f}{9} & -16 + \frac{2f}{3} \end{pmatrix}. \tag{6.194}$$

The explicit calculation of this matrix will be presented next.

The two-loop anomalous dimension matrix $\hat{\gamma}_s^{(1)}$ in the NDR scheme looks truly horrible:

$$\begin{pmatrix} -\frac{21}{2} - \frac{2f}{9} & \frac{7}{2} + \frac{2f}{3} & \frac{79}{9} & -\frac{7}{3} & -\frac{65}{9} & -\frac{7}{3} \\ \frac{7}{2} + \frac{2f}{3} & -\frac{21}{2} - \frac{2f}{9} & -\frac{202}{243} & \frac{1354}{81} & -\frac{1192}{243} & \frac{904}{81} \\ 0 & 0 & -\frac{5911}{486} + \frac{71f}{9} & \frac{5983}{162} + \frac{f}{3} & -\frac{2384}{243} - \frac{71f}{9} & \frac{1808}{81} - \frac{f}{3} \\ 0 & 0 & \frac{379}{18} + \frac{56f}{243} & -\frac{91}{6} + \frac{808f}{81} & -\frac{130}{9} - \frac{502f}{243} & -\frac{14}{3} + \frac{646f}{81} \\ 0 & 0 & \frac{-61f}{9} & \frac{-11f}{3} & \frac{71}{3} + \frac{61f}{9} & -99 + \frac{11f}{3} \\ 0 & 0 & \frac{-682f}{243} & \frac{106f}{81} & -\frac{225}{2} + \frac{1676f}{243} & -\frac{1343}{6} + \frac{1348f}{81} \end{pmatrix}.$$

$$\tag{6.195}$$

The corresponding matrix in the HV scheme can be found in [122, 146]. These two-loop matrices have been first calculated in the latter papers and in [124, 125]. The result in the NDR scheme has been confirmed subsequently in [147].

With all these results at hand one can now evaluate $C_i(\mu_b)$ by using

$$\vec{C}(\mu_b) = \hat{U}_5(\mu_b, M_W)\vec{C}(M_W) \tag{6.196}$$

with $\hat{U}_5(\mu_b, M_W)$ given in (5.113). With the help of Mathematica we can then find

$$C_j(\mu_b) = C_j^{(0)}(\mu_b) + \frac{\alpha_s(\mu_b)}{4\pi}C_j^{(1)}(\mu_b), \tag{6.197}$$

where

$$C_j^{(0)}(\mu_b) = \sum_{i=3}^{8} k_{ji}\eta^{a_i}, \tag{6.198}$$

$$C_j^{(1)}(\mu_b) = \sum_{i=3}^{8}[e_{ji}\eta E_0(x_t) + f_{ji} + g_{ji}\eta]\eta^{a_i} \tag{6.199}$$

with

$$\eta = \left[\frac{\alpha_s(M_W)}{\alpha_s(\mu_b)}\right]. \tag{6.200}$$

The magic numbers a_i, k_{ij}, e_{ij}, f_{ij}, and g_{ij} are collected in tables 6 and 7 of my Les Houches lectures [29], but we will not reproduce them here as eventually the calculation has to be done numerically anyway. But the structure of the preceding formulas gives us some idea about various dependences that clearly cannot be seen by performing numerical calculations.

Matching Conditions for QCD Penguins

It is instructive to derive the matching conditions for QCD penguin operators in (6.186)–(6.189). In particular it is useful to see how the scheme-dependent constant $-2/3$ is generated. After all in Chapter 2 we have briefly discussed what happens to GIM mechanism beyond the tree level. Looking at equation (2.65) it is evident that all mass independent constants in the evaluation of penguin vertices involving m_t-dependent functions like $E_0(x_t)$ can be dropped because of GIM mechanism. Yet as we will see in a moment such statements are valid only in the full theory. In the effective theory the top quark is absent as a dynamical degree of freedom, GIM is no longer valid in certain parts of the theory, and constants like $-2/3$ remain.

To demonstrate this explicitly let us consider first the tree-level Hamiltonian for $\Delta B = 1$ decays:

$$\mathcal{H}_{\text{eff}}^{(0)}(\Delta B = 1) = \frac{G_F}{\sqrt{2}}\left[\lambda_u Q_2^u + \lambda_c Q_2^c + \lambda_t Q_2^t\right], \tag{6.201}$$

in which only current-current operators Q_2^i are present. Note the appearance of the operator Q_2^t.

Next let us include QCD corrections and perform the matching of the full theory to an effective five-quark theory in which the top quark is no longer a dynamical degree of freedom. Because we are only interested in the penguin coefficients we can leave out the QCD corrections to Q_2^q operators and also drop the Q_1^q operators. Calculating then the usual QCD penguin diagram (a) in Figure 6.11 with full $W^\pm$ and internal u, c, t quarks and adding the result to the tree-level matrix element of the Hamiltonian (6.201), we find the amplitude in the full theory:

$$
\mathcal{A}_{\text{full}} = \frac{G_F}{\sqrt{2}} \left(\lambda_u \left[\langle Q_2^u \rangle^0 - \frac{\alpha_s(M_W)}{8\pi} G_u(m_u) \langle Q_P \rangle^0 \right] \right.
$$
$$
\left. + \lambda_c \left[\langle Q_2^c \rangle^0 - \frac{\alpha_s(M_W)}{8\pi} G_c(m_c) \langle Q_P \rangle^0 \right] + \lambda_t \left[-\frac{\alpha_s(M_W)}{8\pi} E_0(x_t) \langle Q_P \rangle^0 \right] \right).
$$
$$
(6.202)
$$

Note that as a preparation for the matching we have already removed the tree-level matrix element of Q_2^t in which the top-quark field is a dynamical degree of freedom. Next

$$
Q_P = Q_4 + Q_6 - \frac{1}{3}(Q_3 + Q_5), \tag{6.203}
$$

where Q_i with $i = 3 - 6$ are the penguin operators defined in (6.181) and (6.182).

The functions $G_i(m_i)$ result from calculating penguin diagrams with internal u and c quarks. They are given explicitly in the appendix of [122]. As we will see in a moment they will cancel out in the process of matching, and their analytic expressions are not needed here. This is the reason why in (6.186)–(6.189) the masses m_u and m_c are absent.

Now the effective theory involves only Q_2^u, Q_2^c, and Q_P. Calculating the insertions of Q_2^u and Q_2^c into QCD penguin diagrams of Figure 6.12 and adding the tree-level contributions of Q_2^q operators as in the full theory, we find

$$
\mathcal{A}_{\text{eff}} = \frac{G_F}{\sqrt{2}} \left(\lambda_u \left[\langle Q_2^u \rangle^0 - \frac{\alpha_s(M_W)}{8\pi} (G_u(m_u) - r) \langle Q_P \rangle^0 \right] \right.
$$
$$
\left. + \lambda_c \left[\langle Q_2^c \rangle^0 - \frac{\alpha_s(M_W)}{8\pi} (G_c(m_c) - r) \langle Q_P \rangle^0 \right] - \lambda_t C_P \langle Q_P \rangle^0 \right), \tag{6.204}
$$

where C_P is the coefficient we are looking for. The minus sign in front of λ_t is a convention that has no impact on physics. Next r is a scheme-dependent constant equal to $2/3$ and 0 for NDR and HV schemes, respectively. Finally, it should be remarked that the insertions of QCD penguin operators into penguin diagrams contribute only at $O(\alpha_s^2)$ to (6.204) and do not contribute to the initial conditions for Wilson coefficients at this order. However, as we will see soon, such insertions play a role in the evaluation of the anomalous dimension matrix involving QCD penguin operators already at this order. This shows again that the RG analysis goes beyond the ordinary perturbation theory.

Comparing (6.202) and (6.204) and using the unitarity relation $\lambda_u + \lambda_c = -\lambda_t$, we determine C_P to be

$$
C_P = \frac{\alpha_s(M_W)}{8\pi} [E_0(x_t) - r]. \tag{6.205}
$$

Inserting this result into (6.204), using the expression for Q_P in (6.203) and comparing the coefficient of λ_t with the one of (6.183), we derive the matching conditions (6.186)–(6.189).

Numerical Values for $C_i(\mu_b)$

With $\Lambda_{\overline{\text{MS}}}^{(5)} = 225\,\text{MeV}$ corresponding to $\alpha_s(M_Z) = 0.118$ and choosing $\mu_b = 4.40\,\text{MeV}$, we find in the NDR scheme

$$C_1(\mu_b) = -0.184, \qquad C_2(\mu_b) = 1.078, \qquad (\text{NDR}), \qquad (6.206)$$

$$C_3(\mu_b) = -0.013, \qquad C_4(\mu_b) = -0.035, \qquad (\text{NDR}), \qquad (6.207)$$

$$C_5(\mu_b) = -0.009, \qquad C_6(\mu_b) = -0.041, \qquad (\text{NDR}). \qquad (6.208)$$

Let us make just a few comments:

- Penguin coefficients are much smaller than C_1 and C_2.
- The largest penguin coefficients are C_4 and C_6.
- A numerical analysis shows that in the range $m_t = (170 \pm 15)\,\text{GeV}$ the m_t dependence of the QCD penguin coefficients can be neglected.

Threshold Effects in the Presence of Penguin Operators

In (5.120) we have given a formula for $\vec{C}(\mu)$ in the presence of flavor thresholds. This formula implies in particular that $\vec{C}_{f-1}(\mu_f) = \vec{C}_f(\mu_f)$ where μ_f is the threshold between an effective f-flavor theory and an effective theory with $f - 1$ flavors. In the presence of penguin operators the matching is more involved. One finds now

$$\vec{C}_{f-1}(\mu_f) = \hat{M}(\mu_f)\vec{C}_f(\mu_f), \qquad (6.209)$$

where $\hat{M}(\mu_f)$ is a matching matrix given by

$$\hat{M}(\mu_f) = \hat{1} + \frac{\alpha_s(\mu_f)}{4\pi}\delta\hat{r}^T. \qquad (6.210)$$

The matrix $\delta\hat{r}^T$ can be found in section VID of [111]. With (6.209) the formula (5.120) generalizes to

$$\vec{C}(\mu) = \hat{U}_3(\mu, \mu_c)\hat{M}(\mu_c)\hat{U}_4(\mu_c, \mu_b)\hat{M}(\mu_b)\hat{U}_5(\mu_b, \mu_W)\vec{C}(\mu_W). \qquad (6.211)$$

6.4.3 Explicit Calculation of 6 × 6 ADM

Preliminaries

It is time to do a real climb by calculating the matrix (6.194). This involves the operator insertions into the penguin diagrams and into the current-current diagrams. Because master formulas for the latter insertions have already been derived and applied for the case of (Q_1, Q_2) in the previous section, we begin this climb by discussing the penguin insertions.

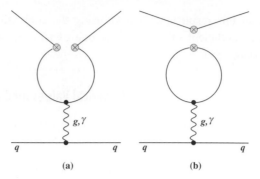

Figure 6.12 One-loop penguin diagrams in the effective theory.

Penguin Insertions: Generalities

The two diagrams contributing to the anomalous dimension matrix of $(Q_1, \ldots Q_6)$ through the penguin insertions are given in Figure 6.12. We observe that two types of insertions of a given operator into a penguin diagram are possible. Type A insertions represented by diagram (a) are constructed by joining two quarks belonging to two different disconnected parts of an operator and attaching the gluon to the resulting internal quark line. For instance, in the case of $(\bar{c}b)_{V-A}(\bar{s}c)_{V-A}$, one can join $\bar{c}$ and c into one line. Type B insertions represented by the diagram (b) are constructed by joining two quarks belonging to the same part of a given operator and attaching the gluon to the resulting quark loop. For instance, in the case of $(\bar{c}c)_{V-A}(\bar{s}b)_{V-A}$ we have a c-quark loop. Because gluons conserve flavor, penguin insertions are only possible if a given operator contains at least two quarks with the same flavor. Note that in the case of $(\bar{s}b)_{V-A}(\bar{s}s)_{V-A}$ both types of insertions are possible and have to be taken into account. Finally, the bottom quark line attached to the lower end of the gluon represents any quark present in the effective theory. In calculating the contribution of a given diagram one has to sum over all quark flavors on this line. In this manner the penguin operators are generated from insertions of operators into penguin diagrams. It is clear from the last statement that the insertion of any operator $(Q_1, \ldots Q_6)$ into the penguin diagrams of Figure 6.12 always results in a linear combination of penguin operators. That is, penguin operators mix under renormalization among themselves, and the current-current operators mix into penguin operators, but the mixing of penguin operators into Q_1 and Q_2 does not take place. This last feature is not affected by the insertions of penguin operators into the current-current diagrams of Figure 5.3 as we will see explicitly later. Consequently, without any calculation we can state that

$$(\hat{\gamma})_{i1} = (\hat{\gamma})_{i2} = 0 \qquad i = 3, ..6, \tag{6.212}$$

and this is also true for the electroweak penguins discussed in the last part of this section as well as other penguins discussed in the rest of this book.

After these general remarks, let us derive two master formulas for penguin insertions of type A and B, which are analogous to the three master formulas for current-current insertions given in (5.150)–(5.152).

As in the case of current-current insertions, we consider an arbitrary operator with the color structure $\hat{V}_1 \otimes \hat{V}_2$ and the Dirac structure $\Gamma_1 \otimes \Gamma_2$. Dropping the external spinors, the insertion of this operator into the penguin vertex in the diagram (a) of Figure 6.12 gives

$$W_\lambda = -ig\mu^\varepsilon \hat{V}_1 t^a \hat{V}_2 I^{\mu\nu} T^\lambda_{\mu\nu}, \tag{6.213}$$

where

$$T^\lambda_{\mu\nu} = \Gamma_1 \gamma_\nu \gamma_\lambda \gamma_\mu \Gamma_2, \tag{6.214}$$

and

$$I^{\mu\nu} = \int \frac{d^D k}{(2\pi)^D} \frac{k^\nu (k-q)^\mu}{k^2 (k-q)^2} = -\frac{i}{16\pi^2} \left[\frac{1}{\varepsilon}\right] \left[\frac{1}{6} q^\mu q^\nu + q^2 \frac{g^{\mu\nu}}{12}\right], \tag{6.215}$$

with q being the gluon momentum. In evaluating $I^{\mu\nu}$ we have kept only the divergent part, as we are only interested in the anomalous dimensions.

The master formula for type A insertions is then obtained by including the gluon propagator and the lower vertex in the penguin diagram (a) of Figure 6.12. We find

$$\mathcal{P}_A = -C_A \frac{\alpha_s}{4\pi} \left[\frac{1}{\varepsilon}\right] \left[\frac{1}{6} \frac{q^\mu q^\nu}{q^2} + \frac{g^{\mu\nu}}{12}\right] \Gamma_1 \gamma_\nu \gamma_\lambda \gamma_\mu \Gamma_2 \otimes \gamma^\lambda, \tag{6.216}$$

where the color factor is given by

$$C_A = \hat{V}_1 t^a \hat{V}_2 \otimes t^a. \tag{6.217}$$

It is understood that the Dirac and color structures on the l.h.s. of $\otimes$ are sandwiched between free spinors belonging to the inserted operator, and $\gamma^\lambda t^a$ standing on the r.h.s. of $\otimes$ are sandwiched between the spinors representing the bottom line of the penguin diagram. Formula (6.216) serves to calculate the coefficients $(b_1)_{ij}$ in the master formula (5.128). We will demonstrate this explicitly later.

The master formula for type B insertions can be derived in an analogous manner. We find

$$\mathcal{P}_B = C_B \frac{\alpha_s}{4\pi} \left[\frac{1}{\varepsilon}\right] \left[\frac{1}{6} \frac{q^\mu q^\nu}{q^2} + \frac{g^{\mu\nu}}{12}\right] \text{Tr}(\Gamma_1 \gamma_\mu \gamma_\lambda \gamma_\nu) \Gamma_2 \otimes \gamma^\lambda, \tag{6.218}$$

where

$$C_B = \text{Tr}(\hat{V}_1 t^a) \hat{V}_2 \otimes t^a, \tag{6.219}$$

with "Tr" in (6.219) standing for the trace in the color space. Note that in this formula we have closed the part $\hat{V}_1 \Gamma_1$ of the inserted operator in the loop. If the part $\hat{V}_2 \Gamma_2$ is closed instead, the indices "1" and "2" in (6.218) and (6.219) should be interchanged. The rules for the incorporation of the external spinors into (6.218) should be evident in view of the comments made after (6.217). The difference in the overall sign compared to (6.216) is a consequence of "-1" for the fermion loop.

Explicit Calculation of Penguin Insertions

Let us apply our master formulas to the case of the operator

$$Q_2 = (\bar{c}_\alpha b_\alpha)_{V-A}(\bar{s}_\beta c_\beta)_{V-A}, \tag{6.220}$$

for which we have

$$\hat{V}_1 \otimes \hat{V}_2 = \mathbf{1} \qquad \Gamma_1 = \Gamma_2 = \gamma_\tau(1 - \gamma_5). \tag{6.221}$$

The flavor structure in (6.220) tells us that only type A insertions are possible. We use therefore the master formula (6.216). As we are only interested in the coefficient of $1/\varepsilon$, we calculate all Dirac structures in $D = 4$ dimensions. This gives

$$\left[\frac{1}{6}\frac{q^\mu q^\nu}{q^2} + \frac{g^{\mu\nu}}{12}\right]\Gamma_1\gamma_\nu\gamma_\lambda\gamma_\mu\Gamma_2 = \frac{4}{3}\left[\gamma_\lambda - \frac{q_\lambda \not{q}}{q^2}\right](1 - \gamma_5), \tag{6.222}$$

where we have used the identity

$$\not{q}\gamma_\lambda \not{q} = 2q_\lambda \not{q} - q^2\gamma_\lambda. \tag{6.223}$$

The term $q_\lambda \not{q}/q^2$ does not contribute here, as using the Dirac equation one has $\bar{s} \not{q} b = 0$ for massless quarks. As discussed in Section 6.2, more care is needed when magnetic penguins corresponding to dipole operators are considered and m_b has to be kept.

Dropping then the second term on the r.h.s. of (6.222), using

$$C_A = t^a \otimes t^a = \frac{1}{2}\left[\tilde{\mathbf{1}} - \frac{1}{N}\mathbf{1}\right], \tag{6.224}$$

with $\tilde{\mathbf{1}}$ defined in (5.143) and inserting the relevant spinors we arrive at

$$\mathcal{P}_A(Q_2) = -\frac{\alpha_s}{4\pi}\left[\frac{1}{\varepsilon}\right]\left[\frac{2}{3}\right]\left[\tilde{\mathbf{1}} - \frac{1}{N}\mathbf{1}\right][\bar{s}_\alpha\gamma_\lambda(1 - \gamma_5)b_\beta] \otimes \sum_q \bar{q}_\gamma\gamma^\lambda q_\delta. \tag{6.225}$$

Next we decompose γ^λ on the r.h.s. of $\otimes$ into $V - A$ and $V + A$ parts

$$\gamma^\lambda = \frac{1}{2}\gamma^\lambda(1 - \gamma_5) + \frac{1}{2}\gamma^\lambda(1 + \gamma_5), \tag{6.226}$$

which allows to express (6.225) in terms of the penguin operators

$$\mathcal{P}_A(Q_2) = -\frac{\alpha_s}{4\pi}\left[\frac{1}{\varepsilon}\right]\left[\frac{1}{3}\right]\left[Q_4 + Q_6 - \frac{1}{N}(Q_3 + Q_5)\right]. \tag{6.227}$$

The contribution of penguin insertions to the coefficients $(b_1)_{2j}$ relevant for the master formula (5.128) are consequently given by

$$\boxed{(b_1)_{23}^P = (b_1)_{25}^P = \frac{1}{3N}, \qquad (b_1)_{24}^P = (b_1)_{26}^P = -\frac{1}{3}.} \tag{6.228}$$

We next consider Q_1 and rewrite it using Fierz reordering as

$$Q_1 = (\bar{c}_\alpha c_\alpha)_{V-A}(\bar{s}_\beta b_\beta)_{V-A}, \tag{6.229}$$

so that the color and Dirac structures are again given by (6.221). This time only type B insertions are possible. However, $\mathrm{Tr}(t^a) = 0$ and consequently the color factor in (6.219) vanishes. Thus

$$\mathcal{P}_B(Q_1) = 0, \tag{6.230}$$

implying

$$\boxed{(b_1)^P_{13} = (b_1)^P_{14} = (b_1)^P_{15} = (b_1)^P_{16} = 0.} \tag{6.231}$$

In the case of Q_3, the type B insertion vanishes as in the case of Q_1, but now two type-A insertions are possible. One involves the internal b quark, the other the s quark. Because gluons are flavor-blind and Q_3 has the same color and Dirac structures as Q_2, we find immediately

$$\mathcal{P}_A(Q_3) = 2\mathcal{P}_A(Q_2). \tag{6.232}$$

Consequently, using (6.228), we find

$$\boxed{(b_1)^P_{33} = (b_1)^P_{35} = \frac{2}{3N}, \qquad (b_1)^P_{34} = (b_1)^P_{36} = -\frac{2}{3}.} \tag{6.233}$$

Next comes Q_4. Performing Fierz reordering, we have

$$Q_4 = \sum_q (\bar{s}_\alpha q_\alpha)_{V-A} (\bar{q}_\beta b_\beta)_{V-A}. \tag{6.234}$$

The type B insertions involving s and b quarks vanish as in the case of Q_1. On the other hand, we have f type A insertions involving all quark flavors. Thus

$$\mathcal{P}_A(Q_4) = f\mathcal{P}_A(Q_2) \tag{6.235}$$

and

$$\boxed{(b_1)^P_{43} = (b_1)^P_{45} = \frac{f}{3N}, \qquad (b_1)^P_{44} = (b_1)^P_{46} = -\frac{f}{3}.} \tag{6.236}$$

The penguin insertions of Q_5 vanish. The type B insertions vanish because of $\mathrm{Tr}(t^a) = 0$. The type A insertions vanish because now

$$\Gamma_1 = \gamma_\tau(1 - \gamma_5), \qquad \Gamma_2 = \gamma_\tau(1 + \gamma_5), \tag{6.237}$$

and the Dirac structure in the master formula (6.216) vanishes. Thus

$$\boxed{(b_1)^P_{53} = (b_1)^P_{54} = (b_1)^P_{55} = (b_1)^P_{56} = 0.} \tag{6.238}$$

Finally, the insertions of Q_6 have to be considered. Performing Fierz reordering, we have

$$Q_6 = -2\sum_q (\bar{s}_\alpha(1 + \gamma_5)q_\alpha)(\bar{q}_\beta(1 - \gamma_5)b_\beta), \tag{6.239}$$

implying

$$\hat{V}_1 \otimes \hat{V}_2 = \mathbf{1}, \qquad \Gamma_1 = (1 + \gamma_5), \qquad \Gamma_2 = (1 - \gamma_5). \tag{6.240}$$

Again, as in the case of Q_4, the type B insertions vanish. The type A insertion of Q_6 is expected at first sight to give a different result than the one of Q_4 because of the different Dirac structure. However, the application of the master formula (6.216) gives

$$\mathcal{P}_A(Q_6) = \mathcal{P}_A(Q_4), \tag{6.241}$$

and consequently

$$(b_1)_{63}^P = (b_1)_{65}^P = \frac{f}{3N}, \qquad (b_1)_{64}^P = (b_1)_{66}^P = -\frac{f}{3}. \tag{6.242}$$

Explicit Calculation of Current-Current Insertions

In the previous chapter we calculated the 2×2 anomalous dimension matrix for the pair (Q_1, Q_2) by inserting these operators into the current-current diagrams of Figure 5.3. Using the master formulas (5.150)–(5.152) for these diagrams together with the basic formula (5.128), we have found the matrix (5.78), which as seen in the left upper corner of (6.194) constitutes a part of the 6×6 matrix we are trying to reproduce.

What remains to be done are the insertions of the penguin operators into the current-current diagrams. The case of the pair (Q_3, Q_4) is simple. From the point of view of current-current insertions the pair (Q_3, Q_4) behaves as (Q_1, Q_2), and we can write immediately

$$(b_1)_{33}^{cc} = 2C_F + \frac{3}{N}, \qquad (b_1)_{34}^{cc} = -3, \tag{6.243}$$

$$(b_1)_{43}^{cc} = -3, \qquad (b_1)_{44}^{cc} = 2C_F + \frac{3}{N}. \tag{6.244}$$

The case of Q_5 and Q_6 operators is different as they have the $(V-A) \otimes (V+A)$ structure. Let us consider Q_5 first. Using master formulas (5.150)–(5.152) for

$$\hat{V}_1 \otimes \hat{V}_2 = \mathbf{1}, \qquad \Gamma_1 = \gamma_\tau(1 - \gamma_5), \qquad \Gamma_2 = \gamma_\tau(1 + \gamma_5), \tag{6.245}$$

we arrive at

$$\sum_i \mathcal{D}_i(Q_5) = \frac{\alpha_s}{4\pi} \frac{1}{\varepsilon} \Gamma_1 \otimes \Gamma_2 \left[C_a^{(1)} + C_a^{(2)} - (C_b^{(1)} + C_b^{(2)}) + 4(C_c^{(1)} + C_c^{(2)}) \right] \tag{6.246}$$

with color factors $C_i^{(j)}$ given in (5.163) and (5.164).

To perform the reduction of Dirac structures in the master formulas (5.150)–(5.152) we had to generalize the Greek method to the $(V - A) \otimes (V + A)$ operators. In this case $\otimes$ should be replaced by 1 as otherwise the Dirac structures would identically vanish. Now the coefficients (4,16,4) in (5.157)–(5.159) are replaced by (4,4,16), respectively, which implies a different weighting of the color factors in (6.246) relative to (5.162). Noting that Q_5 is represented by $\mathbf{1}$ and Q_6 by $\tilde{\mathbf{1}}$ we obtain from (6.246)

$$(b_1)_{55}^{cc} = 2C_F - \frac{3}{N}, \qquad (b_1)_{56}^{cc} = 3. \tag{6.247}$$

Finally, we consider Q_6. Here it is useful to use the form (6.239). As Γ_i are now given by (6.240), one easily finds that the usual Greek method with $\otimes = \gamma_\tau$ applies. The master formulas (5.150)–(5.152) then give

$$\sum_i \mathcal{D}_i(Q_6) = \frac{\alpha_s}{4\pi} \frac{1}{\varepsilon} [-2\Gamma_1 \otimes \Gamma_2] \left[4(C_a^{(1)} + C_a^{(2)}) - (C_b^{(1)} + C_b^{(2)}) + (C_c^{(1)} + C_c^{(2)}) \right] .$$

(6.248)

The weighting of color factors differs from the cases Q_2 and Q_5. In particular, using (5.164) we find that the two last terms in the square bracket cancel each other. Effectively then only the insertions of Q_6 into the diagrams (a) of Figure 5.3 and its symmetric counterpart contribute. However, contrary to the case of the operators Q_{1-5} this contribution will not be canceled by the δ_{ij} term in (5.128) as now $C_a^{(j)}$ are multiplied by 4 instead of 1. Consequently, noting that in this case Q_6 is represented by $\mathbf{1}$ and Q_5 by $\tilde{\mathbf{1}}$ we find

$$(b_1)_{65}^{cc} = 0, \qquad (b_1)_{66}^{cc} = 8C_F.$$

(6.249)

Putting Things Together

Let us add the results for penguin and current-current insertions obtained earlier and listed in white frames. Setting $N = 3$, we find the matrix $\hat{b}_1$ in (5.128):

$$\hat{b}_1 = \begin{pmatrix} 2C_F + 1 & -3 & 0 & 0 & 0 & 0 \\ -3 & 2C_F + 1 & \frac{1}{9} & -\frac{1}{3} & \frac{1}{9} & -\frac{1}{3} \\ 0 & 0 & 2C_F + \frac{11}{9} & -\frac{11}{3} & \frac{2}{9} & -\frac{2}{3} \\ 0 & 0 & -3 + \frac{f}{9} & 2C_F + 1 - \frac{f}{3} & \frac{f}{9} & -\frac{f}{3} \\ 0 & 0 & 0 & 0 & 2C_F - 1 & 3 \\ 0 & 0 & \frac{f}{9} & -\frac{f}{3} & \frac{f}{9} & 8C_F - \frac{f}{3} \end{pmatrix} . \quad (6.250)$$

Inserting this matrix into the one-loop master formula (5.128), we reproduce the full 6×6 matrix in (6.194). Fantastic! We have reproduced all the magic numbers in this matrix. This is almost like reaching the top of Monte Rosa.

We hope that this long exercise and the corresponding exercise for current-current operators in the previous chapter were useful for those readers who have never calculated anomalous dimension matrices. But there is another lesson from these exercises. The corresponding two-loop calculations of current-current and penguin insertions involving many more diagrams, more complicated color factors, and evanescent operators are truly horrible. Moreover, at two-loops we have to use, at least in NDR and HV schemes, the Dirac algebra in $D \neq 4$, and Fierz reordering is not permitted unless we introduce new evanescent operators, the so-called Fierz vanishing evanescent operators, which complicates the calculations further.

Thus two-loop calculations are not like climbing Mont Blanc but rather Mount Everest. They take several months rather than a day or two. The NNLO calculations take sometimes a year or even longer and can be compared to climbing the K2 or flying to the moon. Consequently it is advisable for beginners to take an experienced guide to climb these Himalayas. Fortunately the guides in physics, as opposed to those in the real Himalayas, are doing it for free. On the other hand, the experienced guides these days as opposed to 1990s, when many two-loop calculations were done by hand, are doing such calculations entirely by Mathematica. While this certainly saves time, it prevents, in my view, to feel the details of the calculations and in this manner also to see the beauty of various structures and cancellations of various terms. Moreover, it is often easier to find an error. Yet, these comments apply only to one-loop and two-loop calculations. At three-loops, Mathematica or other computer programs have to be used.

6.4.4 Electroweak Penguins

Operators

The inclusion of the electroweak penguins and box diagrams of Figure 6.11 generates two additional operators, Q_7 and Q_9. With respect to the color structure they are analogous to Q_5 and Q_3 operators, respectively. When QCD effects are also taken into account, two additional operators, Q_8 and Q_{10}, are needed to close the system under renormalization. They are analogous to Q_6 and Q_4, respectively. The full set of operators necessary for the description of $\Delta F = 1$ nonleptonic decays including electroweak effects consists then of ten operators. The four electroweak penguin operators relevant for $\Delta B = 1$ decays with $\Delta S = 1$ are given by

$$Q_7 = \frac{3}{2} (\bar{s}b)_{V-A} \sum_{q=u,d,s,c,b} e_q (\bar{q}q)_{V+A}, \tag{6.251}$$

$$Q_8 = \frac{3}{2} (\bar{s}_\alpha b_\beta)_{V-A} \sum_{q=u,d,s,c,b} e_q (\bar{q}_\beta q_\alpha)_{V+A}, \tag{6.252}$$

$$Q_9 = \frac{3}{2} (\bar{s}b)_{V-A} \sum_{q=u,d,s,c,b} e_q (\bar{q}q)_{V-A}, \tag{6.253}$$

$$Q_{10} = \frac{3}{2} (\bar{s}_\alpha b_\beta)_{V-A} \sum_{q=u,d,s,c,b} e_q (\bar{q}_\beta q_\alpha)_{V-A}. \tag{6.254}$$

Here, α, β denote color indices. The overall factor 3/2 is introduced for convenience. The charge e_q is the charge of the quark coupled to the lower vertex of the photon or Z-propagator.

Wilson Coefficients

To generate the electroweak penguin operators Q_7–Q_{10}, it is sufficient to include the photon penguin together with the relevant QCD renormalization. However, to keep the gauge invariance also Z^0-penguins and box diagrams have to be included at the NLO level. The latter two sets of diagrams involving only heavy fields ($W^\pm, Z^0, t$) contribute only to the Wilson coefficients at $\mu = O(M_W)$ and have no impact on the renormalization group evolution down to low-energy scales, as at these scales being integrated out ($W^\pm, Z^0, t$) do not appear as dynamical degrees of freedom. On the other hand, the inclusion of Z^0-penguins introduces a strong m_t-dependence into Wilson coefficients of the electroweak penguin operators, which in several cases has important phenomenological implications. We will discuss several of them in the phenomenological parts of this book. Here we give some information on the Wilson coefficients of electroweak penguin operators.

The matching at $\mu_W = M_W$ gives in the presence of the electroweak penguin and box diagrams the values of $C_i(M_W)$ with $i = 1, ..10$. In the NDR scheme they are given by:

$$C_1(M_W) = \frac{11}{2} \frac{\alpha_s(M_W)}{4\pi}, \tag{6.255}$$

$$C_2(M_W) = 1 - \frac{11}{6} \frac{\alpha_s(M_W)}{4\pi} - \frac{35}{18} \frac{\alpha}{4\pi}, \tag{6.256}$$

$$C_3(M_W) = -\frac{\alpha_s(M_W)}{24\pi} \widetilde{E}_0(x_t) + \frac{\alpha}{6\pi} \frac{1}{\sin^2\theta_W} [2B_0(x_t) + C_0(x_t)], \tag{6.257}$$

$$C_4(M_W) = \frac{\alpha_s(M_W)}{8\pi} \widetilde{E}_0(x_t), \tag{6.258}$$

$$C_5(M_W) = -\frac{\alpha_s(M_W)}{24\pi} \widetilde{E}_0(x_t), \tag{6.259}$$

$$C_6(M_W) = \frac{\alpha_s(M_W)}{8\pi} \widetilde{E}_0(x_t), \tag{6.260}$$

$$C_7(M_W) = \frac{\alpha}{6\pi} \left[4C_0(x_t) + \widetilde{D}_0(x_t) \right], \tag{6.261}$$

$$C_8(M_W) = 0, \tag{6.262}$$

$$C_9(M_W) = \frac{\alpha}{6\pi} \left[4C_0(x_t) + \widetilde{D}_0(x_t) + \frac{1}{\sin^2\theta_W} (10B_0(x_t) - 4C_0(x_t)) \right], \tag{6.263}$$

$$C_{10}(M_W) = 0. \tag{6.264}$$

The m_t-dependent one-loop functions are known to us from previous sections, and

$$\widetilde{D}_0(x_t) = D_0(x_t) - \frac{4}{9} \tag{6.265}$$

with the constant $-4/9$ characteristic for the NDR scheme. It is absent in the HV scheme. We note that the presence of electroweak effects modifies the values of $C_2(M_W)$ and $C_3(M_W)$ by small $O(\alpha)$ corrections. We also note that $C_8(M_W) = C_{10}(M_W) = 0$. For $\mu \neq M_W$ nonvanishing C_8 and C_{10} are generated through QCD effects.

The anomalous dimension matrices are 10×10:

$$\hat{\gamma}(\alpha_s, \alpha) = \hat{\gamma}_s^{(0)} \frac{\alpha_s}{4\pi} + \hat{\gamma}_e^{(0)} \frac{\alpha}{4\pi} + \hat{\gamma}_s^{(1)} \left(\frac{\alpha_s}{4\pi} \right)^2 + \hat{\gamma}_{se}^{(1)} \frac{\alpha_s}{4\pi} \frac{\alpha}{4\pi} \tag{6.266}$$

with $\gamma_s^{(0)}$ and $\gamma_s^{(1)}$ being 10×10 generalizations of the corresponding 6×6 matrices considered previously. Because now $O(\alpha)$ effects are included in the coefficients at scales $O(M_W)$, the anomalous dimension matrix must also include $O(\alpha)$ contributions, which are represented by $\hat{\gamma}_e^{(0)}$ and $\hat{\gamma}_{se}^{(1)}$ at LO and NLO, respectively. The four matrices in (6.266) can be found in [111], where the references to the original literature are given. See also the review of all NLO and NNLO QCD calculations for weak decays until the summer of 2014 [77].

The calculation of the 6×6 submatrix of $\hat{\gamma}_s^{(0)}$ has been presented in detail earlier. The evaluation of $\hat{\gamma}_e^{(0)}$ proceeds in an analogous manner except that the color factors have to be properly replaced by electric charges and the closed fermion loops coupled to the photon have to be multiplied by $N = 3$. Any reader who succeeded in calculating $\hat{\gamma}_s^{(0)}$ should have no difficulty in calculating within few hours the matrix $\hat{\gamma}_e^{(0)}$. This is a very nice exercise indeed. The calculations of $\hat{\gamma}_s^{(1)}$ and $\hat{\gamma}_{se}^{(1)}$ are even nicer but take more time. Typically six months for $\hat{\gamma}_s^{(1)}$ and then a month for $\hat{\gamma}_{se}^{(1)}$ if the calculation is done by hand. This was in fact the case in 1992, but these days one can do such calculations much faster by using Mathematica.

Due to the simultaneous appearance of α and α_s, the RG analysis is more involved than the one discussed until now. In particular the evolution matrix takes now the general form

$$\hat{U}(m_1, m_2, \alpha) = \hat{U}(m_1, m_2) + \frac{\alpha}{4\pi} \hat{R}(m_1, m_2). \tag{6.267}$$

Here $\hat{U}(m_1, m_2)$ represents the pure QCD evolution matrix given in (5.113). $\hat{R}(m_1, m_2)$ describes the additional evolution in the presence of electromagnetic interactions. It includes both LO and NLO corrections. Let us recall that $\hat{U}(m_1, m_2)$ sums the logarithms $(\alpha_s t)^n$ and $\alpha_s (\alpha_s t)^n$ with $t = \ln(m_2^2/m_1^2)$. On the other hand, $\hat{R}(m_1, m_2)$ sums the logarithms $t(\alpha_s t)^n$ and $(\alpha_s t)^n$. The expression for $\hat{R}(m_1, m_2)$ is rather complicated. It can be found in [111]. The Wilson coefficients are then found by using

$$\vec{C}(\mu_b) = \hat{U}_5(\mu_b, M_W, \alpha)\vec{C}(M_W), \tag{6.268}$$

with $\vec{C}(M_W)$ given in (6.255)–(6.255).

6.4.5 Numerical Values of Wilson Coefficients

With $\Lambda_{\overline{MS}}^{(5)} = 225\,\text{MeV}$ corresponding to $\alpha_s(M_Z) = 0.118$, and choosing $\mu_b = 4.40\,\text{MeV}$ we find in the NDR scheme

$$C_7(\mu_b) = -0.002\,\alpha, \qquad C_8(\mu_b) = 0.054\,\alpha, \qquad \text{(NDR)}, \tag{6.269}$$

$$C_9(\mu_b) = -1.292\,\alpha, \qquad C_{10}(\mu_b) = 0.263\alpha, \qquad \text{(NDR)}. \tag{6.270}$$

Let us make just a few comments:

- Electroweak penguin coefficients being $O(\alpha)$ are smaller than C_{1-6} coefficients. A notable exception is the coefficient C_9, which is in the ballpark of the smallest QCD penguin coefficients C_3 and C_5. It is the operator Q_9 that is the dominant electroweak penguin in nonleptonic B decays [148]. With decreasing μ the coefficient C_8 increases considerably. While its role in B-decays can be fully neglected, it plays an important role in the CP violation in $K \rightarrow \pi\pi$ decays, where also its hadronix matrix element is large. We will return to this operator in the context of the analysis of the ratio ε'/ε in Chapter 10.

- A numerical analysis shows that in contrast to $C_1, \ldots, C_6$, the additional coefficients $C_7, \ldots, C_{10}$ increase strongly with m_t. This strong m_t dependence originates in the Z^0-penguin represented by the function $C_0(x_t)$ in (6.12). Even in the range $m_t = (170 \pm 15)$ GeV with in/decreasing m_t there is a relative variation of $O(\pm 19\%)$ and $O(\pm 10\%)$ for the absolute values of C_8 and $C_{9,10}$, respectively. In obtaining the preceding values, we have just set $m_t = 170$ GeV.

6.5 $B \rightarrow X_s \gamma$ Decay

6.5.1 General Remarks

The inclusive rare decay $B \rightarrow X_s \gamma$ played an important role in constraining NP in the last three decades because both the experimental data and theory have been in good shape for some time. Therefore it is important to discuss this decay in some detail in this book. In particular, we should describe the structure of QCD effects in this decay even if, as we will soon see, this structure is rather complicated. But the QCD effects, as opposed to rare K and B decays with leptons in the final state, are very large, and not including them would totally misrepresent the complete calculation. Therefore, already for almost thirty years several groups of theorists have made a big effort to calculate the branching ratio for $B \rightarrow X_s \gamma$, and as of 2020 this decay is known including NNLO corrections.

The experimentalists study the decays of B^+ and B_d^0 and using isospin symmetry quote the result summarized here. The operators given in (6.274) have an incoming b-quark, which means that the initial particle is really B^- and $\bar{B}_d^0$, and often in theoretical papers $\bar{B} \rightarrow X_s \gamma$ is found. But for the study of branching ratio in question these differences are irrelevant, and to simplify the notation we will just use B in this section. However, when discussing CP-asymmetries in other decays in later chapters, we have to be more explicit to get the correct signs of the asymmetries in question.

Before entering the details let us right away quote the present SM prediction and the most recent data. On the theory side we have a rather precise prediction within the SM [149]

$$\mathcal{B}(B \rightarrow X_s \gamma)_{\text{SM}} = (3.36 \pm 0.23) \times 10^{-4}, \tag{6.271}$$

for $E_\gamma \geq 1.6$ GeV. A detailed historical account of NLO and NNLO calculations leading to this result can be found in [77].

Also experimentalists made impressive progress in measuring this branching ratio, reaching the accuracy of 4.5% [150]

$$\mathcal{B}(B \rightarrow X_s \gamma)_{\text{exp}} = (3.32 \pm 0.15) \times 10^{-4}, \tag{6.272}$$

where again $E_\gamma \geq 1.6\,\text{GeV}$ has been imposed. One expects that in this decade the SuperKEKB will reach the accuracy of 3% so that very precise tests of the SM and its extensions will be possible. In fact, comparing the theory with experiment we observe an impressive agreement of the SM with the data although with present uncertainties there is still some room for NP contributions.

To appreciate these results we will now describe some aspects of the dynamics behind the result in (6.271). We will succeed in describing the leading order in some detail. It will also be possible to present the general structure of the NLO corrections. But any useful description of the calculation of NNLO corrections, which finally lead to (6.271), is beyond the scope of this book.

6.5.2 General Structure

The inclusive decay $B \to X_s\gamma$ with an on-shell γ is governed by the operator $Q_{7\gamma}$, which originates in the photon-penguin vertex, shown on the left in Figure 6.13, with $q^2 = 0$, where q_μ is the momentum of the emitted photon. In addition, the diagrams in which the photon couples to $W^\pm$ are present. They are analogous to the ones involved in the calculation of Z-penguins and are shown in Figure 6.7. To obtain a nonvanishing result, one has to keep the external b-quark mass as well as external momenta. This *mass insertion* together with the expansion to second-order in external momenta generates $Q_{7\gamma}$, which due to the appearance of $\sigma^{\mu\nu}$ is known under the name of *the magnetic photon penguin*. The corresponding gluon-penguin vertex with $q^2 = 0$, shown on the right in Figure 6.13, results in *the magnetic gluon penguin operator* Q_{8G}, which plays the dominant role in the inclusive $B \to X_s$ gluon decay. This operator plays a subleading role in the $B \to X_s\gamma$ decay, but it has to be taken into account when QCD corrections are included.

The renormalization group analysis of $B \to X_s\gamma$ involves in addition to $Q_{7\gamma}$ and Q_{8G} also the operators $Q_1 \ldots Q_6$ discussed previously and given in (6.179)–(6.182). Consequently, the effective Hamiltonian for $B \to X_s\gamma$ at scales $\mu_b = O(m_b)$ reads

$$\mathcal{H}_{\text{eff}}(b \to s\gamma) = -\frac{4G_F}{\sqrt{2}} V_{ts}^* V_{tb} \left[\sum_{i=1}^{6} C_i(\mu_b)Q_i + C_{7\gamma}(\mu_b)Q_{7\gamma} + C_{8G}(\mu_b)Q_{8G} \right], \quad (6.273)$$

where in view of $| V_{us}^* V_{ub}/V_{ts}^* V_{tb} | < 0.02$ we have neglected the term proportional to $V_{us}^* V_{ub}$.

The magnetic-penguin operators are given as follows[1]

$$Q_{7\gamma} = \frac{e}{16\pi^2} m_b \bar{s}_\alpha \sigma^{\mu\nu} P_R b_\alpha F_{\mu\nu}, \qquad Q_{8G} = \frac{g_s}{16\pi^2} m_b \bar{s}_\alpha \sigma^{\mu\nu} P_R t^a_{\alpha\beta} b_\beta G^a_{\mu\nu}. \quad (6.274)$$

---

[1] In some papers a different normalization of these operators is used. It is compensated by the change in Wilson coefficients relative to ours or by changing the overall factor in (6.273).

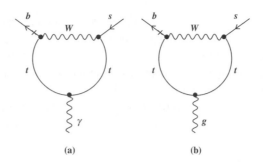

Figure 6.13 Magnetic photon (a) and gluon (b) penguins. The cross on the b signals that m_b has to be kept. The diagrams with photon coupling to $W^\pm$ are not shown. They are obtained from Z-penguin diagrams in Figure 6.7 by replacing Z with γ. We show them in Figure 17.1. From [29].

To derive the contribution of $Q_{7\gamma}$ to the Hamiltonian in (6.273), in the absence of QCD corrections, one multiplies the result of the first diagram in Figure 6.13 by "i" and makes the replacement (q_μ is the four-momentum of the emitted photon)

$$2i\sigma_{\mu\nu}q^\nu \to -\sigma^{\mu\nu}F_{\mu\nu}. \qquad (6.275)$$

Analogous procedure gives the contribution of Q_{8G}.

It is the magnetic γ-penguin that plays the crucial role in this decay. However, the role of the dominant current-current operator Q_2 should not be underestimated. Indeed, the short-distance QCD effects involving in particular the mixing between Q_2 and $Q_{7\gamma}$ are very important in this decay. They are known [151, 152] to enhance $C_{7\gamma}(\mu_b)$ substantially, so that the resulting branching ratio $\mathcal{B}(B \to X_s\gamma)$ turns out to be by a factor of 3 higher than it would be without QCD effects.

A peculiar feature of the renormalization group analysis in $B \to X_s\gamma$ is that the mixing under infinite renormalization between the set $(Q_1 \ldots Q_6)$ and the operators $(Q_{7\gamma}, Q_{8G})$ vanishes at the one-loop level. Consequently, to calculate the coefficients $C_{7\gamma}(\mu_b)$ and $C_{8G}(\mu_b)$ in the leading logarithmic approximation, two-loop calculations of $O(eg_s^2)$ and $O(g_s^3)$ are necessary. The corresponding NLO analysis requires the evaluation of the mixing in question at the three-loop level and four-loop contributions at the NNLO level. This peculiar feature caused that the first fully correct calculation of the leading anomalous dimension matrix relevant for this decay has been obtained only in 1993 [153, 154]. It has been confirmed subsequently in [155–157]. But the result exhibited a very large μ_b dependence in the branching ratios [158], which made the calculation of the NLO corrections mandatory.

The outline of such a calculation has been presented in 1993 [159], but only in 2002 the NLO corrections to $B \to X_s\gamma$ have been completed and summarized in [160]. Many different groups contributed to this result:

- The $O(\alpha_s)$ corrections to $C_{7\gamma}(\mu_W)$ and $C_{8G}(\mu_W)$, involving the gluon corrections to the diagrams in Figure 6.13, have been calculated in [117, 118, 128, 161, 162].

- The two-loop mixing in the sector $(Q_{7\gamma}, Q_{8G})$ has been calculated in [163, 164] and the three-loop mixing between the set $(Q_1 \ldots Q_6)$ and the operators $(Q_{7\gamma}, Q_{8G})$ in [147, 164, 165].
- One-loop matrix elements $\langle s\gamma \text{gluon}|Q_i|b\rangle$ have been calculated in [166, 167], and the dominant contributions to the very difficult two-loop corrections to $\langle s\gamma|Q_i|b\rangle$ have been presented in [168, 169]. Only in 2002 has this result been confirmed in [170, 171], where the evaluation of additional small corrections from QCD penguin operators completed the NLO story.

But then it has been pointed out in [172] that the uncertainty in the scale of the charm-quark mass entering the evaluation of the matrix elements $\langle s\gamma|Q_i|b\rangle$ was significant at the NLO level implying the necessity of NNLO calculations. This means increasing the complexity of each of the three difficult parts just listed by much. In particular the four-loop mixing between the set $(Q_1 \ldots Q_6)$ and the operators $(Q_{7\gamma}, Q_{8G})$ had to be calculated. This heroic effort was basically completed in 2006 in [173], and some refinements have been made since then. We refer to the historical review in [77] and to [149], where further references can be found.

We will now discuss some of these achievements in explicit terms. To demonstrate the importance of higher-order QCD calculations for this decay, it is instructive to discuss first the leading logarithmic approximation.

6.5.3 The Decay $B \to X_s\gamma$ in the Leading Log Approximation

Anomalous Dimension Matrix

In the presence of the magnetic penguin operators the 6×6 anomalous dimension matrix $\hat{\gamma}_s^{(0)}$ involving the operators $Q_1, \ldots, Q_6$ is extended to a 8×8 matrix. The part of this matrix involving $Q_1, \ldots, Q_6$ has been already discussed in Section 6.4. We will therefore concentrate our presentation on the remaining entries in this matrix.

We begin with the mixing between the sets $Q_1, \ldots, Q_6$ and $Q_{7\gamma}, Q_{8G}$ in $\hat{\gamma}_s^{(0)}$. This mixing resulting from two-loop diagrams is generally regularization scheme dependent. This is certainly disturbing because the matrix $\hat{\gamma}_s^{(0)}$, being the first term in the expansion for $\hat{\gamma}_s$, is usually scheme independent. As we will show soon, there is a simple way to circumvent this difficulty [159].

As noticed in [153, 154] the regularization scheme dependence of $\hat{\gamma}_s^{(0)}$ in the case of $b \to s\gamma$ and $b \to sg$ is signaled in the finite parts of the one-loop matrix elements of $Q_1, \ldots, Q_6$ for on-shell photons or gluons. They vanish in any four-dimensional regularization scheme and in the HV scheme, but some of them are nonzero in the NDR scheme. One has

$$\langle Q_i \rangle^\gamma_{\text{one-loop}} = y_i \langle Q_{7\gamma} \rangle_{\text{tree}}, \qquad i = 1, \ldots, 6 \tag{6.276}$$

and

$$\langle Q_i \rangle^G_{\text{one-loop}} = z_i \langle Q_{8G} \rangle_{\text{tree}}, \qquad i = 1, \ldots, 6. \tag{6.277}$$

Here $\langle Q_{7\gamma} \rangle_{\text{tree}}$ and $\langle Q_{8G} \rangle_{\text{tree}}$ denote the operators itself.

In the HV scheme all the y_i and z_i vanish, while in the NDR scheme

$$\vec{y} = \left(0, 0, 0, 0, -\frac{1}{3}, -1\right), \qquad \vec{z} = (0, 0, 0, 0, 1, 0). \tag{6.278}$$

But we know from previous sections that generally the finite parts in one-loop diagrams combined with the coefficients of $1/\varepsilon$ terms in two-loop diagrams give a scheme independent result. Indeed, as first demonstrated in [153, 154], the regularization scheme dependence of y_is and z_is cancels the corresponding regularization scheme dependence found in the mixing of $Q_1, \ldots, Q_6$ and $Q_{7\gamma}, Q_{8G}$ in $\hat{\gamma}_s^{(0)}$. It should be stressed that the numbers y_i and z_i come from purely short-distance parts of the one-loop integrals. So no reference to the hadronic matrix elements is necessary here.

In view of all this it is convenient in the leading order to introduce the so-called effective coefficients [159] for the operators $Q_{7\gamma}$ and Q_{8G}, which are regularization scheme independent. They are given as follows:

$$C_{7\gamma}^{(0)eff}(\mu_b) = C_{7\gamma}^{(0)}(\mu_b) + \sum_{i=1}^{6} y_i C_i^{(0)}(\mu_b) \tag{6.279}$$

and

$$C_{8G}^{(0)eff}(\mu_b) = C_{8G}^{(0)}(\mu_b) + \sum_{i=1}^{6} z_i C_i^{(0)}(\mu_b). \tag{6.280}$$

One can then introduce a scheme-independent vector

$$\vec{C}^{(0)eff}(\mu_b) = \left(C_1^{(0)}(\mu_b), \ldots, C_6^{(0)}(\mu_b), C_{7\gamma}^{(0)eff}(\mu_b), C_{8G}^{(0)eff}(\mu_b)\right). \tag{6.281}$$

From the RGE for $\vec{C}^{(0)}(\mu)$ it is straightforward to derive the RGE for $\vec{C}^{(0)eff}(\mu)$. It has the form

$$\mu \frac{d}{d\mu} C_i^{(0)eff}(\mu) = \frac{\alpha_s}{4\pi} \gamma_{ji}^{(0)eff} C_j^{(0)eff}(\mu), \tag{6.282}$$

where

$$\gamma_{ji}^{(0)eff} = \begin{cases} \gamma_{j7}^{(0)} + \sum_{k=1}^{6} y_k \gamma_{jk}^{(0)} - y_j \gamma_{77}^{(0)} - z_j \gamma_{87}^{(0)} & i=7, j=1,\ldots,6 \\ \gamma_{j8}^{(0)} + \sum_{k=1}^{6} z_k \gamma_{jk}^{(0)} - z_j \gamma_{88}^{(0)} & i=8, j=1,\ldots,6 \\ \gamma_{ji}^{(0)} & \text{otherwise.} \end{cases} \tag{6.283}$$

The matrix $\hat{\gamma}^{(0)eff}$ is a scheme-independent quantity. It equals the matrix that one would directly obtain from two-loop diagrams in the HV scheme. To simplify the notation we will omit the label "eff" in the expressions for the elements of this effective one-loop anomalous dimension matrix given later and keep it only in the Wilson coefficients of the operators $Q_{7\gamma}$ and Q_{8G}.

After the clarification of this rather subtle point we can now write down the leading anomalous dimension matrix relevant for the calculation of the $B \rightarrow X_s \gamma$ rate in the LO approximation. The 6×6 submatrix of $\hat{\gamma}^{(0)}$ involving the operators $Q_1, \ldots, Q_6$ is given in (6.194). Here we only give the remaining nonvanishing entries of $\hat{\gamma}^{(0)}$ [153, 154].

The elements $\gamma_{i7}^{(0)}$ with $i = 1, \ldots, 6$ are

$$\gamma_{17}^{(0)} = 0, \quad \gamma_{27}^{(0)} = \frac{104}{27} C_F, \tag{6.284}$$

$$\gamma_{37}^{(0)} = -\frac{116}{27} C_F, \quad \gamma_{47}^{(0)} = \left(\frac{104}{27} u - \frac{58}{27} d\right) C_F, \tag{6.285}$$

$$\gamma_{57}^{(0)} = \frac{8}{3} C_F, \quad \gamma_{67}^{(0)} = \left(\frac{50}{27} d - \frac{112}{27} u\right) C_F, \tag{6.286}$$

where u and d denote the number of quarks with charge $+2/3$ and $-1/3$, respectively. The elements $\gamma_{i8}^{(0)}$ with $i = 1, \ldots, 6$ are

$$\gamma_{18}^{(0)} = 3, \quad \gamma_{28}^{(0)} = \frac{11}{9} N - \frac{29}{9} \frac{1}{N}, \tag{6.287}$$

$$\gamma_{38}^{(0)} = \frac{22}{9} N - \frac{58}{9} \frac{1}{N} + 3f, \quad \gamma_{48}^{(0)} = 6 + \left(\frac{11}{9} N - \frac{29}{9} \frac{1}{N}\right) f, \tag{6.288}$$

$$\gamma_{58}^{(0)} = -2N + \frac{4}{N} - 3f, \quad \gamma_{68}^{(0)} = -4 - \left(\frac{16}{9} N - \frac{25}{9} \frac{1}{N}\right) f. \tag{6.289}$$

Finally, the 2×2 one-loop anomalous dimension matrix in the sector $Q_{7\gamma}, Q_{8G}$ is given by [174]

$$\gamma_{77}^{(0)} = 8C_F, \quad \gamma_{78}^{(0)} = 0, \tag{6.290}$$

$$\gamma_{87}^{(0)} = -\frac{8}{3} C_F \quad \gamma_{88}^{(0)} = 16C_F - 4N. \tag{6.291?}$$

Renormalization Group Evolution

The coefficients $C_i(\mu_b)$ in (6.273) can be calculated by using

$$\vec{C}(\mu_b) = \hat{U}_5(\mu_b, \mu_W) \vec{C}(\mu_W). \tag{6.291}$$

Here $\hat{U}_5(\mu_b, \mu_W)$ is the 8×8 evolution matrix, which is given in general terms in (5.113) with $\hat{\gamma}$ being this time an 8×8 anomalous dimension matrix. In the leading order $\hat{U}_5(\mu_b, \mu_W)$ is to be replaced by $\hat{U}_5^{(0)}(\mu_b, \mu_W)$ and the initial conditions by $\vec{C}^{(0)}(\mu_W)$ with [174]

$$C_2^{(0)}(\mu_W) = 1, \quad C_{7\gamma}^{(0)}(\mu_W) = -\frac{1}{2} D_0'(x_t), \quad C_{8G}^{(0)}(\mu_W) = -\frac{1}{2} E_0'(x_t). \tag{6.292}$$

In LO all remaining coefficients are set to zero at $\mu = \mu_W$. The two master functions $D_0'(x_t)$ and $E_0'(x_t)$ are given in (6.15) and (6.16): $D_0'(x_t) \approx 0.37$ and $E_0'(x_t) \approx 0.19$.

Using the techniques developed in Section 6.4, the leading order results for the Wilson coefficients of all operators entering the effective Hamiltonian in (6.273) can be written in

Table 6.1 Magic numbers.								
i	1	2	3	4	5	6	7	8
a_i	$\frac{14}{23}$	$\frac{16}{23}$	$\frac{6}{23}$	$-\frac{12}{23}$	0.4086	−0.4230	−0.8994	0.1456
h_i	2.2996	−1.0880	$-\frac{3}{7}$	$-\frac{1}{14}$	−0.6494	−0.0380	−0.0185	−0.0057
$\bar{h}_i$	0.8623	0	0	0	−0.9135	0.0873	−0.0571	0.0209

an analytic form. They are [159]

$$C_j^{(0)}(\mu_b) = \sum_{i=1}^{8} k_{ji}\eta^{a_i}, \qquad (j = 1,\ldots,6) \tag{6.293}$$

$$C_{7\gamma}^{(0)eff}(\mu_b) = \eta^{\frac{16}{23}} C_{7\gamma}^{(0)}(\mu_W) + \frac{8}{3}\left(\eta^{\frac{14}{23}} - \eta^{\frac{16}{23}}\right) C_{8G}^{(0)}(\mu_W) + C_2^{(0)}(\mu_W)\sum_{i=1}^{8} h_i\eta^{a_i}, \tag{6.294}$$

$$C_{8G}^{(0)eff}(\mu_b) = \eta^{\frac{14}{23}} C_{8G}^{(0)}(\mu_W) + C_2^{(0)}(\mu_W)\sum_{i=1}^{8} \bar{h}_i\eta^{a_i}, \tag{6.295}$$

with

$$\eta = \frac{\alpha_s(\mu_W)}{\alpha_s(\mu_b)} \tag{6.296}$$

and $C_{7\gamma}^{(0)}(\mu_W)$ and $C_{8G}^{(0)}(\mu_W)$ given in (6.292). The numbers a_i, h_i, and $\bar{h}_i$ necessary for the calculation of $C_{7\gamma}^{(0)eff}(\mu_b)$ and $C_{8G}^{(0)eff}(\mu_b)$ are given in Table 6.1. k_{ji} have been already discussed in Section 6.4.

It is instructive to perform a quick numerical analysis of (6.294) and (6.295). We use the leading order expression

$$\alpha_s(\mu_b) = \frac{\alpha_s(M_Z)}{1 - \beta_0 \frac{\alpha_s(M_Z)}{2\pi}\ln(M_Z/\mu_b)}. \tag{6.297}$$

We find the results in Table 6.2, where we choose a significantly larger range for $\alpha_s^{(5)}(M_Z)$ than presently known to exhibit the dependence on this important parameter. The central value is close to the present one.

Three features of these results should be emphasized:

- The strong enhancement of the coefficient $C_{7\gamma}^{(0)\mathrm{eff}}$ by short-distance QCD effects, which we illustrate by the relative numerical importance of the three terms in expression (6.294). For instance, for $m_t = 170\,\mathrm{GeV}$, $\mu_b = 5\,\mathrm{GeV}$, and $\alpha_s^{(5)}(M_Z) = 0.118$ one obtains

$$C_{7\gamma}^{(0)\mathrm{eff}}(\mu_b) = 0.695\, C_{7\gamma}^{(0)}(\mu_W) + 0.085\, C_{8G}^{(0)}(\mu_W) - 0.158\, C_2^{(0)}(\mu_W)$$

$$= 0.695\,(-0.193) + 0.085\,(-0.096) - 0.158 = -0.300. \tag{6.298}$$

In the absence of QCD we would have $C_{7\gamma}^{(0)\mathrm{eff}}(\mu_b) = C_{7\gamma}^{(0)}(\mu_W)$ (in that case one has $\eta = 1$). Therefore, the dominant term in the preceding expression (the one proportional

Table 6.2 Wilson coefficients $C_{7\gamma}^{(0)\mathrm{eff}}$ and $C_{8G}^{(0)\mathrm{eff}}$ for $m_t = 163\,\mathrm{GeV}$ and various values of $\alpha_s^{(5)}(M_Z)$ and μ.

$\mu[\mathrm{GeV}]$	$\alpha_s^{(5)}(M_Z) = 0.113$		$\alpha_s^{(5)}(M_Z) = 0.118$		$\alpha_s^{(5)}(M_Z) = 0.123$	
	$C_{7\gamma}^{(0)\mathrm{eff}}$	$C_{8G}^{(0)\mathrm{eff}}$	$C_{7\gamma}^{(0)\mathrm{eff}}$	$C_{8G}^{(0)\mathrm{eff}}$	$C_{7\gamma}^{(0)\mathrm{eff}}$	$C_{8G}^{(0)\mathrm{eff}}$
2.5	−0.328	−0.155	−0.336	−0.158	−0.344	−0.161
5.0	−0.295	−0.142	−0.300	−0.144	−0.306	−0.146
7.5	−0.277	−0.134	−0.282	−0.136	−0.286	−0.138
10.0	−0.265	−0.130	−0.269	−0.131	−0.273	−0.133

to $C_2^{(0)}(\mu_W)$) is the additive QCD correction that causes the enormous QCD enhancement of the $B \to X_s\gamma$ rate [151, 152]. It originates solely from the two-loop diagrams. On the other hand, the multiplicative QCD correction (the factor 0.695) tends to suppress the rate, but fails in the competition with the additive contributions.

In the case of $C_{8G}^{(0)\mathrm{eff}}$ a similar enhancement is observed

$$C_{8G}^{(0)\mathrm{eff}}(\mu_b) = 0.727\, C_{8G}^{(0)}(\mu_W) - 0.074\, C_2^{(0)}(\mu_W)$$

$$= 0.727\,(-0.096) - 0.074 = -0.144. \tag{6.299}$$

- A strong μ_b-dependence of both coefficients as first stressed by Ali and Greub [158] and confirmed in [159]. Because $B \to X_s\gamma$ is dominated by QCD effects, it is not surprising that this scale uncertainty in the leading order is particularly large. We will investigate this scale uncertainty in a moment.
- Relatively small uncertainty in the values of the coefficients from the one in $\alpha_s(M_Z)$, even for the range shown, which is practically negligible in 2020.

Scale Uncertainties at LO

To illustrate the scale uncertainty at LO in $\mathcal{B}(B \to X_s\gamma)$ we just use the spectator model in which the inclusive decay $B \to X_s\gamma$ is approximated by the partonic decay $b \to s\gamma$. In fact, this is the leading term in the heavy quark expansion, an expansion in $1/m_b$, discussed briefly in Chapter 7 with the first corrections appearing at $O(1/m_b^2)$.

We have then the following approximate equality:

$$\frac{\Gamma(B \to X_s\gamma)}{\Gamma(B \to X_c e\bar{\nu}_e)} \simeq \frac{\Gamma(b \to s\gamma)}{\Gamma(b \to c e\bar{\nu}_e)} \equiv R_{\mathrm{quark}}, \tag{6.300}$$

where the quantities on the r.h.s. are calculated in the spectator model corrected for short-distance QCD effects. The normalization to the semileptonic rate is usually introduced to reduce the uncertainties due to the CKM matrix elements and factors of m_b^5 in the r.h.s. of (6.300).

The leading logarithmic calculations can be summarized in a compact form as follows:

$$R_{\mathrm{quark}} = \frac{\mathcal{B}(B \to X_s\gamma)}{\mathcal{B}(B \to X_c e\bar{\nu}_e)} = \frac{|V_{ts}^* V_{tb}|^2}{|V_{cb}|^2} \frac{6\alpha}{\pi f(z)} |C_7^{(0)\mathrm{eff}}(\mu_b)|^2, \tag{6.301}$$

where

$$f(z) = 1 - 8z + 8z^3 - z^4 - 12z^2 \ln z \quad \text{with} \quad z = \frac{m_{c,pole}^2}{m_{b,pole}^2}, \tag{6.302}$$

is the phase space factor in $\mathcal{B}(B \to X_c e \bar{\nu}_e)$ and $\alpha = e^2/4\pi$. To find (6.301) only the tree-level matrix element $< s\gamma|Q_{7\gamma}|B >$ has to be computed. We leave the derivation of (6.301) to motivated readers.

There are three scale uncertainties present in (6.301):

- The low-energy scale $\mu_b = O(m_b)$ at which the Wilson Coefficient $C_7^{(0)\text{eff}}(\mu_b)$ is evaluated.

- The high-energy scale $\mu_W = O(M_W)$ at which the full theory is matched with the effective five-quark theory. In LO this scale enters only η in (6.296). $C_7^{(0)}(\mu_W)$ and $C_8^{(0)}(\mu_W)$ serve in LO as initial conditions to the renormalization group evolution from μ_W down to μ_b. As seen in (6.292), they do not depend on μ_W.

- The scale $\mu_t = O(m_t)$ at which the running top-quark mass is defined. In LO it enters only through $x_t(\mu_t)$. As we stressed in connection with $B^0 - \bar{B}^0$ mixing in Section 6.3, μ_W and μ_t do not have to be equal. Initially when the top quark and the W-boson are integrated out, it is convenient in the process of matching to keep $\mu_t = \mu_W$. Yet one has always the freedom to redefine the top-quark mass and to work with $m_t(\mu_t)$ where $\mu_t \neq \mu_W$.

It is evident from the preceding formulas that in LO the variations of μ_b, μ_W, and μ_t remain uncompensated, which results in potential theoretical uncertainties in the predicted branching ratio. We have seen this feature already in the case of $\Delta F = 2$ transitions. But here this problem is really serious, as far as μ_b uncertainty is concerned, because the renormalization group effects are larger. This is related to the fact that the elements of the anomalous dimension matrix entering $B \to X_s \gamma$ are significantly larger than in the case of $\Delta F = 2$ transitions.

It is customary to estimate the uncertainties due to μ_b by varying it in the range $m_b/2 \leq \mu_b \leq 2m_b$. Similarly one can vary μ_W and μ_t in the ranges $M_W/2 \leq \mu_W \leq 2M_W$ and $m_t/2 \leq \mu_t \leq 2m_t$, respectively. Specifically in our numerical analysis we will consider the ranges

$$2.5 \text{ GeV} \leq \mu_b \leq 10 \text{ GeV} \tag{6.303}$$

and

$$40 \text{ GeV} \leq \mu_W \leq 160 \text{ GeV} \qquad 80 \text{ GeV} \leq \mu_t \leq 320 \text{ GeV}. \tag{6.304}$$

In the LO analysis we use the leading order formula for $\alpha_s(\mu_b)$ in (6.297) with $\alpha_s(M_Z) = 0.118$, set $m_t(m_t) = 163.0$ GeV, and use

$$m_t(\mu_t) = m_t(m_t) \left[\frac{\alpha_s(\mu_t)}{\alpha_s(m_t)} \right]^{\frac{4}{\beta_0}}, \qquad \beta_0 = 23/3. \tag{6.305}$$

$$\gamma_{87}^{(1)} = C_F \left(-\frac{404}{27} N + \frac{32}{3} C_F + \frac{56}{27} f \right),$$

$$\gamma_{88}^{(1)} = -\frac{458}{9} - \frac{12}{N^2} + \frac{214}{9} N^2 + \frac{56}{9} \frac{f}{N} - \frac{13}{9} f N.$$

The generalization of (6.284)–(6.289) to next-to-leading order requires three-loop calculations. The result can be found in [147].

The constants r_i resulting from the calculations of NLO corrections to decay matrix elements [147, 168, 169] are collected in [147]. It should be stressed that the basis of the operators with $i = 1 - 6$ used in [147] differs from the standard basis used in [168, 169] and here. The basis used in [147] has been chosen to avoid γ_5 problems in the three-loop calculations performed in the NDR scheme. This has to be remembered when using formulas of this paper. In particular the constants r_i calculated in [168, 169] have to be transformed to the basis of [147]. We refer to the hep-ph-version of the latter paper and to [164, 178].

For the following discussion it will be useful to have [153]

$$\gamma_{27}^{(0)\mathrm{eff}} = \frac{416}{81}, \qquad \gamma_{28}^{(0)\mathrm{eff}} = \frac{70}{27}, \tag{6.324}$$

which enter (6.320) and (6.321), respectively. They can be obtained from (6.284) and (6.287).

Reduction of Scale Uncertainties at NLO

Let us investigate how much the uncertainties in (6.306) are reduced after including NLO corrections. We begin this discussion by demonstrating analytically that the μ_b, μ_W, and μ_t dependences present in $C_7^{(0)\mathrm{eff}}(\mu_b)$ are indeed cancelled at $O(\alpha_s)$ by the explicit scale-dependent terms in (6.311) and (6.320). The scale-dependent terms in (6.321) do not enter this cancellation at this order in α_s in $B \to X_s \gamma$. On the other hand, they are responsible for the cancellation of the scale dependences in $C_8^{(0)\mathrm{eff}}(\mu_b)$ relevant for the $b \to s$ gluon transition.

Expanding the three terms in (6.294) in α_s and keeping the leading logarithms, we find:

$$\eta^{\frac{16}{23}} C_7^{(0)}(\mu_W) = \left(1 + \frac{\alpha_s}{4\pi} \frac{16}{3} \ln \frac{\mu_b^2}{\mu_W^2} \right) C_7^{(0)}(\mu_W), \tag{6.325}$$

$$\frac{8}{3} \left(\eta^{\frac{14}{23}} - \eta^{\frac{16}{23}} \right) C_8^{(0)}(\mu_W) = -\frac{\alpha_s}{4\pi} \frac{16}{9} \ln \frac{\mu_b^2}{\mu_W^2} C_8^{(0)}(\mu_W), \tag{6.326}$$

$$\sum_{i=1}^{8} h_i \eta^{a_i} = \frac{\alpha_s}{4\pi} \frac{23}{3} \ln \frac{\mu_b^2}{\mu_W^2} \sum_{i=1}^{8} h_i a_i = \frac{208}{81} \frac{\alpha_s}{4\pi} \ln \frac{\mu_b^2}{\mu_W^2}, \tag{6.327}$$

respectively. In (6.327) we have used $\sum h_i = 0$. Inserting these expansions into (6.311), we observe that the μ_W dependences in (6.325), (6.326), and (6.327) are precisely cancelled by the three explicit logarithms in (6.320) involving μ_W, respectively. Similarly, one can convince oneself that the μ_t-dependence of $C_7^{(0)\mathrm{eff}}(\mu_b)$ is cancelled at $O(\alpha_s)$ by the $\ln \mu_t^2/M_W^2$ term in (6.320). Finally and most important, the μ_b dependences in (6.325),

(6.326), and (6.327) are cancelled by the explicit logarithms in (6.311) that result from the calculation of the one-loop matrix elements $< s\gamma|Q_{7\gamma}|B >$ and $< s\gamma|Q_{8G}|B >$ and the two-loop matrix element $< s\gamma|Q_2|B >$ as discussed previously. Interestingly the scale-dependent term in (6.322) does not contribute to any cancellation of the μ_W dependence at this order in α_s due to the relation

$$\sum_{i=1}^{8} \left(\frac{2}{3} e_i + 6 l_i \right) = 0. \tag{6.328}$$

This can be verified by using Table 6.3.

Clearly there remain small μ_b, μ_W, and μ_t dependences in (6.309), which can only be reduced by going beyond the NLO approximation. They constitute the theoretical uncertainty, which should be taken into account in estimating the error in the prediction for $\mathcal{B}(B \to X_s \gamma)$. For this reason also the term $\Delta C_7^{(1)eff}(\mu_b)$ in (6.319) has to be kept as pointed out in [179].

Using the two-loop generalization of (6.305) in (4.153) and varying μ_b, μ_W, and μ_t in the ranges (6.303) and (6.304), one finds [128] the following respective uncertainties in the branching ratio after the inclusion of NLO corrections:

$$\Delta \mathcal{B}(B \to X_s \gamma) = \begin{cases} \pm 4.3\% & (\mu_b) \\ \pm 1.1\% & (\mu_W) \\ \pm 0.4\% & (\mu_t) \end{cases} . \tag{6.329}$$

This reduction of the μ_b-uncertainty by roughly a factor of seven relative to $\pm 22\%$ in LO is impressive. The remaining μ_W and μ_t uncertainties are negligible.

6.5.5 Final Remarks on NLO

We hope that this rather technical part of this book has shown that already the NLO calculations of $\mathcal{B}(B \to X_s \gamma)$ are very involved. It should also be emphasized that there are other uncertainties like the parametric ones related to the elements of the CKM matrix and corrections to the relation (6.300). The latter ones can be calculated using Heavy Quark Expansion (HQE) so that at NLO one finds [147]

$$\mathcal{B}(B \to X_s \gamma) = \mathcal{B}(B \to X_c e \bar{\nu}_e) \cdot R_{\text{quark}} \left(1 - \frac{\delta_{sl}^{NP}}{m_b^2} + \frac{\delta_{rad}^{NP}}{m_b^2} \right), \tag{6.330}$$

where δ_{sl}^{NP} and δ_{rad}^{NP} parametrize nonperturbative corrections to the semileptonic and radiative B-meson decay rates, respectively.

Next, to improve the accuracy also electroweak $O(\alpha)$ corrections to R_{quark} have to be included. In this context an important point has been made by Czarnecki, and Marciano [180] already in 1998 when they made the first analysis of these corrections. It is related to the scale μ in $\alpha_{\text{em}} \equiv e^2(\mu)/4\pi$, which is rather arbitrary if corrections $O(\alpha)$ are not considered. In all calculations prior to their paper $m_b \leq \mu \leq M_W$ has been used, giving $1/\alpha_{\text{em}} = 130.3 \pm 2.3$. The inclusion of fermion loop contributions in the photon propagator indicates [180], however, that α renormalized at $q^2 = 0$, i.e., $\alpha = 1/137.036$ is more appropriate. This reduces the branching ratio by roughly 5%. The fermion loops in the

W-propagator bring a reduction of 2%. Two other reductions, each of roughly 1%, come from short-distance photonic corrections to $b \to s\gamma$ and $b \to ce\nu$. The total reduction of R_{quark} found in [180] amounted then to $(9 \pm 2)\%$. With this reduction and all the NLO corrections discussed by us these authors gave in 1998 the SM prediction

$$\mathcal{B}(B \to X_s \gamma) = (3.28 \pm 0.30) \times 10^{-4}, \tag{6.331}$$

not far from the present NNLO result in (6.271).

There have been other analyses at the end of the previous millennium that addressed other small effects contributing to $\mathcal{B}(B \to X_s \gamma)$. They are discussed in Section 12 in [29], and we will not review them here. These analyses made it clear that to convincingly reduce the error in the branching ratio below 10% and possibly even below 5% a complete calculation of NNLO QCD corrections and NLO electroweak corrections is necessary.

In fact, presently such corrections are known. It is out of the question that we could present the NNLO analysis at the level we have done until now. Therefore we refer to original literature, to the extensive list of references in [77], and to various reports by Mikolaj Misiak.

6.6 $b \to s\ell^+\ell^-$ and $d \to s\ell^+\ell^-$ Transitions

6.6.1 $b \to s\ell^+\ell^-$

General Remarks

The FCNC transitions $b \to s\ell^+\ell^-$ play a very important role in flavor physics. They are responsible for several important decays like $B_d \to K\ell^+\ell^-$, $B_d \to K^*\ell^+\ell^-$, $B \to X_s \mu^+\mu^-$, and $B_s \to \mu^+\mu^-$. Therefore, we will devote this section to the fundamental effective Hamiltonian governing these decays in the SM. The expressions for the branching ratios and in particular for various angular observables related to $B_d \to K^*\ell^+\ell^-$ will be presented in Section 9.1. In fact the formulas presented there are quite general and valid also beyond the SM. But here to get familiar with these decays in a simpler setting we will confine our discussion to the SM.

The starting point in any analysis of $b \to s\ell^+\ell^-$ transitions is the effective Hamiltonian at scales $\mu = O(m_b)$. It is obtained by calculating Z-penguin and photon-penguin diagrams and box diagrams that we discussed at the beginning of this chapter. Subsequently renormalization group effects have to be included to obtain the Wilson coefficients of contributing operators at $\mu = O(m_b)$.

There are different conventions for operators contributing to $b \to s\ell^+\ell^-$ transitions. The effective Hamiltonian used in this book and in many recent publications is given as follows:

$$\mathcal{H}_{\text{eff}}(b \to s\ell\bar{\ell}) = \mathcal{H}_{\text{eff}}(b \to s\gamma) - \frac{4G_F}{\sqrt{2}} \frac{\alpha}{4\pi} V_{ts}^* V_{tb} \left[C_9(\mu) Q_9 + C_{10}(\mu) Q_{10} \right], \tag{6.332}$$

where we have again neglected the term proportional to $V_{us}^* V_{ub}$ and $\mathcal{H}_{eff}(b \to s\gamma)$ is given in (6.273). In addition to the operators relevant for $B \to X_s \gamma$, there are two new operators:

$$Q_9 = (\bar{s}\gamma_\mu P_L b)(\bar{\ell}\gamma^\mu \ell), \qquad Q_{10} = (\bar{s}\gamma_\mu P_L b)(\bar{\ell}\gamma^\mu \gamma_5 \ell). \qquad (6.333)$$

In some papers the factor $\alpha/4\pi$ is included in the definition of these two operators in analogy to what is done in the case of $B \to X_s \gamma$ decay in (6.273), and this difference has to be remembered when calculating the matrix elements of operators and relevant branching ratios. Therefore it is crucial to always check which normalization is used in a given paper. We will stress it whenever for strategical reasons we will make some rescalings of normalization in Wilson coefficients in later chapters of this book. It should also be noticed that using the same notation, namely Q_9 and Q_{10}, for these semileptonic operators and nonleptonic operators in $\Delta B = 1$ transitions is not a good idea, and possibly the notation in (6.339) in the case of $K_L \to \pi^0 \ell^+ \ell^-$, as done originally also in the case of $b \to s\ell^+\ell^-$ in [127], would be better. But presently the literature is dominated by the notation in (6.333), and we will use it here as well with the hope that this will not confuse some readers.

While we do not show explicitly the four-quark operators in (6.332), they play an important role in the calculation of QCD and electroweak corrections as we will see soon.

The Wilson coefficients at $\mu_b = O(m_b)$ are given as follows

$$\sin^2 \vartheta_W C_9(\mu_b) = \sin^2 \vartheta_W P_0^{\text{NDR}} + [\eta_{\text{eff}} Y_0(x_t) - 4 \sin^2 \vartheta_W Z_0(x_t)], \qquad (6.334)$$

$$\sin^2 \vartheta_W C_{10}(\mu_b) = -\eta_{\text{eff}} Y_0(x_t) \qquad (6.335)$$

with functions $Y_0(x_t)$ and $Z_0(x_t)$ calculated already in Section 6.2. The entries P_0^{NDR} and η_{eff} are QCD and electroweak corrections, which we will now discuss in some detail.

QCD and Electroweak Corrections

Wilson Coefficient C_{10}

We begin the discussion with C_{10}, which is simpler because the operator Q_{10} does not renormalize under QCD so that C_{10} has no μ dependence related to the RG evolution to the low scales.[2] But it does not mean that the calculations are simple if one wants to achieve sufficient precision. Indeed, QCD corrections to the relevant penguin and box diagrams have to be included and also electroweak corrections. We will collect all these effects in η_{eff}, which equals unity without these corrections.

The NLO QCD corrections to η_{eff} were calculated a long time ago in [181, 182] with the result $\eta_{\text{eff}} = 1.012$ for $m_t = m_t(m_t)$. Over several years electroweak corrections have been calculated [134, 183–185], but they were incomplete, implying dependence on the renormalization scheme used for electroweak parameters as analyzed in detail in [185, 186]. The complete NLO electroweak corrections have been calculated in [187] and QCD corrections up to NNLO in [188]. The inclusion of these higher-order corrections reduced

---

[2] We write $C_{10}(\mu_b)$ to stress that it can be used at the low scale.

significantly various scale uncertainties so that nonparametric uncertainties in C_{10} have been reduced below 1%.

The calculations performed in [187, 188] are very involved and in analogy to the QCD factors, like η_B and η_{1-3} in $\Delta F = 2$ processes, we find it useful to include all QCD and electroweak corrections into η_{eff} introduced in (6.335) that without these corrections would be equal to unity. Inspecting the analytic formulas in [189] one finds then [190]

$$\eta_{\text{eff}} = 0.9882 \pm 0.0024 . \tag{6.336}$$

One could ask the question whether there was a point in doing such complicated calculations finding at the end such a small correction. But it should be emphasized that it was crucial to perform such involved calculations as these small corrections are only valid for particular definitions of the top-quark mass and of other electroweak parameters involved. In particular one has to use in $Y_0(x_t)$ the $\overline{\text{MS}}$-renormalized top-quark mass $m_t(m_t)$ with respect to QCD but on-shell one with respect to electroweak interactions. This means $m_t(m_t) = 163.5\,\text{GeV}$ as calculated in [189]. Moreover, in using (6.336) to calculate observables like branching ratios, it is important to have the same normalization of the effective Hamiltonian as in the latter paper. There this normalization is expressed in terms of G_F and M_W only. Needless to say one can also use directly the formulas in [189].

On the other hand, if one follows the normalization of an effective Hamiltonian that uses G_F, $\alpha(M_Z)$ and $\sin^2 \vartheta_W$, the consistency with [189] would require $\eta_{\text{eff}} = 0.991$ with $m_t(m_t) = 163.0\,\text{GeV}$. It should be remarked that presently only in the case of the $B_{s,d} \to \mu^+ \mu^-$ decays discussed here and $K^+ \to \pi^+ \nu \bar{\nu}$ and $K_L \to \pi^0 \nu \bar{\nu}$ decays discussed in the next section one has to take such a care about the definition of m_t with respect to electroweak corrections. In most cases such corrections are not known or hadronic uncertainties are too large to worry about such small corrections. But the discussion here shows that at a certain level of precision one has to take into account such small effects.

Wilson Coefficient C_9

While the QCD corrections to C_{10} did not involve any renormalization group running, the case of C_9 is more profound. In particular the structure of these corrections differs from the ones encountered in previous sections. This is related to the appearance of a large logarithm represented by $1/\alpha_s$ in P_0 given in (6.337). Consequently, the renormalization group improved perturbation theory for C_9 has the structure $O(1/\alpha_s) + O(1) + O(\alpha_s)+$ higher-order terms, whereas the corresponding series for $C_{7\gamma}(\mu)$ and C_{10} are $O(1) + O(\alpha_s)+$ higher-order terms. Therefore, to find the next-to-leading $O(1)$ term in various branching ratios the full two-loop renormalization group analysis has to be performed to find C_9, but the coefficients of the remaining operators should be taken in the leading logarithmic approximation. In particular at the NLO level one should include in these decays only the leading term $C_{7\gamma}^{(0)\text{eff}}(\mu)$ in $C_{7\gamma}(\mu)$. The scheme-dependent correction $C_{7\gamma}^{(1)\text{eff}}(\mu)$ should be omitted in a consistent NLO calculation as it is a part of NNLO correction.

The QCD corrections to $C_9(\mu)$ have been calculated over many years with increasing precision by several groups, and presently LO, NLO, and NNLO corrections are known. Here, as in previous sections, we can only discuss the structure of LO and NLO corrections.

This should already give a good idea what is going on. Very motivated readers can study the NNLO analyses listed in table 8 in [77], in particular [183, 184] and references therein.

Collecting NLO corrections into the factor P_0^{NDR} one finds [127] in the NDR scheme

$$P_0^{\mathrm{NDR}} = \frac{\pi}{\alpha_s(M_{\mathrm{W}})}\left(-0.1875 + \sum_{i=1}^{8} p_i \eta^{a_i+1}\right) + 1.2468 + \sum_{i=1}^{8} \eta^{a_i}[r_i^{\mathrm{NDR}} + s_i \eta]. \quad (6.337)$$

The powers a_i are the same as in Table 6.1. The coefficients p_i, r_i^{NDR}, and s_i can be found in Table 6.4. In the HV scheme only the coefficients r_i are changed. They are given in the last row of Table 6.4.

In Table 6.5 we show the constant P_0 in (6.337) for different μ and $\Lambda_{\overline{\mathrm{MS}}}$ in the leading order corresponding to the first term in (6.337) and for the NDR as given by (6.337). In the last row we show the results in the HV scheme. In Table 6.6 we show the corresponding values for $C_9(\mu)$. To this end we set $m_t = 170\,\mathrm{GeV}$ and vary $\Lambda_{\overline{\mathrm{MS}}}^{(5)}$ in a larger range than presently known to better exhibit this dependence.

Table 6.4 Additional magic numbers.

i	1	2	3	4	5	6	7	8
p_i	0,	0,	$-\frac{80}{203}$,	$\frac{8}{33}$,	0.0433	0.1384	0.1648	−0.0073
r_i^{NDR}	0	0	0.8966	−0.1960	−0.2011	0.1328	−0.0292	−0.1858
s_i	0	0	−0.2009	−0.3579	0.0490	−0.3616	−0.3554	0.0072
r_i^{HV}	0	0	−0.1193	0.1003	−0.0473	0.2323	−0.0133	−0.1799

Table 6.5 The coefficient P_0 of C_9 for various values of $\Lambda_{\overline{\mathrm{MS}}}^{(5)}$ and μ.

	$\Lambda_{\overline{\mathrm{MS}}}^{(5)} = 160\,\mathrm{MeV}$			$\Lambda_{\overline{\mathrm{MS}}}^{(5)} = 225\,\mathrm{MeV}$			$\Lambda_{\overline{\mathrm{MS}}}^{(5)} = 290\,\mathrm{MeV}$		
$\mu[\mathrm{GeV}]$	LO	NDR	HV	LO	NDR	HV	LO	NDR	HV
2.5	2.022	2.907	2.787	1.933	2.846	2.759	1.857	2.791	2.734
5.0	1.835	2.616	2.402	1.788	2.591	2.395	1.748	2.568	2.390
7.5	1.663	2.386	2.127	1.632	2.373	2.127	1.605	2.361	2.128
10.0	1.517	2.201	1.913	1.494	2.194	1.917	1.475	2.185	1.920

Table 6.6 Wilson coefficient C_9 for $m_t = 170\mathrm{GeV}$ and various values of $\Lambda_{\overline{\mathrm{MS}}}^{(5)}$ and μ.

	$\Lambda_{\overline{\mathrm{MS}}}^{(5)} = 160\,\mathrm{MeV}$			$\Lambda_{\overline{\mathrm{MS}}}^{(5)} = 225\,\mathrm{MeV}$			$\Lambda_{\overline{\mathrm{MS}}}^{(5)} = 290\,\mathrm{MeV}$		
$\mu[\mathrm{GeV}]$	LO	NDR	HV	LO	NDR	HV	LO	NDR	HV
2.5	2.022	4.472	4.352	1.933	4.410	4.323	1.857	4.355	4.298
5.0	1.835	4.182	3.968	1.788	4.156	3.961	1.748	4.134	3.955
7.5	1.663	3.954	3.694	1.632	3.940	3.694	1.605	3.928	3.695
10.0	1.517	3.769	3.481	1.494	3.761	3.485	1.475	3.754	3.487

Let us briefly discuss these numerical results. We observe:

- The NLO corrections to P_0 enhance this constant relatively to the LO result by roughly 45% and 35% in the NDR and HV schemes, respectively.
- The NLO corrections to C_9, which include also the m_t-dependent contributions, are large as seen in Table 6.6. The results in HV and NDR schemes are by more than a factor of two larger than the leading order result $C_9 = P_0^{\text{LO}}$, which consistently should not include m_t-contributions. This demonstrates very clearly the necessity of NLO calculations, which allow a consistent inclusion of the important m_t-contributions.
- The μ dependence of C_9 is sizable: ~15% in the range of μ considered. On the other hand, its $\Lambda_{\overline{\text{MS}}}^{(5)}$ dependence is weak and can be neglected with the 2020 precision for α_s. Also the m_t dependence of C_9 is weak. Finally, the difference between C_9^{NDR} and C_9^{HV} is small and amounts to roughly 5%.

The phenomenology of $b \to s\ell^+\ell^-$ transitions will be discussed in Section 9.1.

6.6.2 $d \to s\ell^+\ell^-$

The decays $K_L \to \pi^0 e^+ e^-$, $K_{L,S} \to \pi^0 \mu^+ \mu^-$, and $K_{L,S} \to \mu^+ \mu^-$ are the counterparts of the transitions just discussed, but in the K meson system. Presently, they do not play as big role in the phenomenology as the decays governed by $b \to s\ell^+\ell^-$ transitions. There are two reasons for this

- $K_L \to \pi^0 e^+ e^-$ and $K_L \to \pi^0 \mu^+ \mu^-$ have not been observed and there are presently no plans for a dedicated experiment to measure them. $K_L \to \mu^+ \mu^-$ has been measured, but only a small fraction of it can be predicted in the SM, the so-called *short-distance* part. For $K_S \to \mu^+ \mu^-$ only a weak experimental result from LHCb exists.
- All four decays contain significant *long-distance* contributions so that presently precise tests of the SM and of its extensions are not possible. But as we will see in later chapters they still provide useful rough upper bounds on possible new physics contributions. $K_L \to \pi^0 \ell^+ \ell^-$ and $K_S \to \mu^+ \mu^-$ bound in particular new CP-violating contributions while $K_L \to \mu^+ \mu^-$ the corresponding CP-conserving ones. We will see several examples in BSM chapters.

In what follows we will summarize the effective Hamiltonian for the decays in question postponing the phenomenology and the discussion of long-distance contributions to Section 9.7.

The effective Hamiltonian for $d \to s\ell^+\ell^-$ transitions, relevant for the four decays in question, is given at scales $\mu < m_c$ in the SM as follows:

$$\mathcal{H}_{\text{eff}}(d \to \bar{s}\ell^+\ell^-) = 4\frac{G_F}{\sqrt{2}}\frac{\alpha}{4\pi}V_{us}^*V_{ud}\left[\sum_{i=1}^{6,7V}\left[z_i(\mu) + \tau y_i(\mu)\right]Q_i + \tau y_{7A}(M_W)Q_{7A}\right],$$

$$(6.338)$$

where $Q_1, \ldots Q_6$ are the current-current and QCD penguin operators discussed previously, and the new operators Q_{7V} and Q_{7A} are given by

$$Q_{7V} = (\bar{s}\gamma_\mu P_L d)(\bar{\ell}\gamma^\mu \ell), \qquad Q_{7A} = (\bar{s}\gamma_\mu P_L d)(\bar{\ell}\gamma^\mu \gamma_5 \ell). \tag{6.339}$$

At this stage we should warn readers that in the literature these two operators are sometimes denoted by Q_9 and Q_{10}, respectively. Indeed as seen in (6.333), only the flavor-changing current differs from the one in $b \rightarrow s\ell\bar{\ell}$. But as the contributions of the dipole operators Q_7 and Q_8 can be neglected due to the suppression by $m_{s,d}$ in these decays, it appears to be useful to use the notation in (6.339).

As only the coefficients y_i will be of interest to us in this book, we only give the results for them. The renormalization group analysis proceeds as in $b \rightarrow s\ell^+\ell^-$ transitions, except that now the evolution has to be continued down to $\mu < m_c$. One finds including NLO corrections [131]

$$y_{7V} = P_0 + \frac{Y_0(x_t)}{\sin^2 \vartheta_W} - 4Z_0(x_t) + P_E E_0(x_t), \qquad y_{7A} = -\frac{1}{\sin^2 \vartheta_W} Y_0(x_t), \tag{6.340}$$

with Y_0, Z_0 and E_0 given in (6.27), (6.28), and (6.14), respectively. P_E is $O(10^{-2})$, and consequently the last term in (6.340) can be neglected. The next-to-leading QCD corrections to these coefficients enter only P_0. They have been calculated in [131] reducing certain ambiguities present in leading order analyses [191, 192]. It is not possible to show an analytic expression for P_0 as the additional renormalization group evolution from $\mu = m_b$ down to $\mu \leq m_c$ complicates the expressions. We refer to [111, 131] for further details and quote the value of P_0 in the NDR scheme for $\mu = 1.0 \pm 0.2\,\text{GeV}$ using 2019 input parameters

$$P_0 = 2.88 \pm 0.06, \tag{6.341}$$

which is sufficient for our purposes. The error includes also uncertainty in α_s.

6.7 $d \rightarrow sv\bar{v}$ and $b \rightarrow sv\bar{v}$ Transitions

6.7.1 $d \rightarrow sv\bar{v}$

We will now move to discuss the transition $d \rightarrow sv\bar{v}$ that governs the semileptonic rare decays $K^+ \rightarrow \pi^+ v\bar{v}$ and $K_L \rightarrow \pi^0 v\bar{v}$. $K^+ \rightarrow \pi^+ v\bar{v}$ is CP conserving while $K_L \rightarrow \pi^0 v\bar{v}$ is governed by CP violation. Within the SM these decays are loop-induced semileptonic FCNC processes receiving only contributions from Z^0-penguin and box diagrams and are governed by the single-function $X_0(x_t)$ given in (6.26). This function has been calculated in Section 6.2. A particular and very important virtue of $K \rightarrow \pi v\bar{v}$ decays is their clean theoretical character. This is related to the fact that the low-energy hadronic matrix elements required for the calculations of their branching ratios are just the matrix elements of quark currents between hadron states, which can be extracted from the leading (nonrare) semileptonic decays.

The investigation of these low-energy rare decay processes, in conjunction with their theoretical cleanliness, allows to probe, albeit indirectly, high-energy scales of the theory, far beyond the reach of the LHC. They are also very sensitive to the values of the CKM parameters, in particular to V_{td} and $\mathrm{Im}\lambda_t = \mathrm{Im}V_{ts}^* V_{td}$ so that the latter could in principle be extracted from precise measurements of the decay rates for $K^+ \to \pi^+ \nu\bar\nu$ and $K_L \to \pi^0 \nu\bar\nu$, respectively. Moreover, the combination of these two decays offers one of the cleanest measurements of $\sin 2\beta$ [193]. However, the very fact that these processes are based on higher-order electroweak effects implies that their branching ratios are expected to be very small and not easy to access experimentally.

As of 2020 one can look back at four decades of theoretical efforts to calculate the branching ratios for these two decays within the SM. Among early calculations are [5, 132] in which QCD corrections were not considered. The first LO QCD corrections have been calculated in [194, 195] and the NLO ones in the 1990s [181, 182, 196, 197]. Already the NLO calculations reduced significantly various renormalization scale uncertainties present at LO. Yet, in the last twenty years further progress has been made through the following calculations:

- NNLO QCD corrections to the charm contributions: [198–200].
- Isospin breaking effects and nonperturbative effects: [201, 202].
- Complete NLO electroweak corrections to the charm-quark contribution to $K^+ \to \pi^+ \nu\bar\nu$: [203].
- Complete NLO electroweak corrections to the top-quark contribution to $K^+ \to \pi^+ \nu\bar\nu$ and $K_L \to \pi^0 \nu\bar\nu$: [135].

On the experimental side the NA62 experiment at CERN is presently running and is expected to measure the $K^+ \to \pi^+ \nu\bar\nu$ branching ratio with the precision of 10% by 2021, as described in [204], that would improve the accuracy of the previous measurement by a factor of five. The expected measurement of $K_L \to \pi^0 \nu\bar\nu$ by KOTO at J-PARC, [205, 206], should reach the SM level by 2022. Experimental reviews of these two decays can be found in [205, 207–210] and the power of these decays in testing energy scales as high as several hundreds of TeV has been demonstrated in [211]. We will return to this in Chapter 19.

But before we come to discuss all this we have to take a closer look at the effective Hamiltonians for these decays first within the SM.

The effective Hamiltonian for $K^+ \to \pi^+ \nu\bar\nu$ can be written as

$$\mathcal{H}_{\mathrm{eff}}(K^+ \to \pi^+ \nu\bar\nu) = \frac{G_F}{\sqrt{2}} \frac{\alpha}{2\pi \sin^2 \vartheta_W} \sum_{l=e,\mu,\tau} \left(V_{cs}^* V_{cd} X_{\mathrm{NNL}}^l + V_{ts}^* V_{td} X(x_t) \right) (\bar s d)_{V-A} (\bar\nu_l \nu_l)_{V-A}.$$

$$(6.342)$$

The index $l = e$, μ, τ denotes the lepton flavor. The dependence on the charged lepton mass resulting from the box diagram is negligible for the top contribution. In the charm sector this is the case only for the electron and the muon but not for the τ-lepton. We have discussed it and even calculated it in Section 6.2.2.

The function $X(x_t)$ relevant for the top part is given by

$$X(x_t) = X_0(x_t) + \frac{\alpha_s}{4\pi} X_1(x_t) = \eta_X \cdot X_0(x_t), \qquad \eta_X = 0.994 \qquad (6.343)$$

with the QCD correction

$$X_1(x_t) = \tilde{X}_1(x_t) + 8 x_t \frac{\partial X_0(x_t)}{\partial x_t} \ln x_\mu. \qquad (6.344)$$

Here $x_\mu = \mu_t^2/M_W^2$ with $\mu_t = O(m_t)$ and $\tilde{X}_1(x_t)$ is a complicated function given in [181, 182].

The μ_t-dependence of the last term in (6.344) cancels to the considered order the μ_t-dependence of the leading term $X_0(x_t(\mu_t))$. The leftover μ_t-dependence in $X(x_t)$ is below 1%. The factor η_X summarizes the NLO corrections represented by the second term in (6.343). With $m_t \equiv m_t(m_t)$ the QCD factor η_X is practically independent of m_t and α_s and is very close to unity. Moreover, complete NLO electroweak corrections to this function have been calculated in [135]. They reduced significantly the dependence on the definition of electroweak parameters, in particular $\sin^2 \vartheta_W$.

Including all these corrections the present best value is given by [212]

$$\boxed{X(x_t) = 1.481 \pm 0.005_{\text{th}} + 0.008_{\text{exp}} = 1.481 \pm 0.009.} \qquad (6.345)$$

The contribution corresponding to $X(x_t)$ in the charm sector is represented by the functions X_{NNL}^l that are known, including QCD NLO [181, 197] and NNLO corrections [198, 199]. They also include complete two-loop electroweak contributions [203].

The expressions for X_{NNL}^l are very complicated. For our purposes it will be sufficient to collect these results in a parameter that directly enters the branching ratio, which we will derive in Section 9.5. It is

$$P_c(X) = P_c^{\text{SD}}(X) + \delta P_{c,u}, \qquad \delta P_{c,u} = 0.04 \pm 0.02, \qquad (6.346)$$

with the short-distance part given by

$$P_c^{\text{SD}}(X) = \frac{1}{\lambda^4} \left[\frac{2}{3} X_{\text{NNL}}^e + \frac{1}{3} X_{\text{NNL}}^\tau \right], \qquad (6.347)$$

and the long-distance contributions $\delta P_{c,u}$ calculated in [201]. Future lattice calculations could reduce the present error in this part [213], and in fact first steps in this direction have been made in [214, 1337].

An excellent approximation for $P_c^{\text{SD}}(X)$, including QCD and electroweak corrections listed earlier, is given as a function of $\alpha_s(M_Z)$ and $m_c(m_c)$ in (50) of [203]. Using this formula for the most recent input parameters [35, 215]:

$$\lambda = 0.2252(9), \qquad m_c(m_c) = 1.279(13)\,\text{GeV}, \qquad \alpha_s(M_Z) = 0.1181(6), \qquad (6.348)$$

one finds [212]

$$P_c^{\text{SD}}(X) = 0.365 \pm 0.012. \qquad (6.349)$$

Adding the long-distance contribution in (6.346), we finally find

$$P_c(X) = 0.405 \pm 0.024, \tag{6.350}$$

where we have added the errors in quadratures. We will use this value in this book.

The effective Hamiltonian for $K_L \to \pi^0 \nu\bar{\nu}$ is given as follows:

$$\mathcal{H}_{\text{eff}}(K_L \to \pi^0 \nu\bar{\nu}) = \frac{G_F}{\sqrt{2}} \frac{\alpha}{2\pi \sin^2 \vartheta_W} \sum_{l=e,\mu,\tau} V_{ts}^* V_{td} X(x_t) (\bar{s}d)_{V-A} (\bar{\nu}_l \nu_l)_{V-A} + h.c., \tag{6.351}$$

where the function $X(x_t)$, present already in $K^+ \to \pi^+ \nu\bar{\nu}$, is given in (6.345).

The branching ratios for both decays using the Hamiltonians (6.342) and (6.351) will be calculated in Section 9.5, where also the phenomenology of these decays within the SM will be presented in detail. But the most important role of these two decays is in the search for NP. We will witness it in the last part of this book.

6.7.2 $b \to s\nu\bar{\nu}$

These decays are governed in the SM by the penguin and box diagrams considered in the case of $d \to s\nu\bar{\nu}$ except that b has to be replaced by d. The calculation of these loop diagrams is not modified, and as the relevant operators have no anomalous dimensions, also QCD corrections in the effective Hamiltonian are the same. Therefore the effective Hamiltonian for $b \to s\nu\bar{\nu}$ transitions can be easily obtained from the one for $d \to s\nu\bar{\nu}$ by simply changing $b \to d$ both in operators and the CKM factors.

Now these factors are

$$V_{cs}^* V_{cb} \approx V_{ts}^* V_{tb} = 4 \cdot 10^{-2}, \tag{6.352}$$

and consequently the hierarchy of charm and top contributions is governed by the values of the loop functions. These functions to first approximation are given by $X_0(x_c)$ and $X_0(x_t)$, implying that the charm contribution is irrelevant for $b \to s\nu\bar{\nu}$ transitions.

The effective Hamiltonian in the SM is then given by

$$\mathcal{H}_{\text{eff}}(b \to s\nu\bar{\nu}) = \frac{G_F}{\sqrt{2}} \frac{\alpha}{2\pi \sin^2 \vartheta_W} \sum_{l=e,\mu,\tau} V_{ts}^* V_{tb} X(x_t) (\bar{s}b)_{V-A} (\bar{\nu}_l \nu_l)_{V-A}, \tag{6.353}$$

and the decays to be studied in this case are

$$B \to K\nu\bar{\nu}, \qquad B \to K^* \nu\bar{\nu}, \qquad B \to X_s \nu\bar{\nu}. \tag{6.354}$$

As the function $X(x_t)$, including QCD corrections, is precisely the same as the one in (6.345), there are strong correlations between the latter decays and $K \to \pi\nu\bar{\nu}$ in the SM [216]. We will return to this correlation and its breakdown in the BSM part of this book.

The branching ratios of the decays in (6.354), based on the effective Hamiltonian (6.353), will be presented in Section 9.6 where phenomenological virtues of these decays will be discussed in detail.

Nonperturbative Methods in Weak Decays

7.1 General View

We have seen in previous chapters that the calculations of short-distance contributions to various decays of mesons, represented by Wilson coefficients, reached already within the SM a very satisfactory level. But to obtain the values of various decay rates and branching ratios, also long-distance (LD) contributions have to be included. We have seen in Chapter 3 that these effects depend strongly on the decays considered.

In leptonic decays of mesons the dominant LD effects are encoded in weak decay constants, like F_K, F_{B_d}, F_{B_s}. By now they are calculated with respectable precision by lattice QCD (LQCD). We collected their values in Table D.3. See also Table 7.1.

In semileptonic decays various formfactors have to be calculated. For K meson decays, LQCD and chiral perturbation theory (ChPT) appear presently to provide the most accurate results. In particular calculations of strong isospin breaking corrections and of QED effects are dominated by ChPT.

The most difficult are the calculations of LD effects in nonleptonic decays and transitions like $K^0 - K^0$ mixing that are dominantly contained in the matrix elements of local operators. Here still significant progress has to be made.

In $K \to \pi\pi$ decays and $K^0 - \bar{K}^0$ mixing, the so-called Dual QCD (DQCD) approach provided already in the 1980s first results, and these calculations have been improved recently. Among analytic approaches DQCD remains still the leading approach for the calculation of hadronic matrix elements in K meson system in QCD. This is related to the fact that it is presently the only analytic approach that allows to match the LD QCD effects in hadronic matrix elements to SD ones encoded in the Wilson coefficients. Such matching is presently not possible in ChPT, implying very large uncertainties in the calculations of $K \to \pi\pi$ decays in this framework. This deficiency of ChPT precludes also an accurate estimate of strong isospin breaking corrections to $K \to \pi\pi$ within this approach. A better estimate should be available in 2020 from DQCD and in a few years from LQCD.

LQCD provides already now the best results for hadronic matrix elements relevant for $K^0 - \bar{K}^0$, $B^0_{s,d} - \bar{B}^0_{s,d}$, and $D^0 - \bar{D}^0$ mixings with some competition only from DQCD but only for $K^0 - \bar{K}^0$ mixing. It is satisfying that in the latter case, the results from LQCD and DQCD agree well with each other. Moreover, DQCD provides some insight into the dynamics behind the lattice values that are obtained by very demanding numerical simulations.

In fact DQCD turned over three decades by now to be an efficient approximate method for obtaining results for nonleptonic decays, years and even decades before useful results from numerically sophisticated lattice calculations could be obtained. But recently LQCD made impressive progress not only for $K^0 - \bar{K}^0$ mixing but also for $K \to \pi\pi$ decays, and a better comparison of the LQCD and DQCD results is now possible. This is important because also in the case of the latter decays DQCD provides some insight into the LQCD results.

The situation with nonleptonic decays of heavy mesons, in particular very important two-body B decays, is very different. Here DQCD has nothing to say, and LQCD can only help other approaches based on the so-called QCD-factorization (QCDF) approach, heavy quark expansions, and heavy quark effective theory by calculating simpler objects than hadronic matrix elements of four-quark operators present in these methods.

This chapter describes in some detail the DQCD approach (Section 7.2), which is easy for me as I took part in developing the applications of this approach to $K^0 - \bar{K}^0$ mixing and $K \to \pi\pi$ decays. We will also summarize the results of LQCD (Section 7.3) and present the general structure of the QCDF approach to two-body B decays (Section 7.4). Some general statements on heavy quark expansions and heavy quark effective theory will be given in Section 7.5. Finally, we will provide a few references to reviews and lectures in which other nonperturbative methods are presented.

In this chapter we will summarize only the results relevant for the SM. The ones obtained in the extensions of the SM, in which new operators are present, will be discussed in Chapter 13.

7.2 Dual QCD Approach

7.2.1 Preliminaries

This approach is based on the conjecture of 't Hooft [217, 218] and subsequently Witten [219, 220] that QCD (the theory of quarks and gluons) is for a large number of colors N equivalent to a theory of weakly interacting mesons with a quartic meson coupling being $O(1/N)$. This allows us to formulate a dual representation of the strong dynamics in terms of hadronic degrees of freedom. In the large N limit, this representation becomes exact, and a full description of the physics can be achieved using an infinite set of interacting meson fields. For large N QCD simplifies significantly, in particular in the strict large N limit QCD is dual to a free theory of mesons. In this section we will present the application of this dual QCD approach (DQCD) to the calculation of hadronic matrix elements of local operators as proposed in 1986 by Bardeen, Gérard, and myself [221] and developed by us in a series of papers since then. But let us first review the first attempts of using large N ideas in weak decays.

The first attempts to apply $1/N$ expansion to weak decays can be found in [222–224]. However, the first big step forward in the phenomenological applications of this expansion was made in [225] in the context of nonleptonic charm-meson decays, where it was realized that removing the $1/N$ Fierz terms from the so-called vacuum insertion approximation (VIA) softened the disagreement of the theory with both exclusive and inclusive data.

This procedure has been motivated by the analysis in [226]. However, these authors did not attach it with a consistent application of the $1/N$ expansion.

VIA is just a method for evaluating hadronic matrix elements of four-quark operators by just inserting a vacuum or a single meson between the two currents in these operators, and then the product of the resulting matrix elements can easily be calculated. Moreover by making Fierz reordering of quarks, additional $1/N$ corrections to the matrix elements can be obtained.

In the strict large N limit the matrix elements of four-quark operators indeed factorize into a product of matrix elements of quark currents, and as done in [225] one could calculate all hadronic matrix elements for nonleptonic matrix elements in this manner. But it was soon realized that combining these results with the WCs of the contributing operators was problematic as the final result was renormalization-scale and renormalization-scheme dependent. As we have seen in previous chapters these dependences are present in the WCs and must be cancelled by those present in the hadronic matrix elements. But the matrix elements of conserved currents are free of such dependences, and consequently this is also the case of matrix elements calculated in this manner. Moreover, the $1/N$ corrections resulting in VIA from Fierz-reordering do not represent correct $1/N$ QCD corrections to hadronic matrix elements as we will stress later.

In what follows we will discuss the DQCD approach of [221, 227–229], which goes beyond the large N limit and thereby allows to remove the unphysical scale and renormalization-scheme dependences in question by including *nonfactorizable contributions*. We will discuss this approach in the context of K meson flavor physics, and readers not familiar with this subject should refresh the information gained in Sections 6.3 and 6.4 and look first briefly at Section 8.1 and Chapter 10 to get an idea what the following objects mean and why they are important:

$$\hat{B}_K, \quad B_6^{(1/2)}, \quad B_8^{(3/2)}, \quad \Delta M_K, \quad \varepsilon_K, \quad \varepsilon'/\varepsilon, \quad \mathrm{Re}A_0, \quad \mathrm{Re}A_2. \tag{7.1}$$

In particular the isospin amplitudes $A_{0,2}$ are defined in (8.29)–(8.31).

But dropping just $1/N$ terms in the VIA when calculating $\hat{B}_K$, $B_6^{(1/2)}$, and $B_8^{(3/2)}$ still gives correct results in the large N limit, and this simple philosophy of using $1/N$ expansion has been applied by Gérard and myself to $K \to \pi\pi$ decays, ΔM_K, and ε_K in [230]. The first leading order results for the matrix elements of operators relevant for these observables can be found in that paper. The most important results obtained there are $\hat{B}_K = 3/4$, also noticed in [231], and the realization that the removal of $1/N$ Fierz terms from VIA to current-current matrix elements suppresses $\mathrm{Re}A_2$, moving the theory in the direction of the data. In this paper also the first large N result for the matrix elements of the dominant QCD-penguin operator Q_6 can be found. In this case the problematic issue of scale dependence is practically absent because the Q_6 operator has a different structure than current-current operators considered in [225]. Q_6 is a density-density operator, and its factorized matrix element is scale dependent even in the large N limit. In fact, the product $C_6(\mu)\langle Q_6(\mu)\rangle$ is for $\mu \geq 1\,\mathrm{GeV}$ basically scale independent, small μ dependence coming from the mixing with other operators. We will demonstrate this later using the results from previous chapters.

However, just dropping Fierz terms in the VIA results is really not a satisfactory procedure for the calculation of hadronic matrix elements and in particular does not provide the explanation of the $\Delta I = 1/2$ rule as given in (7.28). Moreover, as we will see later, this procedure fails in the case of operators built out of scalar currents, giving totally wrong results. What was still missing in these first papers was an effective Lagrangian describing the weak and strong interactions of mesons in the large N limit and allowing for a systematic evaluation of $1/N$ and higher-order corrections to this limit.

With these goals in mind Bardeen, Gérard, and myself developed in the second half of the 1980s an approach to $K^0 - \bar{K}^0$ mixing and nonleptonic K-meson decays [221, 227–229] based on an effective Lagrangian corresponding to the dual representation of QCD as a theory of weakly interacting mesons for large N [217–220] mentioned earlier. In the first application of this approach in [227, 228] we confirmed the leading order result for $\langle Q_6 \rangle$ [230]. Moreover, we have pointed out that the property of factorization of hadronic matrix elements in the large N limit is automatic in DQCD because in this limit QCD becomes a free theory of mesons. But the main virtue of this approach is the ability to combine consistently, even if approximately, hadronic matrix elements of four-quark operators with the renormalization group improved values of the corresponding WCs.

Our studies of the 1980s culminated in the formulation of the *meson evolution* [221, 229], to be discussed later, and the evaluation in this framework of $1/N$ corrections to $K \to \pi\pi$ amplitudes and to the parameter $\hat{B}_K$ relevant for $K^0 - \bar{K}^0$ mixing. These papers represent the first attempt at a consistent calculation of the weak matrix elements in the continuum field theory. Pedagogical summary of this work has been presented by us in various reviews and lectures [232–238] and more recently in [239, 240].

This approach provided, in particular, first results within QCD for the $K \to \pi\pi$ isospin amplitudes $\mathrm{Re}A_0$ and $\mathrm{Re}A_2$ in the ballpark of experimental values. In this manner, for the first time, the SM dynamics behind the $\Delta I = 1/2$ rule has been identified. In particular, it has been emphasized that at scales $O(1\,\mathrm{GeV})$ long-distance dynamics in hadronic matrix elements of current-current operators and not QCD-penguin operators, as originally proposed in [241], are dominantly responsible for this rule. Moreover, it has been demonstrated analytically why $\mathrm{Re}A_0$ is enhanced and why $\mathrm{Re}A_2$ is suppressed relative to the VIA estimates. In this context, we have emphasized that the Fierz terms in the latter approach totally misrepresent $1/N$ corrections to the strict large N limit for these amplitudes. In particular their sign is wrong.

Our approach, among other applications, allowed us to consistently calculate, for the first time within QCD, the nonperturbative parameters $\hat{B}_K$, $B_6^{(1/2)}$, and $B_8^{(3/2)}$ governing the corresponding matrix elements of the $\Delta S = 2$ SM current-current operator and $K \to \pi\pi$ matrix elements of the dominant QCD-penguin (Q_6) and electroweak penguin (Q_8) operators that we encountered in Section 6.4. As we will see in the next chapters these parameters are crucial for the evaluation of ε_K and ε'/ε within the SM and its various extensions. Also the $K \to \pi\pi\pi$ decays have been analyzed in [242] and the $K_L - K_S$ mass difference ΔM_K including long-distance contributions has been calculated [236, 243] within this approach. During the last two decades some of these calculations have been improved and extended, in particular in [239, 244, 245], which we will briefly review soon.

It is interesting and encouraging that most of our results have been confirmed by several recent LQCD calculations that we will specify in the next section. While the LQCD approach has a better control over the errors than our approach, it does not provide the physical picture of the dynamics behind the obtained numerical results. This is in particular seen in the case of the $\Delta I = 1/2$ rule, where our analytic approach offers a very simple picture of the dynamics behind this rule, as first pointed out in [221], more recently summarized in [239, 240], and described in this section. Moreover, to obtain reliable results for $K \rightarrow \pi\pi$ amplitudes in LQCD required already a quarter of a century, whereas all calculations performed in the DQCD taken together required less than three years and involved teams of two to three physicists compared to large LQCD collaborations. Yet, it should be stressed that LQCD will provide ultimately more precise results than is possible in our approach and has a broader spectrum of quantities for which the calculations can be done than using DQCD.

While until 2010 DQCD was ahead of lattice calculations, this changed in this decade for several quantities in view of the increased computer power and new ideas. Yet, as we will demonstrate in this section and in Chapter 13, DQCD still allows us to obtain an insight in the numerical LQCD results that cannot be gained by LQCD. Moreover, all of the hadronic matrix elements of BSM operators of dimension-six entering $K \rightarrow \pi\pi$ decays have been calculated recently in DQCD [246], while their LQCD values remain still unknown.

Other applications of large N ideas to $K \rightarrow \pi\pi$ and $\hat{B}_K$, but in a different spirit than our original approach, are reviewed in [50, 247]. We refer in particular to [248–256]. A review of SU(N) gauge theories at large N can be found in [257]. Here we present only our approach following closely our published work.

7.2.2 Basic Framework

General Structure

To describe our framework in explicit terms, let us return to $K \rightarrow \pi\pi$ decay amplitudes in the SM [123]

$$A(K \rightarrow \pi\pi) = \frac{G_F}{\sqrt{2}} V_{ud} V_{us}^* \sum_{i=1}^{10} (z_i(\mu) + \tau y_i(\mu)) \langle \pi\pi | Q_i(\mu) | K \rangle, \qquad \tau \equiv -\frac{V_{td} V_{ts}^*}{V_{ud} V_{us}^*},$$

$$(7.2)$$

with the operators Q_i discussed already at length in Section 6.4 and numerical values of the Wilson coefficients $z_i(\mu)$ and $y_i(\mu)$ given in Chapter 10. In our presentation only four of them will be relevant, namely the two current-current operators Q_1, Q_2, the QCD penguin operator Q_6, and the electroweak penguin operator Q_8. The operator Q_4 plays also some role as can be found in our papers, but we will not discuss it here.

In the DQCD it is useful to write these operators in the Fierz transformed form relative to the ones considered in previous chapters so that they are given as products of color singlet currents and densities, as follows

$$Q_1 = 4(\bar{s}_L \gamma_\mu d_L)(\bar{u}_L \gamma_\mu u_L), \qquad Q_2 = 4(\bar{s}_L \gamma_\mu u_L)(\bar{u}_L \gamma_\mu d_L), \qquad (7.3)$$

$$Q_6 = -8(\bar{s}_L q_R)(\bar{q}_R d_L), \qquad Q_8 = -12 e_q(\bar{s}_L q_R)(\bar{q}_R d_L), \qquad (7.4)$$

where e_q is the electric quark charge.

We have already used this representation for Q_6 when calculating anomalous dimensions of QCD penguin operators in Section 6.4. Here $q_{R(L)} = (1/2)(1 \pm \gamma_5)q$ and sums over color indices and $q = u, d, s$ in Q_6 and Q_8 are understood. In this manner the leading, in $1/N$ expansion, contributions can be calculated as products of matrix elements of currents in the case of $Q_{1,2}$ and densities in the case of Q_6 and Q_8.

Because the operators Q_i in (7.3)–(7.4) are constructed from the light quark fields only, the full information about the heavy quark fields (c, b, t) is contained in the Wilson coefficients z_i and y_i. Correspondingly, the normalization scale μ in (7.2) is not completely arbitrary in our approach but must be chosen below the charm-quark mass. In our large N approach the structure of different contributions to physical amplitudes is then as follows. The physics contributions from scales above μ are fully contained in the coefficients $z_i(\mu)$ and $y_i(\mu)$, whereas the remaining contributions from the low-energy physics below μ (i.e., from μ to the factorization scale expected around m_π) are contained in the matrix elements $\langle \pi\pi |Q_i(\mu)|K \rangle$. It follows that for $\mu = O(1\,\text{GeV})$, the coefficients $z_i(\mu)$ and $y_i(\mu)$ can be calculated within a perturbative *quark-gluon picture* by means of renormalization group methods considered in previous chapters.

As far as the meson matrix elements are concerned, the ultimate goal is to compute them in a nonperturbative quark-gluon picture where mesons occur as bound states. This route is followed by lattice computations and in fact, as summarized in the next section, since our work appeared in 1986 impressive progress has been made in this manner. Yet this numerical route is very demanding as even after more than a quarter of a century of hard work by the LQCD community the present results for $K \to \pi\pi$ amplitudes are still not fully satisfactory. Moreover, it is much harder to understand the underlying physics than by means of our analytic approach as we already stressed earlier and will demonstrate in particular in Chapter 13.

Our proposal, summarized most explicitly in [221], was to apply instead the DQCD approach of 't Hooft [217, 218] and Witten [219, 220] to nonleptonic K decays and K^0–$\bar{K}^0$ mixing. The fact that QCD can be formulated both as the theory of quarks and gluons on the one hand and as the theory of mesons on the other hand can now be used for these processes as follows. The main point is that the matrix elements of four-fermion operators governing these transitions can be written at leading order in large N as products of matrix elements of color singlet currents in the case of current-current operators and as products of matrix elements of quark densities in the case of penguin operators. This is evident from (7.3) and (7.4).

At the next-to-leading order one has two classes of contributions:

- $1/N$ corrections to the matrix elements of factorized operators
- Low-energy, nonfactorized matrix elements of two currents or two quark densities

The latter contributions can be written as an integral over the momentum flowing through the currents (densities) in the connected *planar amplitude*. One can then use our knowledge of both the high- and low-energy behavior of the integrand. At high momentum, these are just the short-distance contributions to the coefficient functions of the OPE, which can be computed perturbatively in the quark-gluon picture. While in principle this could also be done in the *meson picture*, such an analysis would be very complex requiring many meson states and complicated interactions. However the long-distance analysis is correspondingly simple as only lowest-lying meson states may be required, and the interactions are largely dictated by the chiral symmetry structure of the effective Lagrangian. Moreover the spontaneous breakdown of chiral symmetry $SU(3)_L \times SU(3)_R \to SU(3)_V$ can be proven to be true in QCD in its large N limit [258].

Our proposal in [221] was to use the meson theory to interpolate to the point where one can match the behavior of the integrand of the short-distance theory. If the amplitude is smooth enough then it may be sufficient to match the meson amplitude to the quark amplitude at an appropriate scale. In this manner one can achieve a consistent unified description of the physics by using the quark-gluon picture at short distances matched to the meson picture at long distances. The accuracy of the method depends on the interpolation of the integrand between short and long distances.

A full AdS/QCD description [259, 260] should be able to interpolate the meson amplitudes to arbitrarily short distance, and first attempts in this direction have been made in [261, 262]. In our approach the matching scale must presently be chosen around 1 GeV, implying approximate treatments in both pictures. In particular, the scheme dependence of the long-distance part comes when one subtracts the short-distance part of the integral using a particular scheme. This scheme dependence can be treated exactly if needed. In this context, calculating Wilson coefficients and the hadronic matrix elements in a momentum scheme in [239] we have made a significant progress relative to our previous papers.

Despite not being exact, this approach has several virtues. Indeed, the simplicity of this formulation lies in the fact that in the strict large N limit QCD becomes a free theory of mesons, and consequently the leading order contributions to any quantity are obtained by calculating tree diagrams with the propagated objects being mesons, not quarks or gluons. In this strict limit, also the factorization of hadronic matrix elements of four-quark operators into the product of matrix elements of quark currents or quark densities follows. Beyond this limit, one obtains $1/N$ expansion represented by a loop expansion in the meson theory. It should be emphasized that in this loop expansion $1/N$ never explicitly appears and is replaced by $1/F_\pi^2$ as discussed later.

Even if naively these loop corrections could be expected to be small, one should notice that one-loop contributions in the meson theory represent, in fact, the leading term in the $1/N$ expansion for observables like the $\pi^+ - \pi^-$ electromagnetic mass difference or the K^0 decay into two neutral pions. In particular, they have to be sizable if one wants to explain why the subleading $K^0 \to \pi^0\pi^0$ decay amplitude turns out to be almost equal to the $K^0 \to \pi^+\pi^-$ leading one, namely the so-called $\Delta I = 1/2$ rule discussed in detail later.

An important issue in this framework is the matching between the quark-gluon and meson theories. In the quark-gluon picture, the scale μ enters naturally as the normalization scale in the renormalization group improved perturbative QCD calculations

$$\mu^2 \frac{d}{d\mu^2} Q_i(\mu^2) = -\frac{1}{2}\hat{\gamma}_{ij} Q_j(\mu^2), \tag{7.5}$$

with $\hat{\gamma}$, the anomalous dimension matrix for the Q_i operators. In our formulation, μ serves as an infrared cutoff below which one should switch to the meson picture unless one wants to perform lattice computations. Now the truncated meson theory, involving a finite set of light pseudoscalar and vector mesons only, appears nonrenormalizable. In particular, if only lowest-lying pseudoscalar mesons are included without ultraviolet QCD completion, it exhibits a quadratic dependence on the cutoff which we will denote by M. This *physical* cutoff must be introduced to restrict the truncated meson theory to the long-distance domain or, in other words, to cutoff the high-mass and high-momentum contributions in the meson loops. Therefore, the physical cutoff introduced here should be distinguished from the usual cutoff regularization procedure in which M could be sent to arbitrarily large values, to disappear from observables after renormalization.

On the other hand, we know that QCD being renormalizable has a logarithmic dependence on the ultraviolet cutoff. While this difference from the quadratic dependence on M in the truncated meson theory has been in the 1980s a subject of criticism of our approach, one should emphasize that these two dependences are not inconsistent with each other. Indeed, the strict logarithmic cutoff dependence of QCD is valid only at short distances, whereas power counting supplemented with chiral symmetry requires quadratic dependence on the cutoff for the long-distance behavior of QCD. For high values of M, after the inclusion of vector mesons and heavier meson states, this quadratic dependence on M should smoothly turn into a logarithmic dependence as expected in the full meson theory. In fact, as demonstrated in [239], already the inclusion of vector mesons shows that this expectation is correct.

In the evaluation of the matrix elements $\langle \pi\pi | Q_i(\mu) | K \rangle$ the simplest choice one can make is $\mu = M$. This identification of μ with M is certainly an idealization in the approximate treatment used in our papers in 1980s, but has been improved in [239]. In particular, to relate μ to M in a meaningful manner, we went beyond the Fermi limit for the W-propagator and calculated at NLO the Wilson coefficients not in the usual NDR-$\overline{\text{MS}}$ scheme but in a momentum scheme. But as the Wilson coefficients in momentum scheme and NDR-$\overline{\text{MS}}$ scheme are related by simple one-loop shifts, one can finally use the latter scheme for phenomenology. Moreover, the inclusion of the lowest-lying vector mesons improved the matching between meson and quark pictures significantly. We refer also to a very nice earlier analysis in [263] in which this improved matching has been discussed first.

Another Look

Let us still formulate these ideas using the language of previous chapters. Our goal is the calculation of a decay amplitude, which one can write generally as follows

$$\mathcal{A} = \langle \vec{Q}^T(\mu) \rangle \vec{C}(\mu) \tag{7.6}$$

with $\mu < m_c$ in our approach but otherwise arbitrary. What is easy to do is

- The calculation of $\vec{C}(M_W)$, which can be done in ordinary perturbation theory.
- The calculation of $\langle \vec{Q}^T(0) \rangle$, corresponding to strict large N limit in which these matrix elements are calculated as products of quark currents or quark densities. Here to simplify the notation, we just replace the factorization scale by $\mu = 0$.

We must next fill the gap between $\mu = 0$ and $\mu = M_W$ to include all physics contributions. Our idea is to divide the renormalization group evolution into a *short-distance evolution* from M_W to $\mu = O(1 \, \text{GeV})$ to be called *quark-gluon evolution* described by

$$\vec{C}(\mu) = \hat{U}(\mu, M_W) \, \vec{C}(M_W), \tag{7.7}$$

and a *long-distance evolution* from scale 0 *up to* μ, to be called *meson evolution* as it is performed in the dual representation of QCD. The latter evolution is described by

$$\langle \vec{Q}^T(\mu) \rangle = \langle \vec{Q}^T(0) \rangle \, \hat{U}(0, \mu), \tag{7.8}$$

with $\hat{U}(0, \mu)$ being the evolution matrix in the long-distance regime. As

$$\hat{U}(0, \mu)\hat{U}(\mu, M_W) = \hat{U}(0, M_W), \tag{7.9}$$

we find inserting (7.7) and (7.8) into (7.6)

$$\mathcal{A} = \langle \vec{Q}^T(0) \rangle \hat{U}(0, M_W) \vec{C}(M_W), \tag{7.10}$$

so that all contributions from scales between $\mu = 0$ and $\mu = M_W$ are included in this manner.

It should be remarked that the relation (7.8) can be generalized to

$$\langle \vec{Q}^T(m_2) \rangle = \langle \vec{Q}^T(m_1) \rangle \, \hat{U}(m_1, m_2), \tag{7.11}$$

with $m_1 < m_2$ so that knowing the hadronic matrix elements at a given scale one can calculate them at a different scale. This is in particular useful if both scales are in perturbative regime and the standard renormalization group technology can be used.

Having these formulas at hand lets us make the following observation [123]. The evolution of $\vec{C}(\mu)$ in (7.7) is governed by $\hat{\gamma}^T$ with $\hat{\gamma}$ being the anomalous dimension matrix. On the other hand, the evolution of $\langle \vec{Q}(\mu) \rangle$ is governed by $\hat{\gamma}$. As this matrix is rather asymmetric, the structure of the evolution in (7.11) is rather different from the evolution in (7.7). Thus, whereas the evolution of $C_{1,2}(\mu)$ is unaffected by the presence of penguin contributions, the evolution of $\langle Q_{1,2}(\mu) \rangle$ depends on the size of $\langle Q_i(\mu) \rangle$, $i \neq 1, 2$. Conversely, whereas the evolution of $C_i(\mu)$, $i \neq 1, 2$ depends on the size of $C_{1,2}(\mu)$, the evolution of the matrix elements of penguin operators $\langle Q_i(\mu) \rangle$, $i \neq 1, 2$ is a sole penguin affair.

Basic Lagrangian of the Truncated Meson Theory

The explicit calculation of the contributions of pseudoscalars to hadronic matrix elements of local operators is based on a truncated chiral Lagrangian describing the low-energy interactions of the lightest mesons [227, 228, 264]

$$L_{\text{tr}} = \frac{F^2}{8} \left[\text{Tr}(D_\mu U D^\mu U^\dagger) + r \text{Tr}(m(U + U^\dagger)) - \frac{r}{\Lambda_\chi^2} \text{Tr}(m(D^2 U + D^2 U^\dagger)) \right], \tag{7.12}$$

where

$$U = \exp\left(i\sqrt{2}\frac{\Pi}{F}\right), \qquad \Pi = \sum_{a=1}^{8} \lambda^a \pi^a \tag{7.13}$$

is the unitary chiral matrix describing the octet of light pseudoscalars and transforming as $U \to g_L U g_R^\dagger$ under the chiral $SU(3)_L \times SU(3)_R$. The parameter F is related to the weak decay constants $F_\pi \approx 130\,\text{MeV}$ and $F_K \approx 156\,\text{MeV}$ through

$$F_\pi = F\left(1 + \frac{m_\pi^2}{\Lambda_\chi^2}\right), \qquad F_K = F\left(1 + \frac{m_K^2}{\Lambda_\chi^2}\right). \tag{7.14}$$

It should be emphasized that to adjust the normalization of F_π to the one used in the rest of this book, we replaced the parameter f_π in our original DQCD papers by $F = \sqrt{2}f_\pi$.

The diagonal mass matrix m involving m_u, m_d, and m_s is such that

$$r(\mu) = \frac{2m_K^2}{m_s(\mu) + m_d(\mu)}, \tag{7.15}$$

with $r(1\,\text{GeV}) \approx 3.75\,\text{GeV}$ for $(m_s + m_d)(1\,\text{GeV}) \approx 132\,\text{MeV}$. The flavor-singlet η_0 meson decouples due to large mass m_0 generated by the nonperturbative $U(1)_A$ anomaly. Consequently the matrix Π reads

$$\Pi = \begin{pmatrix} \pi^0 + \frac{1}{\sqrt{3}}\eta_8 & \sqrt{2}\pi^+ & \sqrt{2}K^+ \\ \sqrt{2}\pi^- & -\pi^0 + \frac{1}{\sqrt{3}}\eta_8 & \sqrt{2}K^0 \\ \sqrt{2}K^- & \sqrt{2}\bar{K}^0 & -\frac{2}{\sqrt{3}}\eta_8 \end{pmatrix}. \tag{7.16}$$

In (7.12), $D_\mu U$ is the usual weak covariant derivative acting on the U field, and m is the real and diagonal quark mass matrix. At $O(p^2)$ and in the isospin limit $m_u = m_d = m_{ud}$,

$$m_\pi^2 = r\,m_{ud}, \quad m_K^2 = \frac{r}{2}(m_s + m_{ud}), \quad m_8^2 = \frac{4}{3}m_K^2 - \frac{1}{3}m_\pi^2. \tag{7.17}$$

It should be emphasized that the chiral Lagrangian in (7.12) must not be viewed as a normal effective tree Lagrangian but instead must be used as a fully interacting field theory including loop effects. In this sense we are providing a bosonization of the fundamental quark theory, where all the quark currents and densities, presented later, have a valid representation in terms of the meson fields. But in the truncated version, the meson representation is valid only for a proper description of long-distance physics.

The parameter Λ_χ in (7.12) sets the scale of higher-order terms, which are always expected in a truncated theory. It should be emphasized that this scale is a hadronic scale different from Λ_{QCD}. As shown in [221, 264] and evident from (7.14), its value can be determined from the physical pseudoscalar masses and decay constants:

$$\Lambda_\chi^2 = F_\pi \frac{m_K^2 - m_\pi^2}{F_K - F_\pi} + O\left(\frac{1}{N}\right) \Rightarrow \Lambda_\chi \approx 1.1\,\text{GeV}, \tag{7.18}$$

where we used the most recent lattice value for the ratio $F_K/F_\pi \approx 1.20$. The $1/N$ correction, calculated in [221], is positive and in the ballpark of 5–10% for the range of

M considered. As this correction is only logarithmically dependent on this scale, Λ_χ is practically independent of M with variation in the range $0.6\,\text{GeV} \le M \le 0.8\,\text{GeV}$ of less than 2%.

As stressed in [221], this cutoff independence of Λ_χ results only if the cutoff dependence of $F(M^2)$ following from our Lagrangian is taken into account. Explicit formula can be found in [221]. As $1/F^2(M^2)$, being $O(1/N)$, is the actual expansion parameter and

$$\frac{\partial}{\partial M^2}\left(\frac{1}{F^2(M^2)}\right) < 0, \tag{7.19}$$

it is the meson picture analog of the QCD running coupling in the quark picture.

The chiral Lagrangian (7.12) contains only terms with a single trace over flavor indices, which reflects the large N structure of QCD. The leading N contributions to any quantity are simply obtained from the tree diagrams, whereas the leading $1/N$ corrections are found by calculating the one-loop contributions. More generally, the $1/N$ expansion corresponds to the loop expansion characterized by inverse powers of $(4\pi F)^2$ ($F^2 \sim N$) with the strong interaction vertices given by the truncated Lagrangian in (7.12). As pointed out by John Donoghue, it is similar to an expansion in inverse powers of M_p^2 ($G_N = 1/M_p^2$) if one treats general relativity as an effective field theory for gravity, which is modified above the Planck scale by new degrees of freedom. Other details on the Lagrangian in (7.12) and explicit results of the relevant loop calculations can be found in [221] and in the lecture notes [233, 236].

The Structure of Hadronic Matrix Elements

The resulting matrix elements of *current-current* operators in this approach have then the structure ($i = 1, 2$)

$$\langle \pi\pi|Q_i(\mu)|K\rangle = A_i\sqrt{N}\left[1 + \frac{K_i(\mu)}{N} + O\left(\frac{1}{N^2}\right)\right], \tag{7.20}$$

where A_i and K_i are N-independent numerical expansion coefficients, which, in our approach, are given in terms of the parameters of the truncated Lagrangian. Note that the μ dependence in the matrix elements of $Q_{1,2}$ appears as a $1/N$ correction. This is consistent with the μ dependence of the Wilson coefficients $z_{1,2}(\mu)$ and reflects the simple fact that the anomalous dimensions of $Q_{1,2}$ vanish in the large N limit. See (7.40).

On the other hand, for *penguin* operators Q_6 and Q_8 the matrix elements have the structure ($i = 6, 8$)

$$\langle \pi\pi|Q_i(\mu)|K\rangle = \tilde{A}_i(\mu)\sqrt{N}\left[1 + \frac{\tilde{K}_i(\mu)}{N} + O\left(\frac{1}{N^2}\right)\right]. \tag{7.21}$$

The important difference relative to (7.20) is the appearance of the μ dependence already in the leading term. Again, this is consistent with the μ dependence of $z_{6,8}(\mu)$ and $y_{6,8}(\mu)$ and reflects the fact that the anomalous dimensions of density-density operators do not vanish in the large N limit but are twice the anomalous dimension of the mass operator.[1]

---

[1] See Section 4.3.

This fact allows a better matching of the truncated meson theory with the short-distance contributions of penguin operators than is possible for the current-current operators in the case of $K \to \pi\pi$ amplitudes. This will be clearer from explicit formulas for the matrix elements of penguin operators given later, and we will return to this point at the end of this section.

To calculate the matrix elements of the local operators in question, we need meson representation of quark currents and quark densities. They are directly obtained from the effective Lagrangian in (7.12) and are given, respectively, as follows

$$
\bar{q}^j_L \gamma_\mu q^i_L = i\frac{F^2}{8} \left\{ (\partial_\mu U)U^\dagger - U(\partial_\mu U^\dagger) - \frac{r}{\Lambda_\chi^2} \left[m(\partial_\mu U^\dagger) - (\partial_\mu U)m^\dagger \right] \right\}_{ij}, \quad (7.22)
$$

$$
\bar{q}^j_R q^i_L = -\frac{F^2}{8} r \left[U - \frac{1}{\Lambda_\chi^2} \partial^2 U \right]_{ij}. \quad (7.23)
$$

At tree level, corresponding to leading order in $1/N$, using these representations one simply expresses the operators in terms of the meson fields and expands the matrix U in powers of $1/F$. For $K^0 - \bar{K}^0$ mixing the relevant contribution to hadronic matrix elements is read off from terms involving only K^0 and $\bar{K}^0$ and for $K \to \pi\pi$ decays from terms involving K^0 or $K^\pm$ and two pions. When expressing derivatives in terms of four-momenta one has to take care that all fields in the amplitude resulting from the Lagrangian are incoming. Therefore for $K^0 - \bar{K}^0$ mixing with $K^0 \to \bar{K}^0$ and general relation $i\partial_\mu = p_\mu$ we have

$$
\partial_\mu K^0 = -ip_{K^0} K^0, \qquad \partial_\mu \bar{K}^0 = ip_{\bar{K}^0} \bar{K}^0, \quad (7.24)
$$

and for $K \to \pi\pi$

$$
\partial_\mu K^0 = -ip_{K^0} K^0, \qquad \partial_\mu \pi^\pm = ip_{\pi^\pm} \pi^\pm, \quad (7.25)
$$

with analogous rules for K^+ and π^0.

Next-to-leading corrections are then obtained from loop diagrams with weak vertices represented by the operators and strong vertices obtained from the effective Lagrangian in (7.12). Examples of the loop calculations with the list of the relevant integrals can be found in [239].

Here we give just one example of the parameter B_K, which is discussed in more detail in Section 7.2.4. Including both pseudoscalar and vector meson contributions, one finds [239]

$$
B_K(M) = \frac{3}{4} \left\{ 1 - \frac{1}{(4\pi F_K)^2} \left[\frac{7}{8}M^2 + \frac{3}{8}m_V^2 \ln\left(1 + \frac{M^2}{m_V^2}\right) + \frac{3}{4} \frac{m_V^2 M^2}{(M^2 + m_V^2)} \right] \right\}, \quad (7.26)
$$

where m_V is the vector meson mass and M the cutoff in the loop integral. If only pseudoscalar meson contributions are included $M \approx 0.7\,\text{GeV}$. With vector mesons taken into account, it can be increased to $M \approx 0.9\,\text{GeV}$. The important virtue of the inclusion of lightest vector mesons is the reduction of the coefficient in front of the quadratic term M^2. In the absence of vector meson contributions the factor 7/8 is increased to 2. This demonstrates that the inclusion of heavier vector mesons would eventually result in a purely logarithmic cutoff dependence for $M > 1\,\text{GeV}$ as expected from QCD at short-distance scales.

Comparison with Chiral Perturbation Theory

It is important to stress two major differences of this approach from the usual chiral perturbative calculations [50, 91, 265].

- First, the large N structure of the basic truncated low-energy Lagrangian provides a simplification over those effective Lagrangians used by chiral perturbation practitioners. In particular, within our ultraviolet quark-gluon completion, no $O(p^4)$ counter-terms are needed to absorb divergences generated by a dimensional regularization.
- More important, our loop calculations employ a cutoff regularization, and consequently our results exhibit a *quadratic* dependence on the *physical* cutoff M. This quadratic dependence is lost in the usual chiral perturbative calculations, which are based on the dimensional regularization. In effect, dimensional regularization makes extra infrared subtractions of quadratically divergent terms. These subtractions are not permitted in the full integration of the loop contributions in the truncated theory. As this quadratic dependence on the physical cutoff is usually a subject of criticism, we want to emphasize that it is an essential ingredient in the matching of the meson and quark-gluon pictures. Once again, it is required by power counting and chiral symmetry. Moreover, it stabilizes the $1/N$ expansion as exemplified through the cutoff independence of the hadronic scale Λ_χ. As discussed soon, it is also at the source of the $\Delta I = 1/2$ rule in our dual approach for QCD.

It is evident from these comments and from the review in [50] that in contrast to our $1/N$ approach, the chiral perturbation theory framework, while being very powerful in the determination of low-energy constants from experiment, cannot by itself address the issue of the dynamics behind the $\Delta I = 1/2$ rule and the evaluation of $\hat{B}_K$, $B_6^{(1/2)}$ and $B_8^{(3/2)}$.

With this brief formulation of DQCD at hand, we are ready to illustrate this approach on the example of the $\Delta I = 1/2$ rule and summarize the most important results obtained by us in [221, 227–230] and recently within the SM in [239, 244, 245]. The DQCD results for hadronic matrix elements of new operators present in the extensions of the SM will be discussed in Chapter 13.

7.2.3 The Dynamics behind the $\Delta I = 1/2$ Rule

Preliminaries

One of the puzzles of the 1950s was a large disparity between the measured values of the real parts of the isospin amplitudes A_0 and A_2 in $K \to \pi\pi$ decays, which on the basis of usual isospin considerations were expected to be of the same order. In 2020 we know the experimental values of the real parts of these amplitudes very precisely [35]

$$\text{Re}A_0 = 27.04(1) \times 10^{-8} \text{ GeV}, \quad \text{Re}A_2 = 1.210(2) \times 10^{-8} \text{ GeV}. \tag{7.27}$$

As $\text{Re}A_2$ is dominated by $\Delta I = 3/2$ transitions but $\text{Re}A_0$ receives contributions also from $\Delta I = 1/2$ transitions, the latter transitions dominate $\text{Re}A_0$, which expresses the so-called $\Delta I = 1/2$ rule [266, 267]

$$R = \frac{\text{Re}A_0}{\text{Re}A_2} = 22.35. \tag{7.28}$$

In the 1950s QCD and Operator Product Expansion did not exist, and clearly one did not know that $W^{\pm}$ bosons exist, in nature, but using the ideas of Fermi [268], Feynman and Gell-Mann [269], and Marshak and Sudarshan [270], one could still evaluate the amplitudes $\text{Re}A_0$ and $\text{Re}A_2$ to find out that such a high value of R is a real puzzle.

In modern times we can reconstruct this puzzle by evaluating the simple $W^{\pm}$ boson exchange between the relevant quark's which after integrating out $W^{\pm}$ generates the current-current operator Q_2:

$$Q_2 = (\bar{s}u)_{V-A} \, (\bar{u}d)_{V-A} \,. \tag{7.29}$$

With only Q_2 contributing we have

$$\text{Re}A_{0,2} = \frac{G_F}{\sqrt{2}} V_{ud} V_{us}^* \langle Q_2 \rangle_{0,2}. \tag{7.30}$$

Calculating the matrix elements $\langle Q_2 \rangle_{0,2}$ in the strict large N limit, which corresponds to factorization of matrix elements of Q_2 into the product of matrix elements of currents, we find

$$\langle Q_2 \rangle_0 = \sqrt{2} \langle Q_2 \rangle_2 = \frac{2}{3} F_{\pi} (m_K^2 - m_{\pi}^2), \tag{7.31}$$

and consequently

$$\text{Re}A_0 = 3.59 \times 10^{-8} \text{ GeV}, \qquad \text{Re}A_2 = 2.54 \times 10^{-8} \text{ GeV}, \qquad R = \sqrt{2}, \tag{7.32}$$

in plain disagreement with the data in (7.27) and (7.28). It should be emphasized that the explanation of the missing enhancement factor of 15.8 in R through some dynamics must simultaneously give the correct values for $\text{Re}A_0$ and $\text{Re}A_2$. This means that this dynamic should suppress $\text{Re}A_2$ by a factor of 2.1, not more, and enhance $\text{Re}A_0$ by a factor of 7.5. This tells us that while the suppression of $\text{Re}A_2$ is an important ingredient in the $\Delta I = 1/2$ rule, it is not the main origin of this rule. It is the enhancement of $\text{Re}A_0$ as already emphasized in [241] even if, in contrast to this paper, as demonstrated next, the current-current operators are responsible dominantly for this rule and not QCD penguins as pointed out first in [221].

We will next describe how the QCD dynamics in the context of the DQCD allow to explain the dominant part of the $\Delta I = 1/2$ rule. We proceed in three steps.

Step 1: Quark-Gluon Evolution

This step involves the calculation of the Wilson coefficients $z_{1,2}(\mu)$ of the current-current operators $Q_{1,2}$ at a low-energy scale $\mu = O(1\,\text{GeV})$ and was discussed in detail in previous chapters. This evolution starts already at scales $O(M_W)$ as opposed to the contribution of the QCD penguin operator Q_6 whose coefficient $z_6(\mu)$ is strongly suppressed by GIM mechanism for $\mu \geq m_c$ and will be included in Step 3. The contributions of the remaining operators, in particular electroweak penguin operators to $\text{Re}A_0$ and $\text{Re}A_2$ in the SM are negligible because of the very small Wilson coefficients $z_i(\mu)$ of these operators.

For the real parts of $K \to \pi\pi$ one finds then

$$\text{Re}A(K \to \pi\pi)_{cc} = \frac{G_F}{\sqrt{2}} V_{ud} V_{us}^* \sum_{i=1,2} z_i(\mu) \langle \pi\pi | Q_i(0) | K \rangle, \qquad (7.33)$$

where we indicate that only current-current operator contributions have been included. We have indicated that in this step the matrix elements are evaluated at $\mu \approx 0$ so that their values calculated in the strict large N limit can be used. As $z_i(\mu)$ are evaluated at $\mu = O(1\,\text{GeV})$, this is clearly inconsistent. But we do it here for pedagogical reasons, and we will improve on this in Step 2.

This first step corresponds in fact to the pioneering 1974 calculations in [119, 120] except that they were done at leading order in the renormalization group improved perturbation theory and now can be done at the NLO level. These 1974 papers have shown that the short-distance QCD effects enhance $\text{Re}A_0$ and suppress $\text{Re}A_2$. However, the inclusion of NLO QCD corrections to $z_{1,2}$ [71, 79] made it clear, as stressed in particular in [71], that the $K \to \pi\pi$ amplitudes without the proper calculation of hadronic matrix elements of Q_i are both scale and renormalization scheme dependent. For instance, setting $\mu = 0.8\,\text{GeV}$ we find for the ratio R in (7.28)[2]

$$R_{cc}(\text{NDR} - \overline{\text{MS}}) \approx 3.0, \qquad R_{cc}(\overline{\text{MOM}}) \approx 4.4, \qquad (7.34)$$

where $\overline{\text{MOM}}$ is a momentum scheme, introduced in [239], which is particularly suited for the calculations of the amplitudes in the DQCD. In this scheme one finds then for $\mu = 0.8\,\text{GeV}$

$$\text{Re}A_0 = 7.1 \times 10^{-8}\,\text{GeV}, \qquad \text{Re}A_2 = 1.6 \times 10^{-8}\,\text{GeV}. \qquad (7.35)$$

This is a significant improvement over the results in (7.32) bringing the theory closer to the data in (7.27) and (7.28). However, this result is scale and renormalization-scheme dependent. For the NDR $- \overline{\text{MS}}$ scheme and $\mu \approx (2-3)\,\text{GeV}$ as used in LQCD calculations, this improvement is much smaller. But, even in the $\overline{\text{MOM}}$ scheme and at $\mu = 0.8\,\text{GeV}$, further enhancement of $\text{Re}A_0$ and further suppression of $\text{Re}A_2$ are needed to be able to understand the $\Delta I = 1/2$ rule. This brings us to Step 2.

---

[2] The subscript cc indicates that only contributions from current-current operators $Q_{1,2}$ have been taken into account.

Step 2: Meson Evolution

The renormalization group evolution down to the scales $O(1\,\text{GeV})$ just performed is continued as a short but fast meson evolution down to zero momentum scales at which the factorization of hadronic matrix elements is at work. Equivalently, starting with factorizable hadronic matrix elements $\langle Q_1 \rangle_{0,2}$ and $\langle Q_2 \rangle_{0,2}$ at $\mu \approx 0$ and evolving them to $\mu = O(1\,\text{GeV})$ at which $z_{1,2}$ are calculated, one is able to calculate the matrix elements of these two operators at $\mu = O(1\,\text{GeV})$ and properly combine them with $z_{1,2}$ calculated in the $\overline{\text{MOM}}$ scheme. Details of these calculations can be found in [221, 239], and there is no space for presenting them here. I just want to make a few comments:

- Our loop calculations in the meson theory with a cutoff $M = O(1\,\text{GeV})$ include the contributions from pseudoscalars and lowest-lying vector mesons, and the result can be cast in the form of evolution equations. It is remarkable that the structure of these evolution equations, in particular the anomalous dimension matrix in the meson theory, is very similar to the one in the quark-gluon picture. This allows to perform an adequate matching between the two evolutions in question thereby removing to a large extent scale and renormalization scheme dependences present in the results of Step 1.
- The inclusion of vector meson contributions in [239] in addition to pseudoscalar contributions calculated in [221] is a significant improvement over our 1986 analysis bringing the theory closer to data.
- The same comment applies to the matching between the quark-gluon and meson theory, which this time has been performed at NLO in QCD. In this manner we could justify equating the physical cutoff M of the truncated meson theory (pseudoscalars and lowest-lying vector mesons) with the renormalization scale μ in the quark-gluon theory.

The resulting values

$$\text{Re}A_0 \approx (13.3 \pm 1.0) \times 10^{-8}\,\text{GeV}, \qquad \text{Re}A_2 \approx (1.1 \pm 0.1) \times 10^{-8}\,\text{GeV} \qquad (7.36)$$

show a very significant improvement over the results in (7.35), bringing the theory closer to the data in (7.27) and (7.28). In particular within the uncertainties of our approach we can claim that the experimental value of $\text{Re}A_2$ has been reproduced. The amplitude $\text{Re}A_0$ has been enhanced in this step by almost a factor of two relative to the result in (7.35), but it is still by a factor of two below the data. But whereas the calculation of $\text{Re}A_2$ has been completed in this step, to complete the calculation of $\text{Re}A_0$, we have to include QCD penguin contributions to this amplitude. This brings us to Step 3.

Step 3: QCD Penguins

As pointed out in [241], QCD penguin operators, of which the dominant is Q_6 in (7.4), could play an important role in enhancing the ratio R as in the isospin limit they do not contribute to A_2 and uniquely enhance the amplitude A_0. However, in 1975 the relevant matrix element $\langle Q_6 \rangle_0$ was unknown within QCD, and its Wilson coefficient z_6 was poorly known. In the strict large N limit in which factorization applies one finds [227, 228, 230] the result in (7.44), where we have introduced the parameter $B_6^{(1/2)}$, which equals unity in the large N limit.

While this matrix element is much larger than the matrix elements of $Q_{1,2}$, its Wilson coefficient z_6 is strongly GIM suppressed at scales $O(m_c)$ due to the fact that it results from the difference of QCD penguin diagrams with charm and up-quark exchanges. If these masses are neglected above $\mu = m_c$ then $z_6(m_c) = 0$ and its value is roughly by an order of magnitude smaller than $z_{1,2}$ at $\mu = 0.8\,\text{GeV}$. In [228] an additional (with respect to previous estimates) enhancement of the QCD penguin contributions to $\text{Re}A_0$ has been identified. It comes from an incomplete GIM cancellation above the charm-quark mass. But as the analyses in [221, 239] show, this enhancement is insufficient to reproduce fully the experimental value of $\text{Re}A_0$. We find that the Q_6 contribution to $\text{Re}A_0$ for $\mu \leq 1\,\text{GeV}$ is relevant as it is by a factor of 3 larger than $\text{Re}A_2$. Yet at $\mu = 0.8\,\text{GeV}$ it contributes at most at the level of 15% of the experimental value of $\text{Re}A_0$.

Summary of Results

Our final results for $K \rightarrow \pi\pi$ amplitudes obtained using the DQCD approach can be summarized as follows

$$\text{Re}A_0 \approx (17.0 \pm 1.5) \times 10^{-8}\,\text{GeV}, \quad \text{Re}A_2 \approx (1.1 \pm 0.1) \times 10^{-8}\,\text{GeV}, \quad R \approx 16.0 \pm 1.5 .$$
$$(7.37)$$

Even if the result for $\text{Re}A_0$ is not satisfactory, it should be noted that the QCD dynamics identified by us were able to enhance the ratio R by an order of magnitude. We therefore conclude that QCD dynamics is dominantly responsible for the $\Delta I = 1/2$ rule. The remaining piece in $\text{Re}A_0$ could come from final state interactions (FSI) between pions as discussed later, neglected $1/N^2$ corrections and/or NP contributions as investigated in [271].

7.2.4 Results for $\hat{B}_K$, $B_6^{(1/2)}$, $B_8^{(3/2)}$, and $B_8^{(1/2)}$

$\hat{B}_K$

We will next discuss the parameter $B_K(\mu)$ that plays an important role in the calculation of the $K^0 - \bar{K}^0$ mass difference ΔM_K and of the parameter ε_K in the SM as discussed in subsequent parts of this book.

$B_K(\mu)$, introduced already in Section 6.3, is scale and renormalization scheme dependent. It is related to the relevant hadronic matrix element of the $\Delta S = 2$ operator

$$Q = (\bar{s}d)_{V-A}(\bar{s}d)_{V-A},$$
$$(7.38)$$

as follows

$$\langle \bar{K}^0 | Q(\mu) | K^0 \rangle = B_K(\mu) \frac{4}{3} F_K^2 m_K.$$
$$(7.39)$$

More useful is the renormalization group invariant parameter $\hat{B}_K$ that is given by [137]

$$\hat{B}_K = B_K(\mu) \left[\alpha_s^{(3)}(\mu) \right]^{-d} \left[1 + \frac{\alpha_s^{(3)}(\mu)}{4\pi} J_3 \right], \qquad d = \frac{9(N-1)}{N(11N-6)}.$$
$$(7.40)$$

We have shown the N-dependence of the exponent d in the leading term to signal that d vanishes in the large N limit. The coefficient J_3 is renormalization-scheme dependent. This dependence cancels the one of $B_K(\mu)$.

As in the strict large N limit the exponent in (7.40) and the NLO term involving J_3 vanish, one finds [230] that independent of any renormalization scale or renormalization scheme for the operator Q

$$\hat{B}_K \to 0.75, \qquad \text{(in large } N \text{ limit, 1986)}. \tag{7.41}$$

It can be shown that including $1/N$ corrections suppresses $\hat{B}_K$ so that [272]

$$\hat{B}_K \leq 0.75, \qquad \text{(in DQCD)}. \tag{7.42}$$

Consistently with this bound our calculations in the 1980s [229] gave $\hat{B}_K = 0.67 \pm 0.07$. The latest analysis in our approach resulted in [239]

$$\hat{B}_K = 0.73 \pm 0.02, \qquad \text{(in DQCD)}, \tag{7.43}$$

where the error should not be considered as a standard deviation. Rather, this result represents the range for $\hat{B}_K$ we expect in our approach after the inclusion of NLO QCD corrections and the contributions of pseudoscalar and vector mesons as discussed in detail in [239]. As we will see in the next section, LQCD confirms this result within present uncertainties after efforts of many groups.

$B_6^{(1/2)}$, $B_8^{(3/2)}$, and $B_8^{(1/2)}$

We discuss next the matrix elements of the penguin operators Q_6 and Q_8, which are crucial for the evaluation of the ratio ε'/ε in the SM and in most extensions of this model. We will see this in Chapter 10 and subsequent chapters. The relevant matrix elements are conveniently parametrized by

$$\langle Q_6(\mu) \rangle_0 = -4h \left[\frac{m_K^2}{m_s(\mu) + m_d(\mu)} \right]^2 (F_K - F_\pi) B_6^{(1/2)} = -0.473 \, h \, B_6^{(1/2)} \, \text{GeV}^3, \tag{7.44}$$

$$\langle Q_8(\mu) \rangle_2 = \sqrt{2}h \left[\frac{m_K^2}{m_s(\mu) + m_d(\mu)} \right]^2 F_\pi B_8^{(3/2)} = 0.862 \, h \, B_8^{(3/2)} \, \text{GeV}^3, \tag{7.45}$$

$$\langle Q_8(\mu) \rangle_0 = \frac{h}{2} \left[\frac{2m_K^2}{m_s(\mu) + m_d(\mu)} \right]^2 F_\pi B_8^{(1/2)} = 1.219 \, h \, B_8^{(1/2)} \, \text{GeV}^3, \tag{7.46}$$

with [230, 273]

$$B_6^{(1/2)} = B_8^{(3/2)} = B_8^{(1/2)} = 1, \tag{7.47}$$

in the large N limit. We have introduced the factor h to emphasize different normalizations of these matrix elements present in the literature. In this book we will set $h = 1$. From (7.44)–(7.46) we obtain

$$
\boxed{\frac{B_8^{(3/2)}(\mu)}{B_6^{(1/2)}(\mu)} = -0.55 \frac{\langle Q_8(\mu)\rangle_2}{\langle Q_6(\mu)\rangle_0}, \qquad \frac{B_8^{(3/2)}(\mu)}{B_8^{(1/2)}(\mu)} = \sqrt{2}\frac{\langle Q_8(\mu)\rangle_2}{\langle Q_8(\mu)\rangle_0}.}
$$

$$(7.48)$$

The standard renormalization group analysis shows that $B_6^{(1/2)}$, $B_8^{(3/2)}$, and $B_8^{(1/2)}$ exhibit a very weak scale dependence for $\mu \geq 1$ GeV [123]. But this is no longer true for $\mu < 1$ GeV in the case of $B_6^{(1/2)}$ and $B_8^{(3/2)}$ as demonstrated within DQCD in [244] with important implications for the ratio ε'/ε. The dimensionful parameters entering (7.44), (7.45) have been calculated at $\mu = m_c$ using [48]

$$
m_K = 497.614\,\text{MeV}, \qquad F_\pi = 130.41(20)\,\text{MeV}, \qquad \frac{F_K}{F_\pi} = 1.194(5), \tag{7.49}
$$

$$
m_s(m_c) = 109.1(2.8)\,\text{MeV}, \qquad m_d(m_c) = 5.44(19)\,\text{MeV}. \tag{7.50}
$$

Before giving the values of $B_6^{(1/2)}$, $B_8^{(3/2)}$, and $B_8^{(1/2)}$ beyond the large N limit, it should be reemphasized that the overall factor h in (7.44)–(7.46) depends on the normalization of the amplitudes $A_{0,2}$. In [123] and recent papers of the RBC-UKQCD collaboration [274–276] $h = \sqrt{3/2}$ is used, whereas in most recent phenomenological papers [50, 212, 239, 271], $h = 1$. Correspondingly, the experimental values quoted for $A_{0,2}$ differ by this factor.

The $1/N$ corrections to these three parameters have been investigated in [252] and more recently in [244]. We will describe now the results of the latter study. In the DQCD the large N limit in (7.47) corresponds to very low scales $O(m_\pi)$, where the factorization of the matrix elements into a product of matrix elements of quark densities is a good approximation. However, to combine these parameters with the Wilson coefficients $z_{6,8}$ and $y_{6,8}$ one has to calculate $B_6^{(1/2)}$, $B_8^{(3/2)}$, and $B_8^{(1/2)}$ at scales $O(1\,\text{GeV})$. This is achieved through the meson evolution from low-energy scales $O(m_\pi)$ to scales $O(1\,\text{GeV})$ as done in [244] with the result

$$
B_6^{(1/2)}(M) = 1 - \frac{3}{2}\left[\frac{F_\pi}{F_K - F_\pi}\right]\frac{(m_K^2 - m_\pi^2)}{(4\pi F_\pi)^2}\ln\left(1 + \frac{M^2}{\tilde{m}_6^2}\right) = 1 - 0.66\ln\left(1 + \frac{M^2}{\tilde{m}_6^2}\right),
$$

$$(7.51)$$

$$
B_8^{(3/2)}(M) = 1 - 2\frac{(m_K^2 - m_\pi^2)}{(4\pi F_\pi)^2}\ln\left(1 + \frac{M^2}{\tilde{m}_8^2}\right) = 1 - 0.17\ln\left(1 + \frac{M^2}{\tilde{m}_8^2}\right), \tag{7.52}
$$

$$
B_8^{(1/2)}(M) = 1 + \frac{(m_K^2 - m_\pi^2)}{(4\pi F_\pi)^2}\ln\left(1 + \frac{M^2}{\tilde{m}_8^2}\right) = 1 + 0.08\ln\left(1 + \frac{M^2}{\tilde{m}_8^2}\right). \tag{7.53}
$$

The pseudoscalar mass scale parameters $\tilde{m}_{6,8}$ are bounded necessarily by the effective cut-off around 1 GeV: $\tilde{m}_{6,8} \leq M$.

We emphasize most important properties of these results:

- For $M = 0$, corresponding to strict large N limit and matrix elements evaluated at zero momentum, $B_6^{(1/2)} = B_8^{(3/2)} = B_8^{(1/2)} = 1$ in accordance with (7.47).
- With increasing M, the parameters $B_6^{(1/2)}$ and $B_8^{(3/2)}$ decrease below unity, and $B_6^{(1/2)}$ decreases faster than $B_8^{(3/2)}$. Consequently, at scales $O(1\,\text{GeV})$ relevant for the phenomenology, both $B_6^{(1/2)}$ and $B_8^{(3/2)}$ are predicted to be below unity and $B_6^{(1/2)} < B_8^{(3/2)}$.
- While the dependence of $B_6^{(1/2)}$ and $B_8^{(3/2)}$ on $M < 1\,\text{GeV}$ is stronger than their dependence on μ in the perturbative regime, these two properties of $B_6^{(1/2)}$ and $B_8^{(3/2)}$ are at the qualitative level consistent with the numerical analysis performed for $B_6^{(1/2)}$ and $B_8^{(3/2)}$ by means of the standard renormalization group running in [123]. Indeed, as seen in figures 11 and 12 of that paper, $B_6^{(1/2)}$ decreases with increasing μ, faster than $B_8^{(3/2)}$, albeit in this perturbative range the dependence of $B_6^{(1/2)}$ and $B_8^{(3/2)}$ on μ is very weak. It should be emphasized that in that paper, similar to other literature in the 1990s, $B_6^{(1/2)}(m_c) = B_8^{(3/2)}(m_c) = 1$ has been chosen, which is not correct as just explained. But the pattern of the μ dependence of these parameters is as given there.

While the analysis in [123] includes NLO QCD and QED corrections, the inspection of the one-loop anomalous dimension matrix allows to see these properties explicitly. In particular Q_6 mixes with the linear combination (Q_4+Q_6), and we find for $\mu_1 \leq \mu_2 \leq m_c$

$$B_6^{(1/2)}(\mu_2) = B_6^{(1/2)}(\mu_1)\left[1 - \frac{\alpha_s(\mu_1)}{2\pi}\ln\left(\frac{\mu_2}{\mu_1}\right)\left(1 + \frac{\langle Q_4(\mu_1)\rangle_0}{\langle Q_6(\mu_1)\rangle_0}\right)\right]. \tag{7.54}$$

As $|\langle Q_6(\mu_1)\rangle_0| > |\langle Q_4(\mu_1)\rangle_0|$, one finds that $B_6^{(1/2)}$ decreases with increasing μ. On the other hand, in the LO Q_8 runs only by itself, and the one-loop anomalous dimension matrix implies

$$B_8^{(1/2,3/2)}(\mu_2) = B_8^{(1/2,3/2)}(\mu_1), \tag{7.55}$$

which follows from exact SU(3) symmetry imposed in SD calculations. The breakdown of SU(3) is only felt in the matrix elements of Q_8 making in the LD range $B_8^{(3/2)}$ dependent weakly on the scales involved. In view of this, the suppression of both $B_6^{(1/2)}$ and $B_8^{(3/2)}$ below the unity can be considered as a solid result, and our explicit calculation as well as different behavior of Q_6 and Q_8 under flavor SU(3) provide a strong support for $B_6^{(1/2)} < B_8^{(3/2)}$.

- $B_8^{(1/2)}$ is only mildly affected by meson evolution.

These results can be then summarized by the upper bounds on both $B_6^{(1/2)}$ and $B_8^{(3/2)}$ [244]

$$B_6^{(1/2)} < B_8^{(3/2)} < 1, \qquad (\text{DQCD}). \tag{7.56}$$

While this approach gives $B_8^{(3/2)}(m_c) = 0.80 \pm 0.10$, the result for $B_6^{(1/2)}$ is less precise, but there is a strong indication that $B_6^{(1/2)} < B_8^{(3/2)}$, with typical values $B_6^{(1/2)} \approx 0.6$ at scales $O(1\,\text{GeV})$, in agreement with the LQCD results discussed in the next section. Therefore the bound on $B_6^{(1/2)}$ in (7.56) should be considered as conservative. On the other hand a number of other large N approaches [255, 256, 277–279] violate strongly the bounds in (7.56) with

$B_6^{(1/2)}$ in the ballpark of 3 and $B_8^{(3/2)} > 1$ in striking disagreement with the DQCD approach and LQCD results.

7.2.5 Final State Interactions

Already a long time ago the chiral perturbation theory practitioners put forward the idea that the amplitude ReA_0, governed by the current-current operator $Q_2 - Q_1$ could be enhanced significantly through FSI [280–286] bringing the values of R in (7.37) close to its experimental value. Some support for this claim comes from the more recent reconsideration of the role of final state interactions (FSI) in the $\Delta I = 1/2$ rule within the DQCD approach [245].

However, the claim by the authors in [280–286] and recently in [247], that FSI would enhance $B_6^{(1/2)}$ above unity in a correlated manner with the ReA_0 enhancement has not been confirmed in [245]. The main reason for the latter conclusion is the absence of the meson evolution of $B_6^{(1/2)}$ in the chiral perturbation analyses in question. According to [245] the suppression of $B_6^{(1/2)}$ through this evolution, seen in (7.51), overcompensates possible enhancements of $B_6^{(1/2)}$ considered in chiral perturbation papers so that the bound $B_6^{(1/2)} < 1$ should be valid in the presence of FSI. We will see in Chapter 10 that the values of $B_6^{(1/2)}$ and $B_8^{(3/2)}$ at scales $O(1\,\text{GeV})$ are crucial for the SM prediction of the CP-violating ratio ε'/ε. Their values obtained by LQCD will be discussed in the next section.

7.2.6 More on the μ Dependence of $\langle Q_{6,8} \rangle$

The μ dependence of $B_6^{(1/2)}$ and $B_8^{(3/2)}$ is related to nonfactorizable contributions to the corresponding hadronic matrix elements in (7.44) and (7.45). But as seen in the latter formulas, these matrix elements depend strongly on μ through the μ dependence of quark masses, and for scales $\mu > 1\,\text{GeV}$ this is by far the dominant μ dependence. This property should be contrasted with the matrix elements of $(V - A) \otimes (V - A)$ operators, which are μ-independent in the large N limit. Only through the inclusion of $1/N$ nonfactorizable corrections, corresponding to meson loops in the effective theory, can the proper μ-dependence of the latter matrix elements be generated.

The μ-dependence in $\langle Q_6 \rangle_0$ and $\langle Q_8 \rangle_2$ related to the μ dependence of quark masses is for $\mu > 1\,\text{GeV}$ exactly canceled in the decay amplitudes by the diagonal evolution (no mixing) of the Wilson coefficients $y_6(\mu)$ and $y_8(\mu)$ taken in the large N limit. Indeed, the μ-dependence of $1/m_{s,d}^2(\mu)$ is governed in LO by $2\gamma_m^{(0)} = 12C_F$. On the other hand, the one-loop anomalous dimensions of $Q_{6,8}$, which govern the diagonal evolution of $y_{6,8}(\mu)$ are given by

$$\gamma_{66}^{(0)} = -2\gamma_m^{(0)} + \frac{2f}{3}, \qquad \gamma_{88}^{(0)} = -2\gamma_m^{(0)}. \tag{7.57}$$

For large N, $\gamma_m^{(0)} \sim O(N)$, we find indeed $\gamma_{66}^{(0)} = \gamma_{88}^{(0)} = -2\gamma_m^{(0)}$ in the large-N limit [273]. Going back to the respective evolutions of $m_{s,d}$ and $y_{7,8}(\mu)$ we indeed confirm the cancellation of the μ-dependence in question. This feature is preserved at the two-loop level as discussed in [122, 146].

7.2.7 The Virtues of DQCD

Let us close this section by listing various virtues of DQCD even if it is really not QCD, the theory of quarks and gluons. But, as we have already seen, and we will see later, it is a successful low-energy approximation of QCD. Moreover, it has several virtues:

- It is an efficient approximate method for obtaining results for nonleptonic decays, years and even decades before useful results from numerically sophisticated and demanding lattice calculations could be obtained.
- It is the only existing method that allows us to study analytically the dominant dynamics below $1\,\text{GeV}$ scale, represented by the meson evolution, which turns out to have the pattern of operator mixing, both for SM and BSM operators, to agree with the one found perturbatively at short-distance scales. This allows for satisfactory, even if approximate, matching between Wilson coefficients and hadronic matrix elements.
- It provides insight into the purely numerical results obtained by LQCD. We will see soon that the results for $B_6^{(1/2)}$, $B_8^{(3/2)}$, and $B_8^{(1/2)}$ in (7.51)–(7.53) explain their numerical values from LQCD given in the next section. Moreover, we will see in Chapter 13 that the pattern of LQCD results for $K^0 - \bar{K}^0$ hadronic matrix elements of BSM operators can also be explained as the result of meson evolution.
- Most important, as discussed in Chapter 14, DQCD allowed already in 2018 to evaluate $K \to \pi\pi$ matrix elements of all BSM operators, providing in this manner for the first time a global view on NP contributions to ε'/ε.

7.3 Lattice QCD Results

7.3.1 Preliminaries

LQCD plays an essential role in weak decays as it allows to calculate various nonperturbative quantities like weak decay constants, formfactors and hadronic matrix elements of local operators from first principles. Moreover, in contrast to DQCD, just discussed, it provides useful results not only for K meson decays but also for D and $B_{s,d}$ decays. However, these calculations are very tedious, and for certain quantities it required the efforts of various collaborations over decades to obtain satisfactory precision. But in the case of nonleptonic decays still the lattice calculations have to be improved by much before one could be satisfied.

It is not the purpose of this section to describe the lattice methods, simply because such a description is far beyond the skills of the present author. What, however, can be done here is the collection of most important results from LQCD that we will use in this book when discussing phenomenology of weak decays. The main source here will be the summary of lattice results from the very recent 2019 FLAG review [287]. They are summarized in Table 7.1. These numbers differ slightly from our input in Appendix D that was used in this book until 2019 FLAG review was available. As they will further change until 2020, when this book will appear, we decided not to update our calculations.

Table 7.1 Values of some theoretical quantities extracted from FLAG 2019 [287].[a]

$F_\pi = *$	$F_K = 155.7(3)\,\text{MeV}$	$\hat{B}_K = 0.717(18)(16)$
$F_{B_d} = 190.0(13)\,\text{MeV}$	$F_{B_s} = 230.3(13)\,\text{MeV}$	$F_{B^+} = (189.4(13)\,\text{MeV}$
$F_\pi = 132(8)$	$F_K = 155.7(7)\,\text{MeV}$	$\hat{B}_K = 0.7625(97)$
$F_{B_d} = 192.0(43)\,\text{MeV}$	$F_{B_s} = 228.4(37)\,\text{MeV}$	$F_{B^+} = (189.4(13)\,\text{MeV}$
$\hat{B}_{B_d} = 1.30(10)$	$\hat{B}_{B_s} = 1.35(6)$	$\hat{B}_{B_s}/\hat{B}_{B_d} = 1.032(38)$
$F_{B_d}\sqrt{\hat{B}_{B_d}} = 225(9)\,\text{MeV}$	$F_{B_s}\sqrt{\hat{B}_{B_s}} = 274(8)\,\text{MeV}$	$\xi = 1.206(17)$

[a] The first two rows correspond to existing 2+1+1 calculations, the subsequent four after the horizontal line to 2+1 calculations. The $*$ means that no result was given in this review.

Indeed, due to significant progress in lattice calculations expected in coming years the numbers collected by us will be changing in the future, and potential readers should always check, by looking up FLAG's webpage, whether some modifications took place. This is similar to experimental data, which also in most cases change with time, in particular the uncertainties decrease. But in the case of many observables the achieved precision, both by LQCD and experiments, is such that any future modifications will only have small impact on phenomenology. In other cases, like hadronic matrix elements of four-quark operators, significant progress has still to be made to be able to reach clear-cut conclusions on the SM predictions and on the room left for NP contributions.

Before entering the details one could ask the question whether the results of DQCD presented in the previous section will become obsolete when the precision of LQCD will become satisfactory. We would like to emphasize once again that this will not be the case for reasons listed at the end of the previous section. In particular LQCD being a purely numerical method does not allow one to understand the dynamics responsible for the numerical values of hadronic matrix elements in $K \to \pi\pi$ decays and $K^0 - \bar{K}^0$ mixing. This insight can be obtained within DQCD as we have already demonstrated in the case of the $\Delta I = 1/2$ rule, the values of $B_6^{(1/2)}$, $B_8^{(3/2)}$, and $B_8^{(3/2)}$ and the values of BSM hadronic matrix elements in $K^0 - \bar{K}^0$ mixing discussed in Chapter 13.

7.3.2 Weak Decay Constants

During the last years very impressive progress has been made by LQCD in evaluating weak decay constants

$$F_\pi, \qquad F_K, \qquad F_{B_d}, \qquad F_{B_s}, \qquad F_{B^+}. \tag{7.58}$$

They play a very important role in the phenomenology, and we will see this explicitly in the next chapters. Their present values are collected in Appendix D and also in Table 7.1.

7.3.3 *B*-Physics Results

LQCD plays an important role in the calculations of form factors. It also provides the best results for hadronic matrix elements entering $B_{s,d}^0 - \bar{B}_{s,d}^0$ mixings both in the SM and beyond. We will encounter them in the phenomenological sections later. However, the multitude of kinematically accessible hadronic final states prevents LQCD from being a useful tool for nonleptonic B decays. Possibly for B physics most useful is presently *QCD Factorization* method, discussed in the next section.

7.3.4 *K*-Physics Results

For K physics significant progress has been made recently by LQCD, and we will list selected results now.

The lattice world average for $\hat{B}_K$ from FLAG based on the calculations of various groups [288–293] reads for $N_f = 2 + 1$ calculations [48],

$$\hat{B}_K = 0.7625(97), \qquad (\text{LQCD}, N_f = 2 + 1). \tag{7.59}$$

While this result violates the bound in (7.42), it should be noted that a number of lattice groups among [288–293] published results with central values satisfying the bound in (7.42), but the errors did not allow for a clear-cut conclusion.

In fact, the most recent update from staggered quarks [294] quotes precisely $\hat{B}_K = 0.738 \pm 0.005$, but the additional systematic error of 0.037 does not allow for a definite conclusion. Similarly, the Rome group [295], the only group working with $N_f = 2 + 1 + 1$, finds basically the result in (7.43):

$$\hat{B}_K = 0.717(18)(16), \qquad (\text{LQCD}, N_f = 2 + 1 + 1). \tag{7.60}$$

We expect therefore that improved lattice calculations will eventually satisfy the bound in (7.42), and in a few years from now the LQCD average for $\hat{B}_K$ will read $\hat{B}_K \approx 0.74$.

It is interesting not only that the values of $\hat{B}_K$ obtained in the DQCD approach and in LQCD are basically equal to each other but also that the resulting value is so close to its large N limit value. While the lattice approach did not provide the explanation why $\hat{B}_K$ is so close to its large N limit 0.75, in DQCD the smallness of $1/N$ corrections follows from the approximate cancellation of negative pseudoscalar meson contributions by the positive vector meson contributions.

Next, the relevant hadronic $K \to \pi\pi$ matrix elements have been evaluated by RBC-UKQCD collaboration [275, 276]. From these results one can extract values of $B_6^{(1/2)}$, $B_8^{(3/2)}$, and $B_8^{(1/2)}$ for $\mu = m_c$. They read in 2015 [296]

$$B_6^{(1/2)}(m_c) = 0.57 \pm 0.19, \qquad B_8^{(3/2)}(m_c) = 0.76 \pm 0.05, \qquad B_8^{(1/2)}(m_c) = 1.0 \pm 0.2, \tag{7.61}$$

in agreement with the bounds from DQCD in (7.56). The improved results should be available in 2020. The implications for the ratio ε'/ε will be presented in Chapter 10.

The present result for the $\Delta I = 1/2$ rule from the RBC-UKQCD collaboration is [276]

$$\left(\frac{\text{Re}A_0}{\text{Re}A_2} \right)_{\text{lattice QCD}} = 31.0 \pm 11.1, \tag{7.62}$$

and, in agreement with results from DQCD, this rule is governed by current-current operators. But the uncertainty is still very large, and it will be interesting to see whether LQCD will be able to come closer to the data than is possible using DQCD. See (7.37).

As far as LQCD results for formfactors are concerned, we will mention them in the phenomenological sections. Lattice QCD is also making progress in calculating long-distance effects in rare decays like $K^+ \to \pi^+ \nu \bar{\nu}$ [214, 297, 1337], $K_L \to \pi^0 \ell^+ \ell^-$ [298, 299], and the mass difference ΔM_K [373, 1330]. Also the progress in calculating QED corrections to decay amplitudes and leptonic decay rates should be mentioned [302–304].

The prospects for LQCD calculations relevant for flavor physics are summarized in [305–308]. In particular the projection on the reduction of errors in LQCD in the coming years is described well in section 11 of [305]. See in particular table 40 therein.

7.4 QCD Factorization for Exclusive B Decays

7.4.1 Introduction

We have seen that the DQCD approach could be successfully applied to the K meson system. In other meson systems one could apply the large N ideas for the calculation of hadronic matrix elements obtaining them as products of matrix elements of currents and quark densities. In fact, the first application of the large N method has been done for two-body charm decays in [225]. But to combine this calculation with the Wilson coefficients evaluated at scales $O(m_c, m_b)$ would require within the DQCD approach the meson evolution through the region of many resonances, which is clearly not doable. Here the so-called *QCD factorization* (QCDF) [309, 310] for exclusive hadronic B decays is much more powerful.

Indeed, the formulation of factorization theorems for exclusive hadronic B-meson decays in 1999 made an entire new class of processes accessible to systematic calculations of higher-order corrections in QCD [309, 310]. These processes include B decays into a pair of light mesons, the prototype of which is $B \to \pi\pi$, but also rare and radiative decays, such as $B \to K^*\gamma$ or $B \to K^* l^+ l^-$. In the heavy-quark limit, that is up to relative corrections of order $\Lambda_{\rm QCD}/m_b$, the problem of computing exclusive hadronic decay amplitudes simplifies considerably. In this limit the decay amplitudes can be written as hard-scattering kernels, which are *process dependent* but perturbatively calculable, multiplied by hadronic quantities such as $B \to \pi$ form factors, meson decay constants, and light-cone wave functions, which are nonperturbative but *process independent*. The decomposition into calculable hard contributions and universal hadronic quantities is in full analogy with the factorization of short-distance and long-distance terms that is the basis of any application of QCD to high-energy processes and has been discussed at length in the context of weak decays in previous chapters in this book. Correspondingly the framework is referred to as QCD factorization for exclusive hadronic B decays, or QCDF for short.

In the present section we review the factorization formula and give an overview of the NLO and NNLO calculations performed for exclusive B decays. To this end we give first the description of this approach by one of its fathers, Gerhard Buchalla, as presented in section 9 of [77]. Subsequently we will briefly describe recent developments.

7.4.2 Factorization Formula

The matrix element of an operator Q_i in the effective weak Hamiltonian for the decay of a $\bar{B}$ meson into a pair of light mesons $M_1 M_2$ is given by [309, 310]

$$\langle M_1 M_2 | Q_i | \bar{B} \rangle = F^{B \to M_1}(m_2^2) \int_0^1 du\, T_i^I(u)\, \Phi_{M_2}(u) + (M_1 \leftrightarrow M_2)$$

$$+ \int_0^1 d\xi\, du\, dv\, T_i^{II}(\xi, u, v)\, \Phi_B(\xi)\, \Phi_{M_1}(v)\, \Phi_{M_2}(u), \qquad (7.63)$$

up to power corrections of order $\Lambda_{\mathrm{QCD}}/m_b$. $F^{B \to M_{1,2}}(m_{2,1}^2)$ are $B \to M_{1,2}$ form factors, where $m_{1,2}$ denote the light-meson masses, and Φ_M is the light-cone distribution amplitude for the quark-antiquark Fock state of the meson M. Here the light-cone distribution amplitudes are understood to include the decay constant F_M of the meson M in their normalization. These quantities define the nonperturbative input needed for the computation of the decay amplitudes in QCDF. They are simpler than the full matrix element on the l.h.s. of (7.63) and universal in the sense that they appear as well in many other processes, which are different from $\bar{B} \to M_1 M_2$. The $T_i^I(u)$ and $T_i^{II}(\xi, u, v)$ are the hard-scattering functions, which are process-specific and depend in particular on the operator Q_i. They are calculable by standard methods in perturbative QCD.

The formula (7.63) exhibits the factorization of the short-distance kernels T_i and the long-distance hadronic quantities $F^{B \to M}$ and Φ_M. The factorization of the latter takes, in general, the form of a convolution over the parton momentum fractions ξ, u, $v \in [0, 1]$. A graphical representation of (7.63) is given in Figure 7.1, where index j accounts for the possibility of more than a single $B \to M_1$ formfactor. The second term ($\sim T^{II}$) is distinguished from the first ($\sim T^I$) by the participation of the B-meson spectator quark in the hard interaction, indicated by the spectator line entering the kernel T^{II}. The spectator interaction requires the exchange of a hard gluon. T^{II} starts therefore at order α_s, whereas T^I is of order unity, schematically

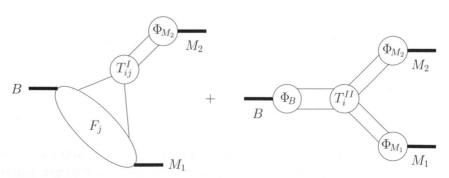

Figure 7.1   Factorization formula.

$$T^I = T^I_{(0)} + \frac{\alpha_s}{4\pi} T^I_{(1)} + \left(\frac{\alpha_s}{4\pi}\right)^2 T^I_{(2)} + \cdots, \qquad T^{II} = \frac{\alpha_s}{4\pi} T^{II}_{(1)} + \left(\frac{\alpha_s}{4\pi}\right)^2 T^{II}_{(2)} + \cdots \quad (7.64)$$

Such a description is only possible because in the two-body decay of the B meson the final-state particles are necessarily very energetic, with light-like four-momenta, in the rest frame of the B. Therefore, a meson emitted from the hard interaction, such as M_2 from T^I in Figure 7.1, is then described by its light-cone distribution amplitude. At leading power in $\Lambda_{\rm QCD}/m_b$ the amplitude is determined by the contribution from the light-cone wave function of leading twist, which corresponds to the simplest, two-particle Fock state. Higher Fock states give power-suppressed contributions and are therefore absent in the heavy-quark limit. For example, an additional energetic gluon, collinear to the light-like quark and antiquark in meson M_2 will generate an additional, far off-shell propagator when attached to the hard process T, which results in a power suppression. The properties of the light-cone wave functions, which vanish at the endpoints ($u = 0, 1$), also imply the suppression of highly asymmetric configurations where one parton carries almost the entire meson momentum and the other parton is soft.

At leading order in QCD, that is at $O(\alpha_s^0)$, the factorized matrix element in (7.63) reduces to a particularly simple result. The second term T^{II} is absent at this order, and $T^I(u)$ becomes a u-independent constant. Taking the matrix element of the operator $Q_2 = (\bar{u}b)_{V-A}(\bar{d}u)_{V-A}$ for $\bar{B} \to \pi^+\pi^-$ as an example, the factorization formula then states that

$$\langle \pi^+\pi^- | (\bar{u}b)_{V-A}(\bar{d}u)_{V-A} | \bar{B} \rangle = \langle \pi^+ | (\bar{u}b)_{V-A} | \bar{B} \rangle \langle \pi^- | (\bar{d}u)_{V-A} | 0 \rangle = i F^{B\to\pi}(0) F_\pi m_B^2. \tag{7.65}$$

This corresponds to the prescription of factorizing the matrix element of the 4-quark operator into a product of matrix elements of bilinear quark currents, which has a long history in phenomenological applications [311]. Here such a prescription receives its proper justification in the context of QCDF. The approximation in (7.65) means that the emission of the π^- is independent of the remaining $\bar{B} \to \pi^+$ transition. The intuitive argument for this, namely that the energetic and highly collinear, color-singlet $\bar{u}d$ pair forming π^- has little interaction with the rest of the process, was described a long time ago in [312]. The factorization theorem (7.63) is the formal implementation of this intuitive idea, and it allows us very importantly to compute corrections systematically.

It is then not surprising that factorization also works for decays of the type $\bar{B} \to D^+\pi^-$ with a heavy and a light meson in the final state, if it is the light meson that is emitted from the hard interaction (meson M_2 in Figure 7.1). In this case spectator scattering is power suppressed, and the factorization theorem in the heavy-quark limit takes the form

$$\langle DM_2 | Q_i | \bar{B} \rangle = F^{B\to D}(m_2^2) \int_0^1 du\, T_i^I(u)\, \Phi_{M_2}(u). \tag{7.66}$$

In fact the expression in (7.66) had already been used in [313] to compute the order-α_s corrections to the ratio of the $\bar{B} \to D\pi^-$ and $\bar{B} \to D^*\pi^-$ decay rates, prior to the systematic development of the QCDF.

The factorization theorem can be formulated using soft-collinear effective theory (SCET) [314, 315]. This formalism is useful for proving factorization [316] and for disentangling the hard and hard-collinear scale in explicit terms. QCDF and SCET are theoretical concepts that are fully compatible with each other, but they refer to different aspects of the problem of B-decay matrix elements. In some sense the relation between QCDF and SCET is similar to the relation between the heavy-quark expansion (HQE) and heavy-quark effective theory (HQET) in their application to *inclusive B* decays.[3] QCDF [309, 310] refers to the separation of the matrix elements into simpler long-distance quantities and calculable hard interactions, where the long-distance form factors are defined in full QCD. SCET, on the other hand, is a general effective field theory formulation for the relevant QCD modes (hard, hard-collinear, collinear, soft) and allows a further separation of scales, for instance in the transition form factors. However, working with form factors in full QCD often seems preferable in practice. An excellent review of SCET can be found in [317]. Some aspects are also discussed in [112].

7.4.3　Beyond NLO and Outlook

Review of NLO and NNLO corrections to the factorization formula by Buchalla can be found in section 9 of [77]. It is clearly beyond the scope of this book to present such details here. Over the years QCDF has been applied to a large number of exclusive B decays, and the associated phenomenology is very rich.

Not all observables accessible in principle to a factorization calculation are equally useful in practice as at a certain level accurate estimates of power corrections are necessary, and these are still beyond our control. One expects typically 10–20% corrections. This makes it difficult to compute direct CP asymmetries because these are sensitive to the relatively small strong phases, of which only the perturbative (though formally leading) part is calculable. However, there are many cases where the level of precision attainable with QCDF provides the basis for accurate predictions and flavor-physics tests.

QCDF can be employed to estimate the size of SU(3) breaking in approaches that rely on flavor symmetries to determine CKM quantities from CP violation in hadronic B decays [318]. Moreover, it can be used to estimate QCD penguin contributions to nonleptonic B decays [319].

In the upcoming era of precision experiments with B mesons, QCDF in the heavy quark limit, at NLO and beyond, provides us with an important tool to control theory predictions at a level adequate for discoveries in flavor physics. A recent summary of the theory status in nonleptonic heavy meson decays can be found in [317]. See also [320]. An interesting application of SCET to decays of new very heavy resonances has been presented in [321, 322].

Finally, we should mention the perturbative QCD approach to exclusive B decays [323–325] with the most recent application presented in [326], where further references can be found.

---

[3] We will describe these approaches briefly at the end of this chapter.

7.5 Heavy Quark Effective Theory (HQET) and Heavy Quark Expansions (HQE)

7.5.1 HQET

Since its advent in 1989 heavy quark effective theory (HQET) [327–331] has developed into an elaborate and well-established formalism, providing a systematic framework for the treatment of hadrons containing a heavy quark. HQET represents a static approximation for the heavy quark, covariantly formulated in the language of an effective field theory. It allows to extract the dependence of hadronic matrix elements on the heavy quark mass and to exploit the simplifications that arise in QCD in the static limit. The most important application of HQET has been to the analysis of exclusive semileptonic transitions involving heavy quarks, where this formalism allows to exploit the consequences of heavy quark symmetry to relate formfactors and provides a basis for systematic corrections to the $m \to \infty$ limit. There are several excellent reviews on this subject [332–335], and as there is no space to cover this topic in a satisfactory manner here, we ask the readers to look at these papers. A simple introduction to this subject can be found in chapter XV of [111].

7.5.2 HQE

The calculation of the decay rate difference $\Delta\Gamma_s$ is more involved than the one of ΔM_s.[4] In the box diagrams we have to take into account now only the internal up- and charm-quarks. Integrating out all heavy particles (in this case only the W boson) we are not left with a local $\Delta B = 2$ operator as in the case of ΔM_s, but with a bilocal object with up and charm quark propagators connecting two local vertices. To obtain a result in terms of local operators, whose matrix elements can be evaluated by lattice QCD, a second OPE is required. It relies on the smallness of the parameter Λ/m_b, where Λ is expected to be of the order of the hadronic scale Λ_{QCD}. This so-called heavy quark expansion (HQE) has been developed in a number of papers [336–342]. The HQE applies in addition to $\Delta\Gamma_s$ also for lifetimes and totally inclusive decay rates of heavy hadrons. Nice reviews with some details are the ones in [335, 343–345], and a nice summary of the present situation including historical development can be found in [346]. It shows that after some difficult periods it is a successful framework. Technical aspects are discussed also in [112].

We cannot cover this topic here in any detail, but let us discuss just few examples. First according to the HQE, the off-diagonal element Γ_{12}^s of the $B_s^0 - \bar{B}_s^0$ mixing matrix, discussed in detail in the next chapter, can be expanded as a power series in the inverse of the heavy b-quark mass m_b and the strong coupling α_s:

$$\Gamma_{12}^s = \frac{\Lambda^3}{m_b^3}\left(\Gamma_3^{s,(0)} + \frac{\alpha_s}{4\pi}\Gamma_3^{s,(1)} + \cdots\right) + \frac{\Lambda^4}{m_b^4}\left(\Gamma_4^{s,(0)} + \cdots\right) + \cdots. \qquad (7.67)$$

---

[4] See Section 8.1 for definitions of these quantities.

Although Λ is assumed to be of the order of Λ_{QCD}, its actual value has to be determined by a nonperturbative calculation. Each of the $\Gamma_i^{s,(j)}$ is a product of perturbative Wilson coefficients and nonperturbative matrix elements. In Γ_3^s these matrix elements arise from dimension 6 four-quark operators, in Γ_4^s from dimension 7 operators, and so on.

The leading term in (7.67), $\Gamma_3^{s,(0)}$, was calculated a long time ago by [347–352]. NLO QCD corrections, i.e. $\Gamma_3^{s,(1)}$ in (7.67); were done for the first time in [353]; they turned out to be quite large. Subleading $1/m_b$ corrections were first calculated by [354], and they also turned out to be quite sizable. Both LQCD and QCD sum rules were used over the years to calculate hadronic matrix elements of the relevant operators. As we will discuss in the next chapter, the present theoretical value for $\Delta\Gamma_s$ agrees well with experiment.

Another known application of HQE is for *inclusive* decays of heavy mesons, like $B_{s,d}$-mesons. In an inclusive decay one sums over all (or over a special class) of accessible final states, and eventually one can show [340, 355–361] that the resulting branching ratio can be calculated in the expansion in inverse powers of m_b with the leading term described by the spectator model in which the B-meson decay is approximated by the decay of the b-quark with corrections $O(1/m_b^2)$.

Because the leading term in this expansion represents the decay of the quark, it can be calculated in perturbation theory or more correctly in the renormalization group improved perturbation theory. It should be realized that also here the basic starting point is the relevant effective Hamiltonian and that the knowledge of the Wilson coefficients $C_i(\mu)$ is essential for the evaluation of the leading term in this expansion. But there is an important difference relative to the exclusive case: The matrix elements of the relevant operators can be "effectively" evaluated in perturbation theory. This means, in particular, that their μ and renormalization scheme dependences can be evaluated and the cancellation of these dependences by those present in $C_i(\mu)$ can be investigated.

Clearly, to complete the evaluation of the rates nonleading terms have to be considered. These terms are of a nonperturbative origin, but fortunately they are suppressed often by two powers of m_b. They have been studied by several authors in the literature with the result that they affect various branching ratios by less than 10% and often by only a few percent. Consequently the inclusive decays were expected for decades to give generally more precise theoretical predictions than the exclusive decays. On the other hand their measurements are harder. Moreover with advances of lattice QCD calculations it should be expected that *exclusive* decays will take the lead in testing the theory as there the issues like hadron duality violations present in inclusive decays and introducing some systematic uncertainties are absent.

7.6 Other Nonperturbative Methods

There are other useful nonperturbative methods, in particular chiral perturbation theory [50, 91, 265, 362–365], QCD sum rules, reviewed in [366], and light-cone sum rules [367–369], but their description is outside the scope of this book.

Particle-Antiparticle Mixing and CP Violation in the Standard Model

8.1 Particle-Antiparticle Mixing

8.1.1 Preliminaries

After the long exposition of the technology, let us begin to use it for the phenomenology. We begin with the phenomenon of particle-antiparticle mixing restricting the discussion to the SM. The impact of NP on this mixing will be discussed in BSM chapters.

The neutral mesons K^0, B^0, and D^0, to be denoted occasionally by P^0, can mix with their corresponding antiparticles. Physically, a K^0 changes with time into $\bar{K}^0$ and subsequently $\bar{K}^0$ changes back into K^0. Similarly $B^0_{d,s}$ can change into $\bar{B}^0_{d,s}$ and subsequently back into $B^0_{d,s}$, and the same applies to neutral D mesons. As the quantum number F describing the flavor content of P^0 differs from the one of $\bar{P}^0$, such oscillations can only be induced by interactions that change the flavor quantum number F. Both strong and electromagnetic interactions are flavor conserving. In the SM such $\Delta F = 2$ processes can only be mediated by weak interactions. Diagrammatically, in the SM, such oscillations appear first at the one-loop level, and as such they are in the case of $K^0 - \bar{K}^0$ and $B^0_{d,s} - \bar{B}^0_{d,s}$ mixings sensitive measures of the top quark couplings $V_{ti} (i = d, s, b)$ and in particular of the CKM phase δ. The *box diagrams* responsible for $K^0 - \bar{K}^0$ mixing have been already discussed and calculated in Section 6.2 and are shown in Figure 6.10. Here we would like to look in detail at the physics behind them and to derive formulas that allow a straightforward phenomenology. These will also include $B^0_{d,s} - \bar{B}^0_{d,s}$ mixings. These processes will play important roles in the construction of the unitarity triangle (UT) as explicitly demonstrated in this chapter.

The fact that in the SM $\Delta F = 2$ processes appear first at one-loop level is due to the GIM mechanism [6], discussed already in Section 6, that forbids the appearance of flavor-changing neutral current processes (FCNCs) at tree-level. As we will see in later chapters beyond the SM this mechanism is not always present, and FCNC processes can receive contributions from tree-level exchanges of new heavy particles, gauge bosons, and scalars.

In this chapter we will discuss the formalism of particle-antiparticle mixing and CP violation. We will begin with the $K^0 - \bar{K}^0$ mixing and the CP violation in this system. It will be useful to discuss in this context also $K_L \to \pi\pi$ decays, in which CP violation related to $K^0 - \bar{K}^0$ mixing, and represented by the parameter ε_K, was discovered in 1964. In fact, historically the K system dominated flavor physics until the 1980s, and it is natural to discuss it first. But then $B_{d,s}$ mesons, with very many decay channels, entered the scene allowing to extract CKM parameters related to the third generation of quarks, directly in leading tree-level decays of these mesons. In K physics the only truly important actor beyond ε_K in the 1980s and 1990s was the parameter ε', describing the direct CP violation in $K_L \to \pi\pi$ decays, that after heroic efforts by experimentalists has been finally measured just at the beginning of this millennium.

After a detail exposition of fundamental aspects of $K^0 - \bar{K}^0$ mixing, ε_K and ε', we will discuss in detail $B^0_{d,s} - \bar{B}^0_{d,s}$ mixings and CP violation in these meson systems, which will bring us to some specific $B_{s,d}$ decays. Having this formalism at hand we will present classification of various types of CP violation. Subsequently, we will show in explicit terms how the UT can be constructed. In this context the determination of the elements of the CKM matrix will play an important role and in particular the determination of the angles α, β and γ in this triangle. We will discuss various strategies for their direct determination from tree-level decays. We end the chapter with the discussion of $B^0_{s,d} \to \pi K$ decays that despite theoretical uncertainties offer an interesting arena for various tests of the SM and its extensions.

8.1.2 $K^0 - \bar{K}^0$ Mixing

$K^0 = (\bar{s}d)$ and $\bar{K}^0 = (s\bar{d})$ are flavor eigenstates that in the SM may mix via weak interactions through the box diagrams in Figure 6.10. We will choose the phase conventions so that

$$CP|K^0\rangle = -|\bar{K}^0\rangle, \qquad CP|\bar{K}^0\rangle = -|K^0\rangle. \tag{8.1}$$

In the absence of mixing the time evolution of $|K^0(t)\rangle$ is given by

$$|K^0(t)\rangle = \exp(-iHt)|K^0(0)\rangle, \qquad H = M - i\frac{\Gamma}{2}, \tag{8.2}$$

where M is the mass and Γ the width of K^0. A similar formula exists for $\bar{K}^0$.

On the other hand, in the presence of flavor mixing the time evolution of the $K^0 - \bar{K}^0$ system is described in the Wigner–Weisskopf approximation [370, 371] by the following Schrödinger equation:

$$i\frac{d\psi(t)}{dt} = \hat{H}\psi(t), \qquad \psi(t) = \begin{pmatrix} |K^0(t)\rangle \\ |\bar{K}^0(t)\rangle, \end{pmatrix}, \tag{8.3}$$

where

$$\hat{H} = \hat{M} - i\frac{\hat{\Gamma}}{2} = \begin{pmatrix} M_{11} - i\frac{\Gamma_{11}}{2} & M_{12} - i\frac{\Gamma_{12}}{2} \\ M_{21} - i\frac{\Gamma_{21}}{2} & M_{22} - i\frac{\Gamma_{22}}{2} \end{pmatrix}. \tag{8.4}$$

Here $\hat{M}$ is a Hermitian mass matrix (dispersive part), $\hat{\Gamma}$ is a Hermitian decay matrix (absorptive part), and $K^0(t)$ ($\bar{K}^0(t)$) describes the time evolution of the K meson that was a K^0 ($\bar{K}^0$) meson at $t = 0$. Note that $\hat{M}$ and $\hat{\Gamma}$ being Hermitian matrices have positive (real) eigenvalues in analogy with M and Γ. M_{ij} and Γ_{ij} are the transition matrix elements from virtual and physical intermediate states, respectively. Using

$$M_{21} = M_{12}^*, \qquad \Gamma_{21} = \Gamma_{12}^*, \qquad \text{(hermiticity)} \tag{8.5}$$

$$M_{11} = M_{22} \equiv M, \qquad \Gamma_{11} = \Gamma_{22} \equiv \Gamma, \quad \text{(CPT)} \tag{8.6}$$

we have

$$\hat{H} = \begin{pmatrix} M - i\frac{\Gamma}{2} & M_{12} - i\frac{\Gamma_{12}}{2} \\ M_{12}^* - i\frac{\Gamma_{12}^*}{2} & M - i\frac{\Gamma}{2} \end{pmatrix}. \tag{8.7}$$

Diagonalizing (8.3) we find:

Eigenstates:

$$K_{L,S} = \frac{(1 + \bar{\varepsilon})K^0 \pm (1 - \bar{\varepsilon})\bar{K}^0}{\sqrt{2(1 + |\bar{\varepsilon}|^2)}}, \tag{8.8}$$

where $\bar{\varepsilon}$ is a small complex parameter given by

$$\frac{1 - \bar{\varepsilon}}{1 + \bar{\varepsilon}} = \sqrt{\frac{M_{12}^* - i\frac{1}{2}\Gamma_{12}^*}{M_{12} - i\frac{1}{2}\Gamma_{12}}} = \frac{2M_{12}^* - i\Gamma_{12}^*}{\Delta M - i\frac{1}{2}\Delta\Gamma} \equiv r \exp(i\kappa) \tag{8.9}$$

with $\Delta\Gamma$ and ΔM given in (8.12).

Eigenvalues:

$$M_{L,S} = M \pm \text{Re}Q, \qquad \Gamma_{L,S} = \Gamma \mp 2\text{Im}Q, \tag{8.10}$$

where

$$Q = \sqrt{\left(M_{12} - i\frac{1}{2}\Gamma_{12}\right)\left(M_{12}^* - i\frac{1}{2}\Gamma_{12}^*\right)}. \tag{8.11}$$

Consequently, we have

$$\boxed{\Delta M = M_L - M_S = 2\text{Re}Q, \qquad \Delta\Gamma = \Gamma_L - \Gamma_S = -4\text{Im}Q.} \tag{8.12}$$

We observe that the mass eigenstates K_S and K_L differ from the CP eigenstates

$$K_1 = \frac{1}{\sqrt{2}}(K^0 - \bar{K}^0), \qquad CP|K_1\rangle = |K_1\rangle, \tag{8.13}$$

$$K_2 = \frac{1}{\sqrt{2}}(K^0 + \bar{K}^0), \qquad CP|K_2\rangle = -|K_2\rangle, \tag{8.14}$$

by a small admixture of the other CP eigenstate:

$$K_S = \frac{K_1 + \bar{\varepsilon}K_2}{\sqrt{1 + |\bar{\varepsilon}|^2}}, \qquad K_L = \frac{K_2 + \bar{\varepsilon}K_1}{\sqrt{1 + |\bar{\varepsilon}|^2}}. \tag{8.15}$$

Because $\bar{\varepsilon}$ turns out to be $O(10^{-3})$, one has to have a very good approximation:

$$\boxed{\Delta M_K = 2\text{Re}M_{12}, \qquad \Delta\Gamma_K = 2\text{Re}\Gamma_{12},} \tag{8.16}$$

where we have introduced the subscript K to stress that these formulas apply only to the $K^0 - \bar{K}^0$ system. They cannot be used in $B_q^0 - \bar{B}_q^0$ systems, in particular because the corresponding parameter ε_B, not used these days, can be $O(1)$, and various approximations made in the derivation of (8.16) are not valid. On the other hand, while $\Delta\Gamma_K \approx -2\Delta M_K$, in $B_q^0 - \bar{B}_q^0$ systems $\Delta M_q \gg \Delta\Gamma_q$. We will demonstrate it soon.

The $K_{\mathrm{L}} - K_{\mathrm{S}}$ mass difference is experimentally measured to be [35]

$$\Delta M_K = M(K_{\mathrm{L}}) - M(K_{\mathrm{S}}) = 3.484(6) \times 10^{-15}\,\mathrm{GeV} = 5.293(9) \times 10^{-3}\,\mathrm{ps}^{-1}. \quad (8.17)$$

Presently the contribution of the SM dynamics to ΔM_K is subject to large theoretical uncertainties. In the SM ΔM_K is described by the real parts of the box diagrams with charm-quark and top-quark exchanges, whereby the contribution of the charm exchanges is by far dominant. Unfortunately, the uncertainties in short-distance QCD corrections to charm contribution amount to roughly ±40% with the central value somewhat below the experimental one [144]. Moreover, there are also nonperturbative long-distance contributions that are known to amount to 20 ± 10% of the measured ΔM_K [239, 243], when calculated using the DQCD approach that we presented in the previous chapter. In the future they should be known more precisely from LQCD [301, 372, 373], and in fact the most recent value from RBC-UKQCD collaboration reads [1330]

$$(\Delta M_K)_{\mathrm{SM}} = 7.7(2.1) \times 10^{-15}\,\mathrm{GeV}, \quad \text{(RBC-UKQCD 2019)} \quad (8.18)$$

hinting for NP at 2σ level. We will return to this issue in Section 15.4 in the context of Z' scenarios. For the time being the rough picture is that box diagrams contribute 80% of the measured ΔM_K, but the long-distance contributions could be large implying through (8.18) possibly new dynamics beyond the SM. It appears that these new dynamics, if present, would be required to suppress the SM prediction to agree with data but in view of large uncertainties ΔM_K does not yet play as important a role in phenomenology as the CP-violation parameter $\varepsilon_K \equiv \varepsilon$ introduced soon. This could change in the future. The situation with $\Delta\Gamma_K$ is even worse. It is fully dominated by long-distance effects. Experimentally one has $\Delta\Gamma_K \approx -2\Delta M_K$, and this relation will be used in what follows.

Generally to observe CP violation one needs an interference between various amplitudes that carry complex phases. As these phases are obviously convention dependent, the CP-violating effects depend only on the differences of these phases. In particular the parameter $\bar{\varepsilon}$ in (8.15) depends on the phase convention chosen for K^0 and $\bar{K}^0$. Therefore it may not be taken as a physical measure of CP violation. But one can define quantities that are independent of phase conventions. In particular

$$r \equiv 1 + \frac{2|\Gamma_{12}|^2}{4|M_{12}|^2 + |\Gamma_{12}|^2}\,\mathrm{Im}\left(\frac{M_{12}}{\Gamma_{12}}\right) \approx 1 - \mathrm{Im}\left(\frac{\Gamma_{12}}{M_{12}}\right) \quad (8.19)$$

depends only on the difference of the phases of Γ_{12} and M_{12}. The departure of r from 1 measures then CP violation in the $K^0 - \bar{K}^0$ mixing.

This type of CP violation can be best isolated in semileptonic decays of the K_L meson. The nonvanishing asymmetry $a_{\mathrm{SL}}(K_L)$:

$$\boxed{a_{\mathrm{SL}}(K_L) = \frac{\Gamma(K_L \to \pi^- e^+ \nu_e) - \Gamma(K_L \to \pi^+ e^- \bar{\nu}_e)}{\Gamma(K_L \to \pi^- e^+ \nu_e) + \Gamma(K_L \to \pi^+ e^- \bar{\nu}_e)} = \left(\mathrm{Im}\frac{\Gamma_{12}}{M_{12}}\right)_K = 2\mathrm{Re}\bar{\varepsilon}} \quad (8.20)$$

signals this type of CP violation. Note that $a_{SL}(K_L)$ is determined purely by the quantities related to $K^0 - \bar{K}^0$ mixing. Specifically, it measures directly the difference between the phases of Γ_{12} and M_{12}.

That a nonvanishing $a_{SL}(K_L)$ is indeed a signal of CP violation can also be understood in the following manner. K_L, that should be a CP eigenstate K_2 in the case of CP conservation, decays into CP conjugate final states with different rates. As $\text{Re}\,\bar{\varepsilon} > 0$, K_L prefers slightly to decay into $\pi^- e^+ \nu_e$ than $\pi^+ e^- \bar{\nu}_e$. This would not be possible in a CP-conserving world. Experimentally we have [374]

$$a_{SL}(K_L) = 3.32(6) \times 10^{-3}. \tag{8.21}$$

8.1.3 The First Look at ε and ε'

Because two pion final states, $\pi^+\pi^-$ and $\pi^0\pi^0$, are CP even while the three pion final state $3\pi^0$ is CP odd, K_S and K_L decay to 2π and $3\pi^0$, respectively, via the following CP-conserving decay modes:

$$K_L \to 3\pi^0 \ \ (\text{via } K_2), \qquad K_S \to 2\pi \ \ (\text{via } K_1). \tag{8.22}$$

Moreover, $K_L \to \pi^+\pi^-\pi^0$ is also CP conserving provided the orbital angular momentum of $\pi^+\pi^-$ is even. This difference between K_L and K_S decays is responsible for the large disparity in their lifetimes:

$$\tau(K_L) = 5.116(21) \times 10^{-8} \ s, \qquad \tau(K_S) = 0.89564(33) \times 10^{-10} \ s, \tag{8.23}$$

a factor of 571.

However, K_L and K_S are not CP eigenstates and may decay with small branching fractions as follows:

$$K_L \to 2\pi \ \ (\text{via } K_1), \qquad K_S \to 3\pi^0 \ \ (\text{via } K_2). \tag{8.24}$$

This violation of CP is called *indirect* as it proceeds not via explicit breaking of the CP symmetry in the decay itself but via the admixture of the CP state with opposite CP parity to the dominant one. The measure for this indirect CP violation is defined by

$$\varepsilon \equiv \frac{A(K_L \to (\pi\pi)_{I=0})}{A(K_S \to (\pi\pi)_{I=0})}, \tag{8.25}$$

I being the *strong isospin*. In the final state one can have $(\pi\pi)_{I=0}$ and $(\pi\pi)_{I=2}$, but only $(\pi\pi)_{I=0}$ enters the definition of ε. The parameter ε is often denoted by ε_K, and we will use this notation in later sections and chapters.

It should be added that the decay $K_S \to \pi^+\pi^-\pi^0$ is CP violating (conserving) if the orbital angular momentum of $\pi^+\pi^-$ is even (odd). Following the derivation in [350] one finds

$$\varepsilon = \bar{\varepsilon} + i\xi = \frac{\exp(i\pi/4)}{\sqrt{2}\Delta M_K} \left(\text{Im}M_{12} + 2\xi \text{Re}M_{12} \right), \qquad \xi = \frac{\text{Im}A_0}{\text{Re}A_0}. \tag{8.26}$$

The phase convention dependence of ξ cancels the one of $\bar{\varepsilon}$ so that ε is free from this dependence. The isospin amplitude A_0 is defined in (8.30). The phase of ε, to be denoted by ϕ_ε, differs slightly from $\pi/4$. We will improve on it later.

The important point in the definition (8.25) is that only the transition to the state $(\pi\pi)_{I=0}$ enters. The transition to $(\pi\pi)_{I=2}$ is absent. This allows to remove a certain type of CP violation that originates in decays only. Yet as $\varepsilon \neq \bar{\varepsilon}$ and only $\mathrm{Re}\varepsilon = \mathrm{Re}\bar{\varepsilon}$, it is clear that ε includes a type of CP violation represented by $\mathrm{Im}\varepsilon$, which is absent in the semileptonic asymmetry (8.20). We will identify this type of CP violation in a more profound manner in Section 8.3, where a more systematic classification of different types of CP violation will be given.

While *indirect* CP violation, characterized by ε, reflects the fact that the mass eigenstates are not CP eigenstates, the so-called direct CP violation is realized via a direct transition of a CP odd to a CP even state: $K_2 \rightarrow \pi\pi$. A measure of such a direct CP violation in $K_L \rightarrow \pi\pi$ is characterized by a complex parameter ε' defined as

$$\varepsilon' \equiv \frac{1}{\sqrt{2}}\left(\frac{A_{2,L}}{A_{0,S}} - \frac{A_{2,S}}{A_{0,S}}\frac{A_{0,L}}{A_{0,S}}\right), \tag{8.27}$$

where $A_{I,L} \equiv A(K_L \rightarrow (\pi\pi)_I)$ and $A_{I,S} \equiv A(K_S \rightarrow (\pi\pi)_I)$.

This time the transitions to $(\pi\pi)_{I=0}$ and $(\pi\pi)_{I=2}$ are included, which allows to study CP violation in the decay itself. We will discuss this issue in general terms in Section 8.3. It is useful to cast (8.27) into

$$\varepsilon' = \frac{1}{\sqrt{2}}\mathrm{Im}\left(\frac{A_2}{A_0}\right)\exp(i\phi_{\varepsilon'}), \qquad \phi_{\varepsilon'} = \frac{\pi}{2} + \delta_2 - \delta_0, \tag{8.28}$$

where the isospin amplitudes A_I in $K \rightarrow \pi\pi$ decays are introduced through

$$A(K^+ \rightarrow \pi^+\pi^0) = \left(\frac{1}{h}\right)\left[\frac{3}{2}A_2 e^{i\delta_2}\right], \tag{8.29}$$

$$A(K^0 \rightarrow \pi^+\pi^-) = \left(\frac{1}{h}\right)\left[A_0 e^{i\delta_0} + \sqrt{\frac{1}{2}}A_2 e^{i\delta_2}\right], \tag{8.30}$$

$$A(K^0 \rightarrow \pi^0\pi^0) = \left(\frac{1}{h}\right)\left[A_0 e^{i\delta_0} - \sqrt{2}A_2 e^{i\delta_2}\right]. \tag{8.31}$$

The parameter h appeared already in (7.44) and (7.45). It distinguishes between various normalizations of $A_{0,2}$ found in the literature. In this book we use $h = 1$, but the RBC-UKQCD lattice collaboration uses $h = \sqrt{3/2}$, implying that their amplitudes $A_{0,2}$ are by a factor of $\sqrt{3/2}$, larger than ours. This difference cancels, of course, in all physical observables.

Here the subscript $I = 0, 2$ denotes $\pi\pi$ states in $K \rightarrow \pi\pi$ with isospin $0, 2$ equivalent to $\Delta I = 1/2$ and $\Delta I = 3/2$ transitions, respectively, and $\delta_{0,2}$ are the corresponding strong phases. The weak CKM phases are contained in A_0 and A_2. The isospin amplitudes A_I are complex quantities that depend on phase conventions. On the other hand, ε' measures the difference between the phases of A_2 and A_0 and is a physical quantity. The strong phases $\delta_{0,2}$ can be extracted from $\pi\pi$ scattering. Then $\phi_{\varepsilon'} \approx \pi/4$. See [375] for more details. Further details on how the isospin amplitudes are extracted from the data can be found in [50]. We will also return to related issues in Chapter 10.

Experimentally ε and ε' can be found by measuring the ratios

$$\eta_{00} = \frac{A(K_L \to \pi^0\pi^0)}{A(K_S \to \pi^0\pi^0)}, \qquad \eta_{+-} = \frac{A(K_L \to \pi^+\pi^-)}{A(K_S \to \pi^+\pi^-)}. \tag{8.32}$$

Indeed, assuming ε and ε' to be small numbers, one finds

$$\eta_{00} = \varepsilon - \frac{2\varepsilon'}{1 - \sqrt{2}\omega}, \quad \eta_{+-} = \varepsilon + \frac{\varepsilon'}{1 + \omega/\sqrt{2}}, \tag{8.33}$$

where

$$\omega = \frac{\mathrm{Re}A_2}{\mathrm{Re}A_0} = 0.045. \tag{8.34}$$

In the absence of direct CP violation $\eta_{00} = \eta_{+-}$. The ratio ε'/ε can then be measured through

$$\mathrm{Re}(\varepsilon'/\varepsilon) = \frac{1}{6(1 + \omega/\sqrt{2})}\left(1 - \left|\frac{\eta_{00}}{\eta_{+-}}\right|^2\right). \tag{8.35}$$

8.1.4 Basic Formula for ε

With all this information at hand, let us derive a formula for ε that can be efficiently used in phenomenological applications. The off-diagonal element M_{12} in the neutral K-meson mass matrix representing K^0–$\bar{K}^0$ mixing is given by

$$M_{12}^* = \langle \bar{K}^0|\mathcal{H}_{\mathrm{eff}}(\Delta S = 2)|K^0\rangle, \tag{8.36}$$

where $\mathcal{H}_{\mathrm{eff}}(\Delta S = 2)$ is the effective Hamiltonian for the $\Delta S = 2$ transitions resulting from diagrams in Figure 6.10 and given in (6.172). That M_{12}^* and not M_{12} stands on the l.h.s. of this formula is evident from (8.7).

The matrix element in (8.36) has been evaluated already in Section 6.3 and is given in (6.177). We find then

$$M_{12} = \frac{G_F^2}{12\pi^2}F_K^2\hat{B}_K m_K M_W^2 \left[\lambda_c^{*2}\eta_1 S_0(x_c) + \lambda_t^{*2}\eta_2 S_0(x_t) + 2\lambda_c^*\lambda_t^*\eta_3 S_0(x_c, x_t)\right]. \tag{8.37}$$

It should be mentioned that in the normalization of external states in which a factor $2m_K$ is present on the l.h.s. in (8.36), the r.h.s. of (6.176) is multiplied by $2m_K$ so that (8.37) remains unchanged. All quantities in (8.37) are known already from Section 6.3.

In the past the last term in (8.26) has been usually neglected as it was estimated to be at most a 5 % correction to ε. This was justified in view of other uncertainties, in particular those connected with $\hat{B}_K$. However, as discussed in the previous chapter, in this decade the latter parameter has been calculated with a precision of 1% by lattice QCD methods, and this term cannot be neglected. Moreover, long-distance contributions to both M_{12} and Γ_{12} have to be taken into account. These effects have been calculated in [376, 377] and can be included in ε by dropping the last term in (8.26) and introducing and overall correcting factor κ_ε in ε that is found to be

$$\boxed{\kappa_\varepsilon = 0.94 \pm 0.02.} \tag{8.38}$$

Inserting then (8.37) into (8.26) and using rather accurate expressions for λ_i in (2.81)–(2.83), we find

$$\varepsilon = C_\varepsilon \kappa_\varepsilon \hat{B}_K \operatorname{Im}\lambda_t \left\{ \operatorname{Re}\lambda_c \left[\eta_1 S_0(x_c) - \eta_3 S_0(x_c, x_t) \right] - \operatorname{Re}\lambda_t \eta_2 S_0(x_t) \right\} e^{i\phi_\varepsilon}, \tag{8.39}$$

where the numerical constant C_ε is given by

$$C_\varepsilon = \frac{G_F^2 F_K^2 m_K M_W^2}{6\sqrt{2}\pi^2 \Delta M_K} = 3.837 \cdot 10^4. \tag{8.40}$$

Comparing (8.39) with the experimental value for ε

$$\varepsilon_{\exp} = (2.228 \pm 0.011) \cdot 10^{-3} \, \exp i\phi_\varepsilon, \qquad \phi_\varepsilon = (43.51 \pm 0.05)^\circ, \tag{8.41}$$

one obtains a constraint on the unitarity triangle in Figure 2.4, which will be discussed in more detail in Section 8.4.

The structure of ε'/ε in the SM is much more complicated, and we postpone its discussion to Chapter 10.

8.1.5 $B_{d,s}^0 - \bar{B}_{d,s}^0$ Mixing

Preliminaries and Basic Definitions

The discussion of $B_{d,s}^0 - \bar{B}_{d,s}^0$ mixing proceeds in an analogous manner but there are differences, and it is important to stress them. In particular CP violation is much larger in this case, and the mass difference $\Delta M_{s,d}$ between two mass eigenstates in each system can be rather precisely calculated already now. Also the corresponding width differences can be calculated using OPE but here the situation has to be improved. In fact in the case of $B_s^0 - \bar{B}_s^0$ system the width effects can have significant impact on the phenomenology as we will see later. They will also play a role in the next chapter in the case of $B_s \to \mu^+ \mu^-$ decay.

There are many lectures and reviews on $B_{d,s}^0 - \bar{B}_{d,s}^0$ mixing which often differ slightly in notations between each other [138, 346, 378–384]. In particular the convention-dependent phases at intermediate stages and the usual sign ambiguity related to square-roots are exhibited in [378, 381]. These ambiguities are absent in the final expressions for physical observables and to simplify our presentation we will fix our conventions as we did in the case of $K^0 - \bar{K}^0$ mixing. But the general presentation in [378, 381] could be useful for readers that use different conventions than used by us here.

The next pages follow to some extent the presentation in [346], although at a few places we have recast a few rather complicated formulas presented by these authors with the hope to increase their transparency. But we adopt the conventions of that paper, which are also used in [383]. The reviews in [378, 381] use occasionally different conventions, and we will spell them out. The most recent short summary of this field, presented at the CKM 2018, is given in [385].

The flavor eigenstates are this time

$$B_d^0 = (\bar{b}d), \qquad \bar{B}_d^0 = (b\bar{d}), \qquad B_s^0 = (\bar{b}s), \qquad \bar{B}_s^0 = (b\bar{s}) . \tag{8.42}$$

We will choose the phase conventions so that $(q = d, s)$

$$\mathcal{CP}|B_q^0\rangle = -|\bar{B}_q^0\rangle, \qquad \mathcal{CP}|\bar{B}_q^0\rangle = -|B_q^0\rangle. \tag{8.43}$$

In the absence of mixing, the time evolution of $|B_q^0(t)\rangle$ is given by

$$|B_q^0(t)\rangle = \exp(-iHt)\,|B_q^0(0)\rangle, \qquad H = M_{B_q} - i\frac{\Gamma_{B_q}}{2} , \tag{8.44}$$

where M_{B_q} is the mass and Γ_{B_q} the width of B_q^0. A similar formula exists for $\bar{B}_q^0$.

These states mix via the box diagrams in Figure 6.10 with s replaced by b in the case of B_d^0–$\bar{B}_d^0$ mixing. In the case of B_s^0–$\bar{B}_s^0$ mixing also d has to be replaced by s. The Wigner–Weisskopf formalism [370, 371] yields now effective Schrödinger equation $(q = d, s)$

$$i\frac{\partial}{\partial t}\left(\begin{array}{c} |B_q^0(t)\rangle \\ |\bar{B}_q^0(t)\rangle \end{array}\right) = \left[\left(\begin{array}{cc} M_0^q & M_{12}^q \\ M_{12}^{q*} & M_0^q \end{array}\right) - \frac{i}{2}\left(\begin{array}{cc} \Gamma_0^q & \Gamma_{12}^q \\ \Gamma_{12}^{q*} & \Gamma_0^q \end{array}\right)\right] \cdot \left(\begin{array}{c} |B_q^0(t)\rangle \\ |\bar{B}_q^0(t)\rangle \end{array}\right), \tag{8.45}$$

describing the time evolution of the $B_q^0 - \bar{B}_q^0$ system.

In writing the mass and decay matrices in (8.45), we have already used relations following from hermiticity of M_{ij}^q and Γ_{ij}^q and invariance under CPT transformations as done in (8.5) and (8.6).

Again the box diagrams include contributions from virtual internal particles, denoted by M_{12}^q and contributions from internal on-shell particles, denoted by Γ_{12}^q. Within the SM M_{12}^q is fully dominated by top-quark contributions, while only internal charm- and up-quarks contribute to Γ_{12}^q. Beyond the SM M_{12}^q can obviously be affected by NP contributions but also Γ_{12}^q can receive NP contributions, and we will briefly mention it at the end of Section 15.1.1.

As M_{12}^q and Γ_{12}^q are generally complex quantities we have

$$\boxed{M_{12}^q = |M_{12}^q|e^{i\phi_M^q}, \qquad \Gamma_{12}^q = |\Gamma_{12}^q|e^{i\phi_\Gamma^q}.} \tag{8.46}$$

The phases ϕ_M^q and ϕ_Γ^q are not physical and depend on the phase convention used in the CKM matrix but as we will see later, only differences in phases will enter observables, and these are phase convention independent.

Diagonalization of this system gives now the physical eigenstates of mesons with a definite mass and decay rate. The meson eigenstates will be denoted by $|B_{q,H}\rangle$ (H = heavy) and $|B_{q,L}\rangle$ (L = light). They are linear combinations of the flavour eigenstates:

$$|B_{q,L}\rangle = p\,|B_q^0\rangle + q\,|\overline{B_q^0}\rangle, \qquad |B_{q,H}\rangle = p\,|B_q^0\rangle - q\,|\overline{B_q^0}\rangle, \qquad |p|^2 + |q|^2 = 1. \quad (8.47)$$

These two states are not orthogonal to each other, and both p and q are complex numbers. Denoting their masses and decay rates by $M_{L,H}^q$ and $\Gamma_{L,H}^q$, respectively, one finds [346]

$$\Delta M_q := M_H^q - M_L^q = 2\left|M_{12}^q\right|\left(1 - \frac{\left|\Gamma_{12}^q\right|^2 \sin^2 \phi_{12}^q}{8\left|M_{12}^q\right|^2} + \cdots\right) \approx 2\left|M_{12}^q\right|, \quad (8.48)$$

$$\Delta \Gamma_q := \Gamma_L^q - \Gamma_H^q = 2\left|\Gamma_{12}^q\right|\cos \phi_{12}^q\left(1 + \frac{\left|\Gamma_{12}^q\right|^2 \sin^2 \phi_{12}^q}{8\left|M_{12}^q\right|^2} + \cdots\right) \approx 2\left|\Gamma_{12}^q\right|\cos \phi_{12}^q, \quad (8.49)$$

with the mixing phase involving only the difference of ϕ_M^q and ϕ_Γ^q

$$\phi_{12}^q := \arg\left(-\frac{M_{12}^q}{\Gamma_{12}^q}\right) = \pi + \phi_M^q - \phi_\Gamma^q, \quad (8.50)$$

and consequently being physical quantity. This phase can be probed by the so-called flavor-specific CP asymmetry [386]

$$a_{\mathrm{fs}}^q = \frac{|\Gamma_{12}^q|}{|M_{12}^q|}\sin \phi_{12}^q = \frac{\Delta M_q}{\Delta \Gamma_q}\tan \phi_{12}^q, \quad (8.51)$$

that we will encounter later on.

It should be emphasized that the approximations made in the preceding formulas are excellent because in the $B_q^0 - \bar{B}_q^0$ systems one has $\Delta \Gamma_q \ll \Delta M_q$. This is in fact the reason why it is more suitable to distinguish the mass eigenstates by their masses than by the corresponding lifetimes as we have done in the K meson system.

Finally an important formula is this one

$$\frac{q}{p} = -e^{-i\phi_M^q}\left[1 - \frac{1}{2}\frac{|\Gamma_{12}^q|}{|M_{12}^q|}\sin \phi_{12}^q + O\left(\frac{|\Gamma_{12}^q|^2}{|M_{12}^q|^2}\right)\right]$$

$$= -e^{-i\phi_M^q}\left[1 - \frac{a_{\mathrm{fs}}^q}{2}\right] + O\left(\frac{|\Gamma_{12}^q|^2}{|M_{12}^q|^2}\right). \quad (8.52)$$

Time Evolution of Neutral Mesons

We move next to discuss the time evolution of the flavor eigenstates following first [346, 380] and generalizing the index s in their formulas to $q = d, s$. We have then

$$|B_q^0(t)\rangle = g_+(t)|B_q^0(t=0)\rangle + \frac{q}{p}g_-(t)|\bar{B}_q^0(t=0)\rangle, \tag{8.53}$$

$$|\bar{B}_q^0(t)\rangle = \frac{p}{q}g_-(t)|B_q^0(t=0)\rangle + g_+(t)|\bar{B}_q^0(t=0)\rangle, \tag{8.54}$$

with the coefficients

$$g_+(t) = e^{-iM_q t}e^{-\frac{1}{2}\Gamma_q t}\left[\cosh\frac{\Delta\Gamma_q t}{4}\cos\frac{\Delta M_q t}{2} - i\sinh\frac{\Delta\Gamma_q t}{4}\sin\frac{\Delta M_q t}{2}\right], \tag{8.55}$$

$$g_-(t) = e^{-iM_q t}e^{-\frac{1}{2}\Gamma_q t}\left[-\sinh\frac{\Delta\Gamma_q t}{4}\cos\frac{\Delta M_q t}{2} + i\cosh\frac{\Delta\Gamma_q t}{4}\sin\frac{\Delta M_q t}{2}\right], \tag{8.56}$$

where

$$M_q = \frac{M_H^q + M_L^q}{2}, \qquad \Gamma_q = \frac{\Gamma_H^q + \Gamma_L^q}{2}. \tag{8.57}$$

Certainly

$$g_+(0) = 1, \qquad g_-(0) = 0, \tag{8.58}$$

but at $t > 0$ the flavor eigenstates are mixtures of $|B_q^0(t=0)\rangle$ and $|\bar{B}_q^0(t=0)\rangle$.

The next goal is to calculate the time evolution of four decay rates:

$$\Gamma\left[B_q^0(t) \to f\right], \qquad \Gamma\left[\bar{B}_q^0(t) \to f\right], \qquad \Gamma\left[B_q^0(t) \to \bar{f}\right], \qquad \Gamma\left[\bar{B}_q^0(t) \to \bar{f}\right], \tag{8.59}$$

where f is an arbitrary final state and $\bar{f}$ its CP conjugate

$$|\bar{f}\rangle = \mathcal{CP}|f\rangle. \tag{8.60}$$

The resulting formulas presented in [346, 380] are rather involved, and it is useful to express them in terms of quantities that enter directly observables. To this end we introduce the amplitudes[1]

$$\mathcal{A}_f = \langle f|\mathcal{H}_{eff}|B_q^0\rangle, \qquad \bar{\mathcal{A}}_f = \langle f|\mathcal{H}_{eff}|\bar{B}_q^0\rangle. \tag{8.61}$$

and

$$\mathcal{A}_{\bar{f}} = \langle\bar{f}|\mathcal{H}_{eff}|B_q^0\rangle, \qquad \bar{\mathcal{A}}_{\bar{f}} = \langle\bar{f}|\mathcal{H}_{eff}|\bar{B}_q^0\rangle, \tag{8.62}$$

with $\mathcal{H}_{eff}$ being effective Hamiltonian relevant for the decays in question. Having these amplitudes, we introduce two important quantities

$$\xi_f = \frac{q}{p}\frac{\bar{\mathcal{A}}_f}{\mathcal{A}_f}, \qquad \xi_{\bar{f}} = \frac{q}{p}\frac{\bar{\mathcal{A}}_{\bar{f}}}{\mathcal{A}_{\bar{f}}}, \tag{8.63}$$

---

[1] To simplify the notation we suppress the index q on the l.h.s. of these definitions.

which in turn allows us to introduce the following six asymmetries

$$\mathcal{A}_{CP}^{dir} = \frac{1 - |\xi_f|^2}{1 + |\xi_f|^2}, \qquad \bar{\mathcal{A}}_{CP}^{dir} = \frac{1 - |\xi_{\bar{f}}|^{-2}}{1 + |\xi_{\bar{f}}|^{-2}}, \tag{8.64}$$

$$\mathcal{A}_{CP}^{mix} = -\frac{2\text{Im}\left(\xi_f\right)}{1 + |\xi_f|^2}, \qquad \bar{\mathcal{A}}_{CP}^{mix} = -\frac{2\text{Im}\left(1/\xi_{\bar{f}}\right)}{1 + |\xi_{\bar{f}}|^{-2}}, \tag{8.65}$$

$$\mathcal{A}_{\Delta\Gamma} = -\frac{2\text{Re}\left(\xi_f\right)}{1 + |\xi_f|^2}, \qquad \bar{\mathcal{A}}_{\Delta\Gamma} = -\frac{2\text{Re}\left(1/\xi_{\bar{f}}\right)}{1 + |\xi_{\bar{f}}|^{-2}}. \tag{8.66}$$

Some readers could be puzzled by the minus signs in $\mathcal{A}_{CP}^{mix}$ and $\mathcal{A}_{\Delta\Gamma}$, which are absent in some other papers. This difference can be traced back by looking at (8.47) and finding that in the other papers q is replaced by $-q$, although the expression for ξ_f in terms of p and q is in that papers as in (8.63). Consequently ξ_f differs by sign in that papers from the one used by us, which translates into additional minus signs in the two asymmetries in question. With these minus signs the final results for all asymmetries listed by us are the same as in papers in which q is replaced by $-q$. However, one has to remember at intermediate stages that ξ_f differ by sign.

As we will see soon, $\mathcal{A}_{CP}^{dir}$ describes effects related to direct CP violation, while $\mathcal{A}_{CP}^{mix}$ encodes effects due to the interference between mixing and decay. Finally, $\mathcal{A}_{\Delta\Gamma}$ is a correction factor, due to a finite value of the decay rate difference $\Delta\Gamma_q$. It enters the definition of the *effective lifetime* τ_q^{eff}

$$\tau_q^{eff} = \tau_{B_q^0} \frac{1}{1 - y_q^2} \left(\frac{1 + 2\mathcal{A}_{\Delta\Gamma} y_q + y_q^2}{1 + \mathcal{A}_{\Delta\Gamma} y_q} \right) \tag{8.67}$$

with

$$\tau_{B_q^0} = \frac{1}{\Gamma_{B_q^0}} = \frac{2}{(\Gamma_H^q + \Gamma_L^q)}, \qquad y_q = \tau_{B_q^0} \frac{\Delta\Gamma_q}{2}. \tag{8.68}$$

In the limit $\Delta\Gamma_q = 0$ the effective lifetime reduces to the total lifetime of $\tau(B_q^0)$. The three quantities in (8.64)–(8.66) are related through

$$\left(\mathcal{A}_{CP}^{dir}\right)^2 + \left(\mathcal{A}_{CP}^{mix}\right)^2 + (\mathcal{A}_{\Delta\Gamma})^2 = 1. \tag{8.69}$$

With all these definitions at hand the formulas in [346, 380] look nicer

$$\Gamma\left[B_q^0(t) \to f\right] = \tilde{\Gamma}_f e^{-\Gamma t} \left\{ \cosh\left(\frac{\Delta\Gamma_q}{2}t\right) + \mathcal{A}_{\Delta\Gamma} \sinh\left(\frac{\Delta\Gamma_q}{2}t\right) \right.$$

$$\left. + \mathcal{A}_{CP}^{dir} \cos\left(\Delta M_q t\right) + \mathcal{A}_{CP}^{mix} \sin\left(\Delta M_q t\right) \right\}. \tag{8.70}$$

$$\Gamma\left[\bar{B}_q^0(t) \to f\right] = \tilde{\Gamma}_f(1 + a_{\mathrm{fs}}^q)e^{-\Gamma t}\left\{\cosh\left(\frac{\Delta\Gamma_q}{2}t\right) + \mathcal{A}_{\Delta\Gamma}\sinh\left(\frac{\Delta\Gamma_q}{2}t\right)\right.$$
$$\left. - \mathcal{A}_{\mathrm{CP}}^{\mathrm{dir}}\cos\left(\Delta M_q t\right) - \mathcal{A}_{\mathrm{CP}}^{\mathrm{mix}}\sin\left(\Delta M_q t\right)\right\}. \tag{8.71}$$

$$\Gamma\left[B_q^0(t) \to \bar{f}\right] = \tilde{\bar{\Gamma}}(1 - a_{\mathrm{fs}}^q)e^{-\Gamma t}\left\{\cosh\left(\frac{\Delta\Gamma_q}{2}t\right) + \bar{\mathcal{A}}_{\Delta\Gamma}\sinh\left(\frac{\Delta\Gamma_q}{2}t\right)\right.$$
$$\left. - \bar{\mathcal{A}}_{\mathrm{CP}}^{\mathrm{dir}}\cos\left(\Delta M_q t\right) - \bar{\mathcal{A}}_{\mathrm{CP}}^{\mathrm{mix}}\sin\left(\Delta M_q t\right)\right\}. \tag{8.72}$$

$$\Gamma\left[\bar{B}_q^0(t) \to \bar{f}\right] = \tilde{\bar{\Gamma}}e^{-\Gamma t}\left\{\cosh\left(\frac{\Delta\Gamma_q}{2}t\right) + \bar{\mathcal{A}}_{\Delta\Gamma}\sinh\left(\frac{\Delta\Gamma_q}{2}t\right)\right.$$
$$\left. + \bar{\mathcal{A}}_{\mathrm{CP}}^{\mathrm{dir}}\cos\left(\Delta M_q t\right) + \bar{\mathcal{A}}_{\mathrm{CP}}^{\mathrm{mix}}\sin\left(\Delta M_q t\right)\right\}. \tag{8.73}$$

The overall time-independent factors $\tilde{\Gamma}_f$ and $\tilde{\bar{\Gamma}}$ cancel out in the asymmetries considered by us. They can be extracted from [346, 380]. Using these expressions we can now calculate the following time-dependent CP asymmetry

$$A_{CP,f}(t) = \frac{\Gamma\left(\bar{B}_q^0(t) \to f\right) - \Gamma\left(B_q^0(t) \to f\right)}{\Gamma\left(\bar{B}_q^0(t) \to f\right) + \Gamma\left(B_q^0(t) \to f\right)}. \tag{8.74}$$

We find [346, 380][2]

$$A_{CP,f}(t) = -\frac{\mathcal{A}_{\mathrm{CP}}^{\mathrm{dir}}\cos(\Delta M_q t) + \mathcal{A}_{\mathrm{CP}}^{\mathrm{mix}}\sin(\Delta M_q t)}{\cosh\left(\frac{\Delta\Gamma_q t}{2}\right) + \mathcal{A}_{\Delta\Gamma}\sinh\left(\frac{\Delta\Gamma_q t}{2}\right)} + O\left(a_{fs}^q\right), \tag{8.75}$$

where the term involving a_{fs}^q is negligible. To be able to isolate this small term and measure it, we have to consider CP asymmetries of the so-called flavor-specific decays, which allow to eliminate large terms in (8.75). These decays are defined as follows

- The decays $\bar{B}_d^0 \to f$ and $B_d^0 \to \bar{f}$ are forbidden, which implies

$$\bar{\mathcal{A}}_f = 0 = \mathcal{A}_{\bar{f}}, \qquad \xi_f = 0 = \frac{1}{\xi_{\bar{f}}}. \tag{8.76}$$

Therefore

$$\mathcal{A}_{\mathrm{CP}}^{\mathrm{mix}} = \bar{\mathcal{A}}_{\mathrm{CP}}^{\mathrm{mix}} = \mathcal{A}_{\Delta\Gamma} = \bar{\mathcal{A}}_{\Delta\Gamma} = 0, \tag{8.77}$$

and the time evolution of these decays is very simple.

[2] The overall minus sign absent in some papers is simply related to the definition of the asymmetry (8.74).

8.1.7 More on $\Delta\Gamma_{d,s}$

We have already stated in Section 7.5 that the calculation of $\Delta\Gamma_{d,s}$ is much more difficult than of $\Delta M_{d,s}$ and relies on HQE with the general expression given in (7.67). We refer to papers listed there and to more recent presentations for more information. In particular in [346] an informative discussion of the 2015 status of $\Delta\Gamma_{d,s}$ is given. A good collection of references can be found in [387, 389, 390]. See also contributions to the CKM 2018 workshop.

Still we would like to get some feeling for $\Delta\Gamma_{d,s}$. First using (8.48), (8.49), and (8.46) we obtain

$$\frac{\Delta\Gamma_q}{\Delta M_q} = -\mathrm{Re}\left(\frac{\Gamma_{12}^{(q)}}{M_{12}^{(q)}}\right). \tag{8.92}$$

Note that the mass difference corresponds to the oscillation frequency as clearly seen in the preceding basic formulas. Final states in which both B and $\bar{B}$ can decay contribute to $\Delta\Gamma$ and correspond to the imaginary parts of the box diagrams.

Indeed, from the absorptive parts of box diagrams one obtains [351, 391],

$$\Gamma_{12}^{(q)} = \frac{G_F^2 m_b^2 \hat{m}_{B_q} \hat{B}_{B_q} F_{B_q}^2}{8\pi}\left[\lambda_t^{(q)2} + \frac{8}{3}\lambda_c^{(q)}\lambda_t^{(q)}\left(z_c + \frac{1}{4}z_c^2 - \frac{1}{2}z_c^3\right) \right. \tag{8.93}$$
$$\left. + \lambda_c^{(q)2}\left\{\sqrt{1-4z_c}\left(1-\frac{2}{3}z_c\right) + \frac{8}{3}z_c + \frac{2}{3}z_c^2 - \frac{4}{3}z_c^3 - 1\right\}\right]^*,$$

where

$$z_c \equiv \frac{m_c^2}{m_b^2}, \qquad \lambda_i^{(q)} \equiv V_{iq}V_{ib}^*. \tag{8.94}$$

Because the expression (8.93) for the off-diagonal element $\Gamma_{12}^{(q)}$ of the decay matrix is similarly to $M_{12}^{(q)}$ dominated by the term proportional to $[\lambda_t^{(q)2}]^*$, we have

$$\frac{\Gamma_{12}^{(q)}}{M_{12}^{(q)}} \approx -\frac{3\pi}{2S_0(x_t)}\frac{m_b^2}{M_W^2}. \tag{8.95}$$

Therefore, $|\Gamma_{12}^{(q)}|/|M_{12}^{(q)}| = O(m_b^2/m_t^2) \ll 1$.

Using (8.92) we obtain

$$\frac{\Delta\Gamma_q}{\Gamma_q} \approx \frac{3\pi}{2S_0(x_t)}\frac{m_b^2}{M_W^2}x_q, \qquad x_q = \frac{\Delta M_q}{\Gamma_q}. \tag{8.96}$$

Consequently $\Delta\Gamma_q$ is positive so that the decay width $\Gamma_H^{(q)}$ of the "heavy" mixing eigenstate is smaller than that of the "light" eigenstate. Because the numerical factor in (8.96) multiplying the mixing parameter x_q is $O(10^{-2})$, the width difference $\Delta\Gamma_d$ is very small within the SM. On the other hand, the large value of x_s in (8.91) implies a sizable $\Delta\Gamma_s$. The dynamical origin of this width difference is related to CKM favored $\bar{b} \to \bar{c}c\bar{s}$ quark-level transitions into final states that are common to B_s^0 and $\bar{B}_s^0$ mesons. Theoretical and experimental analyses of $\Delta\Gamma_s/\Gamma_s$ have advanced over last years far beyond the result in (8.93), with the most recent results in theory [346, 387, 392] and experiment [150] given, respectively,

$$(\Delta\Gamma_s)^{\text{SM}} = 0.088 \pm 0.020\text{ps}^{-1}, \qquad (\Delta\Gamma_s)^{\text{EXP}} = 0.086 \pm 0.006\text{ps}^{-1} \qquad (8.97)$$

and

$$\left(\frac{\Delta\Gamma_s}{\Gamma_s}\right)^{\text{SM}} = 0.132 \pm 0.030, \qquad \left(\frac{\Delta\Gamma_s}{\Gamma_s}\right)^{\text{EXP}} = 0.130 \pm 0.009. \qquad (8.98)$$

We observe a very good agreement of the theory with experiment but it is desirable to decrease the theoretical error. The corresponding values for $B_d^0 - \bar{B}_d$ system are

$$(\Delta\Gamma_d)^{\text{SM}} = 0.0029 \pm 0.0007\text{ps}^{-1}, \qquad (\Delta\Gamma_d)^{\text{EXP}} = -0.001 \pm 0.006\text{ps}^{-1} \qquad (8.99)$$

and

$$\left(\frac{\Delta\Gamma_d}{\Gamma_d}\right)^{\text{SM}} = 0.0040 \pm 0.0009, \qquad \left(\frac{\Delta\Gamma_d}{\Gamma_d}\right)^{\text{EXP}} = -0.002 \pm 0.010, \qquad (8.100)$$

with the SM value for $\Delta\Gamma_d$ from [393].

As was pointed out in [394, 395], $\Delta\Gamma_s$ may lead to interesting CP-violating effects in *untagged* data samples of time-evolved B_s decays, where one does not distinguish between initially present B_s^0 and $\bar{B}_s^0$ mesons. We shall return to detailed discussions of CP-violating asymmetries in the B_d system and of the B_s system in light of significant $\Delta\Gamma_s$, but first let us focus on B_q decays ($q \in \{d, s\}$) into final CP eigenstates.

8.2 B_q Decays into CP Eigenstates

Let us next look in more detail on

$$\boxed{\xi_f^{(q)} = \frac{q}{p} \frac{A(\bar{B}_q^0 \to f)}{A(B_q^0 \to f)}, \qquad \xi_{\bar{f}}^{(q)} = \frac{q}{p} \frac{A(\bar{B}_q^0 \to \bar{f})}{A(B_q^0 \to \bar{f})},} \qquad (8.101)$$

where we now added the index q in the definition in (8.63) and changed the notation to amplitudes as often done in the literature. While in [346] mainly the B_s^0 system has been discussed, all the formulas for time evolution and CP-asymmetries presented by us can also be used for the $B_d^0 - \bar{B}_d^0$ system, but in this case one can set $\Delta\Gamma_d = 0$ for all practical purposes, which simplifies the preceding formulas. We will give them in Section 8.2.3.

Now, a very promising special case, in respect of extracting CKM phases from CP-violating effects in neutral B_q decays, are transitions into final states $|f\rangle$ that are eigenstates of the CP operator and hence satisfy

$$(\mathcal{CP})|f\rangle = \eta_f |f\rangle, \qquad \eta_f = \pm 1. \qquad (8.102)$$

Consequently we have $\xi_f^{(q)} = \xi_{\bar{f}}^{(q)}$ in that case (see (8.101)) and have to deal only with a single observable $\xi_f^{(q)}$ containing essentially all the information that is needed to evaluate the time-dependent decay rates (8.70)–(8.73). Indeed, the six asymmetries in (8.64)–(8.66), that one needs to evaluate these rates, depend for such final states only on $\xi_f^{(q)}$.

But the decays into final states that are not CP eigenstates play also an important role in phenomenology, in particular in the extraction of the UT angle γ from B_s decays. We will discuss them in Section 8.5.4.

8.2.1 Calculation of $\xi_f^{(q)}$

In the SM we have from (8.52) within an excellent approximation

$$\xi_f^{(q)} = -e^{-\phi_M^q} \frac{A(\overline{B_q^0} \to f)}{A(B_q^0 \to f)}, \tag{8.103}$$

with the phases ϕ_M^q given in (8.87) entirely in terms of CKM phases β and β_s. On the other hand the amplitude ratio $A(\overline{B_q^0} \to f)/A(B_q^0 \to f)$ requires the calculation of hadronic matrix elements, which are still poorly known at present. To investigate this amplitude ratio, we shall employ the low-energy effective Hamiltonian for $|\Delta B| = 1$, $\Delta C = \Delta U = 0$ transitions that involve current-current, QCD-penguin and electroweak penguin operators. We met this Hamiltonian in general form in Chapter 6 but now we have to exhibit better various indices. We follow here [378] except that we set the phase ϕ_{CP} in that paper to π. As demonstrated in [378], this phase cancels between two factors in (8.103).

We have then

$$\mathcal{H}_{\text{eff}} = \mathcal{H}_{\text{eff}}(\Delta B = -1) + \mathcal{H}_{\text{eff}}(\Delta B = -1)^\dagger, \tag{8.104}$$

with

$$\mathcal{H}_{\text{eff}}(\Delta B = -1) = \frac{G_F}{\sqrt{2}} \left[\sum_{j=u,c} V_{jr}^* V_{jb} \left\{ \sum_{k=1}^{2} Q_k^{jr} C_k(\mu) + \sum_{k=3}^{10} Q_k^r C_k(\mu) \right\} \right], \tag{8.105}$$

where $\mu = O(m_b)$. In writing this effective Hamiltonian, we have generalized the notation of Chapter 6 to exhibit different cases. We have introduced two quark flavor labels $j = (u,c)$ and $r = (d,s)$ to parametrize $b \to j\bar{j}r$ quark-level transitions with r distinguishing between $b \to d$ and $b \to s$ transitions, respectively. The label $q = (d,s)$ is reserved for B_q so that one can distinguish between the valence quark q in B_q and the one in the final state r. These labels will turn out to be useful for the following discussion. Consequently we have

- Current-current operators:

$$Q_1^{jr} = (\bar{r}_\alpha j_\beta)_{\text{V-A}} (\bar{j}_\beta b_\alpha)_{\text{V-A}},$$
$$Q_2^{jr} = (\bar{r}_\alpha j_\alpha)_{\text{V-A}} (\bar{j}_\beta b_\beta)_{\text{V-A}}. \tag{8.106}$$

- QCD penguin operators:

$$Q_3^r = (\bar{r}_\alpha b_\alpha)_{V-A} \sum_{q'=u,d,s,c,b} (\bar{q}'_\beta q'_\beta)_{V-A},$$

$$Q_4^r = (\bar{r}_\alpha b_\beta)_{V-A} \sum_{q'=u,d,s,c,b} (\bar{q}'_\beta q'_\alpha)_{V-A},$$

$$Q_5^r = (\bar{r}_\alpha b_\alpha)_{V-A} \sum_{q'=u,d,s,c,b} (\bar{q}'_\beta q'_\beta)_{V+A},$$

$$Q_6^r = (\bar{r}_\alpha b_\beta)_{V-A} \sum_{q'=u,d,s,c,b} (\bar{q}'_\beta q'_\alpha)_{V+A}.$$

(8.107)

- EW penguin operators:

$$Q_7^r = \frac{3}{2}(\bar{r}_\alpha b_\alpha)_{V-A} \sum_{q'=u,d,s,c,b} e_{q'} (\bar{q}'_\beta q'_\beta)_{V+A},$$

$$Q_8^r = \frac{3}{2}(\bar{r}_\alpha b_\beta)_{V-A} \sum_{q'=u,d,s,c,b} e_{q'} (\bar{q}'_\beta q'_\alpha)_{V+A},$$

$$Q_9^r = \frac{3}{2}(\bar{r}_\alpha b_\alpha)_{V-A} \sum_{q'=u,d,s,c,b} e_{q'} (\bar{q}'_\beta q'_\beta)_{V-A},$$

$$Q_{10}^r = \frac{3}{2}(\bar{r}_\alpha b_\beta)_{V-A} \sum_{q'=u,d,s,c,b} e_{q'} (\bar{q}'_\beta q'_\alpha)_{V-A}.$$

(8.108)

Consequently

$$A\left(\overline{B_q^0} \to f\right) = \left\langle f \middle| \mathcal{H}_{\text{eff}}(\Delta B = -1) \middle| \overline{B_q^0} \right\rangle$$

(8.109)

$$= \left\langle f \middle| \frac{G_F}{\sqrt{2}} \left[\sum_{j=u,c} V_{jr}^* V_{jb} \left\{ \sum_{k=1}^{2} Q_k^{jr}(\mu) C_k(\mu) + \sum_{k=3}^{10} Q_k^r(\mu) C_k(\mu) \right\} \right] \middle| \overline{B_q^0} \right\rangle.$$

On the other hand, the transition amplitude $A\left(B_q^0 \to f\right)$ is given by

$$A\left(B_q^0 \to f\right) = \left\langle f \middle| \mathcal{H}_{\text{eff}}(\Delta B = -1)^\dagger \middle| B_q^0 \right\rangle$$

(8.110)

$$= \left\langle f \middle| \frac{G_F}{\sqrt{2}} \left[\sum_{j=u,c} V_{jr} V_{jb}^* \left\{ \sum_{k=1}^{2} Q_k^{jr\dagger}(\mu) C_k(\mu) + \sum_{k=3}^{10} Q_k^{r\dagger}(\mu) C_k(\mu) \right\} \right] \middle| B_q^0 \right\rangle,$$

where we assumed that the WCs are real as is the case in the SM. Otherwise they have to be complex conjugated.

Performing appropriate CP transformations in this equation, i.e., inserting the operator $(CP)^\dagger(CP) = \hat{1}$ both after the bra $\langle f|$ and in front of the ket $|B_q^0\rangle$, yields

$$A\left(B_q^0 \to f\right) = -\eta_f$$

(8.111)

$$\times \left\langle f \middle| \frac{G_F}{\sqrt{2}} \left[\sum_{j=u,c} V_{jr} V_{jb}^* \left\{ \sum_{k=1}^{2} Q_k^{jr}(\mu) C_k(\mu) + \sum_{k=3}^{10} Q_k^r(\mu) C_k(\mu) \right\} \right] \middle| \overline{B_q^0} \right\rangle,$$

two penguin diagrams, or one tree diagram and one penguin diagram. Indeed, writing the decay amplitude $\mathcal{A}_{f^+}$ and its CP conjugate $\bar{\mathcal{A}}_{f^-}$ as

$$\mathcal{A}_{f^+} = \sum_{i=1,2} A_i e^{i(\delta_i + \phi_i)}, \qquad \bar{\mathcal{A}}_{f^-} = \sum_{i=1,2} A_i e^{i(\delta_i - \phi_i)}, \tag{8.129}$$

with A_i being real, one finds

$$\mathcal{A}_{\text{CP}}^{\text{dir}}(B^\pm \to f^\pm) = \frac{-2A_1 A_2 \sin(\delta_1 - \delta_2) \sin(\phi_1 - \phi_2)}{A_1^2 + A_2^2 + 2A_1 A_2 \cos(\delta_1 - \delta_2) \cos(\phi_1 - \phi_2)}. \tag{8.130}$$

The sign of the strong phases δ_i is the same for $\mathcal{A}_{f^+}$ and $\bar{\mathcal{A}}_{f^-}$ because CP is conserved by strong interactions. The weak phases have opposite signs.

The presence of hadronic uncertainties in A_i and of strong phases δ_i complicates the extraction of the phases ϕ_i from data. An example of this type of CP violation in K decays is direct CP violation in $K_L \to \pi^+ \pi^-$ described by the parameter ε'. We will demonstrate this later.

8.3.4 CP Violation in the Interference of Mixing and Decay

This type of CP violation can also be called *mixing-induced* CP violation, and we presented already some formulas for it before in connection of the decays to states that are CP eigenstates as defined in (8.102). Here we want to add several observations that will turn out to be useful in next sections. We could denote CP eigenstates by f_{CP}, but as only such states will be considered here we will just call it f as already done in (8.102). The presence of CP-parity η_f and Figure 8.1 will remind us that it is a CP-eigenstate.

A neutral meson M can either directly decay into f or first oscillate into $\overline{M}$ and then decay into f (see Figures 8.1 and 8.2). The interference between these two routes can induce CP violation if

$$\text{Im} \xi_f \neq 0 \Rightarrow \mathcal{A}_{\text{CP}}^{\text{mix}} \neq 0. \tag{8.131}$$

This type of CP violation is only possible in neutral B and K decays, and the basic formula for the asymmetry that involves $\mathcal{A}_{\text{CP}}^{\text{mix}}$ has been given for the B_d system in (8.122), and for the B_s system one has to include $\Delta\Gamma_s$ effects so that here (8.75) has to be used.

Generally several decay mechanisms with different weak and strong phases can contribute to $\mathcal{A}_{\text{CP}}^{\text{mix}}$. These are tree diagram (current-current) contributions, QCD penguin contributions, and electroweak penguin contributions. If they contribute with similar strength to a given decay amplitude, the resulting CP asymmetries suffer from hadronic uncertainties related to matrix elements of the relevant operators Q_i. The situation is

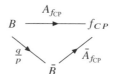

Figure 8.1   Mixing induced CP violation for B mesons.

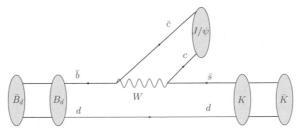

Figure 8.2 The decay $B_d \to \psi K_S$ with possible oscillations before the decay.

then analogous to the case of CP violation in decay just discussed. Indeed then the CP-asymmetry depends on the strong phases $\delta_{T,P}$ and the hadronic matrix elements present in $A_{T,P}$. Thus the measurement of the asymmetry does not allow a clean determination of the weak phases $\phi_{T,P}$.

But as we have seen earlier an interesting case arises when a single mechanism dominates the decay amplitude or the contributing mechanisms have the same weak phases. Then the hadronic matrix elements and strong phases drop out and the quantity $\xi_f^{(q)}$ in (8.120) becomes a pure phase and moreover depends only on CKM phases. We have calculated it for four cases in (8.116)–(8.119). Another important point is that in all these cases $|\xi_f^{(q)}| = 1$ so that the coefficient of $\cos \Delta M_d t$ in $A_{CP,f}(t)$ vanishes. Consequently we find

$$A_{CP,f}(t) = -\mathcal{A}_{CP}^{\text{mix}} \sin(\Delta M_d t) = \text{Im} \xi_f^{(d)} \sin(\Delta M_d t), \qquad (8.132)$$

so that we arrive at an important formula

$$A_{CP,f}(t) = S_f \sin(\Delta M_d t), \qquad S_f = -\eta_f \sin(\phi_M^{(d)} - \phi_D^{(f)}(d,j)). \qquad (8.133)$$

Thus the corresponding measurement of weak phases is free from hadronic uncertainties. A well-known example is the decay $B_d \to \psi K_S$ for which case 1 in (8.116) applies. As $\phi_M^{(d)} = 2\beta$ and $\phi_D^{(f)}(d,j) = 0$ we find with $\eta_f = -1$

$$A_{CP,\psi K_S}(t) = S_{\psi K_S} \sin(\Delta M_d t), \qquad S_{\psi K_S} = \sin\left(\phi_M^{(d)}\right) = \sin(2\beta), \qquad (8.134)$$

which allows a very clean measurement of the angle β in the unitarity triangle.

As this asymmetry measures directly the difference between the phases of the $B_d^0 - \bar{B}_d^0$-mixing $(2\phi_M)$ and of the decay amplitude $(2\phi_D)$, it is impossible to state on the basis of a single asymmetry whether CP violation takes place in the mixing or in the decay. To this end at least two asymmetries for $B_d^0(\bar{B}_d^0)$ decays to different final states f_i have to be measured. As ϕ_M does not depend on the final state, $\text{Im} \xi_{f_1} \neq \text{Im} \xi_{f_2}$ is a signal of CP violation in the decay.

With this in mind let us try as a second example $B_d \to \pi^+\pi^-$, which corresponds to the case 3 in (8.118). With $\eta_{\pi\pi} = 1$ we find now

$$A_{CP,\pi\pi}(t) = S_{\pi\pi}\sin(\Delta M_d t), \qquad S_{\pi\pi} = -\sin(2\beta + 2\gamma) = \sin(2\alpha), \qquad (8.135)$$

which could be in principle a nice measurement of $\beta + \gamma$ or equivalently α in the UT. But unfortunately our assumption of the dominance of tree diagrams in this decay fails because of important QCD penguin contributions. We will demonstrate this in Section 8.5.

Of great interest is also $B_s \to \psi\phi$. We will discuss it in detail in Section 8.5, but as it corresponds to the case 2 in (8.117) we can investigate it already now neglecting $\Delta\Gamma_s$ so that formula (8.133) applies with d replaced by s. With $\eta_{\psi\phi} = 1$ we find with $\phi_M^{(s)} = 2\beta_s$

$$A_{CP,\psi\phi}(t) = S_{\psi\phi}\sin(\Delta M_s t), \qquad S_{\psi\phi} = -\sin\left(\phi_M^{(s)}\right) = -\sin(2\beta_s). \qquad (8.136)$$

Here the assumption of the dominance of tree-diagrams in $b \to sc\bar{c}$ is very good, and this asymmetry allows to measure β_s, which in the SM turns out to be very small: $\beta_s \approx -1°$. As we will discuss in Section 8.5, this SM prediction is consistent with experiment. We will also discuss there what is the impact of QCD penguins and of $\Delta\Gamma_s$ on this determination. But we will find that it is rather small. It should be remarked that in some papers $S_{\psi\phi}$ is defined without the minus sign in (8.136) so that in the SM it is negative. We find it more natural to have the general definition in (8.133), which results in a positive $S_{\psi\phi}$ in the SM. At the end what really counts are the phases ϕ_M and ϕ_D.

We will see in later sections that the ideal situation of the dominance of one decay mechanism does not always take place. We have mentioned it already in the case of $B \to \pi\pi$. Indeed, when two different mechanisms with different weak ($\phi_{1,2}$) and strong ($\delta_{1,2}$) phases contribute to the CP asymmetry, one finds

$$A_{CP,f}(t) = -C_f \cos(\Delta M t) + S_f \sin(\Delta M t), \qquad (8.137)$$

$$C_f = 2r\sin(\phi_1 - \phi_2)\sin(\delta_1 - \delta_2), \qquad (8.138)$$

$$S_f = \eta_f\left[\sin(2\phi_1 - \phi_M) + 2r\cos(2\phi_1 - \phi_M)\sin(\phi_1 - \phi_2)\cos(\delta_1 - \delta_2)\right], \qquad (8.139)$$

where $r = A_2/A_1 \ll 1$ has been assumed. The extraction of weak phases is now polluted by strong phases. For $r = 0$ the previous formulas are obtained.

In the case of K decays, this type of CP violation can be cleanly measured in the rare decay $K_L \to \pi^0\nu\bar{\nu}$. Here the difference between the weak phase in the $K^0 - \bar{K}^0$ mixing and in the decay $\bar{s} \to \bar{d}\nu\bar{\nu}$ matters. We will discuss this decay in Section 9.5.

We can now compare the two classifications of different types of CP violation. CP violation in mixing is a manifestation of indirect CP violation. CP violation in decay is a manifestation of direct CP violation. The third type contains elements of both the indirect and direct CP violation.

It is clear from this discussion that only in the case of the third type of CP violation there are possibilities to measure directly weak phases without hadronic uncertainties and moreover without invoking sophisticated methods. This takes place provided a single mechanism (diagram) is responsible for the decay or the contributing decay mechanisms have the same weak phases. However, we will see in Section 8.5 that there are other strategies, involving also decays to CP noneigenstates, that provide clean measurements of the weak phases.

8.3.5 Another Look at ε and ε'

Let us finally investigate what type of CP violation is represented by ε and ε'. Here instead of different mechanisms, it is sufficient to talk about different isospin amplitudes.

In the case of ε, CP violation in decay is not possible as only the isospin amplitude A_0 is involved. See (8.25). We also know that only Re ε = Re $\bar{\varepsilon}$ is related to CP violation in mixing. Consequently:

- Reε represents CP violation in mixing.
- Imε represents CP violation in the interference of mixing and decay.

To analyze the case of ε' we use the formula (8.28) to find

$$\text{Re } \varepsilon' = -\frac{1}{\sqrt{2}} \left| \frac{A_2}{A_0} \right| \sin(\phi_2 - \phi_0) \sin(\delta_2 - \delta_0), \tag{8.140}$$

$$\text{Im } \varepsilon' = \frac{1}{\sqrt{2}} \left| \frac{A_2}{A_0} \right| \sin(\phi_2 - \phi_0) \cos(\delta_2 - \delta_0) . \tag{8.141}$$

Consequently:

- Re ε' represents CP violation in decay as it is only nonzero provided simultaneously $\phi_2 \neq \phi_0$ and $\delta_2 \neq \delta_0$.
- Im ε' exists even for $\delta_2 = \delta_0$ but as it requires $\phi_2 \neq \phi_0$ it represents CP violation in decay as well.

Experimentally $\delta_2 \neq \delta_0$. Within the SM, ϕ_2 and ϕ_0 are connected with electroweak penguins and QCD penguins, respectively. We will discuss the phenomenology of ε'/ε in Chapter 10.

8.4 Standard Analysis of the Unitarity Triangle (UT)

8.4.1 Preliminaries

With the information collected until now we are ready to determine the UT. Optimally this triangle should be determined by means of tree-level decays because these decays are expected to be subject to very small uncertainties coming from possible NP contributions. Indeed, from semileptonic K and B_d decays and in particular nonleptonic two-body $B_{d,s}$ decays discussed in the next section, it is possible to determine the set

$$|V_{us}|, \qquad |V_{ub}|, \qquad |V_{cb}|, \qquad \gamma \tag{8.142}$$

and consequently both the UT and the full CKM matrix independent of NP contributions. The UT determined in this manner is usually called *the reference unitarity triangle* [396]. In fact, γ is the only angle in the UT that can be determined using only tree-level decays.

Such a determination would allow one to make predictions for various loop-induced processes, like $B_{d,s}^0 - \bar{B}_{d,s}^0$ mixings, ε_K, ε'/ε, and rare K and $B_{s,d}$ decays in the SM independent of whether these processes receive any contributions from NP or not. Due to the progress by the LHCb and soon by Belle II experiment, such a determination with respectable precision should be possible in the coming years.

However, until now the determination of the UT and of the CKM matrix, with the exception of $|V_{us}|$, was dominated by global fits performed in particular by UTfit and CKMfitter collaborations [397, 398] that include all available information. This means also processes that are subject to potential contributions from NP. Such a procedure is self-consistent within the SM, but assumes the absence of NP to observables used in the fit. We should all hope that this assumption is false and that such a procedure will imply various inconsistencies between various determinations of $|V_{ub}|$, $|V_{cb}|$ and γ signaling the presence of NP. In fact, as we will discuss in Section 15.1, there are indeed some hints that this happens when one simultaneously considers present determinations of $|V_{ub}|$, $|V_{cb}|$, γ, and β from tree-level decays together with $\Delta M_{s,d}$ and ε_K.

Yet, a search for NP through UT fits is not very efficient. As we will stress in the next chapters, the optimal procedure is to determine the four parameters in (8.142) and the UT through tree-level decays and having them predict as many flavor observables as possible, in particular various branching ratios for rare decays and CP-asymmetries, within the SM. Comparing them with experimental data and studying the pattern of deviations from SM, predictions would allow one to identify the dynamics behind these *anomalies*. Despite this, we will follow first the traditional route, which involves also loop-induced processes.

8.4.2 General Procedure

Even if these days a UT analysis incorporates numerically constraints from various observables, it is instructive to present the determination of the UT in the form of steps that exhibit the basic uncertainties and also tell us which measurements determine its sides and which determine its angles. The presentation that follows is only meant to be a pedagogical anatomy of the determination of the UT and as such does not use exact parametrization of the CKM matrix. A professional construction has to use exact parametrization and must be done numerically. See frequent updates by UTfitter [397] and CKMfitter [398] collaborations.

An important parameter in this construction is $|V_{us}|$ and the assumption that the CKM matrix is a unitary matrix. Recently the most accurate values for $|V_{us}|$ and for the first-row CKM unitarity relation has been obtained by the ETM LQCD collaboration through the inclusion of leading electromagnetic and strong isospin-breaking corrections to $\pi^+ \to \mu^+ \nu[\gamma]$ and $K^+ \to \mu^+ \nu[\gamma]$ decays. The impressive result reads [399],

$$|V_{us}| = 0.22538(38), \qquad |V_{ud}|^2 + |V_{us}|^2 + |V_{ub}|^2 = 0.99988(44). \tag{8.143}$$

Yet, as we will report in Section 12.2, the fate of the CKM unitarity is unclear at present, in particular because of new measurements of $|V_{ud}|$. For the time being we assume CKM unitarity to be valid.

Setting then $\lambda = |V_{us}| = 0.225$, our simplified analysis uses generalized Wolfenstein parametrization and proceeds in the following five steps:

Step 1

From the $b \rightarrow c$ transition in inclusive and exclusive leading B-meson decays, one finds $|V_{cb}|$ and consequently the scale of the UT:

$$|V_{cb}| \quad \Longrightarrow \quad \lambda|V_{cb}| = \lambda^3 A. \tag{8.144}$$

Step 2

From the $b \rightarrow u$ transition in inclusive and exclusive B meson decays, one finds $|V_{ub}|$ and consequently using (2.100) together with $|V_{cb}|$ from the first step the side $CA = R_b$ of the UT:

$$\left|\frac{V_{ub}}{V_{cb}}\right| \quad \Longrightarrow \quad R_b = \sqrt{\bar{\varrho}^2 + \bar{\eta}^2} = \frac{1}{\lambda}(1 - \frac{\lambda^2}{2})\left|\frac{V_{ub}}{V_{cb}}\right| = 4.35 \cdot \left|\frac{V_{ub}}{V_{cb}}\right|. \tag{8.145}$$

Step 3

From the experimental value of ε_K in (8.41) and the formula (8.39) rewritten in terms of Wolfenstein parameters, one derives the constraint on $(\bar{\varrho}, \bar{\eta})$

$$\bar{\eta}\left[(1 - \bar{\varrho})A^2\eta_2 S_0(x_t) + P_c(\varepsilon)\right]A^2 \hat{B}_K \kappa_\varepsilon = 0.187, \tag{8.146}$$

where

$$P_c(\varepsilon) = [\eta_3 S_0(x_c, x_t) - \eta_1 x_c]\frac{1}{\lambda^4}, \qquad x_i = \frac{m_i^2}{M_W^2}, \tag{8.147}$$

with all symbols defined in previous sections and $P_c(\varepsilon)$ summarizing the contributions of box diagrams with two charm-quark exchanges and the mixed charm-top exchanges.

As seen in Figure 8.3, equation (8.146) specifies a hyperbola in the $(\bar{\varrho}, \bar{\eta})$ plane. The position of the hyperbola depends on m_t, $|V_{cb}| = A\lambda^2$, and $\hat{B}_K$. With decreasing m_t, $|V_{cb}|$ and $\hat{B}_K$, it moves away from the origin of the $(\bar{\varrho}, \bar{\eta})$ plane. Among these three variables only the uncertainty in $|V_{cb}|$ is relevant at present because ε_K is very sensitive to it. It is represented by the A^4 dependence in the leading term in (8.146). The plot in Figure 8.3 shows on the one hand the schematic determination of the UT and on the other hand the situation around the year 2000 when all uncertainties were much larger and $\sin 2\beta$ and γ have not been measured yet. These days the precision is much larger, and the overlapping region is shifted due to direct γ measurements to the right: $\gamma < 90°$. We will discuss this determination later. We show the plot in Figure 8.3 to underscore the dramatic progress that has been made in this millennium by B-factories Babar and Belle, by Tevatron and LHC experiments, as well as theoretical efforts. The latter efforts were very important for next steps.

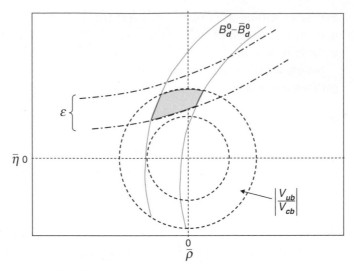

Figure 8.3   Schematic determination of the unitarity triangle: 2000 view. From [29].

Step 4

From the measured ΔM_d and the formula (8.90), the side $AB = R_t$ of the UT can be determined:

$$R_t = \frac{1}{\lambda} \frac{|V_{td}|}{|V_{cb}|} = 0.85 \cdot \left[\frac{|V_{td}|}{8.0 \cdot 10^{-3}} \right] \left[\frac{0.042}{|V_{cb}|} \right], \tag{8.148}$$

$$|V_{td}| = 8.0 \cdot 10^{-3} \left[\frac{227.7\,\text{MeV}}{\sqrt{\hat{B}_{B_d}} F_{B_d}} \right] \sqrt{\frac{2.322}{S(x_t)}} \sqrt{\frac{\Delta M_d}{0.5055/\text{ps}}} \sqrt{\frac{0.5521}{\eta_B}}, \tag{8.149}$$

with all symbols defined in previous sections. Note that R_t suffers from the additional uncertainty in $|V_{cb}|$, which is absent in the determination of $|V_{td}|$ this way. The constraint in the $(\bar{\varrho}, \bar{\eta})$ plane coming from this step is illustrated in Figure 8.3.

Step 5

The measurement of ΔM_s together with ΔM_d, allows to determine R_t in a different manner:

$$R_t = 0.87 \left[\frac{\xi}{1.21} \right] \sqrt{\frac{17.7/\text{ps}}{\Delta M_s}} \sqrt{\frac{\Delta M_d}{0.5055/\text{ps}}}, \qquad \xi = \frac{\sqrt{\hat{B}_{B_s}} F_{B_s}}{\sqrt{\hat{B}_{B_d}} F_{B_d}}. \tag{8.150}$$

One should note that m_t and $|V_{cb}|$ dependences have been eliminated this way and that ξ should in principle contain much smaller theoretical uncertainties than the hadronic matrix elements in ΔM_d and ΔM_s separately.

The main uncertainties in these steps originated for many years in the theoretical uncertainties in $\hat{B}_K$, $\sqrt{\hat{B}_d} F_{B_d}$, to a lesser extent in ξ, and of course in the CKM parameters in the first two steps. But these uncertainties as of 2019 decreased significantly as seen in

Table 8.1 Values for different quantities from the UTfitter and CKMfitter. $\lambda_t = V_{ts}^* V_{td}$.

Quantity	UTfitter	CKMfitter
A	0.826 ± 0.012	$0.8403^{+0.0056}_{-0.0201}$
λ	0.2255 ± 0.0005	$0.224747^{+000254}_{-000059}$
$\bar{\eta}$	0.348 ± 0.010	$0.3493^{+0.0096}_{-0.0071}$
$\bar{\varrho}$	0.148 ± 0.013	$0.1577^{+0.0107}_{-0.0074}$
$\sin 2\beta$	0.699 ± 0.016	$0.708^{+0.013}_{-0.010}$
α	$(91.9 \pm 2.0)°$	$(91.6^{+1.7}_{-1.1})°$
β	$(22.25 \pm 0.65)°$	$(22.51^{+0.55}_{-0.40})°$
γ	$(66.8 \pm 2.0)°$	$(65.81^{+0.99}_{-1.66})°$
R_b	0.379 ± 0.011	$0.3832^{+0.0087}_{-0.0065}$
R_t	0.919 ± 0.013	$0.9128^{+0.0067}_{-0.0124}$
$|V_{td}|\ (10^{-3})$	$8.69 + 0.14$	$8.710^{+0.086}_{-0.246}$
$|V_{ts}|\ (10^{-3})$	41.25 ± 0.55	$41.69^{+0.28}_{-1.08}$
$\mathrm{Im}\lambda_t\ (10^{-4})$	1.419 ± 0.041	1.447 ± 0.041
$\mathrm{Re}\lambda_t\ (10^{-4})$	$-(3.297 \pm 0.096)$	$-(3.329^{+0.170}_{-0.053})$

Table 8.1 and Table D.3. Still the present errors on $|V_{ub}|$ and the angle γ from tree-level decays are substantial. We will return to this issue in Section 15.1.1 in which some tensions between various observables involved in these steps will be discussed.

This five-step procedure was used until the turn of the millennium, but after the discovery of CP-violation in the B_d system one could add a sixth step, a very important one: the determination of the angle β of the UT from $B_d \rightarrow \psi K_S$ and generally $b \rightarrow dc\bar{c}$ transitions. This is such an important step that we devote to it a separate section.

8.4.3 The Angle β from $B_d \rightarrow \psi K_S$

In this decay the final state is a CP eigenstate and the angle β in the UT can be measured by means of the time-dependent mixing induced CP asymmetry in (8.134)

$$A_{CP,\psi K_S}(t) = S_{\psi K_S} \sin(\Delta M_d t), \qquad S_{\psi K_S} = \sin(2\beta). \tag{8.151}$$

The present world average from HFLAV reads

$$(\sin 2\beta)_{\psi K_S} = 0.699 \pm 0.017 . \tag{8.152}$$

This is a milestone in the field of CP violation and in the tests of the SM as we will see in a moment. Not only violation of this symmetry has been confidently established in the B_d

system, but also its size has been measured very accurately. Moreover in contrast to the five constraints listed earlier, the determination of the angle β in this manner is theoretically very clean.

The expression (8.151) assumes that QCD penguin contributions are negligible. While this assumption is very plausible, it is important, in view of the accurate determination of β in this manner, to estimate the error coming from these additional contributions within the SM. Several authors, over many years, expressed their views on the impact of QCD penguin (QCDP) contributions on this determination of β. It appears from present perspective that

$$\Delta(\sin 2\beta)_{\psi K_S}^{\text{QCDP}} \leq 0.01. \tag{8.153}$$

Selected papers on this topic can be found in [400–405]. For the latest summary, see the talk by Martin Jung at CKM 2018.

8.4.4 Unitarity Triangle 2019

General Fits

We are now in the position to combine all these constraints to construct the UT and determine various quantities of interest. In this context the important issue is the error analysis of these formulas, in particular the treatment of theoretical uncertainties. In the literature the most popular are the Bayesian approach [397] and the frequentist approach [398]. Most recent analyses of the UT by the Bayesians ("UTfitter") and frequentists ("CKMfitter") can be found in [397] and [398], respectively.

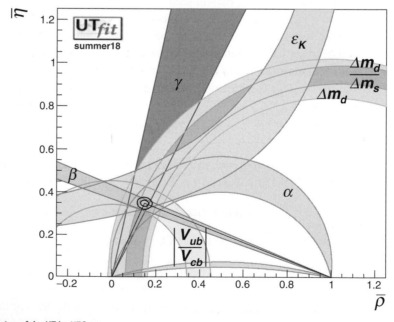

Figure 8.4   Determination of the UT by UTfitter.

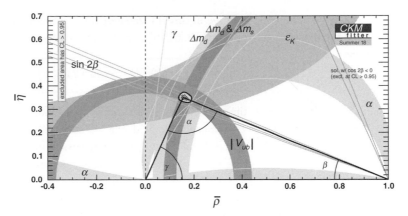

Figure 8.5 Determination of the UT by CKMfitter.

In Figures 8.4 and 8.5 we show the results of the analyses of both groups. In Table 8.1 we show the output for several quantities.[3] This output is changing with time, but we expect that the pattern shown there will not be changed by much in the future.

At first sight there is a good agreement between the direct measurements and the standard analysis of the UT within the SM. This gives a strong indication that the CKM matrix is likely the dominant source of CP violation in the observables used in these fits. However, a closer look shows that this depends on the values of $|V_{cb}|$ and $|V_{ub}|$ determined in the first two steps, and moreover the value of γ from tree-level determinations is significantly larger than from the UT fit. Therefore let us have a closer look at possible deviations.

The Issue of $|V_{ub}|$, $|V_{cb}|$ and the Angle γ

As stressed at the beginning of this section, the CKM matrix can be determined by the tree-level measurements of $|V_{ub}|$, $|V_{cb}|$, $|V_{us}|$, and of the angle γ in the UT. The determinations of $|V_{ub}|$, $|V_{cb}|$, $|V_{us}|$ have been already described in Section 3.4, stressing the difference between the exclusive and inclusive determinations of $|V_{ub}|$ and to a much lesser extent in the case of $|V_{cb}|$. We repeat here our nominal values

$$|V_{ub}|^{\text{nom}} = 3.7 \times 10^{-3}, \qquad |V_{cb}|^{\text{nom}} = 42.0 \times 10^{-3}, \qquad (8.154)$$

with the latter value being the central value favored by Paolo Gambino $|V_{cb}| = (42.0 \pm 0.6) \times 10^{-3}$ [406].

---

[3] The author thanks Sébastien Descotes-Genon, Marcela Bona, and Luca Silvestrini for providing these numbers and the UT-plots.

The current world average of direct measurements of the angle γ from HFLAV [150] and the most recent from the LHCb [407], read, respectively

$$\gamma = \left(71.1^{+4.6}_{-5.3}\right)^{\circ}, \qquad \gamma = \left(74.0^{+5.0}_{-5.8}\right)^{\circ}. \tag{8.155}$$

As seen in Table 8.1, these values are significantly larger than the ones obtained from the UTfit with interesting consequences to be discussed in Section 15.1. The manner in which γ can be extracted from tree-level decays will be discussed in the next section.

On the other hand, the value of $\sin 2\beta$ in Table 8.1 agrees well with the experimental world average in (8.152). In this context it is useful to recall the interplay between $|V_{cb}|$, $|V_{ub}|$, ε_K, and $\sin 2\beta$ with $|V_{cb}|$ and ε_K strongly correlated with each other and $|V_{ub}|$ with $\sin 2\beta$ as one can check analytically.

As realized already in 2008 [376, 408] and also confirmed in [397, 398], this interplay has the following pattern:

- If *exclusive* determinations of $|V_{ub}|$ and $|V_{cb}|$ were used, the resulting $\sin 2\beta$ agreed well with the result in (8.152), while the predicted value of ε_K was visibly below the data.
- If *inclusive* determinations of $|V_{ub}|$ and $|V_{cb}|$ were used, the resulting $\sin 2\beta$ was significantly above the result in (8.152), while the predicted value of ε_K was consistent with the data.
- Optimally for the SM would be $|V_{ub}|$ from the exclusive determinations, while $|V_{cb}|$ from the inclusive ones. This is basically what we have chosen as nominal values for $|V_{ub}|$ and $|V_{cb}|$ in (8.154).

Yet as discussed in Section 3.4, while the issue of $|V_{cb}|$ seems to be solved in favor of the inclusive determination, this is not the case for $|V_{ub}|$. Therefore, if in the future $|V_{ub}|$ from tree-level decays will be found significantly larger than our nominal value, the implied UTfit value of $\sin 2\beta$ will be larger than the experimental one requiring the introduction of new complex phases. We will return to this possibility in Chapter 15.

We will also see in Section 15.1 that although the SM describes the experimental value of ε_K well, new anomalies in $\Delta M_{s,d}$ follow from the inclusive value of $|V_{cb}|$ combined with the most recent LQCD values of hadronic matrix elements entering $\Delta M_{s,d}$. In the case of ΔM_d, this anomaly is further enhanced through the large value of γ in (8.155) so that also the ratio $\Delta M_d/\Delta M_s$ predicted by the SM seems to disagree with data. A different view has been expressed in [1340].

As the UTfitter and CKMfitter analyses are very involved, we list here again real and imaginary parts of λ_t as functions of the four basic parameters of the CKM matrix

$$\mathrm{Re}\lambda_t \simeq |V_{ub}||V_{cb}| \cos\gamma(1-2\lambda^2) + (|V_{ub}|^2 - |V_{cb}|^2)\lambda \left(1 - \frac{\lambda^2}{2}\right), \tag{8.156}$$

$$\mathrm{Im}\lambda_t \simeq |V_{ub}||V_{cb}| \sin\gamma. \tag{8.157}$$

These handy expressions are accurate up to $O(\lambda^4)$ corrections and allow a quick estimate of these two parameters that enter various observables.

8.5 The Angles α, β, and γ from $B_{d,s}$ Decays

8.5.1 Preliminaries

CP violation in B decays has been already for three decades one of the most important targets of B physics dedicated experiments both in e^+e^- machines as B-factories (Belle and BaBar experiments) and hadron facilities with experiments D0 and CDF at the Tevatron and LHCb, CMS, and ATLAS at the LHC. Soon the Belle II experiment at SuperKEKB will also produce new data. It is well known that CP-violating effects are expected to occur in a large number of $B_{s,d}$ decay channels at a level attainable experimentally and in fact as we have seen in previous sections and we will see later, clear signals of CP violation in $B_{s,d}$ decays have already been observed. Moreover there exist channels that offer the determination of the angles in the UT with only small hadronic uncertainties. In fact, the measurements of CP asymmetries in several $B_{s,d}$ decays and the measurements of the angles α, β, and γ by means of various strategies using two-body $B_{s,d}$ decays contributed already substantially to our understanding of CP violation and tests of the KM picture of CP violation. While this framework appears in a rather good shape, we will see that there is still some room left for NP in CP-violating quantities and also in CP-conserving ones.

The various types of CP violation have been already classified in Section 8.3. It turned out that CP violation in the interference of mixing and decay, in a $B_{s,d}$ meson decay into a CP eigenstate, is very suitable for a theoretically clean determination of the angles of the unitarity triangle provided a single complex phase governs the decay. However, as we will see soon several useful strategies for the determination of the angles α, β, and γ have been developed that are effective also in the presence of competing complex phases and when the final state is not a CP eigenstate. The discussion presented later should only be considered as an introduction to this rich field. For more details the references quoted in the course of our presentation should be contacted. In constructing the following pages, we benefited much from [378–381].

8.5.2 Classification of Elementary Processes

Nonleptonic B decays are caused by elementary decays of b quarks that are represented by tree and penguin diagrams in Figure 8.6. Generally we have

$$b \to q_1 \bar{q}_2 d(s), \qquad b \to q\bar{q}d(s) \qquad (8.158)$$

for tree and penguin diagrams, respectively.

There are twelve basic transitions that can be divided into three classes:

Class I: Both tree and penguin diagrams contribute. Here $q_1 = q_2 = q = u, c$, and consequently the basic transitions are

$$b \to c\bar{c}s, \qquad b \to c\bar{c}d, \qquad b \to u\bar{u}s, \qquad b \to u\bar{u}d. \qquad (8.159)$$

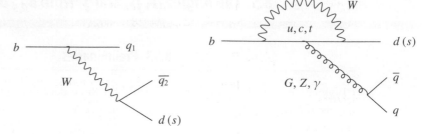

Figure 8.6   Tree and penguin diagrams.

Class II: Only tree diagrams contribute. Here $q_1 \neq q_2 \in \{u, c\}$, and

$$b \rightarrow c\bar{u}s, \qquad b \rightarrow c\bar{u}d, \qquad b \rightarrow u\bar{c}s, \qquad b \rightarrow u\bar{c}d. \tag{8.160}$$

Class III: Only penguin diagrams contribute. Here $q = d, s$, and

$$b \rightarrow s\bar{s}s, \qquad b \rightarrow s\bar{s}d, \qquad b \rightarrow d\bar{d}s, \qquad b \rightarrow d\bar{d}d. \tag{8.161}$$

Now in presenting various decays here, we do not show the corresponding diagrams on purpose. After all this is a textbook and a good exercise for the readers is to draw these diagrams by embedding the elementary diagrams of Figure 8.6 into a given $B_{s,d}$ meson decay. In case of difficulties the reader will find these diagrams in [379, 381, 409].

8.5.3 Neutral B Decays into CP Eigenstates

$B_d^0 \rightarrow \psi K_S$ and β

We have already discussed briefly this decay in the previous section in connection with the determination of $\sin 2\beta$. But let us look at this decay in more detail now. The amplitude for it can be written as follows

$$A(B_d^0 \rightarrow \psi K_S) = V_{cs}V_{cb}^*(A_T + P_c) + V_{us}V_{ub}^*P_u + V_{ts}V_{tb}^*P_t, \tag{8.162}$$

where A_T denotes tree diagram contributions and P_i, with $i = u, c, t$, stand for penguin diagram contributions with internal u, c, and t quarks. Now

$$V_{cs}V_{cb}^* \approx A\lambda^2, \quad V_{us}V_{ub}^* \approx A\lambda^4 R_b e^{i\gamma}, \quad V_{ts}V_{tb}^* = -V_{us}V_{ub}^* - V_{cs}V_{cb}^*, \tag{8.163}$$

with the last relation following from the unitarity of the CKM matrix. Thus

$$A(B_d^0 \rightarrow \psi K_S) = V_{cs}V_{cb}^*(A_T + P_c - P_t) + V_{us}V_{ub}^*(P_u - P_t). \tag{8.164}$$

We next note that

$$\left| \frac{V_{us}V_{ub}^*}{V_{cs}V_{cb}^*} \right| \leq 0.02, \qquad \frac{P_u - P_t}{A_T + P_c - P_t} \ll 1, \tag{8.165}$$

where the last inequality is very plausible because we know from previous chapters that the WCs of the current-current operators responsible for A_T are much larger than the ones of the penguin operators. Consequently, this decay is dominated by a single CKM factor and as discussed in Section 8.3, a clean determination of the relevant CKM phase is possible. Indeed, in this decay $\phi_D = 0$ and $\phi_M = 2\beta$. Using (8.133) we find once more ($\eta_{\psi K_S} = -1$)

$$S_{\psi K_S} = -\eta_{\psi K_S} \sin\left(\phi_M^{(d)} - \phi_D\right) = \sin 2\beta, \qquad C_{\psi K_S} = 0. \tag{8.166}$$

The impact of QCD penguins on this determination is small but has to be taken into account at some level. SU(3) and U-spin methods allow to get some estimate of its size with the result given in (8.153). Eventually, it will be important to measure β in decays to which QCD penguins do not contribute. This is the case of $\bar{B}^0 \to D^0 h^0$, where h^0 is a neutral hadron.

$B_s^0 \to \psi\phi$ and β_s

This decay differs from the previous one by the spectator quark, with $d \to s$ so that the preceding formulas remain unchanged except that now $\phi_M^{(s)} = 2\beta_s = -\lambda^2 \bar{\eta}$. A complication arises as the final state is a mixture of $CP = +$ and $CP = -$ states. This issue can be resolved experimentally [379]. Choosing $\eta_{\psi\phi} = 1$ we then find again

$$S_{\psi\phi} = -\sin\left(\phi_M^{(s)} - \phi_D\right) = -\sin(2\beta_s) = 2\lambda^2\bar{\eta} = 0.0363 \pm 0.0013, \qquad C_{\psi\phi} = 0. \tag{8.167}$$

Thus this asymmetry measures the phase of V_{ts} that is predicted to be very small from the unitarity of the CKM matrix alone. Because of this there was several years ago a lot of room for NP contributions to $\mathcal{A}_{CP}^{mix}(\psi\phi)$. Yet, the most recent measurements from the LHCb show that this asymmetry is indeed very small [410, 411], leading to HFLAV average

$$S_{\psi\phi} = 0.054 \pm 0.020, \tag{8.168}$$

which compared with (8.167) is a significant constraint on NP models. Yet, it should be emphasized that introducing NP parameters one could arrange the $S_{\psi\phi}$ to be close to its SM value, while having significant CP-violating effects in other observables.

The impact of QCD penguins on the determination of $S_{\psi\phi}$ is discussed in [403, 404] and by Martin Jung at the CKM 2018, where further references can be found.

$B_d^0 \to \phi K_S$ and β

This decay proceeds entirely through penguin diagrams and consequently should be much more sensitive to NP contributions than the decay $B_d^0 \to \psi K_S$. Assuming $\phi = (s\bar{s})$, the decay amplitude is given by (8.164) with A_T removed:

$$A(B_d^0 \to \phi K_S) = V_{cs}V_{cb}^*(P_c - P_t) + V_{us}V_{ub}^*(P_u - P_t). \tag{8.169}$$

With

$$\left|\frac{V_{us}V_{ub}^*}{V_{cs}V_{cb}^*}\right| \leq 0.02, \qquad \frac{P_u - P_t}{P_c - P_t} = O(1), \tag{8.170}$$

also in this decay a single CKM phase dominates and as ϕ_D and ϕ_M are the same as in $B_d^0 \to \psi K_S$, we find

$$C_{\phi K_S} = 0, \qquad S_{\phi K_S} = S_{\psi K_S} = \sin 2\beta .\qquad (8.171)$$

The equality of these two asymmetries need not be perfect as the ϕ meson is not entirely a $s\bar{s}$ state, and the approximation of neglecting the second amplitude in (8.169) could be only true within a few percent. However, a detailed analysis in [412] shows that these two asymmetries should be very close to each other within the SM:

$$S_{\phi K_S} - S_{J/\psi K_S} = 0.025 \pm 0.012 \pm 0.010.\qquad (8.172)$$

Any strong violation of this bound would be a signal for NP. It will be interesting to follow the development in the values of $S_{\phi K_s}$ and $C_{\phi K_s}$ and similar values in other channels such as $B \to \eta' K_S$. However, experimental measurements of these channels are difficult. Some recent discussions of these channels can be found in [320, 413].

$B_d^0 \to \pi^+ \pi^-$ and α

This decay receives the contributions from both tree and penguin diagrams. The amplitude can be written as follows

$$A(B_d^0 \to \pi^+ \pi^-) = V_{ud} V_{ub}^* (A_T + P_u) + V_{cd} V_{cb}^* P_c + V_{td} V_{tb}^* P_t \qquad (8.173)$$

where

$$V_{cd} V_{cb}^* \approx A\lambda^3, \quad V_{ud} V_{ub}^* \approx A\lambda^3 R_b e^{i\gamma}, \quad V_{td} V_{tb}^* = -V_{ud} V_{ub}^* - V_{cd} V_{cb}^* .\qquad (8.174)$$

Consequently

$$A\left(B_d^0 \to \pi^+ \pi^-\right) = V_{ud} V_{ub}^* (A_T + P_u - P_t) + V_{cd} V_{cb}^* (P_c - P_t).\qquad (8.175)$$

We next note that

$$\left| \frac{V_{cd} V_{cb}^*}{V_{ud} V_{ub}^*} \right| = \frac{1}{R_b} \approx 2.5, \qquad \frac{P_c - P_t}{A_T + P_u - P_t} \equiv \frac{P_{\pi\pi}}{T_{\pi\pi}}.\qquad (8.176)$$

Now the dominance of a single CKM amplitude in contrast to the cases considered until now is very uncertain and takes place only provided $P_{\pi\pi} \ll T_{\pi\pi}$. Assuming this we have already derived in (8.135) that

$$C_{\pi\pi} = 0, \qquad S_{\pi\pi} = \sin 2\alpha .\qquad (8.177)$$

However, our assumption $P_{\pi\pi} \ll T_{\pi\pi}$ has been invalidated by experimental data a long time ago implying that QCD penguin contributions have to be taken into account or somehow eliminated in the extraction of α. The well-known strategy to deal with this "penguin problem" is the isospin analysis of Gronau and London [414] that uses $B \to \pi\pi$ decays. It has been extended to $B \to \rho\rho$ and $B \to \rho\pi$ modes [415, 416]. Various other strategies are reviewed in [379, 417]. Isospin breaking effects in these methods have been developed in [418]. The most recent analysis in [419] determines

$$\alpha = \left(86.2^{+4.4}_{-4.0} \cup \left(178.4^{+3.9}_{-5.1}\right)\right)^\circ .\qquad (8.178)$$

The result in the first quadrant agrees well with the UT analyses in Table 8.1 but is much less accurate.

8.5.4 Decays to CP Noneigenstates

Preliminaries

The strategies discussed next have the following general properties:

- $B_d^0(B_s^0)$ and their antiparticles $\bar{B}_d^0(\bar{B}_s^0)$ can decay to the same final state.
- Only tree diagrams contribute to the decay amplitudes.
- A full-time dependent analysis of the four processes is required:

$$B_{d,s}^0(t) \to f, \quad \bar{B}_{d,s}^0(t) \to f, \quad B_{d,s}^0(t) \to \bar{f}, \quad \bar{B}_{d,s}^0(t) \to \bar{f}. \tag{8.179}$$

The latter analysis allows to measure

$$\xi_f = \exp(i2\phi_M)\frac{A(\bar{B}^0 \to f)}{A(B^0 \to f)}, \qquad \xi_{\bar{f}} = \exp(i2\phi_M)\frac{A(\bar{B}^0 \to \bar{f})}{A(B^0 \to \bar{f})}. \tag{8.180}$$

It turns out then that

$$\xi_f \cdot \xi_{\bar{f}} = F(\gamma, \phi_M) \tag{8.181}$$

is given in terms of γ and ϕ_M without any hadronic uncertainties, so that determining ϕ_M from other decays as discussed earlier, allows the determination of γ. Let us show this and find an explicit expression for the function F.

$B_d^0 \to D^\pm \pi^\mp, \bar{B}_d^0 \to D^\pm \pi^\mp$ and γ

With $f = D^+\pi^-$ the four decay amplitudes are given by

$$A(B_d^0 \to D^+\pi^-) = M_f A\lambda^4 R_b e^{i\gamma}, \qquad A(\bar{B}_d^0 \to D^+\pi^-) = \bar{M}_f A\lambda^2 \tag{8.182}$$

$$A(\bar{B}_d^0 \to D^-\pi^+) = \bar{M}_{\bar{f}} A\lambda^4 R_b e^{-i\gamma}, \qquad A(B_d^0 \to D^-\pi^+) = M_{\bar{f}} A\lambda^2 \tag{8.183}$$

where we have factored out the CKM parameters, A is one of the Wolfenstein parameters, and M_i stand for the rest of the amplitudes that generally are subject to large hadronic uncertainties. The important point is that each of these transitions receives the contribution from a single phase so that inserting these expressions into (8.180), we find

$$\xi_f^{(d)} = e^{-i(2\beta+\gamma)}\frac{1}{\lambda^2 R_b}\frac{\bar{M}_f}{M_f}, \qquad \xi_{\bar{f}}^{(d)} = e^{-i(2\beta+\gamma)}\lambda^2 R_b \frac{\bar{M}_{\bar{f}}}{M_{\bar{f}}}. \tag{8.184}$$

Now, as CP is conserved in QCD, we simply have

$$M_f = \bar{M}_{\bar{f}}, \qquad \bar{M}_f = M_{\bar{f}} \tag{8.185}$$

and consequently [420],

$$\xi_f^{(d)} \cdot \xi_{\bar{f}}^{(d)} = e^{-i2(2\beta+\gamma)}, \tag{8.186}$$

as promised. The phase β is already known with high precision, and consequently γ can be determined. Unfortunately, as seen in (8.182) and (8.183), the relevant interferences are $O(\lambda^2)$, and the execution of this strategy is a very difficult experimental task. Moreover, NP could enter through β affecting this determination of γ.

$B_s^0 \to D_s^{\pm} K^{\mp}, \bar{B}_s^0 \to D_s^{\pm} K^{\mp},$ and γ

Replacing the d-quark by the s-quark in the strategy just discussed allows to enhance the interference between various contributions. With $f = D_s^+ K^-$, equations (8.182) and (8.183) are replaced by

$$A(B_s^0 \to D_s^+ K^-) = M_f A\lambda^3 R_b e^{i\gamma}, \qquad A(\bar{B}_s^0 \to D_s^+ K^-) = \bar{M}_f A\lambda^3, \tag{8.187}$$

$$A(\bar{B}_s^0 \to D_s^- K^+) = \bar{M}_{\bar{f}} A\lambda^3 R_b e^{-i\gamma}, \qquad A(B_s^0 \to D_s^- K^+) = M_{\bar{f}} A\lambda^3 . \tag{8.188}$$

Proceeding as in the previous strategy one finds [421]

$$\xi_f^{(s)} \cdot \xi_{\bar{f}}^{(s)} = e^{-i2(2\beta_s+\gamma)} \tag{8.189}$$

with $\xi_f^{(s)}$ and $\xi_{\bar{f}}^{(s)}$ being the analogs of $\xi_f^{(d)}$ and $\xi_{\bar{f}}^{(d)}$, respectively. Now, all interfering amplitudes are of a similar size. With β_s extracted from the asymmetry in $B_s^0(\bar{B}_s^0) \to \psi\phi$, the angle γ can be determined. Again, this determination could be affected by NP entering here through β_s.

$B^{\pm} \to D^0 K^{\pm}, B^{\pm} \to \bar{D}^0 K^{\pm},$ and γ

By replacing the spectator s-quark in the last strategy through a u-quark one arrives at decays of $B^{\pm}$ that can be used to extract γ. Also this strategy is unaffected by penguin contributions. Moreover, as particle-antiparticle mixing is absent here, γ can be measured directly without any need for phases in the mixing. Both these features make it plausible that this strategy, not involving to first approximation any loop diagrams, is particularly suited for the determination of γ without any NP pollution.

By considering six decay rates $B^{\pm} \to D_{CP}^0 K^{\pm}$, $B^+ \to D^0 K^+$, $\bar{D}^0 K^+$ and $B^- \to D^0 K^-$, $\bar{D}^0 K^-$ where $D_{CP}^0 = (D^0 + \bar{D}^0)/\sqrt{2}$ is a CP eigenstate, and noting that

$$A(B^+ \to \bar{D}^0 K^+) = A(B^- \to D^0 K^-), \tag{8.190}$$

$$A(B^+ \to D^0 K^+) = A(B^- \to \bar{D}^0 K^-)e^{2i\gamma}, \tag{8.191}$$

the well-known triangle construction due to Gronau and Wyler [422] allows to determine γ. Indeed, the reader can check by drawing the corresponding diagrams that the relations just given are true and that these two processes are governed by $b \to c\bar{u}s$ and $b \to u\bar{c}s$ transitions, respectively.

However, the method is not without problems. The detection of D^0_{CP}, that is necessary for this determination because $K^+ \bar{D}^0 \neq K^+ D^0$ is experimentally challenging. Moreover, the small branching ratios of the color suppressed channels in (8.191) and the absence of this suppression in the two remaining channels in (8.190), evident after the diagrams have been drawn, imply a rather squashed triangle, thereby making the extraction of γ very difficult. More promising variants of this method are discussed in [414, 423–425]. These days these methods carry the names GLW [423], ADS [424], and GGSZ [425], and the fastest way to find out what they are in detail is to look up the talks by Charles, Gerson, Reis, and Resmi in the proceedings of the CKM 2018. The summary of the working group V dealing with γ determinations and also with searches for direct CP violation in B decays can be found in [426] and the analysis of an ultimate error on γ from $B \to DK$ decays in [427] and in Brod's contribution to CKM 2018. See also [428, 429].

In the last years the LHCb collaboration made important progress in measuring γ using combination of these methods and the ones listed later. See, e.g., [430] and references therein. The 2018 LHCb average for γ resulting from these efforts and HFLAV average are given in (8.155).

Other Clean Strategies for γ and β

The three strategies just discussed can be generalized to other decays. In particular [423, 431, 432]

- $2\beta + \gamma$ and γ can be measured in

$$B^0_d \to K_S D^0, \; K_S \bar{D}^0, \qquad B^0_d \to \pi^0 D^0, \; \pi^0 \bar{D}^0, \tag{8.192}$$

and the corresponding CP conjugated channels.
- $2\beta_s + \gamma$ and γ can be measured in

$$B^0_s \to \phi D^0, \; \phi \bar{D}^0, \qquad B^0_s \to K_S D^0, \; K_S \bar{D}^0, \tag{8.193}$$

and the corresponding CP conjugated channels.
- γ can be measured by generalizing the Gronau–Wyler construction to $B^\pm \to D^0 \pi^\pm$, $\bar{D}^0 \pi^\pm$ and to B_c decays [433]:

$$B^\pm_c \to D^0 D^\pm_s, \; \bar{D}^0 D^\pm_s, \qquad B^\pm_c \to D^0 D^\pm, \; \bar{D}^0 D^\pm . \tag{8.194}$$

It appears that these methods may give useful results at later stages of CP-B investigations, in particular at the LHCb and Belle II.

8.5.5 U–Spin Strategies

Preliminaries

Useful strategies for γ using the U-spin symmetry have been proposed by Robert Fleischer in [434, 435]. The first strategy involves the decays $B^0_{d,s} \to \psi K_S$ and $B^0_{d,s} \to D^+_{d,s} D^-_{d,s}$. The second strategy involves $B^0_s \to K^+ K^-$ and $B^0_d \to \pi^+ \pi^-$. They are unaffected by

FSI and are only limited by U-spin breaking effects. They already played a role in the determination of γ at the LHCb.

A method of determining γ, using $B^+ \to K^0\pi^+$ and the U-spin related processes $B_d^0 \to K^+\pi^-$ and $B_s^0 \to \pi^+K^-$ was presented in [436]. A general discussion of U-spin symmetry in charmless B decays and more references to this topic can be found in [379, 437]. I will only briefly discuss the method in [434].

$B_d^0 \to \pi^+\pi^-, B_s^0 \to K^+K^-,$ and (γ, β)

Replacing in $B_d^0 \to \pi^+\pi^-$ the d quark by an s quark we obtain the decay $B_s^0 \to K^+K^-$. The amplitude can then be written in analogy to (8.175) as follows

$$A(B_s^0 \to K^+K^-) = V_{us}V_{ub}^*(A_T' + P_u' - P_t') + V_{cs}V_{cb}^*(P_c' - P_t'). \tag{8.195}$$

This formula differs from (8.175) only by $d \to s$ and the primes on the hadronic matrix elements that in principle can be different in these two decays. As

$$V_{cs}V_{cb}^* \approx A\lambda^2, \qquad V_{us}V_{ub}^* \approx A\lambda^4 R_b e^{i\gamma}, \tag{8.196}$$

the second term in (8.195) is even more important than the corresponding term in the case of $B_d^0 \to \pi^+\pi^-$. Consequently $B_d^0 \to K^+K^-$ taken alone does not offer a useful method for the determination of the CKM phases. On the other hand, with the help of the U-spin symmetry of strong iterations, it allows roughly speaking to determine the penguin contributions in $B_d^0 \to \pi^+\pi^-$ and consequently the extraction of β and γ.

Indeed, from the U-spin symmetry, we have

$$\frac{P_{\pi\pi}}{T_{\pi\pi}} = \frac{P_c - P_t}{A_T + P_u - P_t} = \frac{P_c' - P_t'}{A_T' + P_u' - P_t'} = \frac{P_{KK}}{T_{KK}} \equiv de^{i\delta}, \tag{8.197}$$

where d is a real nonperturbative parameter and δ a strong phase. Measuring S_f and C_f for both decays and extracting β_s from $B_s^0 \to \psi\phi$, we can determine four unknowns: d, δ, β, and γ subject mainly to U-spin breaking corrections.

Using the first measurement of CP violation in $B_s \to K^-K^+$ the LHCb collaboration determined [438]

$$\gamma = \left(63.5_{-6.7}^{+7.2}\right)^\circ, \qquad \phi_M^s = -\left(6.9_{-8.0}^{+9.2}\right)^\circ, \tag{8.198}$$

which agrees with the determinations from pure tree decays within the uncertainties.

Theoretical precision of this method is limited by U-spin breaking corrections. Therefore a new strategy has been proposed in [439, 440]. It uses γ and ϕ_M^d as inputs together with semileptonic decays $B_s^0 \to K^-\ell^+\nu_l$ and $B_d^0 \to \pi^-\ell^+\nu_l$. It is demonstrated that this method could provide the determination of ϕ_M^s with a theoretical precision of $O(0.5^\circ)$.

8.6 $B \to \pi K$ Decays

8.6.1 Preliminaries

The four modes that played an important role already for two decades are

$$B_d^0 \to \pi^- K^+, \qquad B^+ \to \pi^+ K^0, \tag{8.199}$$

$$B_d^0 \to \pi^0 K^0, \qquad B^+ \to \pi^0 K^+ \tag{8.200}$$

and the CP conjugates. They are dominated by QCD penguins with tree-level contributions being small. Also EWP contributions to the first two are color suppressed and practically negligible. On the other hand, the EWP contributions to the last two are not color suppressed so that possible NP could easiest enter this system through them. When these decays were first observed by CLEO, BaBar, and Belle collaborations, they were used on the one hand to obtain some direct information on γ. On the other hand, they gave an interesting insight into the flavor and QCD dynamics and also indicated some departures from SM expectations that led to the so-called $B \to \pi K$ puzzle [441–444] and to the search for NP through them. See also a subsequent series of papers [445–447]. Most recent discussion of the latter authors can be found in [448].

Already at the end of the 1990s there has been a large theoretical activity in this field with the main issues being the final state interactions (FSI), SU(3) symmetry breaking effects, and the importance of electroweak penguin contributions. Several interesting ideas have been put forward to extract the angle γ despite large hadronic uncertainties in $B \to \pi K$ decays [449–456]. Three strategies for bounding and determining γ have been proposed. The "mixed" strategy [449] uses $B_d^0 \to \pi^0 K^0$ and $B^\pm \to \pi^\pm K$. The "charged" strategy [451, 455] involves $B^\pm \to \pi^0 K^\pm$, $\pi^\pm K$, and the "neutral" strategy [453] the modes $B_d^0 \to \pi^\mp K^\pm$, $\pi^0 K^0$. General parametrizations for the study of the FSI, SU(3) symmetry breaking effects, and the electroweak penguin contributions in these channels have been presented in [452–454]. Moreover, general parametrizations by means of Wick contractions [457, 458] have been proposed. The latter can be used for all two-body B-decays. Extensive analyses of the so-called charming-penguins in $B \to \pi K$ decays and generally in B decays can be found in [457, 459, 460].

Parallel to these efforts an important progress has been made by developing approaches for the calculation of the hadronic matrix elements of local operators in QCD beyond the standard factorization method that is QCDF, SCET, and QCD light-cone sum rules, which we encountered in Chapter 7.

From the present perspective the goal of extracting γ from $B \to \pi K$ decays around the year 2000 was a good idea as the data for cleaner tree-level strategies for γ determination discussed in previous sections could not be executed due to the lack of data. It led to the derivation of numerous formulas for the decays in question that allow to study QCD dynamics and possible NP in these decays. As of 2020, with γ determined rather precisely

Table 8.4 Compilation of predictions for the CP-violating $B \to \pi\pi, \pi K$ asymmetries.

Quantity	SM Prediction [461]	Experiment
$\mathcal{A}_{\text{CP}}^{\text{dir}}(B_d \to \pi^0\pi^0)$	0.44 ± 0.21	0.33 ± 0.22
$\mathcal{A}_{\text{CP}}^{\text{mix}}(B_d \to \pi^0\pi^0)$	0.81 ± 0.32	$-----$
$\mathcal{A}_{\text{CP}}^{\text{dir}}(B_d \to \pi^\mp K^\pm)$	-0.085 ± 0.064	-0.082 ± 0.006
$\mathcal{A}_{\text{CP}}^{\text{dir}}(B^\pm \to \pi^0 K^\pm)$	-0.007 ± 0.11	0.037 ± 0.021
$\mathcal{A}_{\text{CP}}^{\text{dir}}(B_d \to \pi^0 K_{\text{S}})$	-0.07 ± 0.15	0.00 ± 0.13
$\mathcal{A}_{\text{CP}}^{\text{dir}}(B^\pm \to \pi^\pm K^0)$	0.003 ± 0.005	-0.017 ± 0.016
$\mathcal{A}_{\text{CP}}^{\text{mix}}(B_d \to \pi^0 K_{\text{S}})$	0.81 ± 0.07	0.58 ± 0.17

appears to depart from the data exhibiting in this manner potential anomaly. Section 4 of [461] discusses these findings in detail. This correlation appears to be the cleanest probe of NP in the $B \to \pi K$ system.

A potential solution to this anomaly would be most naturally modified electroweak penguin contributions, a suggestion proposed already in [443] in the context of the old $B \to \pi K$ anomalies.

Step 3

In turn, the presence of modified EWP with significant CP-violating NP phases can have important implications for rare K and B decays as demonstrated in [441–444]. In this context the modified EWP contribution to $B \to \pi K$ decays should be extracted from future data. To this end the new strategy for the determination of the parameters of electroweak penguins proposed in [461] should be useful and could play an important role when improved data from LHCb and the ones from Belle II will be available. We refer to section 5 of that paper for details.

It is important to emphasize that this strategy as developed in 2003 and 2004 is valid both in the SM and all SM extensions in which NP enters predominantly through the EW penguin sector. Consequently, even if some deviations from the SM seen in 2004 diminished with improved data, this strategy, in particular with the recent improvements in [461], is still useful in correlating the phenomena in $B \to \pi\pi$, $B \to \pi K$, and rare K and B decays within the SM and the NP scenario in question. For a recent analysis see [465], where correlations of $B \to \pi K$ with $b \to s \mu^+ \mu^-$ have been investigated.

If, on the other hand, NP would enter $B \to \pi\pi$ decays, this approach should be properly generalized. The same applies to NP scenarios that enter $B \to \pi K$ decays through other operators than EWP ones.

9   Rare B and K Decays in the Standard Model

9.1 $\bar{B} \to \bar{K}^* \ell^+ \ell^-$ and $\bar{B} \to \bar{K} \ell^+ \ell^-$

9.1.1   Preliminaries

The decays $B_{s,d} \to \ell^+ \ell^-$, being the simplest FCNC processes in the $B_{s,d}$ system, could in principle be discussed first. But the relevant operators responsible for these decays belong to the subset of operators contributing to decays $\bar{B} \to \bar{K}^* \ell^+ \ell^-$ and $\bar{B} \to \bar{K} \ell^+ \ell^-$, and it is strategically useful to discuss the latter first.

Moreover, the latter decays are much more versatile with the possibility to measure, for instance, the differential decay rate in the lepton's invariant mass. One can also construct asymmetries, like the well-known forward-backward asymmetry (A_{FB}), with differing sensitivity to NP effects. Most important, the rare decay $\bar{B} \to \bar{K}^*(\to K^- \pi^+) \mu^+ \mu^-$ is an important channel for B physics as the polarization of the K^* allows a precise angular reconstruction resulting in many observables that offer new important tests of the SM and its extensions [466]. These angular observables can be expressed in terms of CP-conserving and CP-violating quantities, in fact as many as 24. But also the decay $\bar{B} \to \bar{K} \mu^+ \mu^-$ as well as $\bar{B} \to \bar{K}(\bar{K}^*) \ell^+ \ell^-$ with $\ell = e, \tau$ play important roles in the phenomenology, and we will discuss them as well.

Due to the GIM mechanism, the decays $\bar{B} \to \bar{K}^* \ell^+ \ell$ and $\bar{B} \to K \ell^+ \ell^-$ are absent in the SM at tree-level and are governed by electroweak penguin diagrams with the dominant top-quark exchange as seen in Figure 9.1. Beyond the SM they can proceed already at tree-level through the exchange of a heavy Z', leptoquarks, or other heavy particles.

The goal of this section is to define all these observables and express them in terms of the relevant Wilson coefficients (WCs). This will allow us to get a good idea of which of these observables are best suited for testing the SM and its extensions and how measuring several of them would allow to distinguish between different NP scenarios. Reaching this goal is a challenge as it is an exclusive decay and as such requires control not only over short-distance perturbative effects, described by WCs in the relevant effective Hamiltonian, but also control over long-distance nonperturbative effects, described largely, but not completely, by formfactors.

In the first part of this section we recall for reader's convenience the effective Hamiltonian for these decays including all contributions, namely, adding to the SM contributions those coming from operators that become relevant only in BSM scenarios. This we do as a preparation for later chapters where such scenarios will be discussed.

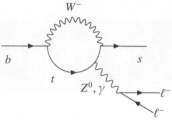

Figure 9.1   Penguin diagrams responsible for $\bar{B} \to \bar{K}^* \ell^+ \ell^-$ and $\bar{B} \to \bar{K} \ell^+ \ell^-$ decays.

Next, we present differential decay distribution for $B \to K^*(\to K\pi)\,\mu^+\mu^-$, introduce transversity amplitudes, and define the basic observables in this decay: twelve CP symmetries and twelve CP asymmetries. Subsequently, we summarize present knowledge of formfactors, which is dominated by light-cone sum rules (LCSR) for low and intermediate q^2 and lattice QCD for large q^2. We focus on the decays of neutral B_d mesons, $\bar{B}_d^0 \to \bar{K}^{*0}(\to K^-\pi^+)\,\mu^+\mu^-$ and its CP-conjugate $B_d^0 \to K^{*0}(\to K^+\pi^-)\,\mu^+\mu^-$, which have the additional advantage that the flavor of the decaying B_d meson (B_d^0 or $\bar{B}_d^0$) is unambiguously tagged by the final state. A nice review of this subject can be found in [467].

9.1.2 Theoretical Framework: $\bar{B} \to \bar{K}^* \mu^+ \mu^-$

Preliminaries

The theoretical framework, which allows one to calculate the decay amplitude of $\bar{B} \to \bar{K}^* \mu^+ \mu^-$, is quite involved and is described in detail in [468]. What follows is a short version of that presentation, which hopefully will give the readers good idea of the basic structure of the calculations and uncertainties involved. This particular paper has been well accepted by the community, and we will follow it rather closely, omitting however some detailed expositions that can always be looked up together with a rich literature collected there.

One comment should be added. In [468] in the text you will find generally $B \to K^* \mu^+ \mu^-$, although the structure of the operators describing $b \to s\mu^+\mu^-$ transitions describes really $\bar{B} \to \bar{K}^* \mu^+ \mu^-$, and we will make this change here. However, the formulas in the latter paper apply to $\bar{B} \to \bar{K}^* \mu^+ \mu^-$, and we could just take them over without any changes.

The three different steps required to obtain any predictions for $\bar{B} \to \bar{K}^* \mu^+ \mu^-$ are as follows:

- The separation of short-distance QCD, electroweak SM interaction, and NP effects from long-distance QCD in an effective Hamiltonian $\mathcal{H}_{\rm eff}$. We know this by now from previous chapters, but in this decay there are many operators, in particular when NP contributions are present, and it is useful to have another look at the relevant $\mathcal{H}_{\rm eff}$.
- The calculation of matrix elements of local quark bilinear operators J of type $\langle \bar{K}^* | J | \bar{B} \rangle$ that are expressed in terms of formfactors.
- The calculation of effects of four-quark operators in $\mathcal{H}_{\rm eff}$, which give rise to so-called non-factorizable corrections and can be calculated using QCDF that we discussed in Section 7.4.

QCDF is only valid for small invariant dilepton mass $q^2 \sim O(1 \text{ GeV}^2)$, or, equivalently, large K^* energy $E \sim O(m_B/2)$, which implies certain cuts on q^2 or E. We will first restrict ourselves to $1 \text{ GeV}^2 < q^2 < 6 \text{ GeV}^2$ for which the LCSR [368, 469, 470] plays an important role. LQCD calculations are more reliable for large q^2, and we will discuss them briefly afterwards.

Effective Hamiltonian

The effective Hamiltonian for $b \to s \mu^+ \mu^-$ transitions is given in [471, 472] as follows

$$\mathcal{H}_{\text{eff}} = -\frac{4 G_F}{\sqrt{2}} \left(\lambda_t \mathcal{H}_{\text{eff}}^{(t)} + \lambda_u \mathcal{H}_{\text{eff}}^{(u)} \right) \tag{9.1}$$

with $\lambda_i = V_{ib} V_{is}^*$ and

$$\mathcal{H}_{\text{eff}}^{(t)} = C_1 Q_1^c + C_2 Q_2^c + \sum_{i=3}^{6} C_i Q_i + \sum_{i=7,8,9,10,P,S} (C_i Q_i + C_i' Q_i'),$$

$$\mathcal{H}_{\text{eff}}^{(u)} = C_1 (Q_1^c - Q_1^u) + C_2 (Q_2^c - Q_2^u).$$

Although the contribution of $\mathcal{H}_{\text{eff}}^{(u)}$ is suppressed by Wolfenstein factor λ^2 (doubly Cabibbo-suppressed) with respect to that of $\mathcal{H}_{\text{eff}}^{(t)}$ and hence often dropped, it proves relevant for certain observables sensitive to complex phases of decay amplitudes, so we keep it. The operators $Q_{i \leq 6}$ are the four-quark operators that we encountered already several times in this book, while the remaining ones are given by

$$Q_7 = \frac{e}{g_s^2} m_b(\mu) (\bar{s} \sigma_{\mu\nu} P_R b) F^{\mu\nu}, \qquad Q_7' = \frac{e}{g_s^2} m_b(\mu) (\bar{s} \sigma_{\mu\nu} P_L b) F^{\mu\nu}, \tag{9.2}$$

$$Q_8 = \frac{1}{g_s} m_b(\mu) (\bar{s} \sigma_{\mu\nu} T^a P_R b) G^{\mu\nu\,a}, \qquad Q_8' = \frac{1}{g_s} m_b(\mu) (\bar{s} \sigma_{\mu\nu} T^a P_L b) G^{\mu\nu\,a}, \tag{9.3}$$

$$Q_9 = \frac{e^2}{g_s^2} (\bar{s} \gamma_\mu P_L b)(\bar{\mu} \gamma^\mu \mu), \qquad Q_9' = \frac{e^2}{g_s^2} (\bar{s} \gamma_\mu P_R b)(\bar{\mu} \gamma^\mu \mu), \tag{9.4}$$

$$Q_{10} = \frac{e^2}{g_s^2} (\bar{s} \gamma_\mu P_L b)(\bar{\mu} \gamma^\mu \gamma_5 \mu), \qquad Q_{10}' = \frac{e^2}{g_s^2} (\bar{s} \gamma_\mu P_R b)(\bar{\mu} \gamma^\mu \gamma_5 \mu), \tag{9.5}$$

$$Q_S = \frac{e^2}{16\pi^2} m_b(\mu) (\bar{s} P_R b)(\bar{\mu} \mu), \qquad Q_S' = \frac{e^2}{16\pi^2} m_b(\mu) (\bar{s} P_L b)(\bar{\mu} \mu), \tag{9.6}$$

$$Q_P = \frac{e^2}{16\pi^2} m_b(\mu) (\bar{s} P_R b)(\bar{\mu} \gamma_5 \mu), \qquad Q_P' = \frac{e^2}{16\pi^2} m_b(\mu) (\bar{s} P_L b)(\bar{\mu} \gamma_5 \mu). \tag{9.7}$$

Here g_s is the strong coupling constant, and $P_{L,R} = (1 \mp \gamma_5)/2$. $m_b(\mu)$ denotes the running b quark mass in the $\overline{\text{MS}}$ scheme. The primed operators with opposite chirality to the unprimed ones vanish or are highly suppressed in the SM, as are $Q_{S,P}$. We neglect the contributions of Q_i' for $1 \leq i \leq 6$. These operators are generated in some NP scenarios, for instance, in left-right symmetric models or through gluino contributions in a general MSSM, but their impact is either heavily constrained or turns out to be very small generically.

The WCs C_i in (9.1) encode short-distance physics and possible NP effects. The technology for calculating them has been explained in detail in previous chapters. In particular, C_i are calculated first at an NP scale, which could be even well above the LHC scales and then evolved down to the matching scale $\mu = M_W$. This evolution can involve new interactions and new operators beyond those present in the SM. Subsequently the evolution to scales $\mu \sim m_b$ takes place. While this evolution is done including QCD and QED interactions, it can contain new operators, generated by NP at very short-distances scales. Moreover, in certain situations the RG effects from top Yukawa couplings and also electroweak loop corrections from scales above the electroweak scales have to be considered. We will return to this important issue in Section 14.1.

Now comes another important point. The operators $Q_7 - Q_{10}$ differ from the ones used in Sections 6.5 and 6.6 by the inclusion of the factors $16\pi^2/g_s^2 = 4\pi/\alpha_s$, and it is crucial to find out the relation of them and of their WCs to the ones used in the latter two sections. It is also important to understand why in [471, 472] such rescalings of operators, known from even earlier QCD calculations, have been made. They allow simply a more transparent organization of the expansion of their Wilson coefficients in perturbation theory, which reads now

$$C_i = C_i^{(0)} + \frac{\alpha_s}{4\pi} C_i^{(1)} + \left(\frac{\alpha_s}{4\pi}\right)^2 C_i^{(2)} + O(\alpha_s^3), \tag{9.8}$$

where $C_i^{(0)}$ is the tree-level contribution, which vanishes for all operators but Q_2. In this normalization of operators also $C_9^{(0)}$ is nonzero. $C_i^{(n)}$ denotes an n-loop contribution. Presently these coefficients are known within the SM at the NNLO [183, 184, 471] corresponding to ($n = 2$). On the other hand, it is sufficient to include NP contributions only with $n = 1$. An explicit calculation of two-loop corrections in the MSSM [472] shows that they are small.

Note that with the operators $Q_{S,P}^{(\prime)}$ defined in this manner, their WCs are scale independent, and their values at $\mu = O(m_b)$ are the same as at $\mu = O(M_W)$. This is related to the fact that the scale dependence of the matrix elements of scalar quark currents is canceled by the scale dependence of $m_b(\mu)$ with $m_b = m_b(\mu_W)$ at the electroweak matching scale but $m_b = m_b(m_b)$ when the rate for this decay is calculated. Q_9 is given by conserved currents, but mixes with $Q_{1,...,6}$, via diagrams with a virtual photon decaying into $\mu^+\mu^-$ as discussed in Section 6.5. Additional scale dependence in C_9 comes from the factor $1/g^2$. The latter dependence is also present in C_{10}, which otherwise would be scale independent as was the case in the normalization in the Section 6.6.

Now comes good news. The coefficients C_7, C_8, C_9, and C_{10} in Sections 6.5 and 6.6 are obtained from the ones used here by multiplying the latter by $4\pi/\alpha_s$. This is good news because as already discussed in Section 6.5, it is convenient to define effective coefficients $C_{7,9}^{(\prime)\mathrm{eff}}$, and also $C_{8,10}^{(\prime)\mathrm{eff}}$, and in these definitions the coefficients C_7, C_8, C_9, and C_{10} as defined in (9.1) must be multiplied precisely by $4\pi/\alpha_s$. Therefore, using the C_i as defined in (9.1), we have [159]

$$C_7^{\text{eff}} = \frac{4\pi}{\alpha_s} C_7 - \frac{1}{3} C_3 - \frac{4}{9} C_4 - \frac{20}{3} C_5 - \frac{80}{9} C_6,$$

$$C_8^{\text{eff}} = \frac{4\pi}{\alpha_s} C_8 + C_3 - \frac{1}{6} C_4 + 20 C_5 - \frac{10}{3} C_6,$$

$$C_9^{\text{eff}} = \frac{4\pi}{\alpha_s} C_9 + Y(q^2),$$

$$C_{10}^{\text{eff}} = \frac{4\pi}{\alpha_s} C_{10}, \qquad C_{7,8,9,10}^{\prime,\text{eff}} = \frac{4\pi}{\alpha_s} C_{7,8,9,10}^{\prime}, \tag{9.9}$$

with

$$Y(q^2) = h(q^2, m_c) \left(\frac{4}{3} C_1 + C_2 + 6 C_3 + 60 C_5 \right)$$

$$- \frac{1}{2} h(q^2, m_b) \left(7 C_3 + \frac{4}{3} C_4 + 76 C_5 + \frac{64}{3} C_6 \right)$$

$$- \frac{1}{2} h(q^2, 0) \left(C_3 + \frac{4}{3} C_4 + 16 C_5 + \frac{64}{3} C_6 \right)$$

$$+ \frac{4}{3} C_3 + \frac{64}{9} C_5 + \frac{64}{27} C_6. \tag{9.10}$$

The function

$$h(q^2, m_q) = -\frac{4}{9} \left(\ln \frac{m_q^2}{\mu^2} - \frac{2}{3} - z \right)$$

$$- \frac{4}{9} (2 + z) \sqrt{|z - 1|} \times \begin{cases} \arctan \dfrac{1}{\sqrt{z-1}} & z > 1 \\[2mm] \ln \dfrac{1 + \sqrt{1-z}}{\sqrt{z}} - \dfrac{i\pi}{2} & z \leq 1 \end{cases} \tag{9.11}$$

with $z = 4m_q^2/q^2$, is related to the basic fermion loop.

We conclude therefore:

The normalization of the coefficients C_7, C_8, C_9, and C_{10} in Sections 6.5 and 6.6 is identical to the one of C_7^{eff}, C_8^{eff}, C_9^{eff}, and C_{10}^{eff} given here. As we will see, only these effective coefficients will appear in the subsequent phenomenological formulas. In the remaining sections of this book the notation C_i will always refer to normalization in Sections 6.5 and 6.6.

The SM values of all WCs to NNLO accuracy based on [183, 184] are given in table 1 of [467]. There also small uncertainties due to the choice of the matching scale $\mu_t = \mu_0$ are listed. Here we give only the values for $\mu_b = 4.2 \, \text{GeV}$ and $\mu_t = 160 \, \text{GeV}$

$$C_7^{\text{eff}} = -0.2957, \quad C_8^{\text{eff}} = -0.1630, \quad C_9 = 4.114, \quad C_{10} = -4.193. \tag{9.12}$$

For C_9 and C_{10} the effective coefficients are q^2-dependent at NLO. We drop this dependence. Again, as stressed in the shaded box, $C_{9,10}$ refer here to those in Section 6.6.

In what follows we will not discuss the coefficient C_7^{eff} as it is already strongly bounded by $B \to X_s \gamma$ decay. We refer to [473], where radiative decays like $B \to V \gamma$ are reviewed.

Similar to $B_{s,d} \to \mu^+ \mu^-$, discussed in the next section, the decay $\bar{B} \to \bar{K}^* (\to \bar{K} \pi) \, \mu^+ \mu^-$ does not allow access to all the preceding coefficients separately: For instance, only the combinations

$$C_9 - C_9', \qquad C_{10} - C_{10}', \qquad C_S - C_S', \qquad C_P - C_P' \tag{9.13}$$

enter the decay amplitude. On the other hand, $B \to K \mu^+ \mu^-$ is sensitive to

$$C_9 + C_9', \qquad C_{10} + C_{10}', \qquad C_S + C_S', \qquad C_P + C_P'. \tag{9.14}$$

This demonstrates that considering simultaneously $\bar{B} \to \bar{K}^* \mu^+ \mu^-$ and $\bar{B} \to \bar{K} \mu^+ \mu^-$ is important if we want to learn in a model independent manner about all these coefficients.

Formfactors

The $\bar{B} \to \bar{K}^*$ matrix elements of the operators $Q_{7,9,10,S,P}^{(\prime)}$ can be expressed in terms of seven formfactors

$$A_i(q^2), \qquad T_i(q^2), \qquad V(q^2), \qquad (i = 1 - 3), \tag{9.15}$$

which depend on the momentum transfer q^2 between the $\bar{B}(p)$ and the $\bar{K}^*(k)$ with

$$q^\mu = p^\mu - k^\mu. \tag{9.16}$$

For Q_9 and Q_{10} the relevant matrix element is

$$\langle \bar{K}^*(k) | \bar{s} \gamma_\mu (1 - \gamma_5) b | \bar{B}(p) \rangle = -i \epsilon_\mu^* (m_B + m_{K^*}) A_1(q^2)$$

$$+ i(2p - q)_\mu (\epsilon^* \cdot q) \frac{A_2(q^2)}{m_B + m_{K^*}}$$

$$+ i q_\mu (\epsilon^* \cdot q) \frac{2m_{K^*}}{q^2} \left[A_3(q^2) - A_0(q^2) \right]$$

$$+ \epsilon_{\mu\nu\rho\sigma} \epsilon^{*\nu} p^\rho k^\sigma \frac{2V(q^2)}{m_B + m_{K^*}}, \tag{9.17}$$

with

$$A_3(q^2) = \frac{m_B + m_{K^*}}{2m_{K^*}} A_1(q^2) - \frac{m_B - m_{K^*}}{2m_{K^*}} A_2(q^2), \qquad A_0(0) = A_3(0). \tag{9.18}$$

For Q_7 the relevant matrix element is

$$\langle \bar{K}^*(k) | \bar{s} \sigma_{\mu\nu} q^\nu (1 + \gamma_5) b | \bar{B}(p) \rangle = i \epsilon_{\mu\nu\rho\sigma} \epsilon^{*\nu} p^\rho k^\sigma \, 2 T_1(q^2)$$

$$+ T_2(q^2) \left[\epsilon_\mu^* (m_B^2 - m_{K^*}^2) - (\epsilon^* \cdot q)(2p - q)_\mu \right]$$

$$+ T_3(q^2)(\epsilon^* \cdot q) \left[q_\mu - \frac{q^2}{m_B^2 - m_{K^*}^2} (2p - q)_\mu \right], \tag{9.19}$$

with $T_1(0) = T_2(0)$. ϵ_μ is the polarization vector of the $\bar{K}^*$. The formfactors A_i and V are observables, i.e., scale independent, while the T_i depend on the renormalization scale μ.

A_0 is also the formfactor of the pseudoscalar current:

$$\langle \bar{K}^* | \partial_\mu A^\mu | \bar{B} \rangle = (m_b + m_s) \langle \bar{K}^* | \bar{s} i \gamma_5 b | \bar{B} \rangle = 2 m_{K^*} (\epsilon^* \cdot q) A_0(q^2). \tag{9.20}$$

The formfactors are hadronic quantities and call for nonperturbative calculations. The status of these calculations by various methods is reviewed in section 7 of [305].

QCD Factorization

In addition to terms proportional to the formfactors, the $\bar{B} \to \bar{K}^* \mu^+ \mu^-$ amplitude also contains certain "nonfactorizable" effects that do not correspond to formfactors. They are related to matrix elements of the purely hadronic operators Q_1 to Q_6 and the chromomagnetic-dipole operator Q_8 with additional (virtual) photon emission. These effects can, in the combined heavy quark and large energy limit, be calculated using QCDF methods [474–476] presented in Section 7.4. Here large energy means large energy of the $\bar{K}^*$, $E \sim O(m_B/2)$. E is related to q^2, the dilepton mass, by

$$2 m_B E = m_B^2 + m_{K^*}^2 - q^2. \tag{9.21}$$

For the phenomenological analysis, one requires $E > 2.1\,\text{GeV}$, which corresponds to $q^2 < 6\,\text{GeV}^2$, well below the charm threshold. We would like to stress here that QCDF *does not work* for large q^2 above the charm resonances – here the only theoretical prediction we have are the contributions to the $\bar{B} \to \bar{K}^* \mu^+ \mu^-$ matrix element given in terms of the formfactors, which is probably a reasonable approximation at the 10 to 20% level.

In the heavy quark and large energy limit, the number of independent formfactors reduces from 7 to 2, which correspond to the polarization of the K^* (transversal or longitudinal) and are usually denoted by $\xi_\perp$ and $\xi_\parallel$. Neglecting $O(\alpha_s)$ corrections, one can define them as [476]

$$\xi_\perp(q^2) = \frac{m_B}{m_B + m_{K^*}} V(q^2), \tag{9.22}$$

$$\xi_\parallel(q^2) = \frac{m_B + m_{K^*}}{2E} A_1(q^2) - \frac{m_B - m_{K^*}}{m_B} A_2(q^2). \tag{9.23}$$

More details can be found in [468, 476].

9.1.3 Differential Decay Distribution and Spin Amplitudes

It is time to discuss the kinematics of the four-body decay $\bar{B} \to \bar{K}^* (\to \bar{K}\pi) \mu^+ \mu^-$, define the angular observables, and derive explicit formulas for them in terms of formfactors and WCs. We begin with *differential decay distribution*. Subsequently we will discuss *transversity amplitudes* and *angular coefficients*, with the latter being used for the construction of twenty-four observables mentioned at the beginning of this section. This subsection is based on Section 3 of [468].

Differential Decay Distribution

The actual decay being observed in experiment is not $\bar{B} \to \bar{K}^* \mu^+ \mu^-$, but $\bar{B} \to \bar{K}^*(\to \bar{K}\pi)\,\mu^+\mu^-$. As emphasized in [466], the additional information provided by the angle between K and π is sensitive to the polarization of the K^* and thus provides an additional probe of the dynamics involved.

The matrix element of the effective Hamiltonian (9.1) for the decay $\bar{B} \to \bar{K}^*(\to \bar{K}\pi)\,\mu^+\mu^-$ can be written, in naïve factorization, as

$$
\begin{aligned}
\mathcal{M} = \frac{G_F \alpha}{\sqrt{2}\pi} V_{tb} V_{ts}^* \Big\{ & \Big[\langle K\pi | \bar{s}\gamma^\mu (C_9^{\text{eff}} P_L + C_9'^{\text{eff}} P_R) b | \bar{B} \rangle \\
& - \frac{2 m_b}{q^2} \langle K\pi | \bar{s} i \sigma^{\mu\nu} q_\nu (C_7^{\text{eff}} P_R + C_7'^{\text{eff}} P_L) b | \bar{B} \rangle \Big] (\bar{\mu}\gamma_\mu \mu) \\
& + \langle K\pi | \bar{s}\gamma^\mu (C_{10}^{\text{eff}} P_L + C_{10}'^{\text{eff}} P_R) b | \bar{B} \rangle (\bar{\mu}\gamma_\mu \gamma_5 \mu) \\
& + \langle K\pi | \bar{s}(C_S P_R + C_S' P_L) b | \bar{B} \rangle (\bar{\mu}\mu) + \langle K\pi | \bar{s}(C_P P_R + C_P' P_L) b | \bar{B} \rangle (\bar{\mu}\gamma_5 \mu) \Big\}.
\end{aligned}
\tag{9.24}
$$

To express the $\bar{B} \to \bar{K}\pi$ matrix elements in terms of the $\bar{B} \to \bar{K}^*$ formfactors discussed earlier, one assumes that the K^* decays resonantly.[1] Then, one can use a narrow-width approximation by making the following replacement in the squared K^* propagator:

$$
\frac{1}{(k^2 - m_{K^*}^2)^2 + (m_{K^*}\Gamma_{K^*})^2} \xrightarrow{\Gamma_{K^*} \ll m_{K^*}} \frac{\pi}{m_{K^*}\Gamma_{K^*}} \delta(k^2 - m_{K^*}^2).
\tag{9.25}
$$

In this way, the formfactors are independent of the $K^* K\pi$ coupling $g_{K^* K\pi}$ [466, 478], because it cancels between the vertex factor and the width

$$
\Gamma_{K^*} = \frac{g_{K^* K\pi}^2}{48\pi} m_{K^*} \beta^3,
\tag{9.26}
$$

where

$$
\beta = \frac{1}{m_{K^*}^2} \left[m_{K^*}^4 + m_K^4 + m_\pi^4 - 2(m_{K^*}^2 m_K^2 + m_K^2 m_\pi^2 + m_{K^*}^2 m_\pi^2) \right]^{1/2}.
\tag{9.27}
$$

Writing the matrix elements as

$$
\langle \bar{K}^*(k) | J_\mu | \bar{B}(p) \rangle = \epsilon^{*\nu} A_{\nu\mu},
\tag{9.28}
$$

where $A_{\nu\mu}$ contains the $\bar{B} \to \bar{K}^*$ formfactors, the corresponding $\bar{B} \to \bar{K}\pi$ matrix element can then be expressed as

$$
\langle \bar{K}(k_1)\pi(k_2) | J_\mu | \bar{B}(p) \rangle = -D_{K^*}(k^2)\, W^\nu A_{\nu\mu},
\tag{9.29}
$$

where [466]

$$
|D_{K^*}(k^2)|^2 = g_{K^* K\pi}^2 \frac{\pi}{m_{K^*}\Gamma_{K^*}} \delta(k^2 - m_{K^*}^2) = \frac{48\pi^2}{\beta^3 m_{K^*}^2} \delta(k^2 - m_{K^*}^2),
\tag{9.30}
$$

$$
W^\mu = K^\mu - \frac{m_K^2 - m_\pi^2}{k^2} k^\mu, \qquad k^\mu = k_1^\mu + k_2^\mu, \qquad K^\mu = k_1^\mu - k_2^\mu.
\tag{9.31}
$$

---

[1] For a study of off-resonance effects, see [477].

With an on-shell K^*, the decay is completely described by four independent kinematical variables:

- The dilepton invariant mass squared q^2
- The three angles θ_{K^*}, θ_l, and ϕ as described after (9.35)

Details on rather involved kinematics of this decay can be found in Appendix A of [468].

Squaring the matrix element, summing over spins of the final state particles, and making use of the kinematical identities sketched in Appendix A of [468], one obtains the full angular decay distribution of $\bar{B}^0 \to \bar{K}^{*0}(\to K^- \pi^+) \, \mu^+ \mu^-$:

$$\frac{d^4\Gamma}{dq^2 \, d\cos\theta_l \, d\cos\theta_{K^*} \, d\phi} = \frac{9}{32\pi} I(q^2, \theta_l, \theta_{K^*}, \phi), \tag{9.32}$$

where

$$\begin{aligned}
I(q^2, \theta_l, \theta_{K^*}, \phi) &= I_1^s \sin^2\theta_{K^*} + I_1^c \cos^2\theta_{K^*} + (I_2^s \sin^2\theta_{K^*} + I_2^c \cos^2\theta_{K^*}) \cos 2\theta_l \\
&\quad + I_3 \sin^2\theta_{K^*} \sin^2\theta_l \cos 2\phi + I_4 \sin 2\theta_{K^*} \sin 2\theta_l \cos\phi \\
&\quad + I_5 \sin 2\theta_{K^*} \sin\theta_l \cos\phi \\
&\quad + (I_6^s \sin^2\theta_{K^*} + I_6^c \cos^2\theta_{K^*}) \cos\theta_l + I_7 \sin 2\theta_{K^*} \sin\theta_l \sin\phi \\
&\quad + I_8 \sin 2\theta_{K^*} \sin 2\theta_l \sin\phi + I_9 \sin^2\theta_{K^*} \sin^2\theta_l \sin 2\phi.
\end{aligned} \tag{9.33}$$

The corresponding expression for the CP-conjugated mode $B^0 \to K^{*0}(\to K^+ \pi^-) \mu^+ \mu^-$ is

$$\frac{d^4\bar{\Gamma}}{dq^2 \, d\cos\theta_l \, d\cos\theta_{K^*} \, d\phi} = \frac{9}{32\pi} \bar{I}(q^2, \theta_l, \theta_{K^*}, \phi). \tag{9.34}$$

The function $\bar{I}(q^2, \theta_l, \theta_{K^*}, \phi)$ is obtained from (9.33) by the replacements [466]

$$I_{1,2,3,4,7}^{(a)} \longrightarrow \bar{I}_{1,2,3,4,7}^{(a)}, \qquad I_{5,6,8,9}^{(a)} \longrightarrow -\bar{I}_{5,6,8,9}^{(a)}, \tag{9.35}$$

where $\bar{I}_i^{(a)}$ equals $I_i^{(a)}$ with all weak phases conjugated. The minus sign in (9.35) is a result of our convention that, while θ_{K^*} is the angle between the $\bar{K}^{*0}$ and the K^- flight direction or between the K^{*0} and the K^+, respectively, the angle θ_l is measured between the $\bar{K}^{*0}$ (K^{*0}) and the lepton μ^- in *both* modes. Thus, a CP transformation interchanging lepton and antilepton leads to the transformations $\theta_l \to \theta_l - \pi$ and $\phi \to -\phi$, as can be verified using formulas in Appendix A of [468]. This convention agrees with the one in [466, 479, 480], but is different from the convention used in some experimental publications of LHCb collaboration, where θ_l is defined as the angle between K^{*0} and μ^+ in the B^0 decay, but between $\bar{K}^{*0}$ and μ^- in the $\bar{B}^0$ decay.

The angular coefficients $I_i^{(a)}$, which are functions of q^2 only, are usually expressed in terms of $\bar{K}^*$ transversity amplitudes, and we will discuss them next.

Transversity Amplitudes

Let us consider for the moment the decay $B \to K^*V^*$, with the B meson decaying to an on-shell K^* and a virtual photon or Z boson denoted here by V^*, which can later decay into a lepton-antilepton pair. The amplitude for this process can be written as

$$\mathcal{M}_{(m,n)}(\bar{B} \to \bar{K}^*V^*) = \epsilon^{*\mu}_{\bar{K}^*}(m) \, M_{\mu\nu} \, \epsilon^{*\nu}_{V^*}(n), \tag{9.36}$$

where $\epsilon^{\mu}_{V^*}(n)$ is the polarization vector of the virtual gauge boson, which can be transverse ($n = \pm$), longitudinal ($n = 0$), or timelike ($n = t$), and $M_{\mu\nu}$ is just defined by this equation. In the B meson rest frame, the four basis vectors can be written as [478]

$$\epsilon^{\mu}_{V^*}(\pm) = (0, 1, \mp i, 0)/\sqrt{2}, \tag{9.37}$$

$$\epsilon^{\mu}_{V^*}(0) = (-q_z, 0, 0, -q_0)/\sqrt{q^2}, \tag{9.38}$$

$$\epsilon^{\mu}_{V^*}(t) = (q_0, 0, 0, q_z)/\sqrt{q^2}, \tag{9.39}$$

where $q^{\mu} = (q_0, 0, 0, q_z)$ is the four-momentum vector of the gauge boson. They satisfy the orthonormality and completeness relations

$$\epsilon^{*\mu}_{V^*}(n)\epsilon_{V^* \, \mu}(n') = g_{nn'}, \tag{9.40}$$

$$\sum_{n,n'} \epsilon^{*\mu}_{V^*}(n)\epsilon^{\nu}_{V^*}(n')g_{nn'} = g^{\mu\nu}, \tag{9.41}$$

where $n, n' = t, \pm, 0$, and $g_{nn'} = \text{diag}(+, -, -, -)$.

The $\bar{K}^*$, on the other hand, is on shell and thus has only three polarization states, $\epsilon^{\mu}_{K^*}(m)$ with $m = \pm, 0$, which read in the $\bar{B}$ rest frame

$$\epsilon^{\mu}_{\bar{K}^*}(\pm) = (0, 1, \pm i, 0)/\sqrt{2}, \tag{9.42}$$

$$\epsilon^{\mu}_{\bar{K}^*}(0) = (k_z, 0, 0, k_0)/m_{K^*}, \tag{9.43}$$

where $k^{\mu} = (k_0, 0, 0, k_z)$ is the four-momentum vector of the K^* (note that $k_z = -q_z$). They satisfy the relations

$$\epsilon^{*\mu}_{\bar{K}^*}(m)\epsilon_{\bar{K}^* \, \mu}(m') = -\delta_{mm'}, \tag{9.44}$$

$$\sum_{m,m'} \epsilon^{*\mu}_{\bar{K}^*}(m)\epsilon^{\nu}_{\bar{K}^*}(m') \, \delta_{mm'} = -g^{\mu\nu} + \frac{k^{\mu}k^{\nu}}{m^2_{K^*}}. \tag{9.45}$$

With this technology at hand one can now construct the *helicity amplitudes* H_0, H_+ and H_-. They can be obtained from $M_{\mu\nu}$ by contracting it with the explicit polarization vectors in (9.36):

$$H_m = \mathcal{M}_{(m,m)}(B \to K^*V^*), \qquad m = 0, +, -. \tag{9.46}$$

Alternatively, one can work with the *transversity amplitudes* defined as [480]

$$A_{\perp,\|} = (H_{+1} \mp H_{-1})/\sqrt{2}, \quad A_0 \equiv H_0. \tag{9.47}$$

In contrast to the decay of B to two (on-shell) vector mesons, to which this formalism can also be applied, there is an additional transversity amplitude in the case of $B \to K^* V^*$ because the gauge boson is virtual, namely

$$A_t = \mathcal{M}_{(0,t)}(B \to K^* V^*). \tag{9.48}$$

It corresponds to a K^* polarization vector, which is longitudinal in the K^* rest frame, and a V^* polarization vector, which is timelike in the V^* rest frame.[2]

If we now consider the subsequent decay of the gauge boson into a lepton-antilepton pair, the amplitude becomes

$$\mathcal{M}(\bar{B} \to \bar{K}^* V^*(\to \mu^+\mu^-))(m) \propto \epsilon_{K^*}^{*\mu}(m)\, M_{\mu\nu} \sum_{n,n'} \epsilon_{V^*}^{*\nu}(n)\epsilon_{V^*}^{\rho}(n')\, g_{nn'}\, (\bar{\mu}\gamma_\rho P_{L,R}\mu).$$

$$\tag{9.49}$$

This amplitude can now be expressed in terms of six transversity amplitudes $A_{\perp,\parallel,0}^L$ and $A_{\perp,\parallel,0}^R$, where L and R refer to the chirality of the leptonic current, as well as the seventh transversity amplitude A_t in (9.48). As the timelike component of the V^* can only couple to an axial-vector current, A_t vanishes in the limit of massless leptons [468].

Now, having shown that the amplitude of the sequential decay $\bar{B} \to \bar{K}^* V^*(\to \mu^+\mu^-)$ can be expressed in terms of seven transversity amplitudes, it is clear that this is true for all contributions of the operators $Q_7^{(\prime)}$, $O_9^{(\prime)}$, and $Q_{10}^{(\prime)}$ to the decay of interest, $\bar{B} \to \bar{K}^*(\to \bar{K}\pi)\mu^+\mu^-$, regardless of whether they originate from virtual gauge boson exchange (i.e., photon or Z penguin diagrams) or from box diagrams.

If the decays are mediated not by a vector, but by a scalar and pseudoscalar operator, a complication arises. Inspecting (9.6) and (9.7) and using current conservation, one finds that the combination $(Q_P - Q'_P)$ can be absorbed into the transversity amplitude A_t because it couples to axial-vector currents, just like the timelike component of a virtual gauge boson. However, this is not possible for the scalar operators $Q_S^{(\prime)}$. Therefore, the inclusion of scalar operators in the decay $\bar{B} \to \bar{K}^*(\to \bar{K}\pi)\mu^+\mu^-$ requires the introduction of an additional, "scalar" transversity amplitude, which we denote by A_S.

To summarize, the treatment of the decay $\bar{B} \to \bar{K}^*(\to \bar{K}\pi)\mu^+\mu^-$ by decomposition of the amplitude into seven transversity amplitudes $A_{\perp,\parallel,0}^{L,R}$ and A_t is sufficient as long as the operators $Q_{7,9,10}^{(\prime)}$ and $Q_P^{(\prime)}$ are considered, but has to be supplemented by an additional, eighth transversity amplitude A_S once contributions from scalar operators are taken into account.

Finally, we give the explicit form of the eight transversity amplitudes in terms of the seven formfactors in (9.15) and the WCs of contributing operators (up to corrections of $O(\alpha_s)$, whose discussion can be found in section 3.4 of [468]). We have

$$A_{\perp L,R} = N\sqrt{2}\lambda^{1/2}\left[\left[(C_9^{\text{eff}} + C_9^{\text{eff}\prime}) \mp (C_{10}^{\text{eff}} + C_{10}^{\text{eff}\prime})\right]\frac{V(q^2)}{m_B + m_{K^*}} + \frac{2m_b}{q^2}(C_7^{\text{eff}} + C_7^{\text{eff}\prime})T_1(q^2)\right],$$

$$\tag{9.50}$$

---

[2] Unlike sometimes stated in the literature, A_t does not correspond to a timelike polarization of the $\bar{K}^*$ meson. We recall that the $\bar{K}^*$ decays on the mass shell and thus has only three polarization states.

$$A_{\parallel L,R} = -N\sqrt{2}(m_B^2 - m_{K^*}^2)\left[B_{\mp}\frac{A_1(q^2)}{m_B - m_{K^*}} + \frac{2m_b}{q^2}(C_7^{\text{eff}} - C_7^{\text{eff}\prime})T_2(q^2)\right], \tag{9.51}$$

$$A_{0L,R} = -\frac{N}{2m_{K^*}\sqrt{q^2}}\left\{B_{\mp}\times\left[(m_B^2 - m_{K^*}^2 - q^2)(m_B + m_{K^*})A_1(q^2) - \lambda\frac{A_2(q^2)}{m_B + m_{K^*}}\right] + T_{23}(q^2)\right\}, \tag{9.52}$$

where

$$B_{\mp} = \left[(C_9^{\text{eff}} - C_9^{\text{eff}\prime}) \mp (C_{10}^{\text{eff}} - C_{10}^{\text{eff}\prime})\right], \tag{9.53}$$

$$T_{23}(q^2) = 2m_b(C_7^{\text{eff}} - C_7^{\text{eff}\prime})\left[(m_B^2 + 3m_{K^*}^2 - q^2)T_2(q^2) - \frac{\lambda}{m_B^2 - m_{K^*}^2}T_3(q^2)\right]. \tag{9.54}$$

$$A_t = \frac{N}{\sqrt{q^2}}\lambda^{1/2}\left[2(C_{10}^{\text{eff}} - C_{10}^{\text{eff}\prime}) + \frac{q^2}{m_\mu}(C_P - C_P')\right]A_0(q^2), \tag{9.55}$$

$$A_S = -2N\lambda^{1/2}(C_S - C_S')A_0(q^2). \tag{9.56}$$

Here

$$N = V_{tb}V_{ts}^*\left[\frac{G_F^2\alpha^2}{3\cdot 2^{10}\pi^5 m_B^3}q^2\lambda^{1/2}\beta_\mu\right]^{1/2}, \tag{9.57}$$

$$\lambda = m_B^4 + m_{K^*}^4 + q^4 - 2(m_B^2 m_{K^*}^2 + m_{K^*}^2 q^2 + m_B^2 q^2), \qquad \beta_\mu = \sqrt{1 - 4m_\mu^2/q^2}. \tag{9.58}$$

Angular Coefficients

With the eight transversity amplitudes just defined, the angular coefficients I_i in (9.33) can be written as

$$I_1^s = \frac{(2+\beta_\mu^2)}{4}\left[|A_\perp^L|^2 + |A_\parallel^L|^2 + (L\to R)\right] + \frac{4m_\mu^2}{q^2}\text{Re}\left(A_\perp^L A_\perp^{R^*} + A_\parallel^L A_\parallel^{R^*}\right), \tag{9.59}$$

$$I_1^c = |A_0^L|^2 + |A_0^R|^2 + \frac{4m_\mu^2}{q^2}\left[|A_t|^2 + 2\text{Re}(A_0^L A_0^{R^*})\right] + \beta_\mu^2|A_S|^2, \tag{9.60}$$

$$I_2^s = \frac{\beta_\mu^2}{4}\left[|A_\perp^L|^2 + |A_\parallel^L|^2 + (L\to R)\right], \tag{9.61}$$

$$I_2^c = -\beta_\mu^2\left[|A_0^L|^2 + (L\to R)\right], \tag{9.62}$$

$$I_3 = \frac{1}{2}\beta_\mu^2\left[|A_\perp^L|^2 - |A_\parallel^L|^2 + (L\to R)\right], \tag{9.63}$$

$$I_4 = \frac{1}{\sqrt{2}}\beta_\mu^2\left[\text{Re}(A_0^L A_\parallel^{L^*}) + (L\to R)\right], \tag{9.64}$$

$$I_5 = \sqrt{2}\beta_\mu\left[\text{Re}(A_0^L A_\perp^{L^*}) - (L\to R) - \frac{m_\mu}{\sqrt{q^2}}\text{Re}(A_\parallel^L A_S^* + A_\parallel^R A_S^*)\right], \tag{9.65}$$

$$I_6^s = 2\beta_\mu \left[\text{Re}(A_\parallel^L A_\perp^{L*}) - (L \to R) \right], \tag{9.66}$$

$$I_6^c = 4\beta_\mu \frac{m_\mu}{\sqrt{q^2}} \text{Re}\left[A_0^L A_S^* + (L \to R) \right], \tag{9.67}$$

$$I_7 = \sqrt{2}\beta_\mu \left[\text{Im}(A_0^L A_\parallel^{L*}) - (L \to R) + \frac{m_\mu}{\sqrt{q^2}} \text{Im}(A_\perp^L A_S^* + A_\perp^R A_S^*) \right], \tag{9.68}$$

$$I_8 = \frac{1}{\sqrt{2}}\beta_\mu^2 \left[\text{Im}(A_0^L A_\perp^{L*}) + (L \to R) \right], \tag{9.69}$$

$$I_9 = \beta_\mu^2 \left[\text{Im}(A_\parallel^{L*} A_\perp^L) + (L \to R) \right]. \tag{9.70}$$

Let us make a few observations:

- In contrast to the transversity amplitudes themselves, the angular coefficients I_i are all physical observables. In fact, they contain the complete information that can be extracted from the measurement of the decay $\bar{B}^0 \to \bar{K}^{*0}(\to K^-\pi^+)\,\mu^+\mu^-$. We will discuss in Section 9.1.4 which combinations of the angular coefficients constitute *theoretically clean* observables.
- In the limit of massless leptons, the following relations hold:

$$I_1^s = 3I_2^s, \qquad I_1^c = -I_2^c. \tag{9.71}$$

- The coefficient I_6^c vanishes unless contributions from scalar operators *and* lepton mass effects are taken into account. However, it is a potentially good observable for scalar currents [468].

9.1.4 Observables in $\bar{B} \to \bar{K}^* \ell^+ \ell^-$

We are now ready to define the twenty-four observables mentioned at the beginning of this section. As just demonstrated, the decay $\bar{B}^0 \to \bar{K}^{*0}(\to K^-\pi^+)\,\mu^+\mu^-$ is completely described in terms of twelve angular coefficient functions $I_i^{(a)}$. The corresponding CP-conjugate mode $B^0 \to K^{*0}(\to K^+\pi^-)\,\mu^+\mu^-$ gives access to twelve additional observables, the CP-conjugate angular coefficient functions $\bar{I}_i^{(a)}$. These quantities have a clear relation to both experiment and theory: Theoretically they are expressed in terms of transversity amplitudes, and experimentally they describe the angular distribution. A physical interpretation of these $I_i^{(a)}$ can be drawn from (9.59) to (9.70). For example, I_6^c depends on scalar operators, and I_7 to I_9 depend on the imaginary part of the transversity amplitudes, and consequently on their phases, which come either from QCD effects and enter the QCD factorization expressions at $O(\alpha_s)$, or are CP-violating SM or NP phases.

To separate CP-conserving and CP-violating NP effects, it is more convenient to consider the twelve CP-averaged angular coefficients $S_i^{(a)}$ and the twelve CP asymmetries $A_i^{(a)}$[3] [468]

$$\boxed{S_i^{(a)} = \left(I_i^{(a)} + \bar{I}_i^{(a)} \right) \Big/ \frac{d(\Gamma + \bar{\Gamma})}{dq^2},} \qquad \boxed{A_i^{(a)} = \left(I_i^{(a)} - \bar{I}_i^{(a)} \right) \Big/ \frac{d(\Gamma + \bar{\Gamma})}{dq^2}.} \tag{9.72}$$

---

[3]   Note that our definition of the CP asymmetries differs from Ref. [479] by a factor of $\frac{3}{2}$.

These decays, similar to $\bar{B} \to \bar{K}^* \ell^+ \ell^-$, give access to a double differential decay spectrum with respect to the invariant mass of the lepton pair q^2 and a lepton charge asymmetry angle $\cos \theta$. The central formula is the normalized angular distribution that can be parametrized as [488, 489]

$$\frac{1}{\Gamma_l} \frac{d\Gamma_l}{d\cos \theta} = \frac{3}{4}(1 - F_H^l)(1 - \cos^2 \theta) + \frac{1}{2}F_H^l + A_{FB}^l \cos \theta, \qquad (9.86)$$

which involves three quantities that will be discussed in more detail in what follows ($i = e, \mu$):

$$\Gamma_l, \qquad F_H^l, \qquad A_{FB}^l. \qquad (9.87)$$

Γ_l is simply decay width and A_{FB}^l the forward-backward asymmetry. F_H^l is an additional quantity needed to describe the full distribution.

The double differential decay rate with respect to q^2 and $\cos \theta$ with lepton flavor l reads as [487]

$$\frac{d^2\Gamma_l}{dq^2\, d\cos \theta} = a_l(q^2) + b_l(q^2) \cos \theta + c_l(q^2) \cos^2 \theta, \qquad (9.88)$$

where

$$\frac{a_l(q^2)}{\Gamma_0 \sqrt{\lambda}\, \beta_l\, \xi_P^2} = q^2 \left(\beta_l^2 |F_S|^2 + |F_P|^2 \right) + \frac{\lambda}{4}(|F_A|^2 + |F_V|^2) \qquad (9.89)$$

$$+ 2m_l(M_B^2 - M_K^2 + q^2)\mathrm{Re}(F_P F_A^*) + 4m_l^2 M_B^2 |F_A|^2, \qquad (9.90)$$

$$\frac{b_l(q^2)}{\Gamma_0 \sqrt{\lambda}\, \beta_l\, \xi_P^2} = 2\left\{ q^2 \left[\beta_l^2 \mathrm{Re}(F_S F_T^*) + \mathrm{Re}(F_P F_{T5}^*) \right] \right. \qquad (9.91)$$

$$\left. + m_l \left[\sqrt{\lambda}\beta_l \mathrm{Re}(F_S F_V^*) + (M_B^2 - M_K^2 + q^2)\mathrm{Re}(F_{T5} F_A^*) \right] \right\}, \qquad (9.92)$$

$$\frac{c_l(q^2)}{\Gamma_0 \sqrt{\lambda}\, \beta_l\, \xi_P^2} = q^2 \left(\beta_l^2 |F_T|^2 + |F_{T5}|^2 \right) - \frac{\lambda}{4}\beta_l^2(|F_A|^2 + |F_V|^2) \qquad (9.93)$$

$$+ 2m_l \sqrt{\lambda}\beta_l \mathrm{Re}(F_T F_V^*) \qquad (9.94)$$

and

$$\Gamma_0 = \frac{G_F^2 \alpha^2 |V_{tb} V_{ts}^*|^2}{512\pi^5 M_B^3}, \quad \lambda = M_B^4 + M_K^4 + q^4 - 2(M_B^2 M_K^2 + M_B^2 q^2 + M_K^2 q^2), \quad \beta_l = \sqrt{1 - 4\frac{m_l^2}{q^2}}. \qquad (9.95)$$

The functions $F_i \equiv F_i(q^2)$, $i = S, P, A, V, T, T5$ are given as

$$F_A = C_{10}, \qquad F_T = \frac{2\sqrt{\lambda}\, \beta_l}{M_B + M_K} \frac{f_T(q^2)}{f_+(q^2)} C_T^l, \qquad F_{T5} = \frac{2\sqrt{\lambda}\, \beta_l}{M_B + M_K} \frac{f_T(q^2)}{f_+(q^2)} C_{T5}^l,$$

$$F_P = \frac{1}{2} \frac{M_B^2 - M_K^2}{m_b - m_s} \frac{f_0(q^2)}{f_+(q^2)}(C_P^l + C_P^{l\prime}) + m_l C_{10} \left[\frac{M_B^2 - M_K^2}{q^2} \left(\frac{f_0(q^2)}{f_+(q^2)} - 1 \right) - 1 \right], \qquad (9.96)$$

$$F_S = \frac{1}{2} \frac{M_B^2 - M_K^2}{m_b - m_s} \frac{f_0(q^2)}{f_+(q^2)} (C_S^l + C_S^{l'}), \quad F_V = C_9 + \frac{2m_b}{M_B} \frac{\mathcal{T}_P(q^2)}{\xi_P(q^2)} + \frac{8m_l}{M_B + M_K} \frac{f_T(q^2)}{f_+(q^2)} C_T^l. \tag{9.97}$$

Here C_i are WCs defined in the context of our discussion of the $\bar{B} \to \bar{K}^* \ell^+ \ell^-$ decay. The coefficients $C_{7,8,9,10}$ are normalized as in the latter case. But the scalar (S) and pseudoscalar (P) include an additional $m_b(\mu)$ factor that was present previously in the definition of the operator. Moreover, contributions of tensor operators are included. They are defined by

$$O_T^l = \frac{e^2}{(4\pi)^2} [\bar{s} \sigma_{\mu\nu} b][\bar{l} \sigma^{\mu\nu} l], \qquad O_{T5}^l = \frac{e^2}{(4\pi)^2} [\bar{s} \sigma_{\mu\nu} b][\bar{l} \sigma^{\mu\nu} \gamma_5 l]. \tag{9.98}$$

The formfactors $\xi_P(q^2) = f_+(q^2)$, $f_T(q^2)$, and $f_0(q^2)$ constitute the main source of theoretical uncertainties. They are defined and discussed in [487]. $\mathcal{T}_P(q^2)$ appearing in the vector coupling to leptons, F_V, takes into account virtual one-photon exchange between the hadrons and the lepton pair and hard scattering contributions. At lowest order it reads as

$$\mathcal{T}_P^{(0)}(q^2) = \xi_P(q^2) \left[C_7^{\text{eff}(0)} + \frac{M_B}{2m_b} Y^{(0)}(q^2) \right]. \tag{9.99}$$

It takes care of the contributions from the $O_{1,...,6}$ matrix elements $\propto Y(q^2)$ that are commonly included in an effective coefficient of the operator O_9 [127].

Integrating over q one finds from (9.88) the angular distribution [487]

$$\frac{d\Gamma_l}{d\cos\theta} = A_l + B_l \cos\theta + C_l \cos^2\theta, \tag{9.100}$$

where

$$A_l = \int_{q_{\min}^2}^{q_{\max}^2} dq^2 \, a_l(q^2), \quad B_l = \int_{q_{\min}^2}^{q_{\max}^2} dq^2 \, b_l(q^2), \quad C_l = \int_{q_{\min}^2}^{q_{\max}^2} dq^2 \, c_l(q^2). \tag{9.101}$$

The values of these coefficients depend on the cuts in q^2. Even if the boundaries of the phase space allow for dilepton masses in the range $4m_l^2 < q^2 \leq (M_B - M_K)^2$, usual calculations are restricted to the range $1\,\text{GeV}^2 \leq q^2 \leq 6\,\text{GeV}^2$ to avoid resonance contributions.

The three basic quantities in (9.87) can now be expressed in terms of A_l, B_l, and C_l

$$\boxed{\Gamma_l = 2 \left(A_l + \frac{1}{3} C_l \right), \qquad A_{\text{FB}}^l = \frac{B_l}{\Gamma_l}, \qquad F_H^l \equiv \frac{2}{\Gamma_l} (A_l + C_l).} \tag{9.102}$$

Detailed numerics of many of these observables can be found in [487]. It includes the SM and various NP scenarios. More recent analyses can be found in papers discussing B physics anomalies that we will mention briefly now. We will encounter them again at various places in this book.

9.1.6 Summary of SM and Experimental Results

During the last years a number of deviations from SM expectations in $b \to s\ell^+\ell^-$ transitions have been observed by the LHCb collaboration [490–492]. They seem to indicate the violation of μ/e universality in $B \to K\ell\bar{\ell}$ and $B \to K^*\ell\bar{\ell}$ decays. In describing them one considers the ratios[4]

$$R(K) = \frac{\mathcal{B}(B \to K\mu^+\mu^-)}{\mathcal{B}(B \to Ke^+e^-)}, \qquad R(K^*) = \frac{\mathcal{B}(B \to K^*\mu^+\mu^-)}{\mathcal{B}(B \to K^*e^+e^-)}, \qquad (9.103)$$

first proposed in [493]. They are useful because a lot of hadronic uncertainties cancel out in them, and consequently they provide a powerful test of μ/e universality [493–495]. LHCb collaboration finds [492]

$$R(K)_{\mathrm{RunI}} = 0.717^{+0.083}_{-0.071}, \qquad R(K)_{\mathrm{RunII}} = 0.928^{+0.089}_{-0.076}, \qquad (9.104)$$

where the first number comes from the revised analysis of Run I [490] and the second from a part of Run II data. The average is [492]

$$R(K)_{\mathrm{exp}} = 0.846^{+0.060\ +0.016}_{-0.054\ -0.014}, \qquad R(K)_{\mathrm{SM}} = 1.00 \pm 0.01 \qquad (9.105)$$

for dilepton invariant mass squared range $1\ \mathrm{GeV}^2 < q^2 < 6\ \mathrm{GeV}^2$, while the SM value of $R(K)$ is very close to unity [487, 496]. The difference is at the level of $2.5\,\sigma$.

In April 2017 LHCb also measured $R(K^*)$, which also hints for violation of LFU [491].

$$R(K^*) = \begin{cases} 0.660^{+0.110}_{-0.070} \pm 0.024 & \text{for } 0.045\ \mathrm{GeV}^2 < q^2 < 1.1\ \mathrm{GeV}^2 \\[2mm] 0.685^{+0.113}_{-0.069} \pm 0.047 & \text{for } 1.1\ \mathrm{GeV}^2 < q^2 < 6\ \mathrm{GeV}^2 \end{cases}. \qquad (9.106)$$

This should be compared with the SM result in [496], which for the second range in q^2 is within 1% from unity and 0.906 ± 0.028 for the first range. The near equality of $R(K^*)$ in (9.106) and of $R(K)$ from Run I indicated the full dominance of NP with left-handed currents. The new result in (9.105) challenges this, implying some amount of NP with right-handed currents responsible for $R(K^*) < R(K)$. But the story is not over because Belle collaboration finds [497]

$$R(K^*) = \begin{cases} 0.52^{+0.36}_{-0.26} \pm 0.05 & \text{for } 0.045\ \mathrm{GeV}^2 < q^2 < 1.1\ \mathrm{GeV}^2 \\[2mm] 0.96^{+0.45}_{-0.29} \pm 0.11 & \text{for } 1.1\ \mathrm{GeV}^2 < q^2 < 6\ \mathrm{GeV}^2 \end{cases}, \qquad (9.107)$$

with the second number fully consistent with the SM, albeit the large error does not allow any firm conclusions.

Moreover, deviation from SM expectations for the angular observable P'_5, shown in Figure 9.2 [498] and the suppression of the branching ratio for $B \to \phi\mu^+\mu$ relative to the SM estimate have been observed by the LHCb [500]. On the other hand, CMS angular analysis in [501] of $B \to K\ell\ell$ is consistent with SM expectations. It should be emphasized that the result on P'_5 hints by itself the presence of NP but not necessarily LFUV. Further tests of the dynamics in $b \to s\mu^+\mu^-$ transitions will be possible through $B_c \to D_s(D_s^*)\mu^+\mu^-$ [502].

Even if the situation in 2019 is unclear, in the last part of this book we will investigate which NP scenarios could be responsible for all these anomalies.

---

[4] The distinction between $K(K^*)$ and $\bar{K}(\bar{K}^*)$ is irrelevant here.

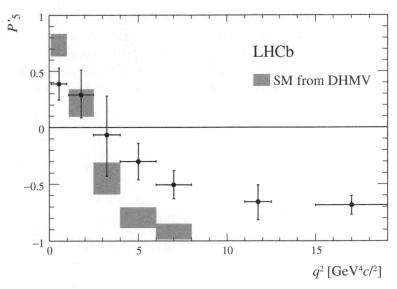

Figure 9.2 P_5' anomaly [498]. The SM prediction is from [499].

9.2 $B_{s,d} \to \mu^+\mu^-$ and $B_{s,d} \to \tau^+\tau^-,\ e^+e^-$

9.2.1 General Remarks

We now turn our attention to two superstars of previous decade in flavor physics: the decays $B_{s,d} \to \mu^+\mu^-$. We will also discuss briefly $B_{s,d} \to \tau^+\tau^-$ and $B_{s,d} \to e^+e^-$, which could become superstars in the future. The decays $B_{s,d} \to \mu^+\mu^-$ are rather special among rare meson decays within the SM. Their branching ratios are not only loop suppressed as other rare decays in the SM. As the final state is purely leptonic and the initial state is a pseudoscalar, the decays in question are strongly helicity suppressed in view of the smallness of m_μ and equally important do not receive photon-mediated one-loop contributions. In contrast to $K^+ \to \pi^+\nu\bar\nu$ decay, discussed in detail in Section 9.5, the internal charm contributions can be neglected removing potential theoretical uncertainty. Even if the relevant hadronic matrix elements of the quark currents cannot be extracted from experiment, they are these days known with high precision from lattice QCD calculations: They just depend on $B_{s,d}$ weak decay constants $F_{B_{s,d}}$ and meson masses. Therefore taking all these properties into account they can compete in theoretical cleanness and in the power in the search for NP with the $K \to \pi\nu\bar\nu$ decays. This also implies that the correlation of $B_{s,d} \to \mu^+\mu^-$ with $K^+ \to \pi^+\nu\bar\nu$ and $K_L \to \pi^0\nu\bar\nu$ will one day play an important role in identifying NP. More about it in Chapter 19.

$B_{s,d} \to \mu^+\mu^-$ decays allow within the SM and constrained MFV models, discussed in Section 15.1, useful measurements of $|V_{ts}|$ and $|V_{td}|$ and in particular their correlations with $\Delta M_{s,d}$ offer very powerful tests of these models [503]. Last but not least, they are very sensitive probes of heavy scalars and pseudoscalars that in various extensions of the SM can contribute already at tree-level [211, 504, 505].

We will discuss all these exciting aspects in final BSM parts of this book. In this chapter we will discuss these decays in detail only in the SM, but to be prepared for NP contributions we will already here, similar to the previous section, consider these decays including contributions from operators strongly suppressed in the SM but playing the role in various extensions of the SM. Therefore we will work with the Hamiltonian

$$\mathcal{H}_{\text{eff}} = -\frac{4G_{\text{F}}}{\sqrt{2}} \frac{\alpha}{4\pi} V_{ts}^* V_{tb} \sum_{i=10,S,P} [C_i(\mu)Q_i(\mu) + C_i'(\mu)Q_i'(\mu)] + h.c., \tag{9.108}$$

where ($\ell = e, \mu, \tau$)

$$Q_{10} = (\bar{s}\gamma_\mu P_L b)(\bar{\ell}\gamma^\mu \gamma_5 \ell), \qquad Q_{10}' = (\bar{s}\gamma_\mu P_R b)(\bar{\ell}\gamma^\mu \gamma_5 \ell), \tag{9.109a}$$

$$Q_S = m_b(\bar{s}P_R b)(\bar{\ell}\ell), \qquad Q_S' = m_b(\bar{s}P_L b)(\bar{\ell}\ell), \tag{9.109b}$$

$$Q_P = m_b(\bar{s}P_R b)(\bar{\ell}\gamma_5 \ell), \qquad Q_P' = m_b(\bar{s}P_L b)(\bar{\ell}\gamma_5 \ell). \tag{9.109c}$$

Including the factors of m_b into the definition of scalar operators makes their matrix elements and their WCs scale independent. Note also that despite a different overall factor, all WCs are the same as in the previous section. It should also be noticed that dipole operators and the operators $Q_9(Q_9')$ are absent as they do not contribute to decays in question. Reviews of these two decays can be found in [208, 506], and the power of these decays in testing energy scales as high as several hundreds of TeV has been demonstrated in [211]. We will discuss it in Chapter 19.

Careful readers will surely notice that the CKM factor in (9.110) and other formulas given subsequently is a complex conjugate of the one in (9.108). Using directly the latter Hamiltonian would result in the formulas for $\bar{B}_{s,d}^0$ and not for $B_{s,d}^0$ as given here, that are derived from the Hermitian conjugate of the Hamiltonian in (9.108). As this is the common practice in the literature, we decided to present formulas for $B_{s,d}^0$ but when correlating the decays $B_{s,d} \to \mu^+\mu^-$ with $\bar{B} \to \bar{K}\mu^+\mu^-$ one has to remember this difference.

9.2.2 Observables for $B_{s,d} \to \mu^+\mu^-$

The central object here is the time-dependent rate for a B_s^0 meson decaying to two muons with a specific helicity $\lambda = L, R$. It is given by [507–509]

$$\Gamma(B_s^0(t) \to \mu_\lambda^+ \mu_\lambda^-) = \frac{G_F^4 M_W^4 \sin^4 \vartheta_W}{16\pi^5} \left| C_{10}^{\text{SM}} V_{ts} V_{tb}^* \right|^2 F_{B_s}^2 m_{B_s} m_\mu^2 \sqrt{1 - \frac{4m_\mu^2}{m_{B_s}^2}} \times \left(|P|^2 + |S|^2 \right)$$

$$\times \left\{ C_{\mu\mu}^\lambda \cos(\Delta M_s t) + \mathcal{S}_{\mu\mu} \sin(\Delta M_s t) \right.$$

$$\left. + \cosh\left(\frac{y_s t}{\tau_{B_s}}\right) + \mathcal{A}_{\Delta\Gamma}^{\mu\mu} \sinh\left(\frac{y_s t}{\tau_{B_s}}\right) \right\} \times e^{-t/\tau_{B_s}}, \tag{9.110}$$

where as in (8.68)

$$\tau_{B_s} \equiv \frac{2}{(\Gamma_H + \Gamma_L)}, \qquad y_s \equiv \tau_{B_s} \frac{\Delta\Gamma_s}{2} = 0.065 \pm 0.005. \qquad (9.111)$$

Next

$$P \equiv \frac{C_{10} - C_{10}'}{C_{10}^{SM}} + \frac{m_{B_s}^2}{2m_\mu} \frac{m_b}{m_b + m_s} \frac{C_P - C_P'}{C_{10}^{SM}} \equiv |P| e^{i\varphi_P}, \qquad (9.112)$$

$$S \equiv \sqrt{1 - \frac{4m_\mu^2}{m_{B_s}^2}} \frac{m_{B_s}^2}{2m_\mu} \frac{m_b}{m_b + m_s} \frac{C_S - C_S'}{C_{10}^{SM}} \equiv |S| e^{i\varphi_S}, \qquad (9.113)$$

and [507][5]

$$C_{\mu\mu}^\lambda = -\eta_\lambda \left[\frac{2|PS| \cos(\varphi_P - \varphi_S)}{|P|^2 + |S|^2} \right], \qquad (9.114)$$

$$S_{\mu\mu} = \frac{|P|^2 \sin(2\varphi_P - 2\varphi_{B_s}) - |S|^2 \sin(2\varphi_S - 2\varphi_{B_s})}{|P|^2 + |S|^2}, \qquad (9.115)$$

where $\eta_\lambda = \eta_{L,R} = \pm 1$.

The time-dependent rate for a $\bar{B}_s^0$ meson is obtained from the preceding expression by replacing $C_{\mu\mu}^\lambda \to -C_{\mu\mu}^\lambda$ and $S_{\mu\mu} \to -S_{\mu\mu}$.

In the general analysis of $B_s \to \mu^+\mu^-$ as presented in [509], the basic four observables are then

$$\overline{R}, \qquad \mathcal{A}_{\Delta\Gamma}^{\mu\mu}, \qquad S_{\mu\mu}, \qquad S_{\psi\phi}. \qquad (9.116)$$

Here, the observable $\overline{R}$, defined in (9.121), is just the ratio of the branching ratio that includes $\Delta\Gamma_s$ effects and of the SM prediction for the branching ratio that also includes them. Following [509] we will denote branching ratios containing $\Delta\Gamma_s$ effects with a *bar* while those without these effects without it.

The next two observables, $\mathcal{A}_{\Delta\Gamma}^{\mu\mu}$ and $S_{\mu\mu}$, can be extracted from flavor untagged and tagged time-dependent measurements of $B_s \to \mu^+\mu^-$, respectively. As $\mathcal{A}_{\Delta\Gamma}^{\mu\mu}$ does not rely on flavor tagging, which is difficult for a rare decay, it will be easier to determine than $S_{\mu\mu}$. Given enough statistics, a full fit to the time-dependent untagged rate gives $\mathcal{A}_{\Delta\Gamma}^{\mu\mu}$ [507]. With limited statistics, an *effective lifetime* measurement may be easier, which corresponds to fitting a single exponential to this rate. For a maximal likelihood fit, the $B_s \to \mu\mu$ effective lifetime is given by [510]:

$$\tau_{\mu\mu} \equiv \frac{\int_0^\infty t \langle \Gamma(B_s(t) \to \mu^+\mu^-) \rangle \, dt}{\int_0^\infty \langle \Gamma(B_s(t) \to \mu^+\mu^-) \rangle \, dt}. \qquad (9.117)$$

The untagged observable $\mathcal{A}_{\Delta\Gamma}^{\mu\mu}$ is then given by

$$\mathcal{A}_{\Delta\Gamma}^{\mu\mu} = \frac{1}{y_s} \left[\frac{(1 - y_s^2)\,\tau_{\mu\mu} - (1 + y_s^2)\tau_{B_s}}{2\tau_{B_s} - (1 - y_s^2)\,\tau_{\mu\mu}} \right], \qquad (9.118)$$

---

[5] In the literature $S_{\mu\mu}$ is sometimes denoted by $S_{\mu\mu}^s$ to distinguish it from $S_{\mu\mu}^d$ in the B_d decays.

to the true values of the SM because of NP contributions. However, proceeding in this manner the following result could be obtained in 2009 [529]

$$\mathcal{B}(B^+ \to \tau^+\nu)_{\text{SM}} = (0.80 \pm 0.12) \times 10^{-4}, \qquad (2009) \tag{9.143}$$

with a similar result found by the UTfit collaboration [528].

These days using the present input one finds

$$\mathcal{B}(B^+ \to \tau^+\nu)_{\text{SM}} = (0.87 \pm 0.01) \left| \frac{V_{ub}}{3.7 \times 10^{-3}} \right|^2 \times 10^{-4}, \qquad (2019). \tag{9.144}$$

Comparison of (9.143) and (9.144) assures us that this SM prediction will not be modified by much in the future. It is compatible with the present experimental world average

$$\mathcal{B}(B^+ \to \tau^+\nu_\tau)_{\text{exp}} = (1.09 \pm 0.24)10^{-4} \tag{9.145}$$

even if somewhat below it.

The full clarification of a possible presence of NP in this decay will have to wait for the data from SuperKEKB. In the meantime hopefully the theoretical advances in the determination of $|V_{ub}|$ from tree-level decays will be made allowing us to make a precise prediction for the branching ratio for this decay. For a recent discussion of NP see [530] and references therein.

It should be emphasized that for low value of $|V_{ub}|$ the increase of F_{B^+}, while enhancing the branching ratio in question, would also enhance ΔM_d, which is not favored by the data. We will show it explicitly in Section 15.1.1. On the other hand, the increase of $|V_{ub}|$ while enhancing $\mathcal{B}(B^+ \to \tau^+\nu)_{\text{SM}}$ would also enhance $S_{\psi K_S}$ shifting it away from the data. This discussion shows clearly that the correlations between

$$F_{B^+}, \qquad |V_{ub}|, \qquad \mathcal{B}(B^+ \to \tau^+\nu), \qquad S_{\psi K_S}, \qquad \Delta M_d \tag{9.146}$$

will turn out to be useful for various tests when all these quantities will be known significantly more precisely than it is the case now. In fact, the decay $B_s \to \mu^+\mu^-$ is presently in better shape than $B^+ \to \tau^+\nu$ as it is governed by $|V_{ts}|$, which is presently better known than $|V_{ub}|$.

Finally, Belle measured [531]

$$\mathcal{B}(B^- \to \mu^-\bar{\nu}_\mu) = (6.46 \pm 2.22 \pm 1.60) \times 10^{-7} \tag{9.147}$$

to be compared with its SM prediction $(3.80 \pm 0.31) \times 10^{-7}$. Possible NP responsible for this enhancement has been discussed in [532]. Belle II should be able to clarify whether this is still another anomaly.

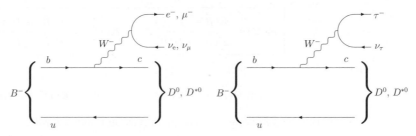

Figure 9.3 $b \to c$ transitions with $\ell = e, \mu$ and $\ell = \tau$.

9.4 $\bar{B} \to D\ell\bar{\nu}_l, \bar{B} \to D^*\ell\bar{\nu}_l, B_c \to J/\psi\ell\bar{\nu}_l$, and $\Lambda_b \to \Lambda_c\ell\bar{\nu}_l$

9.4.1 Preliminaries

In this section we will discuss decays governed by the $b \to c$ transitions with D or D^* meson, a charged lepton, and an antineutrino in the final state. We show two examples in Figure 9.3. As seen there, in contrast to many decays considered in this chapter, these decays proceed in the SM already at tree-level and are governed by the CKM element $|V_{cb}|$ that is roughly by an order of magnitude larger than $|V_{ub}|$ entering $\mathcal{B}(B^+ \to \iota^+\nu)$ just discussed. This means that these decays are not really rare. Still, due to the dynamics involved and the present experimental data, the natural place to describe them in this book is the present chapter.

Experimentalists look also at neutral B decays, and the results for the ratios in (9.148) follow from combining charged and neutral B decays assuming isospin symmetry as given in (9.149). Also $B_c \to J/\psi\ell\bar{\nu}_l$ and $\Lambda_b \to \Lambda_c\ell\bar{\nu}_l$ are of great interest because they can give us additional insights into into $b \to c$ transitions.

Our goal is to give some insight into these decays, in particular to present the basic structure of observables that could be used to test the SM and its extensions. As we will see, similar to the decays $B \to K(K^*)\ell^+\ell^-$ described in Section 9.1, a multitude of observables can be constructed also for these decays. We will discuss several of them, but it will not be possible to derive these expressions here, and we will frequently refer to the literature. Similar to the case of $B \to K(K^*)\ell^+\ell^-$ decays we will include already in this section operators that are irrelevant in the SM but will play an important role in BSM models in later sections.

The study of these decays goes back to [533], where only SM has been considered. More recent studies related to a possible impact of NP on the determination of $|V_{cb}|$ can be found in [51, 53, 55, 59, 60, 534]. The interest in these decays basically exploded in the last years in view of significant departures of the BaBar, Belle, and LHCb data from rather precise SM expectations for the rates for $\bar{B} \to D^*\tau\bar{\nu}_\tau$ and $\bar{B} \to D\tau\bar{\nu}_\tau$ normalized to the corresponding rates with μ and electron, that is for the ratios ($\ell \in \{e, \mu\}$)

$$\boxed{\mathcal{R}(D) = \frac{\mathcal{B}(\bar{B} \to D\tau\bar{\nu})}{\mathcal{B}(\bar{B} \to D\ell\bar{\nu})}, \qquad \mathcal{R}(D^*) = \frac{\mathcal{B}(\bar{B} \to D^*\tau\bar{\nu})}{\mathcal{B}(\bar{B} \to D^*\ell\bar{\nu})}.}$$
(9.148)

Here

$$\mathcal{R}(D) \equiv \mathcal{R}(D^+) = \mathcal{R}(D^0), \qquad \mathcal{R}(D^*) \equiv \mathcal{R}(D^{*+}) = \mathcal{R}(D^{*0}).$$
(9.149)

Similarly, the following ratios begin to play an important role:

$$\boxed{\mathcal{R}(J/\psi) = \frac{\mathcal{B}(B_c \to J/\psi\tau\bar{\nu})}{\mathcal{B}(B_c \to J\psi\ell\bar{\nu})}, \qquad \mathcal{R}(\Lambda_c) = \frac{\mathcal{B}(\Lambda_b \to \Lambda_c\tau\bar{\nu})}{\mathcal{B}(\Lambda_b \to \Lambda_c\ell\bar{\nu})}.}$$
(9.150)

We will summarize the present experimental world averages for these four ratios and corresponding SM predictions at the end of this section. Moreover, we will discuss observables related to the polarizations of τ and D^* as well as other useful quantities. The discussion of possible origin of various anomalies is postponed to several BSM chapters. But first we want to collect formulas that demonstrate there are many observables related to these decays that could allow in the future very efficient tests of the SM and its extensions.

As I have never published any paper on these decays, I made a thorough study of them. In view of numerous papers devoted to this subject this turned out to be a research project by itself. I recommend such a study to motivated readers that could be considered as an advanced exercise relative to exercises present in other books. To this end the papers [60, 535–556] turned out to be very useful. Many angular distributions, analogous to the ones discussed by us in the context of $B \to K(K^*)\ell^+\ell^-$ decays can be found in these papers. We will only present some of them, concentrating on formulas that are particularly useful for phenomenological applications both in the SM and beyond it.

9.4.2 Effective Hamiltonian

The relevant $\text{SU}(2)_L \times \text{U}(1)_Q$ invariant effective Hamiltonian is given as follows [551, 553, 555]

$$\mathcal{H}_{\text{eff}} = 2\sqrt{2}G_F V_{cb}[(1 + C_V^L)O_V^L + +C_V^R O_V^R + C_S^R O_S^R + C_S^L O_S^L + C_T O_T] + h.c.,$$
(9.151)

with

$$O_V^{L,R} = (\bar{c}\gamma^\mu P_{L,R}b)\left(\bar{\tau}\gamma_\mu P_L \nu_\tau\right), \qquad O_S^R = (\bar{c}P_R b)\left(\bar{\tau}P_L \nu_\tau\right),$$
$$O_S^L = (\bar{c}P_L b)\left(\bar{\tau}P_L \nu_\tau\right), \qquad O_T = (\bar{c}\sigma^{\mu\nu}P_L b)\left(\bar{\tau}\sigma_{\mu\nu}P_L \nu_\tau\right),$$
(9.152)

where the absence of both (light) right-handed neutrinos[6] and of NP couplings to the light lepton generations (as studied in [60]) has been assumed. As the SM contribution has been factored out, all coefficients $C_{S,V,T}^{L,R}$ originate from NP only. The vector operator O_V^R with a right-handed coupling to quarks, does not arise at the dimension-6 level in the

---

[6] For studies of right-handed neutrino effects in $\mathcal{R}(D^{(*)})$ see [557–560].

$SU(2)_L$-invariant effective field theory [561–563] if LFU violation is assumed. But it can provide flavor universal contributions [564–566].

9.4.3 $\bar{B} \to D\ell\bar{\nu}_\ell$ decay

Full Decay Distribution

The full spectrum of $\bar{B} \to D\ell\bar{\nu}_\ell$ decay is given as follows,

$$
\frac{d^2\Gamma}{dq^2 d\cos\theta_l} = \frac{1}{32(2\pi)^3 m_B^2} |\vec{q}| \left(1 - \frac{m_\ell^2}{q^2}\right) |\mathcal{M}(\bar{B} \to D\ell\bar{\nu}_\ell)|^2, \tag{9.153}
$$

where $\vec{q}$ stands for the three-momentum of the $\ell\bar{\nu}_\ell$ pair in the B-meson rest frame, and θ_l is the angle between the direction of flight of D and ℓ in the center of mass frame of $\ell\bar{\nu}_\ell$ [533].

The amplitude $\mathcal{M}(\bar{B} \to D\ell\bar{\nu}_\ell)$ depends on the matrix elements of the quark operators involved. The latter are given in terms of formfactors that summarize to a large extent the long-distance QCD effects. To write differential decay rate in an elegant form *helicity amplitudes*, encountered briefly in Section 9.1, are introduced. The expressions for them in terms of formfactors can be found, e.g., in an appendix of [553]. It should be stressed that different papers use different definitions of these amplitudes. Here we use the ones in [553].

We will next list various formulas for the observables of interest.

Differential Decay Rate

Using the effective Hamiltonian in (9.151) one can derive the differential decay rate as a function of the general set of WCs [535, 537], here taken from [553]:

$$
\frac{d\Gamma(\bar{B} \to D\tau\bar{\nu}_\tau)}{dq^2} = \frac{G_F^2 |V_{cb}|^2}{192\pi^3 m_B^3} q^2 \sqrt{\lambda_D(q^2)} \left(1 - \frac{m_\tau^2}{q^2}\right)^2
$$

$$
\times \left\{ |1 + C_V^L + C_V^R|^2 \left[\left(1 + \frac{m_\tau^2}{2q^2}\right) H_{V,0}^{s2} + \frac{3}{2}\frac{m_\tau^2}{q^2} H_{V,t}^{s2}\right] \right.
$$

$$
+ \frac{3}{2} \left|C_S^R + C_S^L\right|^2 H_S^s + 8 \, |C_T|^2 \left(1 + \frac{2m_\tau^2}{q^2}\right) H_T^{s2}
$$

$$
+ 3 \, \mathcal{R}e\left[\left(1 + C_V^L + C_V^R\right)\left(C_S^{R*} + C_S^{L*}\right)\right] \frac{m_\tau}{\sqrt{q^2}} H_S^s H_{V,t}^s \tag{9.154}
$$

$$
\left. - 12 \, \mathcal{R}e\left[\left(1 + C_V^L + C_V^R\right) C_T^*\right] \frac{m_\tau}{\sqrt{q^2}} H_T^s H_{V,0}^s \right\},
$$

from which the full decay width of $\bar{B} \to D\ell\bar{\nu}_\ell$ can be obtained,

$$
\Gamma(\bar{B} \to D\ell\bar{\nu}_\ell) = \int_{m_\ell^2}^{q_{max}^2} \frac{d\Gamma}{dq^2} \, dq^2. \tag{9.155}
$$

Here $q_{max}^2 = (m_B - m_D)^2$, where neutrino masses are neglected. H_i are the helicity amplitudes mentioned earlier. Finally

$$\lambda_D(q^2) \equiv \lambda(m_B^2, m_D^2, q^2) = \left[(m_B - m_D)^2 - q^2\right]\left[(m_B + m_D)^2 - q^2\right]. \qquad (9.156)$$

Forward-Backward Asymmetry

The next one on our list is the forward-backward asymmetry that is given by

$$A_{FB}^D(q^2) = \frac{\displaystyle\int_0^1 \frac{d^2\Gamma}{dq^2 d\cos\theta_l} d\cos\theta_l - \int_{-1}^0 \frac{d^2\Gamma}{dq^2 d\cos\theta_l} d\cos\theta_l}{\dfrac{d\Gamma}{dq^2}}. \qquad (9.157)$$

The important virtue of this asymmetry is its quadratic dependence on the lepton mass, which implies that it should be particularly useful in the presence of τ-lepton in the final state. Moreover, its q^2 dependence is sensitive to scalar and tensor operator contributions.

Lepton-Polarization Asymmetry

One can also decompose the differential branching ratio according to the two possible polarizations of the charged (τ) lepton, giving rise to the τ polarization asymmetry:

$$P_\tau^D(q^2) = \left(\frac{d\Gamma_{\lambda_\tau=1/2}^D}{dq^2} - \frac{d\Gamma_{\lambda_\tau=-1/2}^D}{dq^2}\right) \bigg/ \frac{d\Gamma^D}{dq^2}, \qquad (9.158)$$

where λ_τ is the helicity of the τ lepton, and $d\Gamma_{\lambda_\tau}^D/dq^2$ is the differential decay width of $\bar{B} \to D\tau\bar{\nu}_\tau$ for a given helicity λ_τ.

However, the measurements of q^2 distributions is challenging, and presently one often considers the corresponding integrated quantity

$$P_\tau(D) = \frac{\Gamma(\lambda_\tau = \frac{1}{2}) - \Gamma(\lambda_\tau = -\frac{1}{2})}{\Gamma(\lambda_\tau = \frac{1}{2}) + \Gamma(\lambda_\tau = -\frac{1}{2})}, \qquad (9.159)$$

which is obtained by integrating separately the numerator and the denominator in (9.158). Already $P_\tau(D)$ can help to distinguish between various NP models as stressed, e.g., in [553–555].

9.4.4 $\bar{B} \to D^*\ell\bar{\nu}_\ell$ Decay

Full Decay Distribution

We begin with the fivefold differential decay rate of the $\bar{B} \to D\pi\ell\bar{\nu}_\ell$ decay:

$$\frac{d^5\Gamma}{dq^2 dm_{D\pi}^2 d\cos\theta_D d\cos\theta_l d\chi} = \frac{1}{128(2\pi)^6 m_B^2} |\vec{q}| \left(1 - \frac{m_\ell^2}{q^2}\right) \frac{|\hat{\vec{p}}_D|}{m_{D\pi}} |\mathcal{M}(\bar{B} \to D\pi\ell\bar{\nu}_\ell)|^2,$$

$$(9.160)$$

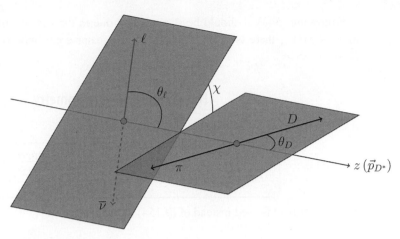

Figure 9.4   Kinematical variables [543].

where $\hat{\vec{p}}_D$ is the three-momentum of D in the rest frame of $D\pi$, and the three angles are specified in Figure 9.4. The details on kinematics can be found, e.g., in the appendices of [543].

As in the case of $\bar{B} \to \bar{K}^*\ell^+\ell^-$, q^2 dependent formfactors parametrize the relevant hadronic matrix elements and similar to $\bar{B} \to D\ell\bar{\nu}_\ell$ decay helicity amplitudes can be introduced. The formulas for them are again collected, e.g., in [553]. Integrating over various angles one can then construct a number of distributions. In analogy to $\bar{B} \to D\ell\bar{\nu}_\ell$ we present here only few of them.

Differential Decay Rate

Following [535, 537, 553] we have

$$
\frac{d\Gamma(\bar{B} \to D^*\tau\bar{\nu}_\tau)}{dq^2} = \frac{G_F^2 \, |V_{cb}|^2}{192\pi^3 m_B^3} \, q^2 \sqrt{\lambda_{D^*}(q^2)} \left(1 - \frac{m_\tau^2}{q^2}\right)^2
$$

$$
\times \left\{ \left(\left|1+C_V^L\right|^2 + \left|C_V^R\right|^2\right)\left[\left(1+\frac{m_\tau^2}{2q^2}\right)\left(H_{V,+}^2 + H_{V,-}^2 + H_{V,0}^2\right) + \frac{3}{2}\frac{m_\tau^2}{q^2} H_{V,t}^2\right] \right.
$$

$$
- 2\,\mathcal{R}e\left[\left(1+C_V^L\right)C_V^{R*}\right]\left[\left(1+\frac{m_\tau^2}{2q^2}\right)\left(H_{V,0}^2 + 2H_{V,+}H_{V,-}\right) + \frac{3}{2}\frac{m_\tau^2}{q^2} H_{V,t}^2\right]
$$

$$
+ \frac{3}{2}\left|C_S^R - C_S^L\right|^2 H_S^2 + 8\,|C_T|^2\left(1 + \frac{2m_\tau^2}{q^2}\right)\left(H_{T,+}^2 + H_{T,-}^2 + H_{T,0}^2\right)
$$

$$\tag{9.161}$$

$$
+ 3\,\mathcal{R}e\left[\left(1+C_V^L - C_V^R\right)\left(C_S^{R*} - C_S^{L*}\right)\right]\frac{m_\tau}{\sqrt{q^2}} H_S H_{V,t}
$$

$$
- 12\,\mathcal{R}e\left[\left(1+C_V^L\right)C_T^*\right]\frac{m_\tau}{\sqrt{q^2}}\left(H_{T,0}H_{V,0} + H_{T,+}H_{V,+} - H_{T,-}H_{V,-}\right)
$$

$$
\left. + 12\,\mathcal{R}e\left[C_V^R C_T^*\right]\frac{m_\tau}{\sqrt{q^2}}\left(H_{T,0}H_{V,0} + H_{T,+}H_{V,-} - H_{T,-}H_{V,+}\right)\right\}.
$$

Following [543], it should be stressed that due to the absence of $H_{V,\pm}(q^2)$ amplitudes in $\overline{B} \to D\ell\overline{\nu}_\ell$, there is a larger sensitivity to scalar contributions in that decay than in $\overline{B} \to D^*\ell\overline{\nu}_\ell$.

Forward-Backward Asymmetry

It is defined as in (9.157), that is

$$A_{FB}^{D^*}(q^2) = \frac{\int_0^1 \frac{d^2\Gamma}{dq^2 d\cos\theta_l} d\cos\theta_l - \int_{-1}^0 \frac{d^2\Gamma}{dq^2 d\cos\theta_l} d\cos\theta_l}{d\Gamma/dq^2}, \tag{9.162}$$

but now (9.161) is used instead of (9.154) to evaluate it.

Lepton-Polarization Asymmetry

We deal here with

$$P_\tau^{D^*}(q^2), \qquad P_\tau(D^*) \tag{9.163}$$

that are defined as in (9.158) and (9.159) but with D replaced by D^*.

Longitudinal Polarization of D^*

Similarly to τ-polarization, one can extract from the angular distribution in the secondary $D^* \to D\pi$ decay the fraction of longitudinally polarized D^* mesons by constructing the following observable (see, e.g., [553]):

$$F_L^{D^*}(q^2) = \frac{d\Gamma_{\lambda_{D^*}=0}}{dq^2} \bigg/ \frac{d\Gamma^{D^*}}{dq^2}. \tag{9.164}$$

Again it is easier to measure integrated quantity

$$F_L(D^*) = \frac{\Gamma(B_d \to D_L^* \tau\nu)}{\Gamma(B_d \to D^* \tau\nu)}. \tag{9.165}$$

The differential decay width into longitudinally polarized ($\lambda_{D^*} = 0$) D^* mesons is given by [553]

$$\frac{d\Gamma_{\lambda_{D^*}=0}^{D^*}}{dq^2} = \frac{G_F^2|V_{cb}|^2}{192\pi^3 m_B^3} q^2 \sqrt{\lambda_{D^*}(q^2)} \left(1 - \frac{m_\tau^2}{q^2}\right)^2 \Bigg\{ |1 + C_V^L - C_V^R|^2 \left[\left(1 + \frac{m_\tau^2}{2q^2}\right) H_{V,0}^2 + \frac{3}{2}\frac{m_\tau^2}{q^2} H_{V,t}^2\right]$$

$$+ \frac{3}{2} |C_S^R - C_S^L|^2 H_S^2 + 8 |C_T|^2 \left(1 + \frac{2m_\tau^2}{q^2}\right) H_{T,0}^2$$

$$+ 3\, \mathcal{R}e[(1 + C_V^L - C_V^R)(C_S^{R*} - C_S^{L*})] \frac{m_\tau}{\sqrt{q^2}} H_S H_{V,t} \tag{9.166}$$

$$- 12\, \mathcal{R}e[(1 + C_V^L - C_V^R) C_T^*] \frac{m_\tau}{\sqrt{q^2}} H_{T,0} H_{V,0} \Bigg\}.$$

9.4.5 $B_c \to J/\psi\ell\bar{\nu}_l$

The differential decay rate for this transition looks rather similar to the one for the $\bar{B} \to D^*\ell\bar{\nu}_l$ in (9.161) [567], here as given in [553]

$$\frac{d\Gamma(B_c \to J/\psi\ell\bar{\nu})}{dq^2} = \frac{G_F^2 |V_{cb}|^2}{192\pi^3 m_{B_c}} q^2 \sqrt{\lambda_{J/\psi}(q^2)} \left(1 - \frac{m_\ell^2}{q^2}\right)$$

$$\times \left\{ (|1 + C_V^L|^2 + |C_V^R|^2) \left[\left(1 + \frac{m_\ell^2}{2q^2}\right)\left(H_{V,+}^2 + H_{V,-}^2 + H_{V,0}^2\right) + \frac{3}{2}\frac{m_\ell^2}{q^2} H_{V,t}^2\right] \right.$$

$$- 2\,\mathcal{R}e\left[(1 + C_V^L)C_V^{R*}\right]\left[\left(1 + \frac{m_\ell^2}{2q^2}\right)\left(H_{V,0}^2 + 2H_{V,+} \cdot H_{V,-}\right) + \frac{3}{2}\frac{m_\ell^2}{q^2} H_{V,t}^2\right]$$

$$+ \frac{3}{2}\left|C_S^L - C_S^R\right|^2 H_S^2 + 8\,|C_T|^2 \left(1 + \frac{2m_\ell^2}{q^2}\right)\left(H_{T,+}^2 + H_{T,-}^2 + H_{T,0}^2\right)$$

$$+ 3\,\mathcal{R}e\left[\left(1 + C_V^L - C_V^R\right)\left(C_S^{L*} - C_S^{R*}\right)\right]\frac{m_\ell}{\sqrt{q^2}} H_S \cdot H_{V,t} \qquad (9.167)$$

$$- 12\,\mathcal{R}e\left[\left(1 + C_V^L\right)C_T^*\right]\frac{m_\ell}{\sqrt{q^2}}\left(H_{T,0} \cdot H_{V,0} + H_{T,+} \cdot H_{V,+} - H_{T,-} \cdot H_{V,-}\right)$$

$$\left. + 12\,\mathcal{R}e\left[C_V^R C_T^*\right]\frac{m_\ell}{\sqrt{q^2}}\left(H_{T,0} \cdot H_{V,0} + H_{T,+} \cdot H_{V,-} - H_{T,-} \cdot H_{V,+}\right)\right\},$$

where

$$\lambda_{J/\psi}(q^2) = \left[(m_{B_c} - m_{J/\psi})^2 - q^2\right]\left[(m_{B_c} + m_{J/\psi})^2 - q^2\right] \qquad (9.168)$$

and H_i are the hadronic helicity amplitudes, that are collected in an appendix of [553]. The corresponding formulas for $\Lambda_b \to \Lambda_c\ell\bar{\nu}_\ell$ given in terms of the helicity amplitudes can be found in [568, 569].

9.4.6 $B_c \to \tau\bar{\nu}_\tau$

Finally, we give the expression for the branching ratio of the pure leptonic decay $B_c \to \tau\bar{\nu}_\tau$, which strongly constrains the axial $(C_V^R - (1 + C_V^L))$ and, especially, the pseudo-scalar $(C_S^R - C_S^L)$ contributions [570, 571]:

$$\mathcal{B}(B_c \to \tau\bar{\nu}_\tau) = \tau_{B_c} \frac{m_{B_c} m_\tau^2 F_{B_c}^2 G_F^2 |V_{cb}|^2}{8\pi}\left(1 - \frac{m_\tau^2}{m_{B_c}^2}\right)^2$$

$$\times \left|1 + C_V^L - C_V^R + \frac{m_{B_c}^2}{m_\tau(m_b + m_c)}\left(C_S^R - C_S^L\right)\right|^2. \qquad (9.169)$$

9.4.7 Useful Formulas for Observables

In [551, 554] a number of useful approximate numerical formulas for the observables just discussed have been presented. They are given directly in terms of WCs of the effective

Hamiltonian in (9.151) setting $C_V^R = 0$ for reasons explained earlier. Consequently these formulas allow to make predictions for the observables in question in any scenario of NP, thereby providing quickly the patterns of correlations between these observables that are characteristic for a given NP scenario. In obtaining the results listed here the formfactors have been set at their central values.

The authors of [551, 554] find then

$$\mathcal{R}(D) \simeq \mathcal{R}_{SM}(D)\{|1 + C_V^L|^2 + 1.54 \operatorname{Re}[(1 + C_V^L)(C_S^{L*} + C_S^{R*})] + 1.09|C_S^L + C_S^R|^2$$
$$+ 1.04 \operatorname{Re}[(1 + C_V^L)C_T^*] + 0.75|C_T|^2\}, \tag{9.170}$$

$$\mathcal{R}(D^*) \simeq \mathcal{R}_{SM}(D^*)\{|1 + C_V^L|^2 + 0.13 \operatorname{Re}[(1 + C_V^L)(C_S^{R*} - C_S^{L*})] + 0.05|C_S^R - C_S^L|^2$$
$$- 5.0 \operatorname{Re}[(1 + C_V^L)C_T^*] + 16.27|C_T|^2\}, \tag{9.171}$$

$$P_\tau(D) \simeq \left(\frac{\mathcal{R}(D)}{\mathcal{R}_{SM}(D)}\right)^{-1}\{0.32|1 + C_V^L|^2 + 1.54 \operatorname{Re}[(1 + C_V^L)(C_S^{L*} + C_S^{R*})]$$
$$1.09|C_S^L + C_S^R|^2 - 0.35 \operatorname{Re}[(1 + C_V^L)C_T^*] + 0.05|C_T|^2\}, \tag{9.172}$$

$$P_\tau(D^*) \simeq \left(\frac{\mathcal{R}(D^*)}{\mathcal{R}_{SM}(D^*)}\right)^{-1}\{-0.49|1 + C_V^L|^2 + 0.13 \operatorname{Re}[(1 + C_V^L)(C_S^{R*} - C_S^{L*})]$$
$$+ 0.05|C_S^R - C_S^L|^2 + 1.67 \operatorname{Re}[(1 + C_V^L)C_T^*] + 0.93|C_T|^2\}, \tag{9.173}$$

$$F_L(D^*) \simeq \left(\frac{\mathcal{R}(D^*)}{\mathcal{R}_{SM}(D^*)}\right)^{-1}\{0.46|1 + C_V^L|^2 + 0.13 \operatorname{Re}[(1 + C_V^L)(C_S^{R*} - C_S^{L*})]$$
$$+ 0.05|C_S^R - C_S^L|^2 - 1.98 \operatorname{Re}[(1 + C_V^L)C_T^*] + 3.2|C_T|^2\}, \tag{9.174}$$

$$\mathcal{R}(\Lambda_c) \simeq \mathcal{R}_{SM}(\Lambda_c)\{|1 + C_V^L|^2 + 0.34 \operatorname{Re}[(1 + C_V^L)C_S^{L*}] + 0.50 \operatorname{Re}[(1 + C_V^L)C_S^{R*}]$$
$$+ 0.53 \operatorname{Re}[C_S^L C_S^{R*}] + 0.33(|C_S^L|^2 + |C_S^R|^2) - 3.10 \operatorname{Re}[(1 + C_V^L)C_T^*]$$
$$+ 10.44|C_T|^2\}, \tag{9.175}$$

$$\mathcal{B}(B_c \to \tau\nu_\tau) \simeq 0.02\left(\frac{F_{B_c}}{0.43 \,\text{GeV}}\right)^2|1 + C_V^L + 4.3\,(C_S^R - C_S^L)|^2, \tag{9.176}$$

in terms of the WCs defined at the low-scale $\mu = m_b$.

From (9.170), (9.171), and (9.176) the following *sum rule* can be derived [551, 554]

$$\boxed{\frac{\mathcal{R}(\Lambda_c)}{\mathcal{R}_{SM}(\Lambda_c)} = 0.262\frac{\mathcal{R}(D)}{\mathcal{R}_{SM}(D)} + 0.738\frac{\mathcal{R}(D^*)}{\mathcal{R}_{SM}(D^*)} - x,} \tag{9.177}$$

where the small remainder x is well approximated by

$$x \simeq -\operatorname{Re}\left[(1 + C_V^L)(0.32\,C_T^* + 0.03\,C_S^{L*})\right] \tag{9.178}$$
$$+ 1.76\,|C_T|^2 - 0.033 \operatorname{Re}(C_S^L C_S^{R*}).$$

This model-independent relation between $\mathcal{R}(D)$, $\mathcal{R}(D^*)$, and $\mathcal{R}(\Lambda_c)$ has been found empirically in [551] by studying the formulas listed. However, as pointed out by Thomas

Mannel, it originates from heavy-quark symmetry: In the heavy-quark limit the inclusive $b \to c\tau\nu$ rate is saturated by the sum of $\bar{B} \to D\tau\nu$ and $\bar{B} \to D^*\tau\nu$ in the mesonic case, and by $\Lambda_b \to \Lambda_c\tau\nu$ in the baryonic case. This sum rule also holds for NP scenarios with right-handed neutrinos, although they are not considered in [551, 554].

The existence of the sum rule in (9.177), which holds in *any* model of NP, implies that a future measurement of $\mathcal{R}(\Lambda_c)$ will serve as a check of the measurements of $\mathcal{R}(D)$ and $\mathcal{R}(D^*)$ and of the formfactor calculations. One predicts then [554]

$$\mathcal{R}(\Lambda_c) = \mathcal{R}_{SM}(\Lambda_c)\,(1.15 \pm 0.06) = 0.38 \pm 0.01. \tag{9.179}$$

Correlations between $\mathcal{R}(D)$, $\mathcal{R}(D^*)$, $P_\tau(D)$, $P_\tau(D^*)$, and $F_L(D^*$, presented in [551, 554], demonstrate the power of polarization observables in distinguishing between various NP scenarios.

9.4.8 Additional Information

It should also be noted that similar to the decays $\bar{B} \to \bar{K}(K^*)\ell^+\ell^-$ also decays considered here can be formulated with the help of the angular coefficients I_i discussed in detail in Section 9.1. The corresponding analyses can be found in [550], and a new analysis of the authors in [543] can be found in [1343].

It should also be emphasized that differential decay distributions and angular coefficients have larger discriminating power than the integrated quantities but to make them more effective the uncertainties in formfactors should be decreased. It could, in fact, turn out that although the ratios $\mathcal{R}(D)$ and $\mathcal{R}(D^*)$ will be found close to the SM ones, these more local quantities will exhibit significant NP effects. For analyses stressing this point see [541, 553, 572–574].

It is also of interest to check whether NP could show up in semileptonic decays governed by $b \to u\ell^-\bar{\nu}_\ell$ transitions like $\bar{B} \to \rho(\pi\pi)\bar{\nu}_\ell$ and $\bar{B} \to a_1(\rho\pi)\bar{\nu}_\ell$. A very detailed analysis of angular distributions and many related observables in these decays can be found in [575]. Even if CKM suppressed relative to $b \to c$ transitions, high-precision measurements of these decays are foreseen in the near future by LHCb and Belle II making this analysis timely.

9.4.9 Summary of SM and Experimental Results

During the last years a number of deviations from SM expectations in $b \to c\tau\bar{\nu}_\tau$ transitions have been observed. They seem to indicate, similar to $b \to s\ell^+\ell^-$ the violation of lepton flavor universality (LFU). In the case at hand these are deviations from $\tau/(\mu, e)$ universality observed in $\bar{B} \to D^*\ell\bar{\nu}$ and $\bar{B} \to D\ell\bar{\nu}$ decays by BaBar [576], Belle [577–579], and LHCb [580] collaborations. See [581] for a review. Moreover also in $B_c^+ \to J/\psi\tau^+\nu_\tau$ anomalous result has been found [582].

Indeed for the ratios $\mathcal{R}(D)$, $\mathcal{R}(D^*)$, and $\mathcal{R}(J/\psi)$ the 2018 experimental world averages from HFLAV based on the results reported by the BaBar [576, 583], Belle [577–579], and LHCb [580, 584] collaborations read as listed in Table 9.1. Taken together they implied a

Consequently neglecting differences in the phase space of these two decays, due to $m_{\pi^+} \neq m_{\pi^0}$ and $m_e \neq 0$, we find

$$\frac{\mathcal{B}(K^+ \to \pi^+ \nu \bar{\nu})}{\mathcal{B}(K^+ \to \pi^0 e^+ \nu)} = \frac{\alpha^2}{|V_{us}|^2 2\pi^2 \sin^4 \vartheta_W} \sum_{l=e,\mu,\tau} \left| V_{cs}^* V_{cd} X_{\mathrm{NL}}^l + V_{ts}^* V_{td} X(x_t) \right|^2 . \quad (9.182)$$

Indeed, we have shown that the relevant hadronic matrix element can be extracted from the data. But to reach sufficient precision one has to go beyond our analysis and include isospin breaking corrections in relating $K^+ \to \pi^+ \nu \bar{\nu}$ to $K^+ \to \pi^0 e^+ \nu$. They include quark mass effects and electroweak radiative corrections and have been calculated in [202]. We will not present these calculations here but we will incorporate these effects in the subsequent phenomenological expressions.

9.5.3 The Branching Ratio

Using (9.182), including isospin breaking corrections and summing over three neutrino flavors, one finds [181, 202]

$$\mathcal{B}(K^+ \to \pi^+ \nu \bar{\nu}) = \kappa_+ (1 + \Delta_{\mathrm{EM}}) \cdot \left[\left(\frac{\mathrm{Im}\lambda_t}{\lambda^5} X(x_t) \right)^2 + \left(\frac{\mathrm{Re}\lambda_c}{\lambda} P_c(X) + \frac{\mathrm{Re}\lambda_t}{\lambda^5} X(x_t) \right)^2 \right]$$

$$(9.183)$$

with $P_c(X)$ defined in (6.347) and

$$\lambda = |V_{us}|, \qquad \kappa_+ = (5.173 \pm 0.025) \cdot 10^{-11} \left[\frac{\lambda}{0.225} \right]^8, \qquad \Delta_{\mathrm{EM}} = -0.003. \quad (9.184)$$

Here $x_t = m_t^2 / M_W^2$, $\lambda_i = V_{is}^* V_{id}$ are the CKM factors, and κ_+ summarizes all the remaining factors, in particular the relevant hadronic matrix element discussed earlier that can be extracted from leading semileptonic decays of K^+, K_L and K_S mesons [202]. In obtaining the numerical value in (9.184)

$$\sin^2 \vartheta_W = 0.23116, \qquad \alpha(M_Z) = \frac{1}{127.9}, \quad (9.185)$$

given in the $\overline{\mathrm{MS}}$ scheme, have been used. Their errors are below 0.1% and can be neglected.

There is an issue related to $\sin^2 \vartheta_W$ that, although very precisely measured in a given renormalization scheme, is a scheme-dependent quantity with the scheme dependence only removed by considering higher-order electroweak effects in $K \to \pi \nu \bar{\nu}$. We have discussed this issue already in Section 6.6 in connection with $B_s \to \mu^+ \mu^-$ decay. The complete analysis of two-loop electroweak contributions to $K \to \pi \bar{\nu} \nu$ in [135] for the top contribution and to the charm contribution in [203] reduced this scheme dependence well below 1%. We recall here the results for $X(x_t)$ and $P_c(X)$ that include all these corrections

$$X(x_t) = 1.481 \pm 0.009, \qquad P_c(X) = 0.405 \pm 0.024 \quad (9.186)$$

with the latter calculated for $\lambda = 0.2252$, which for consistency should be inserted into (9.184).

The apparent large sensitivity of $\mathcal{B}(K^+ \to \pi^+ \nu \bar{\nu})$ to λ is spurious as $P_c(X) \sim \lambda^{-4}$ and the dependence on λ in (9.184) cancels the one in (9.183) to a large extent. However, basically for aesthetic reasons it is useful to write these formulas as given by us. In doing this it is essential to keep track of the λ dependence as it is hidden in $P_c(X)$ (see (6.347)), and changing λ while keeping $P_c(X)$ fixed would give wrong results.

9.5.4 $K^+ \to \pi^+ \nu \bar{\nu}$ and the Unitarity Triangle

The measurement of the branching ratio in (9.183) allows to determine the element $|V_{td}|$ and the unitarity triangle provided additional information on the CKM parameters is available. Let us find out what is still needed to achieve this goal.

Using the improved Wolfenstein parametrization and the approximate formulas (2.81)–(2.83) we can put (9.183) into a more transparent form [27]

$$\mathcal{B}(K^+ \to \pi^+ \nu \bar{\nu}) = \kappa_+ A^4 X^2(x_t) \frac{1}{\sigma} \left[(\sigma \bar{\eta})^2 + (\varrho_0 - \bar{\varrho})^2 \right], \qquad \upsilon = \left(\frac{1}{1 - \frac{\lambda^2}{2}} \right)^2. \quad (9.187)$$

The measured value of $\mathcal{B}(K^+ \to \pi^+ \nu \bar{\nu})$ then determines an ellipse in the $(\bar{\varrho}, \bar{\eta})$ plane centered at $(\varrho_0, 0)$ with

$$\varrho_0 = 1 + \frac{P_c(X)}{A^2 X(x_t)} \quad (9.188)$$

and having the squared axes

$$\bar{\varrho}_1^2 = r_0^2, \qquad \bar{\eta}_1^2 = \left(\frac{r_0}{\sigma} \right)^2, \quad (9.189)$$

where

$$r_0^2 = \frac{1}{A^4 X^2(x_t)} \left[\frac{\sigma \cdot \mathcal{B}(K^+ \to \pi^+ \nu \bar{\nu})}{\kappa_+} \right]. \quad (9.190)$$

Note that r_0 depends only on the top contribution. The departure of ϱ_0 from unity measures the relative importance of the internal charm contributions: $\varrho_0 \approx 1.4$.

The ellipse in question defined by r_0, ϱ_0, and σ intersects with the circle (2.100). This allows to determine $\bar{\varrho}$ and $\bar{\eta}$ with

$$\bar{\varrho} = \frac{1}{1 - \sigma^2} \left(\varrho_0 - \sqrt{\sigma^2 \varrho_0^2 + (1 - \sigma^2)(r_0^2 - \sigma^2 R_b^2)} \right), \qquad \bar{\eta} = \sqrt{R_b^2 - \bar{\varrho}^2} \quad (9.191)$$

and consequently

$$R_t^2 = 1 + R_b^2 - 2\bar{\varrho}, \quad (9.192)$$

where $\bar{\eta}$ is assumed to be positive. Given $\bar{\varrho}$ and $\bar{\eta}$ one can determine V_{td}:

$$V_{td} = A\lambda^3(1 - \bar{\varrho} - i\bar{\eta}), \qquad |V_{td}| = A\lambda^3 R_t. \quad (9.193)$$

We note that in addition to R_b, which can be obtained from $|V_{ub}/V_{cb}|$, the determination of $|V_{td}|$ and of the unitarity triangle requires the knowledge of $|V_{cb}|$ (or A). Both values are subject to theoretical uncertainties present in the existing analyses of tree-level decays as discussed in Section 8.4. Whereas the dependence on $|V_{ub}/V_{cb}|$ is rather weak, the very strong dependence of the branching ratio on A or $|V_{cb}|$ matters in this determination, which of course also requires precise measurement of $\mathcal{B}(K^+ \to \pi^+ \nu \bar{\nu})$ itself. To show these dependences in explicit terms, we will now briefly present the numerical analysis of this decay.

9.5.5 Numerical Analysis of $K^+ \to \pi^+ \nu \bar{\nu}$

For the numerical analysis it is useful to use the parametric formula derived in [212]

$$\mathcal{B}(K^+ \to \pi^+ \nu \bar{\nu})_{\mathrm{SM}} = (8.39 \pm 0.30) \cdot 10^{-11} \cdot \left[\frac{|V_{cb}|}{40.7 \cdot 10^{-3}} \right]^{2.8} \left[\frac{\gamma}{73.2°} \right]^{0.74} . \tag{9.194}$$

This formula exhibits the dominant uncertainties that reside in the values of $|V_{cb}|$ and the angle γ in the UT. The latter dependence is related through the unitarity of the CKM matrix to the dependence on R_t. The error in the first factor represents remaining uncertainties coming dominantly from P_c. It should be emphasized that this formula can be used independent of whether $|V_{cb}|$ and γ have been extracted from tree-level decays or loop-induced processes like ε_K or $B_{s,d} - \bar{B}_{s,d}$ mixings.

The first measurement of $K^+ \to \pi^+ \nu \bar{\nu}$ by E949 at Brookhaven resulted in [594]

$$\mathcal{B}(K^+ \to \pi^+ \nu \bar{\nu})_{\mathrm{exp}} = (17.3^{+11.5}_{-10.5}) \cdot 10^{-11}. \tag{9.195}$$

It has been recently significantly improved by the NA62 collaboration at CERN

$$\mathcal{B}(K^+ \to \pi^+ \nu \bar{\nu})_{\mathrm{exp}} = (4.7^{+7.2}_{-4.7}) \cdot 10^{-11}. \tag{9.196}$$

9.5.6 The Decay $K_L \to \pi^0 \nu \bar{\nu}$

General Remarks

The effective Hamiltonian for $K_L \to \pi^0 \nu \bar{\nu}$ has been given in (6.351). As we will demonstrate shortly, $K_L \to \pi^0 \nu \bar{\nu}$ proceeds in the SM almost entirely through direct CP-violation [595]. The indirectly CP-violating contribution and the CP-conserving contribution analyzed in [596] are fully negligible. Within the SM $K_L \to \pi^0 \nu \bar{\nu}$ is completely dominated by short-distance loop diagrams with top quark exchanges. The charm contribution can be fully neglected, and the theoretical uncertainties present in $K^+ \to \pi^+ \nu \bar{\nu}$ collected in $P_c(X)$ are absent here. Consequently the rare decay $K_L \to \pi^0 \nu \bar{\nu}$ is theoretically even cleaner than $K^+ \to \pi^+ \nu \bar{\nu}$ and is very well suited for the determination of the Wolfenstein parameter $\bar{\eta}$ and in particular $\mathrm{Im}\lambda_t$.

We have stated that the decay $K_L \to \pi^0 \nu \bar{\nu}$ is dominated by *direct* CP violation. Now as discussed in Section 8.1 the standard definition of the direct CP violation requires the presence of strong phases that are absent in $K_L \to \pi^0 \nu \bar{\nu}$. Consequently the violation of CP symmetry in $K_L \to \pi^0 \nu \bar{\nu}$ can only arise through the interference between $K^0 - \bar{K}^0$

mixing and the decay amplitude. This type of CP violation has been discussed already in Section 8.1. However, as already pointed out by [595] and demonstrated explicitly in a moment, the contribution of CP violation to $K_L \rightarrow \pi^0 \nu \bar{\nu}$ via $K^0 - \bar{K}^0$ mixing alone is tiny. It contributes to the branching ratio at the level of $2 \cdot 10^{-15}$ and its interference with the directly CP-violating contribution is $O(10^{-13})$. Consequently, in this sense, CP violation in $K_L \rightarrow \pi^0 \nu \bar{\nu}$ with the branching ratio $O(10^{-11})$ is a manifestation of CP violation in the decay and as such deserves the name of *direct* CP violation. In other words, the difference in the magnitude of CP violation in $K_L \rightarrow \pi\pi$ (ε) and $K_L \rightarrow \pi^0 \nu \bar{\nu}$ is a signal of direct CP violation, and measuring $\mathcal{B}(K_L \rightarrow \pi^0 \nu \bar{\nu})$ at the expected level would be in addition to ε'/ε another signal of this phenomenon. In fact, as we will see later, there are definite correlations between $K_L \rightarrow \pi^0 \nu \bar{\nu}$ and ε'/ε.

Deriving the Branching Ratio

The derivation of the branching ratio for $K_L \rightarrow \pi^0 \nu \bar{\nu}$ is more difficult than for $K^+ \rightarrow \pi^+ \nu \bar{\nu}$. To this end we consider first one neutrino flavor and define the complex function

$$F = \frac{G_F}{\sqrt{2}} \frac{\alpha}{2\pi \sin^2 \theta_W} V_{ts}^* V_{td} X(x_t). \tag{9.197}$$

Then the effective Hamiltonian in (6.351) can be written as

$$\mathcal{H}_{\text{eff}} = F(\bar{s}d)_{V-A}(\bar{\nu}\nu)_{V-A} + F^*(\bar{d}s)_{V-A}(\bar{\nu}\nu)_{V-A}. \tag{9.198}$$

Now, from (8.8) we have

$$K_L = \frac{1}{\sqrt{2}}[(1 + \bar{\varepsilon})K^0 + (1 - \bar{\varepsilon})\bar{K}^0], \tag{9.199}$$

where we have neglected $|\bar{\varepsilon}|^2 \ll 1$. Thus the amplitude for $K_L \rightarrow \pi^0 \nu \bar{\nu}$ is given by

$$A(K_L \rightarrow \pi^0 \nu \bar{\nu}) = \frac{1}{\sqrt{2}} \left[F(1 + \bar{\varepsilon})\langle \pi^0|(\bar{s}d)_{V-A}|K^0\rangle + F^*(1 - \bar{\varepsilon})\langle \pi^0|(\bar{d}s)_{V-A}|\bar{K}^0\rangle \right] (\bar{\nu}\nu)_{V-A}. \tag{9.200}$$

Recalling

$$CP|K^0\rangle = -|\bar{K}^0\rangle, \qquad C|K^0\rangle = |\bar{K}^0\rangle, \tag{9.201}$$

we have

$$\langle \pi^0|(\bar{d}s)_{V-A}|\bar{K}^0\rangle = -\langle \pi^0|(\bar{s}d)_{V-A}|K^0\rangle, \tag{9.202}$$

where the minus sign is crucial for the subsequent steps.

Thus we can write

$$A(K_L \rightarrow \pi^0 \nu \bar{\nu}) = \frac{1}{\sqrt{2}} [F(1 + \bar{\varepsilon}) - F^*(1 - \bar{\varepsilon})] \langle \pi^0|(\bar{s}d)_{V-A}|K^0\rangle (\bar{\nu}\nu)_{V-A}. \tag{9.203}$$

Now the terms $\bar{\varepsilon}$ can be safely neglected in comparison with unity, which implies that the indirect CP violation (CP violation in the $K^0 - \bar{K}^0$ mixing) is negligible in this decay. We have then

$$F(1 + \bar{\varepsilon}) - F^*(1 - \bar{\varepsilon}) = \frac{G_F}{\sqrt{2}} \frac{\alpha}{\pi \sin^2 \theta_W} \text{Im}(V_{ts}^* V_{td}) X(x_t). \tag{9.204}$$

Consequently using the isospin relation

$$\langle \pi^0 | (\bar{d}s)_{V-A} | \bar{K}^0 \rangle = \langle \pi^0 | (\bar{s}u)_{V-A} | K^+ \rangle \tag{9.205}$$

together with (9.180) and taking into account the difference in the lifetimes of K_L and K^+, we have after summation over three neutrino flavors

$$\frac{\mathcal{B}(K_L \to \pi^0 \nu \bar{\nu})}{\mathcal{B}(K^+ \to \pi^0 e^+ \nu)} = 3 \frac{\tau(K_L)}{\tau(K^+)} \frac{\alpha^2}{|V_{us}|^2 2\pi^2 \sin^4 \theta_W} \left[\text{Im}\lambda_t \, X(x_t) \right]^2, \tag{9.206}$$

where $\lambda_t = V_{ts}^* V_{td}$.

The Branching Ratio

Using (9.206) and including again isospin breaking corrections in relating $K_L \to \pi^0 \nu \bar{\nu}$ to $K^+ \to \pi^0 e^+ \nu$, we obtain [111, 458, 597]

$$\mathcal{B}(K_L \to \pi^0 \nu \bar{\nu}) = \kappa_L \cdot \left(\frac{\text{Im}\lambda_t}{\lambda^5} X(x_t) \right)^2, \tag{9.207}$$

where [202]

$$\kappa_L = (2.231 \pm 0.013) \cdot 10^{-10} \left[\frac{\lambda}{0.225} \right]^8. \tag{9.208}$$

Due to the absence of $P_c(X)$ in (9.207), $\mathcal{B}(K_L \to \pi^0 \nu \bar{\nu})$ has essentially no theoretical uncertainties. It is only affected by parametric uncertainties coming from m_t, $\text{Im}\lambda_t$, and κ_L of which only the one due to $\text{Im}\lambda_t$ is important.

Using the Wolfenstein parametrization and (6.343) we can rewrite (9.207) as

$$\mathcal{B}(K_L \to \pi^0 \nu \bar{\nu}) = 3.5 \cdot 10^{-11} \left[\frac{\eta}{0.36} \right]^2 \left[\frac{m_t(m_t)}{163.0 \, \text{GeV}} \right]^{2.3} \left[\frac{|V_{cb}|}{0.042} \right]^4. \tag{9.209}$$

The determination of η using $\mathcal{B}(K_L \to \pi^0 \nu \bar{\nu})$ requires the knowledge of V_{cb} and $m_t(m_t)$. The very strong dependence on V_{cb} or A makes a precise prediction for this branching ratio difficult at present. On the other hand inverting (9.207) and using (6.343) one finds:

$$\text{Im}\lambda_t = 1.4 \cdot 10^{-4} \left[\frac{163 \, \text{GeV}}{m_t(m_t)} \right]^{1.15} \left[\frac{\mathcal{B}(K_L \to \pi^0 \nu \bar{\nu})}{3.5 \cdot 10^{-11}} \right]^{1/2} \tag{9.210}$$

without any uncertainty from $|V_{cb}|$. (9.210) offers the cleanest method to measure $\text{Im}\lambda_t$, even better than the CP asymmetries in *B* decays discussed in Section 8.5 that require the knowledge of $|V_{cb}|$ to determine $\text{Im}\lambda_t$. Yet, the difficulty in measuring this branching ratio will makes this determination useful only in the middle of the 2020s. Moreover, this determination could be polluted by NP contributions, and it appears from present perspective that it is better to determine $\text{Im}\lambda_t$ from B decays and use it for the prediction of the branching ratio for $K_L \to \pi^0 \nu\bar{\nu}$ in the SM.

Numerical Analysis of $K_L \to \pi^0 \nu\bar{\nu}$

Analogous to the formula (9.194) one can derive a formula for $\mathcal{B}(K_L \to \pi^0 \nu\bar{\nu})_{\text{SM}}$. It reads [212]

$$\mathcal{B}(K_L \to \pi^0 \nu\bar{\nu})_{\text{SM}} = (3.36 \pm 0.05) \cdot 10^{-11} \cdot \left[\frac{|V_{ub}|}{3.88 \cdot 10^{-3}}\right]^2 \left[\frac{|V_{cb}|}{40.7 \cdot 10^{-3}}\right]^2 \left[\frac{\sin(\gamma)}{\sin(73.2°)}\right]^2 \cdot \tag{9.211}$$

We note different dependence on the CKM parameters that is the consequence of CP-violating character of this decay in contrast to $K^+ \to \pi^+ \nu\bar{\nu}$.

The 90% C.L. upper bound on $K_L \to \pi^0 \nu\bar{\nu}$ reads [598]

$$\mathcal{B}(K_L \to \pi^0 \nu\bar{\nu})_{\text{exp}} \leq 3.0 \cdot 10^{-9}, \tag{9.212}$$

and the preliminary 2019 status, including theoretical comments can be found in [1338].

Summary of $K_L \to \pi^0 \nu\bar{\nu}$ and Outlook

The KOTO experiment at J-PARC should provide interesting results on $\mathcal{B}(K_L \to \pi^0 \nu\bar{\nu})$ in the coming years [205, 206]. But also the KLEVER experiment at CERN SPS should measure this decay one day [599]. As we will discuss in the BSM chapters the combination of $K^+ \to \pi^+ \nu\bar{\nu}$ and $K_L \to \pi^0 \nu\bar{\nu}$ is particularly powerful in testing NP models and probing very short-distance scales. Assuming that NA62, KOTO, and KLEVER will reach the expected precision and the branching ratios on these decays will be at least as high as the ones predicted in the SM, these two decays are expected to be the superstars of flavor physics in the 2020s. The next pages demonstrate this. In this context we should not forget hyperon decays as at some stage they can also be useful in testing our ideas [600].

9.5.7 Unitarity Triangle and $\sin 2\beta$ from $K \to \pi\nu\bar{\nu}$

As we have just discussed, the branching ratios for $K^+ \to \pi^+ \nu\bar{\nu}$ and $K_L \to \pi^0 \nu\bar{\nu}$ contain still significant parametric uncertainties due to the uncertainties in $|V_{cb}|$, $|V_{ub}|$, and γ and to a lesser extent in m_t. It is then remarkable that within the SM all these uncertainties practically cancel out in the triple correlation between the branching ratios for $K^+ \to \pi^+ \nu\bar{\nu}$ and $K_L \to \pi^0 \nu\bar{\nu}$ and the mixing induced CP asymmetry $S_{\psi K_S}$ as pointed out in [193]. Moreover as we will see in a moment the UT can also be constructed by using these decays.

but by now fully under control. Most important, the decays based on the $b \rightarrow s\nu\bar{\nu}$ transition do not suffer from hadronic uncertainties *beyond* the formfactors, that plague the $b \rightarrow s\ell^+\ell^-$ transitions due to the breaking of factorization caused by photon exchange. For the $B \rightarrow K^{(*)}\nu\bar{\nu}$ transitions, factorization is exact, so a measurement of the decay rates would allow in principle to measure the formfactors.

But to make predictions for these decays, we have to rely on the nonperturbative calculations. Fortunately, in the last years, lattice computations of $B \rightarrow K$ and $B \rightarrow K^*$ formfactors have become available [605, 606] that are valid at large q^2 of the neutrino pair and which complement the existing results within light-cone sum rules (LCSR) [368, 469, 470], valid at low and intermediate q^2. Combining these two sources of information, it is possible to give SM predictions for the observables valid in the entire kinematic range, not relying on model-dependent extrapolations. The inclusive decay $B \rightarrow X_s\nu\bar{\nu}$ is theoretically as clean as $K \rightarrow \pi\nu\bar{\nu}$ decays, but the parametric uncertainties are a bit larger.

We have already mentioned the superiority of these decays over $K^+ \rightarrow \pi^+\nu\bar{\nu}$ and $K_{\mathrm{L}} \rightarrow \pi^0\nu\bar{\nu}$ as far right-handed currents are concerned. Indeed, as we will see in later sections, the latter decays are only sensitive to the sum of the WCs of left-handed and right-handed couplings, whereas $B \rightarrow K^*\nu\bar{\nu}$ is also sensitive to their difference. Also the existence of angular observables in $B \rightarrow K^*\nu\bar{\nu}$ allows a deeper insight into the issue of right-handed currents. Therefore, in contrast to $K \rightarrow \pi\nu\bar{\nu}$ decays we will already in this section present formulas that include contributions from right-handed currents.

But already at this stage we want to stress that simultaneous study of $K^+ \rightarrow \pi^+\nu\bar{\nu}$, $K_{\mathrm{L}} \rightarrow \pi^0\nu\bar{\nu}$, and $b \rightarrow s\nu\bar{\nu}$ transitions discussed here offers a powerful test of flavor symmetries. In the SM and in models with minimal flavor violation (MFV) there are striking correlations between the $d \rightarrow s\nu\bar{\nu}$ and $b \rightarrow s\nu\bar{\nu}$ transitions as they all are governed by the one-loop function $X(x_t)$ [216, 604, 607]. These correlations are generally modified in models with non-MFV interactions.

Moreover, as emphasized in [607] also the correlations between $b \rightarrow s\nu\bar{\nu}$ and $b \rightarrow s\ell^+\ell^-$ processes are important. They are not just relevant in the SM, where they are governed by the same formfactors, but also beyond the SM because the $SU(2)_L$ gauge symmetry relates neutrinos to left-handed charged leptons. The absence of any direct NP signal close to the electroweak scale at the LHC implies that this symmetry should still be reflected approximately in low-energy observables. This fact can be exploited in the so-called SMEFT, which considers dimension-6 operators of SM fields invariant under the SM gauge symmetry. We will discuss it in detail in Chapter 14.

The correlations between $b \rightarrow s\nu\bar{\nu}$ and $b \rightarrow s\ell^+\ell^-$ processes [565, 607–609] are particularly interesting in view of various tensions with the SM predictions observed in exclusive $b \rightarrow s\ell^+\ell^-$ decays. We will elaborate on these tensions or anomalies in Section 12.2. If due to NP, these anomalies might also leave an imprint in $b \rightarrow s\nu\bar{\nu}$ decays. Although on a completely model-independent basis, no general conclusions can be drawn, it turns out that in specific NP models, often only a subset of SMEFT operators are present and, as demonstrated in [607], for several models clear-cut predictions for the size of the effects and various correlations can be obtained. We will present them after presenting general formulas for the transitions in question and in particular after presenting general ideas behind SMEFT in Section 14.1.

Now, detailed analyses of $B \to K^* \nu \bar{\nu}$, $B \to K \nu \bar{\nu}$, and $B \to X_s \nu \bar{\nu}$ have been presented in [604, 607], giving the SM predictions and studying correlations among these decays as well as with $s \to d \nu \bar{\nu}$ and $b \to s \ell^+ \ell^-$ processes. As I am coauthor of these two papers and many analyses in the literature used them [190, 610–616], the present section is based naturally on [604, 607], but we will update the numerics and also include some results from papers written by other authors. For earlier studies of $b \to s \nu \bar{\nu}$ transitions, see in particular [481, 603], where the importance of these decays for the study of right-handed currents has been emphasized.

We will next summarize the effective Hamiltonian for $b \to s \nu \bar{\nu}$ transitions that generalizes (6.353) by the inclusion of right-handed quark currents and collect all observables probing this quark-level transition. While the decay $B \to K^* \nu \bar{\nu}$, due to its additional polarization observable, offers a richer source of information than the two other decays $B \to K \nu \bar{\nu}$ and $B \to X_s \nu \bar{\nu}$, combining all these decays we end up with *four* observables, which are functions of the invariant mass of the neutrino-antineutrino pair. In this manner powerful tests of the short-distance dynamics can be made.

9.6.2 Effective Hamiltonian

The effective Hamiltonian for $b \to s \nu \bar{\nu}$ transitions including right-handed currents reads

$$\mathcal{H}_{\text{eff}} = -\frac{4 G_F}{\sqrt{2}} \frac{\alpha}{4\pi} V_{tb} V_{ts}^* \left(C_L^\nu O_L^\nu + C_R^\nu O_R^\nu \right) + \text{h.c.}, \tag{9.220}$$

with the operators

$$O_L^\nu = (\bar{s} \gamma_\mu P_L b)(\bar{\nu} \gamma^\mu (1 - \gamma_5) \nu), \qquad O_R^\nu = (\bar{s} \gamma_\mu P_R b)(\bar{\nu} \gamma^\mu (1 - \gamma_5) \nu). \tag{9.221}$$

In the SM, C_R^ν is negligible while the coefficient C_L^{SM} is known with a high accuracy, including NLO QCD corrections [181, 182, 196] and two-loop electroweak contributions [135], resulting in

$$C_L^{\text{SM}} = -\frac{X(x_t)}{\sin^2 \vartheta_W}, \qquad\qquad X(x_t) = 1.481 \pm 0.009. \tag{9.222}$$

The formulas given following are derived, similar to $B_{s,d}^0 \to \mu^+ \mu^-$, from the Hermitian conjugate part of the Hamiltonian in (9.220) so that the formulas are for a decaying particle and not an antiparticle as in the case of $\bar{B} \to \bar{K}^* (\bar{K}) \ell^+ \ell^-$ and $\bar{B} \to D^* (D) \ell \bar{\nu}_l$ considered previously.

9.6.3 $B \to K^* \nu \bar{\nu}$

Similar to the decays $\bar{B} \to \bar{K}^* \mu^+ \mu^-$ and $\bar{B} \to D^* \ell \bar{\nu}$ discussed in previous sections, the decay $B \to K^* \nu \bar{\nu}$ has the virtue that the angular distribution of the K^* decay products allows to extract information about the polarization of the K^*. But the number of observables is smaller, and their expressions in terms of the WCs are much simpler because of the presence of only two operators and also because we now deal with neutrinos in the final state and not charged leptons.

Indeed, because the neutrinos escape the detector unmeasured, the experimental information that can be obtained from the process $B \to K^*(\to K\pi)\nu\bar{\nu}$ with an on-shell K^* is completely described by the double differential decay distribution in terms of the two kinematical variables

$$0 \leq s_B = \frac{q^2}{m_B^2} \leq \left(1 - \frac{m_{K^*}}{m_B}\right)^2 \approx 0.69, \qquad 0 \leq \theta \leq 2\pi, \tag{9.223}$$

where q^2 is the invariant mass of the neutrino-antineutrino pair, and θ, the angle between the K^* flight direction in the B rest frame and the K flight direction in the $K\pi$ rest frame. s_B can be called the normalized invariant mass. In what follows, to simplify notation we will use $\widetilde{m}_i = m_i/m_B$. We observe that the kinematics is much simpler than in the case of $\bar{B} \to \bar{K}^* \ell^+ \ell^-$ and $\bar{B} \to D^* \ell \bar{\nu}$.

The spectrum can be expressed in terms of three $B \to K^*$ transversity amplitudes $A_{\perp,\parallel,0}$, which are given in terms of formfactors and WCs as

$$A_\perp(s_B) = 2N\sqrt{2}\lambda^{1/2}(1, \widetilde{m}_{K^*}^2, s_B)(C_L^\nu + C_R^\nu)\frac{V(s_B)}{(1 + \widetilde{m}_{K^*})}, \tag{9.224}$$

$$A_\parallel(s_B) = -2N\sqrt{2}(1 + \widetilde{m}_{K^*})(C_L^\nu - C_R^\nu)A_1(s_B), \tag{9.225}$$

$$A_0(s_B) = -\frac{N(C_L^\nu - C_R^\nu)}{\widetilde{m}_{K^*}\sqrt{s_B}}\left[(1 - \widetilde{m}_{K^*}^2 - s_B)(1 + \widetilde{m}_{K^*})A_1(s_B) - \lambda(1, \widetilde{m}_{K^*}^2, s_B)\frac{A_2(s_B)}{1 + \widetilde{m}_{K^*}}\right]. \tag{9.226}$$

Here

$$\widetilde{m}_{K^*} = \frac{m_{K^*}}{m_B}, \qquad N = V_{tb}V_{ts}^*\left[\frac{G_F^2\alpha^2 m_B^3}{3 \cdot 2^{10}\pi^5}s_B\lambda^{1/2}(1, \widetilde{m}_{K^*}^2, s_B)\right]^{1/2} \tag{9.227}$$

and $\lambda(a, b, c) = a^2 + b^2 + c^2 - 2(ab + bc + ac)$.

It should be noted that whereas $A_\perp(s_B)$ depends on the sum $C_L^\nu + C_R^\nu$ the remaining amplitudes depend on $C_L^\nu - C_R^\nu$. In $K \to \pi\nu\bar{\nu}$ decays only the sum of analogous coefficients is present.

The three formfactors

$$V(q^2), \qquad A_1(q^2), \qquad A_2(q^2) \tag{9.228}$$

entered already our discussion of $\bar{B} \to \bar{K}^*\mu^+\mu^-$ in Section 9.1, and their present status is summarized there.

Defining the invariant mass spectrum with a longitudinally and transversely polarized K^*, respectively, as

$$\frac{d\Gamma_L}{ds_B} = 3m_B^2|A_0|^2, \qquad \frac{d\Gamma_T}{ds_B} = 3m_B^2\left(|A_\perp|^2 + |A_\parallel|^2\right), \tag{9.229}$$

where the factor of 3 stems from the sum over neutrino flavors,[7] the double differential spectrum can be written as

$$\frac{d^2\Gamma}{ds_B d\cos\theta} = \frac{3}{4}\frac{d\Gamma_T}{ds_B}\sin^2\theta + \frac{3}{2}\frac{d\Gamma_L}{ds_B}\cos^2\theta . \qquad (9.230)$$

Thus, $d\Gamma_L/ds_B$ and $d\Gamma_T/ds_B$ can be extracted by an angular analysis of the K^* decay products.

Instead of these two observables, one can choose the following two independent observables accessible from the double differential decay distribution: the dineutrino mass distribution $d\Gamma/ds_B$, where

$$\frac{d\Gamma(B \to K^*\nu\bar{\nu})}{ds_B} = \int_{-1}^{1} d\cos\theta \, \frac{d^2\Gamma}{ds_B d\cos\theta} = \frac{d\Gamma_L}{ds_B} + \frac{d\Gamma_T}{ds_B} = 3m_B^2\left(|A_\perp|^2 + |A_\parallel|^2 + |A_0|^2\right),$$
$$(9.231)$$

and either of the K^* longitudinal and transverse polarization fractions $F_{L,T}$ also used in the studies of $\bar{B} \to \bar{K}^*\ell^+\ell^-$ and $\bar{B} \to D^*\ell\bar{\nu}$ decays and defined as

$$F_{L,T} = \frac{d\Gamma_{L,T}/ds_B}{d\Gamma/ds_B} , \qquad F_L = 1 - F_T . \qquad (9.232)$$

The advantage of this choice of observables is twofold. First, the normalization of $F_{L,T}$ on the total dineutrino spectrum strongly reduces the hadronic uncertainties associated with the formfactors as well as parametric uncertainties associated with CKM elements. Second, in the absence of right-handed currents ($C_R^\nu = 0$), the dependence on the remaining WCs C_L^ν drops out in $F_{L,T}$, making it a perfect observable to probe such right-handed currents.

One can also consider the s_B-integrated form of $F_{L,T}$, which is defined as

$$\langle F_{L,T}\rangle = \frac{\Gamma_{L,T}}{\Gamma}, \qquad \text{where} \qquad \Gamma_{(L,T)} = \int_0^{1-\tilde{m}_{K^*}^2} ds_B \frac{d\Gamma_{(L,T)}}{ds_B} . \qquad (9.233)$$

9.6.4 $B \to K\nu\bar{\nu}$

The dineutrino invariant mass distribution for the exclusive decay $B \to K\nu\bar{\nu}$ can be written as [603]

$$\frac{d\Gamma(B \to K\nu\bar{\nu})}{ds_B} = \frac{G_F^2\alpha^2}{256\pi^5}|V_{ts}^*V_{tb}|^2 m_B^5\lambda^{3/2}(s_B, \tilde{m}_K^2, 1)\left[f_+^K(s_B)\right]^2|C_L^\nu + C_R^\nu|^2 . \qquad (9.234)$$

In this case, similar to $K^+ \to \pi^+\nu\bar{\nu}$ and $K_L \to \pi^0\nu\bar{\nu}$, the rate depends only on $C_L^\nu + C_R^\nu$.

---

[7] Here we assume that the WCs do not depend on the neutrino flavor.

9.6.5 $B \to X_s \nu \bar{\nu}$

The inclusive decay $B \to X_s \nu \bar{\nu}$ offers the theoretically cleanest constraint on the coefficients C_L^ν and C_R^ν as it does not involve any formfactors. Its dineutrino invariant mass distribution is sensitive to yet another combination of C_L^ν and C_R^ν,

$$
\frac{d\Gamma(B \to X_s \nu \bar{\nu})}{ds_b} = m_b^5 \frac{\alpha^2 G_F^2}{128\pi^5} |V_{ts}^* V_{tb}|^2 \kappa(0)(|C_L^\nu|^2 + |C_R^\nu|^2)
$$

$$
\times \sqrt{\lambda(1, \hat{m}_s^2, s_b)} \left[3 s_b (1 + \hat{m}_s^2 - s_b - 4\hat{m}_s \frac{\mathrm{Re}\left(C_L^\nu C_R^{\nu *}\right)}{|C_L^\nu|^2 + |C_R^\nu|^2}) + \lambda(1, \hat{m}_s^2, s_b) \right], \quad (9.235)
$$

where we have defined $\hat{m}_i = m_i/m_b$, and $\kappa(0) = 0.83$ represents the QCD correction to the $b \to s \nu \bar{\nu}$ matrix element [111, 472, 617].

9.6.6 Standard Model Predictions

General Formulas

We summarize next predictions for various observables in the SM. What follows is an overview of the results presented in [607]. We have first

$$
\frac{d\mathcal{B}(B^+ \to K^+ \nu \bar{\nu})_{\mathrm{SM}}}{dq^2} \equiv \mathcal{B}_K^{\mathrm{SM}}(q^2) = \tau_{B^+} 3|N|^2 \frac{X^2(x_t)}{s_W^4} \rho_K(q^2), \quad (9.236)
$$

$$
\frac{d\mathcal{B}(B_d^0 \to K^{*0} \nu \bar{\nu})_{\mathrm{SM}}}{dq^2} \equiv \mathcal{B}_{K^*}^{\mathrm{SM}}(q^2) = \tau_{B^0} 3|N|^2 \frac{X^2(x_t)}{s_W^4} \left[\rho_{A_1}(q^2) + \rho_{A_{12}}(q^2) + \rho_V(q^2) \right], \quad (9.237)
$$

$$
F_L(B \to K^* \nu \bar{\nu})_{\mathrm{SM}} \equiv F_L^{\mathrm{SM}}(q^2) = \frac{\rho_{A_{12}}(q^2)}{\rho_{A_1}(q^2) + \rho_{A_{12}}(q^2) + \rho_V(q^2)}, \quad (9.238)
$$

where the factor of 3 stems from the sum over neutrino flavors. Next

$$
N = V_{tb} V_{ts}^* \frac{G_F \alpha}{16\pi^2} \sqrt{\frac{m_B}{3\pi}} \quad (9.239)
$$

is a normalization factor.

The quantities ρ_i are given in terms of the relevant formfactors as follows

$$
\rho_K(q^2) = \frac{\lambda_K^{3/2}(q^2)}{m_B^4} \left[f_+^K(q^2) \right]^2 \quad (9.240)
$$

$$
\rho_V(q^2) = \frac{2q^2 \lambda_{K^*}^{3/2}(q^2)}{(m_B + m_{K^*})^2 m_B^4} \left[V(q^2) \right]^2, \quad (9.241)
$$

$$\rho_{A_1}(q^2) = \frac{2q^2 \lambda_{K^*}^{1/2}(q^2)(m_B + m_{K^*})^2}{m_B^4} \left[A_1(q^2)\right]^2 ,$$

(9.242)

$$\rho_{A_{12}}(q^2) = \frac{64 m_{K^*}^2 \lambda_{K^*}^{1/2}(q^2)}{m_B^2} \left[A_{12}(q^2)\right]^2 ,$$

(9.243)

where [606]

$$A_{12}(q^2) = \frac{(m_B + m_{K^*})^2 (m_B^2 - m_{K^*}^2 - q^2) A_1(q^2) - \tilde{\lambda} A_2(q^2)}{16 m_B m_{K^*}^2 (m_B + m_{K^*})}$$

(9.244)

with

$$\tilde{\lambda} = (t_+ - q^2)(t_- - q^2), \qquad t_\pm = (m_B \pm m_{K^*})^2 .$$

(9.245)

This combination of A_1 and A_2 is convenient as it is the linear combination that appears in the longitudinal transversity (helicity) amplitude.

As always

$$\lambda(a, b, c) = a^2 + b^2 + c^2 - 2(ab + bc + ac), \qquad \lambda_{K^{(*)}}(q^2) \equiv \lambda(m_B^2, m_{K^{(*)}}^2, q^2).$$ (9.246)

These formulas are given here only to give the reader an idea of what enters the calculation. More details on the formfactors can be found in Appendix A in [607]. Discussion of them is not terribly exciting, so let us not do it here.

In contrast to $\bar{B} \to \bar{K}^{(*)} \ell^+ \ell^-$ decays, the isospin asymmetries of the decays with neutrinos in the final state vanish identically, so the branching ratio of the B_d^0 and $B^\pm$ decays only differ due to the lifetime difference. F_L is equal for charged and neutral B decay.

It is useful to define q^2-binned observables

$$\left\langle \mathcal{B}_{K^{(*)}}^{\rm SM} \right\rangle_{[a,b]} \equiv \int_a^b dq^2 \, \frac{d\mathcal{B}(B \to K^{(*)} \nu \bar{\nu})_{\rm SM}}{dq^2} ,$$

(9.247)

$$\left\langle F_L^{\rm SM} \right\rangle_{[a,b]} \equiv \frac{\int_a^b dq^2 \, \rho_{A_{12}}(q^2)}{\int_a^b dq^2 \left[\rho_{A_1}(q^2) + \rho_{A_{12}}(q^2) + \rho_V(q^2)\right]} ,$$

(9.248)

simply because they give more information than total branching ratios that are obtained by integrating over the full kinematically allowed range of q^2.

Numerical Analysis

The numerical prediction of the observables introduced by us within the SM requires the calculation of the hadronic formfactors. At low q^2 the results from light-cone sum rules [368, 469] are usually taken, while for large q^2 those from lattice QCD [605, 606]. Because the formfactors have to be smooth functions of q^2, one can obtain expressions valid in the whole kinematical range relevant for $B \to K^{(*)} \nu \bar{\nu}$ by performing a combined fit to lattice and LCSR results. Because this approach makes use of theoretical input on both ends of the kinematical range, the results will be very weakly dependent on the parametrization chosen for the formfactors. In the case of $B \to K^*$, such combined fit has been performed in [485] and for $B \to K$ in [618]. The predictions given here were provided in February 2019 by David Straub and are based on these formfactors. They update those given in [607].

In the SM, the branching ratios are then

$$\mathcal{B}(B^+ \to K^+ \nu\bar{\nu})_{\text{SM}} = (4.35 \pm 0.59) \times 10^{-6} \left| \frac{V_{ts} V_{tb}^*}{0.0411} \right|^2 , \tag{9.249}$$

$$\mathcal{B}(B^0 \to K^{0*} \nu\bar{\nu})_{\text{SM}} = (9.44 \pm 0.89) \times 10^{-6} \left| \frac{V_{ts} V_{tb}^*}{0.0411} \right|^2 , \tag{9.250}$$

to be compared with the experimental bounds [374, 619]

$$\mathcal{B}(B^+ \to K^+ \nu\bar{\nu}) \le 1.3 \times 10^{-5} \quad @\ 90\%\ \text{CL} , \tag{9.251}$$

$$\mathcal{B}(B^0 \to K^0 \nu\bar{\nu}) \le 2.6 \times 10^{-5} \quad @\ 90\%\ \text{CL} , \tag{9.252}$$

$$\mathcal{B}(B^+ \to K^{+*} \nu\bar{\nu}) \le 4.0 \times 10^{-5} \quad @\ 90\%\ \text{CL} , \tag{9.253}$$

$$\mathcal{B}(B^0 \to K^{0*} \nu\bar{\nu}) \le 1.8 \times 10^{-5} \quad @\ 90\%\ \text{CL}. \tag{9.254}$$

Moreover,

$$F_L^{\text{SM}} = 0.49 \pm 0.04. \tag{9.255}$$

Note that the value of F_L at the kinematical endpoints is fixed [604, 620].

We also find

$$\mathcal{R}_K^\nu \equiv \frac{\mathcal{B}_K}{\mathcal{B}_K^{\text{SM}}} < 3.5, \qquad \mathcal{R}_{K^*}^\nu \equiv \frac{\mathcal{B}_{K^*}}{\mathcal{B}_{K^*}^{\text{SM}}} < 2.2 \tag{9.256}$$

at 90% C.L.

At the Belle-II experiment, first test of the SM predictions and first signs of NP could in principle be seen already in 2020. Indeed, we will see in later chapters that in several NP models, the experimental upper bounds can be saturated. In this case, a 5σ discovery should be definitely possible at Belle-II.

9.6.7 Model-Independent Constraints on Wilson Coefficients

The four observables accessible in the three different $b \to s\nu\bar{\nu}$ transitions are dependent on the two in principle complex WCs C_L^ν and C_R^ν. However, only two combinations of these complex quantities enter the formulas given by us and are thus observable. These are [617, 621]

$$\epsilon = \frac{\sqrt{|C_L^\nu|^2 + |C_R^\nu|^2}}{|(C_L^\nu)^{\text{SM}}|} , \qquad \eta = \frac{-\text{Re}\left(C_L^\nu C_R^{\nu*}\right)}{|C_L^\nu|^2 + |C_R^\nu|^2} , \tag{9.257}$$

such that $\epsilon > 0$ and η lies in the range $[-\frac{1}{2}, \frac{1}{2}]$. $\epsilon = 1$ in the SM and $\eta \neq 0$ signals the presence of right-handed currents. The ratios $\mathcal{R}_K^\nu$ and $\mathcal{R}_{K^*}^\nu$ in (9.256) can be expressed in terms of ϵ and η as follows

$$\mathcal{R}_K^\nu = (1 - 2\eta)\epsilon^2, \qquad \mathcal{R}_{K^*}^\nu = (1 + \kappa_\eta \eta)\epsilon^2, \qquad \mathcal{R}_{F_L}^\nu \equiv \frac{F_L}{F_L^{SM}} = \frac{1 + 2\eta}{1 + \kappa_\eta \eta}. \tag{9.258}$$

The parameter κ_η depends on the formfactors, and its explicit form is given in an appendix in [607]. Presently $\kappa_\eta = 1.33 \pm 0.05$.

Because the three observables in (9.258) only depend on two combinations of WCs, there is a model-independent prediction [607],

$$F_L = F_L^{SM} \left(\frac{(\kappa_\eta - 2)\mathcal{R}_K + 4\mathcal{R}_{K^*}}{(\kappa_\eta + 2)\mathcal{R}_{K^*}} \right). \tag{9.259}$$

In principle, this relation can be tested experimentally (also on a bin-by-bin basis). A similar relation can be obtained for the modification of the inclusive $B \to X_s \nu \bar{\nu}$ branching ratio [607],

$$\text{BR}(B \to X_s \nu \bar{\nu}) \approx \text{BR}(B \to X_s \nu \bar{\nu})_{SM} \left(\frac{\kappa_\eta \mathcal{R}_K + 2\mathcal{R}_{K^*}}{\kappa_\eta + 2} \right), \tag{9.260}$$

where we have neglected a contribution proportional to η of at most $\pm 5\%$ to the inclusive branching ratio [604]. Following [604] and using our updated numerical input, we obtain

$$\text{BR}(B \to X_s \nu \bar{\nu})_{SM} = (3.0 \pm 0.3) \times 10^{-5}. \tag{9.261}$$

The relations (9.259) and (9.260) hold even in the case of lepton flavor nonuniversality and lepton flavor violation. Consequently, a violation of either of them unambiguously signals the presence of particles other than neutrinos in the final state (as discussed, e.g., in [604, 622]). As ϵ and η can be calculated in any model by means of (9.257), the expressions (9.258), (9.259), and (9.260) can be considered as fundamental formulas for any phenomenological analysis of the decays in question. The experimental bounds on the branching ratios, can then be translated to excluded areas in the ϵ-η-plane, with the SM corresponding to $(\epsilon, \eta) = (1, 0)$.

Because the four observables depend on only two parameters, a measurement of all of them would overconstrain the resulting (ϵ, η) point. Some examples have been presented in [604], and it will be interesting to construct such plots when Belle II will obtain first results [428].

A special role is played by the observable F_L because, as seen in (9.258), it only depends on η, and consequently it leads to a horizontal line in the ϵ-η plane. Although a similar constraint could be obtained by dividing two of the branching ratios to cancel the common factor of ϵ^2, the use of F_L is theoretically much cleaner because in this case, the hadronic uncertainties cancel, while they would add up when using the branching ratios.

Another interesting point about F_L is that because it only depends on η, the distribution $F_L(s_B)$ is universal for all models in which one of the coefficients $C^\nu_{L,R}$ vanishes, such as in the SM and models with constrained minimal flavor violation (CMFV) [520, 623, 624]. Some examples are given in [604].

9.7 $K_{L,S} \to \mu^+\mu^-$ and $K_L \to \pi^0 \ell^+ \ell^-$

9.7.1 $K_L \to \mu^+\mu^-$

Only the so-called short-distance (SD) part of a dispersive contribution to $K_L \to \mu^+\mu^-$ can be reliably calculated. Despite this limitation, as we will see later, already this contribution puts important bounds on certain NP scenarios. The calculation is analogous to the ones for $K^+ \to \pi^+\nu\bar{\nu}$ and $K_L \to \pi^0\nu\bar{\nu}$ except that now instead of the vector current the axial-vector current contributes. The effective Hamiltonian is given in (6.338) with only Q_{7A} contributing. Relating the relevant matrix element $\langle 0|\bar{s}\gamma_\mu P_L d|K_L\rangle$ to the branching ratio $\mathcal{B}(K^+ \to \mu^+\nu_\mu)$, we find ($\lambda = 0.2252$)

$$\mathcal{B}(K_L \to \mu^+\mu^-)^{\text{SM}}_{\text{SD}} = \kappa_\mu \left(\frac{\text{Re}\, Y^{\text{SM}}_{\text{eff}}}{\lambda^5} + \frac{\text{Re}\, \lambda_c}{\lambda} P_c(Y) \right)^2, \qquad (9.262)$$

where

$$\kappa_\mu = \frac{\alpha^2 \mathcal{B}(K^+ \to \mu^+\nu_\mu)}{\pi^2 \sin^4 \vartheta_W} \frac{\tau(K_L)}{\tau(K^+)} \lambda^8 = 2.01 \cdot 10^{-9}. \qquad (9.263)$$

To this end we have used

$$\alpha = \frac{1}{127.9}, \qquad \sin^2 \vartheta_W = 0.2316, \qquad \mathcal{B}(K^+ \to \mu^+\nu_\mu) = 0.635. \qquad (9.264)$$

Next, the charm contribution, represented by $P_c(Y)$, is found at NNLO to be [625]

$$P_c(Y) = 0.115 \pm 0.017. \qquad (9.265)$$

The short-distance contributions are described by

$$Y^{\text{SM}}_{\text{eff}} = V^*_{ts} V_{td}\, Y^{\text{SM}}_L(K), \qquad Y^{\text{SM}}_L(K) = Y(x_t) = 0.942, \qquad (9.266)$$

with the later value extracted from [187]. The structure of this formula is useful for future generalization beyond the SM. We find then

$$\mathcal{B}(K_L \to \mu^+\mu^-)^{\text{SM}}_{\text{SD}} \approx (0.8 \pm 0.1) \cdot 10^{-9}. \qquad (9.267)$$

The extraction of the short-distance part from the data is subject to considerable uncertainties. Here the important issue is the sign of the interference of the SD dispersive part χ_{SD} of the decay amplitude of $K_L \to \mu\bar{\mu}$ with the corresponding LD parts. Allowing for both signs implies a conservative bound $|\chi_{\text{SD}}| \leq 3.1$ [626]. This gives then the known upper bound [626]

$$\mathcal{B}(K_L \to \mu^+\mu^-)_{\text{SD}} \leq 2.5 \cdot 10^{-9}, \tag{9.268}$$

roughly three times as large as the SM value. This bound is also obtained for the sign favored in [627, 628] that implies $-1.7 \leq \chi_{\text{SD}} \leq 3.1$. On the other hand, the opposite sign is favored in [626, 629], giving $-3.1 \leq \chi_{\text{SD}} \leq 1.7$ and consequently approximately

$$\mathcal{B}(K_L \to \mu^+\mu^-)_{\text{SD}} \leq \mathcal{B}(K_L \to \mu^+\mu^-)_{\text{SD}}^{\text{SM}}. \tag{9.269}$$

The implications of these bounds will be discussed in the BSM chapters of this book. We will find there that they do not allow large enhancements of $\mathcal{B}(K^+ \to \pi^+ \nu\bar{\nu})$ for models with NP governed by left-handed currents but are much less important if right-handed currents dominate NP contributions.

9.7.2 $K_S \to \mu^+\mu^-$

The decay $K_S \to \mu\bar{\mu}$ provides another sensitive probe of imaginary parts of short-distance couplings. Its branching fraction receives long-distance (LD) and short-distance (SD) contributions, which are added incoherently in the total rate [626, 630]. This is in contrast to the decay $K_L \to \mu\bar{\mu}$, where LD and SD amplitudes interfere and moreover $\mathcal{B}(K_L \to \mu\bar{\mu})$ is sensitive to real parts of couplings. The SD part of $\mathcal{B}(K_S \to \mu\bar{\mu})$ is given as

$$\mathcal{B}(K_S \to \mu\bar{\mu})_{\text{SD}} = \tau_{K_S} \frac{G_F^2 \alpha^2}{8\pi^3} m_K F_K^2 \sqrt{1 - \frac{m_\mu^2}{m_K^2}} m_\mu^2 \, \text{Im}^2 \left[\lambda_t^{sd} C_{7A} \right]. \tag{9.270}$$

Recently the LHCb collaboration improved the upper bound on $K_S \to \mu\bar{\mu}$ by one order of magnitude [631]

$$\mathcal{B}(K_S \to \mu\bar{\mu})_{\text{LHCb}} < 0.8 \, (1.0) \times 10^{-9} \qquad \text{at } 90\% \, (95\%) \text{ C.L.} \tag{9.271}$$

to be compared with the SM prediction [626, 632]

$$\mathcal{B}(K_S \to \mu\bar{\mu})_{\text{SM}} = (4.99_{\text{LD}} + 0.19_{\text{SD}}) \times 10^{-12} = (5.2 \pm 1.5) \times 10^{-12}. \tag{9.272}$$

While this bound is still by two orders of magnitude above its SM value, it turns out that for several leptoquark models, even the saturation of this bound would barely remove the ε'/ε anomaly, discussed later. There are good future prospects to improve this bound, LHCb expects [305] with 23 fb^{-1} sensitivity to regions $\mathcal{B}(K_S \to \mu\bar{\mu}) \in [4, 200] \times 10^{-12}$, close to the SM prediction.

The description of future prospects for this decay at the LHCb and of decays discussed next including relevant references can be found in [305]. In particular also improvements on $K_S^0 \to \pi^0 \mu^+\mu^-$, $K^+ \to \pi^+ \mu^+\mu^-$, and $K^+ \to \pi^- \mu^+\mu^+$ are expected. These are analogous to τ decays discussed in Section 17.2, with one lepton in the final state replaced by a pion. Moreover, comparison with decays involving electrons will allow the tests of

10 ε'/ε in the Standard Model

10.1 Preliminaries

We will now discuss the ratio ε'/ε that measures the size of the direct CP violation in $K_L \to \pi\pi$ decays relative to the indirect (mixing induced) CP violation in this decay described by $\varepsilon_K \equiv \varepsilon$. After tremendous efforts of experimentalists over two decades, the present experimental world average from NA48 [644] and KTeV [645, 646] collaborations reads

$$(\varepsilon'/\varepsilon)_{\text{exp}} = (16.6 \pm 2.3) \times 10^{-4}. \tag{10.1}$$

The goal of this chapter is to describe in some detail the calculation of ε'/ε within the SM and to compare it with the measured value given here. In the SM ε' is governed by QCD penguins (QCDP) but receives also an important contribution from electroweak penguins (EWP). In the SM QCDP give in total a positive contribution, while the EWP give a negative one. This partial cancellation between these two contributions implies that already for three decades it was not possible to provide a precise prediction for ε'/ε. But as we will see later, significant progress has been made in calculating this ratio in the last years.

Both contributions are subject to the following uncertainties:

- CKM uncertainties that have been significantly reduced in the last years. We have seen this in Chapter 8. What matters here is $\text{Im}\lambda_t^{(K)}$ to be denoted in this chapter simply by $\text{Im}\lambda_t$.
- Perturbative QCD uncertainties that have been significantly reduced at the beginning of the 1990s through complete renormalization group improved calculations of NLO QCD and electroweak corrections to the Wilson coefficients of QCD and EW penguin operators. We have discussed them in Section 6.4. The dominant NNLO corrections to EWP contributions were calculated already in 1999 [647] and the complete ones to QCDP recently [648, 649].
- Hadronic uncertainties originating in the matrix elements of QCDP and EWP operators. These are in fact the largest uncertainties but also here significant progress has been made recently. We have discussed this progress already in Sections 7.2 and 7.3.
- Isospin breaking effects, which decrease the QCD penguin contribution making the cancellation in question stronger.

- Final state interactions, which according to ChPT estimates strongly enhance QCDP contribution and suppress the EWP ones, making the cancellation in question weaker and bringing ε'/ε close to the experimental data. But as discussed in Section 7.2, this claim as far as QCDP contributions are concerned is not supported by DQCD. The verdict from LQCD should be known soon.

The partial cancellation of QCDP contributions and EWP contributions combined with the uncertainties just listed remains even today a challenge for theorists, and as we will see at the end of this chapter, the precision of the SM prediction for ε'/ε cannot presently compete with the experimental one, but it appears that the SM value of ε'/ε is significantly below its experimental value in (10.1), opening speculation on which NP could explain this discrepancy. We will describe various suggestions in the BSM parts of this book, but first we will concentrate on ε'/ε in the SM.

Before, going into details it is appropriate to summarize briefly the history of the calculations of ε'/ε, which started in the second half of the 1970s when one still thought that m_t is much smaller than M_W [650–652]. For $m_t \ll M_W$ only QCDP play a substantial role. Over the 1980s these calculations were refined through the inclusion of isospin braking in the quark masses [273, 653], the inclusion of QED penguin effects for $m_t \ll M_W$ [273, 653, 654], and improved estimates of hadronic matrix elements in the framework of the DQCD [221, 227, 228] presented in Section 7.2. This era of ε'/ε culminated in the analyses in [655, 656], where QCDP, EWP (γ and Z^0 penguins), and the relevant box diagrams were included for arbitrary top quark mass. The strong cancellation between QCDP and EWP for $m_t > 150$ GeV found in these papers was confirmed by other authors [657, 658]. In fact, at that time, due to the aforementioned cancellation between QCDP and EWP contributions, the vanishing of ε'/ε in the SM could not be excluded.

However, all these calculations were done in the leading logarithmic approximation (e.g., one-loop anomalous dimensions of the relevant operators) with the exception of the m_t-dependence, which in the analyses [655–657] has been already included at the NLO level. While such a procedure is not fully consistent, m_t dependence enters ε'/ε first at NLO in the RG improved calculation, it allowed for the first time to exhibit the strong m_t-dependence of the EWP contributions, which is not seen in a strict leading logarithmic approximation.

During the 1990s considerable progress was made by calculating complete NLO corrections to ε' [122–125, 146, 659]. This means both QCD and electroweak corrections. First steps toward the inclusion of NNLO corrections have been already made in [200, 647], in particular in [647] dominant NNLO corrections to EWP have been calculated, but only very recently these calculations have been completed for QCDP [648, 649]. Yet, even more important is the significant improvement in the evaluation of the hadronic matrix elements of QCDP and EWP operators. We have already described this progress in Sections 7.2 and 7.3, but for convenience we will briefly summarize the present status. The reviews of ε'/ε prior to the recent progress can be found in [50, 660–662].

After these general remarks let us enter the details.

with

$$r = \frac{G_F\,\omega}{2\,|\varepsilon_K|\,\mathrm{Re}A_0}, \qquad \omega = \frac{\mathrm{Re}A_2}{\mathrm{Re}A_0}. \tag{10.44}$$

In (10.42) and (10.43) the sums run over all contributing operators. Therefore in $P^{(1/2)}$ in the case of EWP contributions we have to take into account the correction $b \neq 1$ defined in (10.31). It should be mentioned that $P^{(1/2)}$ and $P^{(3/2)}$ do not depend on h, the normalization factor, discussed after (7.45). We stress it because in [296] and in RBC-UKQCD papers $h = \sqrt{3/2}$ has been used and in this book we set $h = 1$.

Writing then

$$P^{(1/2)} = a_0^{(1/2)} + a_6^{(1/2)}\,B_6^{(1/2)} \equiv a_4^{(1/2)} + b a_{\mathrm{EWP}}^{(1/2)} + a_6^{(1/2)}\,B_6^{(1/2)}, \tag{10.45}$$

$$P^{(3/2)} = a_0^{(3/2)} + a_8^{(3/2)}\,B_8^{(3/2)}, \tag{10.46}$$

with the parameters $B_6^{(1/2)}$ and $B_8^{(3/2)}$ taken at $\mu = m_c$ and using the expressions (10.32)–(10.35), we find:

$$a_0^{(1/2)} = r_1 \left[\frac{[4y_4 - b(3y_9 - y_{10})]}{2(1+q)z_-} + b\,\frac{3q(y_9 + y_{10})}{2(1+q)z_+} \right] + r_2\,b\,y_8\,\frac{\langle Q_8 \rangle_0}{\mathrm{Re}A_0}, \tag{10.47}$$

$$a_6^{(1/2)} = r_2\,y_6\,\frac{\langle Q_6 \rangle_0}{B_6^{(1/2)}\,\mathrm{Re}A_0}, \tag{10.48}$$

$$a_0^{(3/2)} = r_1\,\frac{3(y_9 + y_{10})}{2z_+}, \tag{10.49}$$

$$a_8^{(3/2)} = r_2\,y_8^{\mathrm{eff}}\,\frac{\langle Q_8 \rangle_2}{B_8^{(3/2)}\,\mathrm{Re}A_2}, \tag{10.50}$$

where

$$r_1 = \frac{\omega}{\sqrt{2}|\varepsilon_K|}\,\frac{1}{V_{ud}V_{us}^*}, \qquad r_2 = \frac{\omega}{2|\varepsilon_K|}G_F. \tag{10.51}$$

Further details on these formulas can be found in [296], in particular in an Appendix of that paper.

With the values in (10.12)–(10.14), we find at NLO in the NDR-$\overline{\mathrm{MS}}$ scheme [1333]

$$\boxed{a_0^{(1/2)} = -4.19, \qquad a_6^{(1/2)} = 17.68, \qquad a_0^{(3/2)} - a_{\mathrm{EWP}}^{(1/2)} = -2.00, \qquad a_8^{(3/2)} = 8.25.}$$
$$\tag{10.52}$$

We observe that the ratio ε'/ε is governed in the SM by the following four contributions:

i) The term $a_0^{(1/2)}$ in $P^{(1/2)}$ in (10.45). As seen in (10.47) this term is governed by the contribution of the operator Q_4, denoted by $a_4^{(1/2)}$ and includes also smaller contributions from EWP operators Q_8, Q_9 and Q_{10} summarized by $b a_{\mathrm{EWP}}^{(1/2)}$. We find that this term is *negative* and only weakly dependent on q. Also the dependences on α_s and renormalisation scheme (see [123]) are weak. These weak dependences originate from the fact that in our approach the matrix elements entering the first term in $P^{(1/2)}$ cancel out. The dependence on m_t resulting from the contributions of sub-leading EWP operators turns out to be very weak [296].

ii) The contribution of $(V - A) \otimes (V + A)$ QCDP operators to $P^{(1/2)}$, given by the second term in (10.45). This contribution is large and *positive* and is dominated by the operator Q_6.

iii) The contribution of the $(V - A) \otimes (V - A)$ EWP operators Q_9 and Q_{10} to $P^{(3/2)}$, represented by the first term in $P^{(3/2)}$. As in the case of the contribution i), the matrix elements contributing to $a_0^{(3/2)}$ cancel out in the SM. Consequently, the scheme and α_s dependences of $a_0^{(3/2)}$ are weak. As seen in (10.49) the m_t-dependence of $a_0^{(3/2)}$ results from the corresponding dependence of $y_9 + y_{10}$, but the precision on m_t increased by much in the last two decades. $a_0^{(3/2)}$ contributes *positively* to ε'/ε.

iv) The contribution of the $(V - A) \otimes (V + A)$ EWP operators Q_7 and Q_8 to $P^{(3/2)}$, represented by the second term in (10.46). This contribution is dominated by Q_8. It contributes *negatively* to ε'/ε.

The competition between these four contributions is the reason why it is difficult to predict ε'/ε precisely. In this context, one should appreciate the virtue of our approach: The contributions i) and iii) can be determined rather precisely by CP-conserving data so that the dominant uncertainty in our approach in predicting ε'/ε resides in the values of $B_6^{(1/2)}$ and $B_8^{(3/2)}$, which for scales $\mu \geq 1$ GeV are practically μ independent.

The expression (10.41) can be put into a formula that shows explicitly the dependence on m_t and m_s. It can be found in Appendix B of [296]. See also [666, 667]. Here we would like just to insert the values in (10.52) into (10.41), which will allow us to see various effects transparently. The final numerical results in [1333] are based on the formulas listed here with all uncertainties taken into account.

We find then [1333]

$$\left(\frac{\varepsilon'}{\varepsilon} \right)_{\mathrm{SM}} = 10^{-4} \left[\frac{\mathrm{Im}\lambda_t}{1.4 \cdot 10^{-4}} \right] \left[a(1 - \hat{\Omega}_{\mathrm{eff}})(-5.9 + 24.7\, B_6^{(1/2)}) + 2.9 - 11.5\, B_8^{(3/2)} \right].$$

(10.53)

It is an update of a similar formula presented in [296].

Before proceeding, it should be emphasized that although the NLO analysis of ε'/ε reduces renormalization scheme dependence in the QCDP sector, it is still insensitive to the choice of μ_t in $m_t(\mu_t)$. This dependence can only be removed through NNLO calculations but in the QCDP sector it is already weak at the NLO level because of the weak dependence of the QCDP contributions on m_t. On the other hand, as pointed out already in [647], the EWP contributions at the NLO level suffer from a number of unphysical dependences.

- First of all there is the renormalization scheme dependence with ε'/ε in the HV scheme generally smaller than in the NDR scheme.

- The dependence on μ_t, which is much larger than in the QCDP sector because the EWP contributions exhibit much stronger dependence on m_t. Increasing μ_t makes the value of m_t smaller, decreasing the EWP contribution and thereby making ε'/ε larger. At NLO there is no QCD correction that could cancel this effect.

- The dependence on the choice of the matching scale μ_W. It turns out that with increasing μ_W in the EWP contribution, the value of ε'/ε decreases.

Our definitions in (11.8) and (11.9) follow the PDG conventions, also adopted by the HFAG collaboration; for neutral kaons they lead to $\Delta M_K \cdot \Delta \Gamma_K < 0$. Note that if $|\Gamma_{12}^D| \ll |M_{12}^D|$, as appropriate for B^0 mesons, one would recover the familiar expressions $\Delta M \simeq 2|M_{12}^D|$ and $\Delta \Gamma \ll \Delta M$ presented in previous chapters.

We recall that while ΔM_D and $\Delta \Gamma_D$ tell us nothing about the CP symmetry, the ratio q/p and the relative phase between M_{12}^D and Γ_{12}^D,

$$
\boxed{\varphi_{12} = \frac{1}{2} \arg\left(\frac{M_{12}^D}{\Gamma_{12}^D}\right),}
\tag{11.10}
$$

express the CP impurity in the two mass eigenstates through $|q/p| \neq 1$ and/or $2\varphi_{12} \neq \{0, \pm\pi\}$. While the phases of M_{12}^D and Γ_{12}^D depend on the phase conventions chosen, φ_{12} is phase convention independent and consequently an observable.

The experimental world averages as presented in HFAG read [150]

$$
x_D = 0.0046 \pm 0.0013, \qquad y_D = 0.0062 \pm 0.0007, \qquad \frac{x_D^2 + y_D^2}{2} \le (1.3 \pm 2.7) \cdot 10^{-4}.
\tag{11.11}
$$

It appears then that in the D system $x \sim y$ or $\Gamma_{12} \sim M_{12}$, whereas in the B system $|\Gamma_{12}/M_{12}| \ll 1$. For future prospects see [305, 692]. For recent measurements of x_D and y_D by the LHCb see [693].

In the limit of (approximate) CP symmetry x_D, $y_D > 0$ implies the CP *even* state to be slightly heavier and shorter lived than the CP *odd* one (unlike for neutral kaons).

As the $D^0 - \bar{D}^0$ mixing itself is subject to large theoretical uncertainties, we refer to section 3 of [305] for the summary of the present situation and move directly to CP violation, which is cleaner and moreover a good place to search for NP.

11.3 CP Asymmetries in D Decays

11.3.1 General Formalism

For convenience let us apply the technology of Section 8.1.5 to CP asymmetries in the D-meson system with notation properly adjusted to the literature on the D-meson system.

The time evolution of initially pure D^0 and $\bar{D}^0$ states, respectively, can be obtained from solving (11.1) and is given by

$$
|D^0(t)\rangle = f_+(t)|D^0\rangle - \frac{q}{p} f_-(t)|\bar{D}^0\rangle,
\tag{11.12}
$$

$$
|\bar{D}^0(t)\rangle = -\frac{p}{q} f_-(t)|D^0\rangle + f_+(t)|\bar{D}^0\rangle,
\tag{11.13}
$$

where

$$
f_+(t) = e^{-i\bar{M}t} e^{-\bar{\Gamma}t/2} \cos Qt, \qquad f_-(t) = i e^{-i\bar{M}t} e^{-\bar{\Gamma}t/2} \sin Qt,
\tag{11.14}
$$

with q/p given in (11.5), $\bar{M} = (M_1 + M_2)/2$ and

$$Q = \sqrt{(M_{12}^D - \frac{i}{2}\Gamma_{12}^D)(M_{12}^{D*} - \frac{i}{2}\Gamma_{12}^D)^*} = \frac{1}{2}(\Delta M_D - \frac{i}{2}\Delta\Gamma_D). \tag{11.15}$$

From (11.12) and (11.13) we find for the time-dependent decay rates of $D^0(t)$ and $\bar{D}^0(t)$ to a final state f:

$$\Gamma(D^0(t) \to f) = \left|T(D^0 \to f)\right|^2 e^{-\bar{\Gamma}t}\left[\frac{1}{2}\left(1 + |\lambda_f|^2\right)\cosh\frac{\Delta\Gamma_D t}{2}\right. \tag{11.16}$$

$$\left. + \frac{1}{2}\left(1 - |\lambda_f|^2\right)\cos\Delta M_D t - \sinh\frac{\Delta\Gamma_D t}{2}\mathrm{Re}\lambda_f + \sin\Delta M_D t\,\mathrm{Im}\lambda_f\right],$$

$$\Gamma(\bar{D}^0(t) \to f) = \left|T(\bar{D}^0 \to f)\right|^2 e^{-\bar{\Gamma}t}\left[\frac{1}{2}\left(1 + \left|\frac{1}{\lambda_f}\right|^2\right)\cosh\frac{\Delta\Gamma_D t}{2}\right. \tag{11.17}$$

$$\left. + \frac{1}{2}\left(1 - \left|\frac{1}{\lambda_f}\right|^2\right)\cos\Delta M_D t - \sinh\frac{\Delta\Gamma_D t}{2}\mathrm{Re}\frac{1}{\lambda_f} + \sin\Delta M_D t\,\mathrm{Im}\frac{1}{\lambda_f}\right],$$

where we dropped the overall phase space factors and defined

$$\lambda_f = \frac{q}{p}\frac{T(\bar{D}^0 \to f)}{T(D^0 \to f)}. \tag{11.18}$$

These general formulas agree with those of Dunietz and Rosner [694] after their definitions of ΔM and $\Delta\Gamma$ are adjusted to ours in (11.8) and (11.9).

From these results, one can easily obtain the CP asymmetries

$$\frac{\Gamma(D^0(t) \to f) - \Gamma(\bar{D}^0(t) \to f)}{\Gamma(D^0(t) \to f) + \Gamma(\bar{D}^0(t) \to f)}, \tag{11.19}$$

where f is a CP eigenstate

$$C\mathcal{P}|f\rangle = \eta_f|f\rangle, \qquad \eta_f = \pm 1. \tag{11.20}$$

We find

$$\frac{\Gamma(D^0(t) \to f) - \Gamma(\bar{D}^0(t) \to f)}{\Gamma(D^0(t) \to f) + \Gamma(\bar{D}^0(t) \to f)} = \frac{F_-}{F_+ + \cosh\frac{\Delta\Gamma_D t}{2} + \cos\Delta M_D t}, \tag{11.21}$$

where we have introduced the functions

$$F_\pm = \frac{1}{2}\left(\left|\frac{q}{p}\right|^2 \pm \left|\frac{p}{q}\right|^2\right)\left(\cosh\frac{\Delta\Gamma_D t}{2} - \cos\Delta M_D t\right)$$

$$- \left[\left(|\lambda_f| \pm \left|\frac{1}{\lambda_f}\right|\right)\cos 2\varphi_f \sinh\frac{\Delta\Gamma_D t}{2} - \left(|\lambda_f| \mp \left|\frac{1}{\lambda_f}\right|\right)\sin 2\varphi_f \sin\Delta M_D t\right], \tag{11.22}$$

and

$$\varphi_f = \frac{1}{2}\arg\left(\lambda_f\right). \tag{11.23}$$

We emphasize that the phase φ_f is phase convention independent as it depends only on the relative phase between q/p and $T(\bar{D}^0 \to f)/T(D^0 \to f)$ in (11.18).

For practical purposes, as $x_D, y_D \ll 1$, it is sufficient to consider the CP asymmetry in the limit of a small t. Then (11.21) reduces to

$$\frac{\Gamma(D^0(t) \to f) - \Gamma(\bar{D}^0(t) \to f)}{\Gamma(D^0(t) \to f) + \Gamma(\bar{D}^0(t) \to f)}$$

$$= -\left[y_D \left(|\lambda_f| - \left| \frac{1}{\lambda_f} \right| \right) \cos 2\varphi_f - x_D \left(|\lambda_f| + \left| \frac{1}{\lambda_f} \right| \right) \sin 2\varphi_f \right] \frac{t}{2\bar{\tau}_D}, \quad (11.24)$$

where $\bar{\tau}_D = 1/\bar{\Gamma}$.

In the case of a nonnegligible CP phase ξ_f in the decay amplitude $T(D^0 \to f)$, but $|T(D^0 \to f)| = |T(\bar{D}^0 \to f)|$, λ_f simplifies to

$$\lambda_f = \eta_f \frac{q}{p} e^{-i2\xi_f}, \qquad |\lambda_f| = \left| \frac{q}{p} \right|. \quad (11.25)$$

Moreover, if in the adopted phase convention (like CKM convention) the phase ξ_f is negligible as assumed in what follows, we have

$$\lambda_f = \eta_f \frac{q}{p} = \eta_f \left| \frac{q}{p} \right| e^{i2\tilde{\varphi}}, \qquad \tilde{\varphi} = \frac{1}{2} \arg \frac{q}{p}. \quad (11.26)$$

We then find

$$\frac{\Gamma(D^0(t) \to f) - \Gamma(\bar{D}^0(t) \to f)}{\Gamma(D^0(t) \to f) + \Gamma(\bar{D}^0(t) \to f)} \equiv S_f \frac{t}{2\bar{\tau}_D}, \quad (11.27)$$

where we defined in analogy with the B system

$$S_f \simeq -\eta_f \left[y_D \left(\left| \frac{q}{p} \right| - \left| \frac{p}{q} \right| \right) \cos 2\tilde{\varphi} - x_D \left(\left| \frac{q}{p} \right| + \left| \frac{p}{q} \right| \right) \sin 2\tilde{\varphi} \right]. \quad (11.28)$$

Note that in the B system $y \ll x$ and $|q/p| \simeq 1$, so that the preceding result simplifies considerably in the case of the CP asymmetries $S_{\psi K_S}$ and $S_{\psi \phi}$ in the B_d and B_s system, respectively.

Finally, we introduce the semileptonic asymmetry

$$a_{\rm SL}(D^0) \equiv \frac{\Gamma(D^0(t) \to \ell^- \bar{\nu} K^{+(*)}) - \Gamma(\bar{D}^0 \to \ell^+ \nu K^{-(*)})}{\Gamma(D^0(t) \to \ell^- \bar{\nu} K^{+(*)}) + \Gamma(\bar{D}^0 \to \ell^+ \nu K^{-(*)})} = \frac{|q|^4 - |p|^4}{|q|^4 + |p|^4} \approx 2 \left(\left| \frac{q}{p} \right| - 1 \right),$$
$$(11.29)$$

which represents CP violation in $\Delta C = 1$ transitions. In writing the last expression, we assumed that $|q/p| - 1$ is much smaller than unity.

11.3.2 Useful Relations

We next list a number of useful relations that have been derived in [685].

We define

$$x_{12} \equiv \frac{2|M_{12}|}{\Gamma}, \qquad y_{12} \equiv \frac{2|\Gamma_{12}|}{\Gamma}, \qquad \varphi_{12} = \frac{1}{2} \arg \left(\frac{M_{12}^D}{\Gamma_{12}^D} \right). \quad (11.30)$$

Then one finds for x_D and y_D in (11.7)

$$x_D = \text{sign}(\cos\varphi_{12})\left(x_{12}^2 - y_{12}^2 + \sqrt{(x_{12}^2 - y_{12}^2)^2 - 4x_{12}^2 y_{12}^2 \sin^2\varphi_{12}}\right)^{1/2} \tag{11.31}$$

$$y_D = \left(y_{12}^2 - x_{12}^2 + \sqrt{(x_{12}^2 - y_{12}^2)^2 - 4x_{12}^2 y_{12}^2 \sin^2\varphi_{12}}\right)^{1/2}. \tag{11.32}$$

These equations relate the neutral meson mass and width differences to mixing parameters x_{12}, y_{12} and φ_{12}.

Next

$$\tan\varphi = -\frac{\sin 2\varphi_{12}}{\cos 2\varphi_{12} + y_{12}^2/x_{12}^2}, \tag{11.33}$$

$$\tan\varphi = \left(1 - \left|\frac{q}{p}\right|\right)\frac{x_D}{y_D} = -\frac{a_{\text{SL}}}{2}\frac{x_D}{y_D}. \tag{11.34}$$

11.3.3 Correlations

Having all these formulas at hand we can derive two interesting correlations. Following the presentation in [695], we find

$$\sin^2 2\tilde{\varphi} = \frac{x_D^2(1 - |q/p|^2)^2}{x_D^2(1 - |q/p|^2)^2 + y_D^2(1 + |q/p|^2)^2}, \tag{11.35}$$

where in the phase conventions adopted in (11.6) $\tilde{\varphi} = 1/2\arg(q/p)$. In the limit $||q/p|-1| \ll 1$, $x_D \sim y_D$ (11.35) reduces to [695]

$$\sin 2\tilde{\varphi} = \frac{x_D}{y_D}\left(1 - \left|\frac{q}{p}\right|\right), \tag{11.36}$$

where the sign ambiguity in taking the square root of (11.35) can be resolved numerically. Using then (11.28) and (11.29) we find for $\xi_f = 0$

$$S_f = -\eta_f \frac{x_D^2 + y_D^2}{y_D} a_{\text{SL}}(D^0). \tag{11.37}$$

The violation of the relation (11.37) in future experiments would imply the presence of direct CP violation at work [695]. In the presence of a significant phase ξ_f we find

$$S_f = -\eta_f \left[\cos 2\xi_f \frac{x_D^2 + y_D^2}{y_D} a_{\text{SL}}(D^0) + 2x_D \sin 2\xi_f\right]. \tag{11.38}$$

Similar correlation is familiar from the B_s system [695–697], and we recall it using the formulation just presented.

Indeed formulas for CP asymmetries just discussed can be applied to the B_s meson system as well. We recall that in the latter system $|q/p| = 1$ with good accuracy, and in addition $y \ll x$. Using (11.35), we find

$$a_{\text{SL}}^s = -2 \left| \frac{y_s}{x_s} \right| \frac{S_{\psi\phi}}{\sqrt{1 - S_{\psi\phi}^2}}, \tag{11.39}$$

which agrees with [695] and represents an alternative derivation of the correlation found in [696, 697]. Note that we used the definition

$$a_{\text{SL}}^s = \frac{\Gamma(\bar{B}_s(t) \to \ell^+ X) - \Gamma(B_s(t) \to \ell^- X)}{\Gamma(\bar{B}_s(t) \to \ell^+ X) + \Gamma(B_s(t) \to \ell^- X)}, \tag{11.40}$$

and $S_{\psi\phi}$ is the coefficient of $\sin \Delta M_s t$ in

$$\frac{\Gamma(\bar{B}_s(t) \to \psi\phi) - \Gamma(B_s(t) \to \psi\phi)}{\Gamma(\bar{B}_s(t) \to \psi\phi) + \Gamma(B_s(t) \to \psi\phi)}. \tag{11.41}$$

Further

$$x_s = \frac{m_H - m_L}{\bar{\Gamma}_s}, \qquad y_s = \frac{\Gamma_H - \Gamma_L}{2\bar{\Gamma}_s}, \tag{11.42}$$

where we stress that our definition of y_s differs by sign from the HFAG one. Finally, in determining the overall sign of (11.39) we assumed $(y_s)_{\text{SM}} < 0$.

11.4 Connection between D and K Physics

In [683] the connection between $D^0 - \bar{D}^0$ and $K^0 - \bar{K}^0$ mixing has been discussed within the framework of approximately $SU(2)_L$-invariant NP. Due to the connection between up- and down-type quarks in the SM through the CKM matrix, in this scenario the NP contributions to $D^0 - \bar{D}^0$ and $K^0 - \bar{K}^0$ mixing are not independent of each other. This observation has been used in [683] to derive lower bounds on the NP scale in various NP scenarios, emerging if the experimental constraints on $D^0 - \bar{D}^0$ and $K^0 - \bar{K}^0$ mixing are applied to only the $(V - A) \otimes (V - A)$ contribution. One should keep in mind, however, that in models in which new operators contribute to $\Delta F = 2$ processes the power of this approach is limited, as the various contributions interplay with each other and dilute the correlation in question.

On the other hand, the situation is promising in NP models with only SM operators, such as the littlest Higgs model with T-parity (LHT).[1] In fact, this model provides possibly the best example of the physics discussed in [683]. While the following discussion has been triggered by the analysis of $\Delta C = 2$ processes in the LHT model and uses the notations and conventions of [675, 699, 700], it applies as well to all other NP scenarios with only SM operators. Similar to the LHT model, the flavor mixing matrices V_{Hu} and V_{Hd} parameterize the misalignment between the NP and the SM up- and down-type quarks, respectively, and are related via

$$V_{Hu} = V_{Hd} V_{\text{CKM}}^\dagger. \tag{11.43}$$

$D^0 - \bar{D}^0$ oscillations are then governed by the combinations ($i = 1, 2, 3$)

$$\xi_i^{(D)} = V_{Hu}^{iu}{}^* V_{Hu}^{ic}, \tag{11.44}$$

---

[1] Readers not familiar with this model should have a brief look at [698].

while for K, B_d, and B_s physics

$$\xi_i^{(K)} = V_{Hd}^{is}{}^* V_{Hd}^{id}, \qquad \xi_i^{(d)} = V_{Hd}^{ib}{}^* V_{Hd}^{id}, \qquad \xi_i^{(s)} = V_{Hd}^{ib}{}^* V_{Hd}^{is}, \qquad (11.45)$$

respectively, are relevant. By making use of (11.43) we can now express $\xi_i^{(D)}$ through combinations of V_{Hd} and CKM elements. Using the Wolfenstein parameterization for V_{CKM} and expanding in powers of λ, we find

$$\xi_i^{(D)} = \xi_i^{(K)*} + \lambda \left(|V_{Hd}^{is}|^2 - |V_{Hd}^{id}|^2 \right) + \lambda^2 \left(A\xi_i^{(d)*} - 2\mathrm{Re}(\xi_i^{(K)}) \right)$$

$$+ \lambda^3 \left(\frac{1}{2} (|V_{Hd}^{id}|^2 - |V_{Hd}^{is}|^2) + A\xi_i^{(s)*} + A\xi_i^{(s)}(\rho - i\eta) \right) + O(\lambda^4). \qquad (11.46)$$

The following comments are in order:

- At leading order $\xi_i^{(D)} = \xi_i^{(K)*}$, i.e., D and K physics are governed by the *same* NP flavor structure. We note that the complex conjugation arises, as $|D^0\rangle = |\bar{u}c\rangle$ while $|K^0\rangle = |\bar{s}d\rangle$, and it will give rise to a sign difference in CP violating effects in the D and K systems.
- The correction to linear order in λ is real, irrespective of the precise structure of V_{Hd}. However, as $\Delta C = 2$ CP violation is governed by

$$\mathrm{Im}(\xi_i^{(D)})^2 = 2\mathrm{Re}\xi_i^{(D)}\mathrm{Im}\xi_i^{(D)}, \qquad (11.47)$$

corrections to the one-to-one correspondence between $D^0 - \bar{D}^0$ and $K^0 - \bar{K}^0$ mixings will appear already at $O(\lambda)$ in both CP conserving and violating observables. On the other hand, the $\Delta C = 1$ effective Hamiltonian is governed by a single power of $\xi_i^{(D)}$, so that direct CP violation in rare D and K decays will be much more strongly correlated, and deviations from the one-to-one correspondence will arise only at $O(\lambda^2)$.
- The order $O(\lambda^2)$ correction can be complex, provided that $\mathrm{Im}\xi_i^{(d)} \neq 0$, i.e., that there are new CP-violating effects in the B_d system.
- At $O(\lambda^3)$ a complex correction arises due to the CP violation in the CKM matrix, given by $i\eta$. This correction is nonvanishing also in the limit of a real, i.e., CP conserving V_{Hd}, and vanishes only if $\xi_i^{(s)} = 0$.

This discussion shows that when doing NP phenomenology in the K-meson system one should check whether experimental bounds in charm systems are satisfied. For a recent analysis of this issue in the context of the ratio ε'/ε, see [701].

There is no question that charm will play a significant role in the tests of the SM and the search for NP in the coming years.

These expectations have been confirmed at Moriond 2019. LHCb presented the first observation of CP violation in the decays of charm hadrons. For the difference between the CP asymmetries in $D^0 \to K^+K^-$ and $D^0 \to \pi^+\pi^-$, they find [702]

$$\boxed{\Delta A_{CP} = (-15.4 \pm 2.9) \times 10^{-4}.} \qquad (11.48)$$

In this difference the CP violation in mixing cancels out so that this difference measures within excellent approximation direct CP violation. It remains to be seen whether some NP could be responsible for this result [703–705, 1331]. Among these papers I can recommended in particular [705, 1331] and an older paper [706]. A short summary of the LHCb data and the prospects for the future can be found in [707].

of $|V_{us}|$, $|V_{ud}|$, $|V_{cd}|$, and $|V_{cs}|$. Certainly this anomaly should be studied together with the previous one.

Anomaly 10: Possible anomaly in $|V_{ub}|$. With $|V_{ub}|$ much larger than the one implied by UT fits a NP phase would be required to obtain the correct value of the CP asymmetry $S_{\psi K_S}$.

Anomaly 11: The ratio g_τ/g_e in charged currents with 1.031 ± 0.013 exhibits breakdown of lepton universality at 2.4σ, implying g_τ anomaly [715, 716] as $g_\mu/g_e = 1.0003 \pm 0.0012$ [717] is fully consistent with the lepton flavor universality.

Anomaly 12: The correlation between the asymmetries $\mathcal{A}_{CP}^{dir}$ and $\mathcal{A}_{mix}^{dir}$ in $B_d^0 \to \pi^0 K_S$, predicted in the SM, appears to depart from the data exhibiting in this manner potential anomaly. We refer for details to Section 8.6 of this book and in particular to section 4 of [461], which discusses these finding in detail.

The question then arises what these tensions are telling us about possible NP responsible for them. Many authors addressed this issue in the last years and many papers have been published in which a number of model constructions have been presented that aim to find possible NP behind these anomalies. In the rest of the book we will develop the technology that will allow us to address these anomalies and possible future anomalies in a systematic manner. This will also allow us to describe the most interesting NP scenarios behind the anomalies in questions that have been proposed in the literature. A short look at various theoretical reviews could give the readers additional motivations for the final part of this book. A subset of them can be found in [718, 719]. Other will be listed as we proceed. Also section 7 of [305] is very useful in this respect.

Our strategy for Part IV of this book, which consists of eight steps (13–20), has been already described in the Introduction, and I would like to ask the readers to look at short summaries of these steps presented there. This will facilitate the reading of the rest of the book and to follow the logic of this part, the longest part of this expedition. But before starting these steps, let us summarize in a model independent manner what all these anomalies imply for the Wilson coefficients relevant for the $b \to s\ell\ell$ transitions.

12.3 Implications for the Wilson Coefficients

Several groups have performed global fits of the WCs to existing $b \to s\ell\ell$ data. In particular the most extensive analyses have been done by two groups: CCDMV [486, 494] and ANSS [710, 720, 721]. But also the analyses in [722–733] should be mentioned here.

The two most favorite solutions obtained in various global analyses of $b \to s\ell^+\ell^-$ [494, 721] until Moriond 2019, when only scenarios with a single nonvanishing WC or two WCs being equal or of opposite sign are considered, turned out to be

$$i)\ C_{9\mu}^{NP} < 0, \qquad ii)\ C_{9\mu}^{NP} = -C_{10\mu}^{NP} < 0. \tag{12.1}$$

One assumes here that NP is absent in $b \rightarrow se^+e^-$ so that these two coefficients, as indicated by an additional index, apply to $b \rightarrow s\mu^+\mu^-$ transitions only. Here

$$C_{9\mu} = C_{9\mu}^{\mathrm{SM}} + C_{9\mu}^{\mathrm{NP}}, \qquad C_{10\mu} = C_{10\mu}^{\mathrm{SM}} + C_{10\mu}^{\mathrm{NP}}. \tag{12.2}$$

However, as we will discuss next, other possibilities cannot be excluded at present, when several WCs are considered simultaneously and when the information from Moriond 2019 is taken into account. The following new features can be found in the new analyses that we will summarize briefly here:

- As stressed even before Moriond 2019 [734, 735], NP could affect $b \rightarrow se^+e^-$, $b \rightarrow s\mu^+\mu^-$, and $b \rightarrow s\tau^+\tau^-$ channels through two components, one keeping lepton flavour universality and one breaking it.
- The fact that $R(K)$ differs now visibly from $R(K^*)$ increased the room for RH contributions so that now in several papers also solutions with nonvanishing $C'_{9\mu}$ and $C'_{10\mu}$ can be found. We will discuss them briefly now and in more detail in the BSM part of this book.
- The hints for the anomaly in $B_s \rightarrow \mu^+\mu^-$ require the presence of axial muon couplings, that is either $C_{10\mu}^{\mathrm{NP}}$ or $C'_{10\mu}$ or both have to be nonvanishing as stressed in [516]. In this manner the second scenario in (12.1) is now favored.
- Possible NP in $b \rightarrow se^+e^-$, either lepton flavor universal (LFU) as considered in [516, 734, 735], or not lepton flavor universal [736].

New updated global analyses appeared after Moriond 2019 [516, 730, 736–741]. They take some or all new features listed earlier into account. A global analysis dedicated to $b \rightarrow c\tau\nu$ transitions can be found in [553–555]. Numerous tables in these papers show how various scenarios with one, two, or more nonvanishing WCs perform in improving the agreement of the theory with data relative to the SM. The best ones provide the improvement typically by $(5 - 6)\sigma$ dependent which input has been taken into account and how the numerical analysis has been performed.

It would premature to describe all possible solutions here. We pick up here three:

A: A purely muonic solution $C_{9\mu}^{\mathrm{NP}} = -C_{10\mu}^{\mathrm{NP}}$, well suited to UV-complete interpretations discussed in Section 16.4.11, that according to [516] is now favored with respect to a muonic contribution $C_{9\mu}^{\mathrm{NP}}$ only. Considering scenario ii) in (12.1), the authors of [516] find the 1σ range

$$-0.62 \leq C_{9\mu}^{\mathrm{NP}} = -C_{10\mu}^{\mathrm{NP}} \leq -0.45, \tag{12.3}$$

which gives a fit that is by 6.5σ better than the SM one. If $C_{9\mu}^{\mathrm{NP}}$ and $C_{10\mu}^{\mathrm{NP}}$ are treated independently, the best fit point gives

$$C_{9\mu}^{\mathrm{NP}} = -0.72, \qquad C_{10\mu}^{\mathrm{NP}} = 0.40. \tag{12.4}$$

Note that $C_{10\mu}^{\mathrm{NP}} > 0$ suppresses $C_{10\mu}$ as $C_{10\mu}^{\mathrm{SM}} < 0$ and consequently the branching ratio for $B_s \rightarrow \mu^+\mu^-$.

larger than in the SM. Moreover, in models with scalars having both LH and RH scalar couplings also the operators $Q_{1,2}^{LR}$ can contribute.

The anomalous dimensions of new operators including two-loops have been calculated in [744, 745], and a detailed renormalization group analysis with many analytical expressions can be found in [746].

In the literature another basis, the so-called SUSY basis, introduced in [745], is often used. It involves the operators

$$O_1, \quad O_2, \quad O_3, \quad O_4, \quad O_5, \tag{13.10}$$

which are given in the case of $K^0 - \bar{K}^0$ system as follows

$$
\begin{aligned}
O_1 &= \bar{s}^\alpha \gamma_\mu P_L d^\alpha \, \bar{s}^\beta \gamma^\mu P_L d^\beta, \\
O_2 &= (\bar{s}^\alpha P_L d^\alpha)\,(\bar{s}^\beta P_L d^\beta), \\
O_3 &= (\bar{s}^\alpha P_L d^\beta)\,(\bar{s}^\beta P_L d^\alpha), \\
O_4 &= (\bar{s}^\alpha P_L d^\alpha)\,(\bar{s}^\beta P_R d^\beta), \\
O_5 &= (\bar{s}^\alpha P_L d^\beta)\,(\bar{s}^\beta P_R d^\alpha),
\end{aligned}
\tag{13.11}
$$

and for B_q systems with $q = d, s$ by

$$
\begin{aligned}
O_1 &= (\bar{b}^\alpha \gamma_\mu P_L q^\alpha)\,(\bar{b}^\beta \gamma^\mu P_L q^\beta), \\
O_2 &= (\bar{b}^\alpha P_L q^\alpha)\,(\bar{b}^\beta P_L q^\beta), \\
O_3 &= (\bar{b}^\alpha P_L q^\beta)\,(\bar{b}^\beta P_L q^\alpha), \\
O_4 &= (\bar{b}^\alpha P_L q^\alpha)\,(\bar{b}^\beta P_R q^\beta), \\
O_5 &= (\bar{b}^\alpha P_L q^\beta)\,(\bar{b}^\beta P_R q^\alpha).
\end{aligned}
\tag{13.12}
$$

13.2.2 Relation between BMU and SUSY Bases

The operators in BMU and SUSY bases are related through

$$\boxed{Q_1^{VLL} = O_1, \quad Q_1^{LR} = -2O_5, \quad Q_2^{LR} = O_4, \quad Q_1^{SLL} = O_2, \quad Q_2^{SLL} = 4O_2 + 8O_3.}$$

$$\tag{13.13}$$

The structure of one-loop anomalous dimension matrices of the new operators differs significantly from the one of current-current operators. In particular some entries are much larger, which in turn implies much larger RG effects. They are given in the BMU basis of [744] as follows

$$
\hat{\gamma}^{(0)LR} = \begin{pmatrix} \frac{6}{N} & 12 \\ 0 & -6N + \frac{6}{N} \end{pmatrix},
\tag{13.14}
$$

$$
\hat{\gamma}^{(0)SLL} = \begin{pmatrix} -6N + 6 + \frac{6}{N} & \frac{1}{2} - \frac{1}{N} \\ -24 - \frac{48}{N} & 2N + 6 - \frac{2}{N} \end{pmatrix}.
\tag{13.15}
$$

The two operator systems do not mix under renormalization with each other. The corresponding two-loop results can be found in [744].

Using the relations between the two bases just given one finds the one-loop anomalous dimension matrices in the SUSY basis

$$\hat{\gamma}^{(0)}(O_2, O_3) = \begin{pmatrix} -6N + 8 + \frac{2}{N} & 4 - \frac{8}{N} \\ 4N - 4 - \frac{8}{N} & 2N + 4 + \frac{2}{N} \end{pmatrix}, \tag{13.16}$$

$$\hat{\gamma}^{(0)}(O_4, O_5) = \begin{pmatrix} -6N + \frac{6}{N} & 0 \\ -6 & \frac{6}{N} \end{pmatrix}. \tag{13.17}$$

The corresponding two-loop results can be found in [744, 745].

13.2.3 Hadronic Matrix Elements

In some papers the matrix elements of operators are given in terms of B_i parameters with $B_i = 1$ in the vacuum insertion approximation (VIA). This turns out to be useful for getting the insight from DQCD in the values of B_i in the case of $K^0 - \bar{K}^0$ mixing obtained by LQCD. But in the case of $B_{s,d}^0 - \bar{B}_{s,d}^0$ mixing, where there is no reliable analytic method to calculate the B_i, we think that LQCD should provide numerical results for hadronic matrix elements instead of B_i as this avoids additional errors generated in extracting B_i and having directly the matrix elements simplifies the phenomenology.

The expressions for the matrix elements $\langle \bar{K}^0 | Q_i(\mu) | K^0 \rangle \equiv \langle Q_i(\mu) \rangle$ contributing to $K^0 - \bar{K}^0$ mixing are given in the BMU basis as follows

$$\langle Q_1^{\text{VLL}}(\mu) \rangle = \frac{1}{3} m_K F_K^2 B_1^{\text{VLL}}(\mu), \tag{13.18}$$

$$\langle Q_1^{\text{LR}}(\mu) \rangle = -\frac{1}{6} R(\mu) \, m_K F_K^2 B_1^{\text{LR}}(\mu), \tag{13.19}$$

$$\langle Q_2^{\text{LR}}(\mu) \rangle = \frac{1}{4} R(\mu) \, m_K F_K^2 B_2^{\text{LR}}(\mu), \tag{13.20}$$

$$\langle Q_1^{\text{SLL}}(\mu) \rangle = -\frac{5}{24} R(\mu) \, m_K F_K^2 B_1^{\text{SLL}}(\mu), \tag{13.21}$$

$$\langle Q_2^{\text{SLL}}(\mu) \rangle = -\frac{1}{2} R(\mu) \, m_K F_K^2 B_2^{\text{SLL}}(\mu), \tag{13.22}$$

where

$$R(\mu) = \left(\frac{m_K}{m_s(\mu) + m_d(\mu)} \right)^2 = \frac{r^2(\mu)}{4 m_K^2} \tag{13.23}$$

refers to the generic factorized evolution of any density-density operator with μ. F_K is the K-meson decay constant. We have met $r(\mu)$ already in Section 7.2, see (7.15).

Table 13.2 Central values of the scheme-independent hadronic matrix elements evaluated at different values of $M_{Z'}$. $\langle\hat{Q}_1^{\text{VLL}}\rangle^{ij}$ and $\langle\hat{Q}_1^{\text{LR}}\rangle^{ij}$ are in units of GeV^3.

$M_{Z'}$	5 TeV	10 TeV	20 TeV	50 TeV	100 TeV	200 TeV
$\langle\hat{Q}_1^{\text{VLL}}(M_{Z'})\rangle^{sd}$	0.00158	0.00156	0.00153	0.00150	0.00148	0.00146
$\langle\hat{Q}_1^{\text{LR}}(M_{Z'})\rangle^{sd}$	−0.183	−0.197	−0.211	−0.230	−0.244	−0.259
$\kappa_{sd}(M_{Z'})$	−115.46	−126.51	−137.84	−153.24	−165.20	−177.41
$\langle\hat{Q}_1^{\text{VLL}}(M_{Z'})\rangle^{bd}$	0.0423	0.0416	0.0409	0.0401	0.0395	0.0390
$\langle\hat{Q}_1^{\text{LR}}(M_{Z'})\rangle^{bd}$	−0.183	−0.195	−0.206	−0.222	−0.234	−0.246
$\kappa_{bd}(M_{Z'})$	−4.33	−4.68	−5.04	−5.53	−5.92	−6.30
$\langle\hat{Q}_1^{\text{VLL}}(M_{Z'})\rangle^{bs}$	0.0622	0.0611	0.0601	0.0589	0.0581	0.0573
$\langle\hat{Q}_1^{\text{LR}}(M_{Z'})\rangle^{bs}$	−0.268	−0.284	−0.301	−0.323	−0.340	−0.357
$\kappa_{bs}(M_{Z'})$	−4.31	−4.66	−5.01	−5.48	−5.85	−6.23

Table 13.3 Central values of the scheme-independent hadronic matrix elements evaluated at different values of M_H. $\langle\hat{Q}_1^{\text{SLL}}\rangle^{ij}$ and $\langle\hat{Q}_2^{\text{LR}}\rangle^{ij}$ are in units of GeV^3.

M_H	5 TeV	10 TeV	20 TeV	50 TeV	100 TeV	200 TeV
$\langle\hat{Q}_1^{\text{SLL}}(M_H)\rangle^{sd}$	−0.089	−0.093	−0.096	−0.101	−0.105	−0.108
$\langle\hat{Q}_2^{\text{LR}}(M_H)\rangle^{sd}$	0.291	0.312	0.334	0.362	0.384	0.405
$\tilde{\kappa}_{sd}(M_H)$	−3.27	−3.37	−3.46	−3.58	−3.66	−3.75
$\langle\hat{Q}_1^{\text{SLL}}(M_H)\rangle^{bd}$	−0.095	−0.099	−0.103	−0.108	−0.112	−0.116
$\langle\hat{Q}_2^{\text{LR}}(M_H)\rangle^{bd}$	0.245	0.262	0.280	0.304	0.322	0.340
$\tilde{\kappa}_{bd}(M_H)$	−2.57	−2.64	−2.72	−2.81	−2.88	−2.95
$\langle\hat{Q}_1^{\text{SLL}}(M_H)\rangle^{bs}$	−0.140	−0.146	−0.152	−0.159	−0.164	−0.170
$\langle\hat{Q}_2^{\text{LR}}(M_H)\rangle^{bs}$	0.348	0.373	0.399	0.432	0.458	0.484
$\tilde{\kappa}_{bs}(M_H)$	−2.48	−2.56	−2.63	−2.72	−2.79	−2.85

The operators $Q_{1,2}^{\text{LR}}$ were already present in the case of Z' but now, as seen from (13.33), the operator Q_2^{LR} plays the dominant role. In Table 13.3 we give the central values of the renormalization scheme independent matrix elements in (13.32) and (13.33) for the three meson systems and for different values of M_H.

In both tables we show the values of

$$\kappa_{sd}(M_{Z'}) = \frac{\langle\hat{Q}_1^{\text{LR}}(M_{Z'})\rangle^{sd}}{\langle\hat{Q}_1^{\text{VLL}}(M_{Z'})\rangle^{sd}}, \qquad \tilde{\kappa}_{bq}(M_H) = \frac{\langle\hat{Q}_2^{\text{LR}}(M_H)\rangle^{sd}}{\langle\hat{Q}_1^{\text{SLL}}(M_H)\rangle^{sd}} \tag{13.34}$$

that allow to describe the general pattern of these results. Inspecting Tables 13.2 and 13.3 we observe the following pattern:

- The matrix elements of LR operators are in the $K^0 - \bar{K}^0$ system by two orders of magnitude larger than of the SM VLL operator, and this disparity increases with increasing scale. Comparing Tables 13.2 and 13.3, we observe that in particular the matrix elements of the operator $\hat{Q}_2^{\text{LR}}$ are very large with significant impact on phenomenology as we will see in the BSM chapters of this book. This feature has been known already for decades [744, 745, 750, 751].

- The matrix elements of LR operators are in $B^0_{s,d} - \bar{B}^0_{s,d}$ systems also larger than the matrix element of the SM operator but only by factors of 4–6.
- The matrix elements of scalar operators SLL are significantly larger than the ones of the SM operator VLL but smaller than those of LR operators.

13.2.4 DQCD Insight into $K^0 - \bar{K}^0$ Hadronic Matrix Elements from LQCD

During the last years significant progress in the evaluation of $K^0 - \bar{K}^0$ matrix elements by ETM, SWME, and RBC-UKQCD lattice collaborations [295, 747, 752–754] has been made, and we have used their results in constructing the preceding tables.

Let us have a closer look at these results that differ significantly from those obtained in the VIA, which gives for the parameters B_i simply

$$B_1 = B_2 = B_3 = B_4 = B_5 = 1 \qquad \text{(VIA)}. \qquad (13.35)$$

In VIA these parameters do not depend on the renormalization scale μ. Already this property of VIA, which is based on the factorization of matrix elements of four-quark operators into products of quark currents or quark densities, is problematic, as generally these parameters depend on μ.

Now the RBC-UKQCD collaboration working at $\mu = 3$ GeV finds [747, 753, 754]

$$B_1 = 0.523(9)(7), \qquad B_2 = 0.488(7)(17), \qquad B_3 = 0.743(14)(65), \qquad (13.36)$$

and

$$B_4 = 0.920(12)(16), \qquad B_5 = 0.707(8)(44), \qquad (13.37)$$

with the first error being statistical and the second systematic. Similar results are obtained by EMT and SWME collaborations, although the values for B_4 and B_5 from the ETM collaboration are visibly below the ones obtained by RBC-UKQCD collaboration: $B_4 = 0.78(4)(3)$ and $B_5 = 0.49(4)(1)$. Except for B_4 all values differ significantly from unity prohibiting the use of VIA.

To our knowledge no lattice group made an attempt to understand this pattern of values, probably because within LQCD, which works in this case at scale $\mu = 3$ GeV, this pattern cannot be understood. But in Section 7.2.3 we have stressed that the QCD dynamics in $K \to \pi\pi$ decays at low scales, represented in the DQCD by meson evolution, is responsible for the dominant part of the $\Delta I = 1/2$ rule. Let us then see whether DQCD and meson evolution can also be helpful here.

Indeed, it has been demonstrated in [755] that the pattern of the values of the parameters B_i in (13.36) and (13.37) can be understood within DQCD approach through meson evolution from very low energy scales to scales $O(1\,\text{GeV})$ followed by the usual RG QCD evolution to scale $\mu = 3$ GeV at which the parameters B_i have been evaluated in lattice QCD.

The case of B_1 is well known, and we discussed it already in Section 7.2.4. In the large N limit one finds $B_1 = 3/4$ [230]. The meson evolution followed by quark-gluon evolution is $2 \times 40 = 80$. We will next, starting with $N_f = 5$ and ending with $N_f = 3$, list operators corresponding to the numbers in Table 13.4.

13.3.2 $N_f = 5$

We will begin with the SM operators. We will use here the projection $P_{L,R}$, which differs by a factor of 4 from the usual $(V \pm A)$. This has to be remembered when calculating hadronic matrix elements. But this is what these days is done when considering BSM operators, and we can as well do it for SM operators. Despite this difference we will still keep the usual names for SM operators. With this warning no confusion should arise.

We have then two $(sd)(uu)$ current-current (CC)-type operators in the SM from tree-level exchange

$$Q_1 = (\bar{s}^\alpha \gamma_\mu P_L u^\beta)(\bar{u}^\beta \gamma^\mu P_L d^\alpha) = Q_1^{\text{VLL},u},$$
$$Q_2 = (\bar{s}^\alpha \gamma_\mu P_L u^\alpha)(\bar{u}^\beta \gamma^\mu P_L d^\beta) = Q_2^{\text{VLL},u}, \tag{13.45}$$

where the naming given also in [744] will be useful in what follows.

$$\boxed{\text{VLL}, u \Rightarrow 2}$$

The QCD and QED penguin-type diagrams with insertions of CC and P (penguin) operators into CC and P diagrams as already encountered in Section 6.4.3 give rise to mixing into the eight QCD and QED penguin operators, unless the involved Dirac structure does not permit it or vanishing traces of the Dirac structure and/or color structure appear. We have seen this explicitly in Section 6.4.3. They are P-type operators with left-handed $(\bar{s}\gamma_\mu P_L d)$ vector currents and are part of the SM operator basis

$$Q_3 = (\bar{s}^\alpha \gamma_\mu P_L d^\alpha) \sum_{q=u,d,s,c,b} (\bar{q}^\beta \gamma^\mu P_L q^\beta),$$

$$Q_4 = (\bar{s}^\alpha \gamma_\mu P_L d^\beta) \sum_{q=u,d,s,c,b} (\bar{q}^\beta \gamma^\mu P_L q^\alpha),$$

$$Q_5 = (\bar{s}^\alpha \gamma_\mu P_L d^\alpha) \sum_{q=u,d,s,c,b} (\bar{q}^\beta \gamma^\mu P_R q^\beta), \tag{13.46}$$

$$Q_6 = (\bar{s}^\alpha \gamma_\mu P_L d^\beta) \sum_{q=u,d,s,c,b} (\bar{q}^\beta \gamma^\mu P_R q^\alpha),$$

$$Q_7 = \frac{3}{2}(\bar{s}^\alpha \gamma_\mu P_L d^\alpha) \sum_{q=u,d,s,c,b} Q_q (\bar{q}^\beta \gamma^\mu P_R q^\beta),$$

$$Q_8 = \frac{3}{2}(\bar{s}^\alpha \gamma_\mu P_L d^\beta) \sum_{q=u,d,s,c,b} Q_q (\bar{q}^\beta \gamma^\mu P_R q^\alpha),$$

$$Q_9 = \frac{3}{2}(\bar{s}^\alpha \gamma_\mu P_L d^\alpha) \sum_{q=u,d,s,c,b} Q_q (\bar{q}^\beta \gamma^\mu P_L q^\beta), \tag{13.47}$$

$$Q_{10} = \frac{3}{2}(\bar{s}^\alpha \gamma_\mu P_L d^\beta) \sum_{q=u,d,s,c,b} Q_q (\bar{q}^\beta \gamma^\mu P_L q^\alpha).$$

Their chirality-flipped counterparts are not present in the SM.

$$\boxed{\text{Penguins} \Rightarrow 8}$$

Other $(sd)(QQ)$ operators with $Q = d, s$ are [744]

$$Q_{11} = Q_1^{\text{VLL},d+s} = (\bar{s}^\alpha \gamma_\mu P_L d^\alpha) \left[(\bar{d}^\beta \gamma^\mu P_L d^\beta) + (\bar{s}^\beta \gamma^\mu P_L s^\beta) \right],$$
$$Q_{12} = Q_1^{\text{VLR},d+s} = (\bar{s}^\alpha \gamma_\mu P_L d^\beta) \left[(\bar{d}^\beta \gamma^\mu P_R d^\alpha) + (\bar{s}^\beta \gamma^\mu P_R s^\alpha) \right], \qquad (13.48)$$
$$Q_{13} = Q_2^{\text{VLR},d+s} = (\bar{s}^\alpha \gamma_\mu P_L d^\alpha) \left[(\bar{d}^\beta \gamma^\mu P_R d^\beta) + (\bar{s}^\beta \gamma^\mu P_R s^\beta) \right],$$

and

$$Q_{14} = Q_1^{\text{VLL},d-s} = (\bar{s}^\alpha \gamma_\mu P_L d^\alpha) \left[(\bar{d}^\beta \gamma^\mu P_L d^\beta) - (\bar{s}^\beta \gamma^\mu P_L s^\beta) \right],$$
$$Q_{15} = Q_1^{\text{VLR},d-s} = (\bar{s}^\alpha \gamma_\mu P_L d^\beta) \left[(\bar{d}^\beta \gamma^\mu P_R d^\alpha) - (\bar{s}^\beta \gamma^\mu P_R s^\alpha) \right], \qquad (13.49)$$
$$Q_{16} = Q_2^{\text{VLR},d-s} = (\bar{s}^\alpha \gamma_\mu P_L d^\alpha) \left[(\bar{d}^\beta \gamma^\mu P_R d^\beta) - (\bar{s}^\beta \gamma^\mu P_R s^\beta) \right].$$

Only $d + s$ operators mix into QCD- and QED-penguin operators $Q_{3,\dots10}$, whereas $d - s$ ones do not.

$$\boxed{\text{VLL}, d - s, \text{VLR}, d - s, \text{VLL}, d + s, \text{VLR}, d + s \Rightarrow 6}$$

Further $(sd)(uu)$ and $(sd)(cc)$ operators in analogy to $Q_{5,6}$ are

$$Q_{17} = Q_1^{\text{VLR},u-c} = (\bar{s}^\alpha \gamma_\mu P_L d^\beta) \left[(\bar{u}^\beta \gamma^\mu P_R u^\alpha) - (\bar{c}^\beta \gamma^\mu P_R c^\alpha) \right],$$
$$Q_{18} = Q_2^{\text{VLR},u-c} = (\bar{s}^\alpha \gamma_\mu P_L d^\alpha) \left[(\bar{u}^\beta \gamma^\mu P_R u^\beta) - (\bar{c}^\beta \gamma^\mu P_R c^\beta) \right], \qquad (13.50)$$

which do not mix into QCD- and QED-penguin operators $Q_{3,\dots10}$, but give threshold corrections when decoupling the c quark.

$$\boxed{\text{VLR}, u - c \Rightarrow 2}$$

The sector of scalar operators SLL for $(sd)(QQ)$ with $Q = u, c, b$ comprises

$$Q_1^{\text{SLL},Q} = (\bar{s}^\alpha P_L d^\beta)(\bar{Q}^\beta P_L Q^\alpha), \qquad (13.51)$$
$$Q_2^{\text{SLL},Q} = (\bar{s}^\alpha P_L d^\alpha)(\bar{Q}^\beta P_L Q^\beta), \qquad (13.52)$$
$$Q_3^{\text{SLL},Q} = -(\bar{s}^\alpha \sigma_{\mu\nu} P_L d^\beta)(\bar{Q}^\beta \sigma^{\mu\nu} P_L Q^\alpha), \qquad (13.53)$$
$$Q_4^{\text{SLL},Q} = -(\bar{s}^\alpha \sigma_{\mu\nu} P_L d^\alpha)(\bar{Q}^\beta \sigma^{\mu\nu} P_L Q^\beta), \qquad (13.54)$$

with the minus sign resulting from our definition of $\sigma_{\mu\nu}$, which differs by a factor i relative to the one in [744].

Due to Fierz symmetries there are only two operators $(sd)(QQ)$ with $Q = d, s$

$$Q_1^{\text{SLL},Q} = (\bar{s}^\alpha P_L d^\beta)(\bar{Q}^\beta P_L Q^\alpha), \qquad (13.55)$$
$$Q_2^{\text{SLL},Q} = (\bar{s}^\alpha P_L d^\alpha)(\bar{Q}^\beta P_L Q^\beta). \qquad (13.56)$$

$$\boxed{\text{SLL}, Q \ (Q = u, c, b) \text{ and SLL}, Q \ (Q = d, s) \Rightarrow 3 \times 4 + 2 \times 2 = 16}$$

The remaining operators are [744]

$$Q_1^{SLR,Q} = (\bar{s}^\alpha P_L d^\beta)(\bar{Q}^\beta P_R Q^\alpha), \tag{13.57}$$

$$Q_2^{SLR,Q} = (\bar{s}^\alpha P_L d^\alpha)(\bar{Q}^\beta P_R Q^\beta), \tag{13.58}$$

with $Q = u, c, b$.

$$\boxed{SLR, Q \ (Q = u, c, b) \Rightarrow 3 \times 2 = 6}$$

In total there are $8 + 2 + 6 + 2 + 16 + 6 = 40$ operators. Including chirality-flipped operators we end up with 80 operators. Their anomalous dimension matrices will be given in our future paper, and in view of many operators involved we will not list them here.

13.3.3 $N_f = 4$

The minimal basis for $N_f = 4$ contains 30 operators of a given chirality and their chirality-flipped counterparts. It can be found by dropping operators that are exclusively $(sd)(bb)$, which are 6: SLL,b (4) and SLR,b (2). Further, in QCD- and QED-penguin operators the sum runs now only over $q = u, d, s, c$ and in consequence some operators become linearly dependent

$$Q_{10} = Q_9 + Q_4 - Q_3 \tag{13.59}$$

$$Q_1^{VLL,d+s} = \frac{2}{3}(Q_3 - Q_9), \tag{13.60}$$

$$Q_1^{VLR,d+s} = \frac{2}{3}(Q_6 - Q_8), \tag{13.61}$$

$$Q_2^{VLR,d+s} = \frac{2}{3}(Q_5 - Q_7). \tag{13.62}$$

It seems convenient to discard these 4 linearly dependent operators, which also removes one source of mixing into QCD- and QED-penguin operators from the operators VLL,$d + s$ and VLR,$d + s$. This gives a minimal basis for $N_f = 4$.

The operators relevant for $K \to \pi\pi$ are $30 - 6 = 24$ because the SLL,c and SLR,c sectors do not mix into the other operators and have vanishing $K \to \pi\pi$ matrix elements.

13.3.4 $N_f = 3$

The minimal basis for $N_f = 3$ contains 20 operators of a given chirality and their chirality-flipped counterparts. One can start from $N_f = 4$ by dropping operators with exclusively $(sd)(cc)$, which are 6: SLL,c (4) and SLR,c (2). Further, in QCD- and QED-penguin operators the sum runs now only over $q = u, d, s$, and in consequence some operators become linearly dependent

$$Q_4 = Q_3 + Q_2 - Q_1, \tag{13.63}$$

$$Q_9 = \frac{1}{2}(3Q_1 - Q_1). \tag{13.64}$$

By decoupling the c-quark also $Q_{1,2}^{\text{VLR},u-c} \to Q_{1,2}^{\text{VLR},u}$, which are linearly dependent on

$$Q_1^{\text{VLR},u} = \frac{1}{3}(Q_6 + 2Q_8), \tag{13.65}$$

$$Q_2^{\text{VLR},u} = \frac{1}{3}(Q_5 + 2Q_7). \tag{13.66}$$

Thus the operators Q_4, Q_9, $Q_{1,2}^{\text{VLR},u}$ can be discarded from the list of linearly independent operators.

13.3.5 Summary of BSM Operators for $K \to \pi\pi$

As we have seen for $N_f = 3$, relevant for $K \to \pi\pi$ decays, there are thirteen BSM linearly independent four-quark operators that are also linearly independent from the SM ones, in particular none of them mixes into QCD- and QED-penguin operators $Q_{3,\ldots 10}$. Let us group them in classes as done in [246].

Class I

$$Q_1^{\text{VLL},d-s} = (\bar{s}^\alpha \gamma_\mu P_L d^\alpha) \left[(\bar{d}^\beta \gamma^\mu P_L d^\beta) - (\bar{s}^\beta \gamma^\mu P_L s^\beta) \right], \tag{13.67}$$

Class II

$$Q_1^{\text{SLR},u} = (\bar{s}^\alpha P_L d^\beta)(\bar{u}^\beta P_R u^\alpha), \tag{13.68}$$

$$Q_2^{\text{SLR},u} = (\bar{s}^\alpha P_L d^\alpha)(\bar{u}^\beta P_R u^\beta), \tag{13.69}$$

$$Q_1^{\text{VLR},d-s} = (\bar{s}^\alpha \gamma_\mu P_L d^\beta) \left[(\bar{d}^\beta \gamma^\mu P_R d^\alpha) - (\bar{s}^\beta \gamma^\mu P_R s^\alpha) \right], \tag{13.70}$$

$$Q_2^{\text{VLR},d-s} = (\bar{s}^\alpha \gamma_\mu P_L d^\alpha) \left[(\bar{d}^\beta \gamma^\mu P_R d^\beta) - (\bar{s}^\beta \gamma^\mu P_R s^\beta) \right], \tag{13.71}$$

Class III

$$Q_1^{\text{SLL},u} = (\bar{s}^\alpha P_L d^\beta)(\bar{u}^\beta P_L u^\alpha), \tag{13.72}$$

$$Q_2^{\text{SLL},u} = (\bar{s}^\alpha P_L d^\alpha)(\bar{u}^\beta P_L u^\beta), \tag{13.73}$$

$$Q_3^{\text{SLL},u} = -(\bar{s}^\alpha \sigma_{\mu\nu} P_L d^\beta)(\bar{u}^\beta \sigma^{\mu\nu} P_L u^\alpha), \tag{13.74}$$

$$Q_4^{\text{SLL},u} = -(\bar{s}^\alpha \sigma_{\mu\nu} P_L d^\alpha)(\bar{u}^\beta \sigma^{\mu\nu} P_L u^\beta), \tag{13.75}$$

Class IV

$$Q_1^{\text{SLL},d} = (\bar{s}^\alpha P_L d^\beta)(\bar{d}^\beta P_L d^\alpha), \tag{13.76}$$

$$Q_2^{\text{SLL},d} = (\bar{s}^\alpha P_L d^\alpha)(\bar{d}^\beta P_L d^\beta), \tag{13.77}$$

$$Q_1^{\text{SLL},s} = (\bar{s}^\alpha P_L d^\beta)(\bar{s}^\beta P_L s^\alpha), \tag{13.78}$$

$$Q_2^{\text{SLL},s} = (\bar{s}^\alpha P_L d^\alpha)(\bar{s}^\beta P_L s^\beta). \tag{13.79}$$

The one-loop anomalous dimensions for all these BSM operators are then given in units of $\alpha_s/4\pi$) as follows [246, 744, 756]

Class I

$$\gamma^{(0)}(Q_1^{\text{VLL,d-s}}) = 4,$$ (13.80)

Class II

$$\hat{\gamma}^{(0)}(Q_1^{\text{SLR,u}}, Q_2^{\text{SLR,u}}) = \begin{pmatrix} \frac{6}{N} & -6 \\ 0 & -6N + \frac{6}{N} \end{pmatrix} = \begin{pmatrix} 2 & -6 \\ 0 & -16 \end{pmatrix},$$ (13.81)

$$\hat{\gamma}^{(0)}(Q_1^{\text{VLR,d-s}}, Q_2^{\text{VLR,d-s}}) = \begin{pmatrix} -6N + \frac{6}{N} & 0 \\ -6 & \frac{6}{N} \end{pmatrix} = \begin{pmatrix} -16 & 0 \\ -6 & 2 \end{pmatrix},$$ (13.82)

where N denotes the number of colors with $N = 3$.

Class III
In the basis $(Q_1^{\text{SLL,u}}, Q_2^{\text{SLL,u}}, Q_3^{\text{SLL,u}}, Q_4^{\text{SLL,u}})$ we have

$$\hat{\gamma}^{(0)\text{SLL,u}} = \begin{pmatrix} \frac{6}{N} & -6 & \frac{N}{2} - \frac{1}{N} & \frac{1}{2} \\ 0 & -6N + \frac{6}{N} & 1 & -\frac{1}{N} \\ -\frac{48}{N} + 24N & 24 & -\frac{2}{N} - 4N & 6 \\ 48 & -\frac{48}{N} & 0 & 2N - \frac{2}{N} \end{pmatrix}$$ (13.83)

$$= \begin{pmatrix} 2 & -6 & 7/6 & 1/2 \\ 0 & -16 & 1 & -1/3 \\ 56 & 24 & -38/3 & 6 \\ 48 & -16 & 0 & 16/3 \end{pmatrix}.$$ (13.84)

Class IV

$$\hat{\gamma}^{(0)}(Q_1^{\text{SLL,d}}, Q_2^{\text{SLL,d}}) = \begin{pmatrix} 2N + 4 + \frac{2}{N} & 4N - 4 - \frac{8}{N} \\ 4 - \frac{8}{N} & -6N + 8 + \frac{2}{N} \end{pmatrix}$$ (13.85)

$$= \begin{pmatrix} 32/3 & 16/3 \\ 4/3 & -28/3 \end{pmatrix},$$ (13.86)

with the same matrix for the operators $Q_{1,2}^{\text{SLL,s}}$.

We observe that some entries in these matrices are large, implying strong RG effects. A detailed numerical analysis of these effects at the NLO level should be available in the hep-arxiv within 2020. The values of their $K \to \pi\pi$ matrix elements, evaluated in the chiral limit in the DQCD approach, are collected in [246].

IT IS TIME TO LEARN THE BSM LANGUAGE: SMEFT!

14 Standard Model Effective Field Theory

14.1 Basic Framework

If there is a separation of scales between the electroweak scale v and the NP scale Λ, as is suggested by the absence of any new particles close to the electroweak scale in LHC searches so far, one can investigate the implications of the unbroken SM gauge symmetry on the correlations between various observables. To study these correlations in a model-independent manner, one can consider an OPE with dimension-5, dimension-6 or higher-dimension operators built out of SM fields that are invariant under the full SM gauge symmetry. This corresponds to an effective field theory where all the SM degrees of freedom are kept as dynamical degrees of freedom, and only the NP is integrated out.

The application of EFT approach [113, 114] here not only allows to resum large logarithms appearing if NP scales are much higher than the EW scale, but also serves as a convenient intermediate step between an ultraviolet completion formulated at a high NP scale and low-energy phenomenology. The type of EFT considered here is usually called SMEFT because at low-energies, it should reduce to the SM, provided no undiscovered but weakly coupled *light* particles exist, like axions or sterile neutrinos. Yet, as we will see later, the presence of new operators, in particular dimension-6 ones, whose Wilson coefficients can in principle be arbitrary, can introduce very significant modifications of SM predictions for flavor observables and also influence the values of SM parameters. Therefore the name SMEFT is in fact a misnomer because the Wilson coefficients of involved operators contain information about NP beyond the SM, and also there is a multitude of operators that are strongly suppressed in the SM but can be important in SMEFT. But we will use this name in what follows as it is used in the literature. A recent review of SMEFT with applications to Higgs physics can be found in [757]. Here we will concentrate on flavour physics and generally low-energy processes. In any case if no new particles will be found at the LHC in the coming years, the language of the SMEFT will dominate elementary particle physics in the coming decades, and it is crucial to learn it as soon as possible.

The effective Lagrangian of SMEFT at dimension-6 is given as follows

$$\mathcal{L}^{(6)} = \frac{1}{\Lambda^2} \sum_k C_k^{(6)} Q_k^{(6)}, \tag{14.1}$$

with the contributing operators classified in full generality in [561, 562]. The second of these papers removed certain redundant operators present in the first one, and we will use the results of [562] here. The corresponding renormalization group analysis at LO of all these operators has been analyzed in [758–760].

For three generations of fermions there are 2,499 independent operators $Q_k^{(6)}$ (151 irreducible flavour representations) that do not violate baryon and lepton number. This means that the RG analysis involves in full generality a $2,499 \times 2,499$ anomalous matrix with the evolution governed by the Higgs self-coupling λ [758], Yukawa couplings [759], and SM gauge interactions [760]. While several of the entries in this matrix have been known before, not only at LO and already appeared in previous sections, these papers provided a large number of new entries so that RG effects can be investigated, albeit only at LO, in full generality.

The Wilson coefficients $C_k^{(6)}$ contain 1,350 CP-even and 1,149 CP-odd parameters. They contain the information about NP at and above scales $O(\Lambda)$. In this totally model-independent approach, these coefficients should be determined from the pattern of deviations from SM expectations at the electroweak scale and low-energy scales considered in this book. Clearly, this goal cannot be realized in view of very many free parameters, but as we will see later, this classification of operators and the results for RG effects turn out to be very useful in concrete models.

It should also be emphasized that in addition to the 2,499 operators in question, there are 273 D = 6 $\Delta B = \Delta L = 1$ operators (7 irreducible flavor representations) and their Hermitian conjugates, which are the leading operators that permit proton decay in SMEFT [561, 562, 761–764].

The important point stressed already in the past and in particular emphasized in [760] is that RG evolution between the NP scale Λ and the electroweak scale involves operator mixing, and the pattern of deviations from SM expectations observed at the electroweak scale and low-energy scales can differ significantly from the pattern of Wilson coefficients at the NP scale Λ. Here we just quote one example known already for decades in the context of QCD RG evolution. Tiny effects of right-handed currents originating at a very high scale Λ, when accompanied by left-handed currents, also generated by NP or present in the SM, can in $\Delta F = 2$ observables cause very large effects in flavor observables. This is due to the presence of left-right operators with enhanced Wilson coefficients through RG evolution to scales $O(1\,\text{GeV})$ and chirally enhanced hadronic matrix elements, in particular in the K-meson system. See previous chapter.

However, the point made in [760] and in subsequent papers [484, 666, 719, 765–770] is not related to QCD but in particular to top Yukawa couplings. Even in the absence of left-handed currents generated by NP, the presence of right-handed currents in the presence of Yukawa couplings can generate in the process of RG evolution left-right operators at the electroweak scale, because left-handed currents are already present in the SM. Moreover, this effect is further enhanced through QCD RG evolution to low-energy scales as stated earlier. This underlines the importance of RG studies in the search for NP far beyond the LHC scales. As stressed in particular in [770], the bounds from $\Delta F = 2$ processes can affect in turn $\Delta F = 1$ and $\Delta F = 0$ processes.

I find the analyses in [562, 758–760] impressive, but I do not think that they will teach us much about NP at very high scales in the context of a bottom-up approach in which one wants to be totally model independent. Exceptions are observables in which only a very limited number of operators is allowed to contribute because of the special structure of given observables or additional assumptions beyond SM gauge invariance, in particular flavor symmetries and their breakdown. But these analyses form an important basis for a top-down approach in which some model assumptions at high scales are made and in particular are very useful in concrete NP models. While such a strategy towards the identification of NP beyond the LHC requires the study of many models, the number of models studied until now is by roughly two orders of magnitude lower than the numbers of free parameters involved in a general bottom-up approach.

In this section we will list all the operators collected in [562], and we will describe the results of the analyses in [758–760] in general terms. Subsequently, we will present several examples of RG equations in which only few operators will have nonvanishing coefficients at the high scale Λ, but at low-energy new operators will contribute to observables with Wilson coefficients generated through the RG evolution in question. In choosing these examples, we anticipated their application in concrete models to be presented partly in subsequent chapters and partly found in the rich literature. Of particular interest are the following scenarios:

Scenario I

At the high NP scale Λ only neutral flavor-violating right-handed (RH) quark currents are generated by NP with profound implications for $\Delta F = 2$ transitions as advertised earlier and their specific correlations with $\Delta F = 1$ observables. Concrete models with this structure are models with flavor-violating Z couplings considered in Section 15.5 and in particular models with vector-like quarks that we will analyze in Section 16.3.

Scenario II

At the high NP scale Λ only neutral flavor-violating left-handed (LH) quark currents are generated by NP with interesting but different implications for the correlation between $\Delta F = 2$ and $\Delta F = 1$ transitions than found in Scenario I. The corresponding models with vector-like quarks will be analyzed in Section 16.3. These results will also be important for Section 15.5 in which FCNC mediated by the Z boson will be considered.

Scenario III

This scenario is like Scenarios I and II but involving neutral flavor-violating RH and LH lepton currents instead of quark currents with implications through RG evolution on semi-leptonic operators. Concrete models of this scenario are the ones involving vector-like leptons. We will comment on them in Section 16.3.

Scenario IV

At high NP scale Λ only semileptonic operators are present as found in leptoquark models that will be analyzed in Section 16.4. RG evolution in the presence of Yukawa couplings and gauge interactions has then impact on lepton-flavor-violating lepton decays and implies modifications of SM couplings. This case is important in the context of the B physics anomalies that require enhanced contributions from semileptonic operators that in turn can have impact on lepton decays in question implying enhancements of the latter.

The experimental bounds on lepton decays can in turn bound the required enhancements of semileptonic B decays as stressed in particular in [765, 767].

For instance, LFU breaking effects in $\tau \rightarrow \ell\bar{\nu}\nu$ (with $\ell_{1,2} = e, \mu$) are described by the observables

$$R_\tau^{\tau/\ell_{1,2}} = \frac{\mathcal{B}(\tau \rightarrow \ell_{2,1}\nu\bar{\nu})_{\text{exp}}/\mathcal{B}(\tau \rightarrow \ell_{2,1}\nu\bar{\nu})_{\text{SM}}}{\mathcal{B}(\mu \rightarrow e\nu\bar{\nu})_{\text{exp}}/\mathcal{B}(\mu \rightarrow e\nu\bar{\nu})_{\text{SM}}} \tag{14.2}$$

and are experimentally tested at the few permille level [716]

$$R_\tau^{\tau/\mu} = 1.0022 \pm 0.0030, \quad R_\tau^{\tau/e} = 1.0060 \pm 0.0030. \tag{14.3}$$

Similar, modifications of the leptonic Z couplings are constrained by the LEP measurements of the Z decay widths, left-right, and forward-backward asymmetries. The bounds on lepton nonuniversal couplings are according to PDG

$$\frac{v_\tau}{v_e} = 0.959\,(29), \quad \frac{a_\tau}{a_e} = 1.0019\,(15), \tag{14.4}$$

where v_ℓ and a_ℓ are the vector and axial-vector couplings, respectively, defined as $v_\ell = g_{\ell L}^{\ell\ell} + g_{\ell R}^{\ell\ell}$ and $a_\ell = g_{\ell L}^{\ell\ell} - g_{\ell R}^{\ell\ell}$. These few examples show how careful one has to be not to violate the existing bounds. Further examples can be found in [765, 767].

Scenario V

Again, as in Scenario IV, at the high scale only semileptonic operators are present, but this time we will study the impact of RG evolution on $\Delta F = 1$ nonleptonic decays like $K_L \rightarrow \pi\pi$ and the ratio ε'/ε. This case is again interesting in the context of leptoquark models and will be briefly discussed in Section 16.4.13, with more details found in [667].

Scenario VI

In this scenario we will consider only operators relevant for $b \rightarrow s\nu\bar{\nu}$ and $b \rightarrow s\ell^+\ell^-$ transitions with the goal to identify some correlations between them that follow from SM gauge invariance. Indeed the $b \rightarrow s\nu\bar{\nu}$ transition is closely related to the $b \rightarrow s\ell^+\ell^-$ transition because the neutrinos and left-handed charged leptons are related by $\text{SU}(2)_L$ symmetry.

The same comments apply to the $d \rightarrow s\nu\bar{\nu}$ and $d \rightarrow s\ell^+\ell^-$ transitions. While, already at present, the correlations between these transitions imply stringent constraints on parameters of specific NP scenarios, significant LD uncertainties in $d \rightarrow s\ell^+\ell^-$ and no real plans for measuring $K_L \rightarrow \pi^0\ell^+\ell^-$ in the near future make these connections less interesting than the B physics case at present. Yet, there exist dedicated analyses of such correlations in the literature, and we will refer to them as we proceed.

Scenario VII

Of particular interest are scenarios that imply correlations between $b \rightarrow s\nu\bar{\nu}$ transitions studied at BELLE II and $d \rightarrow s\nu\bar{\nu}$ investigated at NA62 and KOTO. In MFV scenarios and simplified NP scenarios, these correlations are very stringent [216, 601]. Involving two meson systems, they test the flavor structure of different models.

Definitely, these are just few simplified scenarios. In reality a given NP model may contain several of these scenarios that are correlated with each other and contain still other scenarios not listed here. The analysis of such models becomes then very involved, and one has to develop efficient computer codes to get useful results. We will list existing codes in the course of our presentation. Yet, we think that already these scenarios will give us some insight in the dynamics involved and even excluding some of them one day will teach us something.

Significant part of this chapter resulted from the collaboration with Christoph Bobeth and benefited from papers and discussions with other colleagues, whose names will appear later.

14.2 Full Set of Dimension-6 Operators

We will now list the operators $Q_k^{(6)}$ classified in [562] and listed in Tables 14.1 and 14.2. The sign conventions for covariant derivatives in that paper differ from ours and correspond to the conventions 2 in Section 1.8, and these are also the conventions used in [758–760]. But it is easy to translate these conventions to ours by simply reversing the signs of all couplings, and we will do it to be consistent with the rest of the book. The reader may check by comparing our expressions with those in the papers that indeed this is what has to be changed. We also warn the reader that our Yukawa matrices, like Y_u, are defined as the Hermitean conjugate w.r.t. [759]. But the most important warning is the following one:

The Wilson coefficients of the operators in the SMEFT are defined by the Lagrangian and not by the Hamiltonian as used in the rest of this book!

Table 14.1 Dimension-6 operators other than the four-fermion ones.					
	X^3		φ^6 and $\varphi^4 D^2$		$\psi^2 \varphi^3$
O_G	$f^{ABC} G_\mu^{A\nu} G_\nu^{B\rho} G_\rho^{C\mu}$	O_φ	$(\varphi^\dagger \varphi)^3$	$O_{e\varphi}$	$(\varphi^\dagger \varphi)(\bar{l}_p e_r \varphi)$
$O_{\widetilde{G}}$	$f^{ABC} \widetilde{G}_\mu^{A\nu} G_\nu^{B\rho} G_\rho^{C\mu}$	$O_{\varphi\Box}$	$(\varphi^\dagger \varphi)\Box(\varphi^\dagger \varphi)$	$O_{u\varphi}$	$(\varphi^\dagger \varphi)(\bar{q}_p u_r \widetilde{\varphi})$
O_W	$\epsilon^{IJK} W_\mu^{I\nu} W_\nu^{J\rho} W_\rho^{K\mu}$	$O_{\varphi D}$	$\left(\varphi^\dagger D^\mu \varphi\right)^\star \left(\varphi^\dagger D_\mu \varphi\right)$	$O_{d\varphi}$	$(\varphi^\dagger \varphi)(\bar{q}_p d_r \varphi)$
$O_{\widetilde{W}}$	$\epsilon^{IJK} \widetilde{W}_\mu^{I\nu} W_\nu^{J\rho} W_\rho^{K\mu}$				
	$X^2 \varphi^2$		$\psi^2 X \varphi$		$\psi^2 \varphi^2 D$
$O_{\varphi G}$	$\varphi^\dagger \varphi\, G_{\mu\nu}^A G^{A\mu\nu}$	O_{eW}	$(\bar{\ell}_p \sigma^{\mu\nu} e_r)\tau^I \varphi W_{\mu\nu}^I$	$O_{\varphi l}^{(1)}$	$(\varphi^\dagger i \overleftrightarrow{D}_\mu \varphi)(\bar{\ell}_p \gamma^\mu \ell_r)$
$O_{\varphi \widetilde{G}}$	$\varphi^\dagger \varphi\, \widetilde{G}_{\mu\nu}^A G^{A\mu\nu}$	O_{eB}	$(\bar{\ell}_p \sigma^{\mu\nu} e_r)\varphi B_{\mu\nu}$	$O_{\varphi l}^{(3)}$	$(\varphi^\dagger i \overleftrightarrow{D}_\mu^I \varphi)(\bar{\ell}_p \tau^I \gamma^\mu \ell_r)$
$O_{\varphi W}$	$\varphi^\dagger \varphi\, W_{\mu\nu}^I W^{I\mu\nu}$	O_{uG}	$(\bar{q}_p \sigma^{\mu\nu} T^A u_r)\widetilde{\varphi}\, G_{\mu\nu}^A$	$O_{\varphi e}$	$(\varphi^\dagger i \overleftrightarrow{D}_\mu \varphi)(\bar{e}_p \gamma^\mu e_r)$
$O_{\varphi \widetilde{W}}$	$\varphi^\dagger \varphi\, \widetilde{W}_{\mu\nu}^I W^{I\mu\nu}$	O_{uW}	$(\bar{q}_p \sigma^{\mu\nu} u_r)\tau^I \widetilde{\varphi}\, W_{\mu\nu}^I$	$O_{\varphi q}^{(1)}$	$(\varphi^\dagger i \overleftrightarrow{D}_\mu \varphi)(\bar{q}_p \gamma^\mu q_r)$
$O_{\varphi B}$	$\varphi^\dagger \varphi\, B_{\mu\nu} B^{\mu\nu}$	O_{uB}	$(\bar{q}_p \sigma^{\mu\nu} u_r)\widetilde{\varphi}\, B_{\mu\nu}$	$O_{\varphi q}^{(3)}$	$(\varphi^\dagger i \overleftrightarrow{D}_\mu^I \varphi)(\bar{q}_p \tau^I \gamma^\mu q_r)$
$O_{\varphi \widetilde{B}}$	$\varphi^\dagger \varphi\, \widetilde{B}_{\mu\nu} B^{\mu\nu}$	O_{dG}	$(\bar{q}_p \sigma^{\mu\nu} T^A d_r)\varphi\, G_{\mu\nu}^A$	$O_{\varphi u}$	$(\varphi^\dagger i \overleftrightarrow{D}_\mu \varphi)(\bar{u}_p \gamma^\mu u_r)$
$O_{\varphi W B}$	$\varphi^\dagger \tau^I \varphi\, W_{\mu\nu}^I B^{\mu\nu}$	O_{dW}	$(\bar{q}_p \sigma^{\mu\nu} d_r)\tau^I \varphi\, W_{\mu\nu}^I$	$O_{\varphi d}$	$(\varphi^\dagger i \overleftrightarrow{D}_\mu \varphi)(\bar{d}_p \gamma^\mu d_r)$
$O_{\varphi \widetilde{W} B}$	$\varphi^\dagger \tau^I \varphi\, \widetilde{W}_{\mu\nu}^I B^{\mu\nu}$	O_{dB}	$(\bar{q}_p \sigma^{\mu\nu} d_r)\varphi\, B_{\mu\nu}$	$O_{\varphi ud}$	$i(\widetilde{\varphi}^\dagger D_\mu \varphi)(\bar{u}_p \gamma^\mu d_r)$

Table 14.2 Four-fermion operators.

$(\bar{L}L)(\bar{L}L)$		$(\bar{R}R)(\bar{R}R)$		$(\bar{L}L)(\bar{R}R)$	
O_{ll}	$(\bar{\ell}_p\gamma_\mu\ell_r)(\bar{\ell}_s\gamma^\mu\ell_t)$	O_{ee}	$(\bar{e}_p\gamma_\mu e_r)(\bar{e}_s\gamma^\mu e_t)$	O_{le}	$(\bar{\ell}_p\gamma_\mu\ell_r)(\bar{e}_s\gamma^\mu e_t)$
$O_{qq}^{(1)}$	$(\bar{q}_p\gamma_\mu q_r)(\bar{q}_s\gamma^\mu q_t)$	O_{uu}	$(\bar{u}_p\gamma_\mu u_r)(\bar{u}_s\gamma^\mu u_t)$	O_{lu}	$(\bar{\ell}_p\gamma_\mu\ell_r)(\bar{u}_s\gamma^\mu u_t)$
$O_{qq}^{(3)}$	$(\bar{q}_p\gamma_\mu\tau^I q_r)(\bar{q}_s\gamma^\mu\tau^I q_t)$	O_{dd}	$(\bar{d}_p\gamma_\mu d_r)(\bar{d}_s\gamma^\mu d_t)$	O_{ld}	$(\bar{\ell}_p\gamma_\mu\ell_r)(\bar{d}_s\gamma^\mu d_t)$
$O_{lq}^{(1)}$	$(\bar{\ell}_p\gamma_\mu\ell_r)(\bar{q}_s\gamma^\mu q_t)$	O_{eu}	$(\bar{e}_p\gamma_\mu e_r)(\bar{u}_s\gamma^\mu u_t)$	O_{qe}	$(\bar{q}_p\gamma_\mu q_r)(\bar{e}_s\gamma^\mu e_t)$
$O_{lq}^{(3)}$	$(\bar{\ell}_p\gamma_\mu\tau^I\ell_r)(\bar{q}_s\gamma^\mu\tau^I q_t)$	O_{ed}	$(\bar{e}_p\gamma_\mu e_r)(\bar{d}_s\gamma^\mu d_t)$	$O_{qu}^{(1)}$	$(\bar{q}_p\gamma_\mu q_r)(\bar{u}_s\gamma^\mu u_t)$
		$O_{ud}^{(1)}$	$(\bar{u}_p\gamma_\mu u_r)(\bar{d}_s\gamma^\mu d_t)$	$O_{qu}^{(8)}$	$(\bar{q}_p\gamma_\mu T^A q_r)(\bar{u}_s\gamma^\mu T^A u_t)$
		$O_{ud}^{(8)}$	$(\bar{u}_p\gamma_\mu T^A u_r)(\bar{d}_s\gamma^\mu T^A d_t)$	$O_{qd}^{(1)}$	$(\bar{q}_p\gamma_\mu q_r)(\bar{d}_s\gamma^\mu d_t)$
				$O_{qd}^{(8)}$	$(\bar{q}_p\gamma_\mu T^A q_r)(\bar{d}_s\gamma^\mu T^A d_t)$

$(\bar{L}R)(\bar{R}L)$ and $(\bar{L}R)(\bar{L}R)$			B-violating	
O_{ledq}	$(\bar{\ell}_p^j e_r)(\bar{d}_s q_t^j)$		$\varepsilon^{\alpha\beta\gamma}\varepsilon_{jk}\left[(d_p^\alpha)^T C u_r^\beta\right]\left[(q_s^{\gamma j})^T C\ell_t^k\right]$	
$O_{quqd}^{(1)}$	$(\bar{q}_p^j u_r)\varepsilon_{jk}(\bar{q}_s^k d_t)$	O_{qqu}	$\varepsilon^{\alpha\beta\gamma}\varepsilon_{jk}\left[(q_p^{\alpha j})^T C q_r^{\beta k}\right]\left[(u_s^\gamma)^T C e_t\right]$	
$O_{quqd}^{(8)}$	$(\bar{q}_p^j T^A u_r)\varepsilon_{jk}(\bar{q}_s^k T^A d_t)$	O_{qqq}	$\varepsilon^{\alpha\beta\gamma}\varepsilon_{jn}\varepsilon_{km}\left[(q_p^{\alpha j})^T C q_r^{\beta k}\right]\left[(q_s^{\gamma m})^T C\ell_t^n\right]$	
$O_{lequ}^{(1)}$	$(\bar{\ell}_p^j e_r)\varepsilon_{jk}(\bar{q}_s^k u_t)$	O_{duu}	$\varepsilon^{\alpha\beta\gamma}\left[(d_p^\alpha)^T C u_r^\beta\right]\left[(u_s^\gamma)^T C e_t\right]$	
$O_{lequ}^{(3)}$	$(\bar{\ell}_p^j\sigma_{\mu\nu}e_r)\varepsilon_{jk}(\bar{q}_s^k\sigma^{\mu\nu}u_t)$			

Yet, it is very straightforward to take this difference (just the sign) into account in the process of the matching of SMEFT Wilson coefficients onto the ones in the low-energy effective theory (LEFT) describing the physics below the electroweak scale. This will be evident from examples presented by us. To make it crystal clear, in contrast to conventions in covariant derivatives and Yukawa couplings, the signs of Wilson coefficients of the operators in the SMEFT presented by us are as in [758–760]. But when matching on LEFT, Wilson coefficients considered by us so far and defined through effective Hamiltonians, we will take this sign difference into account.

The well-known expression for the SM Lagrangian before electroweak spontaneous symmetry breakdown (EWSB) reads

$$\mathcal{L}_{SM}^{(4)} = -\frac{1}{4}G_{\mu\nu}^A G^{A\mu\nu} - \frac{1}{4}W_{\mu\nu}^I W^{I\mu\nu} - \frac{1}{4}B_{\mu\nu}B^{\mu\nu} + \left(D_\mu\varphi\right)^\dagger\left(D^\mu\varphi\right) + m^2\varphi^\dagger\varphi - \frac{1}{2}\lambda\left(\varphi^\dagger\varphi\right)^2$$
$$+ i\left(\bar{l}\slashed{D}l + \bar{e}\slashed{D}e + \bar{q}\slashed{D}q + \bar{u}\slashed{D}u + \bar{d}\slashed{D}d\right) - \left(\bar{l}\,Y_e e\varphi + \bar{q}\,Y_u u\widetilde{\varphi} + \bar{q}\,Y_d d\varphi + \text{h.c.}\right),$$

(14.5)

where the Yukawa couplings $Y_{e,u,d}$ are matrices in the generation space. As stated earlier, our sign convention for covariant derivatives differs from [562] as we use

$$\left(D_\mu q\right)^{\alpha j} = \left(\partial_\mu - ig_s t_{\alpha\beta}^A G_\mu^A - ig S_{jk}^I W_\mu^I - ig' Y_q B_\mu\right)q^{\beta k},$$

(14.6)

so that in front of gauge couplings in the vertices there are no minus signs. Here,

$$t^A = \frac{1}{2}\lambda^A, \qquad S^I = \frac{1}{2}\tau^I$$

(14.7)

are the SU(3) and SU(2) generators, while λ^A and $\tau^I = \sigma^I$ are the Gell-Mann and Pauli matrices, respectively.[1] Complex conjugate of the Higgs field occurs either as $\varphi^\dagger$ or $\widetilde{\varphi}$, where $\widetilde{\varphi}^j = \varepsilon_{jk}(\varphi^k)^\star$, and ε_{jk} is totally antisymmetric with $\varepsilon_{12} = +1$.

It is useful to define Hermitian derivative terms that contain $\varphi^\dagger \overleftarrow{D}_\mu \varphi \equiv (D_\mu \varphi)^\dagger \varphi$ as follows:

$$\varphi^\dagger i \overleftrightarrow{\mathcal{D}}_\mu \varphi \equiv i\varphi^\dagger \left(D_\mu - \overleftarrow{D}_\mu \right) \varphi \quad \text{and} \quad \varphi^\dagger i \overleftrightarrow{\mathcal{D}}^I_\mu \varphi \equiv i\varphi^\dagger \left(\tau^I D_\mu - \overleftarrow{D}_\mu \tau^I \right) \varphi. \quad (14.8)$$

In the following formulas we will often set $\varphi = H$ as this notation is also used in the literature.

The gauge field-strength tensors and their covariant derivatives read

$$\begin{aligned}
G^A_{\mu\nu} &= \partial_\mu G^A_\nu - \partial_\nu G^A_\mu + g_3 f^{ABC} G^B_\mu G^C_\nu, \quad \left(D_\rho G_{\mu\nu} \right)^A = \partial_\rho G^A_{\mu\nu} + g_3 f^{ABC} G^B_\rho G^C_{\mu\nu}, \\
W^I_{\mu\nu} &= \partial_\mu W^I_\nu - \partial_\nu W^I_\mu + g_2 \varepsilon^{IJK} W^J_\mu W^K_\nu, \quad \left(D_\rho W_{\mu\nu} \right)^I = \partial_\rho W^I_{\mu\nu} + g_2 \varepsilon^{IJK} W^J_\rho W^K_{\mu\nu}, \\
B_{\mu\nu} &= \partial_\mu B_\nu - \partial_\nu B_\mu, \quad\quad\quad\quad\quad\quad\quad D_\rho B_{\mu\nu} = \partial_\rho B_{\mu\nu}.
\end{aligned}$$

$$(14.9)$$

Again the signs in front of gauge couplings differ from those in [562].

Dual tensors are defined by

$$\widetilde{X}_{\mu\nu} = \frac{1}{2}\varepsilon_{\mu\nu\rho\sigma} X^{\rho\sigma}, \quad\quad \varepsilon_{0123} = +1, \quad\quad (14.10)$$

where X stands for G^A, W^I, or B.

14.3 Rotations in the Flavor Space

14.3.1 SM Repetition

We have now specified the set of linearly independent operators. Yet this is not the full story. To proceed we have to specify the flavor basis in which we plan to perform calculations, including the RG evolution above the electroweak scale. To this end let us repeat what we already stated in Section 2.5 in the context of the SM.

The SM interaction Lagrangian of (2.3) is invariant under a $[U(3)]^5$ flavor symmetry

$$Q_L \to V^u_L Q, \quad u_R \to V^u_R u_R, \quad d_R \to V^d_R d_R, \quad\quad (14.11)$$

$$L \to V^e_L L, \quad e_R \to V^e_R e_R, \quad\quad (14.12)$$

where $V^u_L, V^u_R, V^d_R, V^e_L$, and V^e_R are unitary 3×3 matrices. This is the consequence of the fact that there is the universality of the gauge couplings in a given family of fermions, left-handed or right-handed. In the SM the Yukawa sector (2.13) breaks flavor universality in a given family, and consequently $[U(3)]^5$ symmetry explicitly simply because the Yukawa

---

[1] In the rest of the book we use σ^I for Pauli matrices but not to confuse the notation with $\sigma_{\mu\nu}$ present in the operators we use the symbol τ^I for them here.

couplings to fermions must be flavor nonuniversal to reproduce the known mass spectrum of quarks and leptons. The preferred basis in which calculations are performed is the mass eigenstate basis in which the Yukawa and consequently mass matrices are diagonalized as explicitly given by

$$(V_L^d)^\dagger Y^D V_R^d = \hat{Y}^D, \quad (V_L^u)^\dagger Y^U V_R^u = \hat{Y}^U, \quad (V_L^e)^\dagger Y^E V_R^e = \hat{Y}^E, \tag{14.13}$$

with $\hat{Y}^i$ being diagonal.

Now because of the universality of gauge couplings and the unitarity of rotation matrices, FCNC are absent and appear only in the charged currents parametrized by CKM and PMNS matrices. It should be stressed that it is irrelevant whether we rotate the down-quarks from flavor to mass eigenstates as in (2.45) and assume flavor and mass eigenstates in the up-quark system to be equal, or rotate the up-quarks only. The physics remains unchanged. The same applies to the lepton sector.

14.3.2 Light Z' Boson with Flavor Nonuniversal Couplings

Let us next add to the SM spectrum a new neutral gauge boson Z' resulting from a $U(1)'$ gauge symmetry.[2] The Lagrangian describing the interaction of Z' with the SM fermions is given by [3]

$$\mathcal{L}(Z') = \sum_{i,j,\psi_L} \Delta_L^{ij}(Z') \, \bar{\psi}_L^i \gamma^\mu P_L \psi_L^j Z'_\mu + \sum_{i,j,\psi_R} \Delta_R^{ij}(Z') \, \bar{\psi}_R^i \gamma^\mu P_R \psi_R^j Z'_\mu, \tag{14.14}$$

where $P_{L,R} = (1 \mp \gamma_5)/2$. Here ψ represent classes of fermions with the same electric charge, i.e., u, d, e, ν, while i, j are generation indices: $1, 2, 3$. For instance, $u_3 = t$ and $e_2 = \mu$. We assume first for simplicity that Z' is not much heavier than the top quark so that it can be integrated out at the electroweak scale.

The couplings $\Delta_{L,R}^{ij}(Z')$ are given in the interaction basis as follows

$$\Delta_L^{ij}(Z') = g_{Z'} z_{\psi_L^i} \delta^{ij}, \qquad \Delta_R^{ij}(Z') = g_{Z'} z_{\psi_R^i} \delta^{ij}, \tag{14.15}$$

with $z_{\psi_L^i}$ and $z_{\psi_R^i}$ being fermion charges under $U(1)'$ corresponding to Z'.

Performing then rotations (14.11) and (14.12) to the mass eigenstate basis we find instead of (14.15) [771]

$$\Delta_L^{ij}(Z') = g_{Z'} B_L^{ij}(\psi_L), \qquad \Delta_R^{ij}(Z') = g_{Z'} B_R^{ij}(\psi_R), \tag{14.16}$$

where

$$B_L^{ij}(\psi_L) = [(V_L^\psi)^\dagger \hat{Z}_L^\psi V_L^\psi]^{ij}, \qquad B_R^{ij}(\psi_R) = [(V_R^\psi)^\dagger \hat{Z}_R^\psi V_R^\psi]^{ij}. \tag{14.17}$$

$\hat{Z}_{\psi,L}$ and $\hat{Z}_{\psi,R}$ are diagonal matrices

$$\hat{Z}_L^\psi = \mathrm{diag}[z_{\psi_L^1}, z_{\psi_L^2}, z_{\psi_L^3}], \qquad \hat{Z}_R^\psi = \mathrm{diag}[z_{\psi_R^1}, z_{\psi_R^2}, z_{\psi_R^3}], \tag{14.18}$$

with the diagonal elements composed of the $U(1)'$ charges of the SM fermions.

---

[2] Based on collaboration with Jason Aebischer, Maria Cerdá-Sevilla, and Fulvia De Fazio.
[3] The physics of Z' will be discussed in a general context in Section 15.4.

Let us next assume that all these charges differ from each other so that the invariance under the $[U(3)]^5$ flavor symmetry is broken explicitly. To see the consequences of this breakdown, let us choose $V_L^u = \hat{1}$ and $V_R^u = \hat{1}$. This means that the Yukawa matrix or equivalently the mass matrix for up-quarks in this flavor basis is diagonal, and the same applies to the interactions of up-quarks with Z'. There is no flavor violation in the up-quark sector mediated by the Z' up to RG effects discussed later. But with $V_L^u = \hat{1}$ we have $V_L^d = V_{CKM}$. Therefore, inserted into (14.16) and (14.17) we find FCNC transitions in the down-quark sector with

$$B_L^{ij}(\psi_L) = [(V_{CKM})^\dagger \hat{Z}_L^d (V_{CKM})]^{ij}, \qquad B_R^{ij}(\psi_R) = [(V_R^d)^\dagger \hat{Z}_R^d V_R^d]^{ij}, \tag{14.19}$$

with $(i, j = d, s, b)$ and $\hat{Z}_{L,R}^d$ being diagonal matrices collecting $U(1)'$ charges of left-handed and right-handed down-quarks.

However, $V_L^u = \hat{1}$ and $V_R^u = \hat{1}$ is an assumption that specifies our model. It assumes that in the basis in which Yukawa matrices for up-quarks are diagonal also the interactions of Z' with the up-quarks are flavor diagonal. In other words, $\hat{Y}_u$ and Z' interactions with the up-quarks are aligned with each other. But we could just as well choose $V_L^d = \hat{1}$ and $V_R^d = \hat{1}$, which would result in FCNC mediated by Z' in the up-quark sector and no FCNC in the down-quark sector up to RG effects.

These simple examples show that in the absence of $[U(3)]^5$ flavor symmetry we have more freedom, and the physics depends on how the Yukawa matrices and matrices describing interactions are oriented in the flavor space. We will soon generalize this discussion to the SMEFT, but let us first have a closer look at the unitary matrix V_R^d. It is present in the SM but does not appear in the SM interactions because of the absence of right-handed charged currents in the SM and flavor universality of right-handed Z^0 couplings. In the Z' case considered here this matrix is responsible for flavor violation in the right-handed down-quark sector. Let us find an expression for it.

Adopting the standard CKM phase convention, where the five relative phases of the quark fields are adjusted to remove five complex phases from the CKM matrix, we have no more freedom to remove the six complex phases from V_R^d. In the standard CKM basis V_R^d can then be parametrized as follows [610, 772]

$$V_R^d = D_U V_R^0 D_D^\dagger, \tag{14.20}$$

where V_R^0 is a "CKM-like" mixing matrix, containing only three real mixing angles and one nontrivial phase. The diagonal matrices $D_{U,D}$ contain the remaining CP-violating phases. Choosing the standard parametrization for V_R^0 we have

$$V_R^0 = \begin{pmatrix} \tilde{c}_{12}\tilde{c}_{13} & \tilde{s}_{12}\tilde{c}_{13} & \tilde{s}_{13}e^{-i\phi} \\ -\tilde{s}_{12}\tilde{c}_{23} - \tilde{c}_{12}\tilde{s}_{23}\tilde{s}_{13}e^{i\phi} & \tilde{c}_{12}\tilde{c}_{23} - \tilde{s}_{12}\tilde{s}_{23}\tilde{s}_{13}e^{i\phi} & \tilde{s}_{23}\tilde{c}_{13} \\ \tilde{s}_{12}\tilde{s}_{23} - \tilde{c}_{12}\tilde{c}_{23}\tilde{s}_{13}e^{i\phi} & -\tilde{s}_{23}\tilde{c}_{12} - \tilde{s}_{12}\tilde{c}_{23}\tilde{s}_{13}e^{i\phi} & \tilde{c}_{23}\tilde{c}_{13} \end{pmatrix}, \tag{14.21}$$

and

$$D_U = \text{diag}(1, e^{i\phi_2^u}, e^{i\phi_3^u}), \qquad D_D = \text{diag}(e^{i\phi_1^d}, e^{i\phi_2^d}, e^{i\phi_3^d}). \tag{14.22}$$

A similar discussion can be made for leptons. We will continue with this simple model in Section 15.4.9. It illustrated rotations in the flavor space in simple settings, but to be general we return now to the SMEFT.

14.3.3 SMEFT Case

In a UV completion, having the complete Lagrangian including the Yukawa matrices along with the Wilson coefficients at the NP scale the flavor basis can be fixed. Eventually at the electroweak scale rotations to the mass eigenstate basis have to be performed. We will list some of such completions in Section 16.4.11.

A subtlety of choosing a basis for flavored SMEFT operators is the choice of the flavor basis in the space of the three fermion generations for each of the quark and lepton fields. Because all fermions are massless in the $SU(2)_L \times U(1)_Y$-symmetric phase, there is no a priori preferred basis, such as the mass basis below the electroweak scale. We will describe here very briefly what is done in numerical codes following [773]. The list of various codes will be made in Section 14.6.4.

The strategy of [773], which works in Warsaw operator basis, is to define a default weak basis in which the fermion mass matrices have a specific form. The running fermion mass matrices at the dimension-6 level, after a general $U(3)^5$ rotation, can be written as

$$M_d = \frac{v}{\sqrt{2}} U_q^\dagger \left(Y_d - \frac{v^2}{2} C_{d\phi} \right) U_d, \tag{14.23}$$

$$M_u = \frac{v}{\sqrt{2}} U_q^\dagger \left(Y_u - \frac{v^2}{2} C_{u\phi} \right) U_u, \tag{14.24}$$

$$M_e = \frac{v}{\sqrt{2}} U_l^\dagger \left(Y_e - \frac{v^2}{2} C_{e\phi} \right) U_e, \tag{14.25}$$

$$M_\nu = -v^2 U_l^T C_{ll\phi\phi} U_l. \tag{14.26}$$

Here $v \approx 246$ GeV is the electroweak vacuum expectation value (VEV) and Y_ψ, with $\psi = u, d, e$, are the SM Yukawa couplings. $C_{\psi\phi}$ is the Wilson coefficients of the SMEFT dimension-6 operator $Q_{\psi\phi} = \left(\phi^\dagger \phi \right) Q_\psi^Y$ with

$$\left(Y_e \bar{l} e\phi + Y_u \bar{q} u\widetilde{\phi} + Y_d \bar{q} d\phi \right) \equiv \sum_\psi Y_\psi Q_\psi^Y. \tag{14.27}$$

$C_{ll\phi\phi}$ is the lepton number-violating Weinberg operator $Q_{ll\phi\phi} = \left(\widetilde{\phi}^\dagger l \right)^T C \left(\widetilde{\phi}^\dagger l \right)$.

Next, using the freedom of $U(3)^5$ field rotations, a basis, called `Warsaw-down`, is chosen where

- M_d and M_e are diagonal with real positive ascending entries.
- M_u has the form $V^\dagger \hat{M}_u$, where $\hat{M}_u$ is diagonal with real positive ascending entries and V has the form of the CKM matrix in the standard phase convention.
- M_ν has the form $U^* \hat{M}_\nu U^\dagger$, where $\hat{M}_\nu$ is diagonal with real positive entries and U has the form of the PMNS neutrino mixing matrix in standard phase convention for the normally ordered neutrino mass spectrum. For inverted hierarchy, see [773].

This basis choice is convenient as it removes all unphysical parameters present in the Yukawa couplings, and it allows to easily translate to the mass basis at the scale of electroweak symmetry breaking. However, one should stress that this diagonality is not invariant under SMEFT renormalization group evolution and only holds at a single scale. Therefore at the electroweak scale additional rotations to the mass eigenstate basis have to be performed. An additional subtlety is the fact that v is scale dependent itself and can even vanish at a high scale. While the overall factor does not affect the rotation matrices, the $O(v^2)$ terms in (14.23)–(14.25) are affected. To avoid this problem, the authors of [773] advocate using the on-shell definition of v in these terms.

As a variant of the default "Warsaw" basis, [773] also defines a basis denoted Warsaw-up that uses a flavor basis where the up-type quark mass matrix M_u rather than M_d is diagonalized, while M_d has the form $V\hat{M}_d$. These two choices are analogous to the ones made in the light Z' scenario discussed earlier, but now this procedure can also be used for a very heavy Z'.

Our presentation was rather brief. Much more details on rotations in flavor space with physical implications can be found in [484, 516, 563, 565, 719, 737, 767, 769, 770, 773, 774].

As we just mentioned, the diagonality of Yukawa couplings at one scale is not guaranteed to be preserved at another scale because of the RG effects. We will discuss them next.

14.4 Renormalization Group Equations

The scale dependence of Wilson coefficients is governed by the RG equations presented in [758–760] for quartic Higgs couplings, Yukawa couplings, and gauge couplings, respectively. They have the general structure

$$\dot{C}_a \equiv (4\pi)^2 \mu \frac{dC_a}{d\mu} = \gamma_{ab}\, C_b, \tag{14.28}$$

with γ_{ab} being the entries of a very big anomalous dimension matrix (ADM). The ADM is known for SMEFT at one-loop and depends on (1) the quartic Higgs coupling λ [758], (2) the fermion Yukawa couplings to the Higgs doublet [759], and (3) the three gauge couplings $g_{1,2,3}$ [760]. For small $\gamma_{ab}/(4\pi)^2 \ll 1$ the approximate solution retains only the first leading logarithm (1stLLA)

$$C_a(\mu_{ew}) = \left[\delta_{ab} - \frac{\gamma_{ab}}{(4\pi)^2} \ln \frac{\mu_\Lambda}{\mu_{ew}}\right] C_b(\mu_\Lambda), \tag{14.29}$$

which is sufficient as long as the logarithm is not too large, so that also $\gamma_{ab}/(4\pi)^2 \ln \frac{\mu_\Lambda}{\mu_{ew}} \ll 1$ holds. Numerically one expects the largest enhancements when the ADM γ_{ab} is proportional to the strong coupling $4\pi\alpha_s \sim 1.4$ or the top-Yukawa coupling squared $y_t^2 \sim 1$. The QCD mixing is flavor-diagonal and hence cannot give rise to new genuine phenomenological effects in $\Delta F = 1, 2$ observables. On the other hand, top-Yukawa couplings are the main source of flavor-off-diagonal interactions responsible for the phenomenology discussed later. The $SU(2)_L$ gauge interactions induce via $\gamma_{ab} \propto g_2^2$ and are parametrically suppressed compared to y_t-induced effects. The suppression is even stronger for the terms $\gamma_{ab} \propto g_1^2$. Yet, in certain situations these effects have to be included, in particular when QCD effects and top-Yukawa contributions are absent. This is, e.g., the case of Scenario V.

The ADMs presented in [758–760] look truly horrible. They imply in principle a large set of coupled differential equations of the RG equations of SM couplings (quartic Higgs, gauge, and Yukawa) and the ones of dim-6 Wilson coefficients. The solution of this system in full generality requires the application of numerical methods, and the imposition of boundary conditions might be highly nontrivial. The 1stLLA neglects all these "secondary mixing" effects that would be present in the general leading logarithmic approximation (LLA), which would also resum large logarithms to all orders in couplings. With "secondary mixing" we refer to the situation where an operator O_1 might not have an ADM entry with operator O_3 (no "direct mixing"), but still contribute to the Wilson coefficient $C_3(\mu_{ew})$, via a direct mixing with some operator O_2 that in turn mixes directly into O_3. We refer to Appendix E for an analytic treatment of such cases.

Here we will first confine our presentation by keeping only contributions proportional to the up-type quark Yukawa coupling Y_u from [759], i.e., neglecting contributions from Yukawa matrices $Y_{d,e}$ of down-quarks and leptons unless the latter contributions are crucial. The same applies to terms involving gauge couplings. For instance, in Scenario V only gauge coupling contributions matter, and they have to be kept. In presenting the results here, we follow here [666, 667, 766] and list the equations only for operators that receive leading logarithmic contributions at the scale μ_{EW} from the initial Wilson coefficients at the scale μ_Λ of the operators in the scenarios defined earlier.

Let us then have a closer look at simplified scenarios in question.

14.4.1 Scenario I

In this scenario only the coefficients $[C_{Hd}]_{ij}$ with $i, j = d, s, b$ are nonvanishing at μ_Λ. Keeping only Y_u couplings the RG evolution from μ_Λ down to μ_{EW} generates $(\bar{L}L)(\bar{R}R)$ $\Delta F = 2$ operators through [766]

$$[\dot{C}_{qd}^{(1)}]_{ijij} = [Y_u Y_u^\dagger]_{ij} [C_{Hd}]_{ij} + \cdots . \tag{14.30}$$

The dots indicate other terms $\propto C_{\psi^2 H^2 D}$ that are not proportional to Y_u. We recall that our Y_u is defined as the Hermitean conjugate w.r.t. [759].

To proceed we have to specify Y_u and transform the quark fields from the flavor to the mass eigenbasis. As Y_u are scale dependent, it is useful to do it at the electroweak scale where EWSB takes place. The quark fields are then rotated by 3×3 unitary rotations in flavor space as done in Section 14.3 such that for $\psi = u, d$

$$(V_L^\psi)^\dagger M_\psi V_R^\psi = M_\psi^{\text{diag}}, \qquad\qquad V \equiv (V_L^u)^\dagger V_L^d, \tag{14.31}$$

with diagonal up- and down-quark mass matrices M_ψ^{diag}. The nondiagonal mass matrices M_ψ include the contributions of dim-6 operators. The quark-mixing matrix V is unitary, similar to the CKM matrix of the SM; however, in the presence of dim-6 contributions the numerical values are different from those obtained in usual SM CKM-fits. We will choose the flavor basis such that down-quarks are already mass eigenstates, which fixes $V_{L,R}^d = \mathbb{1}$, and assume without loss of generality $V_R^u = \mathbb{1}$, yielding $q_L = (V^\dagger u_L, d_L)^T$. It should be noted that this treatment is different from the usual discussion of the CKM matrix, where the down-quarks are rotated. See (2.45).

Neglecting for simplicity dim-6 to M_ψ we have $V \approx V_{\text{CKM}}$, and consequently, we find Y_u as a function of up-type quark masses m_k and the CKM parameters:

$$Y_u \overset{\text{dim-4}}{\approx} \frac{\sqrt{2}}{v} V_L^u M_U^{\text{diag}} V_R^{u\dagger} = \frac{\sqrt{2}}{v} V_{\text{CKM}}^\dagger M_U^{\text{diag}}, \tag{14.32}$$

$$[Y_u Y_u^\dagger]_{ij} = \frac{2}{v^2} \sum_{k=u,c,t} m_k^2 V_{ki}^* V_{kj} \approx \frac{2}{v^2} m_t^2 \lambda_t^{ij}, \tag{14.33}$$

with $v \approx (\sqrt{2} G_F)^{-1/2}$ being the Higgs vacuum expectation value. In the sum over k only the top-quark contribution is relevant ($m_{u,c} \ll m_t$), if one assumes that the unitary matrix V is equal to the CKM matrix up to dim-6 corrections. We expect only tiny contributions from $k = c$ in the case $ij = sd$, for $ij = bd, bs$ such contributions are entirely negligible. We recall that Y_d is diagonal and that all quark masses and CKM parameters are defined at the electroweak scale. When performing RG evolution to low-energy or high-energy scales the modification of their values due to QCD have to be taken into account, as they are significant. See Table 4.2.

Next we want to relate the coefficients on both sides of (14.30) to objects that we know from other parts of this book. The decoupling of heavy SM degrees of freedom at μ_{ew} gives rise to the $\Delta F = 2$ effective Hamiltonian [744, 745]

$$\mathcal{H}_{\Delta F=2}^{ij} = \mathcal{N}_{ij} \sum_a C_a^{ij} O_a^{ij} + \text{h.c.}, \tag{14.34}$$

where the normalization factor and the CKM combinations are

$$\mathcal{N}_{ij} = \frac{G_F^2}{4\pi^2} M_W^2 \left(\lambda_t^{ij}\right)^2, \qquad \lambda_t^{ij} = V_{ti}^* V_{tj}, \tag{14.35}$$

with $ij = sd$ for kaon mixing and $ij = bd, bs$ for B_d and B_s mixing, respectively. The important operators for Scenarios I and II are

$$O_{\text{VLL}}^{ij} = [\bar{d}_i \gamma_\mu P_L d_j][\bar{d}_i \gamma^\mu P_L d_j], \qquad O_{\text{VRR}}^{ij} = [\bar{d}_i \gamma_\mu P_R d_j][\bar{d}_i \gamma^\mu P_R d_j], \tag{14.36}$$

$$O_{\text{LR},1}^{ij} = [\bar{d}_i \gamma_\mu P_L d_j][\bar{d}_i \gamma^\mu P_R d_j], \qquad O_{\text{LR},2}^{ij} = [\bar{d}_i P_L d_j][\bar{d}_i P_R d_j], \tag{14.37}$$

which we met already in the previous chapter. Because of the addition of general flavor indices we had to change the notation of operators slightly.

The SMEFT is then matched to the $\Delta F = 2$–LEFT at μ_{ew} and at tree-level one finds the following modifications of the Wilson coefficients relevant for Scenario I [563]:

$$\Delta C_{\text{LR},1}^{ij} = -\mathcal{N}_{ij}^{-1} \left([C_{qd}^{(1)}]_{ijij} - \frac{[C_{qd}^{(8)}]_{ijij}}{2N_c} \right), \qquad \Delta C_{\text{LR},2}^{ij} = \mathcal{N}_{ij}^{-1} [C_{qd}^{(8)}]_{ijij}, \qquad (14.38)$$

where $\mathcal{N}_{ij}$ is given in (14.35). The minus sign reflects the fact that $[C_{qd}^{(1,8)}]_{ijij}$ are the coefficients in the Lagrangian and $C_{\text{LR},k}^{ij}$ the coefficients in the Hamiltonian. Note that our SMEFT coefficients are $O(1/\Lambda^2)$ as we absorbed the $1/\Lambda^2$ factor in (14.1) into these coefficients. As can be seen, the $\Delta F = 2$ operator $O_{qd}^{(8)}$ does not receive direct leading logarithmic contributions. It is well known though that $C_{qd}^{(8)}$ would be generated via secondary QCD mixing from $C_{qd}^{(1)}$ [744].

From (14.38) and (14.30) we find that the presence of the RH operator O_{Hd} at a short-distance scale μ_Λ, i.e., $[C_{Hd}]_{ij}(\mu_\Lambda) \neq 0$, generates through Yukawa RG effects a leading-logarithmic contribution to the LR operator $O_{\text{LR},1}^{ij}$ at the electroweak scale μ_{ew}, given by [766]

$$\Delta_{\text{1stLLA}} C_{\text{LR},1}^{ij}(\mu_{\text{ew}}) = v^2 \frac{[C_{Hd}]_{ij}(\mu_\Lambda)}{\lambda_t^{ij}} x_t \ln \frac{\mu_\Lambda}{\mu_{\text{ew}}}. \qquad (14.39)$$

The coefficient $[C_{Hd}]_{ij}(\mu_\Lambda)$ is related to the flavor-changing coupling of the Z to down-quarks. Although this relation and the phenomenological implications of (14.39) will be discussed in detail in Section 15.5, we would like to make the following important comment. The RG contribution just given is for μ_Λ sufficiently larger than μ_{ew} not only the most important NP effect in this scenario but has also opposite sign to nonlogarithmic effects at the matching scale.

14.4.2 Scenario II

In this scenario only the coefficients $[C_{Hq}^{(1)}]_{ij}$ and $[C_{Hq}^{(3)}]_{ij}$ are nonvanishing. Keeping only Y_u couplings the RG evolution from μ_Λ down to μ_{EW} generates $(\bar{L}L)(\bar{L}L)$ $\Delta F = 2$ operators through [766]

$$[\dot{C}_{qq}^{(1)}]_{ijij} = +[Y_u Y_u^\dagger]_{ij}[C_{Hq}^{(1)}]_{ij} + \cdots, \qquad (14.40)$$

$$[\dot{C}_{qq}^{(3)}]_{ijij} = -[Y_u Y_u^\dagger]_{ij}[C_{Hq}^{(3)}]_{ij} + \cdots. \qquad (14.41)$$

Matching in this case the SMEFT to the $\Delta F = 2$–LEFT at μ_{ew} one finds the following modifications of the Wilson coefficients at tree-level [563]:

$$\Delta C_{\text{VLL}}^{ij} = -\mathcal{N}_{ij}^{-1} \left([C_{qq}^{(1)}]_{ijij} + [C_{qq}^{(3)}]_{ijij} \right). \qquad (14.42)$$

Again the minus sign reflects the fact that $[C_{qq}^{(1,3)}]_{ijij}$ are the coefficients in the Lagrangian and C_{VLL}^{ij} the coefficients in the Hamiltonian.

Proceeding as in Scenario I, we then find that the presence of two operators $O_{Hq}^{(1)}$ and $O_{Hq}^{(3)}$ generates via (14.40), (14.41), and (14.42)

$$\Delta_{1stLLA} C_{VLL}^{ij}(\mu_{ew}) = \frac{v^2}{\lambda_t^{ij}} \left([C_{Hq}^{(1)}]_{ij}(\mu_\Lambda) - [C_{Hq}^{(3)}]_{ij}(\mu_\Lambda) \right) x_t \ln \frac{\mu_\Lambda}{\mu_{ew}}. \tag{14.43}$$

As we will see in Section 15.5 the coefficient $\sim (C_{Hq}^{(1)} - C_{Hq}^{(3)})$ is related to the flavor-changing coupling of the Z to up-quarks and not down-quarks considered here. The latter involve the linear combination $(C_{Hq}^{(1)} + C_{Hq}^{(3)})$. Important phenomenological consequences of this difference will be discussed in Section 15.5. It demonstrates again that neglecting RG effects driven by top-Yukawa couplings in the presence of $SU(2)_L$ invariance would miss important dynamics.

14.4.3 Self-Mixing in Scenarios I and II

The $\psi^2 H^2 D$-operators governing Scenarios I and II mix between themselves during the evolution from μ_Λ to μ_{ew}. This mixing turns out to be much smaller than the effects considered so far, at the level of a few percent for $\Lambda \leq 10\,\text{TeV}$. For this reason in writing the results in (14.39) and (14.43) this mixing has been neglected. For completeness we give the relevant formulas. They read [766]

$$\dot{C}_{Hq}^{(1)} = 6\,\text{Tr}[Y_u Y_u^\dagger] C_{Hq}^{(1)} + 2\left(Y_u Y_u^\dagger C_{Hq}^{(1)} + C_{Hq}^{(1)} Y_u Y_u^\dagger \right) \tag{14.44}$$
$$- \frac{9}{2}\left(Y_u Y_u^\dagger C_{Hq}^{(3)} + C_{Hq}^{(3)} Y_u Y_u^\dagger \right) - Y_u C_{Hu} Y_u^\dagger,$$

$$\dot{C}_{Hq}^{(3)} = 6\,\text{Tr}[Y_u Y_u^\dagger] C_{Hq}^{(3)} + Y_u Y_u^\dagger C_{Hq}^{(3)} + C_{Hq}^{(3)} Y_u Y_u^\dagger - \frac{3}{2}\left(Y_u Y_u^\dagger C_{Hq}^{(1)} + C_{Hq}^{(1)} Y_u Y_u^\dagger \right), \tag{14.45}$$

$$\dot{C}_{Hd} = 6\,\text{Tr}[Y_u Y_u^\dagger] C_{Hd}, \tag{14.46}$$

$$\dot{C}_{Hu} = -2 Y_u^\dagger C_{Hq}^{(1)} Y_u + 6\,\text{Tr}[Y_u Y_u^\dagger] C_{Hu} + 4\left(Y_u^\dagger Y_u C_{Hu} + C_{Hu} Y_u^\dagger Y_u \right), \tag{14.47}$$

$$\dot{C}_{Hud} = 6\,\text{Tr}[Y_u Y_u^\dagger] C_{Hud} + 3 Y_u^\dagger Y_u C_{Hud}. \tag{14.48}$$

We observe that in SMEFT the LH interactions $C_{Hq}^{(1,3)}$ do not generate RH interactions in the down-type sector (C_{Hd}) and vice versa. However, there is mixing of the LH interaction $C_{Hq}^{(1)}$ into the RH interactions of up-type sector (C_{Hu}) and vice versa. Moreover $C_{Hq}^{(1)}$ and $C_{Hq}^{(3)}$ mix with each other, while C_{Hd} evolves undisturbed by other operators.

14.4.4 Scenario III

Now the operators $O_{H\ell}^{(1,3)}$ and O_{He} are the driving force. The complex-valued coefficients of these operators are denoted by

$$[C_{H\ell}^{(1)}]_{ab}, \qquad [C_{H\ell}^{(3)}]_{ab}, \qquad [C_{He}]_{ab}, \tag{14.49}$$

where the indices $a, b = 1, 2, 3$ denote the different generations of up- and down-type leptons, that is neutrinos and charged leptons.

Through RG Yukawa interactions, these operators generate contributions to semileptonic four-fermion operators. Keeping only Y_u effects we find

$\psi^4 \, (\bar{L}L)(\bar{L}L)$:

$$[\dot{C}_{\ell q}^{(1)}]_{abij} = +[C_{H\ell}^{(1)}]_{ab}[Y_u Y_u^\dagger]_{ij}, \tag{14.50}$$

$$[\dot{C}_{\ell q}^{(3)}]_{abij} = -[C_{H\ell}^{(3)}]_{ab}[Y_u Y_u^\dagger]_{ij}, \tag{14.51}$$

$\psi^4 \, (\bar{R}R)(\bar{R}R)$:

$$[\dot{C}_{eu}]_{abij} = -2\,[C_{He}]_{ab}[Y_u^\dagger Y_u]_{ij}, \tag{14.52}$$

$\psi^4 \, (\bar{L}L)(\bar{R}R)$:

$$[\dot{C}_{\ell u}]_{abij} = -2\,[C_{H\ell}^{(1)}]_{ab}[Y_u^\dagger Y_u]_{ij}, \tag{14.53}$$

$$[\dot{C}_{qe}]_{ijab} = [Y_u Y_u^\dagger]_{ij}[C_{He}]_{ab}. \tag{14.54}$$

Here lepton-flavor indices are ab and quark-flavor indices ij.

This time the self-mixing of $O_{H\ell}^{(1,3)}$ and O_{He} is much simpler than in the first two scenarios. The three operators do not mix with each other ($N_c = 3$):

$$\dot{C}_{H\ell}^{(1)} = 2N_c \, \text{Tr}[Y_u Y_u^\dagger] \, C_{H\ell}^{(1)}, \tag{14.55}$$

$$\dot{C}_{H\ell}^{(3)} = 2N_c \, \text{Tr}[Y_u Y_u^\dagger] \, C_{H\ell}^{(3)}, \tag{14.56}$$

$$\dot{C}_{He} = 2N_c \, \text{Tr}[Y_u Y_u^\dagger] \, C_{He}. \tag{14.57}$$

A detailed analysis of flavor physics in this scenario is in progress and should be available in the hep-arxiv when this book is published.

14.4.5 Scenario IV

We consider in this scenario $(\bar{L}L)(\bar{L}L)$ operators so that the operators $O_{\ell q}^{(1,3)}$ are the driving force. The complex-valued coefficients of these operators are denoted by

$$\boxed{[C_{\ell q}^{(1)}]_{abij}, \qquad [C_{\ell q}^{(3)}]_{abij},} \tag{14.58}$$

where lepton-flavor indices are ab and quark-flavor indices ij.

Keeping only Y_u effects we find (summation over repeated indices ww is understood)

$$[\dot{C}_{\ell q}^{(1)}]_{abij} = +\frac{1}{2}[Y_u Y_u^\dagger]_{iw}[C_{\ell q}^{(1)}]_{abwj} + \frac{1}{2}[C_{\ell q}^{(1)}]_{abiw}[Y_u Y_u^\dagger]_{wj}, \tag{14.59}$$

$$[\dot{C}_{\ell q}^{(3)}]_{abij} = +\frac{1}{2}[Y_u Y_u^\dagger]_{iw}[C_{\ell q}^{(3)}]_{abwj} + \frac{1}{2}[C_{\ell q}^{(3)}]_{abiw}[Y_u Y_u^\dagger]_{wj}, \tag{14.60}$$

which shows that these two Wilson coefficients in this approximation evolve independent of each other. Yet, when gauge couplings are included, they start to mix

$$[\dot{C}_{\ell q}^{(1)}]_{abij} = \frac{2}{9}g_1^2[C_{\ell q}^{(1)}]_{abww}\delta_{ij} + \frac{2}{3}g_1^2[C_{\ell q}^{(1)}]_{wwij}\delta_{ab} - g_1^2[C_{\ell q}^{(1)}]_{abij} + 9g_2^2[C_{\ell q}^{(3)}]_{abij},$$

(14.61)

$$[\dot{C}_{\ell q}^{(3)}]_{abij} = 2g_2^2[C_{\ell q}^{(3)}]_{abww}\delta_{ij} + \frac{2}{3}[C_{\ell q}^{(3)}]_{wwij}\delta_{ab} + 3g_2^2[C_{\ell q}^{(1)}]_{abij} - 6\left(g_2^2 + \frac{1}{6}g_1^2\right)[C_{\ell q}^{(3)}]_{abij}.$$

(14.62)

But what is more important is the generation through electroweak interaction of purely leptonic operators:[4]

$$[\dot{C}_{\ell\ell}]_{abcd} = -\frac{1}{3}g_1^2\left([C_{\ell q}^{(1)}]_{abww}\delta_{cd} + [C_{\ell q}^{(1)}]_{cdww}\delta_{ab}\right)$$
$$- g_2^2\left([C_{\ell q}^{(3)}]_{abww}\delta_{cd} + [C_{\ell q}^{(3)}]_{cdww}\delta_{ab} - 2[C_{\ell q}^{(3)}]_{cbww}\delta_{ad} - 2[C_{\ell q}^{(3)}]_{adww}\delta_{bc}\right),$$

(14.63)

which can contribute to pure leptonic lepton decays thereby putting significant bounds on the coefficients of semileptonic operators as stressed in particular in [765, 767] and mentioned already a few pages before. We refer to these papers for phenomenological implications of these equations.

14.4.6 Scenario V

In what follows we list RG equations, which govern the generation of nonleptonic (NL)-ψ^4 coefficients from semileptonic (SL)-ψ^4 ones. They can be derived from [760] and have been presented in detail in [667]. Only electroweak interactions are involved here. We find

NL-ψ^4 $(\overline{L}L)(\overline{L}L)$:

$$[\dot{C}_{qq}^{(1)}]_{prst} = -\frac{1}{9}g_1^2\left([C_{\ell q}^{(1)}]_{wwst}\delta_{pr} + [C_{\ell q}^{(1)}]_{wwpr}\delta_{st} + [C_{qe}]_{stww}\delta_{pr} + [C_{qe}]_{prww}\delta_{st}\right),$$

(14.64)

$$[\dot{C}_{qq}^{(3)}]_{prst} = +\frac{1}{3}g_2^2\left([C_{\ell q}^{(3)}]_{wwst}\delta_{pr} + [C_{\ell q}^{(3)}]_{wwpr}\delta_{st}\right).$$

(14.65)

NL-ψ^4 $(\overline{R}R)(\overline{R}R)$:

$$[\dot{C}_{uu}]_{prst} = -\frac{4}{9}g_1^2\left([C_{eu}]_{wwst}\delta_{pr} + [C_{eu}]_{wwpr}\delta_{st} + [C_{\ell u}]_{wwst}\delta_{pr} + [C_{\ell u}]_{wwpr}\delta_{st}\right),$$

(14.66)

$$[\dot{C}_{dd}]_{prst} = +\frac{2}{9}g_1^2\left([C_{ed}]_{wwst}\delta_{pr} + [C_{ed}]_{wwpr}\delta_{st} + [C_{\ell d}]_{wwst}\delta_{pr} + [C_{\ell d}]_{wwpr}\delta_{st}\right),$$

(14.67)

$$[\dot{C}_{ud}^{(1)}]_{prst} = +\frac{4}{9}g_1^2\left([C_{\ell u}]_{wwpr}\delta_{st} + [C_{eu}]_{wwpr}\delta_{st} - 2[C_{\ell d}]_{wwst}\delta_{pr} - 2[C_{ed}]_{wwst}\delta_{pr}\right).$$

(14.68)

---

[4] Note that now there are two new lepton indices cd, and summation is only over quark indices.

NL-ψ^4 $(\bar{L}L)(\bar{R}R)$:

$$[\dot{C}_{qu}^{(1)}]_{prst} = -\frac{2}{9}g_1^2 \left(4[C_{\ell q}^{(1)}]_{wwpr}\delta_{st} + 4[C_{qe}]_{prww}\delta_{st} + [C_{\ell u}]_{wwst}\delta_{pr} + [C_{eu}]_{wwst}\delta_{pr}\right),$$
(14.69)

$$[\dot{C}_{qd}^{(1)}]_{prst} = +\frac{2}{9}g_1^2 \left(2[C_{\ell q}^{(1)}]_{wwpr}\delta_{st} + 2[C_{qe}]_{prww}\delta_{st} - [C_{\ell d}]_{wwst}\delta_{pr} - [C_{ed}]_{wwst}\delta_{pr}\right),$$
(14.70)

and finally for all other NL-ψ^4 operators

$$[\dot{C}_{ud}^{(8)}]_{prst} = 0, \quad [\dot{C}_{qu}^{(8)}]_{prst} = 0, \quad [\dot{C}_{quqd}^{(1)}]_{prst} = 0,$$
$$[\dot{C}_{qd}^{(8)}]_{prst} = 0, \quad [\dot{C}_{quqd}^{(8)}]_{prst} = 0.$$
(14.71)

We observe that the SM gauge-mixing of SL-ψ^4 into NL-ψ^4 operators within SMEFT generates in 1stLLA only $(\bar{L}L)(\bar{L}L)$, $(\bar{L}L)(\bar{R}R)$, and $(\bar{R}R)(\bar{R}R)$ NL-ψ^4 operators from the corresponding semileptonic classes. The initial Wilson coefficients of the semileptonic operators at the scale μ_Λ enter only summed over the lepton-flavor-diagonal parts $[C_b^{(a)}]_{ww\cdots}$ (and $[C_{qe}]_{\cdots ww}$), summation over the index $w = 1, 2, 3$ is implied, because all leptons can run inside the loop. In consequence the underlying combination of the couplings is

$$\Sigma_{\chi,\mathrm{LQ}}^{ji} \equiv \sum_a (Y_{ja}^\chi)^* Y_{ia}^\chi.$$
(14.72)

Further details on this scenario can be found in [667], where the ratio ε'/ε was studied in leptoquark models. The results of this analysis are summarized in Section 16.4.13.

Scenario VI has been discussed by many authors, and we devote to it a separate subsection.

14.5　SU(2)$_L$ Correlations between $b \to s\nu\bar{\nu}$ and $b \to s\ell^+\ell^-$

The implication of SM gauge invariance for flavor observables has been discussed in many papers, in particular in [565, 607, 775–779]. Here, as an example, we will just write general formulas, which allow to study correlations on one hand between $b \to s\nu\bar{\nu}$ and $b \to s\ell^+\ell^-$ transitions and on the other hand between $d \to s\nu\bar{\nu}$ and $d \to s\ell^+\ell^-$ transitions.

To this end we recall general $\Delta F = 1$ Hamiltonian for the semileptonic FCNC transition of down-type quarks into leptons and neutrinos below the scale μ_{ew}

$$\mathcal{H}_{d\to d(\ell\ell,\nu\nu)} = -\frac{4G_F}{\sqrt{2}}\lambda_t^{ji}\frac{\alpha_e}{4\pi}\sum_k C_k^{baji}Q_k^{baji} + \mathrm{h.c.},$$
(14.73)

with a, b being lepton indices, i, j down-quark indices and

$$\lambda_u^{ji} \equiv V_{ui}^* V_{uj}, \qquad\qquad u = \{u, c, t\}.$$
(14.74)

There are eight semileptonic operators relevant for $d_i\ell_a \to d_j\ell_b$ when considering UV completions that give rise to SMEFT above the electroweak scale [565]

$$Q_{9(9')}^{baji} = [\bar{d}_j\gamma_\mu P_{L(R)}d_i][\bar{\ell}_b\gamma^\mu\ell_a], \qquad Q_{10(10')}^{baji} = [\bar{d}_j\gamma_\mu P_{L(R)}d_i][\bar{\ell}_b\gamma^\mu\gamma_5\ell_a],$$
$$Q_{S(S')}^{baji} = [\bar{d}_j P_{R(L)}d_i][\bar{\ell}_b\ell_a], \qquad Q_{P(P')}^{baji} = [\bar{d}_j P_{R(L)}d_i][\bar{\ell}_b\gamma_5\ell_a], \qquad (14.75)$$

and two for $d_i\nu_a \to d_j\nu_b$

$$Q_{L(R)}^{baji} = [\bar{d}_j\gamma_\mu P_{L(R)}d_i][\bar{\nu}_b\gamma^\mu(1-\gamma_5)\nu_a]. \qquad (14.76)$$

The SM contribution to these Wilson coefficients is lepton-flavor diagonal

$$C_k^{baji} = C_{k,\mathrm{SM}}\,\delta_{ba} + \frac{\pi}{\alpha_e}\frac{v^2}{\lambda_t^{ji}}C_{k,\mathrm{NP}}^{baji}, \qquad (14.77)$$

where $v = 246\,\mathrm{GeV}$ and a normalization factor has been introduced for the NP contribution that proves convenient for matching with SMEFT in a given model. The nonvanishing SM contributions valid both in the B and in the K system,

$$C_{9,\mathrm{SM}} = \frac{Y_0(x_t)}{s_W^2} - 4Z_0(x_t), \qquad C_{10,\mathrm{SM}} = -\frac{Y_0(x_t)}{s_W^2}, \qquad C_{L,\mathrm{SM}} = -\frac{X_0(x_t)}{s_W^2}, \qquad (14.78)$$

are given by the gauge-independent functions $X_0(x_t)$, $Y_0(x_t)$, and $Z_0(x_t)$ [133] that already appeared many times in this book.

The NP contribution to the Wilson coefficients of the $\Delta F = 1$ semileptonic operators in (14.75) and (14.76) at μ_{ew} in terms of the semileptonic SMEFT Wilson coefficients at μ_{ew} is given as follows [563, 565, 607]

$$C_{9,\mathrm{NP}}^{baji} = [C_{qe} + C_{\ell q}^{(1)} + C_{\ell q}^{(3)}]_{baji}, \qquad C_{9',\mathrm{NP}}^{baji} = [C_{ed} + C_{\ell d}]_{baji},$$
$$C_{10,\mathrm{NP}}^{baji} = [C_{qe} - C_{\ell q}^{(1)} - C_{\ell q}^{(3)}]_{baji}, \qquad C_{10',\mathrm{NP}}^{baji} = [C_{ed} - C_{\ell d}]_{baji},$$
$$C_{L,\mathrm{NP}}^{baji} = [C_{\ell q}^{(1)} - C_{\ell q}^{(3)}]_{baji}, \qquad C_{R,\mathrm{NP}}^{baji} = [C_{\ell d}]_{baji}, \qquad (14.79)$$
$$C_{S,\mathrm{NP}}^{baji} = -C_{P,\mathrm{NP}}^{baji} = [C_{\ell edq}]_{abij}^*, \qquad C_{S',\mathrm{NP}}^{baji} = C_{P',\mathrm{NP}}^{baji} = [C_{\ell edq}]_{baji}.$$

We caution the reader that C_{qe}, used in the literature, should here be written as C_{eq} so that the lepton indices come first as in the remaining WCs in these equations. But as we already stated after (14.73) a, b are lepton indices and i, j down-quark ones, so that no confusion should result from this notation. Here contributions from Z-mediating $\psi^2 H^2 D-$ SMEFT operators $O_{Hq}^{(1,3)}$ to $C_{9,10,L}$ and O_{Hd} to $C_{9',10',R}$, respectively, have been omitted. In rare FCNC Kaon decays scalar and pseudoscalar Wilson coefficients are negligible but are relevant in $B_s \to \mu^+\mu^-$.

We observe that the number of operators in the SMEFT is in general larger than in the low-energy effective Hamiltonian, so on a completely model-independent basis, no general correlations can be derived. Indeed, from (14.79), it is clear that in complete generality, the size of NP effects in $b \to s\nu\bar{\nu}$ is not constrained by the $b \to s\ell^+\ell^-$ measurements.

First, the decays with charged leptons are only sensitive to the combination $C_{\ell q}^{(1)} + C_{\ell q}^{(3)}$, while the decays with neutrinos in the final state probe $C_{\ell q}^{(1)} - C_{\ell q}^{(3)}$. Second, even if the Wilson coefficient $C_{\ell q}^{(3)}$ vanishes, cancellations between the operators with left- and right-handed charged leptons can lead to small deviations from the SM in $b \to s\ell^+\ell^-$ transitions even when large effects are present in $b \to s\nu\bar{\nu}$. However, in concrete NP models, often only a subset of the operators is relevant and in these cases correlations characteristic for these NP scenarios are obtained. We will see later on that the present bounds on $b \to s\nu\bar{\nu}$ transitions can in certain models be problematic for the explanation of the anomalies in $b \to s\ell^+\ell^-$ measurements discussed in Chapter 12.

A solution to the latter problem is the equality $C_{\ell q}^{(1)} = C_{\ell q}^{(3)}$, which removes the contributions not only to $b \to s\nu\bar{\nu}$ transition but if valid for all down-quark flavors, also to $d \to s\nu\bar{\nu}$ transitions, that is $K^+ \to \pi^+\nu\bar{\nu}$ and $K_L \to \pi^0\nu\bar{\nu}$. However, as emphasized in [765, 767], this equality can be valid only at one scale. RG evolution from NP scale to EW scale breaks this relation. See [667] for an explicit analysis of such RG effects in the context of leptoquark models that we will review in Section 16.4.

While such explicit models will be considered also in subsequent chapters, here we just illustrate general correlations on two examples presented in [607] and discussed in detail in Sections 15.4 and 15.5. Readers not familiar with these NP scenarios can skip these examples and return to them after reading these two sections.

We consider first the general case of Z' models in which a single Z' gauge boson with left-handed and right-handed couplings dominates the scene. In this case $C_{\ell q}^{(3)} = 0$ at the NP scale, and at this scale we have the relations

$$C_{L,\text{NP}}^{baji} = \frac{C_{9,\text{NP}}^{baji} - C_{10,\text{NP}}^{baji}}{2}, \qquad C_{R,\text{NP}}^{baji} = \frac{C_{9',\text{NP}}^{baji} - C_{10',\text{NP}}^{baji}}{2}. \tag{14.80}$$

But as we will see in Section 15.5, if NP contributions to the processes considered are fully dominated by induced FCNC couplings of the SM Z boson – this is the case, e.g., in the MSSM, in models with partial compositeness and models with vector-like quarks discussed in Section 16.3 – one finds

$$C_{L,\text{NP}}^{baji} = C_{10,\text{NP}}^{baji}, \qquad C_{9,\text{NP}}^{baji} = -\zeta C_{10}^{\text{NP}}, \qquad C_{R,\text{NP}}^{baji} = C_{10',\text{NP}}^{baji}, \qquad C_9^{\text{NP}} = -\zeta C_{10',\text{NP}}^{baji}, \tag{14.81}$$

where $\zeta = 1 - 4s_W^2 = 0.08$ is the accidentally small vector coupling of the Z to charged leptons. Evidently the correlations in (14.80) and (14.81) differ from each other. In particular the sign in front of $C_{10,\text{NP}}^{baji}$ in $C_{L,\text{NP}}^{baji}$ is different and similar for right-handed coefficients. In this manner the Z' and Z scenarios can be distinguished from each other. The plots in [607] illustrate these differences.

Finally, if the presence of Z' induces FCNC couplings of Z through $Z - Z'$ mixing, the relations just given are modified and depend on the size of $Z - Z'$ mixing. This mixing is clearly model dependent, and the resulting correlations can vary from model to model. We will illustrate this case in Section 16.2 by using the 331 models studied in [615].

The preceding discussion does not include RG effects above the electroweak scale. While such an analysis is missing in the literature at the time of this writing, it should be available when this book is published.

14.6　General Procedure and Useful Results

14.6.1　Preliminaries

We have seen on the examples just presented that specifying general properties about the dynamics at the NP scale, combined with the SMEFT technology, can give us some insight into various patterns of NP effects in low-energy flavor phenomenology. Quite generally, having a specific NP model or some simplified NP scenario with new heavy fields, gauge bosons, scalars, and fermions, the implications of this extension of the SM on low-energy observables can be found systematically by performing the following steps:

Step 1: Matching of NP Model onto the SMEFT
Assuming the NP scale Λ_{NP} to be much larger than the electroweak scale μ_{EW} and integrating out new heavy particles with masses $O(\Lambda_{NP})$ one finds the nonvanishing WCs of the SMEFT at $\mu = \Lambda_{NP}$. Useful results related to this step present in the literature will be briefly described later. It should be emphasized that among the five steps listed here, this is the only one that is model dependent. Once this step has been completed, the remaining steps, even if sometimes tedious, are model independent in the sense that only SM fields and their interactions are involved.

Step 2: Renormalization Group Evolution in the SMEFT
This step has been already discussed earlier and is based on the one-loop ADMs collected in [758–760]. This step generates new operators with nonvanishing WC and modifies the ones we started with at $\mu = \Lambda_{NP}$. If there is a full spectrum of new heavy particles with vastly different masses, this step consists of several steps and corresponding effective theories similar to what we have encountered in the QCD renormalization group evolution with effective theories characterized by the number of flavors $f = 5, 4, 3$. At every threshold between two effective theories matching between the two effective theories in question has to be made. Here, to simplify the presentation, we will assume that the first threshold on our route to low-energy scales is the electroweak scale.

Step 3: Matching of the SMEFT onto the LEFT
In this step the matching of WCs in SMEFT evaluated through Step 2 at $\mu = \Lambda_{EW}$ onto those of the low-energy effective theory below the electroweak scale (LEFT) is performed. Here the issue of the relative signs of SMEFT WCs and LEFT WCs has to be taken into account as we already warned the readers several times and illustrated in a few examples. Also in this case useful results are present in the literature and will be briefly described later.

Step 4: Renormalization Group Evolution in the LEFT
The one-loop anomalous dimensions for the operators of LEFT have been collected in [780]. They include previously known results that are cited in that paper as well as new ones. Complete RG evolution below the electroweak scale down to scales relevant for B physics has been presented in [756]. This evolution involves QCD and QED.

Step 5: Calculation of the Observables

Finally, having evaluated WCs at scales relevant for observables of interest, the latter have to be evaluated. This is often a difficult step, in particular in the case of nonleptonic decays of mesons but also in those semileptonic decays in which good knowledge of formfactors is required.

These five steps should be evident for any reader who studied the first three parts of this book, but we recalled them here as they constitute the generalization of what is usually done within the SM. In the context of electric dipole moments I find the presentation of the preceding steps in [781] very transparent.

During the last years a number of papers were published that contain results on SMEFT and LEFT that are useful for any analysis of NP effects in any model. The next three sections collect the results I found in the literature. I will describe them here very briefly, listing most important references. I hope this collection is useful, in particular for those readers who want to incorporate in their NP studies the full machinery of effective theories involved.

14.6.2 Useful Technology: SMEFT

- The complete matching of any NP model onto the SMEFT at tree-level has been calculated in [782]. Moreover, the classification of the new fields relevant for this work, presented before in a series of papers for new quarks [783], new leptons [784], new vectors [785], and new scalars [786], has been summarized and generalized to non-renormalizable UV-completions. Starting with a general Lagrangian involving scalars, fermions, and vector bosons, the authors have computed WCs of the SMEFT at the NP scale in terms of couplings and masses. In tables 1–3 of this paper one can find the list of new scalar bosons, new vector-like fermions, and new vector bosons contributing to the SMEFT at tree level. In the corresponding tables 7–9 in this paper a list of SMEFT operators generated by each of this field is given. Finally, in Appendix D of that paper the WCs of SMEFT operators as functions of masses and couplings of all new fields in tree-level approximations are given. Even if the formulas in this paper are very complicated, this paper is very useful both for the top-down approach and also for the bottom-up one. But as the authors stress, these results are insufficient for the study of NP that enters first at one-loop level.
- Feynman rules for the SMEFT in R_ξ-gauges can be found in [787, 788] and FEYN-RULES implementation of SMEFT in [789]. In [790] generalization to a generic class of effective theories with operators of arbitrary dimension has been made.
- Complete one-loop matching for a real singlet heavy scalar on the SMEFT has been presented in [791].
- Formulas for Z penguin contributions in generic extensions of the SM that involve arbitrary charged fermions, scalars, and gauge bosons but do not generate FCNC mediated by Z at tree-level, have been presented in [136]. Because of the gauge invariance also the relevant box contributions are given there. This paper is particularly important for models with RH charged currents as in this case one-loop calculations require even more care than in the presence of only LH currents.

- Another useful paper for one-loop calculations is [792], where a generic class of NP models featuring new heavy scalars and fermions that couple to the SM fermions via Yukawa-like interactions has been considered. The extension to a vector-like fourth generation with particular emphasis put on $b \to s\ell^+\ell^-$ has been presented in [793].
- One-loop matching of gauge-invariant dimension-6 operators for $b \to s$ and $b \to c$ transitions has been presented in [563].
- QCD improved matching for semileptonic B decays with leptoquarks has been presented in [794]. The enhancement of the Wilson coefficients by 8%(13%) for scalar (vector) leptoquarks softens the LHC bounds and increases the allowed parameter space of LQ models addressing the flavor anomalies.
- In [795] the complete one-loop contributions in the SMEFT to radiative lepton decays $\mu \to e\gamma$, $\tau \to \mu\gamma$, and $\tau \to e\gamma$ have been calculated as well as to the closely related anomalous magnetic moments and electric dipole moments of charged leptons. Also three body flavor-violating charged lepton decays, which we will consider in some detail in Section 17.2, have been considered, taking into account all tree-level contributions.
- In [796] a complete analysis of $\mu \to e\gamma$, $\mu \to 3e$, and $\mu \to e$ conversion in the framework of LEFT is performed with RG evolution of the relevant WCs between EW scale and low energies at the one-loop level.
- In [768] RG evolution of NP contributions to charged-current leptonic and semileptonic decays has been presented focusing on operators involving chirality flip at the quark level. It has been pointed out that the large mixing of the tensor operators into the (pseudo)scalar ones has important impact on the phenomenology.
- A master formula for neutrinoless double-beta decay rate that takes into account dimension-5, -7, and -9 operators within SMEFT and includes RG evolution down to chiral EFT for nucleons has been presented in [797].
- A master formula for ε'/ε, valid in any extension of the SM, has been presented in [798]. We will discuss it in detail in Section 14.7.
- Generation of $\Delta F = 2$ operators from $\Delta F = 1$ operators O_{Hd}, $O_{Hq}^{(1)}$, and $O_{Hq}^{(3)}$ through RG running in the SMEFT and one-loop matching on LEFT considered in Scenarios I and II (see Sections 14.4.1 and 14.4.2) [766] has been generalized to include all four-fermion Dim=6 $\Delta F = 1$ operators in [799, 800].
- Correlations of lepton-flavor-violating τ decays with $K^+ \to \pi^+\nu\bar{\nu}$ and $K_L \to \pi^0\nu\bar{\nu}$ have been analyzed within the SMEFT in [801]. They will be of interest in the era of NA62, KOTO, and Belle II experiments.
- In the case of $(g - 2)_\mu$ useful general formulas can be found in [802–804].

Several papers in addition to those already cited in other chapters should be mentioned in this context. These are [761, 781, 796, 805–810], where various results related to RG evolutions in the SMEFT and LEFT relevant for various processes are presented.

14.6.3 Matching of SMEFT on LEFT

The SMEFT allows to calculate the WCs of the SMEFT in the Warsaw basis at the electroweak scale. Below the electroweak scale flavor physics is described by an effective theory discussed within the SM in previous chapters of this book. However in the presence of NP there are new operators generated already at the NP scale and/or at the electroweak scale through RG running in the SMEFT and through the matching onto the low-energy theory so that the starting point for the RG evolution down to the low-energy scale differs from the one encountered within the SM. Not only the initial conditions for SM operators are modified by NP but also new WCs are present. The resulting effective theory below the electroweak scale, although as the SM based on the gauge symmetry of QCD × QED, differs then from the SM one, and we have termed it low-energy EFT (LEFT), although these days also the name Weak Effective Theory (WET) is used for it. It should be stressed that this theory does not involve W, Z gauge bosons, the Higgs boson, and the top-quark as dynamical degrees of freedom. Moreover, there are no elementary scalars in this theory.

Therefore the next task is to find all possible gauge-invariant operators of the LEFT. This task has been accomplished in [811]. There are 70 Hermitian D = 5 and 3,631 Hermitian D = 6 operators that conserve baryon and lepton number, as well as $\Delta B = \pm\Delta L = \pm 1$, $\Delta L = \pm 2$, and $\Delta L = \pm 4$ operators. Among the 3,631 operators in question, 1,933 are CP-even and 1,698 CP-odd. In that paper also the matching onto these operators from SMEFT up to order v^2/Λ^2 has been computed at tree-level. Because SMEFT has far fewer D = 6 operators than the LEFT, there are many relations among LEFT coefficients, which for B physics can be found in [563, 565] and more generally in [811].

The situation simplifies for a CP conserving $U(3)^5$-symmetric SMEFT. The full tree and one-loop matching of such a theory on LEFT has been performed in [812].

14.6.4 Numerical Codes

In view of the large number of operators, public computer codes have been generated to calculate many flavor observables and treat the RG equations in full generality both in the context of SMEFT and LEFT. A detailed presentation of these codes goes beyond the scope of this book and beyond author's skills. Their compact description can be found in [773], which defines an exchange format for BSM WCs, called WCxf. It consists of a standard data format for the numerical values of WCs and therefore facilitates the exchange of WCs between different computer codes. Furthermore in [773] a python package is provided, which allows to do the basis change between different bases in LEFT/SMEFT in an automated manner.

As the codes in question are useful for various phenomenological applications, we want to list them here and describe some of them very briefly so that the readers will know what to expect from them. More details can be found in [1332]. A list of different aspects and the corresponding codes is then given as follows:

- Flavor observables: flavio [813], EOS [814], FlavorKit [815], SPheno [816], FormFlavor[817], SuperIso [818], HEPfit [819], FlavBit [820],
- Running and/or Matching: DsixTools [821], wilson [822], MatchingTools [823],

- SMEFT related: SMEFTsim [789], SMEFT Feynman Rules [787],
- Dark matter: DirectDM [824].

Let us describe a few of them in more detail and mention other tools that are very useful.

- DsixTools [821] is a Mathematica package for handling the dimension-6 SMEFT. It allows to perform the full one-loop RG evolution of the Wilson coefficients from the NP scale down to the electroweak scale in the Warsaw basis numerically. It also contains modules devoted to the matching of SMEFT to $\Delta B = \Delta S = 1, 2$ and $\Delta B = \Delta C = 1, 2$ operators of the LEFT and their QCD and QED RG evolution below the electroweak scale. The application of this code to B physics anomalies by these authors can be found in [727]. References to the literature in which the relevant calculations have been done can be found in [821]. Readers interested in them should definitely look at these two papers.

- flavio [813] is an open-source tool for phenomenological analyses in flavor physics and other precision observables in the SM and beyond. It allows to compute predictions for a plethora of observables in quark and lepton-flavor physics and electroweak precision tests. It includes also the neutron electric dipole moment and anomalous magnetic moments of leptons. The list of presently included observables can be found in tables 1 and 2 of that paper. NP effects are parametrized as WCs of dimension-6 operators in the SMEFT or below the electroweak scale, that is in the LEFT. It contains the database of experimental measurements of these observables, a statistical package that allows to construct Bayesian and frequentists likelihoods, and convenient plotting and visualization routines.

- FlavBit [820] is a GAMBIT module for computing flavor observables and likelihoods. See also [825].

- CoDEx [826] is a Mathematica package that allows to calculate WCs of all effective operators at the electroweak scale that are generated from a given BSM theory, including the matching at tree and one-loop level on to the SMEFT and subsequent SMEFT renormalization group evolution down to the electroweak scale.

- MatchingTools [823] is a code that performs the tree-level matching of a given BSM model onto SMEFT in an automated way.

- BasisGen [827] allows for an automatic generation of an operator basis, given the fields and symmetries of any QFT.

- A further match-runner code is wilson [822]. In wilson the complete 1-loop running in the SMEFT, the complete tree-level matching from SMEFT onto LEFT at the EW scale, and the complete 1-loop running below the EW scale is implemented for all possible operators in SMEFT and LEFT. It is based on WCxf [773] mentioned earlier.

- The culmination of all these efforts can be found in [484], where the technology of [773, 813, 822] is combined resulting in a very efficient code, called *SMEFT likelihood (smelli)* that includes 257 observables, among them *all* our main players. Wilson coefficients in any BSM model can be specified at any scale, with one-loop renormalization group running above and below the electroweak scale automatically taken care of. The smelli package allows to compute the global likelihood as a function of the SMEFT WCs.

- In the standard determinations of the CKM parameters, it is usually assumed that the values of these parameters obtained from tree-level decays are not polluted by the presence of NP. To test this assumption and eventually take into account possible shifts in CKM parameters caused by the presence of NP, a strategy has been developed in [828]. It amounts to the redefinition of four Wolfestein parameters to four effective parameters by absorbing in them the effects of dim = 6 operators to four chosen observables. Subsequently, all other observables are expressed in terms of them. Once these four effective parameters are determined, one can insert their values in other observables. In my view it is not obvious whether one gains anything in this manner in the context of the SMEFT because the observables outside the chosen set will still contain additional contributions from dim = 6 operators that are unknown without a concrete model. Moreover, the values of the effective parameters depend on the chosen set of observables. Yet, one cannot exclude at present that one day, when the data improve, one will be able to find four precisely measured observables that will allow to find four optimal effective parameters that will facilitate the usual CKM fits in the presence of NP.

The necessity for efficient computer codes is illustrated by the following problematic, which is also present in the UV completions that do not predict the values of CKM parameters or generally SM parameters. These are usually determined at the electroweak scale or lower scales like m_b. But we would like to know them at the NP scale. If only QCD running is involved, this is straightforward as we have demonstrated in various chapters before. But in the SMEFT one has to take all effects into account. These are EW interactions, Yukawa interactions, and also the fact that the evolution of the SM parameters from EW scale to NP scales depends on the SMEFT Wilson coefficients themselves. This means that starting with a set of WCs at the NP scale, we have to make sure that the values of the SM parameters at this scale for the numerical values of WCs in question gives after RG to low energies the correct values of SM parameters determined in low-energy experiments. An iterative procedure explained in [822] accomplishes this task.

14.6.5 General Comments on the Future

Certainly this impressive technology will be improved in the coming years.

- The complete matching of any tree level model onto SMEFT at $\mu = \Lambda_{NP}$ from [782] should be generalized to one-loop matching.
- The one-loop RG evolution in SMEFT from Λ down to EW scale from [758–760] should be generalized to NLO level to remove scale and renormalization scheme uncertainties present in one-loop matching.
- The tree-level matching of SMEFT operators onto LEFT operators at EW scale from [563, 811] should be generalized to one-loop level.
- The one-loop RG evolution in LEFT from EW scale down to a low-energy scale from [756, 780] should be extended to NLO. In fact, to a large extent this could be done

already now using existing NLO results in the literature, see previous chapter. But one should still investigate whether some elements are missing.

- The numerical codes listed earlier should be extended to include more observables and the NLO contributions listed earlier as soon as they will be available.

Still the existing technology, which is dominantly incorporated in computer codes, allows one already now to compute not only the low-energy Lagrangian relevant for the phenomenology of weak decays of mesons and leptons but also with the help of flavio [813] a multitude of observables outside the flavor physics. In this context the paper [484] is very useful.

14.6.6 Nonlinear Realizations

It should be emphasized that SMEFT assumes the EW symmetry to be broken by a single fundamental Higgs doublet, which is known under the name of a linearly realized effective theory. This assumption can be in principle tested by measuring LEFT parameters from low-energy observables, SMEFT parameters from observables above the electroweak scale, and having the matching between these two theories [566, 811, 829]. I stress that in practice such a test will be very difficult in view of the presence of many operators.

When the breakdown of EW symmetry is realized nonlinearly, one deals with the EW chiral Lagrangians including a light Higgs. In this case the Higgs is a composite Nambu–Goldstone boson as considered in models in which SM gauge symmetry is broken dynamically. As far as Higgs physics and collider physics is concerned, these studies go back to the early 1990s [830–832]. More recent analyses can be found in [833–841]. In particular the complete electroweak chiral Lagrangian with a light Higgs at NLO has been presented in [838, 839] and renormalization group evolution of Higgs effective field theory in [840, 841].

14.7 ε'/ε beyond the SM

14.7.1 Master Formula for ε'/ε beyond the SM

In Chapter 10 we have presented various analytical formulas for ε'/ε in the SM. Writing ε'/ε as a sum of the SM and BSM contributions,

$$\frac{\varepsilon'}{\varepsilon} = \left(\frac{\varepsilon'}{\varepsilon}\right)_{\text{SM}} + \left(\frac{\varepsilon'}{\varepsilon}\right)_{\text{BSM}}, \tag{14.82}$$

we will now present a master formula for the BSM part of this expression, which is valid in *any* theory beyond the SM that is free from nonstandard light degrees of freedom below the electroweak scale. This is in fact possible as we know all BSM $K \to \pi\pi$ hadronic matrix elements calculated in [246] at the low-energy scale μ_{low} and also the RG evolution from μ_{low} to μ_{ew} at which the WCs are calculated.

Such a master formula for $(\varepsilon'/\varepsilon)_{\text{BSM}}$ has been presented in [798]. It exhibits the dependence on each WC at the scale μ_{ew}. Technical details that led to this formula can be found in [798] and in particular in [701].

Before presenting this formula a few comments are in order:

- The WCs are defined by

$$\mathcal{H}_{\Delta S=1} \equiv - \sum_i \left[C_i(\mu_{\text{ew}})O_i + C_i'(\mu_{\text{ew}})O_i' \right], \qquad (14.83)$$

with O_i' obtained from O_i by interchanging $P_L \leftrightarrow P_R$.
- As we have seen in the previous chapter, there are many operators in this sum when all five quarks below the electroweak scale are taken into account. Relating them to the SM basis is very complicated, and the idea in [701, 798] is to avoid this step by using a simpler basis, not involving sums over quarks. This basis is discussed in detail in [701]. As we will see soon, this allows in no time to calculate the BSM part of ε'/ε using numerical tables in both papers.
- For this reason the minus sign in (14.83) is not an issue. In calculating the BSM part, we simply forget about the definitions of SM operators Q_6, Q_8, etc. that we used so far and calculate the WCs defined by (14.83).
- While [798] is very short but allows already to calculate $(\varepsilon'/\varepsilon)_{\text{BSM}}$, we advise motivated readers to study [701] as it contains a lot of information about ε'/ε in the SMEFT, which we can only briefly present here. The complexity of the analysis illustrated in Figure 14.1 will hopefully not stop the readers to look at this paper. The classes A–E of operators indicated in this sketch are explained there.

The master formula of [798] for the BSM part in (14.82) then reads

$$\left(\frac{\varepsilon'}{\varepsilon}\right)_{\text{BSM}} = \sum_i P_i(\mu_{\text{ew}}) \,\text{Im}\left[C_i(\mu_{\text{ew}}) - C_i'(\mu_{\text{ew}}) \right] \times (1\,\text{TeV})^2, \qquad (14.84)$$

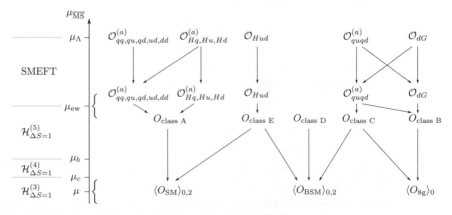

Figure 14.1   Sketch of the different contributions to ε'/ε discussed in section 2 of [701].

where

$$
P_i(\mu_{\text{ew}}) = \sum_j \sum_{I=0,2} p_{ij}^{(I)}(\mu_{\text{ew}}, \mu_{\text{low}}) \left[\frac{\langle O_j(\mu_{\text{low}})\rangle_I}{\text{GeV}^3} \right],
\tag{14.85}
$$

with the sum over i extending over the Wilson coefficients C_i of all operators and their chirality-flipped counterparts, that is $36 + 36'$ linearly independent four-quark operators and $1 + 1'$ chromo-magnetic dipole operators discussed in previous chapter. The C_i' are the WCs of the corresponding chirality-flipped operators as mentioned earlier. The relative minus sign accounts for the fact that their $K \rightarrow \pi\pi$ matrix elements differ by a sign. Among the contributing operators are also operators present already in the SM, but their WCs in (14.84) include only BSM contributions. The list of all contributing operators has been given in the previous chapter, but it is of course also given in [701, 798].

As the coefficients defined by (14.83) carry dimension, to make the coefficients $P_i(\mu_{\text{ew}})$ and $p_{ij}^{(I)}(\mu_{\text{ew}}, \mu_{\text{low}})$ dimensionless, we have included the factor $(1\,\text{TeV})^2$. The coefficients $p_{ij}^{(I)}(\mu_{\text{ew}}, \mu_{\text{low}})$ include the QCD and QED RG evolution from μ_{ew} to μ_{low} for each WC as well as the relative suppression of the contributions to the $I = 0$ amplitude due to $\text{Re}A_2/\text{Re}A_0 \ll 1$ for the matrix elements $\langle O_j(\mu_{\text{low}})\rangle_I$ of all the operators O_j present at the low-energy scale. The index j includes also i so that the effect of self-mixing is included. The $P_i(\mu_{\text{ew}})$ do not depend on μ_{low} to the considered order because the μ_{low}-dependence cancels between matrix elements and the RG evolution operator. Moreover, it should be emphasized that their values are *model-independent* and depend only on the SM dynamics below the electroweak scale, which includes short-distance contributions down to μ_{low} and the long-distance contributions represented by the hadronic matrix elements. The BSM dependence enters our master formula in (14.84) *only* through $C_i(\mu_{\text{ew}})$ and $C_i'(\mu_{\text{ew}})$. That is, even if a given P_i is nonzero, the fate of its contribution depends on the difference of these two coefficients. In particular, in models with exact left-right symmetry this contribution vanishes as first pointed out in [842].

The numerical values of the $P_i(\mu_{\text{ew}})$ are collected in the tables presented in [701, 798]. As seen in (14.85), the P_i depend on the hadronic matrix elements $\langle O_j(\mu_{\text{low}})\rangle_I$ and the RG evolution factors $p_{ij}^{(I)}(\mu_{\text{ew}}, \mu_{\text{low}})$. The numerical values of the hadronic matrix elements rely on LQCD in the case of SM operators and DQCD in the case of BSM operators. Once lattice QCD will provide precise values for the latter, an update for P_i will be possible.

Inspecting the results in the tables of [701] the following observations can be made, here given in a simplified form in terms of three islands of hadronic matrix elements: SM-like island governed by SM operator Q_8, chromo-magnetic one governed by the chromo-magnetic penguin operator O_{8g} and the BSM-island conquered in [246]:

- The largest P_i values on the SM-like island can be traced back to the large values of the matrix element $\langle Q_8 \rangle_2$, the dominant electroweak penguin operator in the SM, and the enhancement by 22 of the $I = 2$ contributions relative to $I = 0$ ones.
- The small P_i values on the chromo-magnetic island are the consequence of the fact that each one is proportional to $\langle O_{8g}\rangle_0$, which has recently been found to be much smaller than previously expected [843, 844]. Moreover, as $\langle O_{8g}\rangle_2 = 0$, all contributions in

this class are suppressed by the factor 22 relative to contributions from other classes. Nevertheless, there are NP scenarios where they play an important role (see, e.g., [845, 846]).

- The large P_i values on the BSM-island can be traced back to the large hadronic matrix elements of scalar and tensor operators calculated in [246]. In particular the largest P_i are found for

$$(\bar{s}^\alpha \sigma_{\mu\nu} P_L d^\alpha)(\bar{u}^\beta \sigma^{\mu\nu} P_L u^\beta), \qquad (\bar{s}^\alpha \sigma_{\mu\nu} P_L d^\beta)(\bar{u}^\beta \sigma^{\mu\nu} P_L u^\alpha), \qquad (14.86)$$

$$(\bar{s}^\alpha \sigma_{\mu\nu} P_L d^\alpha)(\bar{d}^\beta \sigma^{\mu\nu} P_L d^\beta), \qquad (\bar{s}^\alpha P_L d^\alpha)(\bar{u}^\beta P_L u^\beta), \qquad (14.87)$$

where α, β are color indices. The operators in (14.86) are related by Fierz identities to scalar-scalar ones and are allowed by the full gauge symmetry of the SM. This is not the case for the operators in (14.87), which violate hypercharge conservation and are only consistent with $\mathrm{SU}(3)_c \times \mathrm{U}(1)_Q$.

The question then arises of which creatures are living on these islands. To this end the analysis in [782], mentioned previously, turns out to be very useful. Here we provide only a short answer for the SM- and BSM-islands:

- SM-island: Z', W', G', vector-like quarks
- BSM-island: heavy scalars

We will encounter the ones on the SM-island in several concrete models in later chapters, but we will also look briefly at BSM-island as well. This will allow us to illustrate on examples how the master formula in question is used. The detailed phenomenology of ε'/ε in concrete models on the BSM-island is left for the future.

The master formula given in (14.84) tells us what happens below the electroweak scale μ_{ew} with the information collected in the coefficients $P_i(\mu_{\mathrm{ew}})$. While the values of the coefficients C_i and C_i' are certainly model dependent, some general lessons from the extensive SMEFT analysis in [701] can be drawn.

14.7.2 Lessons from the SMEFT Analysis of ε'/ε

In [701] the first model-independent anatomy of the ratio ε'/ε in the context of the SMEFT has been presented. This was only possible thanks to the 2018 calculations of the $K \to \pi\pi$ matrix elements of BSM operators, namely of the chromo-magnetic dipole operators by lattice QCD [843] and DQCD [844] and in particular through the calculation of matrix elements of all four-quark BSM operators, including scalar and tensor operators, by DQCD [246]. Even if the latter calculation has been performed in the chiral limit, it offers for the first time a look into the world of BSM operators contributing to ε'/ε.

The main goal in [701] was to identify those NP scenarios, which are probed by ε'/ε and which could help to explain the emerging anomaly in ε'/ε discussed in Chapter 10.

Since 2015 a number of analyses, addressing the ε'/ε anomaly in concrete models, appeared in the literature. They are collected in Section 14.7.3. We will discuss some of them in the next chapters. But all these analyses concentrated on models in which NP entered exclusively through modifications of the WCs of SM operators, in particular of the WC of the dominant electroweak penguin operator Q_8. Thus all these analyses have been performed on the SM-like island corresponding to the first term in (13.1).

This is a significant limitation if one wants to have a general view of possible BSM scenarios responsible for the ε'/ε anomaly. In particular, in the absence of even approximate values of hadronic matrix elements of BSM operators contributing to the second term in (13.1), no complete model-independent analysis was possible. The 2018 calculations of BSM $K \to \pi\pi$ matrix elements, in particular of those of scalar and tensor operators in [246], allowed to perform the analysis also on the BSM-island. Combined with the LEFT and in particular SMEFT technology in [701] this study widens significantly our view on BSM contributions to ε'/ε. The analysis in [701] has two main virtues:

- It opens the road to the analyses of ε'/ε in any theory beyond the SM and allows with the help of the master formula in (14.84) to search very efficiently for BSM scenarios behind the ε'/ε anomaly. In particular, the values of P_i collected in this paper indicate which routes could be more successful than others both in the context of the LEFT and the SMEFT. By implementing our results in the open source code flavio [813], testing specific BSM theories becomes particularly simple.
- Through this SMEFT analysis it was possible to identify correlations between ε'/ε and various observables that depend sensitively on the operators involved. Here $\Delta S = 2$, $\Delta C = 2$, and electric dipole moments (EDM) play a prominent role but also correlations with $\Delta S = 1$ and $\Delta C = 1$ provide valuable informations. We ask the interested readers to look at [701] for more details. We will return to some of these correlations in the following chapters.

The main messages from [701] are as follows:

- Tree-level vector exchanges, like Z', W', and G' contributions (SM-like island), discussed already by various authors and vector-like quarks [666] can be responsible for the observed anomaly, but generally one has to face important constraints from $\Delta S = 2$ and $\Delta C = 2$ transitions as well as direct searches, and often some fine-tuning is required. Here the main role is played by the electroweak operator Q_8 with its WC significantly modified by NP.
- Models with tree-level exchanges of heavy colorless or colored scalars (BSM-island) are a new avenue, opened with the results for BSM operators from DQCD in [246]. In particular scalar and tensor operators, having chirally enhanced matrix elements and consequently large coefficients P_i, are candidates for the explanation of the anomaly in question. Moreover, some of these models, in contrast to models with tree-level Z' and G' exchanges, are free from both $\Delta S = 2$ and $\Delta C = 2$ constraints. The EDM of the neutron is an important constraint for these models, depending on the couplings, but does not preclude sizable NP effects in ε'/ε.
- Models with modified $W^\pm$ or Z^0 couplings can induce sizable effects in ε'/ε without appreciable constraints from semileptonic decays such as $K^+ \to \pi^+\nu\bar{\nu}$ or $K_L \to \pi^0\ell\bar{\ell}$. In the case of an SM singlet Z' mixing with the Z, sizable Z-mediated contributions are

disfavored by electroweak precision tests. Yet, as discussed in [847] and in references given there, also such models could contribute to our understanding of the role of NP in ε'/ε. This is in particular the case of models with vector-like quarks [666].

Of particular interest in the results of [246] is large mixing of the scalar-scalar operators into tensor-tensor operators that is responsible for the enhancement of the matrix elements of tensor-tensor operators in the process of $O(1/N)$ meson evolution. As pointed out in the latter paper, this feature has some analogy to the observation made in [756, 768], where the QED short-distance RG evolution of NP contributions to charged-current induced leptonic and semileptonic meson decays has been presented, focusing on chirality-flipped operators at the quark level. It has been pointed out that the large mixing of the tensor-tensor operators into the scalar-scalar ones has an important impact on the phenomenology. Recently this aspect has also been discussed in the context of $R(D^{(*)})$ anomalies in [848, 849].

In fact, the one-loop QED diagrams responsible for the latter mixing are the same as the one-loop QCD diagrams with gluon replaced by photon, QCD coupling replaced by the QED one, and color matrices replaced by charge ones. Even if the RG evolution of the QED coupling constant is different from the QCD one, the pattern of mixing analyzed in [756, 768] is very similar to the one found in [246].

The reason why in [756, 768] tensor operators have the impact on the scalar ones, as opposed to the case discussed here, is simply related to the known fact, encountered already in Chapter 5 and Section 7.2, that while the evolution of the matrix elements of operators is governed by the ADM of operators, the evolution of their Wilson coefficients, analyzed in [756, 768], is governed by the corresponding transposed matrix.

We expect then that while in [756, 768, 848, 849] the large mixing in question had impact on the phenomenology of B-meson decays, in the case at hand, it will have significant impact on ε'/ε. But a detailed phenomenological analysis has still to be done.

The future of ε'/ε in the SM and in the context of searches for NP will depend on how accurately it can be calculated. This requires improved lattice calculations not only of the matrix elements of SM operators but also of the BSM ones, which are known presently only from the DQCD approach in the chiral limit. Moreover, the impact of FSI on the values of P_i have to be investigated. As these values are dominated by $I = 2$ matrix elements, we expect that these effects amount to at most 10–20% of the present values. It is also hoped that lattice QCD will be able to take into account isospin breaking corrections and that other lattice collaborations will attempt to calculate hadronic matrix elements of all relevant operators. In this context we hope that the new analysis of the RBC-UKQCD collaboration with improved matrix elements to be expected this year will shed new light on the hinted anomaly. Such future updates can be easily accounted for by the supplementary details on the master formula in the appendices of [701].

On the short-distance side the NNLO results for QCD penguins should be available soon [649]. The dominant NNLO corrections to electroweak penguins were calculated almost 20 years ago [647]. As pointed out in [701] all the estimates of ε'/ε at NLO prior to this paper suffer from short-distance renormalization scheme uncertainties in the electroweak penguin contributions and also scale uncertainties in $m_t(\mu)$ that are removed only in the NNLO matching at the electroweak scale [647]. We have stressed it in Section 10.4.

Table 14.3 Papers studying implications of the ε'/ε anomaly.

NP Scenario	References	Correlations with
LHT	[850]	$K_L \to \pi^0 \nu \bar{\nu}$
Z-FCNC	[766, 799, 851]	$K^+ \to \pi^+ \nu \bar{\nu}$ and $K_L \to \pi^0 \nu \bar{\nu}$
Z′	[851]	$K^+ \to \pi^+ \nu \bar{\nu}$, $K_L \to \pi^0 \nu \bar{\nu}$ and ΔM_K
Simplified models	[601]	$K_L \to \pi^0 \nu \bar{\nu}$
331 models	[852, 853]	$b \to s \ell^+ \ell^-$
Vector-like quarks	[666]	$K^+ \to \pi^+ \nu \bar{\nu}$, $K_L \to \pi^0 \nu \bar{\nu}$ and ΔM_K
Supersymmetry	[854–858]	$K^+ \to \pi^+ \nu \bar{\nu}$ and $K_L \to \pi^0 \nu \bar{\nu}$
2-Higgs doublet model	[846, 859, 860]	$K^+ \to \pi^+ \nu \bar{\nu}$ and $K_L \to \pi^0 \nu \bar{\nu}$
Right-handed currents	[861, 862]	EDMs
Left-right symmetry	[863, 864]	EDMs
Leptoquarks	[667]	all rare Kaon decays
SMEFT	[701]	several processes
SU(8)	[865]	$b \to s \ell^+ \ell^-$, $K^+ \to \pi^+ \nu \bar{\nu}$, $K_L \to \pi^0 \nu \bar{\nu}$
Diquarks	[866, 867]	ε_K, $K^+ \to \pi^+ \nu \bar{\nu}$, $K_L \to \pi^0 \nu \bar{\nu}$
3HDM+ν_R	[868]	$R(K^{(*)})$, $R(D^{(*)})$
Vectorlike compositeness	[869]	$R(K^{(*)})$, $R(D^{(*)})$, ε_K, $K^+ \to \pi^+ \nu \bar{\nu}$, $K_L \to \pi^0 \nu \bar{\nu}$

The fact that these NNLO contributions imply a *negative* shift of about -1.1×10^{-4} in ε'/ε gives another motivation for the search for new physics responsible for it, and thus for the analyses just described and those listed soon.

As far as BSM operators are concerned, an NLO analysis of their Wilson coefficients should be completed in 2020, but its importance is not as high as of hadronic matrix elements due to significant additional parametric uncertainties residing in any NP model. In any case, in the coming years the ratio ε'/ε is expected to play a significant role in the search for NP. In this respect, the results presented in [701, 798] will be helpful in disentangling potential models of new CP violating sources beyond the SM as well as constraining the magnitude of their effects.

14.7.3 ε'/ε in Specific Models

A number of authors investigated what kind of NP could give sufficient upward shift in ε'/ε and what would then be the implications for $K^+ \to \pi^+ \nu \bar{\nu}$ and $K_L \to \pi^0 \nu \bar{\nu}$. There is no space to discuss them all here but some of them will be discussed in later chapters. The list of papers known to the author in May 2019 is collected in Table 14.3. In these models ε'/ε can be enhanced significantly without violating existing constraints. The exceptions are leptoquark models, which we will discuss in Section 16.4.13.

It should be mentioned that in most models listed in Table 14.3, only modifications of the WCs of SM operators by NP contributions have been considered. The exceptions are the SMEFT analyses discussed earlier and the inclusion of chromomagnetic penguins in [846]. Finally, I can recommend an interesting paper in [1326] that demonstrates the correlation of ε'/ε with hadronic B decays via $U(2)^3$ flavor symmetry. This symmetry will be discussed in the following chapter.

Simplest Extensions of the SM

15.1 Minimal Flavor Violation

15.1.1 Constrained Minimal Flavor Violation (CMFV)

Formulation

This is possibly the simplest class of extensions of the SM. It is defined pragmatically as follows [520]:

- The only source of flavor and CP violation is the CKM matrix. This implies that the only CP-violating phase is the KM phase. Moreover, it is assumed that CP-violating flavor blind phases, to be discussed later on, are absent.
- The only relevant operators in the effective Hamiltonian *below* the electroweak scale are the ones present within the SM.

Detailed expositions of phenomenological consequences of this NP scenario have been given already a long time ago in [521, 697, 870] and more recently in [388]. Here we will collect most important properties of this class of models and summarize their status.

We have seen in Chapter 6 that an elegant formulation of FCNC processes within the SM is the PBE in which the basic object is the set of seven gauge independent master functions in (6.29). They result from calculations of penguin and box diagrams within the SM and depend only on the top quark mass. This description of FCNC processes can be generalized in a straightforward manner to CMFV models by simply replacing in all SM formulas the master functions in (6.29) by the following seven master functions [521]

$$S(\omega), \ X(\omega), \ Y(\omega), \ Z(\omega), \ E(\omega), \ D'(\omega), \ E'(\omega), \tag{15.1}$$

where the variable ω collects the parameters of a given model. Two properties of these functions should be emphasized.

- They are *real* valued as the only complex phases reside in CKM factors that multiply these functions in effective Hamiltonians and flavor observables.
- They are flavor universal, as the full flavor dependence comes from the CKM factors.

These properties imply relations between various observables not only within a given meson system but also in particular between observables in different meson systems that are valid for the whole class of CMFV models and most importantly do not depend on any NP parameters. These relations are in fact the same as in the SM and as such do not allow the distinction between various CMFV models that is only possible by means of observables that explicitly depend on the functions in (15.1). On the other hand violation of any of these relations by experimental data would automatically signal new sources of flavor and CP violation beyond CMFV framework and would be a problem for all models of this class.

As the SM contributions are included in the functions in (15.1) it is often useful to write

$$S(\omega) = S_0(x_t) + \Delta S, \qquad X(\omega) = X_0(x_t) + \Delta X, \qquad Y(\omega) = Y_0 + \Delta Y, \qquad (15.2)$$

with analogous decomposition for the remaining four functions. The shifts ΔS, ΔX, and ΔY summarize the modifications of SM master functions by NP contributions.

This formulation tells us right away which master functions contribute to a given decay. We have, similar to the SM, the following correspondence between the most interesting FCNC processes and the master functions in the CMFV models

$K^0 - \bar{K}^0$-mixing (ε_K)	$S(\omega)$
$B_{d,s}^0 - \bar{B}_{d,s}^0$-mixing ($\Delta M_{s,d}$)	$S(\omega)$
$K \to \pi\nu\bar{\nu}, B \to X_{d,s}\nu\bar{\nu}$	$X(\omega)$
$K_L \to \mu\bar{\mu}, B_{d,s} \to \ell^+\ell^-$	$Y(\omega)$
$K_L \to \pi^0 e^+ e^-$	$Y(\omega), Z(\omega), E(\omega)$
ε', Nonleptonic $\Delta B = 1, \Delta S = 1$	$X(\omega), Y(\omega), Z(\omega), E(\omega)$
$B \to X_s\gamma$	$D'(\omega), E'(\omega)$
$B \to X_s$ gluon	$E'(\omega)$
$B \to X_s\ell^+\ell^-$	$Y(\omega), Z(\omega), E(\omega), D'(\omega), E'(\omega)$

At this point, the impact of QCD corrections on this formulation should be mentioned. As the operator structure is the same as in the SM, the renormalization group evolution from scale $O(m_t)$ down to low energies is precisely the same as in the SM. That is, the functions in (15.1) enter the Wilson coefficients at the matching scale $\mu = O(\mu_{ew})$ with μ_{ew} being any scale like m_t, M_W, or M_Z. But the QCD corrections to the shifts in (15.2) at the matching scale and at all scales above it differ from the ones in the SM and vary from model to model. This model dependence of QCD corrections can be significant between models in which NP enters at vastly different scales as then the size of renormalization group effects in these models is different. But the important feature of these QCD corrections is that they are flavor blind. Therefore, even if the effect of QCD corrections on the shifts in (15.2), as the shifts themselves, is model dependent, the flavor universality of these functions is retained. The breakdown of flavor universality of the master functions could be caused in principle by electroweak corrections and Yukawa couplings that are not flavor blind. But these corrections are not considered in [521, 697, 870] and will be assumed to be small until we discuss the formulation of MFV as an effective theory that is based on flavor symmetries and that goes beyond CMFV.

Under these assumptions the preceding table means that the observables like branching ratios for various decays of K and $B_{s,d}$ mesons, mass differences $\Delta M_{d,s}$ in $B_{d,s}^0 - \bar{B}_{d,s}^0$-mixing, and the CP violation parameters ε and ε' within CMFV models, all can be to a very good approximation entirely expressed in terms of the corresponding master functions and the relevant CKM factors.

CMFV Relations as Standard Candles of Flavor Physics

The implications of this framework are so stringent that they could be considered as standard candles of flavor physics. Of particular importance are the relations that do not depend on the master functions and involve only CKM parameters and nonperturbative parameters. A review of these relations is given in [521]. Some of them appeared already in previous chapters but it is useful to collect them here at one place. We will see that several more involved NP scenarios violate these relations.

We have:

1. $S_{\psi K_S}$ and $S_{\psi \phi}$ are as in the SM and therefore given by

$$S_{\psi K_S} = \sin(2\beta), \qquad S_{\psi \phi} = \sin(2|\beta_s|), \tag{15.3}$$

with

$$V_{td} = |V_{td}|e^{-i\beta}, \qquad V_{ts} = -|V_{ts}|e^{-i\beta_s}. \tag{15.4}$$

2. While ΔM_d and ΔM_s can differ from the SM values, their ratio is as in the SM

$$\left(\frac{\Delta M_d}{\Delta M_s}\right)_{\text{CMFV}} = \left(\frac{\Delta M_d}{\Delta M_s}\right)_{\text{SM}}. \tag{15.5}$$

   Moreover, this ratio is given entirely in terms of CKM parameters and the non-perturbative parameter ξ:

$$\frac{\Delta M_d}{\Delta M_s} = \frac{m_{B_d}}{m_{B_s}} \frac{1}{\xi^2} \left|\frac{V_{td}}{V_{ts}}\right|^2, \qquad \xi = \frac{\sqrt{\hat{B}_s} F_{B_s}}{\sqrt{\hat{B}_d} F_{B_d}}. \tag{15.6}$$

3. These two properties allow the construction of the *Universal Unitarity Triangle* (UUT) of models with CMFV that uses as inputs the measured values of $S_{\psi K_S}$ and $\Delta M_s / \Delta M_d$ [520]. We will present this construction later. In this manner the angle γ and the ratio $|V_{ub}|/|V_{cb}|$ can be determined and compared with their tree-level determinations.

4. The flavor universality of $S(\omega)$ allows to derive universal relations between ε_K and $\Delta M_{s,d}$ that depend only on $|V_{us}|$, $|V_{cb}|$, known from tree-level decays, and nonperturbative parameters entering the evaluation of ε_K and $\Delta M_{s,d}$ [27, 697, 871].

5. For fixed CKM parameters determined in tree-level decays, $|\varepsilon_K|$, ΔM_s and ΔM_d, if modified, can only be *enhanced* relative to SM predictions [872]. Moreover, this happens in a correlated manner [871]. This follows from the lower bound [872]

$$S(\omega) \geq S_0(x_t) = 2.32 \tag{15.7}$$

with $S_0(x_t)$ given in (6.17). We will see later that this particular correlation between ε_K, ΔM_s, and ΔM_d, appears to be in variance with the data hinting at the presence of NP beyond the CMFV framework in $\Delta F = 2$ transitions.

6. Two other interesting universal relations in models with CMFV are

$$\frac{\mathcal{B}(B \to X_d \nu \bar{\nu})}{\mathcal{B}(B \to X_s \nu \bar{\nu})} = \left| \frac{V_{td}}{V_{ts}} \right|^2, \tag{15.8}$$

$$\frac{\mathcal{B}(B_d \to \mu^+ \mu^-)}{\mathcal{B}(B_s \to \mu^+ \mu^-)} = \frac{\tau(B_d)}{\tau(B_s)} \frac{m_{B_d}}{m_{B_s}} \frac{F_{B_d}^2}{F_{B_s}^2} \left| \frac{V_{td}}{V_{ts}} \right|^2. \tag{15.9}$$

7. Eliminating $|V_{td}/V_{ts}|$ from (15.6) and (15.9) allows to obtain another universal relation within the CMFV models [503]

$$\frac{\mathcal{B}(B_s \to \mu^+ \mu^-)}{\mathcal{B}(B_d \to \mu^+ \mu^-)} = \frac{\hat{B}_d}{\hat{B}_s} \frac{\tau(B_s)}{\tau(B_d)} \frac{\Delta M_s}{\Delta M_d} \tag{15.10}$$

that does not involve F_{B_q} and CKM parameters and consequently contains smaller hadronic and parametric uncertainties than the preceding formulas. It involves only measurable quantities except for the ratio $\hat{B}_s/\hat{B}_d$ that is known from lattice calculations already with precision of roughly $\pm 1\%$ [522]. Consequently the r.h.s. of this equation is already very precisely known. This should allow to identify possible NP in $B_{s,d} \to \mu^+ \mu^-$ decays and also in $\Delta M_{s,d}$ beyond the CMFV framework even if it was only at the level of 20% of the SM contributions. Therefore the relation (15.10) should allow a precision test of the CMFV framework. Unfortunately, because of difficulties in measuring precisely $\mathcal{B}(B_d \to \mu^+ \mu^-)$ it could take even a decade to perform this test.

8. The fact that all amplitudes for FCNC processes within the CMFV framework can be expressed in terms of seven *real* and *universal* master loop functions listed in (15.1) implies numerous correlations between various observables. See Chapter 19.

Construction of the UUT

There are many ways to construct UT but the construction of the UUT must make sure that it does not involve any NP parameters. This is achieved by using

$$\frac{\Delta M_d}{\Delta M_s}, \qquad S_{\psi K_S}. \tag{15.11}$$

We present here the most recent construction of the UUT, which is an update of the one in [388]. From (15.6) we obtain first

$$\frac{|V_{td}|}{|V_{ts}|} = \xi \sqrt{\frac{m_{B_s}}{m_{B_d}}} \sqrt{\frac{\Delta M_d}{\Delta M_s}} = 0.2046 \pm 0.0033, \tag{15.12}$$

where we have used [873]

$$\xi = 1.206 \pm 0.019. \tag{15.13}$$

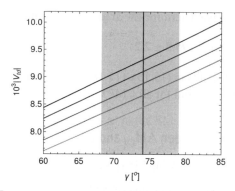 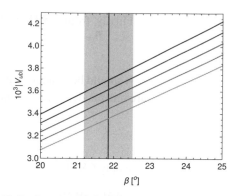

Figure 15.1 Left: $|V_{td}|$ as function of γ, for different values of $|V_{cb}|$. Right: $|V_{ub}|$ as function of β, for different values of $|V_{cb}|$. From bottom to top: $|V_{cb}| = 39 \cdot 10^{-3}, 40 \cdot 10^{-3}, 41 \cdot 10^{-3}, 42 \cdot 10^{-3}, 43 \cdot 10^{-3}$. From [874].

Emerging ΔM_d-Anomaly, the Angle γ, and $|V_{cb}|$

As we discussed already in Chapter 3, most recent tree-level measurements of $|V_{cb}|$ favor its high value and consequently no ε_K anomaly. As then NP is expected in $\Delta M_{s,d}$, it appears as a better strategy to replace their ratio by the angle γ and instead treat $\Delta M_{s,d}$ as outputs being functions of γ, β, and $|V_{cb}|$ [874]. The four fundamental CKM parameters used to determine the remaining entries in the CKM matrix are then

$$|V_{us}|, \qquad |V_{cb}|, \qquad \gamma, \qquad \beta. \tag{15.27}$$

In Figure 15.1 we show $|V_{td}|$ as a function of γ and $|V_{ub}|$ as a function of β for different values of $|V_{cb}|$. The dependences of $|V_{td}|$ on β and of $|V_{ub}|$ on γ are very small. These plots will allow to monitor the values of $|V_{td}|$ and $|V_{ub}|$ that enter various observables as the uncertainties of γ, β, and $|V_{cb}|$ will shrink with time.

In Figure 15.2 we show in the left panel ΔM_d and ΔM_s normalized to their experimental values. Evidently, for central values of all parameters, ΔM_d differs by roughly 30% from the data while in the case of ΔM_s the corresponding difference amounts only to 12%. But the uncertainties in other parameters like $|V_{cb}|$ and the hadronic parameters are still significant. However, we expect that in the coming years these uncertainties will be reduced by much so that a possible ΔM_d anomaly will be better seen.

In the right panel of Figure 15.2 we show the ratio $\Delta M_s/\Delta M_d$ as a function of γ. The dependence on $|V_{cb}|$ cancels in this ratio, and the error on ξ in (15.13) is much smaller than the hadronic uncertainties affecting the plot in the left panel of Figure 15.2. Consequently the disagreement of the ratio in question with the data, shown as a horizontal line at 35.1, is clearly visible and expresses the problem of CMFV models and those based on the $U(2)^3$ symmetry as discussed later in this book.

As discussed in [874], in the presence of only left-handed currents, the required suppression of ΔM_d and to a lesser extent ΔM_s implies large new CP-violating phases and significant deviations from the SM in CP-asymmetries of radiative and rare $b \to d$ and $b \to s$ decays.

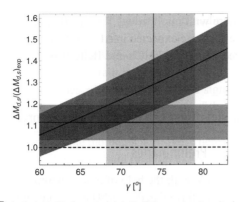
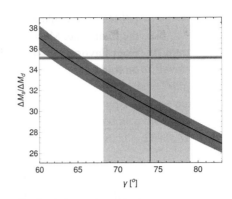

Figure 15.2   Left: ΔM_d (dark gray) and ΔM_s (light gray) as functions of γ, normalized to their experimental values. The 1σ-band includes all other uncertainties. Right: $\Delta M_s / \Delta M_d$ as function of γ. From [874].

In the presence of both left- and right-handed couplings, on the other hand, the suppression of ΔM_d is much easier to achieve without introducing large CP-violating phases. In this context probably most interesting are models in which the SMEFT operator O_{Hd} involving right-handed flavor-violating couplings to down-quarks is generated at the NP scale, that is the Scenario I discussed in Section 14.4.1. We saw there that the renormalization group evolution to low-energy scales involving also left-handed currents present already within the SM generates left-right $\Delta F = 2$ operators representing FCNC mediated by the Z boson.

An explicit realization of such an NP scenario is provided by models with vector-like quarks with an additional U(1) gauge symmetry so that both tree-level Z and Z' exchanges are present, and in some models of this type also box diagram contributions with vector-like quarks, Higgs and other scalar and pseudoscalar exchanges are important [666]. We will discuss models with vector-like quarks in Section 16.3.2.

In any case, this anomaly, if confirmed by improved measurements of the angle γ, will have implications for observables sensitive to $b \to d$ transitions like $b \to d\ell^+\ell^-$ and $b \to d\nu\bar{\nu}$ which will be explored by Belle II. It will open a new oasis of NP, analogous to the one related to the recent anomalies in $b \to s\ell^+\ell^-$ and their implications for $b \to s\nu\bar{\nu}$ transitions. Depending on the NP flavor structure, it could also have implications for $K^+ \to \pi^+\nu\bar{\nu}$ and $K_L \to \pi^0\nu\bar{\nu}$.

New Physics in $\Delta\Gamma_{d,s}$

In this book, as far as particle-antiparticle mixing is concerned we will mainly discuss NP contributions to ΔM_K and $\Delta M_{d,s}$ but NP could also affect $\Delta\Gamma_{d,s}$. Good references here are [387, 393], where collection of references to earlier papers can be found. One finds that NP can affect $\Delta\Gamma_d$ by even 100%, as pointed out in [875] and analyzed in detail in [393]. NP related to the violation of CKM unitarity and automatically beyond the MFV framework could be responsible for sizable modifications of $\Delta\Gamma_d$ that could be measured through semileptonic CP asymmetries in B_d^0 decays. Also effective operators of the form corresponding to the independent unitary rotations in flavor space of the five fermion fields in (15.29). The nonabelian factors are

$$G_q = SU(3)_{Q_L} \times SU(3)_{U_R} \times SU(3)_{D_R}, \qquad G_\ell = SU(3)_{L_L} \otimes SU(3)_{E_R}. \qquad (15.31)$$

The five U(1) subgroups can be identified as follows

$$U(1)^5 = U(1)_B \times U(1)_L \times U(1)_Y \times U(1)_{PQ} \times U(1)_E \qquad (15.32)$$

that is with baryon number (B), lepton number (L), and hypercharge (Y), which are respected by the Yukawa interactions. The two remaining U(1) groups can be identified with the Peccei-Quinn symmetry whereby the Higgs and D_R, E_R fields have opposite charges, and with a global rotation of E_R only.

While $U(1)_B$, $U(1)_L$, and $U(1)_Y$ are unaffected by Yukawa interactions, the nonabelian groups G_q and G_ℓ are explicitly broken by them. Indeed

$$-\mathcal{L}_{Yukawa}^{SM} = Y_d^{ij} \bar{Q}_L^i \phi D_R^j + Y_u^{ij} \bar{Q}_L^i \tilde{\phi} U_R^j + Y_e^{ij} \bar{L}_L^i \phi E_R^j + \text{h.c.} \qquad (\tilde{\phi} = i\tau_2 \phi^\dagger) \qquad (15.33)$$

breaks explicitly G_q and G_ℓ because the Yukawa couplings $Y_{d,u,e}$ are not proportional to a unit matrix because of the known mass spectrum of quarks and leptons.

The MFV hypothesis as formulated in [624] is then based on the assumption that Y_d and Y_u are the only sources of flavor symmetry breaking not only in the SM but also beyond it. On the technical side to implement this hypothesis in a consistent manner, one assumes that G_q is a good symmetry and promotes first $Y_{u,d}$ to be nondynamical fields (*spurions*) with nontrivial transformation properties under G_q:

$$Y_u \sim (3, \bar{3}, 1), \qquad Y_d \sim (3, 1, \bar{3}). \qquad (15.34)$$

In this manner also the Yukawa Lagrangian is invariant under G_q. Once these Yukawas take their background values, that is their SM values, the symmetry G_q is broken.

Within an effective-theory approach to physics beyond the SM, an effective theory satisfies the criterion of MFV in the quark sector if all higher-dimensional operators, constructed from SM and Y fields, are invariant under CP and (formally) under the flavor group G_q [624].

Strictly speaking one should then consider operators with arbitrary powers of the (dimensionless) Yukawa fields. But fortunately except for top Yukawa coupling all the eigenvalues of the Yukawa matrices are small and also off-diagonal elements of the CKM matrix are strongly suppressed so that

$$\left[Y_u (Y_u)^\dagger\right]_{i \neq j}^n \approx y_t^n V_{it}^* V_{tj}. \qquad (15.35)$$

Then, in the limit where we neglect light quark masses, the leading $\Delta F = 2$ and $\Delta F = 1$ FCNC amplitudes get exactly the same CKM suppression as in the SM [883]:

$$\mathcal{A}(d^i \to d^j)_{MFV} = (V_{tj}^* V_{ti}) \ \mathcal{A}_{SM}^{(\Delta F=1)} \left[1 + a_1 \frac{16\pi^2 M_W^2}{\Lambda^2}\right], \qquad (15.36)$$

$$\mathcal{A}(M_{ij} - \bar{M}_{ij})_{MFV} = (V_{tj}^* V_{ti})^2 \mathcal{A}_{SM}^{(\Delta F=2)} \left[1 + a_2 \frac{16\pi^2 M_W^2}{\Lambda^2}\right], \qquad (15.37)$$

where the $\mathcal{A}_{\text{SM}}^{(i)}$ are the SM loop amplitudes and the a_i are $O(1)$ *real* parameters. The a_i depend on the specific operator considered but are *flavor independent*. This implies the same relative correction in $s \to d$, $b \to d$, and $b \to s$ transitions, and the structure is very similar to the one in CMFV models.

Yet, one should stress that MFV as formulated here is, in contrast to CMFV, based on a renormalization-group-invariant symmetry argument, which can easily be extended above the electroweak scale where new degrees of freedoms, such as extra Higgs doublets or SUSY partners of the SM fields are included. It can also be extended to strongly coupled gauge theories, although in this case the expansion in powers of the Yukawa spurions is not necessarily a rapidly convergent series. In this case, a resummation of the terms involving the top-quark Yukawa coupling needs to be performed [898]

This model-independent structure does not hold in CMFV in which one assumes that the effective FCNC operators playing a significant role within the SM are the only relevant ones also beyond the SM. This condition is realized only in weakly coupled theories at the TeV scale with only one light Higgs doublet. In the MSSM, where two Higgs doublets are present, it still works for small $\tan\beta$ but does not work for large $\tan\beta$ as then new scalar operators become important. A very nice summary of this situation can be found in [883], and further details are presented in [624].

Despite these differences between CMFV and MFV, many phenomenological implications listed by us in the CMFV section are the same. In particular the CP asymmetries $S_{\psi K_S}$ and $S_{\psi\phi}$ are the same, and this applies also to relations (15.8) and (15.9). But (15.10) could be violated in MFV if new operators to $B_{s,d} \to \mu^+\mu^-$ and $\Delta M_{s,d}$ become important.

Personally, I think that CMFV relations being more stringent than the MFV ones are a useful starting point. If one day significant violation of them will be found, one can then see whether they can be accommodated within MFV.

15.1.3 Further Implications of CMFV for Weak Decays

$$B^+ \to \tau^+ \nu_\tau$$

Let us then see what MFV can tell us about this simple decay. Here we would like to point out that in this class of models the branching ratio for this decay is enhanced (suppressed) for the same (opposite) sign of the lepton coupling of the new charged gauge boson relative to the SM one. Indeed, the only possibility to modify the SM result up to loop corrections in CMFV is through a tree-level exchange of a new charged gauge boson, whose flavor interactions with quarks are governed by the CKM matrix. In particular the operator structure is the same.

Denoting this gauge boson by $\tilde{W}$ and the corresponding gauge coupling by $\tilde{g}_2$, one has

$$\frac{\mathcal{B}(B^+ \to \tau^+\nu)}{\mathcal{B}(B^+ \to \tau^+\nu)^{\text{SM}}} = \left(1 + r\frac{\tilde{g}_2^2}{g_2^2}\frac{M_W^2}{M_{\tilde{W}}^2}\right)^2, \tag{15.38}$$

where we introduced a factor r allowing a modification in the lepton couplings relative to the SM ones, in particular of its sign. Which sign is favored will be known once the data and SM prediction improve.

If $\tilde{W}$ with these properties is absent, the branching ratio in this framework is not modified with respect to the SM up to loop corrections that could involve new particles but are expected to be small. A $H^{\pm}$ exchange generates new operators and is outside this framework. The same comment applies to gauge bosons with right-handed couplings (W') that we will discuss later.

15.1.4 Comparison with Other Suppressions of FCNC

Still in the context of natural flavor conservation, as present in the MFV framework, one should mention that FCNC can be avoided at tree-level without fine-tuning through a special arrangement of the couplings of the Higgs system with two or more doublets to SM fermions [899, 900], so that there are no FCNC mediated by neutral scalars at tree-level. This is, for instance, the case of the MSSM and many 2HDM models in which one Higgs doublet is responsible for the up-quark masses and a second for down-quark ones.

However, there are still other ways of suppressing FCNC proposed in the literature, discussed in in the context of two-Higgs doublet models in [891, 896, 901–903]. They have been compared critically with the MFV framework in [896]. We describe here briefly only some of these models.

In the so-called BGL models [901], six models in total, the strength of FCNCs in the up- or down-type sector, is unambiguously related to the off-diagonal elements of the CKM matrix. While all the six BGL models are interesting, only one of them is compatible with the MFV principle. It is the one in which $d_i \rightarrow d_j$ FCNC transitions are proportional to $V_{3j}^* V_{3i}$. Recently, motivated by the $b \rightarrow s\mu^+\mu^-$ anomalies, these ideas have been used in the context of Z' models in which Z' couples flavor nonuniversally to SM quarks [904]. Other recent analyses of FCNC processes in this framework can be found in [905, 906].

On the other hand, the *Yukawa alignment model* [891] is a limiting case of the general MFV construction, where the higher-order powers in Y_u and Y_d are not included. It includes also flavor-blind CP-violating phases mentioned earlier. Neutral-Higgs phenomenology of this model is presented in [891] and the one for charged Higgs in [907]. Electric dipole moments are discussed in [908].

15.2 2HDM$_{\overline{\text{MFV}}}$

15.2.1 Preliminaries

We will next discuss a specific 2HDM model, namely 2HDM with MFV accompanied by flavor-blind CP phases that is called for short 2HDM$_{\overline{\text{MFV}}}$ [896] with the "bar" on MFV indicating the presence of FBPs. We present this model mainly to illustrate how flavor-blind phases can affect various observables.

Let us first list a few important points of the 2HDM$_{\overline{\text{MFV}}}$ framework.

• The presence of FBPs in this MFV framework modifies through their interplay with the standard CKM flavor violation the usual characteristic relations of the MFV framework.

In particular the mixing induced CP asymmetries in $B_d^0 \to \psi K_S$ and $B_s^0 \to \psi\phi$ take the form known from non-MFV frameworks like LHT, RSc, and in particular Z' models discussed later

$$S_{\psi K_S} = \sin(2\beta + 2\varphi_{B_d}), \qquad S_{\psi\phi} = \sin(2|\beta_s| - 2\varphi_{B_s}), \qquad (15.39)$$

where φ_{B_q} are NP phases in $B_q^0 - \bar{B}_q^0$ mixings. Thus in the presence of nonvanishing φ_{B_d} and φ_{B_s}, originating here in nonvanishing FBPs, these two asymmetries do not measure β and β_s but $(\beta + \varphi_{B_d})$ and $(|\beta_s| - \varphi_{B_s})$, respectively.

- The FBPs in the 2HDM$_{\overline{\text{MFV}}}$ can appear both in Yukawa interactions and in the Higgs potential. While in [896] only the case of FBPs in Yukawa interactions has been considered, in [909] these considerations have been extended to include also the FBPs in the Higgs potential. The two flavor-blind CPV mechanisms can be distinguished through the correlation between $S_{\psi K_S}$ and $S_{\psi\phi}$ that is strikingly different if only one of them is relevant. In fact the relations between generated new phases are very different in each case:

$$\varphi_{B_d} = \frac{m_d}{m_s} \varphi_{B_s} \quad \text{and} \quad \varphi_{B_d} = \varphi_{B_s} \qquad (15.40)$$

for FBPs in Yukawa couplings and Higgs potential, respectively.

- The heavy Higgs contributions to ε_K are negligible, and consequently this model, similar to CMFV, favors the high value of $|V_{cb}|$ for which the SM is consistent with the data on ε_K. But in contrast to CMFV if one day $|V_{ub}|$ will turn out to be high, the presence of a negative phase φ_{B_d} will allow in principle to remove the tension with the experimental value of $S_{\psi K_S}$. Simultaneously $S_{\psi\phi}$ will be enhanced over its SM value with the size of enhancement depending on whether FBPs in Yukawas or Higgs potential are at work.

- The selection of the large value of $|V_{ub}|$ would improve the agreement between the experimental value of $\mathcal{B}(B \to \tau^+\nu_\tau)$ and its SM value that is somewhat lower than experimentally found, see Section 9.3.

- Further implications of this NP scenario, in particular in connection with EDMs, can be found in [896, 909].

What is nice about this model is that while having new sources of CP violation it has a small number of free parameters and a number of definite predictions and correlations between various flavor observables that provide very important tests of this model. The question then arises how this simple model would face the $\Delta F = 2$ tensions and related anomalies if they turned out more significant than presently known. The lessons listed next are based on [909] and updates that happened in this decade.

1. The removal of the $|V_{ub}| - S_{\psi K_S}$ anomaly, in case of a large $|V_{ub}|$, which is achieved through the negative phase φ_{B_d}, is only possible with the help of FBPs in the Higgs potential so that optimally $\varphi_{B_s} = \varphi_{B_d}$ implying the full dominance of the $Q_{1,2}^{\text{SLL}}$ operators as far as CP-violating contributions are concerned.

2. The negative value of φ_{B_d} would through (15.39) automatically increase $S_{\psi\phi}$ relative to the SM value. Finding in the future that nature chooses a *negative* value of $S_{\psi\phi}$ and simultaneously large value of $|V_{ub}|$ would practically rule out 2HDM$_{\overline{\text{MFV}}}$.

3. In the case of the full dominance of NP effects from the Higgs potential, represented by the operators $Q_{1,2}^{\mathrm{SLL}}$,[1] also in the case of $\Delta M_{s,d}$ the CMFV relation in (15.6) is valid. Yet, the CMFV correlation between ε_K and $\Delta M_{s,d}$ is absent, and $\Delta M_{s,d}$ can be both suppressed and enhanced if necessary. In view of the tension discussed in Section 15.1.1 this could be a problem. The presence of large contributions of operators $Q_{1,2}^{\mathrm{LR}}$ would not help. They suppress ΔM_s with basically no effect on ΔM_d.

15.3 Beyond MFV: Models with U(2)^3 Symmetry

Possibly the simplest solution to the possible tensions between ε_K and $\Delta M_{s,d}$ in models with MFV is to reduce the flavor symmetry U(3)^3 down to U(2)^3 [910–916]. As pointed out in [917], in this case NP effects in ε_K and $B_{s,d}^0 - \bar{B}_{s,d}^0$ are not correlated with each other so that the enhancement of ε_K and suppression of $\Delta M_{s,d}$ can be achieved if necessary for the values of $|V_{cb}|$, $F_{B_s}\sqrt{\hat{B}_{B_s}}$, and $F_{B_d}\sqrt{\hat{B}_{B_d}}$ for which MFV has problems as discussed in Section 15.1.1.

A pragmatic definition of these models in the case of $\Delta F = 2$ transitions could be as follows:

- Flavor and CP-violation in $K^0 - \bar{K}^0$ mixing is governed by CMFV.
- The dominant source of flavor and CP violation in $B_{d,s} - \bar{B}_{d,s}$ mixings is the CKM matrix. Yet, new universal, with respect to B_d and B_s, flavor-violating and CP-violating effects in these transitions are possible. The universality in question is a direct consequence of the U(2)^3 symmetry imposed on the quark doublets of the first two generations. Moreover, only SM operators are relevant in $B_{d,s} - \bar{B}_{d,s}$ mixings. We comment on the possible small nonuniversal corrections and contributions of new operators later.
- Very importantly NP effects in K physics and $B_{d,s}$ observables are uncorrelated with each other, although in specific models such correlation could be forced by the underlying theory and the data.

In the grander formulation by means of the effective theory [910], these models are governed by a global flavor symmetry

$$\boxed{G_F = \mathrm{U}(2)_Q \times \mathrm{U}(2)_u \times \mathrm{U}(2)_d} \tag{15.41}$$

broken *minimally*[2] by three spurions transforming under G_F as follows

$$\Delta Y_u = (2, \bar{2}, 1), \quad \Delta Y_d = (2, 1, \bar{2}), \quad V = (2, 1, 1). \tag{15.42}$$

As demonstrated by means of a spurion analysis in section 5 of [910], the phenomenological consequences of this framework for $\Delta F = 2$ transitions as summarized in the pragmatic definition are general consequences of U(2)^3 symmetry and its breaking pattern.

---

[1] Not a likely situation because these operators are forbidden by SMEFT.
[2] More complicated breakdown is discussed in [910].

In particular, if one considers leading flavor-changing amplitudes no assumption of the dominance of SM operators has to be made. Moreover, the universality of NP effects in $B_{d,s} - \bar{B}_{d,s}$ systems, up to the overall usual CKM factors and the hermicity of the $\Delta F = 2$ Hamiltonian implies that the function S_K is real [910]. This result in combination with the dominance of SM operators in this framework implies the CMFV structure of $\Delta S = 2$ transitions.

In this context the following remark should be made. As in the MFV case the leading flavor-changing amplitudes in the framework of [910] are of a left-handed type and to a very good approximation can be evaluated neglecting the effects of light-quark masses. To generate dimension-6 LR operators contributing to $\Delta F = 2$ transitions, one needs at least two extra insertions of the down-type spurion ΔY_d, which implies an extra suppression of amplitudes proportional to down-quark masses. Such effects being proportional to light quark masses ($m_{s,d}$) break the flavor universality between B_d and B_s systems. They can be at best relevant for B_s-mixing, for instance in 2HDM models or supersymmetric models. However, unless one goes to a specific regime of large $\tan \beta$ and small Higgs masses, such effects are always subleading, and we will neglect them.

Let us compare then the general properties of the $\Delta F = 2$ functions S_i ($i = K, d, s$) in this scenario with the ones in CMFV models.

1. In CMFV we have

$$S_K = S_d = S_s \geq S_0(x_t), \qquad \varphi_K = \varphi_{B_d} = \varphi_{B_s} = 0 \qquad \text{(CMFV)}, \qquad (15.43)$$

Consequently,

$$S_{\psi K_S} = \sin(2\beta), \qquad S_{\psi \phi} = \sin(2|\beta_s|) \qquad \text{(CMFV)}. \qquad (15.44)$$

$|\varepsilon_K|$, ΔM_d and ΔM_s can only be enhanced in CMFV models. Moreover, this happens in a correlated manner. The enhancement of one of these observables implies automatically and uniquely the enhancement of the other two observables [697, 871].

2. On the other hand in U(2)^3 models in question (15.43) and (15.3) are replaced by

$$S_K = r_K S_0(x_t), \quad |S_d| = |S_s| = r_B S_0(x_t), \quad \varphi_K = 0, \quad \varphi_{B_d} = \varphi_{B_s} \equiv \varphi_{\text{new}}, \quad \text{(MU(2)}^3) \qquad (15.45)$$

with r_K and r_B unrelated to each other. The last equality implies

$$S_{\psi K_S} = \sin(2\beta + 2\varphi_{\text{new}}), \qquad S_{\psi \phi} = \sin(2|\beta_s| - 2\varphi_{\text{new}}), \qquad \text{(U(2)}^3), \qquad (15.46)$$

that is an anticorrelation between these two asymmetries. As there is no relation of $S_d = S_s$ to S_K in these models, ΔM_d and ΔM_s can be suppressed or enhanced with respect to the SM values, and there is no direct correlation between $|\varepsilon_K|$ and $\Delta M_{s,d}$.

In short, with respect to $\Delta F = 2$ processes there are only three new parameters in this class of models

$$r_K \geq 1, \qquad r_B, \qquad \varphi_{\text{new}}, \qquad (15.47)$$

with r_K and r_B being real and positive definite.

The first inequality in (15.47) is a direct consequence of the CMFV structure in the $\Delta S = 2$ transitions in this framework [872]. Therefore in the U(2)^3 framework $|\varepsilon_K|$ can only

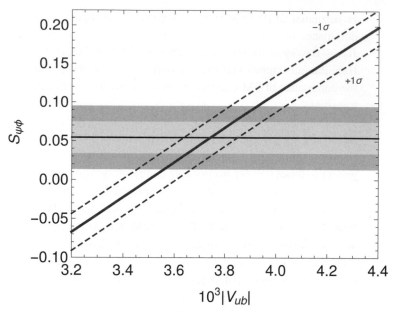

$S_{\psi\phi}$

$10^3|V_{ub}|$

Figure 15.3 $S_{\psi\phi}$ vs. $|V_{ub}|$ in models with $U(2)^3$ symmetry for different values of $|V_{ub}|$. From top to bottom: $S_{\psi K_S} = 0.682, 0.699, 0.716$. Light/dark gray: experimental $1\sigma/2\sigma$ region.

be increased over the SM value, but as stated here, this property is generally uncorrelated with $B_{s,d}$ systems.

3. Flavor universality of the functions S_q in CMFV and $U(2)^3$ models implies

$$\left(\frac{\Delta M_d}{\Delta M_s}\right)_{\mathrm{CMFV}} = \left(\frac{\Delta M_d}{\Delta M_s}\right)_{U(2)^3} = \left(\frac{\Delta M_d}{\Delta M_s}\right)_{\mathrm{SM}} = \frac{m_{B_d}}{m_{B_s}} \frac{\hat{B}_d}{\hat{B}_s} \frac{F_{B_d}^2}{F_{B_s}^2} \left|\frac{V_{td}}{V_{ts}}\right|^2 \equiv \frac{m_{B_d}}{m_{B_s}} \frac{1}{\xi^2} \left|\frac{V_{td}}{V_{ts}}\right|^2 . \tag{15.48}$$

This could be problematic in view of tensions between ΔM_d and ΔM_s and γ pointed out in [874] and discussed in Section 15.1.

4. As pointed out in [917] the relation $\varphi_{B_d} = \varphi_{B_s}$ implies for fixed $|V_{cb}|$ and γ a triple $S_{\psi K_S} - S_{\psi\phi} - |V_{ub}|$ correlation that constitutes an important test of this NP scenario. The dependence of $S_{\psi K_S}$ on $S_{\psi\phi}$ for different $|V_{ub}|$ can be found in figure 1 of [917]. In view of the precise experimental value of $S_{\psi K_S}$ we show this time in Figure 15.3 only $S_{\psi\phi}$ as a function of $|V_{ub}|$ for three values of $S_{\psi K_S}$ which is an update of figure 2 in [917].[3] It is a good exercise to check this plot, which was obtained for $\gamma = 74°$ and $|V_{cb}| = 42.0 \times 10^{-3}$. The γ dependence is fully negligible. The value of $|V_{ub}|$ is proportional to $|V_{cb}|$ as one can check performing the standard construction of the UT. Note that this correlation is independent of the values of $F_{B_s}\sqrt{\hat{B}_{B_s}}$ and $F_{B_d}\sqrt{\hat{B}_{B_d}}$. As seen in this figure the important

---

[3] The author thanks Monika Blanke for providing this plot.

advantage of U(2)^3 models over 2HDM$_{\overline{\text{MFV}}}$ is that in the case of $S_{\psi\phi}$ being very small or even having opposite sign to SM prediction, this framework can survive with concrete prediction for $|V_{ub}|$. Other implications of this NP scenario that could turn out to be relevant when the data improve can be found in [917]. In particular the impact of U(2)^3 symmetry on tree-level FCNC due to gauge boson and scalar exchanges has been analyzed in [611] and [504], respectively.

15.4 Beyond MFV: Z' Boson

15.4.1 Preliminaries

In the following sections we will generalize the discussion of the previous section to NP models, which go beyond the framework of MFV. In these models new sources of flavor violation are present, and in some of them new operators strongly suppressed in the SM become important. Even if on the fundamental level the WCs are the ones that play the crucial role, it turns out that also in this case the formulation of FCNC processes in terms of master functions is useful, simply because then the effects of non-MFV interactions are transparently seen. Indeed, we will see that beyond MFV

- The master functions become complex quantities.
- Their flavor universality is broken.
- The presence of right-handed neutral and charged vector currents will require the introduction of new master functions.
- The presence of scalar and pseudoscalar interactions will require the introduction of still other master functions.

In this section we will first introduce the notation for the master functions in this more general framework. Subsequently we will calculate them in a number of very simple NP models:

- A model with an additional U(1)$'$ gauge symmetry and related heavy neutral gauge boson Z' with flavor-violating couplings and implied tree-level contributions to FCNC processes. This scenario is very useful for the illustration of the new formulas in explicit terms. Reviews of these models can be found in [918, 919], and more references will be given later.
- A model in which the SM Z boson mediates FCNC already at tree level. We will see that the pattern of flavor violation in this case differs significantly from the one caused by tree-level Z' exchanges. We mentioned it already at the end of Section 14.5.
- A model with tree-level FCNC mediated by heavy scalars and pseudoscalars. The pattern of flavor violation in this case will not only differ from the Z' and Z cases, but will also depend on whether a scalar or pseudoscalar is considered.
- While Z' and the heavy scalars and pseudoscalars just mentioned will be colorless, it will be of interest to consider in a separate section simple models with heavy gauge

bosons, scalars, and pseudoscalars that carry color as in these cases the patterns of flavor violation will still be different from the ones encountered in models with new colorless gauge bosons, scalars, and pseudoscalars.

The different NP scenarios just listed and presented in this chapter can be considered as *simplified models* because in concrete NP models that we will discuss subsequently in the next chapter, several of these tree-level exchanges can contribute to a given flavor observable simultaneously. Moreover, in certain models also new one-loop diagrams will turn out to be important. But the technology developed in the context of the simplified models in question will turn out to be very useful for the discussion of more complicated NP scenarios. We will also see that although some of these specific models will contain more free parameters than a single simplified model, the couplings of the gauge bosons, scalars, and pseudoscalars, rather arbitrary in simplified models, will be given in terms of fundamental parameters of a given model. This will imply correlations between observables in different meson systems and also the correlations with flavor-conserving processes studied both at high-energy colliders and low-energy facilities. Moreover, it will be possible to find out whether the contributions from tree-level diagrams or one-loop diagrams to a given observable are most important.

15.4.2 Introducing Generalized Master Functions

The formulation given next in the case of models having the same operator structure as the SM but including new sources of flavor violation has been first presented in the context of the Littlest Higgs model with T-parity in [700]. It has been generalized to include in addition to left-handed (LH) currents also the right-handed (RH) ones in an analysis of Randall–Sundrum models in [920]. While in the latter paper it turned out to be useful to work with $V - A$ and V currents, we will here follow the formulation in terms of L, R, V, and A currents presented in [611]. The indices L and R in the functions following correspond to LH and RH currents, respectively.

For our purposes it will be sufficient to consider the following functions:

- For $\Delta F = 2$ processes

$$\boxed{S(K), \qquad S(B_d), \qquad S(B_s),} \tag{15.49}$$

where we will include in the definitions of these functions the contributions of operators with LL, RR, and LR Dirac structures.

- For decays with a meson and $\nu\bar{\nu}$ in the final state

$$\boxed{X_{L,R}(K), \qquad X_{L,R}(B_d), \qquad X_{L,R}(B_s).} \tag{15.50}$$

- For decays with $\mu\bar{\mu}$ or generally $\ell^+\ell^-$ in the final state

$$\boxed{Y_A(K), \qquad Y_A(B_d), \qquad Y_A(B_s).} \tag{15.51}$$

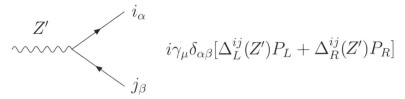

Figure 15.4 Feynman rule for the coupling of a colorless neutral gauge boson Z' to quarks, where i, j denote different quark flavors and α, β the colors. $P_{L,R} = (1 \mp \gamma_5)/2$.

- For decays with $P\ell^+\ell^-$ in the final state, where $P = K, B_{s,d}$

$$Z_V(K), \qquad Z_V(B_d), \qquad Z_V(B_s). \tag{15.52}$$

All these functions, in contrast to the SM and more generally CMFV models, depend on the meson considered and moreover are complex valued. In the absence of new sources of flavor violation and the absence of RH currents, the functions with the index R vanish and the remaining ones reduce to S_0, X_0, Y_0, and Z_0 within the SM and to S, X, Y, and Z within MFV models.

15.4.3 Master Functions in Z' Models

We will next illustrate how these functions look like in Z' models with arbitrary LH and RH complex quark couplings

$$\Delta_L^{sd}(Z'), \qquad \Delta_R^{sd}(Z') \tag{15.53}$$

that are defined in Figure 15.4.

Moreover, we will also work with real lepton couplings

$$\Delta_V^{\mu\bar\mu}(Z') = \Delta_R^{\mu\bar\mu}(Z') + \Delta_L^{\mu\bar\mu}(Z'), \qquad \Delta_A^{\mu\bar\mu}(Z') = \Delta_R^{\mu\bar\mu}(Z') - \Delta_L^{\mu\bar\mu}(Z'). \tag{15.54}$$

We recall the relevant CKM factors

$$\lambda_i^{(K)} = V_{is}^* V_{id}, \qquad \lambda_t^{(d)} = V_{tb}^* V_{td}, \qquad \lambda_t^{(s)} = V_{tb}^* V_{ts}, \tag{15.55}$$

which will be kept as overall factors in front of the master functions to have a better comparison with MFV scenario. We also introduce

$$g_{\rm SM}^2 = 4\frac{G_F}{\sqrt2}\frac{\alpha}{2\pi \sin^2\theta_W} = 4\frac{G_F^2 M_W^2}{2\pi^2} = 1.78137 \times 10^{-7}\,{\rm GeV}^{-2}. \tag{15.56}$$

$$T(B_q) = \frac{G_F^2}{12\pi^2} F_{B_q}^2 \hat{B}_{B_q} m_{B_q} M_W^2 \left(\lambda_t^{(q)}\right)^2 \eta_B, \tag{15.65}$$

$$T(K) = \frac{G_F^2}{12\pi^2} F_K^2 \hat{B}_K m_K M_W^2 \left(\lambda_t^{(K)}\right)^2 \eta_2, \tag{15.66}$$

where η_i are QCD corrections and $\hat{B}_i$ known SM nonperturbative factors.

The first difference from VLL and VRR cases is the appearance of two operators that mix with each other under renormalization, but even more important is the increased size of hadronic matrix elements of these new operators and the increased role of renormalization group effects due to the presence of large anomalous dimensions of these operators. We have seen this already in Section 13.2.

When constructing the effective Hamiltonian one has to remember that with two different currents, two different diagrams giving the same result can be constructed so that the combinatorial factor 1/2 included in VLL and VRR cases is canceled. We have then

$$T(K)[\Delta S(K)]_{\mathrm{LR}} = \frac{\Delta_L^{sd}(Z')\Delta_R^{sd}(Z')}{M_{Z'}^2}\left[C_1^{\mathrm{LR}}(\mu_{Z'})\langle Q_1^{\mathrm{LR}}(\mu_{Z'},K)\rangle + C_2^{\mathrm{LR}}(\mu_{Z'})\langle Q_2^{\mathrm{LR}}(\mu_{Z'},K)\rangle\right].$$
$$\tag{15.67}$$

To expose NP present in the couplings $\Delta_{L,R}^{sd}(Z')$, we have chosen as the renormalization scale, not the low-energy scale at which hadronic matrix elements of $Q_{1,2}^{\mathrm{LR}}$ are evaluated by LQCD, but a high scale $\mu_{Z'} = O(M_{Z'})$ at which Z' is integrated out. These matrix elements are given by

$$\langle Q_i^{\mathrm{LR}}(\mu_{Z'},K)\rangle \equiv \frac{m_K F_K^2}{3} P_i^{\mathrm{LR}}(\mu_{Z'},K) \tag{15.68}$$

with the coefficients P_i^{LR}, introduced in [746], relating these matrix elements to the ones calculated by LQCD. Including NLO QCD corrections [749] the Wilson coefficients of LR operators are given by

$$C_1^{\mathrm{LR}}(\mu_{Z'}) = 1 + \frac{\alpha_s}{4\pi}\left(-\log\frac{M_{Z'}^2}{\mu_{Z'}^2} - \frac{1}{6}\right), \qquad C_2^{\mathrm{LR}}(\mu_{Z'}) = \frac{\alpha_s}{4\pi}\left(-6\log\frac{M_{Z'}^2}{\mu_{Z'}^2} - 1\right). \tag{15.69}$$

The $\mu_{Z'}$ dependence of $C_i^a(\mu_{Z'})$ cancels the one of $P_i^a(\mu_{Z'})$ so that $\Delta S(K)$ does not depend on $\mu_{Z'}$. Similarly for B_q systems we have

$$T(B_q)[\Delta S(B_q)]_{\mathrm{LR}} = \frac{\Delta_L^{bq}(Z')\Delta_R^{bq}(Z')}{M_{Z'}^2}\left[C_1^{\mathrm{LR}}(\mu_{Z'})\langle Q_1^{\mathrm{LR}}(\mu_{Z'},B_q)\rangle\right.$$
$$\left. + C_2^{\mathrm{LR}}(\mu_{Z'})\langle Q_2^{\mathrm{LR}}(\mu_{Z'},B_q)\rangle\right], \tag{15.70}$$

where the Wilson coefficients $C_i^a(\mu_{Z'})$ are as in the K system. Information on hadronic matrix elements has been given in Section 13.2, in particular in Table 13.2.

Combinatoric Factors in $\Delta F = 2$ Calculations

In calculating the box diagrams in the SM, MFV, and Z' models, we argued that combinatoric factor of $1/2$ has to be included in the VLL and VRR cases but none in the LR case. While this rule can be derived diagrammatically, it is much faster to find it by integrating out Z' by means of equations of motion given in (1.2) with ϕ replaced by Z'.

Consider then the Lagrangian

$$\mathcal{L}(Z') = \frac{1}{2}M_{Z'}^2 Z'^{,\mu} Z'_\mu + \Delta_L^{sd}(Z')\bar{s}_L\gamma^\mu d_L Z'_\mu + \Delta_R^{sd}(Z')\bar{s}_R\gamma^\mu d_R Z'_\mu. \tag{15.71}$$

Inserting it into the Euler–Lagrange equation in (1.2), we find

$$Z'_\mu = -\frac{1}{M_{Z'}^2}\left[\Delta_L^{sd}(Z')\bar{s}_L\gamma_\mu d_L + \Delta_R^{sd}(Z')\bar{s}_R\gamma_\mu d_R\right]. \tag{15.72}$$

Inserting it back into (15.71) and reversing the overall sign to obtain the Hamiltonian, a simple algebra convinces us that our rule for combinatoric factors is correct.

$\Delta F = 1$ Master Functions ($\nu\bar{\nu}$ and $\mu\bar{\mu}$)

Proceeding in the same manner for $\Delta F = 1$ transitions and calculating the diagrams in Figure 15.5 with leptons in the final state, we readily find[4]

$$X_{\mathrm{L}}(K) = \eta_X X_0(x_t) + \frac{\Delta_L^{\nu\bar{\nu}}(Z')\,\Delta_L^{sd}(Z')}{g_{\mathrm{SM}}^2 M_{Z'}^2}\,\frac{1}{V_{ts}^* V_{td}} \equiv |X_L(K)|e^{i\theta_X^K}, \tag{15.73}$$

$$X_{\mathrm{R}}(K) = \frac{\Delta_L^{\nu\bar{\nu}}(Z')\,\Delta_R^{sd}(Z')}{g_{\mathrm{SM}}^2 M_{Z'}^2}\,\frac{1}{V_{ts}^* V_{td}}, \tag{15.74}$$

$$X_{\mathrm{L}}(B_q) = \eta_X X_0(x_t) + \left[\frac{\Delta_L^{\nu\nu}(Z')}{M_{Z'}^2\, g_{\mathrm{SM}}^2}\right]\frac{\Delta_L^{qb}(Z')}{V_{tq}^* V_{tb}} \equiv |X_L(B_q)|e^{i\theta_X^{B_q}}, \tag{15.75}$$

$$X_{\mathrm{R}}(B_q) = \left[\frac{\Delta_L^{\nu\nu}(Z')}{M_{Z'}^2\, g_{\mathrm{SM}}^2}\right]\frac{\Delta_R^{qb}(Z')}{V_{tq}^* V_{tb}}, \tag{15.76}$$

---

[4] Obviously no combinatoric factors are present in $\Delta F = 1$ transitions.

$$Y_A(K) = \eta_Y Y_0(x_t) + \frac{\left[\Delta_A^{\mu\bar{\mu}}(Z')\right]}{M_{Z'}^2 g_{\mathrm{SM}}^2} \left[\frac{\Delta_L^{sd}(Z') - \Delta_R^{sd}(Z')}{V_{ts}^* V_{td}}\right] \equiv |Y_A(K)| e^{i\theta_Y^K}, \tag{15.77}$$

$$Y_A(B_q) = \eta_Y Y_0(x_t) + \frac{\left[\Delta_A^{\mu\bar{\mu}}(Z')\right]}{M_{Z'}^2 g_{\mathrm{SM}}^2} \left[\frac{\Delta_L^{qb}(Z') - \Delta_R^{qb}(Z')}{V_{tq}^* V_{tb}}\right] \equiv |Y_A(B_q)| e^{i\theta_Y^{B_q}}. \tag{15.78}$$

Here $\eta_{X,Y}$ are QCD factors, which for $m_t = m_t(m_t)$ are close to unity [181, 182]. In 2019 after the inclusion of electroweak NLO corrections and NNLO QCD corrections they read [187, 203]

$$\eta_X = 0.994, \qquad \eta_Y = 0.988. \tag{15.79}$$

15.4.4 Effective Hamiltonian for $b \to s\ell^+\ell^-$ and $d \to s\ell^+\ell^-$

In the case of $b \to s\ell^+\ell^-$ and $d \to s\ell^+\ell^-$ transitions, the functions Y_A describe only axial-vector couplings, and it is necessary to look at the general effective Hamiltonian given in (9.1). In the case of Z' models this Hamiltonian involves only the Wilson coefficients

$$C_9, \qquad C_{10}, \qquad C_9', \qquad C_{10}' \tag{15.80}$$

with only the first two present in the SM. The new coefficients signal the presence of flavor-violating right-handed currents. The effective Hamiltonian for Z' models reads then

$$\mathcal{H}_{\mathrm{eff}}(b \to s\ell\bar{\ell}) = \mathcal{H}_{\mathrm{eff}}(b \to s\gamma) - \frac{4G_F}{\sqrt{2}} \frac{\alpha}{4\pi} V_{ts}^* V_{tb} \sum_{i=9,10} [C_i(\mu)Q_i(\mu) + C_i'(\mu)Q_i'(\mu)], \tag{15.81}$$

where

$$Q_9 = (\bar{s}\gamma_\mu P_L b)(\bar{\ell}\gamma^\mu \ell), \qquad Q_{10} = (\bar{s}\gamma_\mu P_L b)(\bar{\ell}\gamma^\mu \gamma_5 \ell), \tag{15.82}$$

$$Q_9' = (\bar{s}\gamma_\mu P_R b)(\bar{\ell}\gamma^\mu \ell), \qquad Q_{10}' = (\bar{s}\gamma_\mu P_R b)(\bar{\ell}\gamma^\mu \gamma_5 \ell). \tag{15.83}$$

Here $\mathcal{H}_{\mathrm{eff}}(b \to s\gamma)$ stands for the effective Hamiltonian for the $b \to s\gamma$ transition that involves the dipole operators. In the SM this Hamiltonian is given in (6.273). An explicit formula for this Hamiltonian in the presence of Z' contributions can be found in [611], but as NP contributions to this part turn out to be negligible, we do not present them here.

The Wilson coefficients (15.80) in Z' models are found in the case of $\mu\bar{\mu}$ to be

$$\sin^2\vartheta_W C_9 = [\eta_Y Y_0(x_t) - 4\sin^2\vartheta_W Z_0(x_t)] - \frac{1}{g_{SM}^2}\frac{1}{M_{Z'}^2}\frac{\Delta_L^{sb}(Z')\Delta_V^{\mu\bar{\mu}}(Z')}{V_{ts}^*V_{tb}}, \tag{15.84}$$

$$\sin^2\vartheta_W C_{10} = -\eta_Y Y_0(x_t) - \frac{1}{g_{SM}^2}\frac{1}{M_{Z'}^2}\frac{\Delta_L^{sb}(Z')\Delta_A^{\mu\bar{\mu}}(Z')}{V_{ts}^*V_{tb}}, \tag{15.85}$$

$$\sin^2\vartheta_W C_9' = -\frac{1}{g_{SM}^2}\frac{1}{M_{Z'}^2}\frac{\Delta_R^{sb}(Z')\Delta_V^{\mu\bar{\mu}}(Z')}{V_{ts}^*V_{tb}}, \tag{15.86}$$

$$\sin^2\vartheta_W C_{10}' = -\frac{1}{g_{SM}^2}\frac{1}{M_{Z'}^2}\frac{\Delta_R^{sb}(Z')\Delta_A^{\mu\bar{\mu}}(Z')}{V_{ts}^*V_{tb}}, \tag{15.87}$$

with $\Delta_{V,A}^{\mu\bar{\mu}}(Z')$ defined in (15.54) and explicit expressions for the SM one-loop functions X_0, Y_0, and Z_0 given in Chapter 6.

The presence of several couplings introduces new parameters and consequently significant freedom in phenomenological analyses of general Z' scenarios. But as we will see later on, in concrete models these couplings are given in terms of fundamental parameters of a given theory implying more stringent constraints.

It should be stressed that the preceding formulas do not include QCD renormalization group effects, which influence only C_9 and C_9'. As Z' is integrated out at a scale $O(M_{Z'})$, the RG QCD evolution has to be performed in the Z' part down to EW scale. Strictly speaking, also other RG effects present in the SMEFT have to be included to obtain the full picture. Such a complete analysis should be available in 2020.

15.4.5 Basic Formulas for Observables

We will next generalize the formulas for various observables presented already for the SM and models with MFV in the previous sections. The formulation of NP effects as modifications of the master functions makes this generalization straightforward. The main changes originate in the presence of new phases, and consequently one has to remember that the master functions are now complex quantities. The presence of new operators induced by the right-handed currents is another modification that forces us to distinguish functions with indices L and R and take care of the related minus sign in certain decays, but this is straightforward as well.

$\Delta F = 2$ Observables

The $\Delta B = 2$ mass differences are given as follows:

$$\Delta M_d = \frac{G_F^2}{6\pi^2}M_W^2 m_{B_d}|\lambda_t^{(d)}|^2 F_{B_d}^2 \hat{B}_{B_d}\eta_B|S(B_d)|, \tag{15.88}$$

$$\Delta M_s = \frac{G_F^2}{6\pi^2} M_W^2 m_{B_s} |\lambda_t^{(s)}|^2 F_{B_s}^2 \hat{B}_{B_s} \eta_B |S(B_s)|. \tag{15.89}$$

The absolute values of $S(B_q)$ should be noticed.

The corresponding mixing induced CP-asymmetries are modified with respect to the ones in MFV models because $S(B_q)$ carry now nonvanishing phases. We have

$$S_{\psi K_S} = \sin(2\beta + 2\varphi_{B_d}), \qquad S_{\psi\phi} = \sin(2|\beta_s| - 2\varphi_{B_s}), \tag{15.90}$$

where the phases β and β_s are defined in (15.4) and $\beta_s \simeq -1°$. The new phases φ_{B_q} are directly related to the phases of the functions $S(B_q)$ defined in (15.57)

$$2\varphi_{B_q} = -\theta_S^{B_q}. \tag{15.91}$$

Our phase conventions have been summarized in Section 8.1. Different signs in front of new phases in $S_{\psi K_S}$ and $S_{\psi\phi}$ in (15.90) should be noticed. This difference becomes important in concrete NP models.

For the CP-violating parameter ε_K and ΔM_K, we have, respectively,

$$\varepsilon_K = \frac{\kappa_\varepsilon e^{i\varphi_\varepsilon}}{\sqrt{2}(\Delta M_K)_{\exp}} \left[\mathrm{Im}\left(M_{12}^K\right) \right], \qquad \Delta M_K = 2\mathrm{Re}\left(M_{12}^K\right), \tag{15.92}$$

where

$$\left(M_{12}^K\right)^* = \frac{G_F^2}{12\pi^2} F_K^2 \hat{B}_K m_K M_W^2 \left[\lambda_c^2 \eta_1 x_c + \lambda_t^2 \eta_2 S(K) + 2\lambda_c \lambda_t \eta_3 S_0(x_c, x_t)\right]. \tag{15.93}$$

Relative to the SM the only difference is the new function $S(K)$, which has been evaluated including QCD corrections few pages before.

$B_{d,s} \to \mu^+ \mu^-$

With the assumption that the CKM parameters have been determined independent of NP and are therefore the same in all models considered, we find

$$\frac{\overline{\mathcal{B}}(B_s \to \mu^+\mu^-)}{\overline{\mathcal{B}}(B_s \to \mu^+\mu^-)^{\mathrm{SM}}} = \frac{r^{\mathrm{SM}}(y_s)}{r(y_s)} \left| \frac{Y_A(B_s)}{\eta_Y Y_0(x_t)} \right|^2, \tag{15.94}$$

and

$$\frac{\mathcal{B}(B_d \to \mu^+\mu^-)}{\mathcal{B}(B_d \to \mu^+\mu^-)^{\mathrm{SM}}} = \left| \frac{Y_A(B_d)}{\eta_Y Y_0(x_t)} \right|^2, \tag{15.95}$$

where $Y_A(B_q)$ is given in (15.78) and in (15.94), following the discussion in Section 9.2, we included $\Delta\Gamma_s$ effects, which are summarized in

$$r(y_s) \equiv \frac{1 - y_s^2}{1 + \mathcal{A}_{\Delta\Gamma}^\lambda y_s} \approx 1 - \mathcal{A}_{\Delta\Gamma}^\lambda y_s \tag{15.96}$$

with

$$y_s \equiv \tau_{B_s} \frac{\Delta\Gamma_s}{2} = 0.065 \pm 0.005 \tag{15.97}$$

and general formula for $\mathcal{A}_{\Delta\Gamma}^\lambda$ given in (9.122). Of particular interest now is also the CP-asymmetry $S_{\mu^+\mu^-}$ for which general formula in terms of Wilson coefficients is given in (9.115). Using these formulas we find in Z' models very simple expressions

$$\mathcal{A}_{\Delta\Gamma}^\lambda = \cos(2\theta_Y^{B_s} - 2\varphi_{B_s}), \quad S_{\mu^+\mu^-} = \sin(2\theta_Y^{B_s} - 2\varphi_{B_s}) \tag{15.98}$$

$\theta_Y^{B_s}$ is defined in (15.78). Both $\mathcal{A}_{\Delta\Gamma}^\lambda$ and $S_{\mu^+\mu^-}^s$ are theoretically clean observables.

We stress again the presence of new phases in the $B_q - \bar{B}_q$ mixings as we deal here with the mixing-induced CP violation. Their presence is also required to cancel any phase convention dependences. The SM phases cancel in this asymmetry [507, 508], and in the SM and CMFV models one has

$$\mathcal{A}_{\Delta\Gamma}^\lambda = 1, \quad S_{\mu^+\mu^-} = 0, \quad r(y_s) = 0.935 \pm 0.007 \tag{15.99}$$

independent of NP parameters.

While $\Delta\Gamma_d$ is very small and y_d can be set to zero, in the case of $B_d \to \mu^+\mu^-$ one can still consider the CP asymmetry $S_{\mu^+\mu^-}^d$ [508], for which in the case of Z' models we simply find

$$S_{\mu^+\mu^-}^d = \sin(2\theta_Y^{B_d} - 2\varphi_{B_d}). \tag{15.100}$$

$K_L \to \mu^+\mu^-$

As discussed in Section 9.7.1 only the so-called short-distance (SD) part of the dispersive contribution to $K_L \to \mu^+\mu^-$ can be reliably calculated. Therefore usually this decay is treated only as an additional constraint with the rough upper bound given in (9.268). The relevant formula in the models considered is easily found by simply taking into account that $Y_A(K)$ is complex. We find then ($\lambda = 0.2252$)

$$\mathcal{B}(K_L \to \mu^+\mu^-)_{\text{SD}} = 2.01 \cdot 10^{-9} \left[\frac{\text{Re}Y_{\text{eff}}}{\lambda^5} + \frac{\text{Re}\lambda_c}{\lambda} P_c\left(Y(K)\right) \right]^2, \tag{15.101}$$

where $P_c\left(Y(K)\right) = 0.115 \pm 0.017$ [625] and

$$Y_{\text{eff}} = V_{ts}^* V_{td} Y_A(K). \tag{15.102}$$

As LH and RH couplings in $Y_A(K)$ in (15.77) enter with opposite signs, the impact of the upper bound in (9.268) on the observables in which these signs are equal is markedly different. This is the case of decays $K^+ \to \pi^+\nu\bar{\nu}$ and $K_L \to \pi^0\nu\bar{\nu}$, which we will discuss next.

with

$$\omega_{7V} = \frac{1}{2\pi} \left[P_0 + \frac{|\tilde{Y}(K)|}{\sin^2 \theta_W} \frac{\sin \beta_Y^K}{\sin(\beta - \beta_s)} - 4|\tilde{Z}(K)| \frac{\sin \beta_Z^K}{\sin(\beta - \beta_s)} \right] \left[\frac{\operatorname{Im} \lambda_t}{1.4 \cdot 10^{-4}} \right], \quad (15.114)$$

$$\omega_{7A} = -\frac{1}{2\pi} \frac{|\tilde{Y}(K)|}{\sin^2 \theta_W} \frac{\sin \beta_Y^K}{\sin(\beta - \beta_s)} \left[\frac{\operatorname{Im} \lambda_t}{1.4 \cdot 10^{-4}} \right], \quad (15.115)$$

where $P_0 = 2.88 \pm 0.06$

$$\beta_Y^K = \beta - \beta_s - \theta_Y^K, \qquad \beta_Z^K = \beta - \beta_s - \theta_Z^K. \quad (15.116)$$

The expressions for ω_{7V} and ω_{7A} in terms of WC of the SMEFT can be found in [667].

Finally, as derived in [611] from the formulas in [920],

$$\tilde{Y}(K) = \eta_Y Y_0(x_t) + \left[\frac{\Delta_A^{\mu\bar{\mu}}(Z')}{M_{Z'}^2 g_{SM}^2} \right] \frac{\Delta_V^{sd}(Z')}{V_{ts}^* V_{td}}, \quad (15.117)$$

$$\tilde{Z}(K) = Z_0(x_t) + \frac{1}{4 \sin^2 \vartheta_W} \left[\frac{2\Delta_R^{\mu\bar{\mu}}(Z')}{M_{Z'}^2 g_{SM}^2} \right] \frac{\Delta_V^{sd}(Z')}{V_{ts}^* V_{td}}. \quad (15.118)$$

$\Delta_{V,A}^{sd}(Z')$ are defined in (15.54). We indeed observe that $\tilde{Y}(K)$ differs from $Y(K)$ in (15.77).

The presence of additional coupling $\Delta_R^{\mu\bar{\mu}}(Z')$, in addition to $\Delta_A^{\mu\bar{\mu}}(Z')$, introduces as in $B \to K^* \ell^+ \ell^-$, $B \to K \ell^+ \ell^-$ and $B \to X_s \ell^+ \ell^-$ two new parameters and allows thereby to avoid present constraints if necessary. In the case of FCNC processes mediated by Z, which will be discussed in Section 15.5, all leptonic couplings are known, and the predictions for $K_L \to \pi^0 e^+ e^-$ and $K_L \to \pi^0 \mu^+ \mu^-$ are more specific.

15.4.6 Lessons on NP Patterns in Z' Scenarios

A detailed phenomenology of tree-level Z' exchanges has been presented in [601, 611, 851]. We will now summarize the highlights of these papers in the form of lessons gained through these analyses. Numerous plots related to these lessons can be found in these papers.

We begin with the K-meson system and concentrate on the results obtained in [851]. The summary of the lessons is rather brief. On the other hand, the presentation in [851] is very detailed with numerous analytic expressions that could be useful for nonexperts who want to understand the details of the patterns of various effects. In the latter paper the main goal was to investigate what are the implications of the ε'/ε anomaly discussed in Chapter 10 on rare decays $K^+ \to \pi^+ \nu \bar{\nu}$ and $K_L \to \pi^0 \nu \bar{\nu}$ in various scenarios for the Z' couplings to quarks.

We begin with the K-meson system. Z' models exhibit quite different pattern of NP effects in the K-meson system than the Z scenarios discussed later in Section 15.5. In Z scenarios only electroweak penguin (EWP) Q_8 and Q_8' operators can contribute in an important manner to ε'/ε because of flavor-dependent diagonal Z coupling to quarks. But in Z' models the diagonal quark couplings can be flavor universal so that QCD penguin operators (QCDP) (Q_6, Q_6') can dominate NP contributions to ε'/ε. Interestingly, the pattern of NP in rare K decays depends on whether NP in ε'/ε is dominated by QCDP or EWP operators [851]. Let us discuss this briefly.

In the strategy in [851] the central role is played by ε'/ε and ε_K for which in the presence of NP contributions we have

$$\frac{\varepsilon'}{\varepsilon} = \left(\frac{\varepsilon'}{\varepsilon}\right)^{\text{SM}} + \left(\frac{\varepsilon'}{\varepsilon}\right)^{\text{NP}}, \qquad \varepsilon_K \equiv e^{i\varphi_\varepsilon}\left[\varepsilon_K^{\text{SM}} + \varepsilon_K^{\text{NP}}\right]. \qquad (15.119)$$

As the size of NP contributions is presently not precisely known, the strategy of [851] is to parametrize this contributions as

$$\left(\frac{\varepsilon'}{\varepsilon}\right)^{\text{NP}} = \kappa_{\varepsilon'} \cdot 10^{-3}, \qquad 0.5 \le \kappa_{\varepsilon'} \le 1.5 \qquad (15.120)$$

and

$$(\varepsilon_K)^{\text{NP}} = \kappa_\varepsilon \cdot 10^{-3}, \qquad 0.1 \le \kappa_\varepsilon \le 0.4. \qquad (15.121)$$

The ranges for $\kappa_{\varepsilon'}$ and κ_ε only indicate possible size of NP contributions as argued in [851] in 2016. In 2019 κ_ε is more likely in the range $-0.2 \le \kappa_\varepsilon \le 0.2$ [874]. Anyway, using the formulas in [851] $\kappa_{\varepsilon'}$ and κ_ε can also be treated as free parameters. Once $|V_{cb}|$ will be precisely known the size and sign of the latter will be determined. The results summarized following are based on (15.121).

To simplify the discussion we assume that only LH flavor-violating coupling of Z' are nonvanishing. The first general findings of [851] are then as follows:

- For a given κ_ε that determines the imaginary part of $\Delta_L^{sd}(Z')$, the real part of this coupling is determined for a not too large κ_ε from the ε_K.
- There is a large hierarchy between real and imaginary parts of the flavor-violating couplings implied by anomalies in QCDP and EWP scenarios. As shown in [851] in the case of QCDP imaginary parts dominate over the real ones, while in the case of EWP this hierarchy is opposite unless the κ_ε is very small. This is related to the fact that strong suppression of QCDP to ε'/ε by the factor $1/22$ coming from $\Delta I = 1/2$ rule requires a large imaginary coupling in QCDP scenario to enhance significantly this ratio. This suppression is absent in the case of EWP, and this coupling can be smaller.

Because of this important difference in the manner QCDP and EWP enter ε'/ε, there are striking differences in the implications for the correlation between $K^+ \to \pi^+\nu\bar{\nu}$ and $K_L \to \pi^0\nu\bar{\nu}$ in these two NP scenarios if significant NP contributions to ε'/ε are required. Various plots in [851] show clearly the differences between QCDP and EWP scenarios. We refer to [851] for more details, in particular analytic derivation of all these results. We extract from these results the following lessons:

Lesson 1: In the case of QCDP scenario the correlation between $\mathcal{B}(K_L \to \pi^0\nu\bar{\nu})$ and $\mathcal{B}(K^+ \to \pi^+\nu\bar{\nu})$ takes place along the branch parallel to the Grossman–Nir (GN) bound

[602] in (9.218) that is illustrated in Figure 15.6. Moreover, this feature is independent of $M_{Z'}$. Also the dependence on κ_ε is negligible.

Lesson 2: In the EWP scenario the correlation between $\mathcal{B}(K_L \to \pi^0 \nu \bar{\nu})$ and $\mathcal{B}(K^+ \to \pi^+ \nu \bar{\nu})$ proceeds away from this branch for diagonal quark couplings $O(1)$ if NP in ε_K is present and it is very different from the one of the QCDP case as seen in the plots in [851]. NP effects are also much smaller than in QCDP scenario.

Lesson 3: For fixed values of the neutrino and diagonal quark couplings in ε'/ε, the predicted enhancements of $\mathcal{B}(K_L \to \pi^0 \nu \bar{\nu})$ and $\mathcal{B}(K^+ \to \pi^+ \nu \bar{\nu})$ are much larger when NP in QCDP is required to remove the ε'/ε anomaly than it is in the case of EWP. This is simply related to the fact, as mentioned earlier, that the $\Delta I = 1/2$ rule suppresses QCDP contributions to ε'/ε so that QCDP operators are less efficient in enhancing ε'/ε than EWP operators. Consequently, the imaginary parts of the flavor-violating couplings are required to be larger, implying then larger effects in rare K decays. Only for the diagonal quark couplings $O(10^{-2})$ the requirement of shifting upwards ε'/ε implies large effects in $K^+ \to \pi^+ \nu \bar{\nu}$ and $K_L \to \pi^0 \nu \bar{\nu}$ in EWP scenario. See [851] for a detail discussion of this point and a more detailed analysis to appear in 2020.

Lesson 4: In QCDP scenario ΔM_K is *suppressed*, and this effect increases with increasing $M_{Z'}$ whereas in the EWP scenario ΔM_K is *enhanced*, and this effect decreases with increasing $M_{Z'}$ as long as real couplings dominate. Already on the basis of this property one could differentiate between these two scenarios when the SM prediction for ΔM_K improves. Interestingly, the recent RBC-UKQCD result in (8.18), if confirmed, would then at first sight favor QCDP scenario. Yet, this conclusion is only true if the $Z' \bar{q} q$ couplings are $O(1)$ and RH FCNC couplings are absent. In LH scenario with couplings $Z' \bar{q} q = O(10^{-2})$ also EWP scenario would lead to the suppression of ΔM_K due to enhanced imaginary parts of FCNC couplings required for the explanation of the ε'/ε anomaly. This is clearly seen in the relevant formulas in [851]. But the presence of both LH and RH FCNC couplings, introducing more parameters, would enrich the analysis and only correlations with other observables, in particular with the branching ratios for $K^+ \to \pi^+ \nu \bar{\nu}$ and $K_L \to \pi^0 \nu \bar{\nu}$ could lead to clear cut conclusions.

In summary once the results from NA62 and KOPIO are available, it should be possible to select between various scenarios just presented.

Yet, one should realize that the analysis in [851] included only QCD evolution, and it would be of interest to see how the lessons just listed are modified when the full machinery of the SMEFT is switched on. In particular the effects of Yukawa couplings could play here a significant role. We expect that this issue should be clarified in 2020.

15.4.7 $K_L \to \pi^0 \nu \bar{\nu}, K^+ \to \pi^+ \nu \bar{\nu}$ and ε_K beyond the SM

It should be stressed that the correlation between $K_L \to \pi^0 \nu \bar{\nu}$ with $K^+ \to \pi^+ \nu \bar{\nu}$ gives us generally more information about NP than the correlation between these two decays and ε'/ε. The reason is very simple. While the flavor-violating couplings in $K_L \to \pi^0 \nu \bar{\nu}, K^+ \to \pi^+ \nu \bar{\nu}$, and ε'/ε are the same in a given model, the flavor conserving ones in ε'/ε involving quarks are generally different from neutrino ones, and some assumptions about them have to be made as done in [851] and also in a number of simplified scenarios considered in [601]. Table 3 in [851] illustrates this point.

However, as pointed out by Monika Blanke [924], more information about the nature of NP can be obtained by looking simultaneously at $K_L \to \pi^0 \nu \bar\nu$, $K^+ \to \pi^+ \nu \bar\nu$, and ε_K, simply because the latter depends only on the flavor-violating couplings, which equal to those in $K_L \to \pi^0 \nu \bar\nu$ and $K^+ \to \pi^+ \nu \bar\nu$. We will now summarize the insight gained from [924] that have been verified in Z' models considered in [611] and also in [601].

To this end let us parametrize X_{eff} in (15.105) as follows [601]

$$X_{\text{eff}} = V_{ts}^* V_{td} (X_L(K) + X_R(K)) \equiv V_{ts}^* V_{td} X_L^{\text{SM}}(K)(1 + \xi e^{i\theta}), \qquad (15.122)$$

so that the short-distance dynamics is encapsulated in two real parameters ξ and θ that vanish in the SM. Measuring these branching ratios one day will allow to determine those parameters and, comparing them with their expectations in concrete models, to obtain insight into the flavor structure of the NP contributions. Those, as we have seen earlier, can be dominated by left-handed currents, by right-handed currents, or by both with similar magnitudes and phases. In general one can then distinguish between three classes of models [924]:

1. Models with a CKM-like structure of flavour interactions. If based on flavor symmetries only, they include MFV and $U(2)^3$ models. In this case the function $X_L(K)$ is real, and $X_R(K) = 0$. There is then only one variable to our disposal, the value of $X_L(K)$, and the only allowed values of both branching ratios are on the two broad light gray branches in Figure 15.6 that touch the origin of the plot. But due to stringent correlations with other observables present in this class of models, only certain ranges for $\mathcal{B}(K^+ \to \pi^+ \nu \bar\nu)$ and $\mathcal{B}(K_L \to \pi^0 \nu \bar\nu)$ are still allowed, which have been determined in simplified models in [601].

2. Models with new flavor- and CP-violating interactions in which either left-handed currents or right-handed currents fully dominate, implying that left-right operator contributions to ε_K can be neglected. In this case there is a strong correlation between NP contributions to ε_K and $K \to \pi \nu \bar\nu$ and the ε_K constraint implies the thin light gray branch structure shown in Figure 15.6. On the horizontal branch NP contribution to $K \to \pi \nu \bar\nu$ is real and therefore vanishes in the case of $K_L \to \pi^0 \nu \bar\nu$. On the second branch NP contribution is purely imaginary, and this branch is parallel to the Grossman–Nir (GN) bound [602] shown in the plot. In practice, due to uncertainties in ε_K, there are moderate deviations from this structure, which is characteristic for the LHT model [921], or Z or Z' FCNC scenarios with either pure LH or RH couplings [211, 611].

3. If left-right operators have significant contribution to ε_K or generally if the correlation between ε_K and $K \to \pi \nu \bar\nu$ is weak or absent, the two-branch structure is also absent. Dependent on the values of ξ or θ, any value of $\mathcal{B}(K^+ \to \pi^+ \nu \bar\nu)$ and $\mathcal{B}(K_L \to \pi^0 \nu \bar\nu)$ is in principle possible. The dark gray region in Figure 15.6 shows the resulting structure for a fixed value of ξ and $0 \le \theta \le 2\pi$. Randall–Sundrum models with custodial protection belong to this class of models [920]. However, it should be kept in mind that usually the removal of the correlation with ε_K requires subtle cancellations between different contributions to ε_K and consequently some tuning of parameters [211, 920].

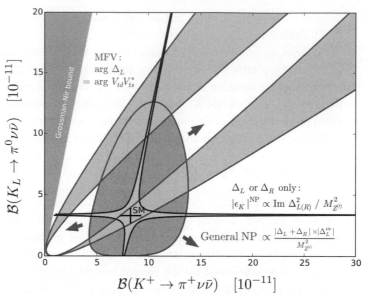

Figure 15.6 Correlations in the $\mathcal{B}(K^+ \to \pi^+ \nu \bar{\nu})$ versus $\mathcal{B}(K_L \to \pi^0 \nu \bar{\nu})$ plane beyond the SM [601]. See text for explanations.

More details with colorful plots can be found in [601, 924] and also in figure 35 of [208], where the dependence of both branching ratios on ξ for various values of θ is shown.

15.4.8 *Z′* Facing *B* Physics Anomalies

In Chapter 12 we mentioned various *B* physics anomalies, and we will return to them again in Chapter 18, where we will compare various explanations of them. Here we would like to concentrate on *Z′* models.

In fact, general *Z′* models can easily explain the $b \to s\ell^+\ell^-$ anomalies but cannot explain the charged current anomalies without the participation of other new particles. Beginning with $b \to s\ell^+\ell^-$ anomalies, the most successful *Z′* models are the following ones.

i) *Z′* models with vector couplings to muons are free of gauge anomalies and consequently offer a natural C_9^{NP}-like solution. The most extensively studied models of this type are $L_\mu - L_\tau$ models [925–931] in which NP is absent in electron channels. Unfortunately in these models, unless some additional heavy particles are added, only very small NP effects in $B_{s,d} \to \mu^+\mu^-$ and $K \to \pi\nu\bar{\nu}$ transitions are possible. On the other hand, they can explain the $(g-2)_\mu$ anomaly, which will be discussed in Section 17.4.

ii) Besides $L_\mu - L_\tau$, various other combinations of gauged flavor symmetries have been used to construct *Z′* models that can address the $b \to s\ell^+\ell^-$ anomalies [904, 927, 928, 932–942].

iii) Also models with loop-induced *Z′* couplings [943] and *Z′* models with heavy vector-like fermions [933] can work. Yet when choosing properly *Z′* couplings to explain the

$b \to s\ell^+\ell^-$ anomalies, one has to be careful that no gauge anomalies are introduced. We will discuss this point soon.

As far as $R(D)$ and $R(D^*)$ anomalies are concerned, they cannot be explained in Z' models without the addition of charged current processes or new particles generating them. Here we briefly describe a few papers that tried to address these anomalies in addition to $b \to s\mu^+\mu^-$ ones. The leptoquarks and vector-like fermions that are mentioned next are discussed in detail in Chapter 16.

In [931] the $R(K^{(*)})$ and $R(D^{(*)})$ anomalies have been addressed in a model with the $U(1)_{L_\mu - L_\tau}$ gauge symmetry. To be able to achieve this goal, one vector-like lepton (VLL) doublet and one singlet scalar leptoquark (LQ) S_1 are considered. The $b \to sZ'$ effective interaction relevant for the $R(K^{(*)})$ is generated at one-loop from VLL and LQ, and $R(D^{(*)})$ is governed by LQ. While this model can, as claimed by the authors, explain $R(K^{(*)})$ and $R(D^{(*)})$ anomalies, it will likely fail in the case of ε'/ε unless it can be generated at tree-level. We will discuss this issue in Section 16.4.

Next models in which the Z' is part of a $SU(2)_L$ triplet have been suggested as a simultaneous explanation of the $b \to s\ell^+\ell^-$ anomalies and of $R(D)$ and $R(D^*)$ ones [719, 769, 777, 933, 934, 944, 945].

In [946] Z' model with MFV has been considered and $b \to s\ell^+\ell^-$ anomalies analyzed. Constraints from $\mu \to e\gamma$, $l_i \to \ell_j\ell_k\ell_l$, $B^0 - \bar{B}^0$ mixing that all appear at tree-level have been taken into account. Results for $\tau \to 3\mu$, $B \to Ke\mu$, $K_L \to e\mu$, $Z \to \ell\ell'$ are found close to the experimental bounds.

In [942] flavor symmetry group $SU(3)_L \times SU(3)_R$ has been considered with the goal to connect NP responsible for the $R(K^{(*)})$ anomalies with the understanding of the origin of flavor. As this symmetry is gauged and broken spontaneously, we are dealing eventually with 16 new heavy gauged bosons, one of them being Z', light enough to explain $R(K^{(*)})$ anomalies.

As far as a light Z' is concerned, it has been considered in [947] with the goal to explain low q^2 region of the $R(K^*)$ anomaly. In [948] the role of Z' in $R(K^{(*)})$ and $(g - 2)_\mu$ anomalies has been considered. Finally, in [949, 950] lepton-flavor violating Z' has been used to explain $(g - 2)_\mu$ anomaly. A light Z' is again required. See also [951, 952].

One important constraint on Z' explanations of $b \to s\ell^+\ell^-$ anomalies comes from $B_s - \bar{B}_s$-mixing [953–955]. The Z' mass can be at most several TeV; otherwise an explanation of the anomalies requires couplings to leptons that are nonperturbatively large. This constraint will increase in importance, when lattice QCD calculations increase the precision on hadronic matrix elements. Most recent analysis of collider constraints on simple Z' models can be found in [955]. A combination of $B_s - \bar{B}_s$-mixing constraint and collider data implies further reduction of the parameter space in simple Z' models, but the effect is clearly model dependent.

We should also mention that certain concrete models, like 331 models, in which LFUV is absent at tree-level, will be disfavored [612, 956] if LFUV will be confirmed in the future. On the other hand, as discussed in Section 16.2, some of them can explain the suppression of C_9 or C_{10}.

As described in detail in [957], neutrino trident production can provide stringent constraints on Z' models. It is a process by which a neutrino, scattering off the Coulomb field of a heavy nucleus, generates a pair of charged leptons: $\nu_\mu \to \nu_\mu \mu^+ \mu^-$. Future measurements of this process at DUNE will be highly sensitive to NP in general. But in particular, as found in this paper, they will cover almost entirely the parameter space of the $L_\mu - L_\tau$ models. Finally in an interesting paper [1327] the ε'/ε anomaly, the $B \to K\pi$ puzzle, and $b \to s\ell^+\ell^-$ anomalies have been addressed simultaneously in an $U(2)^3$ model.

15.4.9 Gauge (triangle) Anomalies and Z'

Preliminaries

One of the important issues often not considered in papers on Z' models is the cancellation of the gauge anomalies, called also triangle anomalies, that naturally are generated in the presence of an additional $U(1)'$ gauge group. This issue has been studied in [730, 904, 935, 938, 958–964] but in most of these papers, exception being [938, 959], flavor universality has been assumed. Also in most of these papers Z' couplings to the first generation have been set to zero and similar for complex phases. For recent studies of anomaly-free models see [1323–1325].

Here we would like to list all anomaly equations without any assumptions on $U(1)'$ charges and present a simple solution that has been found in collaboration with Jason Aebischer, Maria Cerdá-Sevilla, and Fulvia De Fazio. We assume that only SM fermions, except for heavy right-handed neutrinos, are present.

As Z' is a singlet under the SM gauge group, the left-handed leptons in a given doublet ℓ_i with $i = 1, 2, 3$ must have the same $U(1)'$ charges and similar for the members of left-handed quark doublets q_i. On the other hand, the right-handed leptons ν_i and e_i can have different $U(1)'$ charges and the same for right-handed quarks u_i and d_i. We will allow all these charges to be generation dependent. This is required to generate flavor-violating neutral currents after rotating from flavor to mass eigenstate basis. The right-handed neutrinos will sometimes turn out to be relevant for the cancellation of the gauge anomalies.

Denoting by q_i and ℓ_i left-handed doublets and by u_i, d_i, ν_i, and e_i right-handed singlets allows us to drop the subscripts L and R on the fields. For convenience we recall their hypercharges:

$$y_{q_i} = 1/6, \qquad y_{u_i} = 2/3, \qquad y_{d_i} = -1/3, \qquad y_{\ell_i} = -1/2, \qquad y_{e_i} = -1. \qquad (15.123)$$

General Formulas

The rules for calculating triangle anomalies can be found in any textbook on QFT. For a left-handed fermion in the loop, one just multiplies the couplings (charges) of this fermion to a given gauge boson in the vertex. For a right-handed fermion an additional overall minus sign is included. This means that for fermions with left-handed and right-handed charges being equal, that is, for vector-like fermions, there are no gauge anomalies. The SM fermions are not vector-like, and the cancellation of gauge anomalies related to the SM gauge group takes place between quarks and leptons. To this end one has to remember that a color factor of 3 has to be introduced when a quark is running in the loop. As the gauge anomalies must be cancelled, the values of the gauge couplings, as well as the result of integrating over momentum in the triangle diagram, are irrelevant. Moreover, being mass independent, this integral is universal for all fermions. Only the charges under a given gauge group matter. In the case of $U(1)'$ we will denote these charges by

$$z_{q_i}, \quad z_{u_i}, \quad z_{d_i}, \quad z_{l_i}, \quad z_{v_i}, \quad z_{e_i}, \tag{15.124}$$

with $i = 1, 2, 3$ labeling different generations.

There are six gauge anomalies generated by the presence of Z' [958]. Four of them are linear in $U(1)'$ charges, one quadratic in them, and one cubic one. It is useful to begin with linear equations as they are simpler and teach us already something.

The structure of the four linear equations can be simplified by defining

$$z_q = z_{q_1} + z_{q_2} + z_{q_3}, \qquad z_u = z_{u_1} + z_{u_2} + z_{u_3}, \qquad z_d = z_{d_1} + z_{d_2} + z_{d_3}, \tag{15.125}$$

$$z_l = z_{l_1} + z_{l_2} + z_{l_3}, \qquad z_v = z_{v_1} + z_{v_2} + z_{v_3}, \qquad z_e = z_{e_1} + z_{e_2} + z_{e_3}. \tag{15.126}$$

Then the $[SU(3)_c]^2 U(1)'$, $[SU(2)]^2 U(1)'$, and $[U(1)_Y]^2 U(1)'$ anomaly cancellation conditions can be written, respectively, as

$$\boxed{A_{33z} = 2z_q - z_u - z_d = 0,} \tag{15.127}$$

$$\boxed{A_{22z} = 3z_q + z_l = 0,} \tag{15.128}$$

$$\boxed{A_{11z} = \frac{1}{6} z_q - \frac{4}{3} z_u - \frac{1}{3} z_d + \frac{1}{2} z_l - z_e = 0.} \tag{15.129}$$

The fourth linear condition involves two gravitons and Z' and is given by

$$A_{GGz} = 3[2z_q - z_u - z_d] + 2z_l - z_e - z_v = 0, \tag{15.130}$$

but using (15.127) it reduces to

$$\boxed{A_{GGz} = 2z_l - z_e - z_v = 0.} \tag{15.131}$$

To write the remaining two anomaly cancellation conditions in a transparent form, we define

$$z_f^{(3)} = \sum_{i=1,2,3} z_{f_i}^3, \qquad z_f^{(2)} = \sum_{i=1,2,3} z_{f_i}^2. \tag{15.132}$$

Then the $U(1)_Y [U(1)']^2$ anomaly cancellation condition is given by

$$\boxed{A_{1zz} = [z_q^{(2)} - 2z_u^{(2)} + z_d^{(2)}] - [z_l^{(2)} - z_e^{(2)}] = 0,} \tag{15.133}$$

and the one for $[U(1)']^3$ anomaly by

$$\boxed{A_{zzz} = 3[2z_q^{(3)} - z_u^{(3)} - z_d^{(3)}] + [2z_l^{(3)} - z_v^{(3)} - z_e^{(3)}] = 0.} \tag{15.134}$$

It is a good exercise to check these equations.

Solution to Linear Equations

There are four linear equations with six unknowns. This implies that we can express four of the charges in (15.125) and (15.126) in terms of the two remaining ones. We give here two examples:

d-case

We choose as free parameters z_q and z_d, then

$$z_u = 2z_q - z_d, \qquad z_l = -3z_q, \qquad z_e = z_d - 4z_q, \qquad z_v = -2z_q - z_d. \qquad (15.135)$$

u-case

We choose as free parameters z_q and z_u, then

$$z_d = 2z_q - z_u, \qquad z_l = -3z_q, \qquad z_e = -z_u - 2z_q, \qquad z_v = -4z_q + z_u. \qquad (15.136)$$

We emphasize that the simple relations (15.135) and (15.136) are model-independent. Moreover, it turns out, that the solutions in both cases solve in the case of flavor universality the quadratic and cubic equations (15.133)–(15.134). Furthermore, the SM case for the hypercharges can be retrieved from these relations. This is not surprising because the same equations as (15.127), (15.128), (15.131), and (15.134) are obtained in the case of the $U(1)_Y$ gauge boson, by simply replacing $z_f \to y_f$. Indeed, the anomaly equations are solved by choosing $z_{q_i} = 1/6$ independent of i as well as $z_{d_i} = -1/3$ in the d-case and $z_{u_i} = 2/3$ in the u-case.

A Simple Gauge Anomaly Free Model

To obtain phenomenologically interesting scenario we break flavor universality in all four families. But not to end up with many free parameters we break the flavor universality in a rather special manner. This simple model has been constructed by Fulvia De Fazio and analyzed by us four.

To this end let us denote by f one of the following fermions: $f = q, u, d, \ell, e, v$, and let us add a generation index: f_i for $i = 1, 2, 3$. We know that the SM hypercharge is generation universal so that $y_{f_i} = y_f \ \forall i = 1, 2, 3$. The solution of all equations is provided by the charges

$$z_{f_i} = y_f + \epsilon_i, \qquad \sum_{i=1}^{3} \epsilon_i = 0 \qquad (15.137)$$

so that we have

$$z_f = \sum_{i=1}^{3} z_{f_i} = \sum_{i=1}^{3} (y_f + \epsilon_i) = 3y_f + \sum_{i=1}^{3} \epsilon_i = 3y_f + \epsilon, \qquad \epsilon = \sum_{i=1}^{3} \epsilon_i. \qquad (15.138)$$

It should be noted that the breakdown of flavor universality has been made in a very special manner:

- The parameters ϵ_i, while generation dependent, are universal within a given generation. For instance, ϵ_1 is the same not only for the members of the left-handed doublets q_1, ℓ_1 but also for right-handed singlets u_1, d_1, ν_1, and e_1. This implies that cancellation of gauge anomalies involves all generations as opposed to the SM, where it takes place within one generation. See (15.137).
- This implies that the shifts ϵ_i are vector-like, which allows for a straightforward solution of all anomaly cancellation equations. But as hypercharges of left-handed and right-handed fields, as seen in (15.123), differ from each other, the flavor-conserving Z' couplings are not vector-like.

It is a good exercise to check that indeed the charges in (15.137) satisfy six anomaly equations listed earlier. They are only satisfied if $\epsilon = 0$. Having only few parameters, the model implies correlations not only between various quark observables and between various lepton observables, but also in particular quark and lepton observables. The derivation of the relevant couplings proceeds as illustrated in Section 14.3.2. Details of this analysis should appear in 2020.

In [938] an additional $U(1)_X$ gauge symmetry with $X = B_3 - L_\mu$, has been considered. Unlike other alternative schemes, which typically require the addition of new charged fermions or leptoquarks, all anomalies induced by this gauge symmetry, including gravitational ones, are canceled. This model then can explain $R(K^{(*)})$ anomalies with the help of Z'.

The cancellation of triangle anomalies in the preceding simple example can be verified by hand. In more complicated cases, and in particular when constructing models involving many new particles, computer programs turn out to be very useful. We refer to one recent analysis of this type [965], where references to earlier literature can be found. In particular in [966] the $U(1)'$ charges of the third generation equal their hypercharges, with the first two generations being uncharged under $U(1)'$. Even if hypercharge plays a role in this model, the model differs significantly from our model. In particular, ϵ'/ϵ and K-meson physics cannot be successfully addressed in the model in [966].

15.5 Beyond MFV: *Z* Boson with FCNC

15.5.1 Preliminaries

At first sight the discussion of FCNC mediated by Z' in the previous section and the numerous formulas presented there could be used for models in which FCNC are mediated by the SM Z boson already at tree-level. See Figures 15.7 and 15.8. In particular the following changes in all formulas could be made right away

- The flavor-violating couplings $\Delta_{L,R}^{ij}(Z')$ are replaced by $\Delta_{L,R}^{ij}(Z)$.
- Flavor-diagonal couplings of Z to quarks and leptons are replaced by SM ones, which reduces the number of free parameters. Explicit expressions are collected in Appendix B.
- $M_{Z'}$ is replaced by M_Z and $\mu_{Z'}$ by $m_t(m_t)$.

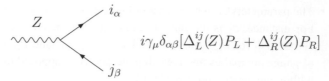

$$i\gamma_\mu \delta_{\alpha\beta}[\Delta_L^{ij}(Z)P_L + \Delta_R^{ij}(Z)P_R]$$

Figure 15.7   Feynman rule for the coupling of the Z to quarks, where i, j denote different quark flavors and α, β the colors. $P_{L,R} = (1 \mp \gamma_5)/2$.

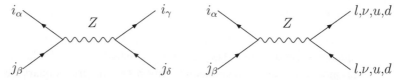

Figure 15.8   Tree-level Z contribution to $\Delta F = 2$ (left) and $\Delta F = 1$ (right) processes. Labels are explained in Figure 15.7.

- The hadronic matrix elements are evaluated at $m_t(m_t)$ as the inclusion of NLO QCD corrections allows us to choose any scale of $O(M_Z)$ without changing physical results. They are collected in Table 13.1.
- The QCD correction $\tilde{r}$ in (15.61) takes now the value $\tilde{r} = 1.068$.

From the point of view of SMEFT, this procedure is correct for $\Delta F = 1$ processes but misses important contributions to $\Delta F = 2$ processes as pointed out in [766] and also analyzed partially in [799]. The present section is based on [766, 851].

We will first address this NP scenario from the point of view of SMEFT, expressing $\Delta_{L,R}^{ij}(Z)$ in terms of Wilson coefficients of SMEFT and identifying the missing contributions. Subsequently we will collect the formulas for the master functions in these NP scenarios. Concrete examples are models with vector-like quarks (VLQs) in which the mixing of SM quarks with these new heavy quarks generates flavor-violating Z couplings. We will discuss these models in Section 16.3.

In should be remarked that in concrete models there could be still other contributions to $\Delta F = 2$ processes that in certain ranges of parameters could be more important than the ones considered here. This is the case of box diagrams with VLQs and Higgs exchanges, which we will discuss in Section 16.3.

15.5.2 The SMEFT View

The operators that induce FC quark couplings of the Z are the ones of Scenarios I and II of the previous chapter. The ones with LH quark currents are[5]

$$O_{Hq}^{(1)} = (H^\dagger i \overleftrightarrow{D}_\mu H)[\bar{q}_L^i \gamma^\mu q_L^j], \qquad O_{Hq}^{(3)} = (H^\dagger i \overleftrightarrow{D}_\mu^a H)[\bar{q}_L^i \sigma^a \gamma^\mu q_L^j], \qquad (15.139)$$

---

[5] To simplify notations we suppress flavour indices on the operators.

including also modified LH $W^{\pm}$ couplings. The ones with RH quark currents are

$$O_{Hd} = (H^{\dagger} i \overleftrightarrow{\mathcal{D}}_{\mu} H)[\bar{d}_R^i \gamma^{\mu} d_R^j], \qquad O_{Hu} = (H^{\dagger} i \overleftrightarrow{\mathcal{D}}_{\mu} H)[\bar{u}_R^i \gamma^{\mu} u_R^j]. \qquad (15.140)$$

Finally, there is one operator with charged RH quark currents:

$$O_{Hud} = (\widetilde{H}^{\dagger} i \mathcal{D}_{\mu} H)[\bar{u}_R^i \gamma^{\mu} d_R^j]. \qquad (15.141)$$

Here $\widetilde{H} \equiv i\sigma_2 H^*$. The complex-valued coefficients of these operators are denoted by

$$[C_{Hq}^{(1)}]_{ij}, \qquad [C_{Hq}^{(3)}]_{ij}, \qquad [C_{Hd}]_{ij}, \qquad [C_{Hu}]_{ij}, \qquad [C_{Hud}]_{ij}, \qquad (15.142)$$

where the indices $ij = 1, 2, 3$ denote different generations of up- and down-type quarks.

After EWSB, the transition from a weak to the mass eigenbasis takes place for gauge and quark fields. The quark fields are rotated by 3×3 unitary rotations in flavor space as explicitly shown in Section 14.4.1. We end up then with the important formulas (14.32) and (14.33), which give us Y_u as a function of up-quark masses and the CKM parameters. Y_d are diagonal as explained when deriving formulas (14.32) and (14.33).

The $\psi^2 H^2 D$ operators lead to modifications of the couplings of quarks to the weak gauge bosons ($V = W^{\pm}, Z$ and $g_Z \equiv \sqrt{g_1^2 + g_2^2}$):

$$\mathcal{L}_{\psi\bar{\psi}V}^{\text{dim-6}} = -\frac{g_Z}{2} v^2 Z_{\mu} \Big(\big[V_L^{d\dagger}(C_{Hq}^{(1)} + C_{Hq}^{(3)})V_L^d \big]_{ij} [\bar{d}_i \gamma^{\mu} P_L d_j] + \big[V_R^{d\dagger} C_{Hd} V_R^d \big]_{ij} [\bar{d}_i \gamma^{\mu} P_R d_j]$$

$$+ \big[V_L^{u\dagger}(C_{Hq}^{(1)} - C_{Hq}^{(3)})V_L^u \big]_{ij} [\bar{u}_i \gamma^{\mu} P_L u_j] + \big[V_R^{u\dagger} C_{Hu} V_R^u \big]_{ij} [\bar{u}_i \gamma^{\mu} P_R u_j] \Big),$$

$$+ \frac{g_2}{\sqrt{2}} v^2 \Big(\big[V_L^{u\dagger} C_{Hq}^{(3)} V_L^d \big]_{ij} [\bar{u}_i \gamma^{\mu} P_L d_j] W_{\mu}^+ + \big[V_R^{u\dagger} \frac{C_{Hud}}{2} V_R^d \big]_{ij} [\bar{u}_i \gamma^{\mu} P_R d_j] W_{\mu}^+ + \text{h.c.} \Big),$$

$$(15.143)$$

where we display all rotation matrices for completeness.

Note that in our notation fermion fields with an index for their handedness correspond to weak eigenstates, whereas mass eigenstates – like in this equation – do not carry this index. The values for v and the gauge couplings $g_{1,2}$ differ from the SM ones by dim-6 contributions. However, because the couplings in (15.143) are already at the level of dim-6, such corrections would count as dim-8 contributions, which is of higher order than considered here. From this equation it is apparent that our definition of the Wilson coefficients $[C_a]_{ij}$ (at μ_{ew}) for our special choice of the weak quark basis – $V_{L,R}^d = \mathbb{1}$ and $V_R^u = \mathbb{1}$ – is particularly convenient for the study of down-type quark $\Delta F = 1, 2$ transitions, see also [563], because additional CKM factors appear only in couplings of operators involving left-handed up-type quarks. Thus the associated Wilson coefficients $C_{Hq}^{(1,3)}$ enter down- *and* up-type-quark processes, leading to correlations between the affected processes that depend on the appearing CKM factors.

If one deals with down-type quark $\Delta F = 1, 2$ processes, it is justified to omit O_{Hu} and O_{Hud} in the phenomenology. This is mainly because their RG flow does not induce leading logarithmic contributions to down-type quark $\Delta F = 1, 2$ processes. The Wilson coefficients

$C_{Hq}^{(1,3)}$, C_{Hd} (and C_{Hu}) are complex-valued matrices in flavor space with a symmetric real part and antisymmetric imaginary part, such that each contains $6 + 3 = 9$ real degrees of freedom.

We can now express the usual FC quark couplings of the Z in terms of $C_{Hq}^{(1,3)}$, C_{Hd}. We have first

$$\mathcal{L}_{\psi\bar{\psi}Z}^{\text{NP}} = Z_\mu \sum_{\psi=u,d} \bar{\psi}_i \gamma^\mu \left([\Delta_L^\psi(Z)]_{ij} \, P_L + [\Delta_R^\psi(Z)]_{ij} \, P_R \right) \psi_j, \tag{15.144}$$

with $\psi = u, d$ distinguishing between *up*- and *down*-quark couplings. These complex-valued couplings are related to the WCs of SMEFT operators of Chapter 14 through

$$[\Delta_L^u(Z)]_{ij} = -\frac{g_Z}{2} v^2 \left[C_{Hq}^{(1)} - C_{Hq}^{(3)} \right]_{ij}, \qquad [\Delta_R^u(Z)]_{ij} = -\frac{g_Z}{2} v^2 [C_{Hu}]_{ij}, \tag{15.145}$$

$$[\Delta_L^d(Z)]_{ij} = -\frac{g_Z}{2} v^2 \left[C_{Hq}^{(1)} + C_{Hq}^{(3)} \right]_{ij}, \qquad [\Delta_R^d(Z)]_{ij} = -\frac{g_Z}{2} v^2 [C_{Hd}]_{ij}. \tag{15.146}$$

As all these vertices are $O(v^2/\Lambda^2)$, it is evident that a simple tree-level Z exchange with FC couplings corresponds to dimension-6 contribution in the case of $\Delta F = 1$ processes while to dimension-8 in the case of $\Delta F = 2$ transitions. Fortunately our results from Section 14.4 can tell us what the dimension-6 contributions to $\Delta F = 2$ are. In the LO they are simply given by the expressions in (14.39) and (14.43) for RH and LH scenarios, respectively.

A detailed analysis of these contributions has been presented in [766]. As this paper uses our notations and conventions, we will not repeat this analysis here and ask interested readers to consult this paper. Instead we will summarize here the main lessons from this study, which demonstrate that the simplified approach used for Z' contributions cannot be used in Z scenarios for $\Delta F = 2$ contributions. These lessons are as follows.

1. In the presence of right-handed FC Z couplings, i.e., $C_{Hd} \neq 0$ or $[\Delta_R^d(Z)]_{ij}$, the RG analysis of Section 14.4 yields that at μ_{ew} the left-right $\Delta F = 2$ operators $O_{\text{LR},1}$ in (14.37) are generated, and as seen in (14.39) their contributions are enhanced by the large leading logarithm $\ln \mu_\Lambda/\mu_{\text{ew}}$. Such operators are known to provide very important contributions to $\Delta F = 2$ observables because of their enhanced hadronic matrix elements and an additional enhancement from QCD RG effects below μ_{ew}, in particular in the K-meson system. As a result *these* operators – and not O_{VRR}^{ij} in (14.36), as used in [601, 611, 851] – dominate $\Delta F = 2$ processes. The results in [759] allow the calculation of this dominant contribution including only leading logarithms, but this is sufficient for our purposes, and even for scales μ_Λ as high as 20 TeV a good approximation is to keep only leading logarithm.

2. Because of the usual μ_{ew} scale ambiguity present at leading order (LO) the next-to-leading order (NLO) matching corrections of O_{Hd} to $\Delta F = 2$ processes at μ_{ew} within SMEFT have to be calculated. One NLO contribution is obtained by replacing the flavor-diagonal lepton vertex in the SM Z-penguin diagram by $[C_{Hd}]_{ij}$, which again generates the operator $O_{\text{LR},1}^{ij}$ simply because the flavor-changing part of the SM penguin

diagram is LH. In fact, this contribution has been first pointed out in [799]. But such contributions are by themselves gauge dependent, simply because the function $C_0(x_t)$ present in the SM vertex is gauge dependent. In [766] the missing contributions have been calculated using SMEFT, obtaining a gauge-independent contribution, which was later confirmed in [799]. However, the LO contribution is not only more important due to the large logarithm $\ln \mu_\Lambda/\mu_{ew}$, but has also opposite sign to the NLO term, allowing to remove the LO μ_{ew} dependence. Moreover, being strongly enhanced with respect to the contributions considered in [601, 611, 851], it has a very large impact on phenomenology, in particular as discussed below the correlations between $\Delta F = 2$ and $\Delta F = 1$ observables are drastically changed.

3. The situation for LH FC Z couplings is different from the RH case both qualitatively and quantitatively. The result in (14.43) tells us that the two operators $O_{Hq}^{(1)}$ and $O_{Hq}^{(3)}$ in SMEFT generate only the $\Delta F = 2$-operator O_{VLL} in (14.36) that is dominant already in the SM. The operator structure is then the same as in [611]. The resulting NP effects are then much smaller than in the RH case because no LR operators are present. But now comes an important difference from [611]. The correlations between $\Delta F = 1$ and $\Delta F = 2$ processes are weakened very significantly: While $\Delta F = 1$ transition amplitudes are proportional to the sum $C_{Hq}^{(1)} + C_{Hq}^{(3)}$, the leading RG contribution to $\Delta F = 2$ processes is proportional to the difference of these couplings. Correlations between $\Delta F = 1$ and $\Delta F = 2$ processes are hence only present in specific scenarios, e.g., when the couplings are given in terms of the fundamental parameters of a given model that can be determined in other processes. This is in stark contrast to the contributions considered in [601, 611, 851], where the same couplings enter both classes of processes, and no involvement of specific models was necessary. Of course, correlations remain in each sector separately because both are governed by two complex couplings, but as previously only one complex coupling was present, one needs more observables to determine them model independently. Moreover, in models where $\Delta F = 2$ and $\Delta F = 1$ observables are correlated, the constraints become weaker, allowing for larger NP effects in rare decays.

4. Also for the operators $O_{Hq}^{(1,3)}$ the NLO contributions to $\Delta F = 2$ corresponding to the replacement of the flavor-diagonal lepton vertex in the SM Z-penguin diagram by $C_{Hq}^{(1,3)}$ are gauge dependent. Including the remaining contributions to remove this gauge dependence one finds two gauge-independent functions of x_t, analogous to $X_0(x_t)$, $Y_0(x_t)$, and $Z_0(x_t)$ known from the SM. Because the NLO contributions are different for $C_{Hq}^{(1)}$ and $C_{Hq}^{(3)}$, it is not just their difference contributing to O_{VLL} anymore, but also their sum.

5. At NLO also new gauge-independent contributions are generated that are unrelated to tree-level Z exchanges and only proportional to $C_{Hq}^{(3)}$, analogous to the usual box diagrams with $W^\pm$ and quark exchanges. They turn out to be important for gauge-independence and depend not only on the coefficients for the quark transition under consideration, but also on additional couplings to the possible intermediate quarks in the box diagrams. But when the hierarchies in CKM elements are taken into account, $C_{Hq}^{(3)}$ for the quark transition under consideration is the only free entry in this part.

It should be stressed in this context that the contributions to $\Delta F = 2$ transitions from FC quark couplings of the Z could be less relevant in NP scenarios with other sources of $\Delta F = 2$ contributions. Most important, $\Delta F = 2$ operators could receive a direct contribution at tree-level at the scale μ_Λ, but also in models where this does not happen Z contributions could be subdominant. Examples are models in which the only new particles are vector-like quarks (VLQs), where box diagrams with VLQ and Higgs exchanges generate $\Delta F = 2$ operators at one-loop level [666, 967], which were found in these papers to be larger than the Z contributions at tree-level. However, in [967] the new effects just listed have not been included. As shown in [666] for right-handed FC Z couplings these box contributions are dwarfed by the LR operator contributions mentioned in point **1**, whereas in B-mixing they are comparable. We will discuss VLQ models in Section 16.3.

15.5.3 Master Functions in Z-Models

$\Delta F = 2$ Master Functions

The $\Delta F = 2$ master functions for $M = K, B_q$ are given as follows

$$S(M) = S_0(x_t) + \Delta S(M) \equiv |S(M)|e^{i\theta_S^M}, \tag{15.147}$$

with $\Delta S(M)$ receiving contributions from various operators so that it is useful to write

$$\Delta S(M) = [\Delta S(M)]_{\text{VLL}} + [\Delta S(M)]_{\text{VRR}} + [\Delta S(M)]_{\text{LR}} + \overline{[\Delta S(M)]}_{\text{VLL}} + \overline{[\Delta S(M)]}_{\text{LR}}. \tag{15.148}$$

The first three contributions are directly obtained from the previous section by making the substitutions listed earlier. They represent dimension-8 contributions coming from the simplified approach. The last two represent new dimension-6 contributions obtained by means of SMEFT as discussed earlier. Explicit formulas for them can be found in [766].

$\Delta F = 1$ Master Functions

These master functions can be directly obtained from the ones in Z' scenarios. They are given as follows

$$X_{\text{L}}(K) = \eta_X X_0(x_t) + \frac{\Delta_L^{v\bar{v}}(Z)}{g_{\text{SM}}^2 M_Z^2} \frac{\Delta_L^{sd}(Z)}{V_{ts}^* V_{td}}, \tag{15.149}$$

$$X_{\text{R}}(K) = \frac{\Delta_L^{v\bar{v}}(Z)}{g_{\text{SM}}^2 M_Z^2} \frac{\Delta_R^{sd}(Z)}{V_{ts}^* V_{td}}, \tag{15.150}$$

$$X_{\text{L}}(B_q) = \eta_X X_0(x_t) + \left[\frac{\Delta_L^{vv}(Z)}{M_Z^2 g_{\text{SM}}^2}\right] \frac{\Delta_L^{qb}(Z)}{V_{tq}^* V_{tb}}, \tag{15.151}$$

$$X_{\text{R}}(B_q) = \left[\frac{\Delta_L^{vv}(Z)}{M_Z^2 g_{\text{SM}}^2}\right] \frac{\Delta_R^{qb}(Z)}{V_{tq}^* V_{tb}}, \tag{15.152}$$

$$Y_A(K) = \eta_Y Y_0(x_t) + \frac{\left[\Delta_A^{\mu\bar{\mu}}(Z)\right]}{M_Z^2 g_{SM}^2} \left[\frac{\Delta_L^{sd}(Z) - \Delta_R^{sd}(Z)}{V_{ts}^\star V_{td}}\right] \equiv |Y_A(K)| e^{i\theta_Y^K}, \qquad (15.153)$$

$$Y_A(B_q) = \eta_Y Y_0(x_t) + \frac{\left[\Delta_A^{\mu\bar{\mu}}(Z)\right]}{M_Z^2 g_{SM}^2} \left[\frac{\Delta_L^{qb}(Z) - \Delta_R^{qb}(Z)}{V_{tq}^\star V_{tb}}\right] \equiv |Y_A(B_q)| e^{i\theta_Y^{B_q}}. \qquad (15.154)$$

Here $\eta_{X,Y}$ are QCD factors given in (15.79).

The Wilson coefficients (15.80) in Z-models read

$$\sin^2\vartheta_W C_9 = [\eta_Y Y_0(x_t) - 4\sin^2\vartheta_W Z_0(x_t)] - \frac{1}{g_{SM}^2}\frac{1}{M_Z^2}\frac{\Delta_L^{sb}(Z)\Delta_V^{\mu\bar{\mu}}(Z)}{V_{ts}^* V_{tb}}, \qquad (15.155)$$

$$\sin^2\vartheta_W C_{10} = -\eta_Y Y_0(x_t) - \frac{1}{g_{SM}^2}\frac{1}{M_Z^2}\frac{\Delta_L^{sb}(Z)\Delta_A^{\mu\bar{\mu}}(Z)}{V_{ts}^* V_{tb}}, \qquad (15.156)$$

$$\sin^2\vartheta_W C_9' = -\frac{1}{g_{SM}^2}\frac{1}{M_Z^2}\frac{\Delta_R^{sb}(Z)\Delta_V^{\mu\bar{\mu}}(Z)}{V_{ts}^* V_{tb}}, \qquad (15.157)$$

$$\sin^2\vartheta_W C_{10}' = -\frac{1}{g_{SM}^2}\frac{1}{M_Z^2}\frac{\Delta_R^{sb}(Z)\Delta_A^{\mu\bar{\mu}}(Z)}{V_{ts}^* V_{tb}}. \qquad (15.158)$$

Finally, in the case of $K_L \to \pi^0 \mu^+ \mu^-$ decays, one can use formulas of Section 15.4.5 with Z' replaced by Z. Note that the flavor-conserving couplings are SM ones.

15.5.4 Lessons on NP Patterns in Z Scenarios

A detailed phenomenology of tree-level Z exchanges has been presented in [601, 851] and in particular in [766]. We will now summarize the highlights of these papers in the form of lessons gained through these analyses. Numerous plots related to these lessons can be found in these papers.

We begin with the K-meson system and concentrate on the results obtained in [851] including the insights on $\Delta F = 2$ processes from [766]. The summary of the lessons is rather brief. On the other hand, the presentation in [851] is very detailed with numerous analytic expressions that could be useful for nonexperts who want to understand the details of the patterns of various effects. In the latter paper the main goal was to investigate, similar to the Z' case in Section 15.4, what are the implications of the ε'/ε anomaly discussed in Chapter 10 on rare decays $K^+ \to \pi^+ \nu\bar{\nu}$ and $K_L \to \pi^0 \nu\bar{\nu}$ in various scenarios for the Z couplings to quarks: left-handed scenario (LHS) and right-handed scenario (RHS).

Lesson 1: In the LHS, a given request for the enhancement of ε'/ε determines the coupling $\text{Im}\Delta_L^{sd}(Z)$. Similar in the RHS the coupling $\text{Im}\Delta_R^{sd}(Z)$ is determined.

Lesson 2: In LHS there is a direct unique implication of an enhanced ε'/ε on $K_L \to \pi^0 \nu\bar{\nu}$: *suppression* of $\mathcal{B}(K_L \to \pi^0 \nu\bar{\nu})$. This property is known from NP scenarios in which NP to $K_L \to \pi^0 \nu\bar{\nu}$ and ε'/ε enters dominantly through the modification of Z-penguins. The known flavor-diagonal Z couplings to quarks and leptons and the sign of the matrix element $\langle Q_8 \rangle_2$ determines this anticorrelation, which has been verified in all models with only LH flavor-violating Z couplings.

Lesson 3: The imposition of the $K_L \to \mu^+\mu^-$ constraint in LHS determines the range for $\mathrm{Re}\Delta_L^{sd}(Z)$, which with the already fixed $\mathrm{Im}\Delta_L^{sd}(Z)$ would allow to calculate the shifts in ε_K and ΔM_K if not for new contributions identified in [766], which were not included in [851]. There it was concluded that these shifts are very small for ε_K and negligible for ΔM_K. But this conclusion is not valid in the presence of these new contributions. Moreover, in concrete models new contributions beyond Z exchange are possible. For instance, in VLQ models box diagrams with VLQs can indeed provide contributions to ε_K and ΔM_K that are larger than coming from tree-level Z-exchange provided the masses of VLQs are far above 3 TeV [967]. We will discuss this in the context of the analysis in [666]. In any case $K_L \to \mu^+\mu^-$ determines the allowed range for $\mathrm{Re}\Delta_L^{sd}(Z)$.

Lesson 4: With fixed $\mathrm{Im}\Delta_L^{sd}(Z)$ and the allowed range for $\mathrm{Re}\Delta_L^{sd}(Z)$, the range for $\mathcal{B}(K^+ \to \pi^+\nu\bar{\nu})$ can be obtained. But in view of uncertainties in the $K_L \to \mu^+\mu^-$ constraint both an enhancement and a suppression of $\mathcal{B}(K^+ \to \pi^+\nu\bar{\nu})$ are possible, and no specific pattern of correlation between $\mathcal{B}(K_L \to \pi^0\nu\bar{\nu})$ and $\mathcal{B}(K^+ \to \pi^+\nu\bar{\nu})$ is found. In the absence of a relevant ε_K constraint, this is consistent with the general analysis in [924]. $\mathcal{B}(K^+ \to \pi^+\nu\bar{\nu})$ can be enhanced by a factor of 2 at most if conservative bound on $K_L \to \mu^+\mu^-$ is used and only suppressed if stricter bound in (9.268) is used. See [766] and discussion at the end of Section 9.7.1.

Lesson 5: As far as the correlation of ε'/ε with $K_L \to \pi^0\nu\bar{\nu}$ is concerned analogous pattern is found in RHS, although the numerics is different. See plots in [851]. But the new contributions from LR operators to ε_K found in [766] have dramatic impact on the results for $K^+ \to \pi^+\nu\bar{\nu}$ presented in [851]. Now not $K_L \to \mu^+\mu^-$ but the constraint from ε_K determines the allowed enhancement of $\mathcal{B}(K^+ \to \pi^+\nu\bar{\nu})$. While in [851] an enhancement of $\mathcal{B}(K^+ \to \pi^+\nu\bar{\nu})$ up to a factor of 5.7 was possible, now only an enhancement up to a factor of 1.5 is possible.

Lesson 6: In a general Z scenario in which the underling theory contains all the operators in (15.139) and (15.140) and simultaneously dimension-8 LR operators are present, the pattern of NP effects can change relative to LH and RH scenarios because of many parameters involved independent of whether new contributions considered in [766] are taken into account or not. As demonstrated in [851] the main virtue of the general scenario is the possibility of enhancing simultaneously ε'/ε, ε_K, $\mathcal{B}(K^+ \to \pi^+\nu\bar{\nu})$, and $\mathcal{B}(K_L \to \pi^0\nu\bar{\nu})$, which is not possible in LHS and RHS. Thus the presence of both LH and RH flavor-violating currents is essential for obtaining simultaneously the enhancements in question when NP is dominated by tree-level Z exchanges. We refer to examples in [851]. Then the main message from this analysis is that in the presence of both LH and RH new flavour-violating couplings of Z to quarks, large departures from SM predictions for $K^+ \to \pi^+\nu\bar{\nu}$ and $K_L \to \pi^0\nu\bar{\nu}$ are still possible. Similar conclusions have been reached in [799].

Concerning B physics observables we have the following lessons [766].

Lesson 7: In the $b \to s, d$ sector $\Delta F = 1$ constraints remain dominant, but allow still for sizable NP contributions, e.g., in $B_{d,s} \to \mu^+\mu^-$. In particular $B_d \to \mu^+\mu^-$ can be enhanced to the present upper bound. Nevertheless, the strong correlations in this scenario will allow for distinguishing it from other NP models with coming data from the LHC (LHCb, CMS, ATLAS) and Belle II.

Lesson 8: The correlations between R_K^ν and $R_{K^*}^\nu$ of Section 9.6 and between $\mathcal{B}(B_s \to \mu^+\mu^-)$ and $\mathcal{B}(B_d \to \mu^+\mu^-)$ allow to distinguish LH and RH scenarios for a large part of the parameter space. Because in the LH scenario the interference with the SM contribution is the same in $B \to K\nu\bar\nu$ and $B \to K^*\nu\bar\nu$, there is a very strict prediction $R_K^\nu/R_{K^*}^\nu \equiv 1$ [607]. An enhancement of each of the ratios is possible only up to $\sim 10\%$ in this scenario, but a strong suppression is possible, in contrast to RH models, where NP effects are tiny. For $\mathcal{B}(B_d \to \mu^+\mu^-)$ a moderate enhancement up to $\sim 2 \times 10^{-10}$ is possible, but also a strong suppression, in contrast with the RH case.

Lesson 9: Possible tensions in $\Delta M_{s,d}$ and ε_K can be successfully addressed in these NP scenarios.

Lesson 10: Very importantly this NP scenario fails in explaining the anomalies in $b \to s\ell^+\ell^-$ transitions like $R(K)$ and $R(K^*)$ and in $b \to c\tau\bar\nu_\tau$ like $R(D)$ and $R(D^*)$. In particular, due to small vector-couplings of Z to muons, the observed suppression of C_9 cannot be explained.

In summary, while the tree-level Z exchanges could be responsible for a hinted ε'/ε anomaly and could provide sizable departures from SM expectations in $K^+ \to \pi^+\nu\bar\nu$, $K_L \to \pi^0\nu\bar\nu$ and $B_{s,d} \to \mu^+\mu^-$, NP behind possible violation of lepton flavor universality must be different. Here Z' models of Section 15.4 in $b \to s\ell^+\ell^-$ transitions and leptoquark models of Section 16.4 in both $b \to s\ell^+\ell^-$ and $b \to c\tau\bar\nu_\tau$ transitions are performing much better.

Finally, let us mention that in models containing Z', FCNC processes mediated by the Z boson can be generated by $Z - Z'$ mixing. We will present the relevant expressions in the context of the presentation of 331 models in Section 16.2. FCNC processes mediated by the Z boson can also be generated in the process of EW symmetry breaking through mixing of the SM fermions with vector-like fermions. We will discuss this case in Section 16.3.

15.6 Beyond MFV: Right-Handed *W'*

A very important property of the SM regarding flavor-violating processes is the LH structure of the charged current interactions reflecting the maximal violation of parity observed in low-energy processes. LH charged currents encode at the level of the Lagrangian the full information about flavor mixing and CP violation represented compactly by the CKM matrix. As we have seen, due to the GIM [6] mechanism this structure has automatically profound implications for the pattern of FCNC processes that seems to be in remarkable accordance with the present data within theoretical and experimental uncertainties, bearing in mind the anomalies listed in Section 12.2.

As the SM is expected to be only the low-energy limit of a more fundamental theory, it is conceivable that at very short distance scales parity could be a good symmetry implying the existence of RH charged currents. Prominent examples of such fundamental theories are left-right (LR) symmetric models based on $SU(2)_L \times SU(2)_R \times U(1)_{B-L}$ gauge group on which a rich literature exists.

Indeed, LR symmetric models were born 45 years ago [968–972]. Early papers mainly cover the examinations of two special cases, known as *manifest* scenario [972] and *pseudo-manifest* scenario [842, 973–975], which are characterized by no spontaneous and fully spontaneous CP violation, respectively. The right-handed counterpart of the CKM matrix appears then in a special form being either identical to or the complex-conjugate of the CKM matrix up to certain phases as, e.g., summarized in [976]. The phenomenology of both scenarios has widely been studied in the literature [750, 977–979]. The book [7] contains a lot of information on these models as well.

By now in both scenarios strong constraints on the heavy charged gauge boson mass $M_{W_R} \geq 4\,\text{TeV}$ have been obtained [750, 980, 981] from the constraints on the $K_L - K_S$ mass difference, CP violation in kaon decays, and the neutron electric dipole moment, making these scenarios difficult to access in direct searches at the LHC. In addition the pseudo-manifest scenario has been ruled out by the correlation of ε_K and $\sin(2\beta)$ [982]. This means that the right-handed mixing matrix must be different from the CKM matrix to reach agreement with experiment. Motivated by this fact more general studies of CP violation have been performed in [976, 983, 984] and more recently in [772, 981, 985–988]. For a very recent analysis with many references see [989].

Moreover, the phenomenological interest in having another look at the RH currents in general originated from tensions between inclusive and exclusive determinations of the elements of the CKM matrix $|V_{ub}|$ and $|V_{cb}|$. In this context it has been pointed out in [990–992] that the presence of RH currents could either remove or significantly weaken some of these tensions, especially in the case of $|V_{ub}|$. The implications of these findings for many observables within an effective theory approach have been studied in [610].

As of 2019, it does not seem that RH currents play a significant role in the explanation of the tension between inclusive and exclusive determinations of $|V_{ub}|$. But they could still play a significant role in the flavor phenomenology, as in the presence of both LH currents in the SM and RH currents beyond it, new operators are generated, in particular LR operators with enhanced $K^0 - \bar{K}^0$ matrix elements. But also in rare decays they could be relevant.

It should also be stressed that in such models not only heavy charged gauge bosons with couplings to RH quarks and leptons are present but also charged and neutral scalars. The charged scalars are most important for $B \to X_s \gamma$ decays, while the neutral ones for $\Delta F = 2$ processes. We refer to [772] for a detailed analysis of these processes. In this paper also a list of Feynman rules for a general model with $SU(2)_L \times SU(2)_R \times U(1)_{B-L}$ gauge group is given.

Among recent studies let me just mention [861, 862, 993, 994], where further references can be found. While, until today, no clear signal of the presence of RH-charged currents have been identified, they could still play a significant role in the future. On the theoretical side, the study of FCNC at one-loop in the presence of RH-charged currents has recently been improved [136].

As a simple example let us write the effective Hamiltonian for the exchange of a charged gauge bosons W'^+ contributing to $B^+ \to \tau^+ \nu_\tau$ as follows

$$\mathcal{H}_{\text{eff}} = C_L O_L + C_R O_R, \qquad (15.159)$$

where

$$O_L = (\bar{b}\gamma_\mu P_L u)(\bar{\nu}_\tau \gamma^\mu P_L \tau^-), \quad O_R = (\bar{b}\gamma_\mu P_R u)(\bar{\nu}_\tau \gamma^\mu P_L \tau^-) \qquad (15.160)$$

and

$$C_L = C_L^{SM} + \frac{\Delta_L^{ub*}(W'^+)\Delta_L^{\tau\nu}(W'^+)}{M_{W'^+}^2}, \quad C_R = \frac{\Delta_R^{ub*}(W'^+)\Delta_L^{\tau\nu}(W'^+)}{M_{W'^+}^2}, \quad (15.161)$$

with C_L^{SM} having the same structure as the correction from W'^+ with $M_{W'^+}$ replaced $M_{W'}$, and

$$\Delta_L^{ub} = \frac{g}{\sqrt{2}}V_{ub}, \qquad \Delta_L^{\nu\tau} = \frac{g}{\sqrt{2}}, \qquad \Delta_R^{ub} = 0, \qquad (SM). \qquad (15.162)$$

The couplings $\Delta_{L,R}^{ub*}(W'^+)$ could be complex numbers and contain new sources of flavor violation. Then

$$\mathcal{B}(B^+ \to \tau^+\nu_\tau)_{W'^+} = \frac{1}{64\pi}m_{B^+}m_\tau^2\left(1 - \frac{m_\tau^2}{m_{B^+}^2}\right)^2 F_{B^+}^2\tau_{B^+}|C_R - C_L|^2. \qquad (15.163)$$

Evidently in a model like this it is possible to improve the agreement with the data by choosing appropriately the couplings of W'^+.

15.7 Beyond MFV: Neutral Scalars and Pseudoscalars

15.7.1 Preliminaries

In BSM scenarios FCNC processes can be mediated by tree-level heavy neutral scalars and/or pseudoscalars $H^0(A^0)$.[6] This generally introduces new sources of flavor violation and CP violation as well as LH and RH *scalar* $(1 \mp \gamma_5)$ currents as opposed to vector LH and RH currents that we studied in previous sections in connection with tree-level Z' and Z exchanges.

As already mentioned in Section 15.1.4, there are ways to avoid such transitions at tree-level without fine-tuning through a special arrangement of the couplings of the Higgs system with two or more doublets to SM fermions [899, 900] so that there are no FCNC mediated by neutral scalars at tree-level. This is, for instance, the case of the MSSM and many 2HDM models in which one Higgs doublet is responsible for up-quark masses and a second for down-quark ones. An excellent review of 2HDMs discussing this topic is the review in [897], and some comments can also be found in Section 15.1.4.

However, even if in the SM and in many of its extensions there are no fundamental flavor-violating couplings of scalars to quarks and leptons, such couplings can be generated through loop corrections leading in the case of $\Delta F = 1$ transitions to Higgs-penguins (HP) and in $\Delta F = 2$ transitions to double Higgs-penguins (DHP). Yet, when the masses of the scalar particles are significantly lower than the heavy new particles exchanged in the loops, the HP and DHP look at the electroweak scale as flavor-violating tree diagrams with the couplings determined by the fundamental parameters of a given model.

---

[6] Unless otherwise specified, we will use the name *scalar* for both scalars and pseudoscalars.

But there exist models in which tree-level scalar exchanges take place at a fundamental level. That is, neutral scalar particles have flavor-violating couplings to quarks and leptons. Important examples are left-right symmetric models just discussed. From the point of view of low-energy theory there is no distinction between these possibilities as long as the vertices involving heavy particles in a HP cannot be resolved and to first approximation what really matters is the mass of the exchanged scalar and its flavor-violating couplings, either fundamental or generated at one-loop level.

Therefore, as the first step, similar to Z' scenarios presented in Section 15.4, we will study the patterns of NP contributions to FCNC processes that depend only on the couplings of $H^0(A^0)$ to fermions and on their masses. In situations in which a single H^0 or A^0 dominates NP contributions, we will find stringent correlations between $\Delta F = 2$ and $\Delta F = 1$ observables that will be markedly different from the ones caused by tree-level Z' and Z FCNC. This section is dominantly based on [504, 509].

Thus the goal of the present section is to analyze how SM predictions for FCNC processes are modified by tree-level scalar exchanges and to demonstrate that these modifications can be distinguished through correlations between quark flavor observables from the ones identified in Sections 15.4 and 15.5. Such a distinction would be crucial in case no new particles would be discovered directly at the LHC.

Before going into details, it will be useful to mention several differences between NP governed by Z' and scalar particles that are related to the existing lower bounds on their masses and the fact that these particle have different spins.

- The lower bounds on masses of Z' gauge bosons from the LHC are by now in the ballpark of 3 TeV in case their couplings to quarks are SM-like,[7] while new neutral scalars with masses below 1 TeV are not excluded.
- In the Z' scenarios in addition to new operators involving RH vector currents also SM operators with modified Wilson coefficients can be present. On the other hand in the case of tree-level scalar exchanges all effective low-energy operators are new.
- Even if after the inclusion of QCD corrections, there is some overlap between operators contributing to $\Delta F = 2$ processes in Z' and scalar scenarios, their Wilson coefficients are differently affected by QCD corrections. We will see this explicitly later. But in $\Delta F = 1$ transitions there is no overlap between the operators present in these two scenarios.
- We have seen that $\Delta F = 1$ FCNC can be significantly modified by Z exchanges still satisfying $\Delta F = 2$ constraints. On the other hand the smallness of the SM Higgs couplings to muons and electrons precludes any visible effects from tree-level Higgs exchanges in rare K and B_d decays with muon or electron pair in the final state once constraints from $\Delta F = 2$ processes are taken into account.
- It should be emphasized that the couplings of new scalars to neutrinos could be sizable if the masses of neutrinos are generated by a different mechanism than coupling to scalars, like in the case of the see-saw mechanism. We will assume that this is not the case so that NP effects of scalars in $K^+ \to \pi^+ \nu \bar{\nu}$, $K_{\rm L} \to \pi^0 \nu \bar{\nu}$, and $b \to s \nu \bar{\nu}$ transitions will be negligible in contrast to Z' models, where NP effects in these decays can still be significant.

---

[7] In models in which Z' couples only to leptons Z' masses can be much lower.

Now comes a warning so that the readers are not disappointed at the end of this section. We will present many formulas that can be used for phenomenology but there is no space to present here numerical analysis. Such analysis with numerous plots can be found in [504, 509].

15.7.2 Model Assumptions and Notations

To proceed we have to specify our notations that will be analogous to the ones for Z' scenarios except that now we have to distinguish between a heavy neutral scalar and a heavy pseudoscalar. We will use a common name, H^0, for them unless otherwise specified. When a distinction will have to be made, we will either use H^0 and A^0 for scalar and pseudoscalar, respectively, or to distinguish SM Higgs from additional spin-0 particles we will use the familiar two-Higgs doublet model (2HDM) and MSSM notation: (H, A, h) with h denoting the SM Higgs particle.

Our strategy will be to consider in this section the simplest model in which the only new particle in the low-energy effective theory is a single neutral particle with spin-0, a scalar, or pseudoscalar. As in most specific extensions of the SM, like 2HDMs, of various sorts, and the MSSM, several spin-0 particles are present, the question arises of whether this minimal construction is possible from the point of view of an underlying fundamental theory.

The first question to be investigated are the transformation properties of this particle under the $SU(2)_L$ of the SM. If the scalar in question is not a $SU(2)_L$ singlet, then it must be placed in a complete $SU(2)_L$ multiplet, e.g., a second doublet, as is the case of 2HDM or the MSSM. However, this implies the existence of its $SU(2)_L$ partners in a given multiplet with masses close to the masses of our scalar. In fact, in the decoupling regime in 2HDM and MSSM the masses of $(H^\pm, H^0, A^0)$ are approximately the same. While $SU(2)_L$ breaking effects in the Higgs potential allow for mass splittings, they must be of $O(v)$ at most with $v = 246$ GeV being the electroweak symmetry breaking scale. Consequently we are lead to the conclusion that the case of the dominance of a single scalar with nontrivial transformations under $SU(2)_L$ is rather unlikely.

It follows then that our scalar should be a $SU(2)_L$ singlet. This does not preclude its couplings to SM fermions. For instance, the scalar-quark couplings of H to quarks can originate in the following low-energy effective operator

$$L = \lambda_L^{ij} \frac{H^0}{\Lambda} \bar{q}_R^i q_L^j h_{\text{SM}} + h.c., \tag{15.164}$$

with Λ denoting the cutoff scale of the low-energy theory and h_{SM} being the Higgs particle. After the spontaneous breakdown of $SU(2)_L$ the scalar left-handed coupling is given by

$$\Delta_L^{ij}(H^0) = \frac{1}{\sqrt{2}} \frac{v}{\Lambda} \lambda_L^{ij}, \tag{15.165}$$

with analogous expression for the right-handed coupling $\Delta_R^{ij}(H^0)$.

All the FCNC discussed in this section will be governed by the couplings $\Delta_{L,R}^{ij}(H^0)$ and the scalar mass M_H. The Feynman rule for the coupling of H^0 to quarks is shown

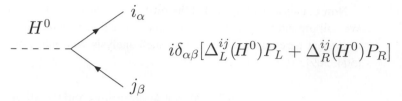

Figure 15.9 Feynman rules for a neutral colorless scalar particle H^0 with mass M_H, where i, j denote different quark flavors and α, β the colors. $P_{L,R} = (1 \mp \gamma_5)/2$.

in Figure 15.9 with (i, j) denoting quark flavors. But in contrast to the couplings of Z', for which there is generally no relation between LH and RH couplings, we have now the following important relation

$$\Delta_L^{ij}(H^0) = [\Delta_R^{ji}(H^0)]^*. \tag{15.166}$$

This difference is, of course, related to the fact that in the case at hand there is a chirality flip, while there is no chirality flip in the case of Z' and any gauge boson. In turn, as we will see later, the low-energy operators governing FCNC are quite different from the ones encountered in previous sections.

The couplings $\Delta_{L,R}^{ij}(H^0)$, as defined in Figure 15.9, are dimensionless quantities, but whereas in the case of gauge bosons they are built out of gauge couplings and some charges, in the present case we are used from the SM, 2HDM, and MSSM to deal with Yukawa couplings and eventually with the ratios of quark masses and the electroweak vacuum expectation value v or other mass scales. But it should be emphasized that the proportionality of the couplings in question to the masses of the participating quarks is not a general property. It applies only if the scalar and the SM Higgs, responsible for SU(2) breakdown, are in the same SU(2) multiplet or a multiplet of a larger gauge group G. Then after the breakdown of G to SU(2), the scalar appears as a singlet of SU(2) symmetry, with couplings to quarks involving their masses after SU(2) breakdown. While this is the case in several models, in our simple extension of the SM, we will assume that the involved scalar couplings are unrelated to the generation of quark masses and the couplings $\Delta_{L,R}^{ij}(H^0)$ are simply arbitrary complex numbers to be determined from experimental data or calculated from an explicit model. This will allow us to derive general expressions for FCNC observables that can be used in specific models constructed by the readers themselves.

15.7.3 Scalar versus Pseudoscalar

We will see in the next subsections that the pattern of flavor violations caused by the spin-0 particle in question depends, even at the qualitative level, on whether it is a scalar or pseudoscalar. This distinction is often made through the study of angular distributions in collider processes, but as we will demonstrate soon also flavor-violating processes can be helpful in distinguishing between a scalar and a pseudoscalar.

When listing the differences between the scalar and the pseudoscalar spin-0 particles, one should emphasize that in the presence of CP violation, the mass eigenstate H^0 propagating in a tree-diagram is not necessarily a CP eigenstate. Therefore, generally the coupling to $\mu^+\mu^-$ appearing at many places in this section can have the general structure

$$L = \frac{1}{2}\bar{\mu}(\Delta_S^{\mu\bar{\mu}}(H) + \gamma_5\Delta_P^{\mu\bar{\mu}}(H))H\mu, \tag{15.167}$$

with analogous expressions for other leptons and quarks. We drop the charge index as in this section only neutral scalars are present. The couplings $\Delta_{S,P}^{\mu\bar{\mu}}$ are defined in analogy to (15.54) by

$$\Delta_S^{\mu\bar{\mu}}(H) = \Delta_R^{\mu\bar{\mu}}(H) + \Delta_L^{\mu\bar{\mu}}(H), \qquad \Delta_P^{\mu\bar{\mu}}(H) = \Delta_R^{\mu\bar{\mu}}(H) - \Delta_L^{\mu\bar{\mu}}(H). \tag{15.168}$$

$\Delta_S^{\mu\bar{\mu}}$ is real and $\Delta_P^{\mu\bar{\mu}}$ purely imaginary as required by the hermiticity of the Hamiltonian, which can be verified by means of (15.166).

We will give the expressions for various observables in terms of the couplings $\Delta_{L,R}^{ij}(H)$ and $\Delta_{S,P}^{\mu\bar{\mu}}(H)$. They can be directly used in the case of the scalar particle being CP-even eigenstate, like (H^0, h) in 2HDM or MSSM setting $\Delta_P^{\mu\bar{\mu}}(H) = 0$. However, when the mass eigenstate is a pseudoscalar A, implying $\Delta_S^{\mu\bar{\mu}} = 0$, it will be useful to exhibit the i, which we illustrate here for the B_s^0 system:

$$\Delta_L^{bs}(A) = -i\tilde{\Delta}_L^{bs}(A), \qquad \Lambda_R^{bs}(A) = +i\tilde{\Delta}_R^{bs}(A), \qquad \Delta_P^{\mu\bar{\mu}}(A) = i\tilde{\Delta}_P^{\mu\bar{\mu}}(A). \tag{15.169}$$

Here the flavor-violating couplings $\tilde{\Delta}_{L,R}^{bs}(A)$ are still complex, while $\tilde{\Delta}_P^{\mu\bar{\mu}}(A)$ is real.

The following useful relations follow from (15.166) and (15.169):

$$\Delta_R^{sb}(A) = i[\tilde{\Delta}_L^{bs}(A)]^*, \qquad \Delta_L^{sb}(A) = -i[\tilde{\Delta}_R^{bs}(A)]^*. \tag{15.170}$$

Interestingly, if the couplings defined here are treated as arbitrary complex numbers to be bounded by $\Delta F = 2$ observables, and there is only one particle contributing, it will not be possible to distinguish a scalar from a pseudoscalar on the basis of $\Delta F = 2$ transitions alone. On the other hand, when rare decays, in particular $B_{s,d} \to \mu^+\mu^-$, are considered, there is a difference between these two cases as the pseudoscalar contributions interfere with SM contribution, while the scalar ones do not. Consequently, the allowed values for $\tilde{\Delta}_P^{\mu\bar{\mu}}(A)$ and $\Delta_S^{\mu\bar{\mu}}(A)$ will differ from each other, and we will find other differences. But, if both scalar and pseudoscalar contribute to tree-level decays and have approximately the same mass as well as couplings related by symmetries, also their contributions to $\Delta F = 2$ processes differ.

After these introductory remarks we will present in the next subsections formulas for contributions of scalar particles to various flavor observables as functions of the preceding couplings. On several occasions the formulas will turn out to be very similar to the ones presented for Z', but the building blocks in these formulas will be different, and we will stress it whenever this will be necessary to avoid confusion.

15.7.4 $\Delta F = 2$ Processes

Master Functions Including H Contributions

In the presence of H tree-level contributions, similar to the Z' case, we will work with three different master functions

$$S(K), \qquad S(B_d), \qquad S(B_s), \tag{15.171}$$

for $K^0 - \bar{K}^0$ and $B^0_{s,d} - \bar{B}^0_{s,d}$ systems, respectively. In introducing these functions we will include in their definitions the contributions of all operators involved.

The derivation of the formulas listed soon proceeds as in Section 15.4 and is left as an exercise for the reader.

The $\Delta F = 2$ master functions for $M = K, B_q$ are then given as follows

$$S(M) = S_0(x_t) + \Delta S(M) \equiv |S(M)|e^{i\theta_S^M}, \tag{15.172}$$

with $\Delta S(M)$ receiving contributions from various operators so that it is useful to write

$$\Delta S(M) = [\Delta S(M)]_{\text{SLL}} + [\Delta S(M)]_{\text{SRR}} + [\Delta S(M)]_{\text{LR}}. \tag{15.173}$$

We have encountered the contributing new operators already in Section 7.3 but we recall them here for convenience. For the K system we have [746, 749]

$$Q_1^{\text{LR}} = \left(\bar{s}\gamma_\mu P_L d\right)\left(\bar{s}\gamma^\mu P_R d\right), \tag{15.174a}$$

$$Q_2^{\text{LR}} = (\bar{s}P_L d)(\bar{s}P_R d). \tag{15.174b}$$

$$Q_1^{\text{SLL}} = (\bar{s}P_L d)(\bar{s}P_L d), \tag{15.175a}$$

$$Q_1^{\text{SRR}} = (\bar{s}P_R d)(\bar{s}P_R d), \tag{15.175b}$$

$$Q_2^{\text{SLL}} = \left(\bar{s}\sigma_{\mu\nu} P_L d\right)\left(\bar{s}\sigma^{\mu\nu} P_L d\right), \tag{15.175c}$$

$$Q_2^{\text{SRR}} = \left(\bar{s}\sigma_{\mu\nu} P_R d\right)\left(\bar{s}\sigma^{\mu\nu} P_R d\right), \tag{15.175d}$$

where $P_{R,L} = (1 \pm \gamma_5)/2$, and we suppressed color indices as they are summed up in each factor. For instance, $\bar{s}\gamma_\mu P_L d$ stands for $\bar{s}_\alpha \gamma_\mu P_L d_\alpha$ and similarly for other factors. For $B^0_q - \bar{B}^0_q$ mixing our conventions for new operators are

$$Q_1^{\text{LR}} = \left(\bar{b}\gamma_\mu P_L q\right)\left(\bar{b}\gamma^\mu P_R q\right), \tag{15.176a}$$

$$Q_2^{\text{LR}} = \left(\bar{b}P_L q\right)\left(\bar{b}P_R q\right), \tag{15.176b}$$

$$Q_1^{\text{SLL}} = \left(\bar{b}P_L q\right)\left(\bar{b}P_L q\right), \tag{15.177a}$$

$$Q_1^{\text{SRR}} = \left(\bar{b}P_R q\right)\left(\bar{b}P_R q\right), \tag{15.177b}$$

$$Q_2^{\text{SLL}} = \left(\bar{b}\sigma_{\mu\nu}P_L q\right)\left(\bar{b}\sigma^{\mu\nu}P_L q\right), \tag{15.177c}$$

$$Q_2^{\text{SRR}} = \left(\bar{b}\sigma_{\mu\nu}P_R q\right)\left(\bar{b}\sigma^{\mu\nu}P_R q\right). \tag{15.177d}$$

To calculate the SLL, SRR, and LR contributions to $\Delta S(M)$, we use again the expressions $T(B_q)$ and $T(K)$, which we recall here for convenience

$$T(B_q) = \frac{G_F^2}{12\pi^2} F_{B_q}^2 \hat{B}_{B_q} m_{B_q} M_W^2 \left(\lambda_t^{(q)}\right)^2 \eta_B, \tag{15.178}$$

$$T(K) = \frac{G_F^2}{12\pi^2} F_K^2 \hat{B}_K m_K M_W^2 \left(\lambda_t^{(K)}\right)^2 \eta_2, \tag{15.179}$$

with all entries defined in Sections 8.1 and 7.3. We have then

$$T(K)[\Delta S(K)]_{\text{SLL}} = -\frac{(\Delta_L^{sd}(H))^2}{2M_H^2}\left[C_1^{\text{SLL}}(\mu_H)\langle Q_1^{\text{SLL}}(\mu_H,K)\rangle + C_2^{\text{SLL}}(\mu_H)\langle Q_2^{\text{SLL}}(\mu_H,K)\rangle,\right] \tag{15.180}$$

with the SRR contribution obtained by replacing L by R. Note that this replacement only affects the coupling $\Delta_L^{sd}(H)$ as the hadronic matrix elements being evaluated in QCD remain unchanged and the Wilson coefficients have been so defined that they also remain unchanged. See (15.182) and (15.183). For LR contributions we find

$$T(K)[\Delta S(K)]_{\text{LR}} = -\frac{\Delta_L^{sd}(H)\Delta_R^{sd}(H)}{M_H^2}\left[C_1^{\text{LR}}(\mu_H)\langle Q_1^{\text{LR}}(\mu_H,K)\rangle + C_2^{\text{LR}}(\mu_H)\langle Q_2^{\text{LR}}(\mu_H,K)\rangle\right]. \tag{15.181}$$

Including NLO QCD corrections [749] the Wilson coefficients of the involved operators are given by

$$C_1^{\text{SLL}}(\mu) = C_1^{\text{SRR}}(\mu) = 1 + \frac{\alpha_s}{4\pi}\left(-3\log\frac{M_H^2}{\mu^2} + \frac{9}{2}\right), \tag{15.182}$$

$$C_2^{\text{SLL}}(\mu) = C_2^{\text{SRR}}(\mu) = \frac{\alpha_s}{4\pi}\left(-\frac{1}{12}\log\frac{M_H^2}{\mu^2} + \frac{1}{8}\right), \tag{15.183}$$

$$C_1^{\text{LR}}(\mu) = -\frac{3}{2}\frac{\alpha_s}{4\pi}, \tag{15.184}$$

$$C_2^{\text{LR}}(\mu) = 1 - \frac{\alpha_s}{4\pi}\frac{3}{N} = 1 - \frac{\alpha_s}{4\pi}. \tag{15.185}$$

It should be noticed that, similar to Z' case, also here the operators $Q_{1,2}^{\text{LR}}$ contribute. But their Wilson coefficients are totally different. In particular, while in the Z' case the operator

Q_1^{LR} was the dominant one, it is Q_2^{LR} in the scalar case. Yet, as we will see in the context of our presentation also the operators Q_1^{SLL} and Q_1^{SRR} will play an important role, in particular when only one of the couplings $\Delta_L^{sd}(H)$ and $\Delta_R^{sd}(H)$ is nonvanishing as then LR operators do not contribute. Next

$$\langle Q_i^a(\mu_H, K) \rangle \equiv \frac{m_K F_K^2}{3} P_i^a(\mu_H, K) \tag{15.186}$$

are the matrix elements of operators evaluated at the matching scale $\mu_H = O(M_H)$, and P_i^a are the coefficients introduced and evaluated in [746]. The μ_H dependence of $P_i^a(\mu_H)$ cancels the one of $\Delta_{L,R}(H)$ and of $C_i^a(\mu_H)$ so that $S(K)$ does not depend on μ_H. It should be emphasized at this point that in contrast to gauge boson couplings $\Delta_{L,R}(Z')$ and $\Delta_{L,R}(Z)$ that are scale independent as far as leading QCD corrections are concerned, the couplings $\Delta_{L,R}(H)$ are scale dependent. Therefore consistent with (15.182) and (15.183), they are defined at $\mu_H = O(M_H)$. In numerical calculations one can simply set $\mu_H = M_H$. Similarly for B_q systems we have

$$T(B_q)[\Delta S(B_q)]_{\text{SLL}} = -\frac{(\Delta_L^{bq}(H))^2}{2M_H^2}\left[C_1^{\text{SLL}}(\mu_H)\langle Q_1^{\text{SLL}}(\mu_H, B_q)\rangle\right.$$
$$\left. + C_2^{\text{SLL}}(\mu_H)\langle Q_2^{\text{SLL}}(\mu_H, B_q)\rangle\right], \tag{15.187}$$

$$T(B_q)[\Delta S(B_q)]_{\text{LR}} = -\frac{\Delta_L^{bq}(H)\Delta_R^{bq}(H)}{M_H^2}\left[C_1^{\text{LR}}(\mu_H)\langle Q_1^{\text{LR}}(\mu_H, B_q)\rangle\right.$$
$$\left. + C_2^{\text{LR}}(\mu_H)\langle Q_2^{\text{LR}}(\mu_H, B_q)\rangle\right], \tag{15.188}$$

where the Wilson coefficients $C_i^a(\mu_H)$ are as in the K system, and the matrix elements are given by

$$\langle Q_i^a(\mu_H, B_q)\rangle \equiv \frac{m_{B_q} F_{B_q}^2}{3} P_i^a(\mu_H, B_q). \tag{15.189}$$

For SRR contributions one proceeds as in the K system.

The values of $\langle Q_i^a(\mu_H)\rangle$ obtained using the lattice results are collected in Table 13.3. They are given in the $\overline{\text{MS}}$-NDR scheme. For simplicity we choose this scale to be M_H, but any scale of this order would give the same results for the physical quantities up to NNLO QCD corrections that are negligible at these high scales. The renormalization scheme dependence of the matrix elements is canceled by the one of the Wilson coefficients.

In the case of tree-level SM Higgs exchanges, we evaluate the matrix elements at $m_t(m_t)$ as the inclusion of NLO QCD corrections allows us to choose any scale of $O(M_H)$ without changing physical results. Then in the preceding formulas one should replace M_H by the SM Higgs mass and μ_H by $m_t(m_t)$. This also means that the flavor-violating couplings of

SM Higgs are defined here at $m_t(m_t)$. The values of hadronic matrix elements at $m_t(m_t)$ in the $\overline{\text{MS}}$-NDR scheme are given in Table 13.1.

We do not repeat the basic formulas for $\Delta F = 2$ Observables as they are identical to the ones in Section 15.4. Only the master functions in (15.171), given in (15.173) are now different.

15.7.5 Rare B Decays

These decays, in particular $B_{s,d} \to \mu^+\mu^-$, played already for many years a significant role in bounding scalar contributions in the framework of the MSSM and 2HDM. While the precise branching ratios of these two decays will constitute an important test of the SM and CMFV models and of several other extensions, in the future time-dependent studies proposed in [507] and analyzed in detail in [504, 509] will help in distinguishing between various NP scenarios. Indeed, the correlations between $B_{s,d} \to \mu^+\mu^-$ observables to be measured in time-dependent studies with the CP-asymmetries $S_{\psi\phi}$ and $S_{\psi K_S}$ will be helpful in this respect. General formulas for such time-dependent studies can be found in Section 9.2 and in the original papers.

Effective Hamiltonian for $b \to s\ell^+\ell^-$

To describe the decays $B_{d,s} \to \mu^+\mu^-$, $B \to K^*\ell^+\ell^-$, $B \to K\ell^+\ell^-$, and $B \to X_s\ell^+\ell^-$ and the related observables in scalar NP scenarios we begin with recalling the effective Hamiltonian that we already encountered in Section 9.1. We do it not only for convenience but also to assure the reader about the normalization of Wilson coefficients.

$$\mathcal{H}_{\text{eff}}(b \to s\ell\bar{\ell}) = \mathcal{H}_{\text{eff}}(b \to s\gamma) - \frac{4G_{\text{F}}}{\sqrt{2}}\frac{\alpha}{4\pi}V_{ts}^*V_{tb}\sum_{i=9,10,S,P}[C_i(\mu)Q_i(\mu) + C_i'(\mu)Q_i'(\mu)],$$

$$(15.190)$$

where

$$Q_9 = (\bar{s}\gamma_\mu P_L b)(\bar{\ell}\gamma^\mu\ell), \qquad Q_9' = (\bar{s}\gamma_\mu P_R b)(\bar{\ell}\gamma^\mu\ell), \qquad (15.191a)$$

$$Q_{10} = (\bar{s}\gamma_\mu P_L b)(\bar{\ell}\gamma^\mu\gamma_5\ell), \qquad Q_{10}' = (\bar{s}\gamma_\mu P_R b)(\bar{\ell}\gamma^\mu\gamma_5\ell), \qquad (15.191b)$$

$$Q_S = m_b(\bar{s}P_R b)(\bar{\ell}\ell), \qquad Q_S' = m_b(\bar{s}P_L b)(\bar{\ell}\ell), \qquad (15.191c)$$

$$Q_P = m_b(\bar{s}P_R b)(\bar{\ell}\gamma_5\ell), \qquad Q_P' = m_b(\bar{s}P_L b)(\bar{\ell}\gamma_5\ell). \qquad (15.191d)$$

Including the factors of m_b into the definition of scalar operators makes their matrix elements and their Wilson coefficients scale independent. $\mathcal{H}_{\text{eff}}(b \to s\gamma)$ stands for the effective Hamiltonian for the $b \to s\gamma$ transition that involves the dipole operators. As already mentioned in Section 15.4, Z' contributions to $b \to s\gamma$ can be neglected. We will

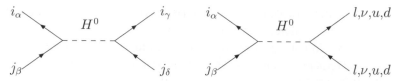

Figure 15.10 Tree-level H^0 contribution to $\Delta F = 2$ (left) and $\Delta F = 1$ (right) processes. Labels are explained in Figure 15.9.

comment on scalar contributions to this decay later, but for the time being we assume that they can be neglected as well.

It is important to notice the difference of ordering of flavors relatively to $\Delta F = 2$, which in view of the relation (15.166) implies the difference in the projections $P_{L,R}$ entering the operators $Q_{9,10}$ and $Q_{S,P}$. Therefore, the unprimed operators Q_S and Q_P represent the LHS scenario and the primed ones Q'_S and Q'_P the RHS scenario. We neglect effects proportional to m_s in each case but keep m_s and m_d different from zero when they are shown explicitly.

The Wilson coefficients C_9 and C_{10} do not receive any new contributions from scalar exchanges and take SM values as given in Section 9.1. A straightforward calculation of tree-level scalar exchanges in Figure 15.10 gives

$$m_b(\mu_H) \sin^2 \theta_W C_S = \frac{1}{g_{SM}^2} \frac{1}{M_H^2} \frac{\Delta_R^{sb}(H)\Delta_S^{\mu\bar{\mu}}(H)}{V_{ts}^* V_{tb}}, \tag{15.192}$$

$$m_b(\mu_H) \sin^2 \theta_W C'_S = \frac{1}{g_{SM}^2} \frac{1}{M_H^2} \frac{\Delta_L^{sb}(H)\Delta_S^{\mu\bar{\mu}}(H)}{V_{ts}^* V_{tb}}, \tag{15.193}$$

$$m_b(\mu_H) \sin^2 \theta_W C_P = \frac{1}{g_{SM}^2} \frac{1}{M_H^2} \frac{\Delta_R^{sb}(H)\Delta_P^{\mu\bar{\mu}}(H)}{V_{ts}^* V_{tb}}, \tag{15.194}$$

$$m_b(\mu_H) \sin^2 \theta_W C'_P = \frac{1}{g_{SM}^2} \frac{1}{M_H^2} \frac{\Delta_L^{sb}(H)\Delta_P^{\mu\bar{\mu}}(H)}{V_{ts}^* V_{tb}}, \tag{15.195}$$

where $\Delta_{S,P}^{\mu\bar{\mu}}(H)$ are defined in (15.168). It should be emphasized at this point that the couplings $\Delta_{L,R}^{sb}(H)$ extracted from ΔM_s and $S_{\psi\phi}$ are defined at $\mu_H = M_H$, therefore, as shown explicitly, m_b has to be evaluated also at this scale to keep these coefficients scale independent. In the case of the SM Higgs m_b has to be evaluated at $m_t(m_t)$ as at this scale the flavor-violating SM Higgs couplings in $\Delta F = 2$ processes are defined. In what follows we will not show this dependence explicitly. The values of m_b at different scales are collected in Table 4.2.

Next we recall that in terms of the couplings used in the analysis of $B^0_{s,d} - \bar{B}^0_{s,d}$ mixings we have according to (15.166)

$$\Delta_R^{sb}(H) = [\Delta_L^{bs}(H)]^*, \qquad \Delta_L^{sb}(H) = [\Delta_R^{bs}(H)]^*, \tag{15.196}$$

which should be kept in mind when studying correlations between $\Delta F = 1$ and $\Delta F = 2$ transitions. These relations can be directly used in the case of C_S and C'_S but in the case of

C_P and C'_P, as discussed in Section 15.7.3, it is useful to use in this context the following relations:

$$\Delta_R^{sb}\Delta_P^{\mu\bar{\mu}} = -[\tilde{\Delta}_L^{bs}]^*\tilde{\Delta}_P^{\mu\bar{\mu}}, \qquad \Delta_L^{sb}\Delta_P^{\mu\bar{\mu}} = [\tilde{\Delta}_R^{bs}]^*\tilde{\Delta}_P^{\mu\bar{\mu}}, \qquad (15.197)$$

with $\Delta_P^{\mu\bar{\mu}}$ being imaginary but $\tilde{\Delta}_P^{\mu\bar{\mu}}$ real.

15.7.6 Rare K Decays

Effective Hamiltonian for $d \to s\ell^+\ell^-$

For the study of $K_{L,S} \to \mu^+\mu^-$ and $K_L \to \pi^0\ell\ell^-$ decays we will need the relevant effective Hamiltonian. It can be obtained from the formulas of subsection 15.4.4. For completeness we list here explicit formulas for operators and Wilson coefficients:

$$Q_9 = (\bar{s}\gamma_\mu P_L d)(\bar{\ell}\gamma^\mu \ell), \qquad\qquad Q'_9 = (\bar{s}\gamma_\mu P_R d)(\bar{\ell}\gamma^\mu \ell), \qquad (15.198\text{a})$$

$$Q_{10} = (\bar{s}\gamma_\mu P_L d)(\bar{\ell}\gamma^\mu \gamma_5 \ell), \qquad Q'_{10} = (\bar{s}\gamma_\mu P_R d)(\bar{\ell}\gamma^\mu \gamma_5 \ell), \qquad (15.198\text{b})$$

$$Q_S = m_s(\bar{s}P_L d)(\bar{\ell}\ell), \qquad\qquad Q'_S = m_s(\bar{s}P_R d)(\bar{\ell}\ell), \qquad (15.198\text{c})$$

$$Q_P = m_s(\bar{s}P_L d)(\bar{\ell}\gamma_5 \ell), \qquad\quad Q'_P = m_s(\bar{s}P_R d)(\bar{\ell}\gamma_5 \ell). \qquad (15.198\text{d})$$

Note that because of the sd ordering instead of qb, scalar operators have L and R interchanged with respect to $b \to s, d$ transitions.

The Wilson coefficients C_9 and C_{10} do not receive any new contributions from scalar exchange and take SM values as given in (6.334). However, to include charm component in $K_L \to \mu^+\mu^-$ we make replacement:

$$\eta_Y Y_0(x_t) \longrightarrow \eta_Y Y_0(x_t) + \frac{V_{cs}^* V_{cd}}{V_{ts}^* V_{td}} Y_{\text{NNL}}, \qquad (15.199)$$

where at NNLO [625]

$$Y_{\text{NNL}} = \lambda^4 P_c(Y), \qquad P_c(Y) = 0.115 \pm 0.017 . \qquad (15.200)$$

The coefficients of scalar operators are

$$m_s \sin^2 \vartheta_W C_S = \frac{1}{g_{\text{SM}}^2} \frac{1}{M_H^2} \frac{\Delta_L^{sd}(H)\Delta_S^{\mu\bar{\mu}}(H)}{V_{ts}^* V_{td}}, \qquad (15.201)$$

$$m_s \sin^2 \vartheta_W C'_S = \frac{1}{g_{\text{SM}}^2} \frac{1}{M_H^2} \frac{\Delta_R^{sd}(H)\Delta_S^{\mu\bar{\mu}}(H)}{V_{ts}^* V_{td}}, \qquad (15.202)$$

$$m_s \sin^2 \vartheta_W C_P = \frac{1}{g_{\text{SM}}^2} \frac{1}{M_H^2} \frac{\Delta_L^{sd}(H)\Delta_P^{\mu\bar{\mu}}(H)}{V_{ts}^* V_{td}}, \qquad (15.203)$$

$$m_s \sin^2 \vartheta_W C'_P = \frac{1}{g_{\text{SM}}^2} \frac{1}{M_H^2} \frac{\Delta_R^{sd}(H)\Delta_P^{\mu\bar{\mu}}(H)}{V_{ts}^* V_{td}}. \qquad (15.204)$$

$K_L \to \mu^+\mu^-$ and $K_S \to \mu^+\mu^-$

We know from previous material that only the so-called short-distance (SD) part to a dispersive contribution to $K_L \to \mu^+\mu^-$ can be reliably calculated. Therefore, in what follows, as in other NP scenarios, this decay can be presently only treated as an additional constraint with the rough upper bound given in (9.268).

The relevant branching ratio can be obtained by first introducing:

$$\hat{P}(K) \equiv C_{10} - C'_{10} + \frac{m_K^2}{2m_\mu} \frac{m_s}{m_d + m_s}(C_P - C'_P) \tag{15.205}$$

and

$$\hat{S}(K) \equiv \sqrt{1 - \frac{4m_\mu^2}{m_K^2}} \frac{m_K^2}{2m_\mu} \frac{m_s}{m_d + m_s}(C_S - C'_S). \tag{15.206}$$

We then find [504]

$$\mathcal{B}(K_L \to \mu^+\mu^-)_{\text{SD}} = \frac{G_F^4 M_W^4}{4\pi^5} F_K^2 m_K \tau_{K_L} m_\mu^2 \sqrt{1 - \frac{4m_\mu^2}{m_K^2}} \sin^4 \vartheta_W \tag{15.207}$$
$$\times \left\{ \left[\text{Re}\left(V_{ts}^* V_{td} \hat{P}\right)\right]^2 + \left[\text{Im}\left(V_{ts}^* V_{td} \hat{S}\right)\right]^2 \right\}.$$

We recall that C_{10} does not receive any contribution from scalar exchanges and includes also SM charm contribution as given in (15.199). $C'_{10} = 0$ for scalar exchanges.

Equivalently we can write

$$\mathcal{B}(K_L \to \mu^+\mu^-)_{\text{SD}} = \kappa_\mu \left\{ \left[\text{Re}\left(V_{ts}^* V_{td} \hat{P}\right)\right]^2 + \left[\text{Im}\left(V_{ts}^* V_{td} \hat{S}\right)\right]^2 \right\}, \tag{15.208}$$

where

$$\kappa_\mu = \frac{\alpha^2 \mathcal{B}(K^+ \to \mu^+\nu)}{\lambda^2 \pi^2} \frac{\tau(K_L)}{\tau(K^+)}. \tag{15.209}$$

Finally, for $K_S \to \mu^+\mu^-$ decay, we just interchange Re and Im in all preceding expressions so that

$$\mathcal{B}(K_S \to \mu^+\mu^-)_{\text{SD}} = \kappa_\mu \left\{ \left[\text{Im}\left(V_{ts}^* V_{td} \hat{P}\right)\right]^2 + \left[\text{Re}\left(V_{ts}^* V_{td} \hat{S}\right)\right]^2 \right\}. \tag{15.210}$$

$K_L \to \pi^0 \ell^+ \ell^-$

While in the case of Z' models large enhancements of branching ratios were not possible due to constraints from data on $K^+ \to \pi^+ \nu\bar{\nu}$ [611], this constraint is absent in the case of scalar contributions, and it is of interest to see by how much the branching ratios can be enhanced in the models considered here still being consistent with all data, in particular with the bound on $K_L \to \mu^+ \mu^-$ in (9.268).

Probably the most extensive model independent analysis of decays in question has been performed in [637], where formulas for branching ratios for both decays in the presence of new operators have been presented. We have already elaborated on these formulas in the case of Z' exchanges in Section 15.4.5. The case of scalar contributions is more involved.

To use the formulas of [637] for scalar contributions, we introduce the following quantities [504]:

$$\omega_{7A} = -\frac{1}{2\pi} \frac{\eta_Y Y_0(x_t)}{\sin^2 \vartheta_W} \frac{\text{Im}(\lambda_t^{(K)})}{1.4 \cdot 10^{-4}}, \tag{15.211}$$

$$\bar{y}_P = \frac{y_P + y_P'}{2}, \qquad \bar{y}_S = \frac{y_S + y_S'}{2}, \tag{15.212}$$

with y_i related to the Wilson coefficients $C_{P,S}$ as follows:

$$y_P = -\frac{M_W^2 \sin^2 \vartheta_W}{m_l} V_{ts}^* V_{td} C_P, \qquad y_S = -\frac{M_W^2 \sin^2 \vartheta_W}{m_l} V_{ts}^* V_{td} C_S, \tag{15.213}$$

with analogous formulas for primed coefficients. Here m_l stands for m_e and m_μ as the authors of [637] anticipating helicity suppression included these masses already in the effective Hamiltonian.

Using [637] the authors of [504] find then corrections from tree-level A^0 and H^0 exchanges to the branching ratios that should be added directly to SM results in (9.282) and (9.283):

$$\Delta \mathcal{B}_P^{e^+ e^-} = \left(1.9 \, \omega_{7A} \text{Im}(\bar{y}_P) + 0.038 \, (\text{Im}(\bar{y}_P))^2\right) \cdot 10^{-17}, \tag{15.214}$$

$$\Delta \mathcal{B}_P^{\mu^+ \mu^-} = \left(0.26 \, \omega_{7A} \text{Im}(\bar{y}_P) + 0.0085 \, (\text{Im}(\bar{y}_P))^2\right) \cdot 10^{-12}, \tag{15.215}$$

$$\Delta \mathcal{B}_S^{e^+ e^-} = \left(1.5 \, \text{Re}(\bar{y}_S) + 0.0039 \, (\text{Re}(\bar{y}_S))^2\right) \cdot 10^{-16}, \tag{15.216}$$

$$\Delta \mathcal{B}_S^{\mu^+ \mu^-} = \left(0.04 \, \text{Re}(\bar{y}_S) + 0.0041 \, (\text{Re}(\bar{y}_S))^2\right) \cdot 10^{-12}. \tag{15.217}$$

Note that in the absence of helicity suppression, the large suppression factors are canceled by the conversion factors in (15.213). The numerical results for these new contributions can be found in [504].

15.7.7 ε'/ε

To get an idea of the impact of tree-level neutral scalar exchange on ε'/ε, it is useful to return to the master formula for ε'/ε in Section 14.7. The effective Hamiltonian for $K \to \pi\pi$ obtained from tree-level neutral scalar exchange in Figure 15.10 reads

$$\mathcal{H}_{\text{eff}} = -\frac{1}{M_H^2} \sum_i \sum_{A,B=L,R} \Delta_A^{sd}(H) \Delta_B^{ii}(H) (\bar{s} P_A d)(\bar{q}_i P_B q_i), \tag{15.218}$$

with $i = d, s, b, u, c$.

The contributing scalar-scalar operators can be compactly denoted by

$$O^q_{SAB} = (\bar{s}P_A d)(\bar{q}P_B q), \qquad q = d, s, b, u, c \tag{15.219}$$

and, with the definition (14.83), their coefficients and P_i factors in (14.84) are given by

$$C^q_{SAB} = \frac{1}{M_H^2}\Delta^{sd}_A(H)\Delta^{qq}_B(H), \qquad P^q_{SLR} = -P^q_{SRL}, \qquad P^q_{SLL} = -P^q_{SRR}. \tag{15.220}$$

It should be noted that generally C^q_{SLL} and C^q_{SRR} differ from each other, and this is also the case of C^q_{SLR} and C^q_{SRL}.

From the tables in the appendix of [798] one finds then the largest values of P^q_{SAB}. These are

$$P^u_{SLR} = -266 \pm 20, \quad P^d_{SLR} = 214 \pm 18, \quad P^d_{SLL} = -87 \pm 16, \quad P^u_{SLL} = 74 \pm 16.$$

$$\tag{15.221}$$

All other P_i factors are below 1, and the corresponding contributions can be neglected, unless there is a strong enhancement of the corresponding Wilson coefficients.

In the latter context, it should be emphasized, that these results, as they stand, cannot be used for phenomenology because the P_i factors, representing hadronic matrix elements, have been calculated at the electroweak scale, while the Wilson coefficients are evaluated at a much higher scale $O(M_H)$. To complete the analysis the gap between these two scales has to be closed by including RG effects between these scales in the framework of the SMEFT. Here contributions from top Yukawa couplings could turn out to be important. Such an analysis should be available in the literature when this book is published.

15.8 Beyond MFV: Charged Scalar Exchanges

We have already mentioned charged scalars at various places in this book. They naturally are present in 2HDM and in supersymmetric models, which we decided not to discuss in detail in this book as there is a very rich literature on them. In particular I recommend strongly the general review in [897] and the review of flavor-phenomenology of 2HDM models with generic Yukawa structure in [995].

Even before $R(D)$ and $R(D^*)$ anomalies became popular, the role of charged Higgs bosons in $B^+ \rightarrow D^0\tau^+\nu$ decays has been studied. A thorough analysis of this decay is presented in [996], where further references to older papers can be found. However, the $R(D)$ and $R(D^*)$ anomalies gave motivation for several newer theoretical analyses of which we just quote four. First, the study of these decays in 2HDM of type III [995, 997] and in NP models with general flavor structure in [775]. Moreover, in [998] 2HDM and 3HDM models with the nonminimal flavor violations originating from flavor-dependent gauge interactions have been analyzed. A recent summary of the situation can be found in

[999]. In particular 2HDM of type II cannot simultaneously describe the data on $\mathcal{R}(D)$ and $\mathcal{R}(D^*)$, but this is possible in 2HDM of type III.

It is evident that $B^+ \to \tau^+ \nu_\tau$, $\bar{B} \to D\tau\nu$, and $\bar{B} \to D^*\tau\nu$ can play a potential role in constraining the charged scalar contributions. Yet, due to the fact that the data in the case of $B^+ \to \tau^+ \nu_\tau$ moved significantly toward the SM and because of large uncertainty in $|V_{ub}|$, the identification of a concrete NP in this decay appears to us presently as a big challenge. The decays $B \to D\tau\nu$ and $B \to D^*\tau\nu$ seem to be more promising, and new insight in their role in charged scalar physics will be gained in the Belle II era.

15.9 Beyond MFV: Colored Gauge Bosons and Scalars

15.9.1 Colored Neutral Gauge Bosons G'

In various NP scenarios neutral gauge bosons with color (G') are present. One of the prominent examples of this type are Kaluza–Klein gluons in Randal–Sundrum scenarios that belong to the adjoint representation of the color $SU(3)_C$. In what follows we will assume that these gauge bosons carry a common mass $M_{G'}$ and being in the octet representation of $SU(3)_C$ couple to fermions in the same manner as gluons do. However, we will allow for different values of their left-handed and right-handed couplings. Therefore, up to the color matrix t^a, the couplings to quarks will be again parametrized by:

$$\Delta_L^{sd}(G'), \qquad \Delta_R^{sd}(G'), \qquad \Delta_L^{qq}(G'), \qquad \Delta_R^{qq}(G'). \tag{15.222}$$

General couplings are defined in the Feynman rules in Figure 15.11. Calculating then the tree-diagrams with G' gauge boson exchanges and expressing the result in terms of the operators encountered in previous sections, we find that the initial conditions at $\mu = M_{G'}$ are modified. As G' do not couple to leptons, they can only influence $\Delta F = 2$ and non-leptonic decays. Therefore, we collect here only the initial conditions for these two classes of processes.

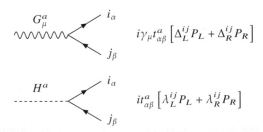

Figure 15.11   Feynman rules for the colored neutral gauge boson G_μ with mass M_G and colored neutral scalar particle H with mass M_H, where i, j denote down-type quarks.

$\Delta F = 2$ Transitions

The initial conditions for $\Delta F = 2$ transition change relative to Z' case. In LO they are given as follows [1007]

$$C_1^{\text{VLL}}(M_{G'}) = \left[\frac{1}{3}\right]\frac{\Delta_L^{sd}(G')^2}{2M_{G'}^2}, \qquad C_1^{\text{VRR}}(M_{G'}) = \left[\frac{1}{3}\right]\frac{\Delta_R^{sd}(G')^2}{2M_{G'}^2}, \qquad (15.223)$$

$$C_1^{\text{LR}}(M_{G'}) = \left[-\frac{1}{6}\right]\frac{\Delta_L^{sd}(G')\Delta_R^{sd}(G')}{M_{G'}^2}, \qquad C_2^{\text{LR}}(M_{G'}) = [-1]\frac{\Delta_L^{sd}(G')\Delta_R^{sd}(G')}{M_{G'}^2}. \qquad (15.224)$$

In particular we note that the coefficient $C_2^{\text{LR}}(M_{G'})$ is now the largest as opposed to Z' case, where it was vanishing. As the matrix element of Q_2^{LR} is large, Kaluza–Klein gluons were problematic ten years ago in the context of RS scenarios. Deriving the preceding conditions is a good exercise. One will then notice that these coefficients are not normalized to unity and are not dimensionless. They are just coefficients of the operators in the relevant effective Hamiltonian.

The NLO QCD corrections to tree-level colored gauge boson exchanges at $\mu = M_{G'}$ to $\Delta S = 2$ are not known. They are expected to be small due to small QCD coupling at this high scale and serve mainly to remove certain renormalization scheme and matching scale uncertainties. More important is the RG evolution from low-energy scales to $\mu = M_{G'}$ necessary to evaluate $\langle Q_1^{\text{VLL}}(M_{G'})\rangle$ and $\langle Q_{1,2}^{\text{LR}}(M_{G'})\rangle$. Here NLO QCD corrections can be included using the technology in [746].

$K \to \pi\pi$ Decays

The new initial conditions for the operators entering $K \to \pi\pi$ read now at LO as follows [271]

$$C_3(M_{G'}) = \left[-\frac{1}{6}\right]\frac{\Delta_L^{sd}(G')\Delta_L^{qq}(G')}{4M_{G'}^2}, \qquad C_3'(M_{G'}) = \left[-\frac{1}{6}\right]\frac{\Delta_R^{sd}(G')\Delta_R^{qq}(G')}{4M_{G'}^2}, \qquad (15.225)$$

$$C_4(M_{G'}) = \left[\frac{1}{2}\right]\frac{\Delta_L^{sd}(G')\Delta_L^{qq}(G')}{4M_{G'}^2}, \qquad C_4'(M_{G'}) = \left[\frac{1}{2}\right]\frac{\Delta_R^{sd}(G')\Delta_R^{qq}(G')}{4M_{G'}^2}, \qquad (15.226)$$

$$C_5(M_{G'}) = \left[-\frac{1}{6}\right]\frac{\Delta_L^{sd}(G')\Delta_R^{qq}(G')}{4M_{G'}^2}, \qquad C_5'(M_{G'}) = \left[-\frac{1}{6}\right]\frac{\Delta_R^{sd}(G')\Delta_L^{qq}(G')}{4M_{G'}^2}, \qquad (15.227)$$

$$C_6(M_{G'}) = \left[\frac{1}{2}\right]\frac{\Delta_L^{sd}(G')\Delta_R^{qq}(G')}{4M_{G'}^2}, \qquad C_6'(M_{G'}) = \left[\frac{1}{2}\right]\frac{\Delta_R^{sd}(G')\Delta_L^{qq}(G')}{4M_{G'}^2}. \qquad (15.228)$$

Due the nonvanishing value of $C_6(M_{G'})$ the dominance of the operator Q_6 is more pronounced than in the case of a colorless Z'.

Phenomenology of ε_K, ε'/ε, ΔM_K, and $\Delta I = 1/2$ rule in this scenario can be found in [271].

15.9.2 Colored Neutral Scalars (H^c)

$\Delta F = 2$ Transitions

Denoting the couplings to quarks by $\lambda_{L,R}^{ij}$, we find for $\Delta F = 2$ transitions

$$C_1^{\text{SLL}}(M_{H^c}) = \left[\frac{5}{12}\right]\frac{\lambda_L^{sd}(H^c)^2}{2M_{H^c}^2}, \qquad C_2^{\text{SLL}}(M_{H^c}) = \left[-\frac{1}{16}\right]\frac{\lambda_L^{sd}(H^c)^2}{2M_{H^c}^2}, \quad (15.229)$$

$$C_1^{\text{SRR}}(M_{H^c}) = \left[\frac{5}{12}\right]\frac{\lambda_R^{sd}(H^c)^2}{2M_{H^c}^2}, \qquad C_2^{\text{SRR}}(M_{H^c}) = \left[-\frac{1}{16}\right]\frac{\lambda_R^{sd}(H^c)^2}{2M_{H^c}^2}, \quad (15.230)$$

$$C_1^{\text{LR}}(M_{H^c}) = \left[\frac{1}{4}\right]\frac{\lambda_L^{sd}(H^c)\lambda_R^{sd}(H^c)}{M_{H^c}^2}, \qquad C_2^{\text{LR}}(M_{H^c}) = \left[\frac{1}{6}\right]\frac{\lambda_L^{sd}(H^c)\lambda_R^{sd}(H^c)}{M_{H^c}^2}.$$
$$(15.231)$$

In fact, these results have been obtained in collaboration with Jennifer Girrbach-Noe.

ε'/ε

Here we proceed as in the case of the colorless neutral scalar but including appropriate color factors. The effective Hamiltonian for $K \to \pi\pi$ obtained from tree-level neutral scalar exchange in Figure 15.11 reads

$$\mathcal{H}_{\text{eff}} = -\frac{1}{M_H^2}\sum_i\sum_{A,B=L,R}\lambda_A^{sd}(H)\lambda_B^{ll}(H)(\bar{s}_\alpha P_A t_{\alpha\beta}^a d_\beta)((\bar{q}_i)_\gamma P_B t_{\gamma\delta}^a(q_i)_\delta), \quad (15.232)$$

with $i = d, s, b, u, c$ and $\alpha, \beta, \gamma, \delta$ being color indices.

The contributing scalar-scalar operators can be compactly denoted by

$$[O_{SAB}^q]^c = (\bar{s}_\alpha P_A t_{\alpha\beta}^a d_\beta)(\bar{q}_\gamma P_B t_{\gamma\delta}^a q_\delta), \qquad q = d, s, b, u, c \quad (15.233)$$

and their coefficients and P_i factors in (14.84) by

$$[C_{SAB}^q]^c = \frac{1}{M_{H^c}^2}\lambda_A^{sd}(H^c)\lambda_B^{qq}(H^c), \qquad [P_{SLR}^q]^c = -[P_{SRL}^q]^c, \qquad [P_{SLL}^q]^c = -[P_{SRR}^q]^c,$$
$$(15.234)$$

with the superscript "c" denoting the special color structure. Again it should be noted that generally $[C_{SLL}^q]^c$ and $[C_{SRR}^q]^c$ differ from each other and this is also the case of $[C_{SLR}^q]^c$ and $[C_{SRL}^q]^c$.

To use (14.84) and calculate colored Higgs contribution to ε'/ε we need the factors $[P_{SAB}^q]^c$. But the operators in (15.233) do not belong to the operator basis used in [798], and these factors cannot be found in the tables of that paper.

To find $[P_{SAB}^q]^c$ we have to express the operators in (15.233) in terms of the ones used in [798]. To this end we use the relation (5.145):

$$\Pi = \frac{1}{2}\left[\tilde{1} - \frac{1}{N}1\right], \quad (15.235)$$

where

$$\mathbf{\Pi}_{\alpha\beta,\gamma\delta} = (t^a)_{\alpha\beta} \otimes (t^a)_{\gamma\delta}, \quad \tilde{\mathbf{1}}_{\alpha\beta,\gamma\delta} = \delta_{\alpha\delta} \otimes \delta_{\gamma\beta}, \quad \mathbf{1}_{\alpha\beta,\gamma\delta} = \delta_{\alpha\beta} \otimes \delta_{\gamma\delta}. \quad (15.236)$$

This allows one to write

$$[P^q_{SAB}]^c = \frac{1}{2}\left[\tilde{P}^q_{SAB} - \frac{1}{N}P^q_{SAB}\right], \qquad q = u, c, b, \quad (15.237)$$

and using tables in [798], to calculate $[P^q_{SAB}]^c$ for $q = u, c, b$. But for $q = d, s$ we have to use in addition the following relations obtained by means of Fierz transformations

$$\tilde{O}^{d,s}_{SLR} = -\frac{1}{2}O^{d,s}_{VRL}, \qquad \tilde{O}^{d,s}_{SLL} = -\frac{1}{2}O^{d,s}_{SLL} - \frac{1}{8}O^{d,s}_{TLL} \quad (15.238)$$

and likewise for their chirality-flipped counterparts. $\tilde{P}^q_{SAB}$ for $q = d, s$ can now be expressed it terms of those found in the tables of [798],

$$\tilde{P}^{d,s}_{SLR} = -\frac{1}{2}P^{d,s}_{VRL}, \qquad \tilde{P}^{d,s}_{SLL} = -\frac{1}{2}P^{d,s}_{SLL} - \frac{1}{8}P^{d,s}_{TLL}. \quad (15.239)$$

We find finally for $N = 3$

$$[P^{d,s}_{SLR}]^c = -\frac{1}{4}P^{d,s}_{VRL} - \frac{1}{6}P^{d,s}_{SLR}, \qquad [P^{d,s}_{SLL}]^c = -\frac{5}{12}P^{d,s}_{SLL} - \frac{1}{16}P^{d,s}_{TLL}. \quad (15.240)$$

Having all these expressions at hand and using the tables in the appendix of [798], we find then the largest values of $[P^q_{SAB}]^c$. These are[8]

$$[P^u_{SLR}]^c = 14 \pm 4, \qquad [P^d_{SLR}]^c = -5 \pm 4 \qquad [P^d_{SLL}]^c = 24 \pm 7, \qquad [P^u_{SLL}]^c = -20 \pm 3 \quad (15.241)$$

All other P_i factors are far below 1, and the corresponding contributions can be neglected, unless there is a strong enhancement of the corresponding Wilson coefficients. Comparing these P_i factors with the ones for a colorless scalar in (15.221), we observe that they are much lower. But the fate of these contributions depends also on the size of the couplings, and without an analysis in a concrete model it is not possible to decide whether colorless scalars or colored one will be more relevant for ε'/ε. Yet, for colored scalars to be able to compete with the colorless ones, their couplings must be sufficiently larger or their masses smaller. This could in principle be forbidden by the bounds on other processes.

Finally, as in the case of colorless scalars, RG effects between scales $O(M_{H^c})$ and the electroweak scale have to be taken into account in the framework of the SMEFT. Such analysis should be available in the literature when this book is published.

---

[8] The large errors result from cancellations between the two terms in (15.240).

16 Specific Models

16.1 Preliminaries

There have been many studies of weak decays in concrete NP models. Here we want to describe in some detail only three classes of them:

- The so-called 331 models in which FCNC transitions are governed by a heavy Z' gauge boson with couplings that are more specific than in a general Z' scenario considered in Section 15.4.
- Models with vector-like quarks and leptons, that is new heavy fermions of which left-handed and right-handed components transform in the same manner under the SM gauge group. In simplest models the FCNC processes are dominated by tree-level Z exchanges and in certain models in addition by box diagrams with vector-like fermions and the Higgs boson. In more complicated models also other new heavy particles, in particular heavy gauge bosons and scalars can be present.
- Leptoquark models in which FCNC processes are dominated by the exchanges of heavy scalar or vector leptoquarks that directly couple quarks to leptons.

The reason for choosing these three classes of models is that they exhibit three different patterns of BSM flavor violation with different particles mediating FCNC processes. Moreover, they play some role in the explanation of the existing anomalies. But there are several other models that in principle deserve to be described here. Yet, in view of space limitations, we will only refer to three reviews [208, 870, 1000] in which I participated, and I will list the models and corresponding FCNC analyses of those that have been briefly described there. In particular in each case flavor properties of these models are listed, and in most cases Feynman rules (FR) for a given model are given. Also references to the relevant original papers can be found there. I would recommend to start with [208] and read the other two subsequently to appreciate that this field is very rich and changing with time. We have then:

- 2HDM$_{\overline{\text{MFV}}}$ model [896], which we briefly described in Section 15.2.
- Littlest Higgs Model with T-parity (LHT) [700] and without it (LH) [1001]. In both papers FR can be found.
- The SM with sequential fourth generation [1002].
- Supersymmetric flavor models that contain fewer parameters than the general MSSM [529].
- Supersymmetric SU(5) GUT model with right-handed neutrinos [1003].

- Supersymmetric SO(10) GUT model [1004].
- Flavor-blind MSSM [890].
- The minimal effective model with right-handed currents [610].
- Universal extra dimensions [1005, 1006], where FR are given.
- Randall–Sundrum model with custodial protection [920, 1007, 1008], where FR are given. See also [922, 1009–1011].
- Left-right symmetric models [772], where FR are given.
- Gauged flavor models [1012], where FR are given.
- Minimal theory of fermion masses with vector-like fermions [1013].
- Composite Higgs and partial compositeness. See [614, 1014–1017] and references to the original papers therein.
- Part III of the book [7] contains a lot of information on multi-Higgs doublet models, spontaneous CP violation, massive neutrinos and CPV in the leptonic sector, and strong CP problem, which we did not cover here. Also some additional information about vector-like quarks and some basics of left-right symmetric models can be found there.

16.2 331 Models

16.2.1 Introduction

Until now we have considered simplified models in which a single new heavy-gauge boson or a single scalar was added to the SM spectrum. In this section we will consider the so-called 331 models based on the gauge group $SU(3)_C \times SU(3)_L \times U(1)_X$ [1018, 1019], that is at first spontaneously broken to the SM group $SU(3)_C \times SU(2)_L \times U(1)_Y$ and then undergoes the spontaneous symmetry breaking to $SU(3)_C \times U(1)_Q$. The extension of the gauge group with respect to SM leads to a number of interesting consequences. The first one is that the requirement of anomaly cancellation together with that of asymptotic freedom of QCD implies that the number of generations must necessarily be equal to the number of colors, hence giving an explanation for the existence of three generations. But this is only possible if quark generations transform differently under the action of $SU(3)_L$. In particular, two quark generations should transform as triplets, one as an antitriplet. Choosing the latter to be the third generation, this different treatment could be at the origin of the large top quark mass. This choice imposes that the leptons should transform as antitriplets. However, one could choose a different scenario in which the role of triplets and antitriplets is exchanged, provided that the number of triplets equals that of antitriplets, to fulfill the anomaly cancellation requirement. Therefore, different versions of the model are obtained according to the way one fixes the fermion representations. Here we will consider two such representations called F_1 and F_2, which we will specify in the next subsection. This section is based on a series of papers [190, 613, 615, 852, 853] where more details can be found.

In these models a fundamental relation holds among the diagonal generators of the gauge group:

$$Q = T_3 + \beta T_8 + X, \tag{16.1}$$

where Q is the electric charge generator, T_3 and T_8 are two diagonal generators of the SU(3)$_L$, and X is the generator of U(1)$_X$. β is a key parameter that defines a specific variant of the model. The 331 models comprise several new particles, in particular a heavy Z' that is related to the generator of U(1)$_X$. There are moreover new heavy neutral and charged gauge bosons Y and V related to new generators present in SU(3)$_L$ beyond those of the SM SU(2)$_L$, a subgroup of SU(3)$_L$. Moreover, there are new heavy fermions, all with electric charges depending on β. Also the Higgs system is extended. But our discussion will concentrate dominantly on Z' as we would like to apply the technology developed in the previous chapter to a concrete model. Various properties of the remaining new particles are discussed in [613].

As described in detail in [1020], β could be in principle arbitrary. But as stressed in [613] only values

$$|\beta| \leq \sqrt{3} \tag{16.2}$$

are allowed. This follows from the fact that in 331 models

$$M_{Z'}^2 = \frac{g_2^2 u^2 c_W^2}{3[1 - (1 + \beta^2)s_W^2]} \, , \tag{16.3}$$

where u is the vacuum expectation value related to the first symmetry breaking and $s_W = \sin \vartheta_W$. It is then evident that only values of β satisfying

$$[1 - (1 + \beta^2)s_W^2] > 0 \tag{16.4}$$

are allowed. With the known value of s_W^2 this implies (16.2), and in fact the only explicit models analyzed in the literature are the ones with

$$\beta = \pm \frac{n}{\sqrt{3}}, \qquad n = 0, 1, 2, 3. \tag{16.5}$$

For $n = 1$ there are singly charged $Y^\pm$ bosons and neutral ones $V^0(\bar{V}^0)$, while for $n = 3$ one finds instead two new singly charged bosons $V^\pm$ and two doubly charged ones $Y^{\pm\pm}$. For $n = 2$ exotic charges $\pm 1/2$ and $\pm 3/2$ for gauge bosons are found. From table 1 in [613] we also find that while for $n = 1$ no exotic charges for heavy fermions are present, for $n = 2$ heavy quarks carry exotic electric charges $\pm 5/6$ and $\pm 7/6$, while heavy leptons $\pm 1/2$ and $\pm 3/2$. Discovering such fermions at the LHC would be a spectacular event. We refer to [613] for further details. In principle β could be a continuous variable satisfying (16.2), but in the present section we will only consider the cases $n = 1, 2$. The case $n = 3$ implies Landau singularities already at the LHC scales [190, 1021], while the model with $n = 0$, proposed in [1022], turns out to imply very small NP effects [853].

Most important for our discussion for all β, a new neutral gauge boson Z' is present, which as discussed in Section 15.4 and also in this section could help to explain various anomalies in several observables in $b \rightarrow s\mu^+\mu^-$ transitions. Compared with the general case of a Z' boson, the flavor-violating and flavor-conserving couplings are given in terms of fundamental parameters of a given 331 model, and consequently there is much less freedom in describing the data than in general Z' models. We will list these couplings in the next subsection.

Finally, a full description of flavor violation should take into account $Z - Z'$ mixing and implied flavor-violating contributions mediated by tree-level Z exchanges. We will describe these effects soon.

16.2.2 Parameters and Couplings

As in the SM, quark mass eigenstates are defined upon rotation of flavor eigenstates through two unitary matrices U_L (for up-type quarks) and V_L (for down-type quarks). The relation

$$V_{\text{CKM}} = U_L^\dagger V_L \tag{16.6}$$

holds in analogy with the SM case. However, while in the SM V_{CKM} appears only in charged current interactions and the two rotation matrices never appear individually, this is not the case in this model, and both U_L and V_L can generate tree-level FCNCs mediated by Z' in the up-quark and down-quark sector, respectively. But these two matrices have to satisfy the relation (16.6). A useful parametrization for V_L proposed in [613] is

$$V_L = \begin{pmatrix} \tilde{c}_{12}\tilde{c}_{13} & \tilde{s}_{12}\tilde{c}_{23}e^{i\delta_3} - \tilde{c}_{12}\tilde{s}_{13}\tilde{s}_{23}e^{i(\delta_1-\delta_2)} & \tilde{c}_{12}\tilde{c}_{23}\tilde{s}_{13}e^{i\delta_1} + \tilde{s}_{12}\tilde{s}_{23}e^{i(\delta_2+\delta_3)} \\ -\tilde{c}_{13}\tilde{s}_{12}e^{-i\delta_3} & \tilde{c}_{12}\tilde{c}_{23} + \tilde{s}_{12}\tilde{s}_{13}\tilde{s}_{23}e^{i(\delta_1-\delta_2-\delta_3)} & -\tilde{s}_{12}\tilde{s}_{13}\tilde{c}_{23}e^{i(\delta_1-\delta_3)} - \tilde{c}_{12}\tilde{s}_{23}e^{i\delta_2} \\ -\tilde{s}_{13}e^{-i\delta_1} & -\tilde{c}_{13}\tilde{s}_{23}e^{-i\delta_2} & \tilde{c}_{13}\tilde{c}_{23} \end{pmatrix}. \tag{16.7}$$

This matrix implies through (16.6) new sources of flavor violation in the up-sector, but they will not be discussed here. However, when $U_L = \mathbb{1}$, as used in [1021], $V_L = V_{\text{CKM}}$, and we deal with a particular simple CMFV model. In this book we would like to be more general, and V_L is given by (16.7).

With this parametrization, denoting the elements of the matrix V_L in (16.7) by v_{ij}, the Z' couplings to quarks, for the three meson systems, K, B_d, and B_s

$$\Delta_L^{sd}(Z'), \qquad \Delta_L^{bd}(Z'), \qquad \Delta_L^{bs}(Z'), \tag{16.8}$$

being proportional to $v_{32}^* v_{31}$, $v_{33}^* v_{31}$, and $v_{33}^* v_{32}$, respectively, depend only on four new parameters:

$$\boxed{\tilde{s}_{13}, \quad \tilde{s}_{23}, \quad \delta_1, \quad \delta_2.} \tag{16.9}$$

Here $\tilde{s}_{13}$ and $\tilde{s}_{23}$ are positive definite and δ_i in the range $[0, 2\pi]$. As $\tilde{s}_{13}$ and $\tilde{s}_{23}$ must be very small to satisfy flavor constraints, we have within an excellent approximation [613]

$$\boxed{v_{32}^* v_{31} = \tilde{s}_{13}\tilde{s}_{23}e^{i(\delta_2-\delta_1)}, \qquad v_{33}^* v_{31} = -\tilde{s}_{13}e^{-i\delta_1}, \qquad v_{33}^* v_{32} = -\tilde{s}_{23}e^{-i\delta_2}.} \tag{16.10}$$

Therefore, for fixed $M_{Z'}$ and β, the Z' contributions to all processes in the down-quark sector depend only on the four parameters in (16.9), implying very strong correlations between NP contributions to various observables. Indeed, as seen in (16.10), the B_d system involves only the parameters $\tilde{s}_{13}$ and δ_1 while the B_s system depends on $\tilde{s}_{23}$ and δ_2. Moreover, stringent correlations between observables in $B_{d,s}$ sectors and in the kaon sector are found because kaon physics depends on $\tilde{s}_{13}$, $\tilde{s}_{23}$, and $\delta_2 - \delta_1$. A very constraining feature of this model is that the diagonal couplings of Z' to quarks and leptons are fixed for a given β, except for a weak dependence on $M_{Z'}$ due to running of $\sin^2 \vartheta_W$ provided β differs significantly from $\pm\sqrt{3}$, which is satisfied for $n = 1, 2$ in (16.5).

It follows that $\tilde{s}_{13}$ and δ_1 can be determined from ΔM_d and CP-asymmetry $S_{\psi K_S}$ while $\tilde{s}_{23}$ and δ_2 from ΔM_s and CP-asymmetry $S_{\psi\phi}$. Then the parameters in the K system are fixed.

We will now recall Z' couplings to SM fermions for arbitrary β concentrating on those that are relevant for this book. The expressions for other couplings and masses of new gauge bosons and fermions as well as expressions for their electric charges that depend on β can be found in [613].

The fundamental quantity that enters all couplings is the function

$$f(\beta) = \frac{1}{1 - (1 + \beta^2)s_W^2} > 0, \tag{16.11}$$

where the positivity of this function results from the reality of $M_{Z'}$ as stressed earlier.

The following properties should be noted:

- For $\beta \approx \sqrt{3}$ there is a Landau singularity for $s_W^2 = 0.25$. As at M_W one has $s_W^2 \approx 0.23$ (with exact number depending on its definition considered) and renormalization group evolution of weak couplings increases s_W^2 with increasing scale, $s_W^2(M_{Z'})$ reaches 0.25, and the singularity in question for $M_{Z'} \approx 4\,\mathrm{TeV}$.
- For $|\beta| \leq \sqrt{3} - 0.20$ this problem does not arise even up to the GUT scales.

While we will specifically consider only the cases $\beta = \pm n/\sqrt{3}$ with $n = 1, 2$, we list here the formulas for the relevant couplings for arbitrary real $\beta \neq \sqrt{3}$ satisfying (16.2). The case $\beta = \sqrt{3}$ is considered in an appendix in [190].

We stress that the couplings of Z' to quarks and leptons have to be evaluated at the scale μ at which Z' is integrated out, that is at $\mu = O(M_{Z'})$ and not at M_W. For $n = 1$ this difference is irrelevant. For $n = 2$ it plays a role if acceptable precision is required, and it is crucial for $n = 3$. The numerical values of couplings listed following are collected in an appendix in [190].

The relevant Z' couplings to quarks are then given as follows:

$$\Delta_L^{ij}(Z') = \frac{g}{\sqrt{3}} c_W \sqrt{f(\beta)} v_{3i}^* v_{3j}, \tag{16.12a}$$

$$\Delta_L^{ji}(Z') = \left[\Delta_L^{ij}(Z')\right]^*, \qquad \Delta_L^{u\bar{u}}(Z') = \Delta_L^{d\bar{d}}(Z'), \tag{16.12b}$$

$$\Delta_L^{d\bar{d}}(Z') = \frac{g}{2\sqrt{3}c_W} \sqrt{f(\beta)} \left[-1 + \left(1 + \frac{\beta}{\sqrt{3}}\right) s_W^2\right], \tag{16.12c}$$

$$\Delta_R^{u\bar{u}}(Z') = \frac{g}{2\sqrt{3}c_W} \sqrt{f(\beta)} \frac{4}{\sqrt{3}} \beta s_W^2 = -2\Delta_R^{d\bar{d}}(Z'), \tag{16.12d}$$

$$\Delta_V^{d\bar{d}}(Z') = \frac{g}{2\sqrt{3}c_W} \sqrt{f(\beta)} \left[-1 + \left(1 - \frac{\beta}{\sqrt{3}}\right) s_W^2\right], \tag{16.12e}$$

$$\Delta_A^{d\bar{d}}(Z') = \frac{g}{2\sqrt{3}c_W} \sqrt{f(\beta)} \left[1 - \left(1 + \sqrt{3}\beta\right) s_W^2\right], \tag{16.12f}$$

$$\Delta_V^{u\bar{u}}(Z') = \frac{g}{2\sqrt{3}c_W} \sqrt{f(\beta)} \left[-1 + \left(1 + \frac{5}{\sqrt{3}}\beta\right) s_W^2\right], \tag{16.12g}$$

$$\Delta_A^{u\bar{u}}(Z') = \frac{g}{2\sqrt{3}c_W} \sqrt{f(\beta)} \left[1 - \left(1 - \sqrt{3}\beta\right) s_W^2\right], \tag{16.12h}$$

where v_{ij} are the elements of the matrix V_L in (16.7). The diagonal couplings are valid for the first two generations of quarks neglecting the very small nondiagonal contributions in the matrices V_L and U_L. For the third generation there is an additional term that can be found in [613].

For leptons we have

$$\Delta_L^{v\bar{v}}(Z') = \frac{g}{2\sqrt{3}c_W} \sqrt{f(\beta)} \left[1 - \left(1 + \sqrt{3}\beta\right) s_W^2\right], \tag{16.13a}$$

$$\Delta_L^{\mu\bar{\mu}}(Z') = \Delta_L^{v\bar{v}}(Z'), \tag{16.13b}$$

$$\Delta_R^{\mu\bar{\mu}}(Z') = \frac{-g\,\beta\,s_W^2}{c_W} \sqrt{f(\beta)}, \tag{16.13c}$$

$$\Delta_V^{\mu\bar{\mu}}(Z') = \frac{g}{2\sqrt{3}c_W} \sqrt{f(\beta)} \left[1 - \left(1 + 3\sqrt{3}\beta\right) s_W^2\right], \tag{16.13d}$$

$$\Delta_A^{\mu\bar{\mu}}(Z') = \frac{g}{2\sqrt{3}c_W} \sqrt{f(\beta)} \left[-1 + \left(1 - \sqrt{3}\beta\right) s_W^2\right]. \tag{16.13e}$$

Until now we discussed only flavor-violating couplings of Z', but in the process of electroweak symmetry breaking $Z - Z'$ mixing takes place implying that there is a small mixing between Z and Z' so that the heavy mass eigenstates are really

$$Z_\mu^1 = \cos\xi Z_\mu + \sin\xi Z_\mu', \qquad Z_\mu^2 = -\sin\xi Z_\mu + \cos\xi Z_\mu'. \tag{16.14}$$

Here Z_μ is the mass eigenstate in the SM, and Z' is the gauge boson corresponding to the generator of $U(1)_X$. For $\sin\xi = 0$ the two mass eigenstates are simply $Z_\mu^1 = Z_\mu$ and $Z_\mu^2 = Z_\mu'$. Consequently only Z_μ' has flavor-violating couplings in the mass eigenstate basis for quarks as a result of different transformation properties of the third generation under the extended gauge group.

When the small but nonvanishing mixing represented by $\sin\xi$ is taken into account, not only the flavor-violating couplings of the mass eigenstate Z_μ^1 to quarks are generated but also its flavor-diagonal couplings to SM fermions differ from the ones of the SM Z boson. Explicitly we have for $i \neq j$

$$\Delta_L^{ij}(Z^1) = \sin\xi\,\Delta_L^{ij}(Z') \equiv \Delta_L^{ij}(Z), \qquad \Delta_L^{ij}(Z^2) = \cos\xi\,\Delta_L^{ij}(Z') \approx \Delta_L^{ij}(Z'), \tag{16.15}$$

where in order not to modify the notation in flavor-violating observables relative to the previous chapter, we will still use Z for Z^1 and Z' for Z^2 with masses M_Z and $M_{Z'}$, respectively. The small shifts in the masses of these gauge bosons relative to the case $\sin\xi = 0$ are irrelevant in flavor-violating processes.

For flavor-diagonal couplings to fermions (generically denoted with f), we have with $k = L, R, A, V$

$$\Delta_k^{ff}(Z^1) = \cos\xi\,\Delta_k^{ff}(Z) + \sin\xi\,\Delta_k^{ff}(Z'), \qquad (16.16)$$

$$\Delta_k^{ff}(Z^2) = \cos\xi\,\Delta_k^{ff}(Z') - \sin\xi\,\Delta_k^{ff}(Z). \qquad (16.17)$$

In the calculations of flavor-violating effects we can neglect the mixing effects in these couplings so that we can simply set

$$\Delta_k^{ff}(Z^1) = \Delta_k^{ff}(Z), \qquad \Delta_k^{ff}(Z^2) = \Delta_k^{ff}(Z'), \qquad (16.18)$$

but in the analysis of electroweak precision tests, performed in [615], one has to keep mixing effects in (16.16).

As the mass M_Z and flavor-diagonal Z-couplings to all SM fermions are known, the model is also predictive after the inclusion of $Z - Z'$ mixing, although one additional parameter, $\tan\bar\beta$, enters the game. This mixing has been calculated correctly first in [615] with the result

$$\boxed{\sin\xi = \frac{c_W^2}{3}\sqrt{f(\beta)}\left(3\beta\frac{s_W^2}{c_W^2} + \sqrt{3}a\right)\left[\frac{M_Z^2}{M_{Z'}^2}\right],} \qquad (16.19)$$

where $f(\beta)$ is given in (16.11) and

$$-1 < a = \frac{v_-^2}{v_+^2} < 1, \qquad (16.20)$$

with $v_\pm^2$ given in terms of the vacuum expectation values of two Higgs triplets ρ and η present in these models as follows

$$v_+^2 = v_\eta^2 + v_\rho^2, \qquad v_-^2 = v_\eta^2 - v_\rho^2. \qquad (16.21)$$

As the Higgs system responsible for the breakdown of the SM group has the structure of a two Higgs doublet model and the triplets ρ and η are responsible for the masses of up-quarks and down-quarks, respectively, one can, express the parameter a in terms of the usual $\tan\bar\beta$ where we introduced a *bar* to distinguish the usual angle β from the parameter β in 331 models. We have then

$$a = \frac{1 - \tan^2\bar\beta}{1 + \tan^2\bar\beta}, \qquad \tan\bar\beta = \frac{v_\rho}{v_\eta}. \qquad (16.22)$$

Thus for $\tan\bar\beta = 1$ the parameter $a = 0$, which simplifies the formula for $\sin\xi$ relating uniquely its sign to the sign of β. On the other hand, in the large $\tan\bar\beta$ limit we find $a = -1$ and in the low $\tan\bar\beta$ limit one has $a = 1$.

The expression in (16.19) tells us definitely that $\sin\xi$ is very small but one should remember that the propagator suppression of FCNC transitions in the case of Z' is by a factor of $M_{Z'}^2/M_Z^2$ stronger than in the case of Z at the amplitude level. Therefore a more detailed numerical analysis of the values of $\sin\xi$ and Z' couplings to leptons as functions of β and $\tan\bar\beta$ and the comparison with the known Z couplings to fermions is required to decide whether Z boson contributions to FCNC processes can be neglected or not. Such an analysis has been performed in [615], and we will report on its outcome later. However, first we have to elaborate on the choice of fermion representations.

16.2.3 Choice of Fermion Representations

It has been emphasized in [1020, 1023] that the choice of β does not uniquely specify the phenomenology of the 331 model considered, which further depends on the choice of fermion representations under $SU(3)_L$. Here we will discuss two choices for which a detailed analysis has been presented in [615].

In the first choice, to be denoted by F_1, and used exclusively in [190, 613], the first two generations of quarks are put into triplets (3), while the third one into the antitriplet (3^*):

$$
\begin{pmatrix} u \\ d \\ D \end{pmatrix}_L, \qquad
\begin{pmatrix} c \\ s \\ S \end{pmatrix}_L, \qquad
\begin{pmatrix} b \\ -t \\ T \end{pmatrix}_L. \tag{16.23}
$$

The corresponding right-handed quarks are singlets. The anomaly cancellation then requires that leptons are put into antitriplets:

$$
\begin{pmatrix} e \\ -\nu_e \\ E_e \end{pmatrix}_L, \qquad
\begin{pmatrix} \mu \\ -\nu_\mu \\ E_\mu \end{pmatrix}_L, \qquad
\begin{pmatrix} \tau \\ -\nu_\tau \\ E_\tau \end{pmatrix}_L. \tag{16.24}
$$

On the other hand, as done in [1020, 1023], the triplets and antitriplets are interchanged relative to F_1. That is, the first two quark generations are in antitriplets while the third one in a triplet. Therefore leptons are also in triplets. We will call this fermion assignment F_2. In [1020, 1023] still two other quark assignments are discussed in which the first or the second quark generation transforms differently under $SU(3)_L$ than the remaining two. But we find the ones just listed more natural due to large top-quark mass, and we do not discuss these two additional possibilities.

The important two features are the following ones. For a given β:

- The expression for $\sin\xi$ in (16.19) is independent of whether F_1 or F_2 is used.
- On the other hand the signs in front of β in Z' couplings to quarks and leptons are changed when going from F_1 to F_2. This property can be derived from the action of the relevant operator $\hat{Q}_W$ on triplet and antitriplet. See formulas in section 2 of [613]. This can also be seen by comparing Z' couplings to fermions, given for F_1, with table 4 of [1023], where the couplings are given for F_2.

As pointed out in [615] these properties have the following important phenomenological implications listed here first without FCNC mediated by the Z boson:

- In F_1 scenario the models with $\beta = -2/\sqrt{3}$ and $\beta = -1/\sqrt{3}$ are useful for the explanation of the anomalies in $B_d \to K^* \mu^+ \mu^-$ because with F_1 representations the coupling $\Delta_V^{\mu\bar{\mu}}(Z')$ is large. On the other hand, the models with $\beta = 2/\sqrt{3}$ and $\beta = 1/\sqrt{3}$ having significant $\Delta_A^{\mu\bar{\mu}}(Z')$ coupling provide interesting NP effects in $B_{s,d} \to \mu^+ \mu^-$.

- In F_2 scenario the situation is reversed. The models with $\beta = 2/\sqrt{3}$ and $\beta = 1/\sqrt{3}$ are useful for the explanation of the anomalies in $B_d \to K^* \mu^+ \mu^-$, while the ones with $\beta = -2/\sqrt{3}$ and $\beta = -1/\sqrt{3}$ for $B_{s,d} \to \mu^+ \mu^-$.

- While these two scenarios cannot be distinguished by flavor observables when only Z' contributions are considered, they can be distinguished when Z boson contributions are taken into account. This originates in the fact that the $\sin \xi$ entering the $\Delta_L^{ij}(Z)$ couplings in (16.15) *does depend* on the sign of β but *does not depend* on whether F_1 scenario or F_2 scenario is considered. In other words, the invariance in flavor observables under the transformations

$$\beta \to -\beta, \qquad F_1 \to F_2 \tag{16.25}$$

present in the absence of $Z - Z'$ mixing is broken by this mixing. We will see this explicitly in our following discussion.

- As a particular sign of β could be favored by flavor-conserving observables, in particular electroweak precision tests, this feature allows in principle to determine whether the representation F_1 or F_2 is favored by nature. We will see this explicitly in the next subsection.

16.2.4 Distinguishing 331 Models through Electroweak Precision Tests

A detailed analysis of 24 models in question corresponding to different values of β and $\tan \bar{\beta}$ for the representations F_1 and F_2 has been presented in [615]. They are collected in Table 16.1. In particular a detailed analysis of electroweak precision tests in these models has been performed there. Interested readers are asked to look at section 5 of that paper. Here we just summarize the main outcome of that study.

Requiring that the 24 models in question perform at least as well as the SM in these tests, only seven models passed this test and have been selected for a more detailed study of FCNC processes. These are

$$\text{M9,} \quad \text{M8,} \quad \text{M6,} \quad \text{M11,} \quad \text{M3,} \quad \text{M16,} \quad \text{M14,} \qquad (\text{favored}), \tag{16.26}$$

Table 16.1 Definition of the various 331 models.

MI	scen.	β	$\tan \bar{\beta}$	MI	scen.	β	$\tan \bar{\beta}$	MI	scen.	β	$\tan \bar{\beta}$
M1	F_1	$-2/\sqrt{3}$	1	M9	F_2	$-2/\sqrt{3}$	1	M17	F_1	$-2/\sqrt{3}$	0.2
M2	F_1	$-2/\sqrt{3}$	5	M10	F_2	$-2/\sqrt{3}$	5	M18	F_2	$-2/\sqrt{3}$	0.2
M3	F_1	$-1/\sqrt{3}$	1	M11	F_2	$-1/\sqrt{3}$	1	M19	F_1	$-1/\sqrt{3}$	0.2
M4	F_1	$-1/\sqrt{3}$	5	M12	F_2	$-1/\sqrt{3}$	5	M20	F_2	$-1/\sqrt{3}$	0.2
M5	F_1	$1/\sqrt{3}$	1	M13	F_2	$1/\sqrt{3}$	1	M21	F_1	$1/\sqrt{3}$	0.2
M6	F_1	$1/\sqrt{3}$	5	M14	F_2	$1/\sqrt{3}$	5	M22	F_2	$1/\sqrt{3}$	0.2
M7	F_1	$2/\sqrt{3}$	1	M15	F_2	$2/\sqrt{3}$	1	M23	F_1	$2/\sqrt{3}$	0.2
M8	F_1	$2/\sqrt{3}$	5	M16	F_2	$2/\sqrt{3}$	5	M24	F_2	$2/\sqrt{3}$	0.2

with the first five performing better than the SM, while the last two basically as the SM. The models with *odd* index I correspond to $\tan \bar{\beta} = 1.0$ and the ones with *even* one to $\tan \bar{\beta} = 5.0$. The models with $\tan \bar{\beta} = 0.2$ did not pass these tests, which is unfortunate, as only these models among all models considered could give significant contributions to rare decays $K^+ \to \pi^+ \nu \bar{\nu}$ and $K_L \to \pi^0 \nu \bar{\nu}$.

16.2.5 Summary of FCNC Results

Extensive flavor analyses in 331 models can be found in [190, 613, 615, 852, 853]. References to earlier analysis of flavor physics in 331 models can be found there and in [1020, 1023].

The most recent updated analyses in [852, 853] concentrated on the ratio ε'/ε and its correlation with ε_K and B-physics observables like $\Delta M_{s,d}$, $B_s \to \mu^+ \mu^-$, and the Wilson coefficient C_9. They were motivated by the anomalies in ε'/ε [244, 275, 276, 296], tension between ε_K and $\Delta M_{s,d}$ within the SM [388] implied by the lattice data in [873], and in the case of C_9 by the LHCb anomalies in the angular observables in $B \to K^*(K)\mu^+\mu^-$ mentioned in Section 12.2.

We briefly summarize the main results of [852, 853]. These analyses show that the requirement of an enhancement of ε'/ε has a significant impact on other flavor observables. Moreover, in [853] it has also been shown that the results are rather sensitive to the value of $|V_{cb}|$, as has been illustrated there by choosing two values: $|V_{cb}| = 0.040$ and $|V_{cb}| = 0.042$. The main lessons extracted from [852, 853] for $M_{Z'} = 3$ TeV are as follows:

Lesson 1: Among the seven 331 models in (16.26), selected on the basis of the electroweak precision study, only three (M8, M9, M16) can provide for both choices of $|V_{cb}|$, significant shift of ε'/ε, even though not larger than 8×10^{-4}.

Lesson 2: The tensions between $\Delta M_{s,d}$ and ε_K pointed out in [388], see Section 15.1, can be removed in these models (M8, M9, M16) for both values of $|V_{cb}|$.

Lesson 3: In agreement with our previous expectations, two of them (M8 and M9) can simultaneously with the enhancement of ε'/ε suppress $B_s \to \mu^+\mu^-$ by at most 10% and 20% for $|V_{cb}| = 0.042$ and $|V_{cb}| = 0.040$, respectively. This is presently sufficient for bringing the theory close to the central experimental value. On the other hand, the maximal deviations from SM in the Wilson coefficient C_9 are $C_9^{NP} = -0.1$ and $C_9^{NP} = -0.2$ for these two $|V_{cb}|$ values, respectively. Due to this moderate shift, these models do not really help in the case of $B_d \to K^*\mu^+\mu^-$ anomalies that require deviations as high as $C_9^{NP} = -1.0$.

Lesson 4: In M16 the situation is reversed. It is possible to reduce the rate for $B_s \to \mu^+\mu^-$ for $M_{Z'} = 3$ TeV for the two $|V_{cb}|$ values by at most 3% and 10%, respectively. With the corresponding values $C_9^{NP} = -0.3$ and -0.5, the anomaly in $B_d \to K^*\mu^+\mu^-$ can be significantly reduced.

Lesson 5: The maximal shifts in ε'/ε decrease fast with increasing $M_{Z'}$ in the case of $|V_{cb}| = 0.042$ but are practically unchanged for $M_{Z'} = 10$ TeV when $|V_{cb}| = 0.040$ is used.

Lesson 6: On the other hand, for larger values of $M_{Z'}$ the effects in $B_s \to \mu^+\mu^-$ and $B_d \to K^*\mu^+\mu^-$ are much smaller. NP effects in rare K decays and $B \to K(K^*)\nu\bar{\nu}$ remain small in all 331 models even for $M_{Z'}$ of few TeV. This could be challenged by NA62, KOTO, and Belle II experiments in this decade.

All these results are illustrated in numerous plots in [852, 853].

Finally, we return to the outcome of general analyses of B-physics anomalies from which, as we saw in Section 12.2, the most favorite relations between NP contributions to C_9 and C_{10} were before Moriond 2019

$$C_9^{\rm NP} = -C_{10}^{\rm NP} = -0.62 \pm 0.12 \quad \text{or} \quad C_9^{\rm NP} = -1.0 \pm 0.2, \; C_{10}^{\rm NP} = 0. \qquad (16.27)$$

These relations should be contrasted with models M8 and M9 for which one finds [852]

$$C_9^{\rm NP} = 0.49 \, C_{10}^{\rm NP} \quad (\text{M8}), \qquad C_9^{\rm NP} = 0.42 \, C_{10}^{\rm NP} \quad (\text{M9}). \qquad (16.28)$$

Thus as already seen earlier, models M8 and M9 have significant problems with the observed pattern of these anomalies. On the other hand,

$$C_9^{\rm NP} = -4.59 \, C_{10}^{\rm NP} \quad (\text{M16}) \qquad (16.29)$$

is close to the second relation in (16.27) where NP resides dominantly in the coefficient C_9. Thus already on the basis of B physics observables, we should be able to distinguish between the three favorite 331 models M8, M9, and M16.

The messages from Moriond 2019 did not change this pattern, and presently M16 appears to be in the best shape, but we have to wait for improved data to have a clear-cut conclusion. In particular as the RH FCNC are absent in 331 models; this could be another problem for these models if the presence of RH currents, proposed by several authors as a solution to the anomalies in question, will be established.

Thus the main message from [852, 853] is that NP contributions in 331 models can simultaneously solve $\Delta F = 2$ tensions, enhance ε'/ε, and suppress either the rate for $B_s \to \mu^+\mu$ or C_9 Wilson coefficient without any significant effect on $K^+ \to \pi^+\nu\bar{\nu}$ and $K_{\rm L} \to \pi^0\nu\bar{\nu}$ and $b \to s\nu\bar{\nu}$ transitions. While sizable NP effects in $\Delta F = 2$ observables and ε'/ε can persist for $M_{Z'}$ outside the reach of the LHC, such effects in $B_s \to \mu^+\mu^-$ will only be detectable provided Z' will be discovered soon. The future of 331 models will also depend on the fate of the violation of flavor lepton universality hinted at by the data as these models cannot explain it, although they can provide modifications in C_9 and C_{10} as just discussed.

We did not review lepton physics in 331 models. We refer to [1024], where a good collection of relevant references and phenomenology of lepton decays in a specific 331 model can be found.

16.3 Vector-Like Quarks and Leptons

16.3.1 Introduction

Presently we know that three generations of quarks and leptons exist. Their left-handed and right-handed components transform differently under the electroweak part of the SM gauge group, and only after this symmetry group is spontaneously broken can their masses be generated through their Yukawa couplings to the SM Higgs. We also know that the fourth generation of quarks and leptons does not exist because otherwise LHC would have already discovered them, and also the required large Yukawa coupling to Higgs would imply much larger Higgs width than measured. But heavy fermions with left-handed and right-handed components transforming identically under the SM gauge group can exist because the generation of their masses does not require spontaneous breakdown of electroweak symmetry (SSB). They can even be so heavy that the LHC cannot discover them directly in collisions. Yet, we should still hope that the LHC will see them in the coming years.

The masses of these vector-like fermions can be generated by some mechanism at very short-distance scales that we do not have to specify at the time being. We can just add the corresponding mass term to the SM Lagrangian by hand and specify how these new heavy fermions interact with SM particles. Such fermions will be termed vector-like fermions (VLFs), vector-like quarks (VLQs), and vector-like leptons (VLLs). This section deals with flavor physics of these new heavy particles, which this time carry spin-1/2 as opposed to the spin-1 gauge boson Z' and heavy scalars with spin-0.

In fact, among the simplest renormalizable extensions of the SM that do not introduce any additional fine-tunings of parameters are models in which the only new particles are VLFs. As already stated such fermions can be much heavier than the SM fermions as they can acquire masses in the absence of electroweak SSB. But if in the process of this breaking they can mix with ordinary fermions, the most natural implication is the generation of FCNC processes mediated by the SM Z boson. We will first discuss this simplest situation concentrating on VLQs, postponing the extension of the SM group by a U(1) factor to the second part of this section. With a new heavy gauge boson Z' and new heavy scalars necessary to provide Z' mass and to break the extended gauge symmetry group down to the SM gauge group, the models with VLFs are more involved but are also interesting.

There is a rich literature on FCNC implied by the presence of VLQs. See in particular [783, 792, 925, 967, 1025–1032]. Here we will be guided by the analysis in [666], which analyzed the patterns of flavor violation in a number of VLQ models classified in [783, 784, 967]. The latter authors identified all renormalizable models with additional fermions residing in a single vector-like complex representation of the SM gauge group with a mass M. It turns out that there are 11 models where new fermions have proper quantum numbers so that they can couple in a renormalizable manner to the SM Higgs and SM fermions, thereby implying new sources of flavor violation.

In this section our discussion will be confined to FCNC in K, B_d, and B_s systems, and only five of these models will be relevant for us. They will be specified in Section 16.3.2. We will see that in these models Yukawa interactions of the SM scalar doublet H involving ordinary quarks and VLQs imply flavor-violating Z couplings to ordinary quarks. In this manner VLQ models are explicit realizations of Z models discussed in Section 15.5 with various flavor-violating couplings of Z given in terms of Yukawa couplings and VLQ masses.

NP contributions to $\Delta F = 1$ FCNC transitions are then dominated by tree-level Z exchanges. The pattern of NP contributions to $\Delta F = 2$ processes is more involved. Indeed, as pointed out in [967], for the masses of VLQs $M \geq 5\,\text{TeV}$ NP contributions to $\Delta F = 2$ transitions from box diagrams with VLQs and Higgs become very important. This property will modify the correlation between $\Delta F = 1$ and $\Delta F = 2$ processes encountered in

Z-models discussed in Section 15.5. But in [967] the contributions from Yukawa RG effects discussed in Section 15.5 have not been taken into account. As we have seen, there these effects depend on whether RH or LH flavor-violating quark couplings to the Z are present. If they are RH, the effects of RG evolution from M (the common VLQ mass) down to the electroweak scale, μ_{ew}, generate left-right operators [766] via top-Yukawa induced mixing. These operators are strongly enhanced through QCD RG effects below the electroweak scale and in the case of the K system through chirally enhanced hadronic matrix elements. They dominate then NP contributions to ε_K, but in the $B_{s,d}$ meson systems for VLQ-masses above 5 TeV they have to compete with contributions from box diagrams with VLQs mentioned earlier. If they are LH the Yukawa enhancement is less important because left-right operators are not present, and box diagrams play an important role both in the $B_{s,d}$ and K systems.

Now NP contributions to flavor observables in K, B_d, and B_s systems depend in each model, respectively, on the products of complex Yukawa couplings $\lambda_s^* \lambda_d$, $\lambda_b^* \lambda_d$, and $\lambda_b^* \lambda_s$ and the common VLQ mass M. This structure allows to set one of the phases to zero so that each model depends on five Yukawa parameters and M, implying a number of correlations between flavor observables not only within a given meson system but also between those in different meson systems. This goes beyond the Z scenario of Section 15.5, where there was no relation between different meson systems.

The outline of this section is as follows. In the next subsection we will summarize the particle content of the five VLQ models considered by us together with gauge interactions and Yukawa interactions. Subsequently we will discuss the decoupling of VLQs both at tree-level and one-loop level. Next phenomenology of these models will be presented. We will close this section by describing briefly what happens in VLQ models in which the gauge group is extended by a U(1) factor, a heavy gauge boson Z' is present, and the scalar sector contains new heavy scalars beyond the SM Higgs. This section is dominantly based on [666], where further details and references to rich literature can be found.

16.3.2 The VLQ Models

VLQ Representations

We will focus on models with VLQs residing in complex representations of the SM gauge group. We adapt the usual SM fermion content of the three generations ($i = 1, 2, 3$) of quarks ($q_L^i = (u_L^i, d_L^i)^T, u_R^i, d_R^i$) and leptons ($L_L^i = (\nu_i, \ell_L^i)^T, \ell_R^i$), which acquire masses via spontaneous symmetry breaking from the standard scalar SU(2)$_{\mathrm{L}}$ doublet H.

As we are mainly interested in the phenomenlogy of down-quark physics, we will restrict our analysis to SU(3)$_C$ triplets and consider the following five models with SU(2)$_{\mathrm{L}}$ singlets, doublets, and triplets

$$
\begin{aligned}
\text{Singlets:} \quad & D(1, -1/3), & & \text{(V)} \;\; \text{(LH)} \\
\text{Doublets:} \quad & Q_V(2, +1/6), \;\; Q_d(2, -5/6), & & \text{(IX, XI)} \;\; \text{(RH)} \\
\text{Triplets:} \quad & T_d(3, -1/3), \;\; T_u(3, +2/3), & & \text{(VII, VIII)} \;\; \text{(LII)},
\end{aligned}
\tag{16.30}
$$

where the transformation properties are indicated as $(SU(2)_L, U(1)_Y)$. The representations D, Q_V, Q_d, T_d, and T_u correspond to the models V, IX, XI, VII, and VIII introduced in [967]. Already at this stage we indicated which models bring in new LH currents and which RH currents. We will be more explicit about it later. The kinetic and gauge interactions of the new VLQs are given by

$$\mathcal{L}_{\text{kin}} = \overline{D}(i\mathcal{D}\!\!\!/ - M_D)D + \sum_{a=V,d} \overline{Q}_a(i\mathcal{D}\!\!\!/ - M_{Q_a})Q_a + \sum_{a=d,u} \text{Tr}\left[\overline{T}_a(i\mathcal{D}\!\!\!/ - M_{T_a})T_a\right], \quad (16.31)$$

with appropriate covariant derivatives $\mathcal{D}_\mu$, and we follow [967] for the triplet representations as given in (2.13) and (2.14) of that paper. The masses M of the VLQs introduce a new scale, which we will assume throughout to be significantly larger than the electroweak scale. Typical at least $O(1\text{ TeV})$ as required by the LHC bounds. The covariant derivative is, omitting the $SU(3)_C$ part,

$$\mathcal{D}_\mu = \partial_\mu - ig_2\frac{\sigma^a}{2}W_\mu^a - ig_1 Y B_\mu, \quad (16.32)$$

with the SM gauge couplings $g_{1,2}$ and σ^a being Pauli matrices.

Yukawa Interactions of VLQs

The scalar sector contains only the standard doublet $H(2, +1/2)$, with its usual scalar potential. It provides masses to gauge bosons and standard fermions in the course of electroweak SSB via the VEV $v \simeq 246$ GeV, where

$$\langle H \rangle = (0, v/\sqrt{2})^T. \quad (16.33)$$

The VLQs interact with SM quarks (q_L, u_R, d_R) via Yukawa interactions

$$-\mathcal{L}_{\text{Yuk}}(H) = \left(\lambda_i^D H^\dagger \overline{D}_R + \lambda_i^{T_d} H^\dagger \overline{T}_{dR} + \lambda_i^{T_u} \widetilde{H}^\dagger \overline{T}_{uR}\right) q_L^i$$
$$+ \lambda_i^{V_u} \overline{u}_R^i \widetilde{H}^\dagger Q_{VL} + \overline{d}_R^i \left(\lambda_i^{V_d} H^\dagger Q_{VL} + \lambda_i^{Q_d} \widetilde{H}^\dagger Q_{dL}\right) + \text{h.c.,} \quad (16.34)$$

where $\widetilde{H} \equiv i\sigma_2 H^*$. The complex-valued Yukawa couplings λ_i^{VLQ} with $i = 1, 2, 3$ or equivalently in the models considered $i = d, s, b$ give rise to mixing with the SM quarks and flavor-changing Z-couplings, which have been worked out in detail [967] and confirmed up to one sign in T_u model in [666]. We will be more explicit in Section 16.3.3.

16.3.3 Decoupling of VLQs

Preliminaries

The VLQ models are characterized by the masses M of the VLQs and the various Yukawa couplings λ_i^{VLQ} ($i = 1, 2, 3$) listed earlier. The present lower bound on M from the LHC is in the ballpark of 1 TeV.

We have thus the hierarchy

$$m_b \ll v \ll M. \tag{16.35}$$

It is then natural to decouple first VLQs and to consider an effective theory valid between scales $\mu_M \sim M$ and $\mu_{\text{EW}} \sim v$ and subsequently match it to the effective $SU(3)_C \otimes U(1)_{em}$ theory valid between μ_{EW} and $\mu_b \sim m_b$. Here m_b denotes the mass of the bottom quark, which is the heaviest of the light standard particles. This will indicate which operators are most important. The evolution to lower scales relevant for K meson physics is done as in previous chapters.

In this subsection we present the results from the decoupling of the VLQs that are important for phenomenological applications within the framework of the LEFT relevant for weak decays of mesons. The decoupling proceeds either by explicit matching calculations starting at tree-level and including subsequently higher orders or by integrating VLQs out in the path integral method [783]. The tree-level decoupling is known for a long time [783]. The one-loop decoupling has been presented in [967] and confirmed up to small differences in [666].

The VLQs have a very limited set of couplings to light fields, which are either via gauge interactions (16.31) to the gauge bosons or via Yukawa interactions (16.34) to light – w.r.t. to VLQ mass M – SM quarks. At tree-level, this particular structure of interactions can give rise only to flavor-changing (FC) Z couplings, whereas all other decoupling effects are loop-suppressed.

The decoupling of the VLQs proceeds in the unbroken phase of $SU(2)_L \otimes U(1)_Y$, hence quark fields are flavor-eigenstates, and neutral components of SM Higgs doublet are without VEV at this stage. After the RG-evolution from μ_M to μ_{EW}, SSB will take place, and the transformation from flavor- to mass-eigenstates for fermions and gauge bosons can be performed.

A detailed discussion of this decoupling has been presented in [666, 783, 967], but we will only present the results following the presentation in [666].

Tree-Level Decoupling and Z Interactions

As discussed in detail in [666] the relation of quark masses to Yukawa interactions includes now also $1/M^2$ contributions, and their diagonalization proceeds as usual for the quark fields with the help of 3×3 unitary rotations

$$\psi_L \to V_L^\psi \psi_L, \qquad\qquad \psi_R \to V_R^\psi \psi_R \tag{16.36}$$

in flavor space, implying

$$V_L^{\psi\dagger} m_\psi V_R^\psi = m_\psi^{\text{diag}}, \qquad\qquad V = (V_L^u)^\dagger V_L^d \tag{16.37}$$

that is diagonal up- and down-quark masses m_ψ^{diag} and the quark-mixing matrix V (CKM matrix). Throughout our discussion we will assume that down quarks are already mass eigenstates, which fixes also the basis for the VLQ Yukawa couplings λ_i^{VLQ} and implies $q_L = (V^\dagger u_L, d_L)^T$.

Coupling	D	Q_V	Q_d	T_d	T_u
Table 16.2 $\Delta_{LR}^{ij}(Z)$ for $i,j = d,s,b$ in VLQ models.					
$\Delta_L^{ij}(Z)$	Δ^{ij}	0	0	$\Delta^{ij}/2$	$-\Delta^{ij}$
$\Delta_R^{ij}(Z)$	0	$-(\Delta^{ij})^*$	$(\Delta^{ij})^*$	0	0

After SSB flavor-changing Z interactions for fermions ($f = \ell, d, u$) are generated. We parametrize them as in Section 15.5 through

$$\mathcal{L}_{\text{VLQ}}^{(Z)} = \bar{f}^i \left[\Delta_L^{ij}(Z)\, \gamma^\mu P_L + \Delta_R^{ij}(Z)\, \gamma^\mu P_R \right] f^j Z_\mu. \tag{16.38}$$

The flavor-diagonal couplings of Z to quarks and leptons will be set to the ones of the SM as corrections from NP to them are fully negligible. Thus flavor-diagonal Z couplings to fermions are universal for all models considered. On the other hand, flavor-violating couplings are model dependent.

The decoupling of VLQs, presented in details in [666], gives the results for $\Delta_{L,R}(Z)$ couplings collected for down-quarks in Table 16.2, where

$$\Delta^{ij} \equiv \frac{\lambda_i^* \lambda_j}{g_Z} \frac{M_Z^2}{M^2}, \qquad\qquad g_Z \equiv \sqrt{g_1^2 + g_2^2}. \tag{16.39}$$

Except for the sign in the case of T_u, the results of [666] agree with those in [967]. Further, also nonzero couplings to up-type quarks arise [967], but they will not be investigated in this book.

Decoupling at One-Loop Level

All other decoupling processes proceed via loops. They turn out to be relevant only for $\Delta F = 2$ processes. VLQs contribute then to the Wilson coefficients of the by-now-familiar $|\Delta F| = 2$ operators O_a^{ij} for $a = \text{VLL, VRR, LR1}$ via box diagrams that contain two heavy VLQ propagators with representations F_m and F_n and massless components of the standard doublet $H = (H^+, H^0)^T$. The resulting Wilson coefficients have the following general structure

$$C_a^{ij} = \frac{\eta_{mn}}{(4\pi)^2} \frac{\Lambda_{ij}^m \Lambda_{ij}^n}{\mathcal{N}_{ij}} f_1(M_m, M_n), \qquad \mathcal{N}_{ij} = \frac{G_F^2}{4\pi^2} M_W^2 (\lambda_t^{ji})^2 \tag{16.40}$$

at the scale μ_M. Here the prefactor $\mathcal{N}_{ij}$ corresponds to the SM normalization of the $|\Delta F| = 2$ Hamiltonian. The function

$$f_1(M_m, M_n) = \frac{\ln(M_m^2/M_n^2)}{M_m^2 - M_n^2} \qquad \text{with} \qquad f_1(M_m, M_m) = \frac{1}{M_m^2}, \tag{16.41}$$

Table 16.3 The index $a = $ VLL, VRR, LR1 appearing in (16.40) for representations (F_m, F_n), followed by corresponding η_{mn}.

(F_m, F_n)	D	Q_d	Q_V	T_d	T_u
D	VLL, $+1/8$	LR1, $+1/4$	LR1, $-1/4$	VLL, $+1/16$	VLL, $-1/8$
Q_d		VRR, $+1/4$	VRR, $-1/4$	LR1, $+3/8$	LR1, $-3/8$
Q_V			VRR, $+1/4$	LR1, $-3/8$	LR1, $+3/8$
T_d				VLL, $+5/32$	VLL, $-1/8$
T_u					VLL, $+5/32$

depends on the VLQ masses of representations $F_{m,n}$. The couplings Λ_{ij}^m are

$$\Lambda_{ij}^m = (\lambda_i^m)^* \lambda_j^m \qquad \text{for} \qquad F_m = D, T_d, T_u, \qquad (16.42)$$

$$\Lambda_{ij}^m = \lambda_i^m (\lambda_j^m)^* \qquad \text{for} \qquad F_m = Q_d, Q_V. \qquad (16.43)$$

The index a of the operator and the numerical factors η_{mn} are collected in Table 16.3. Note that $a = $ VLL for $F_{m,n} = D, T_d, T_u$, and $a = $ VRR for $F_{m,n} = Q_d, Q_V$, whereas $a = $ LR1 for $F_m = D, T_d, T_u$ and $F_n = Q_d, Q_V$. The factors η_{mn} are positive except for interference of $F_m = D, Q_d, T_d$ with $F_n = Q_V, T_u$, because in this case the scalar propagators are crossed, which gives rise to an additional sign w.r.t. the diagram with noncrossed scalar propagators. For $F_m = F_n$, these results agree with [967] for D, T_u, T_d, but for Q_d (model XI), we find an additional factor of 2. Concerning Q_V (model IX), we find a contribution to $\sim O_{\text{VRR}}$ instead of $\sim O_{\text{VLL}}$ and also opposite sign. For completeness we provide also the results for $F_m \neq F_n$.

Before discussing the Yukawa RG effects, let us compare the relative size of box-to-Z exchange contribution to the box function S that we encountered in previous chapters. With the information collected so far and the formulas for tree-level Z contributions of Section 15.5, we readily find

$$\frac{(\Delta S)_{\text{Box}}}{(\Delta S)_Z} = a \, \eta_{LL} \frac{g_Z^2}{8\pi^2} \left[\frac{\tilde{r}_{\text{Box}}}{\tilde{r}_Z} \right] \frac{M^2}{M_Z^2} \qquad (16.44)$$

with $a = 4$ for T_d and unity otherwise. $r_{\text{Box}} \approx \tilde{r}_Z \approx 1$. Moreover, as seen in Table 16.3 $\eta_{LL} = (1/8, 1/4, 5/32)$ for D, (Q_d, Q_V) and (T_d, T_u), respectively. While the Z contribution is comparable to the box contribution for $M \approx 1 - 2$ TeV, it amounts only to a few percent for $M = 10$ TeV. This is not surprising. While to box contributions are dimension-6 contributions to the effective Hamiltonian, the tree-level Z contributions correspond to dimension-8 operators.

The presence of box diagrams modifies the correlation between $|\Delta F| = 2$ and $|\Delta F| = 1$ observables relative to Section 15.5. But to complete the analysis, we have to discuss the RG effects. The ones from QCD alone have been already included in the estimates earlier but we will recall them next. The ones related to Yukawa couplings are more involved, and we will discuss them in more detail.

QCD Renormalization Group Evolution

For $|\Delta F| = 2$ the QCD evolution from μ_M to μ_{EW} involves only $N_f = 6$ and is given at LO as follows [744]

$$C_{\text{VLL,VRR}}(\mu_{\text{EW}}) = \eta_6^{2/7} C_{\text{VLL,VRR}}(\mu_M),$$

$$C_{\text{LR,1}}(\mu_{\text{EW}}) = \eta_6^{1/7} C_{\text{LR,1}}(\mu_M), \tag{16.45}$$

$$C_{\text{LR,2}}(\mu_{\text{EW}}) = \frac{2}{3}\left(\eta_6^{1/7} - \eta_6^{-8/7}\right) C_{\text{LR,1}}(\mu_M) + \eta_6^{-8/7} C_{\text{LR,2}}(\mu_M),$$

with $\eta_6 = \alpha_s^{(6)}(\mu_M)/\alpha_s^{(6)}(\mu_{\text{EW}})$. Initial conditions of $C_a^{ij}(\mu_M)$ resulting from box diagrams are collected in (16.40). Note that $C_{\text{LR,1}}(\mu_M) \neq 0$ only in the presence of several VLQ representations, but always $C_{\text{LR,2}}(\mu_M) = 0$. Renormalization group evolution for tree-level Z contributions starts at μ_{EW}.

Yukawa Renormalization Group Effects

These effects have been already formulated in the framework of SMEFT in Section 14.4. To use these results we have to calculate the coefficients of dimension-6 operators in VLQ models in question at the scale μ_M of order of the VLQ masses. Such a calculation has been done in [783] and confirmed in [666]. We have then

$$D: \quad [C_{Hq}^{(1)}]_{ij} = [C_{Hq}^{(3)}]_{ij} = -\frac{1}{4}\frac{\lambda_i^* \lambda_j}{M^2},$$

$$T_d: \quad [C_{Hq}^{(1)}]_{ij} = -3\,[C_{Hq}^{(3)}]_{ij} = -\frac{3}{8}\frac{\lambda_i^* \lambda_j}{M^2},$$

$$T_u: \quad [C_{Hq}^{(1)}]_{ij} = 3\,[C_{Hq}^{(3)}]_{ij} = \frac{3}{8}\frac{\lambda_i^* \lambda_j}{M^2},$$

$$Q_d: \quad [C_{Hd}]_{ij} = -\frac{1}{2}\frac{\lambda_i \lambda_j^*}{M^2},$$

$$Q_V: \quad [C_{Hd}]_{ij} = \frac{1}{2}\frac{\lambda_i^{V_d} \lambda_j^{V_d*}}{M^2}, \quad [C_{Hu}]_{ij} = -\frac{1}{2}\frac{\lambda_i^{V_u} \lambda_j^{V_u*}}{M^2}, \quad [C_{Hud}]_{ij} = \frac{\lambda_i^{V_u} \lambda_j^{V_d*}}{M^2}. \tag{16.46}$$

Using these results and the technology of Section 14.4, we find the contributions to the Wilson coefficients of $\Delta F = 2$ operators from the Yukawa RG effects

$$C_a^{ij}(\mu_{\text{ew}}) = \frac{\kappa_m}{(4\pi)^2}\frac{\Lambda_{ij}^m \lambda_{ij}^{(t)}}{N_{ij}}\frac{1}{M^2}\frac{2m_t^2}{v^2}\ln\frac{\mu_M}{\mu_{\text{ew}}}, \tag{16.47}$$

with Λ_{ij}^m in (16.42) and (16.43), the chirality of the $|\Delta F| = 2$ operator

$$
\begin{aligned}
a &= \text{VLL} &\quad \text{for} &\quad F_m = D, T_d, T_u, \\
a &= \text{LR}, 1 &\quad \text{for} &\quad F_m = Q_d, Q_V,
\end{aligned}
\tag{16.48}
$$

and the VLQ-model-dependent factor

$$
\kappa_m = \left(0, \ -\frac{1}{2}, \ +\frac{1}{2}, \ -\frac{1}{2}, \ +\frac{1}{4}\right) \quad \text{for} \quad F_m = (D, \ Q_d, \ Q_V, \ T_d, \ T_u). \tag{16.49}
$$

We note the relations

$$
\kappa_m \frac{\Lambda_{ij}^m}{M^2} = [C_{Hq}^{(1)} - C_{Hq}^{(3)}]_{ij}, \qquad (F_m = D, T_d, T_u), \tag{16.50}
$$

and

$$
\kappa_m \frac{\Lambda_{ij}^m}{M^2} = [C_{Hd}]_{ij}, \qquad (F_m = Q_d, Q_V). \tag{16.51}
$$

We point out the different flavor structure of the 1stLLA contribution (16.47) compared to the one of the direct box contribution (16.40) discussed previously:

$$
C_a^{ij}|_{\text{1stLLA}} \ \sim \ \Lambda_{ij} \times \lambda_{ij}^{(t)}, \qquad\qquad C_b^{ij}|_{\text{Box}} \ \sim \ (\Lambda_{ij})^2, \tag{16.52}
$$

showing linear versus quadratic dependence on the product of VLQ Yukawa couplings Λ_{ij}.

Yukawa RG Effects versus Box Contributions

With only the box and tree-level Z contributions taken into account the $|\Delta F| = 2$ observables are not sensitive to the chirality of the VLQ interactions as long as only one VLQ representation is present. However, the inclusion of RG Yukawa effects just presented and NLO contributions discussed in [766] changes this picture drastically in the case of VLQ models with flavor-changing RH currents (Q_d, Q_V) and has also significant impact in the remaining three models with LH currents.

In the case of D, T_d, and T_u models, we find

$$
\left[\frac{(\Delta S)_{\text{RG}}}{(\Delta S)_{\text{Box}}}\right]^{ij} = \frac{\kappa_m}{\eta_{mm}} \frac{\lambda_{ij}^t}{\Lambda_{ij}^m} \frac{2m_t^2}{v^2 \eta_6^{2/7}} \left[\ln\frac{\mu_M}{\mu_{\text{EW}}} + \frac{F_{\text{NLO}}(x_t, \mu_{\text{EW}})}{\kappa_m \Lambda_{ij}^m}\right], \tag{16.53}
$$

with κ_m given in (16.49) and η_{mm} in Table 16.3. The NLO correction

$$
\begin{aligned}
F_{\text{NLO}}(x_t, \mu_{\text{EW}}) = {}&[C_{Hq}^{(1)}]_{ij} H_1(x_t, \mu_{\text{ew}}) - [C_{Hq}^{(3)}]_{ij} H_2(x_t, \mu_{\text{ew}}) \\
&+ \frac{2S_0(x_t)}{x_t} \sum_m \left(\lambda_t^{im}[C_{Hq}^{(3)}]_{mj} + [C_{Hq}^{(3)}]_{im}\lambda_t^{mj}\right)
\end{aligned}
\tag{16.54}
$$

has been calculated in [766], where also the x_t-dependent functions $H_{1,2}$ can be found.

In the case of Q_d and Q_V models, the box and RG contributions yield coefficients to different operators, hence a meaningful comparison of their impact on observables has to include their QCD running between μ_{ew} and the light flavor scales (we choose 3 GeV for Kaons and M_B for $B_{d,s}$), as well as the corresponding matrix elements. We find

$$\left[\frac{(M_{12}^*)_{\text{RG}}}{(M_{12}^*)_{\text{Box}}}\right]^{ij} = \left[\frac{(M_{12}^*)_{\text{LR}}}{(M_{12}^*)_{\text{VRR}}^{\text{box}}}\right]^{ij} = \frac{\kappa_m}{\eta_{mm}} \frac{\lambda_{ij}^t}{\Lambda_{ij}^m} \frac{2m_t^2}{v^2 \eta_6^{2/7}} \left[\ln\frac{\mu_M}{\mu_{\text{EW}}} + H_1(x_t, \mu_{\text{EW}})\right] R^{ij}, \quad (16.55)$$

with R^{ij} including RG factors and the ratio of the hadronic matrix elements. From Eqs. (60) and (61) in [766], we obtain

$$R^{sd} \approx -80 \qquad \text{and} \qquad R^{b(d,s)} \approx -3. \qquad (16.56)$$

This large chiral enhancement in the Kaon system renders the RG contribution dominant, while in the $B_{d,s}$ systems the contribution remains comparable with the box contribution.

16.3.4　Structure of the Phenomenology

The free parameters relevant for the study of K, B_s, and B_d meson systems in a given VLQ model are the complex Yukawa couplings:

$$\lambda_d, \qquad \lambda_s, \qquad (K), \qquad\qquad\qquad (16.57)$$

$$\lambda_b, \qquad \lambda_s, \qquad (B_s), \qquad\qquad\qquad (16.58)$$

$$\lambda_b, \qquad \lambda_d, \qquad (B_d) \qquad\qquad\qquad (16.59)$$

and the mass M of the VLQs, which we assume to be common for all members of the VLQ multiplet.

Thus every model contains three complex Yukawa couplings:

$$\boxed{\lambda_d = |\lambda_d|e^{i\phi_d}, \qquad \lambda_s = |\lambda_s|e^{i\phi_s}, \qquad \lambda_b = |\lambda_b|e^{i\phi_b},} \qquad (16.60)$$

which enter flavor observables in K, B_d, and B_s systems in the case of models with LH currents through three products

$$\Lambda_{sd} = \lambda_s^* \lambda_d = A_{sd}e^{-i\phi_{sd}}, \ \Lambda_{bd} = \lambda_b^* \lambda_d = A_{bd}e^{-i\phi_{bd}}, \ \Lambda_{bs} = \lambda_b^* \lambda_s = A_{bs}e^{-i\phi_{bs}}, \qquad (16.61)$$

respectively. Here

$$\boxed{\phi_{ij} = \phi_i - \phi_j \quad \Rightarrow \quad \phi_{bs} = \phi_{bd} - \phi_{sd}.} \qquad (16.62)$$

For models with RH currents just complex conjugation in the definition of Λ_{ij} has to be made. See (16.42) and (16.43).

The A_{ij} are positive definite real parameters, and the phases ϕ_{ij} can vary in the full range $[0, 2\pi]$. The latter fact implies discrete phase ambiguities when these phases are determined from experiment as explicitly seen in the plots in [611, 666]. As only the differences of phases are involved, without lost of generality, we can set one of them to zero. Setting then $\phi_d = 0$ we find

$$|\lambda_d| = \sqrt{\frac{A_{bd}A_{sd}}{A_{bs}}}, \qquad |\lambda_s| = \sqrt{\frac{A_{bs}A_{sd}}{A_{bd}}}, \qquad |\lambda_b| = \sqrt{\frac{A_{bs}A_{bd}}{A_{sd}}} \qquad (16.63)$$

and

$$\phi_d = 0, \qquad\qquad \phi_s = \phi_{sd}, \qquad\qquad \phi_b = \phi_{bd}. \qquad (16.64)$$

These couplings enter $|\Delta F| = 2$ observables, and one might expect strongest constraints on A_{sd} because of the strong suppression of the SM contribution by $V_{td}V_{ts}^*$, whereas λ_b in turn is constrained by $|\Delta B| = 2$ measurements.

In a sense the flavor structure of VLQ models has some parallels to the one in 331 models discussed in the previous section. However, in 331 models the NP contributions are dominated by Z' tree-level exchanges, and once the constraints from $B_{s,d}$ observables are taken into account, NP effects in the K system are found to be small, with the exception of ε'/ε. In VLQ models important Z boson contributions are present, and this allows for more interesting NP effects than in 331 models in $K^+ \to \pi^+ \nu \bar{\nu}$ and $K_L \to \pi^0 \nu \bar{\nu}$, as we will see soon. Furthermore, the presence of important box diagram contributions to $|\Delta F| = 2$ processes and of Yukawa RG effects in VLQ models discussed earlier increases the impact of $|\Delta F| = 2$ constraints on $|\Delta F| = 1$ processes relative to the ones found in 331 models.

16.3.5 Patterns of NP in VLQ Models

We will now summarize various lessons gained from the extensive analysis of VLQ models in [666]. The interested readers will find numerous plots in this paper that show visually the patterns listed next. The patterns are similar to the ones encountered in Z scenarios of Section 15.5, but the models here are more concrete and allow some correlations between different meson systems.

Lesson 1: The VLQ models allow to enhance ε'/ε significantly, thereby addressing the apparent gap between the SM prediction and data, at the expense of suppressing $\mathcal{B}(K_L \to \pi^0 \nu \bar{\nu})$. This suppression is significantly weaker for Q_V and Q_d models (RH currents) than for D, T_d, and T_u (LH currents), in accordance with the general study in [851]. Simultaneous agreement with the data for ε_K and ε'/ε can be obtained without fine-tuning of parameters.

Lesson 2: While the impact of ε'/ε on $K_L \to \pi^0 \nu \bar{\nu}$ is large as just stated, $K^+ \to \pi^+ \nu \bar{\nu}$ and ε'/ε are only weakly correlated. However, in RH models ϵ_K prevents large enhancements of $\mathcal{B}(K^+ \to \pi^+ \nu \bar{\nu})$, the maximal enhancement is about 50% of its SM value. In models with LH currents, a strong suppression is possible, and the SM value corresponds to an upper bound in this case when a stricter bound from $K_L \to \mu \bar{\mu}$ is used. This implies that a measurement of a significantly enhanced $\mathcal{B}(K^+ \to \pi^+ \nu \bar{\nu})$, as presently still allowed by data, could exclude all VLQ models with a single VLQ representation, although in models with LH currents a more conservative bound from $K_L \to \mu \bar{\mu}$ would presently still allow the enhancement of $\mathcal{B}(K^+ \to \pi^+ \nu \bar{\nu})$ up to a factor of two. In this context it should be again emphasized that the modes $K^+ \to \pi^+ \nu \bar{\nu}$ and $K_L \to \mu \bar{\mu}$ are strongly correlated in VLQ models, however, again differently so for LH and RH currents. While for RH currents one can easily infer the allowed range in one mode from a determination of the other, within the limited range allowed by ε'/ε and ϵ_K, LH-current models are more strongly constrained from $K_L \to \mu \bar{\mu}$. But there is basically no correlation between ε'/ε and $K_L \rightarrow \mu\bar\mu$, as they are governed by imaginary and real parts of the corresponding couplings, respectively.

Lesson 3: The VLQ mass does not have a large impact on all these correlations, as can be seen in the plots in [666].

Moving then to $b \rightarrow s$ processes one finds.

Lesson 4: Because NP effects in all three quark transitions are governed by different parameters, the slight tensions in $|\Delta F| = 2$ observables hinted at by lattice data [873] can easily be removed in VLQ models. This is in contrast to constrained-MFV models, where ε_K prohibits large effects in $\Delta M_{d,s}$ [388].

Lesson 5: $\mathcal{B}(B_s \rightarrow \mu\bar\mu)$ can be strongly suppressed below its SM value, as slightly favored by experiment, while still allowing for sizable NP effects in $\sin(2\beta_s)$, in particular in the case of models with LH currents. For $M_{\mathrm{VLQ}} = 1$ TeV $|\Delta F| = 1$ observables constrain the NP effects in ϕ_s to be smaller than for larger VLQ masses.

Lesson 6: Sizable deviations from the SM prediction are still possible for the mass-eigenstate rate asymmetry $A_{\Delta\Gamma}(B_s \rightarrow \mu\bar\mu)$ and the mixing-induced CP-asymmetry $S^s_{\mu\mu}$. Indeed, both can essentially vary in the full range $[-1, 1]$ for LH models for $M_{\mathrm{VLQ}} = 1$ TeV. For RH models, $A_{\Delta\Gamma}(B_s \rightarrow \mu\bar\mu) \geq 50\%$ for $M_{\mathrm{VLQ}} = 1$ TeV, but still $|S^s_{\mu\mu}|$ can reach up to 80%. For $M_{\mathrm{VLQ}} = 10$ TeV, the former is restricted to positive values in both LH and RH models, the latter slightly stronger constrained in RH models, but not in LH ones.

Lesson 7: CP-violating quantities are almost 100% correlated in $b \rightarrow s$ transitions as long as only one representation is considered. The reason is that the SM predictions are tiny, and all NP contributions therefore directly proportional to the imaginary part of Λ_{bs}, which hence cancels in the ratio of two CP-violating quantities. For small NP contributions, the asymmetries are simply proportional to each other, for larger effects the relation depends on the normalization of the asymmetry. These statements hold not only in VLQ models, but in all models that provide only a single new phase in $b \rightarrow s$ transitions, only the proportionality constant changes in other models.

Lesson 8: Interesting CP-violating effects in various asymmetries due to imaginary parts of $b \rightarrow s\mu\bar\mu$ Wilson coefficients $C_{9,9',10,10'}$ have been identified. The correlations between these asymmetries provide transparent distinction between LH and RH currents.

Lesson 9: NP effects in $B \rightarrow K^{(*)}\nu\bar\nu$ turn out to be small in RH scenario. In LH scenario basically only suppression of branching ratios is possible.

Lesson 10: Similar to the case of Z models the anomalies in $b \rightarrow \ell^+\ell^-$ and $b \rightarrow c\tau\bar\nu_\tau$ cannot be explained in these models.

Having the latter anomalies in mind the paper [666] considered also VLQ models with an additional Z' related to $\mathrm{U}(1)_{L_\mu - L_\tau}$. Here the results depend on the Higgs system necessary to generate the mass of Z'. The main lessons are as follows.

Lesson 11: In a model with an additional singlet, as considered in [925], $b \rightarrow s\mu^+\mu^-$ anomalies can be partly explained by providing sufficient suppression of the coefficient C_9, but NP effects in $B_{s,d} \rightarrow \mu\bar\mu$ and $K_L \rightarrow \mu\bar\mu$ are absent, those in $b \rightarrow s\nu\bar\nu$ transitions small and the ones in $K^+ \rightarrow \pi^+\nu\bar\nu$ and $K_L \rightarrow \pi^0\nu\bar\nu$ much smaller than in the VLQ models discussed earlier. Most important these models fail badly in explaining the ε'/ε anomaly. This pattern is related to the absence of FCNC processes mediated by Z and to pure vector coupling of Z' to charged leptons.

However, adding an additional scalar doublet Φ FCNC processes mediated by Z can be generated [666], and the patterns of NP are modified as we describe now.

Lesson 12: While the explanation of $b \rightarrow s\mu^+\mu^-$ anomalies is more difficult than in models just discussed, due to the presence of Z contributions interesting effects in other observables can be found. In particular, NP effects in ε'/ε *and* $K^+ \rightarrow \pi^+\nu\bar{\nu}$ can be large, in contrast to VLQ models based on the SM gauge group.

Lesson 13: NP effects in ΔM_K can be larger and correlated with $K^+ \rightarrow \pi^+\nu\bar{\nu}$. In particular if in the future the ΔM_K constraint will be improved, large enhancements of $\mathcal{B}(K^+ \rightarrow \pi^+\nu\bar{\nu})$ are likely to be excluded. On the other hand, NP effects in $K_L \rightarrow \pi^0\nu\bar{\nu}$, $K_L \rightarrow \mu\bar{\mu}$, $B \rightarrow K(K^*)\nu\bar{\nu}$, and $B_{d,s} \rightarrow \mu\bar{\mu}$ are very small and beyond the reach of even presently planned future facilities.

Thus if NP will be found in $B_{s,d} \rightarrow \mu\bar{\mu}$ and the ε'/ε-anomaly will be fully established in the future, the VLQ models based on the SM gauge group would offer the best explanation among VLQ models. If, on the other hand, the B physics anomalies will be confirmed in the future and no visible NP will be found in rare K decays, VLQ models with $U(1)_{L_\mu - L_\tau}$ would be favored.

Lesson 14: A large enhancement of $\mathcal{B}(K^+ \rightarrow \pi^+\nu\bar{\nu})$ would uniquely select RH VLQ models with a second doublet Φ subject to the future status of ΔM_K, although LH models with SM gauge group and the ones with $U(1)_{L_\mu - L_\tau}$ and Φ could provide a moderate enhancement, in case of the latter depending on the theoretical treatment of $K_L \rightarrow \mu\bar{\mu}$. On the other hand, a large enhancement of $\mathcal{B}(B \rightarrow K^{(*)}\nu\bar{\nu})$ would disfavor all considered models, at least with only one VLQ representation. Some improvements are possible in the presence of several representations.

While the discovery of VLQs at the LHC would give a strong impetus to the models considered here, nonobservation of them at the LHC would not preclude their importance for flavor physics. In fact, as has been shown in [666], large NP effects in flavor observables can be present for $M_{\mathrm{VLQ}} = 10$ TeV, and in the flavor-precision era one is sensitive to even higher scales. In this context one finds that the combination of $|\Delta F| = 2$ and $|\Delta F| = 1$ observables in a given meson system generally allows to determine the masses of VLQs in a given representation independent of the size of Yukawa couplings.

Finally, those readers who got motivated to construct new UV completions with VLQs should first read [1033].

16.3.6 Vector-Like Leptons

According to the classification in [967] there are six renormalizable models with vector-like leptons (VLL) in a single complex representation of the SM gauge group. The literature on these models is not as rich as on VLQ and is dominated by the desire to explain the $(g-2)_\mu$ anomaly [943, 1034–1038]. For collider signatures see, e.g., [1039]. We can recommend also the analysis in [1040], where a model with a complete vector-like fourth family and additional $U(1)'$ symmetry has been proposed. One can then see VLQs, VLLs, and Z' together in action with the goal to explain the $(g-2)_\mu$ anomaly and $b \rightarrow s\ell^+\ell^-$ anomalies in the context of a rich flavor structure. We hope to describe the physics of VLLs in the second edition of this book.

16.4 Leptoquark Models

16.4.1 Introduction

Leptoquarks (LQs) are very special new particles as they directly couple quarks to leptons and consequently carry both baryon and lepton number, B and L. Moreover, they are strongly interacting and carry fractional electric charges, but in contrast to quarks and leptons they are bosons: either scalars or vectors [1041–1044]. Because of these rather special properties of LQs the phenomenological implications of models containing them are markedly different from the ones where new scalars and gauge bosons couple directly only leptons to leptons and quarks to quarks. They also differ from those just discussed in VLQ models.

Indeed quite generally the pattern of flavor violations within LQ models is as follows.

- The semileptonic and leptonic decays of mesons are privileged as in these models they can naturally appear already at tree-level. This applies in particular to many rare decays which are loop suppressed within the SM. Therefore, the presence of departures from SM expectations for such decays can naturally be explained in some LQ models.
- On the other hand, all nonleptonic decays of mesons and also purely leptonic processes are loop-suppressed within LQ models.
- In consequence, large LQ effects in semileptonic and leptonic decays of mesons do not necessarily imply large modifications of SM predictions for nonleptonic observables for which the SM, with few notable exceptions like ε'/ε and $\Delta M_{d,s}$, offers a good description of the data. On the other hand, in view of a very strong suppression of FCNC in purely leptonic processes within the SM, still large LQ effects in these processes can be found despite their loop suppression providing thereby strong constraints on LQ models.
- Moreover, it should be emphasized that in LQ models in which the only new particles are LQs, the fermions exchanged in the loops are leptons in nonleptonic meson decays but SM quarks in the case of purely leptonic processes. Therefore Z-penguin diagrams that are important for several nonleptonic FCNC processes in the SM because of the large top-quark mass are very strongly suppressed in LQ models by the ratios of lepton masses and LQ masses. In consequence, in these models Z-penguin diagrams do not play any role in nonleptonic processes but can be important in purely leptonic processes in which quarks, in particular the top-quark, can be exchanged in the loop.
- Nonleptonic decays in LQ models are then expected at first sight to be governed by gluon-penguins rather than Z-penguins but these effects turn out to be subleading. Indeed, as pointed out in [667], in the context of ε'/ε the dominant contributions to this ratio originate from the electroweak effects in the renormalization group evolution between LQ scale and electroweak scale. This evolution generates through mixing four-quark operators contributing to ε'/ε from semileptonic operators, characteristic for LQ models. In addition in models with both left-handed and right-handed couplings enhanced box diagram contributions to ε'/ε are possible. However, as demonstrated in [667] and presented at the end of this section, when constraints from other processes

are taken into account, these effects are unlikely to be able to explain the ε'/ε-anomaly discussed in Section 10.5.

The suppression of LQ contributions to nonleptonic transitions in LQ models is certainly useful for those nonleptonic observables for which the SM predictions agree well with data, as strong constraints on LQ parameters from them can be avoided, making the explanation of B physics anomalies in semileptonic decays easier. On the other hand, as mentioned already in the context of ε'/ε, in cases in which the SM would have problems with explaining the data on nonleptonic decays and transitions, LQs are expected to be challenged because bringing the theory to agree with data will likely imply too large deviations in at least some tree-level mediated decays.

There is a very rich literature on LQs with extensive analyses of collider, electroweak, and flavor-violating processes among others. In particular, a model-independent Lagrangian of LQs that is invariant under the SM gauge group $\mathrm{SU}(3)_C \otimes \mathrm{SU}(2)_L \otimes \mathrm{U}(1)_Y$ and conserves B and L was given first in [1041] (see also [1042]). Extensive phenomenology based on this Lagrangian can be found in [1043]; however, many aspects were already discussed before throughout the literature. The analyses in [1045–1048] are particularly useful and informative. For reviews see [1044, 1049] and section 7 of [305].

The goal of the present section is to summarize the properties of leptoquarks, five scalar leptoquarks, and five vector leptoquarks. We give the transformation properties of them under the SM gauge group and present the model-independent, as far as Yukawa couplings are concerned, Lagrangian from which general flavor properties of the 10 leptoquarks can be deduced. In this context we allow leptoquarks to couple to all generations of ordinary quarks and leptons. Even if this increases the number of free parameters, it is useful to present the general formulas for various observables without any particular assumptions on Yukawa couplings. On the other hand, we will confine a detailed discussion of various observables to five models that were studied extensively in the literature. These are three scalar leptoquark models with leptoquarks transforming as doublets (R_2, $\tilde{R}_2$) and triplet (S_3) under the $\mathrm{SU}(2)_L$ and two vector leptoquark models with leptoquarks transforming as singlet (U_1) and a triplet (U_3) under $\mathrm{SU}(2)_L$. The interest in $\tilde{R}_2$, S_3, and U_3 models is that all three provide NP contributions to most interesting rare decays $K^+ \to \pi^+ \nu \bar{\nu}$, $K_L \to \pi^0 \nu \bar{\nu}$, and $K_{L,S} \to \mu^+ \mu^-$ and those governed by the transitions $b \to s \mu^+ \mu^-$, $b \to s \nu \bar{\nu}$, and $b \to c \ell \nu$, in particular the decays $B_d \to D(D^*) \tau \nu_\tau$. As we have seen in previous sections, all these transitions are of great interest in view of new measurements to be performed in this decade and expected advances in the theory.

The remaining two models, R_2 and U_1, are special as in these models LQs do not contribute to $K_L \to \pi^0 \nu \bar{\nu}$, $K^+ \to \pi^+ \nu \bar{\nu}$, and $b \to s \nu \bar{\nu}$ transitions at tree-level. Even if RG effects generate such transitions and they can be generated through one-loop matching, this feature allows to face easier the constraints from decays with $\nu \bar{\nu}$ in the final state than it is possible in other models.

Having explicit formulas at hand will allow us to have another look at the anomalies observed in the $b \to s \mu^+ \mu^-$ transitions but this time in LQ models. $K^+ \to \pi^+ \nu \bar{\nu}$ and $K_L \to \pi^0 \nu \bar{\nu}$ are of great interest in view of new measurements of their branching ratios by NA62 and KOPIO and $b \to s \nu \bar{\nu}$ transitions in view of Belle II experiments. NP in these models enters also $\Delta F = 2$ transitions and the ratio ε'/ε. Moreover, in some of these observables departures from SM expectations have been found, and in fact an increased interest in LQs arose from the fact that some of these anomalies could be attributed to these new particles.

We will only briefly discuss the remaining models and hybrid scenarios in which two or more leptoquarks in different representations are present exhibiting new features relative to the five models listed earlier. An important issue in LQ models is the proton decay that in some models takes place already at tree-level excluding such models for LQ masses of $O(1 \text{ TeV})$ unless some discrete symmetries are imposed.

The material presented in this section originated in an extensive collaboration with Christoph Bobeth, which includes not only the aspects related to [667] but also beyond it.

16.4.2 Model-Independent Leptoquark Lagrangian

Quantum Numbers and the Basic Lagrangian

The LQs are singlets, doublets, or triplets w.r.t. to $SU(2)_L$. There quantum numbers and transformation properties under the SM gauge group are collected in Table 16.4 with our convention

$$Q = T_3 + Y. \tag{16.65}$$

The superscript $i = (1, 2, 3)$ corresponds to the $SU(2)_L$ multiplet (singlet, doublet, triplet) to which the LQs belong. All color-antitriplets (S_1, $\tilde{S}_1$, S_3, V_2, $\tilde{V}_2$) carry fermion number

$$F = 3B + L = -2, \tag{16.66}$$

whereas all color-triplets (U_1, $\tilde{U}_1$, U_3, R_2, $\tilde{R}_2$) carry $F = 0$.

The phenomenological LQ Lagrangian that describes the interactions of spin-0 (S) and spin-1 (V) LQs with quarks and leptons is given as follows

$$\mathcal{L}_{\text{LQ}} = \mathcal{L}_S + \mathcal{L}_V + \text{h.c.,} \tag{16.67}$$

where adopting the conventions from [1050] we have

$$
\begin{aligned}
\mathcal{L}_S = {} & \left(g^{1L} \left[\overline{Q_L^c} i\sigma_2 L_L \right] + g^{1R} \left[\overline{u_R^c} \ell_R \right] \right) S_1 + \tilde{g}^{1R} \left[\overline{d_R^c} \ell_R \right] \tilde{S}_1 \\
& + h^{2L} \left[\overline{u_R} R_2^T i\sigma_2 L_L \right] + h^{2R} \left[\overline{Q_L} R_2 \ell_R \right] + \tilde{h}^{2L} \left[\overline{d_R} \tilde{R}_2^T i\sigma_2 L_L \right] \\
& + g^{3L} \left[\overline{Q_L^c} i\sigma_2 \vec{\tau} L_L \right] \cdot \vec{S}_3,
\end{aligned}
\tag{16.68}
$$

$$
\begin{aligned}
\mathcal{L}_V = {} & \left(h^{1L} \left[\overline{Q_L} \gamma_\mu L_L \right] + h^{1R} \left[\overline{d_R} \gamma_\mu \ell_R \right] \right) U_1^\mu + \tilde{h}^{1R} \left[\overline{u_R} \gamma_\mu \ell_R \right] \tilde{U}_1^\mu \\
& + g^{2L} \left[\overline{d_R^c} \gamma_\mu (V_2^\mu)^T i\sigma_2 L_L \right] + g^{2R} \left[\overline{Q_L^c} \gamma_\mu i\sigma_2 V_2^\mu \ell_R \right] + \tilde{g}^{2L} \left[\overline{u_R^c} \gamma_\mu (\tilde{V}_2^\mu)^T i\sigma_2 L_L \right] \\
& + h^{3L} \left[\overline{Q_L} \gamma_\mu \vec{\tau} L_L \right] \cdot \vec{U}_3^\mu .
\end{aligned}
\tag{16.69}
$$

Table 16.4 Quantum numbers of interactions of LQs.[a]

LQ	$SU(3)_c$	$SU(2)_L$	$U(1)_Y$	T_3	Q_{em}	$d_i, \ell^+\ell^-$	$d_i, \nu\bar{\nu}$	$u_i, \ell^+\ell^-$	$u_i, \nu\bar{\nu}$
S_1	3^*	1	1/3	0	1/3		*	*	
$\tilde{S}_1$	3^*	1	4/3	0	4/3	*			
R_2	3	2	7/6	+1/2	5/3			*	
				−1/2	2/3	*			*
$\tilde{R}_2$	3	2	1/6	+1/2	2/3	*			
				−1/2	−1/3		*		
S_3	3^*	3	1/3	+1	4/3	*			
				0	1/3		*	*	
				−1	−2/3				*
U_1	3	1	2/3	0	2/3	*			*
$\tilde{U}_1$	3	1	5/3	0	5/3			*	
V_2	3^*	2	5/6	+1/2	4/3	*			
				−1/2	1/3		*	*	
$\tilde{V}_2$	3^*	2	−1/6	+1/2	1/3			*	
				−1/2	−2/3				*
U_3	3	3	2/3	+1	5/3			*	
				0	2/3	*			*
				−1	−1/3		*		

[a] The stars indicate to which procesess a given LQ contributes to at tree-level. Note that antitriplets 3^* interact with charge conjugated quark fields.

The $A = (L, R)$-superscript in the couplings y^{iA} ($y = g, \tilde{g}, h, \tilde{h}$) indicates the chirality of the leptons in the coupling, and σ_i are Pauli matrices given again in (16.86). The $\vec{\tau}$, relevant for triplet representations, are defined in (16.87). Q^c, u^c, d^c are charge-conjugated fields.

The fermion fields are the usual quark and lepton left-handed $SU(2)_L$ doublets Q_L and L_L as well as the right-handed singlets u_R, d_R, and ℓ_R. For convenience we recall their hypercharges

$$Y(Q_L) = +1/6, \quad Y(L_L) = -1/2, \quad Y(u_R) = +2/3, \quad Y(d_R) = -1/3, \quad Y(\ell_R) = -1. \tag{16.70}$$

If ν_R is included additional terms appear in the Lagrangians [1042].

We will now expand the $SU(2)_L$ structure to make explicit the interactions of up- and down-type quarks and leptons with $SU(2)_L$ components of the LQ fields. For this purpose, we denote components of the LQ doublet $X_2 = (R_2, \tilde{R}_2, V_2, \tilde{V}_2)$ and triplet $X_3 = (S_3, U_3)$ fields as

$$(X_2)_a = X_{2,a}, \qquad\qquad a = (+, -), \tag{16.71}$$
$$(X_3)_a = X_{3,a}, \qquad\qquad a = (1, 2, 3). \tag{16.72}$$

Note that the triplet fields correspond to a Cartesian representation with complex fields $X_{3,a}$. To have a representation where the components correspond to a definite isospin, one has to transform the fields to the spherical basis $Y_{3,i}$:

$$Y_{3,\pm} = \frac{1}{\sqrt{2}}(X_{3,1} \mp iX_{3,2}), \qquad\qquad Y_{3,0} = X_{3,3}, \qquad (16.73)$$

where $i = (+, 0, -)$ denotes isospin $(+1, 0, -1)$ for $Y_{3,i}$. The Cartesian and spherical bases are related via the unitary transformation

$$Y = U^\dagger X, \qquad\qquad U = \frac{1}{\sqrt{2}} \begin{pmatrix} 1 & 0 & 1 \\ i & 0 & -i \\ 0 & \sqrt{2} & 0 \end{pmatrix}. \qquad (16.74)$$

Lagrangian in SU(2)$_L$ Components

Rewriting the Lagrangian in (16.68) in terms of SU(2)$_L$ components we find

$$\mathcal{L}_{S_1} = \left\{ g^{1L} \left([\overline{u^c}P_L\ell] - [\overline{d^c}P_L\nu] \right) + g^{1R} \ [\overline{u^c}P_R\ell] \right\} S_1 \\ + \left\{ (g^{1L})^\dagger \left([\bar{\ell}P_R u^c] - [\bar{\nu}P_R d^c] \right) + (g^{1R})^\dagger [\bar{\ell}P_L u^c] \right\} S_1^*, \qquad (16.75)$$

$$\mathcal{L}_{\tilde{S}_1} = \tilde{g}^{1R}[\overline{d^c}P_R\ell]\tilde{S}_1 + (\tilde{g}^{1R})^\dagger[\bar{\ell}P_L d^c]\tilde{S}_1^*, \qquad (16.76)$$

$$\mathcal{L}_{R_2} = h^{2L} \left([\bar{u}P_L\ell]R_{2,+} - [\bar{u}P_L\nu]R_{2,-} \right) + h^{2R} \left([\bar{u}P_R\ell]R_{2,+} + [\bar{d}P_R\ell]R_{2,-} \right) \\ + (h^{2L})^\dagger \left([\bar{\ell}P_R u]R_{2,+}^* - [\bar{\nu}P_R u]R_{2,-}^* \right) + (h^{2R})^\dagger \left([\bar{\ell}P_L u]R_{2,+}^* + [\bar{\ell}P_L d]R_{2,-}^* \right), \qquad (16.77)$$

$$\mathcal{L}_{\tilde{R}_2} = \tilde{h}^{2L} \left([\bar{d}P_L\ell]\tilde{R}_{2,+} - [\bar{d}P_L\nu]\tilde{R}_{2,-} \right) + (\tilde{h}^{2L})^\dagger \left([\bar{\ell}P_R d]\tilde{R}_{2,+}^* - [\bar{\nu}P_R d]\tilde{R}_{2,-}^* \right), \qquad (16.78)$$

$$\mathcal{L}_{S_3} = g^{3L} \left[\sqrt{2}\,[\overline{u^c}P_L\nu]S_{3,-} - \sqrt{2}\,[\overline{d^c}P_L\ell]S_{3,+} - (\overline{u^c}P_L\ell + \overline{d^c}P_L\nu)S_{3,0} \right] \\ + (g^{3L})^\dagger \left[\sqrt{2}\,[\bar{\nu}P_R u^c]S_{3,-}^* - \sqrt{2}\,[\bar{\ell}P_R d^c]S_{3,+}^* - (\bar{\ell}P_R u^c + \bar{\nu}P_R d^c)S_{3,0}^* \right]. \qquad (16.79)$$

For the vector leptoquarks we obtain from (16.69)

$$\mathcal{L}_{U_1} = \left\{ h^{1L} \left([\bar{u}\gamma_\mu P_L\nu] + [\bar{d}\gamma_\mu P_L\ell] \right) + h^{1R} \ [\bar{d}\gamma_\mu P_R\ell] \right\} U_1^\mu \\ + \left\{ (h^{1L})^\dagger \left([\bar{\nu}\gamma_\mu P_L u] + [\bar{\ell}\gamma_\mu P_L d] \right) + (h^{1R})^\dagger [\bar{\ell}\gamma_\mu P_R d] \right\} U_1^{\mu,*}, \qquad (16.80)$$

$$\mathcal{L}_{\tilde{U}_1} = \tilde{h}^{1R}[\bar{u}\gamma_\mu P_R \ell]\tilde{U}_1^\mu + (\tilde{h}^{1R})^\dagger[\bar{\ell}\gamma_\mu P_R u]\tilde{U}_1^{\mu,*}, \tag{16.81}$$

$$\mathcal{L}_{V_2} = g^{2L}\left([\overline{d^c}\gamma_\mu P_L \ell]V_{2,+}^\mu - [\overline{d^c}\gamma_\mu P_L \nu]V_{2,-}^\mu\right) + g^{2R}\left([\overline{u^c}\gamma_\mu P_R \ell]V_{2,-}^\mu - [\overline{d^c}\gamma_\mu P_R \ell]V_{2,+}^\mu\right)$$
$$+ (g^{2L})^\dagger\left([\bar{\ell}\gamma_\mu P_L d^c]V_{2,+}^{\mu,*} - [\bar{\nu}\gamma_\mu P_L d^c]V_{2,-}^{\mu,*}\right) + (g^{2R})^\dagger\left([\bar{\ell}\gamma_\mu P_R u^c]V_{2,-}^{\mu,*} - [\bar{\ell}\gamma_\mu P_R d^c]V_{2,+}^{\mu,*}\right), \tag{16.82}$$

$$\mathcal{L}_{\tilde{V}_2} = \tilde{g}^{2L}\left([\overline{u^c}\gamma_\mu P_L \ell]\tilde{V}_{2,+}^\mu - [\overline{u^c}\gamma_\mu P_L \nu]\tilde{V}_{2,-}^\mu\right) + (\tilde{g}^{2L})^\dagger\left([\bar{\ell}\gamma_\mu P_L u^c]\tilde{V}_{2,+}^{\mu,*} - [\bar{\nu}\gamma_\mu P_L u^c]\tilde{V}_{2,-}^{\mu,*}\right), \tag{16.83}$$

$$\mathcal{L}_{U_3} = h^{3L}\left[\sqrt{2}\,[\bar{u}\gamma_\mu P_L \ell]U_{3,+}^\mu + \sqrt{2}\,[\bar{d}\gamma_\mu P_L \nu]U_{3,-}^\mu + (\bar{u}\gamma_\mu P_L \nu - \bar{d}\gamma_\mu P_L \ell)U_{3,0}^\mu\right]$$
$$+ (h^{3L})^\dagger\left[\sqrt{2}\,[\bar{\ell}\gamma_\mu P_L u]U_{3,+}^{\mu,*} + \sqrt{2}\,[\bar{\nu}\gamma_\mu P_L d]U_{3,-}^{\mu,*} + (\bar{\nu}\gamma_\mu P_L u - \bar{\ell}\gamma_\mu P_L d)U_{3,0}^{\mu,*}\right]. \tag{16.84}$$

We summarize all interactions in Table 16.4. Their impact on flavor observables in the case of R_2, $\tilde{R}_2$, S_3, U_1, and U_3 models will be worked out in Section 16.4.4. General properties of the remaining models will be summarized at the end of this section.

It should be emphasized that the preceding formulas are in the flavor eigenbasis of quarks and leptons, and rotations to the mass eigenbasis have to be performed, depending on the mass generation mechanism of quarks and leptons, i.e., for example, in the SM from diagonalization of Yukawa couplings with one Higgs doublet and via SSB – bringing in potential factors from the CKM and PMNS matrices. See, for example, eqs. (2.2) and (2.6) in [1046]. This issue has been already discussed in the context of vector-like quarks with explicit formulas given in (16.36) and (16.37). As there, we will take the freedom to choose the weak basis such that down-type quarks are already mass eigenstates, which fixes $V_{L,R}^d = \mathbb{1}$, and assume without loss of generality $V_R^u = \mathbb{1}$, yielding $q_L = (V^\dagger u_L, d_L)^T$. Analogously, we choose also the down-type lepton mass matrix to be diagonal and leave the neutrinos in the flavor eigenbasis. This defines the SMEFT Wilson coefficients below the LQ scale unambiguously and avoids the appearance of the PMNS lepton-mixing matrix in interactions involving neutrinos.

LQ Kinetic and Gauge Interactions

The kinetic and gauge interactions for a LQ Φ are fixed from the quantum numbers given in Table 16.4. The covariant derivative under $SU(2)_L \otimes U(1)_Y$ for the components Φ_α of a LQ multiplet under $SU(2)_L$ is given by

$$D_\mu \Phi_\alpha = \left[\partial_\mu \delta_{\alpha\beta} - ig_2 \tau_{\alpha\beta}^a W_\mu^a - ig_1 Y_\Phi \delta_{\alpha\beta} B_\mu\right]\Phi_\beta, \tag{16.85}$$

where the $SU(2)_L$ generators τ^a depend on the representation of Φ under $SU(2)_L$. For doublets $\tau^a = \sigma_a/2$ is given by the Pauli matrices

$$\sigma_1 = \begin{pmatrix} 0 & 1 \\ 1 & 0 \end{pmatrix}, \qquad \sigma_2 = \begin{pmatrix} 0 & -i \\ i & 0 \end{pmatrix}, \qquad \sigma_3 = \begin{pmatrix} 1 & 0 \\ 0 & -1 \end{pmatrix}. \tag{16.86}$$

For triplets in the spherical representation Y in (16.73) one has $\tau^a = t^a$ with t^a being the generators of $SU(2)_L$ given in the adjoint representation as follows

$$t^1 = \frac{1}{\sqrt{2}} \begin{pmatrix} 0 & -1 & 0 \\ -1 & 0 & 1 \\ 0 & 1 & 0 \end{pmatrix}, \qquad t^2 = \frac{1}{\sqrt{2}} \begin{pmatrix} 0 & i & 0 \\ -i & 0 & -i \\ 0 & i & 0 \end{pmatrix}, \qquad t^3 = \begin{pmatrix} 1 & 0 & 0 \\ 0 & 0 & 0 \\ 0 & 0 & -1 \end{pmatrix}.$$
(16.87)

The diagonalization of the $SU(2)_L \otimes U(1)_Y$ gauge sector is the usual one of the SM, but we repeat it here for convenience. We have

$$\sqrt{2} W_\mu^\pm = W_\mu^1 \mp i W_\mu^2, \qquad A_\mu = c_W B_\mu + s_W W_\mu^3, \qquad Z_\mu = -s_W B_\mu + c_W W_\mu^3. \quad (16.88)$$

The weak mixing angle ϑ_W and the $U(1)_{\mathrm{em}}$ coupling are related to $g_{1,2}$ as

$$s_W \equiv \sin(\vartheta_W) = \frac{g_1}{\sqrt{g_1^2 + g_2^2}}, \qquad c_W \equiv \cos(\vartheta_W) = \frac{g_2}{\sqrt{g_1^2 + g_2^2}}, \qquad e = \frac{g_1 g_2}{\sqrt{g_1^2 + g_2^2}}.$$
(16.89)

Moreover, with the vacuum expectation value $v = \sqrt{\mu^2/\lambda} = 246\,\mathrm{GeV}$ one has

$$M_W = \frac{g_2}{2} v, \qquad M_Z = \frac{\sqrt{g_1^2 + g_2^2}}{2} v, \qquad e = g_1 c_W = g_2 s_W, \qquad M_W = c_W M_Z. \quad (16.90)$$

The covariant derivative (16.85) will be then

$$D_\mu = \partial_\mu - i \frac{g_2}{\sqrt{2}} \left(\tau^+ W_\mu^+ + \tau^- W_\mu^- \right) - i \frac{g_2}{c_W} (\tau^3 - s_W^2 Q) Z_\mu - i e Q A_\mu.$$
(16.91)

Here

$$\tau^\pm = \tau^1 \pm i \tau^2, \qquad Q = Y \mathbb{1} + \tau^3.$$
(16.92)

The corresponding expression for $SU(2)_L$-singlet LQs is easily obtained by omitting terms with $\tau^{+,-,3}$. For doublets

$$\tau^+ = \begin{pmatrix} 0 & 1 \\ 0 & 0 \end{pmatrix}, \qquad \tau^- = \begin{pmatrix} 0 & 0 \\ 1 & 0 \end{pmatrix}, \qquad \tau^3 = \frac{1}{2} \begin{pmatrix} 1 & 0 \\ 0 & -1 \end{pmatrix}, \qquad (16.93)$$

and for triplets using (16.87)

$$\tau^+ = \frac{2}{\sqrt{2}} \begin{pmatrix} 0 & -1 & 0 \\ 0 & 0 & 1 \\ 0 & 0 & 0 \end{pmatrix}, \qquad \tau^- = \frac{2}{\sqrt{2}} \begin{pmatrix} 0 & 0 & 0 \\ -1 & 0 & 0 \\ 0 & 1 & 0 \end{pmatrix}, \qquad \tau^3 = \begin{pmatrix} 1 & 0 & 0 \\ 0 & 0 & 0 \\ 0 & 0 & -1 \end{pmatrix}.$$
(16.94)

The kinetic and gauge terms are given by

$$\mathcal{L}_{\mathrm{kin}} = (D_\mu \Phi)^\dagger (D^\mu \Phi) + \text{LQ mass term}$$
(16.95)

using (16.91).

Additional interactions of LQs with the SM Higgs doublet were introduced in [1051]. In consequence the LQs of different $SU(2)_L$ multiplets start to mix and phenomenology becomes more involved. See, for example, the generation of tensor operators for $b \to s\bar{\ell}\ell$ [487, 1045].

16.4.3 Selection of Models

It should be stressed that until now we did not specify the structure of Yukawa couplings, which even within one of these 10 scenarios for LQs contains a large model dependence unless some fundamental theory predicting or constraining these couplings has been constructed. One should remember that these Yukawa couplings y^{iA} are general complex 3×3 matrices, and this introduces many free parameters. Moreover, within a given theory, there could be several generations of LQs having the same transformation properties under $SU(2)_L$. Consequently, one could construct several models depending on which of this situations is realized. Some of the obvious possibilities are

1. The couplings y^{iA} are diagonal in generation space of quarks and leptons, and there are several generations for each LQ representation.
2. The couplings y^{iA} are general 3×3 matrices in generation space, and there is just one LQ.
3. Or a combined version of the two aforementioned variants, i.e., the couplings y^{iA} are general 3×3 matrices in generation space that carry also a generation index for several generations of LQs. That is, every leptoquark has its own 3×3 Yukawa coupling.
4. Several LQ representations are considered simultaneously.

Most phenomenological studies found in the literature are restricted to one generation of LQs with general 3×3 couplings y^{iA} in the generation space. But also models with different LQ representations have been considered with the goal to avoid some experimental constraints. See, e.g., [719, 1052, 1053].

Here, to reduce the numbers of free parameters we will first consider models in which

- Only one LQ representation is present.
- Only one generation of LQs is present, which means that we have to deal in most models with one Yukawa coupling describing either left-handed or right-handed interactions. But in models S_1, R_2, U_1, and V_2 both interactions (without taking h.c. in scalar models) can be present. Only R_2 and U_1 will be discussed in some detail here.
- A given LQ couples to both charged leptons and neutrinos as this increases the number of constraints and makes our analysis more interesting.

The inspection of Table 16.4 shows that this selects the first three models among the five models, which we will discuss in some detail first,

$$S_3, \qquad U_3, \qquad \tilde{R}_2, \qquad U_1, \qquad R_2, \qquad (16.96)$$

and these three models will be analyzed by us in some detail first. U_1 and R_2 LQs do not couple to neutrinos at tree-level, which presently is an advantage over the other three models as it allows to avoid constraints from $b \to s\nu\bar{\nu}$ processes. As R_2, $\tilde{R}_2$, and S_3 involve scalar LQs and U_1 and U_3 vector LQs, we will also see explicitly the differences arising from the different spins of these particles. But the problem is that U_1 and U_3 models as they stand are non-renormalizable, and we will only be able to present the results from tree-level diagrams and leading RG effects for which no loops with internal LQs have to be evaluated. This situation can be improved by providing ultraviolet (UV) completion for these models. On the other hand, we will be able to do loop calculations for scalar LQs.

16.4.4 General Structure of Basic Formulas

It will be useful to use the technology of previous chapter and calculate the impact of LQs on various processes as shifts in the master functions known from CMFV models. Similar to models with Z' tree-level FCNC and VLQ models discussed in the previous section, also in the case of LQ models flavor universality of master functions is broken, and they depend on the meson considered. Moreover, they become complex quantities bringing new CP-violating effects. The use of this technology will facilitate the comparison with other models and will avoid the repetition of the formulas for observables in which the master functions listed here should be inserted. For our purposes it will be sufficient to consider the following functions, which we recall here for convenience.

For $\Delta F = 2$ transitions we have

$$S(K), \qquad S(B_d), \qquad S(B_s), \tag{16.97}$$

which describe among others ε_K, ΔM_d, and ΔM_s, respectively. The latter two allow to calculate NP effects in the CP-asymmetries $S_{\psi K_S}$ and $S_{\psi\phi}$. For decays with $\mu\bar{\mu}$ in the final state we have

$$Y_A(K), \qquad Y_A(B_d), \qquad Y_A(B_s) \tag{16.98}$$

and for decays with $\nu\bar{\nu}$ in the final state

$$X(K), \qquad X(B_d), \qquad X(B_s). \tag{16.99}$$

In this particular case, when distinguishing between different neutrinos, we will have to generalize these functions to X^{ij} with the indices denoting different neutrinos.

The relevant CKM factors are again

$$\lambda_i^{(K)} = V_{is}^* V_{id}, \qquad \lambda_t^{(d)} = V_{tb}^* V_{td}, \qquad \lambda_t^{(s)} = V_{tb}^* V_{ts}, \tag{16.100}$$

and

$$g_{SM}^2 = 4\frac{M_W^2 G_F^2}{2\pi^2} = 1.78137 \times 10^{-7}\,\text{GeV}^{-2}. \tag{16.101}$$

This allows to introduce for each LQ the dimensionless quantity

$$\boxed{\kappa(\text{LQ}) = \frac{1}{g_{SM}^2 m_{LQ}^2} = 1.403, \qquad (m_{LQ} = 2\,\text{TeV}),} \tag{16.102}$$

which makes the following formulas more transparent.

In the case of the shifts ΔS it is useful to introduce the notation for sums of products of LQ couplings ($q = d, s$)

$$Y^i_{Qq} \equiv \sum_a y^i_{Qa} \, (y^i_{qa})^*,$$

(16.103)

where y stands for all the LQ couplings g, $\tilde{g}$, h, $\tilde{h}$ in (16.68) and (16.69), and Y are accordingly for G, $\tilde{G}$, H, $\tilde{H}$, with $i = 1L$, $1R$, $2L$, $2R$, $3L$. The summation is either over the neutrinos or charged leptons, which are exchanged in the box diagrams together with LQs. In the case of models considered next, and as explained earlier, working with down-quarks and charged leptons without any rotations from flavor basis and choosing the flavor basis for neutrinos, these sums are the same for neutrinos and charged leptons. We will therefore encounter only

$$G^{3L}_{Qq} = \sum_a g^{3L}_{Qa} \left(g^{3L}_{qa}\right)^*, \qquad H^{3L}_{Qq} = \sum_a h^{3L}_{Qa} \left(h^{3L}_{qa}\right)^*, \qquad \tilde{H}^{2L}_{Qq} = \sum_a \tilde{h}^{2L}_{Qa} \left(\tilde{h}^{2L}_{qa}\right)^*,$$

(16.104)

for S_3, U_3 and $\tilde{R}_2$, respectively. But the difference between neutrino and charged lepton contributions will be indicated in the following formulas through the masses of different LQs coupling to them.

On the other hand, in the U_1 model we will encounter

$$H^{1L}_{Qq} = \sum_a h^{1L}_{Qa} \left(h^{1L}_{qa}\right)^*, \qquad H^{1R}_{Qq} = \sum_a h^{1R}_{Qa} \left(h^{1R}_{qa}\right)^*,$$

(16.105)

and for R_2

$$H^{2L}_{Qq} = \sum_a h^{2L}_{Qa} \left(h^{2L}_{qa}\right)^*, \qquad H^{2R}_{Qq} = \sum_a h^{2R}_{Qa} \left(h^{2R}_{qa}\right)^*.$$

(16.106)

We will just quote the results as except for box diagrams the calculations are tree-level ones and very easy. In quoting the results for box diagrams we take into account that the masses of LQs are much larger than those of leptons so that lepton masses can be set to zero when evaluating these diagrams. The result for a box diagram is just a constant multiplied by some couplings. As there is no GIM mechanism in LQ models this constant remains at the end of the calculation. We advise motivated readers to check the following results. It is easier than many other calculations in this book.

Yet, a few comments should be made. While box diagram calculations result directly in the operators we encountered already in this book, tree diagrams of the type shown in Figure 16.2 result in operators like

$$(\bar{d}^i \gamma_\mu P_L l^j)(\bar{l}^l \gamma^\mu P_L d^k), \qquad (\bar{d}^i \gamma_\mu P_R l^j)(\bar{l}^l \gamma^\mu P_R d^k), \qquad (\bar{d}^i \gamma_\mu P_L l^j)(\bar{l}^l \gamma^\mu P_R d^k).$$

(16.107)

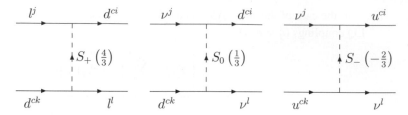

Figure 16.1 Tree-level contributions of S_+, S_0, and S_- in the S_3 model.

To find the Wilson coefficients of the operators consisting of product of quark and lepton currents, as encountered in this book, Fierz transformations have to be performed. One finds then, respectively,

$$(\bar{d}^i \gamma_\mu P_L d^k)(\bar{l}^l \gamma^\mu P_L l^j), \qquad (\bar{d}^i \gamma_\mu P_R d^k)(\bar{l}^l \gamma^\mu P_R l^j), \qquad -2(\bar{d}^i P_R d^k)(\bar{l}^l P_L l^j).$$
$$(16.108)$$

This means that in the presence of both LH and RH vector couplings, in LQ models scalar-scalar operators contribute to various observables. On the other hand in some models, like S_3, operators with LH and RH scalar couplings and charge conjugated fields, shown in Figure 16.1, are present. The corresponding Fierz transformations, which can be found in Appendix A.3, imply LR vector-vector operators. In case of the difficulties with charge conjugated fields the help comes from [1054].

Before entering the details, another important comment should be made. When listing the following results, we did not include any renormalization group effects that were discussed generally in Chapter 14. As the scales involving LQs are much higher than the EW scale, the shifts in the master functions listed following must be corrected by including these effects along the lines discussed in Chapter 14. Then it turns out that the results listed here do not necessarily dominate a given process because RG effects enhanced by large logarithms could be more important. This is, for instance, the case of LQ contributions to ε'/ε as analyzed in [667] and summarized in Section 16.4.13.

Finally, in the case of difficulties, the following formulas can be derived from the general SMEFT formulas, like the ones in (14.79), and the expressions for SMEFT coefficients in terms of LQ couplings and masses given in Appendix B of [667]. But to find the NP shifts in the usual WCs like C_9 and C_{10}, one has to remember to multiply the SMEFT coefficients in the latter paper and the ones in (14.79) by

$$\frac{\pi}{\alpha_e} \frac{v^2}{\lambda_t^{ji}} = \frac{1}{g_{SM}^2 \sin^2 \vartheta_W} \frac{1}{\lambda_t^{ji}} \tag{16.109}$$

as explicitly stated in (14.77).

After these introductory comments we are ready to list results for the chosen LQ models.

16.4.5 S_3

For the shifts ΔS we find

$$\Delta S(K) = \frac{\tilde{r}}{16\pi^2} \left[\frac{G_{sd}^{3L}}{\lambda_t^{(K)}} \right]^2 [4\,\kappa(S_+) + \kappa(S_0)] \tag{16.110}$$

and

$$\Delta S(B_q) = \frac{\tilde{r}}{16\pi^2} \left[\frac{G_{bq}^{3L}}{\lambda_t^{(q)}} \right]^2 [4 \kappa(S_+) + \kappa(S_0)], \tag{16.111}$$

where $\tilde{r} \approx 1$ is a QCD correction that takes into account RG effects between $\mu = m_{\mathrm{LQ}}$ and $\mu = m_t$ with the latter used as matching scale in the SM. The contributions of S_+ and S_0 involve charged leptons and neutrinos, respectively.

Calculating the first diagram in Figure 16.1, we find first

$$Y_A(K) = Y_{\mathrm{SM}} + \kappa(S_+) \left[\frac{g_{s\mu}^{3L}(g_{d\mu}^{3L})^*}{\lambda_t^{(K)}} \right], \qquad Y_A(B_q) = Y_{\mathrm{SM}} + \kappa(S_+) \left[\frac{g_{q\mu}^{3L}(g_{b\mu}^{3L})^*}{\lambda_t^{(q)}} \right]. \tag{16.112}$$

The same diagram implies also NP contributions to the Wilson coefficients C_9 and through Y_A to C_{10} giving

$$C_9^{\mathrm{NP}} = \frac{\kappa(S_+)}{\sin^2 \vartheta_W} \left[\frac{g_{s\mu}^{3L}(g_{b\mu}^{3L})^*}{\lambda_t^{(s)}} \right], \qquad C_{10}^{\mathrm{NP}} = -C_9^{\mathrm{NP}}. \tag{16.113}$$

The relevant functions X^{ij} are found from the second diagram in Figure 16.1:

$$X^{ij}(K) = X_{\mathrm{SM}} \delta_{ij} - \frac{\kappa(S_0)}{2} \left[\frac{g_{s\nu_i}^{3L}(g_{d\nu_j}^{3L})^*}{\lambda_t^{(K)}} \right] \tag{16.114}$$

and

$$X^{ij}(B_q) = X_{\mathrm{SM}} \delta_{ij} - \frac{\kappa(S_0)}{2} \left[\frac{g_{q\nu_i}^{3L}(g_{b\nu_j}^{3L})^*}{\lambda_t^{(q)}} \right]. \tag{16.115}$$

Defining then

$$X_{\mathrm{eff}}^{ij} = \lambda_t^{(K)} X^{ij}(K) \tag{16.116}$$

the usual expressions for the branching ratios for $K^+ \to \pi^+ \nu \bar{\nu}$ and $K_{\mathrm{L}} \to \pi^0 \nu \bar{\nu}$ are generalized as follows

$$\mathcal{B}(K^+ \to \pi^+ \nu \bar{\nu}) = \frac{\kappa_+}{3} \sum_{i,j} \left[\left(\frac{\mathrm{Im}\, X_{\mathrm{eff}}^{ij}}{\lambda^5} \right)^2 + \left(\frac{\mathrm{Re}\, \lambda_c}{\lambda} P_c(X) \delta_{ij} + \frac{\mathrm{Re}\, X_{\mathrm{eff}}^{ij}}{\lambda^5} \right)^2 \right], \tag{16.117}$$

$$\mathcal{B}(K_{\mathrm{L}} \to \pi^0 \nu \bar{\nu}) = \frac{\kappa_L}{3} \sum_{ij} \left(\frac{\mathrm{Im}\, X_{\mathrm{eff}}^{ij}}{\lambda^5} \right)^2, \tag{16.118}$$

where all the quantities κ_+, κ_L and $P_c(X)$ are the same as in the SM. The last diagram in Figure 16.1 is relevant for charm meson decays.

16.4.6 U_3

As the model is not renormalizable, we cannot reliably calculate the box diagrams and correspondingly the shifts ΔS, but the structure of various couplings implies

$$\Delta S(K) = \frac{c_K}{4\pi^2} \left[\frac{H_{sd}^{3L}}{\lambda_t^{(K)}} \right]^2 \left[\kappa(U_{3,0}) + 4\,\kappa(U_{3,-}) \right] \tag{16.119}$$

and

$$\Delta S(B_q) = \frac{c_q}{4\pi^2} \left[\frac{H_{bq}^{3L}}{\lambda_t^{(q)}} \right]^2 \left[\kappa(U_{3,0}) + 4\,\kappa(U_{3,-}) \right] \tag{16.120}$$

with c_i dependent on the UV completion considered. Without it the box diagram is quadratically divergent. The contributions of $U_{3,0}$ and $U_{3,-}$ involve charged leptons and neutrinos, respectively. We give this formula only for curiosity, but it should not be used in phenomenological applications. On the other hand, no loops have to be evaluated to obtain the following formulas. They just result from tree diagrams.

For Y_A we find

$$Y_A(K) = Y_{SM} - \kappa(U_{3,0}) \left[\frac{h_{s\mu}^{3L}(h_{d\mu}^{3L})^*}{\lambda_t^{(K)}} \right], \qquad Y_A(B_q) = Y_{SM} - \kappa(U_{3,0}) \left[\frac{h_{q\mu}^{3L}(h_{b\mu}^{3L})^*}{\lambda_t^{(q)}} \right], \tag{16.121}$$

and consequently NP contributions to the Wilson coefficients C_9 and C_{10}

$$C_9^{NP} = -\frac{\kappa(U_{3,0})}{\sin^2 \vartheta_W} \left[\frac{h_{s\mu}^{3L}(h_{b\mu}^{3L})^*}{\lambda_t^{(s)}} \right], \qquad C_{10}^{NP} = -C_9^{NP}. \tag{16.122}$$

The relevant functions X^{ij} are this time given by

$$X^{ij}(K) = X_{SM}\delta_{ij} + 2\,\kappa(U_{3,-}) \left[\frac{h_{s\nu_i}^{3L}(h_{d\nu_j}^{3L})^*}{\lambda_t^{(K)}} \right] \tag{16.123}$$

and

$$X^{ij}(B_q) = X_{SM}\delta_{ij} + 2\,\kappa(U_{3,-}) \left[\frac{h_{q\nu_i}^{3L}(h_{b\nu_j}^{3L})^*}{\lambda_t^{(q)}} \right]. \tag{16.124}$$

16.4.7 $\tilde{R}_2$

For the shifts ΔS we find

$$\Delta S(K) = \frac{\tilde{r}}{16\pi^2} \left[\frac{\tilde{H}_{sd}^{2L}}{\lambda_t^{(K)}} \right]^2 \left[\kappa(\tilde{R}_2^+) + \kappa(\tilde{R}_2^-) \right] \tag{16.125}$$

and

$$\Delta S(B_q) = \frac{\tilde{r}}{16\pi^2} \left[\frac{\tilde{H}_{bq}^{2L}}{\lambda_t^{(q)}}\right]^2 \left[\kappa(\tilde{R}_2^+) + \kappa(\tilde{R}_2^-)\right]. \tag{16.126}$$

The contributions of $\tilde{R}_2^+$ and $\tilde{R}_2^-$ involve charged leptons and neutrinos, respectively.

For Y_A we find

$$Y_A(K) = Y_{SM} - \frac{\kappa(\tilde{R}_2^+)}{2}\left[\frac{\tilde{h}_{s\mu}^{2L}(\tilde{h}_{d\mu}^{2L})^*}{\lambda_t^{(K)}}\right], \qquad Y_A(B_q) = Y_{SM} - \frac{\kappa(\tilde{R}_2^+)}{2}\left[\frac{\tilde{h}_{q\mu}^{2L}(\tilde{h}_{b\mu}^{2L})^*}{\lambda_t^{(q)}}\right]. \tag{16.127}$$

Equivalently we find for the Wilson coefficients C_9' and C_{10}'

$$C_9' = -\frac{\kappa(\tilde{R}_2^+)}{2\sin^2\vartheta_W}\left[\frac{\tilde{h}_{s\mu}^{2L}(\tilde{h}_{b\mu}^{2L})^*}{\lambda_t^{(s)}}\right], \qquad C_{10}' = -C_9'. \tag{16.128}$$

There are no NP contributions to C_9 and C_{10}. In writing the shifts in (16.127), we took into account that NP contributions come from right-handed currents, which implies the minus sign in the shift in Y_A if it is inserted in the SM expressions. This flip of the sign should not be made in evaluating C_9' and C_{10}' as it is taken into account already in the operators.

The relevant functions X^{ij} are

$$X^{ij}(K) = X_{SM}\delta_{ij} - \frac{\kappa(\tilde{R}_2^-)}{2}\left[\frac{\tilde{h}_{sv_i}^{2L}(\tilde{h}_{dv_j}^{2L})^*}{\lambda_t^{(K)}}\right] \tag{16.129}$$

and

$$X^{ij}(B_q) = X_{SM}\delta_{ij} - \frac{\kappa(\tilde{R}_2^-)}{2}\left[\frac{\tilde{h}_{qv_i}^{2L}(\tilde{h}_{bv_j}^{2L})^*}{\lambda_t^{(q)}}\right]. \tag{16.130}$$

We now turn our attention to two models in which, as seen in Table 16.4, there are no tree-level contributions to processes with neutrinos in the final state. This avoids constraints from such processes but could be problematic if one day NA62 experiment will find NP in $K^+ \to \pi^+ \nu\bar{\nu}$, KOTO in $K_L \to \pi^0 \nu\bar{\nu}$, and Belle II experiment in $b \to s\nu\bar{\nu}$ transitions.

16.4.8 U_1

Proceeding as in the case of the U_3 model, we first find from the box diagram in Figure 16.2, involving either only left-handed or only right-handed couplings, the shifts in $\Delta S(B_q)$

$$\Delta S(B_q) = \frac{c}{4\pi^2}\kappa(U_1)\left(\left[\frac{H_{bq}^{1L}}{\lambda_t^{(q)}}\right]^2 + \left[\frac{H_{bq}^{1R}}{\lambda_t^{(q)}}\right]^2\right), \tag{16.131}$$

with c dependent on the UV completion considered

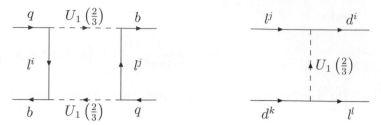

Figure 16.2 Left: Contribution to $B_q^0 - \bar{B}_q^0$ mixing in the U_1 model. Right: Contribution to semileptonic decay.

But the dominant is the box diagram in Figure 16.2 when both LH and RH couplings are present. It contributes to the Wilson coefficient of the operator Q_2^{LR} that has enhanced matrix elements, in particular in the K system. For $B_s^0 - \bar{B}_s^0$ mixing we find

$$\Delta H_{\text{eff}} = -\frac{1}{2\pi^2} \frac{c}{m_{LQ}^2} H_{bs}^{1L} H_{bs}^{1R} Q_2^{LR}, \tag{16.132}$$

where the operator Q_2^{LR} is defined in (13.8). Again c depends on the UV completion. Analogous expressions are found for $B_d^0 - \bar{B}_d^0$ and $K^0 - \bar{K}^0$ mixings.

Next, the second diagram in Figure 16.2, with (LH,LH), (RH,RH), (LH,RH), and (RH,LH) combinations of couplings, generates four diagrams and after proper Fierz transformations the following contributions to the Wilson coefficients relevant for $b \to s\mu^+\mu^-$ transitions:

$$C_9^{NP} = -\frac{\kappa(U_1)}{\sin^2 \vartheta_W} \left[\frac{h_{s\mu}^{1L}(h_{b\mu}^{1L})^*}{\lambda_t^{(s)}} \right], \qquad C_{10}^{NP} = -C_9^{NP}, \tag{16.133}$$

$$C_9' = -\frac{\kappa(U_1)}{\sin^2 \vartheta_W} \left[\frac{h_{s\mu}^{1R}(h_{b\mu}^{1R})^*}{\lambda_t^{(s)}} \right], \qquad C_{10}' = C_9', \tag{16.134}$$

$$m_b C_S = 2\frac{\kappa(U_1)}{\sin^2 \vartheta_W} \left[\frac{h_{s\mu}^{1L}(h_{b\mu}^{1R})^*}{\lambda_t^{(s)}} \right], \qquad C_P = -C_S, \tag{16.135}$$

$$m_b C_S' = 2\frac{\kappa(U_1)}{\sin^2 \vartheta_W} \left[\frac{h_{s\mu}^{1R}(h_{b\mu}^{1L})^*}{\lambda_t^{(s)}} \right], \qquad C_P' = C_S'. \tag{16.136}$$

As we will see later, the U_1 model appears presently to be the favorite LQ model. This motivated various authors to perform detailed analyses of collider signatures of this LQ that in a large class of ultraviolet complete models appears together with Z' and color-octet G'. We refer to a very recent analysis in [1055], where references to other analyses can be found. We will still encounter this model at several stages of our expedition.

16.4.9 R_2

For the shifts ΔS we find

$$\Delta S(K) = \frac{\tilde{r}}{16\pi^2} \left[\frac{H_{sd}^{2R}}{\lambda_t^{(K)}} \right]^2 \kappa(R_2^-) \tag{16.137}$$

and

$$\Delta S(B_q) = \frac{\tilde{r}}{16\pi^2} \left[\frac{H_{bq}^{2R}}{\lambda_t^{(q)}} \right]^2 \kappa(R_2^-). \tag{16.138}$$

For Y_A function

$$Y_A(K) = Y_{\text{SM}} + \frac{\kappa(R_2^-)}{2} \left[\frac{h_{s\mu}^{2R}(h_{d\mu}^{2R})^*}{\lambda_t^{(K)}} \right], \qquad Y_A(B_q) = Y_{\text{SM}} + \frac{\kappa(R_2^-)}{2} \left[\frac{h_{q\mu}^{2R}(h_{b\mu}^{2R})^*}{\lambda_t^{(q)}} \right]. \tag{16.139}$$

Moreover,

$$C_9^{\text{NP}} = -\frac{\kappa(R_2^-)}{2\sin^2\vartheta_W} \left[\frac{h_{s\mu}^{2R}(h_{b\mu}^{2R})^*}{\lambda_t^{(s)}} \right], \qquad C_{10}^{\text{NP}} = C_9^{\text{NP}}, \qquad C_{10}' = C_9' = 0. \tag{16.140}$$

Note that RH and not LH couplings enter $C_{9,10}^{\text{NP}}$. It is a good practice to verify this and also to calculate LQ contributions to $D^0 - \bar{D}^0$ mixing in this model, which are dominated by the operator Q_2^{LR}.

As we will see soon, this model at first sight cannot help in explaining the B physics anomalies because at tree-level it implies the relation between the shifts in the Wilson coefficients given in (16.140), which disagrees with the successful one in (12.1).

The solution to this problem proposed in [1056] is to simply set the relevant Yukawa coupling to zero so that the transition $b \to s\ell^+\ell^-$ is absent at tree-level, and it takes place first at one-loop level. The box diagram with the exchange of the W boson, top-quark, and the leptoquark with charge $+5/3$ allows then to satisfy the second relation in (12.1). In this manner the $R(K)$ and $R(K^*)$ anomalies can be explained. A similar idea was used in [1057] in the case of the S_1 leptoquark, which does not contribute to $b \to s\ell^+\ell^-$ at tree-level. While at first sight this model looked like being useful for the explanation of the $R(K)$ anomaly, the induced problems elsewhere made it phenomenologically unviable [1048].

The model suggested in [1056] is still viable as far as $R(K)$ and $R(K^*)$ are concerned, but in contrast to the U_1 model cannot explain $R(D)$ and $R(D^*)$ anomalies [538]. We refer to [1056] for detailed calculations in this model. Subsequently, it has been demonstrated in [1058] that R_2 model with flavor anomalies generated at one-loop level can not only explain the $R(K)$ and $R(K^*)$ anomalies but also the $(g-2)_\mu$ anomaly and the IceCube data [1059] on the sterile neutrinos. For a recent analysis of IceCube data in S_1 and U_1 models, see [1060].

16.4.10 Remaining LQ Models

For completeness we want to list some general properties of the remaining five models and refer to the literature where detailed phenomenological analyses of most of them can be found. The collection of some of these properties can be found in Table 16.4, and others can be deduced from the Lagrangians listed earlier. They are

- The leptoquarks S_1, and V_2, similar to U_1 and R_2, have both left-handed and right-handed couplings and can provide through left-right operators more important contributions to

$\Delta F = 2$ observables than is the case of other models in which such operators are absent. In this context U_1 and V_2 are relevant for $K^0 - \bar{K}^0$ and $B_{s,d} - \bar{B}_{s,d}$ mixings, while S_1 and R_2 for $D^0 - \bar{D}^0$ mixing.

- These four models can also be distinguished from each other by their couplings to leptons. S_1 does not contribute to $b \to s\ell^+\ell^-$ transitions but has an impact on decays of K and $B_{s,d}$ with neutrinos in the final state. The opposite is true for R_2 and U_1. On the other hand, V_2 provides contributions to all these decays.

- The leptoquark $\tilde{S}_1$ is rather special as it only couples to charged leptons and down-quarks. As such it predicts no NP contributions to $K^+ \to \pi^+ \nu\bar{\nu}$, $K_L \to \pi^0 \nu\bar{\nu}$, $b \to s\nu\bar{\nu}$ transitions and charm physics. On the other hand, it contributes to $b \to s\ell^+\ell^-$. Having only one coupling it is more predictive than the remaining models not considered by us in detail but at the end, it is not useful for B physics anomalies and decays with neutrinos in the final state.

- The leptoquarks $\tilde{U}_1$ and $\tilde{V}_2$ are less interesting than other models as they only give contributions to charm observables. Therefore they cannot help with B physics anomalies nor with the ε'/ε anomaly. But possibly they could address the CP violation in charm decays mentioned at the end of Chapter 11.

16.4.11 LQ Models and B Physics Anomalies

General View

In Chapter 12 we have mentioned various anomalies, and we will return to them again in Chapter 18, where we will compare various explanations of them. In particular LQs provide a very natural description of lepton flavor universality violation (LFUV), as due to $\mu \to e\gamma$ constraint they should have small couplings to electrons.

Here we would like to concentrate on LQ models recalling that the two most favorite solutions obtained in various global analyses of $b \to s\ell^+\ell^-$ [494, 721] turned out before Moriond 2019 to be

$$i)\ C_9^{\text{NP}} < 0, \qquad ii)\ C_9^{\text{NP}} = -C_{10}^{\text{NP}} < 0. \qquad (16.141)$$

While these solutions, as reported in Chapter 12, are even after Moriond 2019 among the favored ones, other possible solutions involving several WCs become competitive and even superior to them. We will return to them later, but let us first find out how different LQ models face the solutions in (16.141). We identify the following relations

$$C_9^{\text{NP}} = -C_{10}^{\text{NP}}, \qquad (S_3, U_1, U_3) \qquad (16.142)$$

$$C_9' = -C_{10}', \qquad (\tilde{R}_2, V_2) \qquad (16.143)$$

$$C_9^{\text{NP}} = C_{10}^{\text{NP}}, \quad (R_2, V_2) \qquad C_9' = C_{10}', \quad (\tilde{S}_1, U_1) \qquad (16.144)$$

$$C_P = -C_S \quad (U_1, V_2) \qquad C_P' = C_S', \quad (U_1, V_2) \qquad (16.145)$$

and recall that in the S_1 model there are no contributions to $b \to s\ell^+\ell^-$ at tree-level. The following observations should be made:

- The first solution in (12.1) in which NP is only present in C_9 cannot be generated with a single leptoquark representation. But it can be generated, e.g., by adding scalar representations R_2 and S_3 [1053].
- The second solution $C_9^{\mathrm{NP}} = -C_{10}^{\mathrm{NP}}$ is the most natural solution for LQs and can be obtained with a single representation: either S_3 or U_1 [778, 779, 1061–1064]. Also U_3 satisfies this relation.
- On the other hand, LQ models with only right-handed currents, like $\tilde{R}_2$, considered in detail in [1047, 1065] before $R(K^*)$ was measured, are disfavored because in these models $R(K) < 1$ implies $R(K^*) > 1$ or $R(K^*) < 1$ implies $R(K) > 1$ [609, 1066] in disagreement with data, even after Moriond 2019.

Thus as far as $b \to s\ell^+\ell^-$ anomalies are concerned, in the context of LQ models the solution $C_9^{\mathrm{NP}} = -C_{10}^{\mathrm{NP}}$ is the favorite one. But the question arises how these models face the $R(D)$ and $R(D^*)$ anomalies. Here a solution to these anomalies can be achieved with NP contribution to the operator $(\bar{c}_L \gamma_\mu b_L)(\bar{\tau}_L \gamma_\mu \nu_L)$. This is consistent with B_c lifetime [571] and q^2 distributions [541, 549, 574] in these decays in contrast to models like 2HDM, in which one attempts to explain the anomalies with scalar and pseudoscalar mediators. Assuming then $SU(2)_L$ gauge invariance, these anomalies are not only correlated with the ones in $b \to s\ell^+\ell^-$ but have also implications for $b \to s\tau^+\tau^-$, which must be significantly enhanced, and this is also the case of $b \to s\nu\bar{\nu}$. We have discussed this in detail in Section 14.5.

In a general analysis the bounds from $B \to K^{(*)}\nu\bar{\nu}$ can be avoided if NP couples dominantly to third generation. But this disagrees with direct LHC searches [1067] and electroweak precision tests [765]. Introducing more operators and assuming special flavor structure, one can with some tuning avoid these constraints as demonstrated in [719]. But within the LQ models the difficulty with $b \to s\nu\bar{\nu}$ transitions can be avoided easily in the U_1 model in which the coefficients $C_{lq}^{(1)}$ and $C_{lq}^{(3)}$ in the SMEFT satisfy the relation $C_{lq}^{(1)} = C_{lq}^{(3)}$. In this case, as seen in (14.79), at the LQ scale there are no contributions to $b \to s\nu\bar{\nu}$ at the tree-level [778, 779]. Also in the R_2 model tree-level contributions to processes with neutrinos in the final state are absent. This is not the case of S_3 and U_3, which are very constrained by these transitions.

The absence of tree-level contributions to $b \to s\nu\bar{\nu}$ and also $K^+ \to \pi^+\nu\bar{\nu}$ and $K_L \to \pi^0\nu\bar{\nu}$ at tree-level in the U_1 and R_2 models at the matching scale allows to avoid the constraints from these decays, which are generally problematic for LQ models. But if future data will require significant NP contributions to them, the fate of these models will depend on whether the RG effects generating such contributions will be able to fit the data [667]. We will return to this issue in Section 16.4.13. The fate of the U_1 model, as far as $\Delta F = 2$ observables are concerned, can only be answered after some UV completion is constructed. But as in this model both left-handed and right-handed couplings are present, possibly future anomalies in nonleptonic observables could be easier solved than in S_3, U_3, and $\tilde{R}_2$ models. One of the reasons is the increase of free-parameters. Another is the

LQ representation	References	Comments (see text)
Table 16.5 Selected references to specific LQ models.		
S_1	[1057, 1068–1070]	R_K, $R(D)$, $R(D^*)$, $(g-2)_\mu$
S_1 and $\tilde{R}_2$	[1071]	Charged and neutral, ν-masses
R_2	[1072]	Charged and neutral
R_2, loop	[1056, 1058, 1068]	R_K, R_{K^*}, $(g-2)_\mu$, IceCube event.
$\tilde{R}_2$, loop	[1047, 1073]	R_K and $R(D)$
S_1 and S_3	[593, 719, 1052, 1098]	Charged and neutral
S_3 and $\tilde{R}_2$	[1053, 1071]	Charged and neutral
R_2 and $\tilde{R}_2$	[1074, 1075]	Charged and neutral
S_3 and R_2	[849]	Charged and neutral
S_1, R_2, S_3	[1076]	EDMs
U_1	[719, 945, 1052, 1064, 1077–1079]	R_K and $R(D)$ and $R(D^*)$
U_1	[1080–1082]	Gauge LQ
U_1	[1083]	Composite LQ
V_2	[1084]	Neutral
U_3	[944, 1063, 1078]	General analysis
U_3	[1061]	Strong dynamics
(U_1, U_3)	[945, 1085]	R_K and $R(D)$ and $R(D^*)$
(U_1, U_3, S_3)	[729]	$R(K)$ and $R(K^*)$
$\tilde{U}_1$ and R_2	[1086]	Neutral

generation of left-right operators, which as we will see later could help to explain the ε'/ε-anomaly. On the other hand left-right operators provide enhanced NP contributions to ΔM_K and ε_K so that possibly some tuning of parameters will be required to satisfy all constraints. In the U_1 model also new operators to $b \to s\mu^+\mu^-$ contribute, in particular $Q_S^{(\prime)}$ and $Q_P^{(\prime)}$, but their Wilson coefficients must be small for the model to agree with $B_s \to \mu^+\mu^-$ data. Similar comments, as far as left-right operators are concerned, apply also to the R_2 model. We will return to some of these issues in the context of UV completions later.

In Table 16.5 we collect selected papers on various LQ models. As I cannot discuss them here in detail, but some of these papers could turn out to be useful for future model building, I will just make a few comments on them. One should realize that some anomalies could go away, and the pattern of anomalies could change in the future, opening thereby the road for models that are not favored in 2019. The ideas presented in the papers listed in Table 16.5 could, however, still turn out to be useful for the construction of new models. The description of these papers presented here has also a historical value, showing a big excitement of theorists in the presence of deviations from the SM predictions.

The neutral anomaly ($b \to s\mu^+\mu^-$) has been discussed extensively in models with several LQ representations in [729]. Tables I and II in this paper show relations between C_9 and C_{10} and also between $R(K)$ and $R(K^*)$. The authors favor (U_1, U_3, S_3) as expected from our comments earlier.

I found the paper [945] very useful. The authors look at $R(K)$, $R(D)$, and $R(D^*)$ anomalies, which they studied already in [777] and compare a model with a vector boson triplet with the U_1 LQ model giving arguments for the superiority of U_1. In this context they review previous analyses in SMEFT [778, 779], in which it is shown that among S_3, U_1, and U_3 only U_1 can simultaneously explain neutral and charged anomalies while satisfying other constraints, in particular from $b \to s\nu\bar{\nu}$ transitions.

One should, however, note that [945] appeared before the $R(K^*)$ anomaly has been announced. Taking the latter anomaly into account, it has been found in [1078] that U_1, U_3, and S_3 models cannot simultaneously explain $R(K)$, $R(K^*)$, $R(D)$, and $R(D^*)$ anomalies within 1σ. But with the new data on $R(D)$ and $R(D^*)$ from Belle the situation changed, and in particular the U_1 model performs better than previously.

Next, it has been demonstrated in [1049] that none of the *scalar* LQs of mass $m_{\rm LQ} \approx 1$ TeV can alone accommodate simultaneously $R(K)$, $R(K^*)$, $R(D)$, and $R(D^*)$ anomalies when constraints arising from both low-energy observables and from direct searches at the LHC are taken into account. It appears then that the only single LQ scenario that works for $m_{\rm LQ} \approx 1$–2 TeV is the vector LQ U_1. Otherwise models with two scalar LQs, as listed in Table 16.5, have to be considered. A general analysis of constraints on scalar LQs from lepton and kaon physics can be found in [1346]. Effects of scalar LQs in $B \to \mu\bar{\nu}$ are discussed in [1345].

In [1087] a general LQ analysis correlating neutral anomalies with LFV in pure leptonic decays and semileptonic B decays has been performed. Also implications for $\mu \to e\gamma$ have been presented. This paper contains useful formulas for a number of observables. A solution to B physics anomalies is provided by a combination of two scalar LQs S_1 and S_3 [719, 1052], which allows to avoid $b \to s\nu\bar{\nu}$ constraints. Large effects in $b \to s\tau^+\tau^-$, like $\mathcal{B}(B \to \tau^+\tau^-)$ of $O(10^{-3})$, are predicted to be compared with $O(10^{-6})$ in the SM.

In [849] another solution to B physics anomalies with two scalar representations has been proposed. This time S_3 and R_2 explain $R(K^{(*)})$ and $R(D^{(*)})$ anomalies, respectively. It is a UV complete model based on SU(5) Grand Unified Theory. Certainly, an interesting paper.

In [1064] a study of flavor symmetry U(2)^5 in LQ models has been investigated. It is assumed that LQs couple first only to third generation. The remaining couplings are generated in the process of the breakdown of the symmetry. Considering U_1, S_3, U_3 it is concluded in agreement with [778, 779] that U_1 should be favored. In [719] both charged and neutral anomalies in an EFT-framework based on U(2)$_q$ × U(2)$_l$ symmetry have been studied with couplings predominantly to third generation of left-handed quarks and leptons. LQs in different representations under the SU(2)$_L$ and colorless vector SU(2)$_L$-triplets (W', B') have been considered, and again the U_1 model has been declared as a winner, although the model with S_1 and S_3, discussed in [1052, 1098], provides also the explanation of all B physics anomalies while satisfying other constraints without fine-tuning.

In models considered in [719] it is shown that the constraints from direct LHC searches and electroweak precision tests pointed out in [1067] and [765], respectively, can be presently avoided through a particular flavor structure that allows to explain anomalies with higher LQ masses than required in the construction in [765, 1067]. The paper [719], known in the flavor community as the "Zürich guide," can be strongly recommended. See also [1319].

At this stage it should be again stressed that having explicit models like LQ models, Z' models, and other simplified models discussed in the previous chapter has some advantage over pure EFT approach. Indeed, complementing EFT approaches with appropriate simplified models with new heavy mediators is essential [944, 1064], if one wants to address the compatibility of models addressing present anomalies with other low-energy constraints and with high-p_T data. Due to the relatively low scale of NP required for the explanations of the charged-current anomalies, high-p_T constraints are very relevant [1044, 1055, 1067, 1088–1093].

UV Completions

The fact that the U_1 model with a vector LQ seems to be the favorite LQ model, but is not renormalizable, motivated a number of authors to look for UV completions in which the vector LQ is a gauge boson. In this context most popular are models involving Pati–Salam group $SU(4) \times SU(2)_L \times SU(2)_R$. In this construction proton remains stable in contrast to several LQ models. We refer to [935, 1080–1082, 1094] for details. See also [849, 1095–1104].

Although the Pati–Salam group is present in all models of this type, some papers invoke larger groups and include additional new particles. Here we provide an express review of some of these papers.

- In [1080] Pati–Salam gauge group $SU(4) \times SU(2)_L \times SU(2)_R$ is considered and both charged and neutral anomalies are explained as well as $(g - 2)_\mu$ anomaly. The model contains also vector-like fermions.
- In [1081] the authors study the gauge group $SU(4) \times SU(3)' \times SU(2)_L \times U(1)'$. See also the papers [1055, 1089, 1100, 1105]. In fact, as stressed in [1055, 1081] this is the minimal gauge group containing the U_1 as a gauge boson, which fulfills the necessary requirements to provide a successful explanation of the anomalies while remaining consistent with high-p_T data.
- A larger gauge group, that involves the product of three Pati–Salam groups was proposed in [1082] with the goal to explain not only flavor anomalies but also the pattern of fermion masses and CKM parameters (see also [1095, 1099]). In contrast to [1055, 1089, 1100, 1105] in this model the gauge group is nonuniversal among the different SM-like families. For an earlier paper by these authors on a general EFT with $U(2)^n$ flavor symmetry, see [1106], where it has been pointed out that the correlation between $K \to \pi\nu\bar{\nu}$, $B \to \pi\nu\bar{\nu}$, and $R(D^*)$ could distinguish between various NP scenarios. A very interesting paper in this context is also [1107], the generalization of the model in [1082] to include RH couplings of the U_1 LQ. This is motivated by the Moriond 2019 news that possibly $R(K) \neq R(K^*)$ and distinguishes this analysis from other analyses of B physics anomalies in which only LH couplings of the U_1 LQ have been kept.[1] I think this is an interesting direction even if the number of gauge bosons and free parameters is large in these models. See also [1316].

---

[1] The presence of both LH and RH couplings in the U_1 model is crucial for addressing the ε'/ε anomaly in this model as pointed out earlier in [667] and discussed in Section 16.4.13.

- In [1100] the Pati–Salam group has been incorporated into the Randall–Sundrum scenario. Breaking the gauge group by boundary conditions the resulting massive vector LQ has the same mass scale as the vector-like fermions and other resonances present in this model. $R(K)$ and $R(K^*)$ anomalies can easily be explained in this model. One also finds sizable effects in $R(D)$, $R(D^*)$, and $R(J/\psi)$, but the model could not account for their central values prior to Moriond 2019 due to stringent constraints from $D^0 - \bar{D}^0$ mixing and $\tau \to 3\mu$. The situation improved after Moriond 2019 so that within 2σ the model is consistent with new $R(D)$ and $R(D^*)$ values in Table 9.1. Also, addition of RH couplings, increasing the effect by scalar operators in $R(D)$ and $R(D^*)$, would help.

- In [933, 934] solutions to B physics anomalies within some other nonabelian groups have been proposed.

While, the virtue of concrete simplified Z' and LQ models over EFT has already been mentioned earlier, the most important consequence following from the requirement of a consistent UV completion is the necessity of extra particles, with interesting high-p_T signatures that cannot be deduced within the simplified model [1107]. In the latter paper these new states include both a color-octet (G') and a color singlet (Z') vector field, as extensively discussed in [1081, 1082, 1095, 1099], and a pair of vector-like quarks and leptons. Also as pointed out in [1097], the $\Delta F = 2$ constraints imply that the vector-like leptons are likely to be the lightest exotic states. In my view this is an interesting direction, as it is hard to imagine that NP consists of a single LQ, even as successful as the U_1.

Still another idea is to explain B physics anomalies via composite Higgs model, in which both Higgs and the S_3 LQ arise as pseudo-Goldstone bosons of a strong dynamics [1061]. Fermion masses are assumed to be generated via the mechanism of partial compositeness, which largely determines the LQ couplings increasing the predictive power of the model. Also nonuniversal lepton interactions are implied naturally in this manner.

The impact of flavor symmetries on the explanations of B physics anomalies in general context and in LQ models has been studied in [1062]. Analyses as the ones in [774, 1061, 1062, 1108] and also in [1082, 1095, 1099, 1106]), mentioned already earlier, could provide valuable information on flavor symmetries beyond the SM when the data improves. This is also evident from an interesting analysis in [774], which demonstrates how some assumptions about flavor dynamics at short-distance scales could provide correlations between B, K, and D physics as well as collider physics, which go beyond MFV and $U(2)^3$ symmetry.

On the other hand, the investigation of scalar leptoquarks accommodating the B-physics anomalies in the context of GUTS is another interesting avenue. See [849] and references to earlier papers therein. Vector leptoquarks in the context of GUTS are discussed in [1109].

As we have seen, some of LQ scenarios are able to address simultaneously the $b \to s\ell\ell$ anomalies as well as the hints for LFUV in $R(D^{(*)})$ [719, 769, 849, 945, 1052, 1057, 1081–1083, 1095, 1100, 1110, 1111]. However, such models are strongly constrained by measurements of the $\tau^+\tau^-$ invariant mass spectrum at the LHC as already pointed out in [1067] and by existing bounds on $B \to K\nu\nu$ and $B \to K^*\nu\nu$ from BaBar and Belle, by existing bounds on LFV tau decays like $\tau \to 3\mu$, from precision measurements of the leptonic couplings of the Z at LEP [765, 767, 1112], and by lepton-universality tests in leptonic tau decays $\tau \to \ell \nu_\tau \bar{\nu}_\ell$ [765, 767, 1112].

It appears to me that not all of these constraints have been taken into account in several papers listed earlier because such analysis is very involved, in particular if RG effects within SMEFT are taken into account. Therefore the codes listed in Chapter 14 turn out to be an invaluable help here.

In this context the analysis in [516], which uses these codes extensively, is important. Considering a multitude of observables this analysis demonstrates again the importance of taking into account loop-effects, both in the RG running and in the matching, as emphasized already in [743, 765, 767]. In particular analyzing the U_1 model the authors of [516] demonstrated that an additional flavor universal shift in C_9, as considered by several authors, can be generated through RG effects from four-fermion operators above the electroweak scale. In this context, as pointed out already in [743], semitautonic operators involving $b \to c$ transitions are most interesting as they allow a simultaneous explanation of $R(D)$ and $R(D^*)$ anomalies. Such effect has been pointed out earlier in [742].

Finally, we recommend a number of analyses devoted to $b \to c\tau\bar{\nu}_\tau$ transitions after Moriond 2019 data that we discussed already in Section 9.4. These are [552–556], which again are dominated by LQ models.

16.4.12 LQ Models and MFV

There is an interesting question whether imposing MFV on LQ models could still allow for the explanation of $R(K)$ and $R(K^*)$. This question has been addressed in [1113]. To this end the charged lepton Yukawa matrix has been taken as the only spurion that violates lepton flavor universality. It has been found that a combination of constraints from $b \to s\mu^+\mu^-$, $b \to s\tau^+\tau^-$, $b \to s\nu\bar{\nu}$, $b\bar{b} \to \tau^+\tau^-$, $b \to c\tau\nu$ transitions, and for scalar LQs from $B_s - \bar{B}_s$ mixing excludes MFV in these models. A similar conclusion has been reached in [1114].

It should be noted that B physics anomalies do not require new complex phases but as shown in these papers already the MFV flavor structure cannot be imposed in LQ models if one wants to understand these anomalies in these models. But MFV is quite generally much stronger violated through the ε'/ε-anomaly. Let us then turn our attention to the ratio ε'/ε within the LQ models.

16.4.13 LQ Models and ε'/ε Anomaly

We have seen in the previous chapter and also in this one that several NP scenarios are able to provide sufficient upward shift in ε'/ε and obtain agreement with experiment. These include in particular tree-level Z' exchanges with explicit realization in 331 models [852, 853] or models with tree-level Z exchanges [766, 799] with explicit realization in models with mixing of heavy vector-like fermions with ordinary fermions [666] and Littlest Higgs model with T-parity [850]. Also simplified Z' scenarios [601, 851] and the MSSM [857] are of help here. As we have just discussed, LQ models are rather successful in addressing B-physics anomalies, which some of the scenarios listed earlier are not able to do. The question then arises how LQs face the ε'/ε anomaly.

Already from the beginning one can expect that the ε'/ε anomaly will be a challenge for those LQ analyses of B-physics anomalies in which all NP couplings have been chosen to be real and those to the first generation set to zero. It should also be realized that the anomalies $R(D)$ and $R(D^*)$, although being very significant, can still be explained in some LQ models through a tree-level LQ exchange. On the other hand, the ε'/ε anomaly, being even larger, if the bound on ε'/ε in [244, 245] is assumed, can only be addressed in these models at one-loop level. This shows that the hinted ε'/ε anomaly is a big challenge for LQ models.

These expectations have been confirmed by a very detailed analysis in [667]. Assuming a mass gap to the EW scale, the main mechanism for LQs to contribute to ε'/ε turns out to be EW gauge-mixing of semileptonic into nonleptonic operators, which is the Scenario V in the context of the SMEFT in Chapter 14. The detailed analysis can be found in [667]. There also one-loop decoupling for scalar LQs has been performed, finding that in all models with both left-handed and right-handed LQ couplings, that is S_1, R_2, and V_2 and U_1, box diagrams generate numerically strongly enhanced EWP operators Q_8 and Q_8' already at the LQ scale. This behavior is rather special for LQs as in most models Q_8 and Q_8' operators cannot be generated at high scale even at NLO and are generated only in QCD RG running to low-energy scales from the operators Q_7 and Q_7', respectively. A good example is the SM and all NP models discussed by us until now.

Investigating correlations of ε'/ε with rare Kaon processes ($K_L \to \pi^0 \nu \bar{\nu}$, $K^+ \to \pi^+ \nu \bar{\nu}$, $K_L \to \pi^0 \ell \bar{\ell}$, $K_S \to \mu \bar{\mu}$, ΔM_K, and ε_K), one finds then that even imposing only a moderate enhancement of $(\varepsilon'/\varepsilon)_{\text{NP}} = 5 \times 10^{-4}$ to explain the current anomaly leads to conflicts with experimental upper bounds on rare Kaon processes. They exclude all LQ models with only a single coupling as an explanation of the ε'/ε anomaly and put serious constraints on parameter spaces of the models S_1, R_2, and V_2 and U_1, where the box diagrams can in principle provide a rescue to LQ models provided both left-handed and right-handed couplings are nonvanishing. However, then the presence of left-right operators contributing not only to ε'/ε but also to $D^0 - \bar{D}^0$ and $K^0 - \bar{K}^0$ mixings requires some fine-tuning of parameters to satisfy all constraints. In the case of V_2 and U_1 the analysis of box diagrams can only be done in a UV completion.

Future results on $K^+ \to \pi^+ \nu \bar{\nu}$ from the NA62 collaboration, $K_L \to \pi^0 \nu \bar{\nu}$ from the KOTO experiment and $K_S \to \mu \bar{\mu}$ from the LHCb will even stronger exhibit the difficulty of LQ models in explaining the measured ε'/ε, in case the ε'/ε anomaly will be confirmed by improved LQCD calculations. Hopefully also improved measurements of $K_L \to \pi^0 \ell \bar{\ell}$ decays will one day help in this context.

The main messages of [667] are then the following ones. If the future improved LQCD calculations will confirm the ε'/ε anomaly at the level $(\varepsilon'/\varepsilon)_{\text{NP}} \geq 5 \times 10^{-4}$ LQs are likely not responsible for it. But if the ε'/ε anomaly will disappear one day, large NP effects in rare K decays that are still consistent with present bounds will be allowed. The analysis in [667] is rather involved, and we will not present it here. But it is an excellent arena to practice the technology of SMEFT, which we introduced in Chapter 14, and anybody who wants to test her (his) skills gained in that chapter should study [667] in detail.

17 Beyond Quark Flavor Physics

17.1 General View

17.1.1 Preliminaries

Until now this book has been dominated by quark flavor-violating processes. Yet in the search for NP it is crucial to consider other phenomena and observables. Here in particular

- Charged lepton violation (LFV),
- Electric dipole moments of the nucleons, atoms, molecules, and leptons (EDMs),
- Anomalous magnetic moments of leptons $a_\ell = \frac{1}{2}(g-2)_\ell$ with $\ell = e, \mu, \tau$, and
- Neutrino oscillations and neutrinoless double β decay

play important roles. Therefore, this chapter will be devoted to the these topics, although our presentation will not be as detailed as was the case of previous chapters not only because the author worked less on lepton flavor physics than quark flavor physics but also because some of the relevant experiments are not as advanced as is the case of those studying meson decays. But as we will see in the case of LFV and EDMs, the contributions from the SM dynamics being very small cannot be measured even in the next decade, and the observation of any positive signal would be a clear evidence for NP. In this respect they differ profoundly from all processes discussed by us until now, which suffer from a large background coming from the SM, and one needs precise theory and precise experiment to identify NP. In the case of LFV and EDMs, theoretical uncertainties by a factor of two still do not matter even if present upper experimental bounds on various observables put already now significant constraints on the parameters of NP models.

The case of $(g-2)_\mu$ is different. On the one hand, the Brookhaven measurement [1115], at the beginning of this millennium, exhibited a significant departure from SM expectations. But the presence of hadronic uncertainties and the fact that only one experiment has seen this anomaly do not allow to make yet firm conclusions whether NP has been seen here or not. The new experiment at Fermilab and persistent efforts by theorists should improve this situation in the coming years. An experiment at J-PARC should provide an independent measurement in the coming years.

Concerning the last item on our list, the observation of neutrino oscillations is a clear signal of physics beyond the SM and so far together with dark matter and matter-antimatter asymmetry observed in our universe the only clear signal of NP. In principle, we could devote one section to this topic in this book. But as in this case, there are already books devoted entirely to this topic, in particular [23], we will only make a few comments and refer to recent reviews.

In the course of our presentation we will frequently refer to the literature. As these four topics are often discussed together, it is useful to refer already now to selected reviews on them, which can be found in [803, 1116–1121]. Many references to the original literature can be found there. Moreover, the study of correlations between LFV, $(g-2)_\mu$, and EDMs in supersymmetric flavor models and SUSY GUTS can be found in [529, 1003, 1122, 1123]. Analogous correlations in models with vector-like leptons have been presented in [1039], and general expressions for these observables in terms of Wilson coefficients of dimension-6 operators can be found in [795]. We will encounter some of these expressions later.

17.1.2 General Considerations

Before going into mathematical details in the next sections, it is useful to discuss certain properties of the topics in question, as in concrete models there exist definite correlations between the observables representing them. But fortunately for us these correlations are model dependent, and studying them one can learn about the dynamics at very short distance scales in a manner we did already in the case of quark flavor physics.

To illustrate it on a concrete example, let us consider first

$$\mathcal{B}(\mu \to e\gamma), \qquad \mathcal{B}(\mu \to 3e), \qquad T(\mu N \to eN), \qquad d_e, \qquad a_e, \qquad (17.1)$$

with the last three representing, respectively, the $\mu \to e$ conversion in nuclei, electric dipole moment of the electron, and its anomalous magnetic moment.

In general, a given interaction can be CP-violating, CP-conserving, flavor-violating, and flavor-conserving. Moreover, the operator structure governing the observables just listed can be model dependent. We note in particular the following features that will be seen explicitly in the formulas presented in the next sections.

- The decay $\mu \to e\gamma$ requires lepton flavor-violation, which could be accompanied by CP-violating phases but could also take place without them. Moreover, similar to $b \to s\gamma$ decay, this transition is fully dominated by dipole operators.
- The decay $\mu \to 3e$ and the transition $\mu N \to eN$ require, similar to $\mu \to e\gamma$, LFV with or without CP-violating phases but are not necessarily fully dominated by dipole operators. Box diagrams and penguin diagrams of various sorts can also give important contributions in certain NP scenarios.
- The electric dipole moment d_e requires CP-violating sources but can proceed without flavor violation. The dipole operator is the prime source here.
- The anomalous magnetic moment a_e does not require any CPV and LFV in contrast to the remaining observables considered here. Similar to $\mu \to e\gamma$ and d_e, it is governed by dipole operators.

These different properties demonstrate that the correlations between observables in (17.1) must be model dependent. This is clearly seen in the correlation between d_e and a_e, both governed by dipole operators and both being flavor conserving [1124]

In the SM enriched by light neutrino masses, lepton-flavor-violating decays $\ell_j \to \ell_i \gamma$ occur at unobservable small rates because the transition amplitudes are suppressed by a factor of $(m_{\nu_j}^2 - m_{\nu_i}^2)/M_W^2$. We will demonstrate this later.

On the other hand, in many extensions of the SM, like supersymmetric models, littlest Higgs model with T-parity (LHT), or models with vector-like leptons (VLLs) measurable in the coming years, branching ratios are predicted, in particular when the masses of involved new particles are in the LHC reach. However, it should be stressed that in principle LFV can even be sensitive to energy scales as high as 1000 TeV. We will return to this point in Chapter 19.

Additional interest in LFV arose in the last years due to indications of the violation of *lepton flavor universality* (LFU) in B decays, discussed in previous chapters. In some BSM models it implies measurable LFV branching ratios. We will discuss these intriguing implications only very briefly at the end of this section. For the time being we will assume LFU.

A very prominent role in the LFV studies is played by the radiative decays

$$\mu \to e\gamma, \qquad \tau \to \mu\gamma, \qquad \tau \to e\gamma, \tag{17.14}$$

which are analogs of $b \to s\gamma$ considered already by us. But the full picture of possible NP can only be obtained through a simultaneous study of the decays

$$\mu^- \to e^- e^+ e^-, \qquad \tau^- \to \mu^- \mu^+ \mu^-, \qquad \tau^- \to e^- e^+ e^-, \tag{17.15}$$

$$\tau^- \to e^- \mu^+ \mu^-, \qquad \tau^- \to \mu^- e^+ e^-, \qquad \tau^- \to \mu^- e^+ \mu^-, \qquad \tau^- \to e^- \mu^+ e^-, \tag{17.16}$$

as well as of the $\mu - e$ conversion in nuclei. Also processes like $e^+ \mu^- \to e^+ e^-$ are of interest.

In this context let us mention already now how a clear-cut distinction between supersymmetric models, LHT model, and a model with a fourth sequential generation (SM4) is possible on the basis of all these decays. While it is not possible to distinguish the LHT model from the supersymmetric models on the basis of $\mu \to e\gamma$ alone, it has been pointed out in [1125] that such a distinction can be made by measuring any of the ratios $\mathcal{B}(\mu \to 3e)/\mathcal{B}(\mu \to e\gamma)$, $\mathcal{B}(\tau \to 3\mu)/\mathcal{B}(\tau \to \mu\gamma)$, etc. In supersymmetric models all these decays are governed by dipole operators so that these ratios are $O(\alpha)$ [1123, 1126–1131]. In the LHT model the LFV decays with three leptons in the final state are not governed by dipole operators but by Z-penguins and box diagrams, and the ratios in question turn out to be by almost an order of magnitude larger than in supersymmetric models. Other analyses of LFV in the LHT model can be found in [1132, 1133] and in the MSSM in [1123]. In the latter paper $(g - 2)_e$ was used to probe lepton flavor-violating couplings that are correlated with $\tau \to e\gamma$. For a detailed analysis of the correlation between $\mu \to 3e$ and $(g - 2)_\mu$, see [804]. A recent review of LFV including experimental prospects for coming years can be found in [1134].

Similarly, as pointed out in [1135], the pattern of the LFV branching ratios in the SM4 differs significantly from the one encountered in the MSSM, allowing to distinguish these

Table 17.1 Experimental upper limits on the branching ratios of the radiative lepton decays and experimental prospects.

Process	Experimental bound	Experimental prospects
$\mathcal{B}(\tau \to \mu\gamma)$	4.4×10^{-8} [1150]	4.4×10^{-9} [1151]
$\mathcal{B}(\tau \to e\gamma)$	3.3×10^{-8} [1150]	3.3×10^{-9} [1151]
$\mathcal{B}(\mu \to e\gamma)$	4.2×10^{-13} [1152]	4×10^{-14} [1139]
$\mathcal{B}(\mu \to e)$	$O(10^{-12})$ [1153]	10^{-17} [1154, 1155]

Table 17.2 Experimental upper limits on the branching ratios of the three body charged lepton decays and experimental prospects.

Process	Experimental bound	Experimental prospects
$\mathcal{B}(\mu^- \to e^- e^+ e^-)$	1.0×10^{-12} [1156]	10^{-16} [1140]
$\mathcal{B}(\tau^- \to \mu^- \mu^+ \mu^-)$	1.2×10^{-8} [150]	10^{-9} [1157]
$\mathcal{B}(\tau^- \to e^- e^+ e^-)$	1.4×10^{-8} [150]	10^{-9} [1157]
$\mathcal{B}(\tau^- \to e^- \mu^+ \mu^-)$	1.6×10^{-8} [150]	10^{-9} [1157]
$\mathcal{B}(\tau^- \to \mu^- e^+ e^-)$	1.1×10^{-8} [150]	10^{-9} [1157]
$\mathcal{B}(\tau^- \to \mu^- e^+ \mu^-)$	9.8×10^{-9} [150]	10^{-9} [1157]
$\mathcal{B}(\tau^- \to e^- \mu^+ e^-)$	8.4×10^{-9} [150]	10^{-9} [1157]

two models with the help of LFV processes in a transparent manner. Also differences from the LHT model were identified. Even if fourth generation has been already ruled out, these studies demonstrate the power of a combined analysis of the decays in (17.14)–(17.16).

Of particular interest is the study of τ decays because as opposed to muon decays, we have at our disposal many more channels involving also pseudoscalar mesons in the final state. In fact hadronic τ decays offer powerful probes of NP in the LHC era [1136]. The physics reach and model-discriminating power of LFV tau decays is most efficiently analyzed above the electroweak scale using SMEFT and in the corresponding LEFT [1137]. For an earlier analysis, see [1138]. We will give the relevant formulas in Section 17.2.3.

A detailed analysis of LFV in various extensions of the SM is also motivated by the prospects (see Tables 17.1 and 17.2) for the measurements of LFV processes with much higher sensitivity than presently available. In particular, the MEG experiment at PSI is already testing $\mathcal{B}(\mu \to e\gamma)$ at the level of $O(10^{-13})$ putting severe constraints on the parameters of a number of models to be mentioned later. An upgrade for MEG is also already approved [1139] so that the sensitivity down to $4 \cdot 10^{-14}$ after three years of running is expected. There is also an approved proposal at PSI to do $\mu \to eee$ [1140]. The planned accuracy of SuperKEKB of $O(10^{-8})$ for $\tau \to \mu\gamma$ is also of great interest. This decay can also be studied by the LHCb experiment at CERN.

An improved upper bound on $\mu - e$ conversion in titanium will also be very important. In this context the dedicated J-PARC experiment PRISM/PRIME [1141] should reach the sensitivity of $O(10^{-18})$, i.e., an improvement by six orders of magnitude relative to the present upper bound from SINDRUM-II at PSI [1142]. Similar comments apply to COMET at J-PARC [1143, 1144]. Mu2e collaboration will measure $\mu - e$ conversion on aluminium to $6 \cdot 10^{-17}$ at 90% CL around the year 2020 [1145], which is a factor of 10^4 better than SINDRUM-II. In [1146] the model discriminating power of a combined phenomenological analysis of $\mu \to e\gamma$ and $\mu \to e$ conversion on different nuclei targets is discussed. A nice summary of all these efforts is given in [1147].

For further detailed reviews of LFV, see [1117, 1148, 1149]. An experimenter's guide for charged LFV can be found in [1121]. In what follows we will first discuss these decays in the SM. Subsequently we will collect formulas for them that will be sufficiently general to allow motivated readers to perform analyses in specific NP scenarios. Having these formulas we will illustrate in Section 17.2.9 their application in the context of Z' models.

17.2.2 $\mu \to e\gamma$ in the SM

To calculate the rate for $\mu \to e\gamma$ in the SM, let us compare the diagrams contributing to the $B \to X_s\gamma$ and $\mu \to e\gamma$ decays. They are shown in Figures 17.1 and 17.2, respectively. In R_ξ gauges also the corresponding diagrams with Goldstone bosons contribute. The diagram with the photon coupled directly to the internal fermion line, present in $B \to X_s\gamma$, is absent in the case of $\mu \to e\gamma$ due to the vanishing neutrino charge.

In the case of the $B \to X_s\gamma$ decay the function $D_0'(x)$ resulting from the diagrams in Figure 17.1 can be decomposed as follows

$$D_0'(x) = \left[D_0'(x)\right]_{\text{Abelian}} + \left[D_0'(x)\right]_{\text{triple}}, \tag{17.17}$$

where the first term on the r.h.s. represents the sum of the first and the last diagram in Figure 17.1 and the second term represents the second diagram.

The inspection of an explicit calculation of the $b \to s\gamma$ transition in the 't Hooft–Feynman gauge gives

$$\left[D_0'(x)\right]_{\text{Abelian}}^{b \to s\gamma} = (2Q_d - Q_u)E_0'(x), \tag{17.18}$$

where $E_0'(x)$ is one of the one-loop master functions introduced in Chapter 6. Q_d and Q_u are the electric charges of external and internal fermions, respectively. Setting $Q_d = -1/3$ and $Q_u = 2/3$ in (17.18) and using (17.17), we find

$$\left[D_0'(x)\right]_{\text{triple}} = D_0'(x) + \frac{4}{3}E_0'(x). \tag{17.19}$$

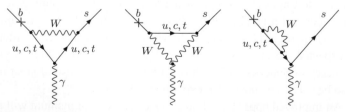

Figure 17.1 Diagrams contributing to $b \to s\gamma$ in the SM.

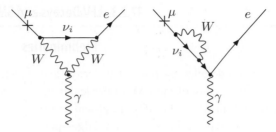

Figure 17.2 Diagrams contributing to $\mu \to e\gamma$ in the SM.

This contribution is independent of fermion electric charges and can be directly used in the $\mu \to e\gamma$ decay.

Turning now our attention to the latter decay, we find first from (17.18)

$$\left[D_0'(x)\right]_{\text{Abelian}}^{\mu \to e\gamma} = -2E_0'(x), \tag{17.20}$$

as $Q_u = Q_\nu = 0$ and $Q_d = Q_l = -1$ in this case. The final result for the relevant short-distance function in the case of $\mu \to e\gamma$ is then given by the sum of (17.19) and (17.20). Denoting this function by $H(x_\nu^i)$, we find

$$H(x_\nu^i) = D_0'(x_\nu^i) - \frac{2}{3}E_0'(x_\nu^i), \qquad x_\nu^i = \left(\frac{m_\nu^i}{M_W}\right)^2. \tag{17.21}$$

The generalization of the known SM result to arbitrary neutrino masses reads then

$$\mathcal{B}(\mu \to e\gamma)_{\text{SM}} = \frac{3\alpha}{2\pi}\left|\sum_i U_{ei}U_{\mu i}^* H(x_\nu^i)\right|^2, \tag{17.22}$$

with U_{ij} being the elements of the PMNS matrix.

Now, in the limit of small neutrino masses,

$$H(x_\nu^i) \longrightarrow \frac{x_\nu^i}{4} \qquad \text{as} \quad x_\nu^i \to 0, \tag{17.23}$$

and we confirm the known result

$$\mathcal{B}(\mu \to e\gamma)_{\text{SM}} = \frac{3\alpha}{32\pi}\left|\sum_i U_{ei}U_{\mu i}^* x_\nu^i\right|^2. \tag{17.24}$$

With the input in (2.112), we find

$$\mathcal{B}(\mu \to e\gamma)_{\text{SM}} \simeq 0.015\,\alpha\, s_{13}^2 \left[\frac{\Delta m_{32}^2}{M_W^2}\right]^2 < 10^{-54}. \tag{17.25}$$

Certainly such branching ratios cannot be measured in this century.

17.2.3 LFV Decays and SMEFT

Preliminaries

Over several decades by now LFV decays and transitions have been studied in details in many extensions of the SM, in particular in supersymmetric models and models like LHT model, in which FCNCs are absent at tree-level. In such models LFV decays are governed by γ-penguin, Z-penguin, Higgs-penguin, and box diagrams with heavy new particles in the loop. Useful general expressions for branching ratios used in the MSSM have been derived in [1127, 1158] and used in the context of the LHT model in [1125]. The interested reader can look at these papers that in principle can be used in any extension of the SM. However, for models, like Z' models, LQ models, models with FCNC mediated by heavy scalars, where FCNC are entering already at tree-level, it is more efficient to use the SMEFT formulation. Yet, also in this case, as we will see soon, contributions of γ-penguins and their correlations with four-fermion operators have to be included.

For LFV in charged lepton decays such an analysis has been done in [795], and we will collect the main results of this paper here. Some of these expressions will also give us automatically the results for EDMs of leptons and also $(g-2)_{e,\mu}$ that will be discussed in two subsequent sections.

In Table 17.3 we collect the independent dimension-6 operators relevant for our discussion. It is a subset of the operators listed in Tables 14.1 and 14.2. These are all operators that can contribute to LFV processes in the charged lepton sector at the tree-level or at the 1-loop level. The operators that could give LFV effects only via the interference with the dimension-4 SM vertices containing the PMNS matrix are neglected because such effects are suppressed by the small neutrino masses. To simplify the notation, the flavor indices in the name of operators are suppressed in this table, but they will be exposed whenever it will be necessary. Here we just give one example that specifies the order of the flavor indices:

$$Q_{\ell\ell}^{jikl} = (\bar{\ell}_j\gamma_\mu\ell_i)(\bar{\ell}_k\gamma^\mu\ell_l) \tag{17.26}$$

with i, j, k, l running from 1 to 3. For instance, $i = 2 = \mu$, $i = 3 = \tau$, and $k = 1 = e$.

Table 17.3 Complete list of the dimension-6 operators (invariant under the SM gauge group) that contribute to the LFV observables under consideration at the tree- or at the one-loop level [795].

$\ell\ell\ell\ell$		$\ell\ell X\varphi$		$\ell\ell\varphi^2 D$ and $\ell\ell\varphi^3$	
$Q_{\ell\ell}$	$(\bar{\ell}_i\gamma_\mu\ell_j)(\bar{\ell}_k\gamma^\mu\ell_l)$	Q_{ew}	$(\bar{\ell}_o\sigma^{\mu\nu}e_j)\tau^I\varphi W^I_{\mu\nu}$	$Q_{\varphi\ell}^{(1)}$	$(\varphi^\dagger i\overleftrightarrow{D}_\mu\varphi)(\bar{\ell}_i\gamma^\mu\ell_j)$
Q_{ee}	$(\bar{e}_i\gamma_\mu e_j)(\bar{e}_k\gamma^\mu e_l)$	Q_{eB}	$(\bar{\ell}_i\sigma^{\mu\nu}e_j)\varphi B_{\mu\nu}$	$Q_{\varphi\ell}^{(3)}$	$(\varphi^\dagger i\overleftrightarrow{D}^I_\mu\varphi)(\bar{\ell}_i\tau^I\gamma^\mu\ell_j)$
$Q_{\ell e}$	$(\bar{\ell}_i\gamma_\mu\ell_j)(\bar{e}_k\gamma^\mu e_l)$			$Q_{\varphi e}$	$(\varphi^\dagger i\overleftrightarrow{D}_\mu\varphi)(\bar{e}_i\gamma^\mu e_j)$
				$Q_{e\varphi 3}$	$(\varphi^\dagger\varphi)(\bar{\ell}_i e_j\varphi)$
		$\ell\ell qq$			
$Q_{\ell q}^{(1)}$	$(\bar{\ell}_i\gamma_\mu\ell_j)(\bar{q}_k\gamma^\mu q_l)$	$Q_{\ell d}$	$(\bar{\ell}_i\gamma_\mu\ell_j)(\bar{d}_k\gamma^\mu d_l)$	$Q_{\ell u}$	$(\bar{\ell}_i\gamma_\mu l_j)(\bar{u}_k\gamma^\mu u_l)$
$Q_{\ell q}^{(3)}$	$(\bar{\ell}_i\gamma_\mu\tau^I\ell_j)(\bar{q}_k\gamma^\mu\tau^I q_l)$	Q_{ed}	$(\bar{e}_i\gamma_\mu e_j)(\bar{d}_k\gamma^\mu d_l)$	Q_{eu}	$(\bar{e}_i\gamma_\mu e_j)(\bar{u}_k\gamma^\mu u_l)$
Q_{eq}	$(\bar{e}_i\gamma^\mu e_j)(\bar{q}_k\gamma_\mu q_l)$	$Q_{\ell edq}$	$(\bar{\ell}_i^a e_j)(\bar{d}_k q_l^a)$	$Q_{\ell equ}^{(1)}$	$(\bar{\ell}_i^a e_j)\varepsilon_{ab}(\bar{q}_k^b u_l)$
				$Q_{\ell equ}^{(3)}$	$(\bar{\ell}_i^a\sigma_{\mu\nu}e_a)\varepsilon_{ab}(\bar{q}_k^b\sigma^{\mu\nu}u_l)$

One could interchange, of course, j and i to have alphabetic ordering $ijkl$, but as the authors of [795] made this labeling, it is safer to keep it in this section. In case a reader will be lost in the indices, this formula should be useful as well as the discussion of the indices in section II of the latter paper.

In [795] the full one-loop predictions for the $\ell_i \rightarrow \ell_f \gamma$ decays and all tree-level contributions for $\ell_i \rightarrow \ell_j \bar{\ell}_l \ell_k$ decays in terms of the Wilson coefficients of the dimension-6 operators have been calculated. Motivated readers should have a look at this paper. In particular it has been found that the contributions of some of the operators in Table 17.3 can be neglected, and their Wilson coefficients will be absent in the final formulas for branching ratios.

17.2.4 General Formulas for $l_j^- \rightarrow l_i^- \gamma$ beyond the SM

We will first present the formulas in terms of WCs of more familiar operators, and subsequently we will relate them to the WCs of the SMEFT. We begin with the general form of the flavor-violating photon-lepton vertex that can be written in terms of vector, scalar, and dipole formfactors as follows

$$
V_{\ell\ell\gamma}^{fi\,\mu} = \frac{i}{\Lambda^2} \left[\gamma^\mu (F_{VL}^{fi} P_L + F_{VR}^{fi} P_R) + (F_{SL}^{fi} P_L + F_{SR}^{fi} P_R) q^\mu + (F_{TL}^{fi} i\sigma^{\mu\nu} P_L + F_{TR}^{fi} i\sigma^{\mu\nu} P_R) q_\nu \right].
$$

$$(17.27)$$

Once the formfactors entering this equation have been expressed in terms of WCs of the SMEFT operators, the branching ratios for the $\ell_i \rightarrow \ell_f \gamma$ decays (with $i > f$) at the 1-loop level up to the order $1/\Lambda^2$ can be directly given in terms of these WCs and used for any model. In addition, the obtained results are directly related to the EDMs and anomalous magnetic moments of leptons after setting $f = i$. We will use them in the next two sections.

Now, gauge-invariance requires that F_{VL} and F_{VR} must vanish for on-shell external particles. Moreover, the formfactors F_{SL} and F_{SR} do not contribute to the $\ell_i \rightarrow \ell_f \gamma$ decay amplitude. Therefore, finally the branching ratios can be expressed in terms of F_{TL}^{fi} and F_{TR}^{fi} only:

$$
\mathcal{B}\left(\ell_i \rightarrow \ell_f \gamma\right) = \frac{m_{\ell_i}^3}{16\pi\Lambda^4 \Gamma_{\ell_i}} \left(\left|F_{TR}^{fi}\right|^2 + \left|F_{TL}^{fi}\right|^2 \right).
$$

$$(17.28)$$

Here Γ_{ℓ_i} denotes the total decay width of the decaying lepton that was discussed in Chapter 3. Comparing with (17.13), evidently

$$
\frac{F_{TR}}{\Lambda^2} = F_{M1} + F_{E1}, \qquad \frac{F_{TL}}{\Lambda^2} = F_{M1} - F_{E1}.
$$

$$(17.29)$$

Now comes an important point. Among the operators in Table 17.3 only the operators Q_{eW} and Q_{eB} can contribute to $F_{TL,R}^{fi}$ at the tree-level. If their coefficients are comparable to other Wilson coefficients of the dimension-6 operators, they dominate the effective photon-lepton vertex, with the formfactors simply given by ($v = 246\,\text{GeV}$):

$$F_{TR}^{fi} = F_{TL}^{if*} = v\sqrt{2}\left(c_W C_{eB}^{fi} - s_W C_{eW}^{fi}\right) \equiv v\sqrt{2}C_\gamma^{fi}. \qquad (17.30)$$

Yet, as emphasized in [795], in a renormalizable theory of NP the operators Q_{eW} and Q_{eB} can only be generated at the loop-level while other operators, like the effective four-lepton couplings, can already be generated at the tree-level. In some extensions of the SM C_{eW} and C_{eB} may even not be generated at all [1159]. Thus, comparable (or even dominant) contributions to the flavor-violating lepton-lepton-photon vertex can come from other dimension-6 operators, which for consistency, as done in [795], should be included at the one-loop level.

The final expressions for the nonvanishing one-loop contributions are collected in Table 17.4, where the first order of expansions in $1/\Lambda^2$ and m_l has been kept. In this excellent approximation only the five Wilson coefficients $C_{\ell e}^{fjji}$, $C_{\ell equ}^{(3)fijj}$, $C_{\varphi\ell}^{(3)fi}$, $C_{\varphi e}^{fi}$, and $C_{\varphi\ell}^{(1)fi}$ contribute.

We note that the terms proportional to $C_{\ell equ}^{(3)}$ contain a divergence Δ. This divergence must be canceled by a counterterm to Q_{eW} and/or Q_{eB}. We refer to [795] for the discussion of this issue.

Table 17.4 One-loop contributions to formfactors F_{TL}^{fi} and F_{TR}^{fi} giving rise to $\ell_i \to \ell_f \gamma$ up to order $1/\Lambda^2$ [795].

Different contributions	Tensor formfactors
Z	$F_{TL}^{Z\,fi} = \dfrac{4e\left[\left(C_{\varphi\ell}^{(1)fi} + C_{\varphi\ell}^{(3)fi}\right)m_f(1 + s_W^2) - C_{\varphi e}^{fi}m_i\left(\frac{3}{2} - s_W^2\right)\right]}{3(4\pi)^2}$
	$F_{TR}^{Z\,fi} = \dfrac{4e\left[\left(C_{\varphi\ell}^{(1)fi} + C_{\varphi l}^{(3)fi}\right)m_i(1 + s_W^2) - C_{\varphi e}^{fi}m_f\left(\frac{3}{2} - s_W^2\right)\right]}{3(4\pi)^2}$
W	$F_{TL}^{W\,fi} = -\dfrac{10em_f C_{\varphi\ell}^{(3)fi}}{3(4\pi)^2}$
	$F_{TR}^{W\,fi} = -\dfrac{10em_i C_{\varphi\ell}^{(3)fi}}{3(4\pi)^2}$
Contact 4-fermion	$F_{TL}^{4f\,fi} = -\dfrac{16e}{3(4\pi)^2}\sum_{j=1}^3 C_{\ell equ}^{(3)fijj\star}m_{u_j}\left(\Delta - \log\dfrac{m_{u_j}^2}{\mu^2}\right)$
	$F_{TR}^{4f\,fi} = -\dfrac{16e}{3(4\pi)^2}\sum_{j=1}^3 C_{\ell equ}^{(3)fijj}m_{u_j}\left(\Delta - \log\dfrac{m_{u_j}^2}{\mu^2}\right)$
Contact 4-lepton	$F_{TL}^{4\ell\,fi} = \dfrac{2e}{(4\pi)^2}\sum_{j=1}^3 C_{\ell e}^{fjji}m_j$
	$F_{TR}^{4\ell\,fi} = \dfrac{2e}{(4\pi)^2}\sum_{j=1}^3 C_{\ell e}^{jifj}m_j$

17.2.5 General Formulas for $\ell_i \to \ell_j \bar{\ell}_l \ell_k$ beyond the SM

We will next present general formulas for the branching ratios of the seven decays listed in (17.15) and (17.16). The authors of [795] split the expressions for the $\ell_i \to \ell_j \bar{\ell}_l \ell_k$ decays into three groups, depending on the composition of the final state leptons:

(A) Three leptons of the same flavor: $\mu^- \to e^- e^+ e^-$, $\tau^- \to e^- e^+ e^-$, and $\tau^- \to \mu^- \mu^+ \mu^-$.
(B) Three distinguishable leptons: $\tau^- \to e^- \mu^+ \mu^-$ and $\tau^- \to \mu^- e^+ e^-$.
(C) Two leptons of the same flavor and charge and one with different flavor and opposite charge: $\tau^- \to \mu^- e^+ \mu^-$ and $\tau^- \to e^- \mu^+ e^-$.

To calculate the branching ratios the amplitude A for the decay $\ell_i \to \ell_j \bar{\ell}_l \ell_k$ is decomposed as

$$A = A_0 + A_\gamma, \tag{17.31}$$

where A_0 contains all operators for which one can neglect the momenta of the external leptons and A_γ is the photon contribution generated by Q_{eW} and Q_{eB} only. The amplitude A_0 can then be written as:

$$A_0 = \frac{1}{\Lambda^2} \sum_I C_I [\bar{u}(p_j) Q_I u(p_i)][\bar{u}(p_k) Q'_I v(p_l)], \tag{17.32}$$

where u and v are usual Dirac spinors, and the momenta assignments are indicated by the indices and are shown explicitly in figure 4 of [795].

For a given $(ji; kl)$ the various operators contributing to A_0 are

$$[Q_{VXY}]^{ji}_{kl} = (\bar{\psi}_j \gamma^\mu P_X \psi_i)(\bar{\psi}_k \gamma_\mu P_Y \psi_l), \qquad [Q_{SXY}]^{ji}_{kl} = (\bar{\psi}_j P_X \psi_i)(\bar{\psi}_k P_Y \psi_l) \tag{17.33}$$

$$[Q_{TXX}]^{ji}_{kl} = (\bar{\psi}_j \sigma^{\mu\nu} P_X \psi_i)(\bar{\psi}_k \sigma_{\mu\nu} P_X \psi_l), \tag{17.34}$$

with X, Y standing for the chiralities L and R. The corresponding effective Hamiltonian is then simply given as

$$[\Delta\mathcal{H}_{\text{eff}}]^{ji}_{kl} = -\frac{1}{\Lambda^2} \sum_{X,Y} \left([C_{VXY}]^{ji}_{kl}[Q_{VXY}]^{ji}_{kl} + [C_{SXY}]^{ji}_{kl}[Q_{SXY}]^{ji}_{kl} + [C_{TXX}]^{ji}_{kl}[Q_{TXX}]^{ji}_{kl} \right), \tag{17.35}$$

so that in the branching ratios that follow we will encounter the coefficients C_{VLL}, C_{VLR}, C_{SLL}, C_{SLR}, C_{TLL}, and those with L and R interchanged.

The complication arises when two identical leptons are present in the final state because then one has to include crossed diagrams in which a different ordering of spinors appears. Using Fierz transformations collected in Appendix A.3, one can always bring the sum of these diagrams into the form in (17.32) but the presence of two diagrams has an effect on the Wilson coefficients of the relevant operators. In addition, when calculating branching ratios a factor 1/2 has to be included.

The contributions from photon exchange for various types of decays (A), (B), (C) read (retaining only $1/\Lambda^2$ terms) [795] ($p_{ij} = p_i - p_j$)

$$A_\gamma^{(A)} = \frac{ev}{\Lambda^2}\left(\frac{1}{p_{ij}^2}[\bar{u}(p_j)i\sigma^{\mu\nu}(C_{\gamma L}P_L + C_{\gamma R}P_R)(p_{ij})_\nu u(p_i)][\bar{u}(p_k)\gamma_\mu v(p_l)] - (p_j \leftrightarrow p_k)\right),$$

$$A_\gamma^{(B)} = \frac{ev}{\Lambda^2}\frac{1}{p_{ij}^2}[\bar{u}(p_j)i\sigma^{\mu\nu}(C_{\gamma L}P_L + C_{\gamma R}P_R)(p_{ij})_\nu u(p_i)][\bar{u}(p_k)\gamma_\mu v(p_l)],$$

$$A_\gamma^{(C)} = 0. \tag{17.36}$$

They bring in two additional coefficients $C_{\gamma L}$ and $C_{\gamma R}$.

The general expression for the spin averaged square matrix element $\mathcal{M} = \frac{1}{2}\sum_{pol}|A|^2$ can be significantly simplified by using the hierarchy of the charged lepton masses: $m_i \equiv M \gg m_j, m_k, m_l$. That is, $M = m_\mu$ or $M = m_\tau$. One finds then [795]

$$\mathcal{B}(\ell_i \to \ell_j\bar{\ell}_l\ell_k) = \frac{N_c M^5}{6144\pi^3\Lambda^4\Gamma_{\ell_i}}\left(4\left(|C_{VLL}|^2 + |C_{VRR}|^2 + |C_{VLR}|^2 + |C_{VRL}|^2\right)\right.$$

$$+ |C_{SLL}|^2 + |C_{SRR}|^2 + |C_{SLR}|^2 + |C_{SRL}|^2$$

$$\left. + 48\left(|C_{TLL}|^2 + |C_{TRR}|^2\right) + X_\gamma\right), \tag{17.37}$$

where $N_c = 1/2$ if two of the final state leptons are identical, $N_c = 1$ in all other cases, and Γ_{ℓ_i} is the total decay width of the initial lepton. The photon penguin contribution is a bit more complicated and reads

$$X_\gamma^{(A)} = -\frac{16ev}{M}\text{Re}\left[\left(2C_{VLL} + C_{VLR} - \frac{1}{2}C_{SLR}\right)C_{\gamma R}^\star + \left(2C_{VRR} + C_{VRL} - \frac{1}{2}C_{SRL}\right)C_{\gamma L}^\star\right]$$

$$+ \frac{64e^2v^2}{M^2}\left(\log\frac{M^2}{m^2} - \frac{11}{4}\right)(|C_{\gamma L}|^2 + |C_{\gamma R}|^2),$$

$$X_\gamma^{(B)} = -\frac{16ev}{M}\text{Re}\left[(C_{VLL} + C_{VLR})C_{\gamma R}^\star + (C_{VRR} + C_{VRL})C_{\gamma L}^\star\right]$$

$$+ \frac{32e^2v^2}{M^2}\left(\log\frac{M^2}{m^2} - 3\right)(|C_{\gamma L}|^2 + |C_{\gamma R}|^2),$$

$$X_\gamma^{(C)} = 0. \tag{17.38}$$

Having these general formulas at hand, the next step is to express the coefficients C_I appearing in these formulas in terms of the WCs of the SMEFT operators. This has been done in [795], and we will now collect their results.

17.2.6 $\mu^- \to e^- e^+ e^-, \tau^- \to \mu^- \mu^+ \mu^-, \tau^- \to e^- e^+ e^-$ in the SMFT

$$
\begin{aligned}
C_{VLL} &= 2\left((2s_W^2 - 1)\left(C_{\varphi\ell}^{(1)ji} + C_{\varphi\ell}^{(3)ji}\right) + C_{\ell\ell}^{jijj}\right), \\
C_{VRR} &= 2\left(2s_W^2 C_{\varphi e}^{ji} + C_{ee}^{jijj}\right), \\
C_{VLR} &= -\frac{1}{2}C_{SRL} = 2s_W^2\left(C_{\varphi\ell}^{(1)ji} + C_{\varphi\ell}^{(3)ji}\right) + C_{\ell e}^{jijj}, \\
C_{VRL} &= -\frac{1}{2}C_{SLR} = (2s_W^2 - 1)C_{\varphi e}^{ji} + C_{\ell e}^{jjji}, \\
C_{SLL} &= C_{SRR} = C_{TLL} = C_{TRR} = 0, \\
C_{\gamma L} &= \sqrt{2}C_{\gamma}^{ij*}, \\
C_{\gamma R} &= \sqrt{2}C_{\gamma}^{ji}, \qquad \text{with } C_{\gamma}^{ji} \text{ in (17.30)}. \tag{17.39}
\end{aligned}
$$

17.2.7 $\tau^- \to e^- \mu^+ \mu^-$ and $\tau^- \to \mu^- e^+ e^-$ in the SMFT

Before giving the results for these decays let us make a few comments to distinguish them from the previous ones. They do not have identical leptons in the final state and have two types of contributions. First of all, they proceed as in $\tau^- \to \mu^- \mu^+ \mu^-$ and $\tau^- \to e^- e^+ e^-$ through $\Delta L = 1$ tree, penguin, and box diagrams. As there are no identical particles in the final state, the effective Hamiltonians for these contributions can be directly obtained from the decay $B \to X_s \ell^+ \ell^-$, and this is a good exercise for any reader that reached so far. In case of difficulties an explicit derivation of the effective Hamiltonian for $\tau^- \to \mu^- e^+ e^-$ is presented in the context of the LHT model in [1125]. The generalization to $\tau^- \to e^- \mu^+ \mu^-$ is automatic.

But these decays can also receive $\Delta L = 2$ contributions and also in this case an explicit derivation of the branching ratio in the context of the LHT model can be found in [1125]. Which of the contributions dominates the branching ratios depends to some extent on the pattern of flavor-violating couplings. But quite generally $\Delta L = 1$ contributions dominate.

For these decays the WCs of the most familiar operators are related to the SMEFT ones as follows [795]

$$
\begin{aligned}
C_{VLL} &= (2s_W^2 - 1)\left(C_{\varphi\ell}^{(1)ji} + C_{\varphi l}^{(3)ji}\right) + C_{\ell\ell}^{jikk}, \\
C_{VRR} &= 2s_W^2 C_{\varphi e}^{ji} + C_{ee}^{jikk}, \\
C_{VLR} &= 2s_W^2\left(C_{\varphi\ell}^{(1)ji} + C_{\varphi\ell}^{(3)ji}\right) + C_{\ell e}^{jikk}, \\
C_{VRL} &= (2s_W^2 - 1)C_{\varphi e}^{ji} + C_{\ell e}^{jkki}, \\
C_{SLR} &= -2C_{\ell e}^{jkki}, \\
C_{SRL} &= -2C_{\ell e}^{jikk}, \\
C_{SLL} &= C_{SRR} = C_{TLL} = C_{TRR} = 0, \\
C_{\gamma L} &= \sqrt{2}C_{\gamma}^{ij*}, \\
C_{\gamma R} &= \sqrt{2}C_{\gamma}^{ji}. \tag{17.40}
\end{aligned}
$$

17.2.8 $\tau^- \to \mu^- e^+ \mu^-$ and $\tau^- \to e^- \mu^+ e^-$ in the SMFT

In these decays either there are two muons or two electrons in the final state with the third lepton being e^+ and μ^+, respectively. These decays can proceed at tree-level. In models in which FCNC appear first at one-loop level, like the LHT model, they are dominated by box-diagrams as Z-penguins and γ-penguins cannot contribute. As these two decays are of $\Delta L = 2$ type, they are generally strongly suppressed. The relevant effective Hamiltonian can be obtained from $\Delta F = 2$ processes in the quark sector by properly changing flavors and taking into account that two identical particles are present in the final state. Interested readers can find the construction of the effective Hamiltonian and the calculation of branching ratios for the case of the LHT model in [1125]. The resulting branching ratio is very strongly suppressed. Here we just quote the result from [795]

$$C_{VLL} = 2C_{\ell\ell}^{kikj},$$

$$C_{VRR} = 2C_{ee}^{kikj},$$

$$C_{VLR} = -\frac{1}{2}C_{SRL} = C_{\ell e}^{kikj},$$

$$C_{VRL} = -\frac{1}{2}C_{SLR} = C_{\ell e}^{kjki},$$

$$C_{SLL} = C_{SRR} = C_{TLL} = C_{TRR} = 0,$$

$$C_{\gamma L} = C_{\gamma R} = 0. \tag{17.41}$$

17.2.9 Three-Body Decays in Z' Models

Calculating in Z' models directly tree diagrams for the three classes (A) – (C) just discussed, one finds for each of them the Wilson coefficient

$$[C_{VXY}]_{kl}^{ji} = -\frac{\Delta_X^{ji}(Z')\Delta_Y^{kl}(Z')}{M_{Z'}^2}, \tag{17.42}$$

where the minus sign is the consequence of the definition of WCs by the Lagrangian and not the Hamiltonian.

In calculating the branching ratios for the three classes of decays just discussed, we note that always $i \neq j$, but dependently on the class considered $k = l$ and/or $k \neq l$.

Class A

As defined earlier, these are decays with $k = l$ only. The presence of two identical leptons in the final state requires the introduction of a factor $1/2$ at the level of the branching ratio. Moreover, always two diagrams, differing by the interchange of identical leptons, contribute. They interfere with each other for VLL and VRR cases but not for VLR and VRL. We find then ($\Gamma_\tau = 1/\tau_\tau$)

$$\mathcal{B}(\tau^- \to \mu^- \mu^+ \mu^-) = \frac{m_\tau^5}{1536\pi^3\Gamma_\tau} \left[2|C_{VLL}|^2 + 2|C_{VRR}|^2 + |C_{VLR}|^2 + |C_{VRL}|^2 \right]_{\mu\mu}^{\mu\tau},$$

(17.43)

with analogous expressions for $\mu^- \to e^- e^+ e^-$ and $\tau^- \to e^- e^+ e^-$. Γ_τ is the total τ decay width.

Class B

Here both $k = l$ and $k \neq l$ are possible, but there are no identical leptons in the final state so that the branching ratio can easily be obtained from the formula for μ or τ decay in the limit of vanishing final lepton masses. We find then for the $\Delta L = 1$ contribution

$$\mathcal{B}(\tau^- \to \mu^- e^+ e^-) = \frac{m_\tau^5}{1536\pi^3\Gamma_\tau} \left[|C_{VLL}|^2 + |C_{VRR}|^2 + |C_{VLR}|^2 + |C_{VRL}|^2 \right]_{ee}^{\mu\tau},$$

(17.44)

with an analogous expression for $\tau^- \to e^- \mu^+ \mu^-$. To obtain $\Delta L = 2$ contribution μ^- and e^- should be interchanged.

Class C

Here we only have $k \neq l$ but two identical leptons in the final state. We find similar to Class A

$$\mathcal{B}(\tau^- \to \mu^- e^+ \mu^-) = \frac{m_\tau^5}{1536\pi^3\Gamma_\tau} \left[2|C_{VLL}|^2 + 2|C_{VRR}|^2 + |C_{VLR}|^2 + |C_{VRL}|^2 \right]_{\mu e}^{\mu\tau},$$

(17.45)

with an analogous expression for $\tau^- \to e^- \mu^+ e^-$.

These formulas agree with the ones present in FLAVIO [813] and the first two with [1128], where the result for class C has not been given. They can also be derived from (17.37), but the route to them is a bit different, which explains the additional factors in the formulas (17.43) and (17.45) relative to (17.37).

The reason for this difference is that the authors in [795] look first at all diagrams and making Fierz transformations in the case of Classes A and C, they bring the amplitude into the form (17.32). In the VLL and VRR cases the Fierz transformation for fields does not change the Dirac structure of the operator so that the coefficient VLL or VRR in that paper gets a factor of two. Squaring it to obtain the branching ratio and multiplying by 1/2, because of identical fermions, one reproduces our factors of 2 in (17.43) and (17.45).

More interesting is the case VLR as the Fierz transformation results in the operator SRL multiplied by -2. But as seen in (17.37) the contribution of this operator to the branching ratio misses the factor 4 present in VLR contribution. This is compensated by $(-2)^2$ when squaring the Wilson coefficient in the evaluation of the branching ratio. At the end these two contributions are equal to each other, and multiplying their sum by 1/2 gives the last two terms in (17.43) and (17.45).

Table 17.5 Experimental upper limits (95 % CL) on the lepton flavor-violating Z^0 decay rates.

Process	Experimental bound
$\mathcal{B}\left[Z^0 \to \mu^\pm e^\mp\right]$	7.5×10^{-7} [35]
$\mathcal{B}\left[Z^0 \to \tau^\pm e^\mp\right]$	9.8×10^{-6} [35]
$\mathcal{B}\left[Z^0 \to \tau^\pm \mu^\mp\right]$	1.2×10^{-5} [35]

17.2.10 Lepton Flavor-Violating Z Decays

In [795] also lepton flavor-violating decays of the SM Z boson $Z \to \ell_f^- \ell_i^+$ have been calculated, and we list the results here. They will constitute an important constraint on NP couplings when the data improve. One finds

$$\mathcal{B}\left(Z \to \ell_f^\pm \ell_i^\mp\right) = \frac{M_Z}{24\pi\Gamma_Z}\left[\frac{M_Z^2}{2}\left(\left|C_{fi}^{ZR}\right|^2 + \left|C_{fi}^{ZL}\right|^2\right) + \left|\Gamma_{fi}^{ZL}\right|^2 + \left|\Gamma_{fi}^{ZR}\right|^2\right], \qquad (17.46)$$

where $\Gamma_Z \approx 2.495$ GeV is the total decay width of the Z boson. Including all tree-level contributions, the authors of [795] find

$$\Gamma_{fi}^{ZL} = \frac{e}{2s_W c_W}\left(\frac{v^2}{\Lambda^2}\left(C_{\varphi l}^{(1)fi} + C_{\varphi l}^{(3)fi}\right) + \left(1 - 2s_W^2\right)\delta_{fi}\right), \qquad (17.47)$$

$$\Gamma_{fi}^{ZR} = \frac{e}{2s_W c_W}\left(\frac{v^2}{\Lambda^2}C_{\varphi e}^{fi} - 2s_W^2\delta_{fi}\right), \qquad (17.48)$$

$$C_{fi}^{ZR} = C_{if}^{ZL\star} = -\frac{v}{\sqrt{2}\Lambda^2}C_Z^{fi}, \qquad (17.49)$$

where C_Z^{fi} is defined as

$$C_Z^{fi} = \left(s_W C_{eB}^{fi} + c_W C_{eW}^{fi}\right). \qquad (17.50)$$

The experimental bounds on these decays are given in Table 17.5. It should be kept in mind that that theoretical prediction in (17.46) is for the decay $Z \to \ell_f^- \ell_i^+$ or $Z \to \ell_f^+ \ell_i^-$ while the experimental values are for the sum $Z \to \ell_f^- \ell_i^+ + \ell_i^- \ell_f^+$. Therefore, (17.46) must be multiplied by a factor of 2 to compare it with the experimental values.

Z' contributions to $Z \to \ell_f^\pm \ell_i^\mp$ decays are generated by $Z - Z'$ mixing and given by

$$\mathcal{B}(Z \to \ell_i^+ \ell_j^-) = \sin^2 \xi \frac{M_Z}{24\pi\Gamma_Z}\left[|\Delta_L^{ij}(Z')|^2 + |\Delta_R^{ij}(Z')|^2\right]. \qquad (17.51)$$

17.2.11 $\mu - e$ Conversion in Nuclei

We give next the formulas for the μ-e conversion in nuclei, that is

$$\mu + (A, Z) \to e + (A, Z), \qquad (17.52)$$

where Z and A denote the proton and atomic numbers in a nucleus, respectively.

Here we follow [1158], where general formulas can be found. The effective Lagrangian relevant to this process at the quark level as given in that paper includes photon-penguin, Z-penguin, and box-type diagrams. But the part involving the Z-penguin and box-diagrams can be combined and rewritten just in terms four fermion operators so that it can be used for operators contributing also at tree-level. Writing then the effective Lagrangian as

$$
\mathcal{L}_{eff} = -\frac{e^2}{q^2}\bar{e}\left[q^2\gamma_\alpha(A_1^L P_L + A_1^R P_R) + m_\mu i\sigma_{\alpha\beta}q^\beta(A_2^L P_L + A_2^R P_R)\right]\mu
$$
$$
\times \sum_{q=u,d} Q_{em}^q \bar{q}\gamma^\alpha q
$$
$$
+ e^2 \sum_{q=u,d} \bar{q}\gamma_\alpha q \, \bar{e}\gamma^\alpha (D_q^L P_L + D_q^R P_R)\mu, \tag{17.53}
$$

where the first term comes from the photon penguin-type diagrams and the second from any four-fermion operator having the appropriate flavors. The coefficient Q_{em}^q denotes the electric charge of the quark q.

Comparing this effective Lagrangian with the one in [1158] and using their results we find the rate

$$
\Gamma(\mu \to e) = 4\alpha^5 \frac{Z_{eff}^4}{Z}|F(q)|^2 m_\mu^5 \left[|Z(A_1^L - A_2^R) - (2Z + N)D_u^L - (Z + 2N)D_d^L|^2 \right.
$$
$$
\left. + |Z(A_1^R - A_2^L) - (2Z + N)D_u^R - (Z + 2N)D_d^R|^2\right], \tag{17.54}
$$

where Z and N denote the proton and neutron numbers in a nucleus, respectively. Z_{eff} is an effective parameter and $F(q^2)$ the nuclear formfactor. In $^{48}_{22}\text{Ti}$, one finds $Z_{eff} = 17.6$ and $F(q^2 \simeq -m_\mu^2) \simeq 0.54$ [1161].

Renormalization group analysis below the EW scale including correlations with $\mu \to e\gamma$ and $\mu \to 3e$ has been presented in [796].

17.2.12 $K_{L,S} \to \mu e$ and $K_{L,S} \to \pi^0 \mu e$

In the SM the decay $K_L \to \mu e$ is forbidden at the tree-level but can proceed through box diagrams in the case of nondegenerate neutrino masses. Similar to $\mu \to e\gamma$ it is too small to be measured.

It is also known from the studies of the Pati–Salam (PS) model [968, 1162] where it proceeds through a tree-level LQ exchange in the t-channel. The stringent upper bound on its branching ratio[1163],

$$
\mathcal{B}(K_L \to \mu e) = \mathcal{B}(K_L \to \mu^+ e^-) + \mathcal{B}(K_L \to \mu^- e^+) < 4.7 \cdot 10^{-12}, \tag{17.55}
$$

implies the mass of the leptoquark gauge boson to be above 10^3 TeV [1164]. By increasing the weak gauge group of the PS model in the context of the so-called Petite Unification [1164, 1165] and placing ordinary fermions in multiplets with heavy new fermions, and not with the ordinary SM fermions as in the PS model, it is possible to avoid tree-level contributions to $K_L \to \mu e$ so that the process is dominated by the box diagrams with new heavy-gauge bosons, heavy quarks, and heavy leptons exchanged. The relevant masses of these particles can then be decreased to $O(1\,\text{TeV})$ without violating the bound in (17.55).

More recently the studies of this decay and related decays like $K_S^0 \to \pi^0 \mu^+ e^-$ intensified in view of hints for LFUV in B decays [995, 1160, 1166, 1167]. As discussed extensively in the last paper in this list and in section 4 of [305], both NA62 and LHCb Upgrade II phase will allow to study this issue. In this context one should also mention the analysis in [1168], where the impact of LFUV effects on $K \to \pi \nu \bar{\nu}$ has been investigated.

17.2.13 Lepton Flavor-Violating B Decays

These decays are useful for tests of the violation of flavor universality. We give here only references to papers in which these decays have been analyzed. These papers contain useful formulas. In [779] these decays and also those with $K^{(*)}\ell^+\ell'^-$ in the final state have been studied. In [1125] these decays have been studied in the LHT model. Specific application to Z' models is given in [1160] and in 2HDM III model in [995]. The last two papers discuss simultaneously charged lepton flavor violation. In particular in [1160] LFV B decays in generic Z' models have been discussed. Constraints from LFV in charged lepton decays have been considered. It has been found that $\tau \to 3\mu$ and $\tau \to \mu\nu\bar{\nu}$ but also $B_s^0 - \bar{B}_s^0$ mixing provide very strong constraints. Finally, the recent improved bounds on $B_d^0 \to \mu^\pm\tau^\mp$ $B_s^0 \to \mu^\pm\tau^\mp$ and $B^+ \to K^+\mu^\pm e^\mp$ from the LHCb put some pressure on models explaining B-physics anomalies.

The experimental bounds on the sums $\ell^\pm\ell'^\mp = \ell^-\ell'^+ + \ell^+\ell'^-$ are given in Table 17.6.

Table 17.6	Current limits on branching ratios for LFV decay channels.	
Channel	**Branching ratio**	**Reference**
$K^+ \to \pi^+\mu^+e^-$	$\leq 1.3 \times 10^{-11}$	E865, E777 [1169]
$K^+ \to \pi^+\mu^-e^+$	$\leq 5.2 \times 10^{-10}$	E865 [1170]
$K_L \to \pi^0\mu^\pm e^\mp$	$\leq 7.6 \times 10^{-11}$	KTeV [1171]
$K_L \to \mu^\pm e^\mp$	$\leq 4.7 \times 10^{-12}$	E871 [1163]
$B_d^0 \to \mu^\pm e^\mp$	$\leq 1.3 \times 10^{-9}$	LHCb [1172]
$B_s^0 \to \mu^\pm e^\mp$	$\leq 6.3 \times 10^{-9}$	LHCb [1172]
$B_d^0 \to \mu^\pm\tau^\mp$	$\leq 1.2 \times 10^{-5}$	LHCb [1173]
$B_s^0 \to \mu^\pm\tau^\mp$	$\leq 3.4 \times 10^{-5}$	LHCb [1173]
$B^+ \to K^+\tau^\pm\mu^\mp$	$\leq 4.8 \times 10^{-5}$	[35]
$B^+ \to K^+\mu^\pm e^\mp$	$\leq 7.1 \times 10^{-9}$	LHCb
$B \to K^*\mu^\pm e^\mp$	$\leq 1.4 \times 10^{-6}$	[35]

17.3 Electric Dipole Moments (EDMs)

17.3.1 Preliminaries

Even though the experimental sensitivities have improved a lot no EDM of a fundamental particle has been observed so far, and we have only upper bounds on them to our disposal that we collected in Table 17.7. Nevertheless, EDM experiments have already put strong limits on NP models. A permanent EDM of a fundamental particle violates both T and P, and thus – assuming CPT symmetry – is another way to measure CP violation. In the SM the individual quark EDMs vanish at two-loop order [1174, 1175] so that the only CP-violating phase of the CKM matrix enters quark EDMs first at three loop, which results in negligibly small SM EDMs. EDMs of leptons are in the SM even stronger suppressed than quark EDMs and enter first at four-loop level [1176]. Consequently EDMs are excellent probes of new CP-violating phases of NP models, especially flavor-blind phases, and of strong CP violation related to the so-called θ parameter.

The fact that EDMs are so strongly suppressed in the SM makes them very special compared to most observables considered in this book. A measurement of a nonvanishing EDM of any particle including nucleons, nuclei, atoms, and molecules will be a clear signal of NP similar to lepton flavor-violating processes. But whereas the latter are in most cases theoretically clean, the case of EDMs is very different. In addition to the short-distance dynamics present in the Wilson coefficients of contributing operators, not only hadronic physics as in other sections, but often also nuclear physics including nuclear many-body calculations and atomic physics is responsible for the final value of a given EDM. Therefore the identification of NP responsible through a measurement of a nonvanishing EDM is much more challenging than is the case of remaining observables considered in this book. Similar the determination of NP scale responsible is harder, but as in the case of rare K and B decays and LFV, EDMs can in principle probe short-distance scales far beyond the reach of the LHC.

As we will see later there are different sources for EDMs. For hadronic EDMs there is the θ term of QCD, which is very much constrained due to the nonobservation of permanent EDMs of the mercury atom (^{199}Hg) and neutron. Apart from the θ term, the SM CKM induced EDMs would be far smaller in magnitude than the next-generation EDM sensitivities. Consequently, one does not need the same kind of refined

Table 17.7 Current experimental upper bounds (at 90% C.L.) on electric dipole moments.

EDM	$|d_e|$	$|d_\mu|$	d_τ
Bound (e cm)	1.1×10^{-29} [1177]	1.5×10^{-19} [1178]	3.4×10^{-17} [1179]
EDM	$|d_n|$	$|d_{Hg}|$	d_{Tl}
Bound (e cm)	3.0×10^{-26} [1180]	6.3×10^{-30} [1181]	9.4×10^{-25} [1182]
EDM	$|d_{Xe}|$	$|d_{Ra}|$	
Bound (e cm)	1.4×10^{-27} [1183, 1334]	1.2×10^{-23} [1184, 1185]	

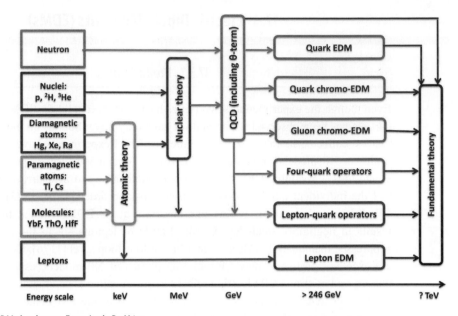

Figure 17.3   EDMs landscape. From Jordy De Vries.

hadronic structure computations as one often needs in flavor physics to interpret the EDM results in terms of NP. That being said, the hadronic matrix element problem remains a considerable challenge. At dimension-6 one encounters several different operators for the first generation fermions that could give rise to EDMs: pure gauge operators $\tilde{G}GG$, four-fermion operators (semileptonic and nonleptonic), gauge-Higgs operators $\varphi^\dagger \varphi \tilde{G}G$ and gauge-Higgs-fermion operators $(\bar{Q}T^A q_R)\varphi G$. We will be more specific about them later.

In experiments one often deals with composite systems, and thus nuclear physics is important in determining the EDMs of neutral atoms. Nuclear structure can also provide an amplifier of atomic EDMs. In heavier neutral systems there is the shielding of the EDMs of constituents of one charge by those of the other (e.g., protons and electrons). The transmission of CP violation through a nucleus into an atom must overcome this shielding. Its effectiveness in doing so is expressed by a nuclear Schiff moment. In nuclei with asymmetric shapes Schiff moments can be enhanced by two or even three orders of magnitude. For example, an octupole deformed nuclei such as ^{225}Ra give enhanced nuclear Schiff moments and, thus, enhanced atomic EDMs in a diamagnetic system.

Because of many sources of EDMs and significant theoretical uncertainties, it is a challenge to present this topic in a transparent manner. But this topic is very important, and it is mandatory to include it in this book and face this challenge. As seen in Figure 17.3, the connection between the fundamental theory and EDMs in measurable systems is rather involved. Our main goal for next pages is to present this problematic in explicit terms and to illuminate the connections between various elements of theoretical calculations that involve several different scales. The chart in Figure 17.3 and also the one in Figure 1 in [1120] offer a general view of such analyses. Indeed, a useful review about EDMs can be found in [1120], which updates the review in [1186]. See also [1187, 1188, 1347] and in particular in the case of EDMs of diamagnetic atoms the review in [1189]. As despite a few publications involving EDMs I do not consider myself to be an expert in this field, the present section benefited in a very significant manner from these reviews, in particular from [1120, 1186, 1188] and a transparent summary in section 2 of [908]. Also the very recent analysis in [1076] can be strongly recommended. It contains up-to-date hadronic, nuclear, and atomic matrix elements necessary for the phenomenology.

Many details not presented here can be found in these papers and the ones quoted later. I hope that potential readers will find this section to be a useful guide to these reviews and to the rich literature.

Before going into details let us recall the following elementary facts [1120, 1186]. The interaction of a neutral nonrelativistic particle of spin S with magnetic and electric fields can be described by the following Hamiltonian,

$$H = -\mu \vec{B} \cdot \frac{\vec{S}}{S} - d\vec{E} \cdot \frac{\vec{S}}{S} . \tag{17.56}$$

Now under the reflection of spatial coordinates, we have

$$P(\vec{B} \cdot \vec{S}) = \vec{B} \cdot \vec{S}, \qquad P(\vec{E} \cdot \vec{S}) = -\vec{E} \cdot \vec{S}. \tag{17.57}$$

Similar under time reflection,

$$T(\vec{B} \cdot \vec{S}) = \vec{B} \cdot \vec{S}, \qquad T(\vec{E} \cdot \vec{S}) = -\vec{E} \cdot \vec{S}. \tag{17.58}$$

Therefore a nonzero electric dipole moment d may exist if and only if both parity and time reversal invariance are broken. Assuming then the CPT theorem, this also implies that the violation of CP is required. The profound difference from other CP-violating phenomena considered in this book is that nonvanishing EDMs do not require simultaneous flavor violation so that flavor-conserving CP-violating interactions are prime sources of them. But as we will discuss later also flavor-violating interactions can generate EDMs.

In the case of a spin-1/2 particle the relativistic generalization of its interaction with the electric field implies the familiar Lagrangian

$$\mathcal{L} = -d\frac{i}{2}\overline{\psi}\sigma^{\mu\nu}\gamma_5\psi F_{\mu\nu}. \tag{17.59}$$

17.3.2 Basic Structure: Nuclear Scale

To see the basic structure of interactions involved eventually in the calculation of EDMs, we do not begin our presentation with high-energy NP or electroweak scale but with the nuclear scale. Subsequently, in Section 17.3.3 we will proceed as in the rest of the book by starting with a fundamental Lagrangian involving elementary fields. Finally, with the help of renormalization group and SMEFT, we will find out how the objects encountered at the nuclear scale are related to the fundamental parameters of a given theory. We follow then [1186] and write down first the relevant CP-odd terms at the nuclear scale:

$$\mathcal{L}_{eff}^{\text{nuclear}} = \mathcal{L}_{\text{edm}} + \mathcal{L}_{\pi NN} + \mathcal{L}_{eN}. \tag{17.60}$$

The different contributions represent the nucleon (and electron) EDMs,

$$\mathcal{L}_{\text{edm}} = -\frac{i}{2} \sum_{i=e,p,n} d_i \, \bar{\psi}_i (F\sigma) \gamma_5 \psi_i, \tag{17.61}$$

the CP-odd pion nucleon interactions

$$\mathcal{L}_{\pi NN} = \bar{g}_{\pi NN}^{(0)} \bar{N} \tau^a N \pi^a + \bar{g}_{\pi NN}^{(1)} \bar{N} N \pi^0 + \bar{g}_{\pi NN}^{(2)} (\bar{N} \tau^a N \pi^a - 3 \bar{N} \tau^3 N \pi^0), \tag{17.62}$$

and CP-odd electron-nucleon interactions

$$\begin{aligned}
\mathcal{L}_{eN} &= C_S^{(0)} \bar{e} i \gamma_5 e \bar{N} N + C_P^{(0)} \bar{e} e \bar{N} i \gamma_5 N + C_T^{(0)} \epsilon_{\mu\nu\alpha\beta} \bar{e} \sigma^{\mu\nu} e \bar{N} \sigma^{\alpha\beta} N \\
&+ C_S^{(1)} \bar{e} i \gamma_5 e \bar{N} \tau^3 N + C_P^{(1)} \bar{e} e \bar{N} i \gamma_5 \tau^3 N + C_T^{(1)} \epsilon_{\mu\nu\alpha\beta} \bar{e} \sigma^{\mu\nu} e \bar{N} \sigma^{\alpha\beta} \tau^3 N.
\end{aligned} \tag{17.63}$$

We should warn the reader that at a few places the notation of couplings in [1186] and [1120], that we use heavily here, differs from each other. In particular,

$$\bar{g}_{\pi NN}^{(0)} = \bar{g}_{\pi}^{(0)}, \qquad \bar{g}_{\pi NN}^{(1)} = \bar{g}_{\pi}^{(1)}, \qquad \bar{g}_{\pi NN}^{(2)} = -\bar{g}_{\pi}^{(2)} \tag{17.64}$$

for [1186] and [1120], respectively. The *bar* stresses the CP-odd character of the couplings. Relations to notations of couplings in other reviews are collected in table 4 of [1120].

Even if at first sight one would think that only elementary d_i would contribute to EDMs of atoms and molecules, this is in reality not the case as also the CP-odd pion nucleon iterations and CP-odd electron-nucleon couplings can contribute. This makes the extraction of the elementary EDMs d_e and $d_{n,p}$ a challenge.

The basic entries in these formulas are then

$$d_e, \quad d_{n,p}, \quad \bar{g}_{\pi NN}^{(0)}, \quad \bar{g}_{\pi NN}^{(1)}, \quad \bar{g}_{\pi NN}^{(2)}, \quad C_S^{(0)}, \quad C_S^{(1)}, \quad C_P^{(0)}, \quad C_P^{(1)}, \quad C_T^{(0)}, \quad C_T^{(1)}. \tag{17.65}$$

While the neutron EDM d_n is already an observable, the other preceding quantities enter the formulas for measureable EDMs of atoms and molecules. Our goal will be the presentation of a general formula for the EDM d_A of an arbitrary atom in terms of the objects in (17.65) as given in [1120]. But as seen in Section 17.3.4, this formula is rather involved, and it is useful to give first some background behind it and to stress which parts can be calculated in perturbation theory and for which nonperturbative methods have to be used.

Observable EDMs

As discussed in [1186] it is convenient to classify the EDM searches into three main categories, distinguished by the dominant physics that would induce the EDMs. These categories are

- the EDMs of *paramagnetic* atoms and molecules involving electron EDM, d_e, with the main actor being the atomic EDM of the thallium atom (Tl),
- the EDMs of *diamagnetic* atoms with the EDM of the mercury atom (^{199}Hg) playing presently the main role, and
- the EDMs of hadrons with particular emphasize put on EDMs of nucleons: d_n and d_p.

EDMs of Paramagnetic Atoms

The EDMs of paramagnetic atoms are governed by the EDM of one unpaired electron, that is d_e. At the nonrelativistic level, this would not be possible due to the Schiff shielding theorem. Indeed, because the atom is neutral, any applied electric field will be shielded and so an EDM of the unpaired electron will not induce an atomic EDM. But this theorem is violated by relativistic effects, in particular for atoms with a large atomic number and molecules. As an example, for the paramagnetic atoms one finds [1190]

$$d_{\text{para}}(d_e) \sim 10\frac{Z^3\alpha^2}{J(J+1/2)(J+1)^2}d_e, \tag{17.66}$$

up to numerical $O(1)$ factors, with J the angular momentum and Z the atomic number.

The large enhancement factor for large Z allows to probe the electron EDM very efficiently. Beyond d_e CP-odd electron-nucleon interactions may also play a significant role, in particular the couplings $C_S^{(0)}$ and $C_S^{(1)}$, because they involve the spin of the electron and are enhanced by the large nucleon number in heavy atoms. As an example, we quote the expression for the EDM of the thallium atom that currently provides the best constraints on fundamental flavor-diagonal CP violation. A number of atomic calculations established the relation between the EDM of thallium, d_e, and the CP-odd electron-nucleon couplings $C_S^{(0)}$ and $C_S^{(1)}$ [1190]

$$d_{\text{Tl}} = -585d_e - e\,43\,\text{GeV} \times (C_S^{(0)} - 0.2\,C_S^{(1)}). \tag{17.67}$$

The relevant atomic numerical factors are known to within 10–20% [1190]. Further examples will be given later. But we already see that to extract the electron EDM d_e from d_{Tl} requires the knowledge of numerical factors resulting from atomic calculations and also of the coefficients $C_S^{(0)}$ and $C_S^{(1)}$ that can be generated by NP but also involve some nonperturbative uncertainties.

EDMs of Diamagnetic Atoms

The total electron angular momentum in diamagnetic atoms equals to zero. This time the Schiff theorem is violated through a net misalignment between the distribution of charge and EDM (i.e., first and second moments) in the nucleus. But unfortunately the induced atomic EDM is considerably suppressed relative to the underlying EDM of the nucleus. This time also the calculations are much more involved requiring input from QCD as well as nuclear and atomic physics. Thus generally the EDMs of the diamagnetic atoms receive contributions from many interactions

$$d_{\text{dia}} = d_{\text{dia}}(S[\bar{g}_{\pi NN}], d_N, C_S, C_P, C_T, d_e). \tag{17.68}$$

Here S is the Schiff moment. Its computation is a nontrivial nuclear many-body problem. It can be written generally as follows

$$S[\bar{g}_{\pi NN}] = a_0\, g\bar{g}^{(0)} + a_1 g\bar{g}^{(1)} + a_2 g\bar{g}^{(2)} \tag{17.69}$$

where $g \equiv g_{\pi NN}$ is the CP-even pion-nucleon coupling, and $\bar{g}^{(i)} \equiv \bar{g}_{\pi NN}^{(i)}$ denote the CP-odd couplings entering (17.62). As an example, one finds for the mercury atom [1191]

$$S(^{199}\text{Hg}) = -0.0004g\bar{g}^{(0)} - 0.055g\bar{g}^{(1)} + 0.009g\bar{g}^{(2)}\ e\ \text{fm}^3. \tag{17.70}$$

We note very strong suppression factors.

Keeping only the isovector pion-nucleon coupling and $C_S^{(0)}$ one finds then for the mercury EDM [1186]

$$d_{\text{Hg}} = -(1.8 \times 10^{-4}\,\text{GeV}^{-1})e\,\bar{g}_{\pi NN}^{(1)} + 10^{-2}d_e + (3.5 \times 10^{-3}\,\text{GeV})e\,C_S^{(0)}. \tag{17.71}$$

Neutron and Proton EDMs

Neutron EDM can be searched for directly with ultracold neutron technology, and its calculation does not involve any atomic or nuclear physics. But similar to many observables considered in this book, it requires good control over nonperturbative QCD effects to find the dependence of d_n on CP-odd sources at the quark-gluon level. As we deal now with a baryon instead of a meson such calculations are much more demanding than is the case of K and B physics. Yet, currently d_n provides one of the strongest constraints on new CP-violating physics, and one should make all efforts to improve the precision of the theory.

17.3.3 Electroweak Scale, SMEFT, and RG Evolution

We have just described the basic structure of EDMs at the nuclear scales. While in principle we could next discuss the *hadronic* scale, I find it strategically better to present first what happens at the electroweak scale because this was the starting point for all processes discussed in this book. Thus at the electroweak scale of $O(M_W)$ after $W^\pm$, Z^0, t, Higgs particle, and all heavy new particles have been integrated out, we will have an effective Hamiltonian written as the sum of products of WCs and local operators. The construction of this Hamiltonian proceeds as in previous chapters, and we make only few comments:

- In a given NP model one integrates out heavy particles at a given scale, which could be many orders of magnitude higher than the EW scale. This then implies the presence of local operators multiplied by corresponding WCs. Which operators are relevant and which not depends on the NP model considered and on the next step.

- This next step is the RG evolution to the EW scale crossing generally various thresholds at which some of new particles are integrated out. The SMEFT technology should be used here so that in addition to QCD effects top Yukawa effects, and EW effects play some role. The RG evolution generates new operators through mixing, and in addition threshold corrections can have an impact on WCs. There is an extensive literature on the relevant calculations of anomalous dimensions that was already given in Section 14.1. In the context of EDMs a very transparent presentation is given in [781], where references to the relevant papers can be found. See also table 5 in [1120].
- Next after the integration of heavy SM particles that is $W^{\pm}$, Z^0, t, and the Higgs particle one continues the RG evolution to hadronic scales at which the relevant hadronic matrix elements of resulting operators are evaluated in QCD. The RG evolution involves now QCD and QED effects.
- Finally, in many cases nuclear and atomic calculations have to be performed to connect fundamental parameters of a given theory to the low-energy parameters in (17.65). Various methods are critically reviewed in sections 3 and 4 in [1120].

If a given NP model is specified this analysis is rather straightforward even if often tedious. In particular only certain operators are relevant in a given model, and their WCs can be calculated in terms of the fundamental parameters. If the number of these parameters is small, some relations between these coefficients characteristic for a given model are present.

On the other hand, if a model is not specified it is a common practice these days to use the SMEFT that we presented in Section 14.1 and applied it already to several processes. The advantage of this approach over a specific model is its generality. The disadvantage is the appearance of many operators with arbitrary WCs even if in certain situations some relations between them are implied by the invariance under the SM gauge symmetry. The list of operators relevant for EDMs will be given soon.

The main sources of CPV are then encoded in

$$\mathcal{L}_{\text{CPV}} = \mathcal{L}_{\text{CKM}} + \mathcal{L}_{\bar{\theta}} + \mathcal{L}_{\text{BSM}}^{\text{eff}}, \qquad (17.72)$$

where the first term on the r.h.s. is the standard CKM contribution and the second term is the so-called QCD θ-term

$$\mathcal{L}_{\bar{\theta}} = -\frac{g_3^2}{16\pi^2} \bar{\theta} \, \text{Tr}\left(G^{\mu\nu}\tilde{G}^{\mu\nu}\right), \qquad \tilde{G}^{\mu\nu} = \frac{1}{2}\epsilon^{\mu\nu\alpha\beta}G_{\alpha\beta}, \qquad (17.73)$$

where $\tilde{G}^{\mu\nu}$ is the dual to the gluon field strength $\tilde{G}^{\mu\nu}$.

The last term in (17.72) representing dimension-6 CPV operators is given generally as follows

$$\mathcal{L}_{\text{BSM}}^{\text{eff}} = \frac{1}{\Lambda^2}\sum_i \alpha_i^{(n)} O_i^{(6)}. \qquad (17.74)$$

The coefficients C_i are the WCs of the four-fermion operators listed earlier.

It should be emphasized that in left-right symmetric models the chirality flip can be generated in a different manner and still other contributions with different flavor structure are possible:

$$\mathcal{L}_{\text{LR,CPV}}^{\text{eff}} = i\frac{\text{Im}C_{\phi ud}}{3\Lambda^2}\{\bar{u}u\,\bar{d}\gamma_5 d - \bar{u}\gamma_5 u\,\bar{d}d + 3[\bar{u}T^A u\,\bar{u}\gamma_5 T^A d - \bar{u}\gamma_5 T^A u\,\bar{d}T^A d]\}. \tag{17.96}$$

θ Term

It is time to discuss briefly the θ-term, which enters the Lagrangian in (17.73). This dim-4 term is allowed by all symmetries of the SM but violates both CP and T. It is by now constrained to be at most 10^{-10}, precisely because of the nonobservation of the permanent EDM of the mercury atom.[1] The unnatural small value of this term signals the so-called strong CP-problem. One of its solutions is the increase of the Higgs system with additional symmetry, Peccei–Quinn (PQ) symmetry $U(1)_{PQ}$ [1193, 1194], which allows to eliminate this term. The spontaneous breakdown of this symmetry implies the existence of a pseudoscalar Goldstone boson, the axion [1195, 1196], that has not been observed. Other solutions to strong CP problem without the PQ symmetry have been presented in [1197, 1198]. A recent review of this subject with implications for cosmology can be found in [1199]. Also the updates in PDG are useful in this respect.

Actually, what really enters physical expressions is not θ but

$$\bar{\theta} = \theta + \arg\det(Y_u Y_d), \tag{17.97}$$

with Y_i being Yukawa couplings. Thus this term is connected with quark masses, and the second term arises after the phases of all quark fields have been redefined. However, in some BSM scenarios radiative corrections to quark masses can generate unacceptably large contributions to $\bar{\theta}$ implying severe constraints on model parameters unless some symmetries like the PQ one is present.

The presence of the $\bar{\theta}$ term complicates the search for NP in the case of nucleons and diamagnetic systems, but measuring deviations in several systems one could disentangle NP from this term [1200].

Additional Remarks on EDMs in SMEFT

The formfactors listed in Table 17.4 for $f = i$ can be directly used to calculate EDMs of charged leptons [795]

$$\boxed{d_{l_i} = -\frac{1}{\Lambda^2}\text{Im}[F_{TR}^{ii}].} \tag{17.98}$$

---

[1] Apparently as stressed in [1188] this bound can be relaxed in a global analysis by a few orders of magnitude.

A very clear presentation of the application of SMEFT and RG ideas to EDMs is given in [1076]. Although it is given in the context of specific LQ models, it shows transparently many different aspects of the physics of EDMs and can be strongly recommended.

17.3.4 General Formula

With the information about various contributions at hand, we are ready to write down a formula for a general atomic or molecular EDM d_A in terms of d_e, d_N, the nuclear Schiff moment S, and the coefficients $C_{S,P,T}^{(0)}$ and $C_{S,P,T}^{(1)}$. These are the basic entries in (17.65) with $\bar{g}_{\pi NN}^{(i)}$ hidden in S as given in (17.69). Rewriting (5.171) and (5.172) of [1120] in terms of these coefficients we have then

$$d_A = \rho_A^e d_e + \sum_{N=p,n} \rho_Z^N d_N + \kappa_S S - \left[k_S^{(0)} C_S^{(0)} + k_P^{(1)} C_P^{(1)} \right] \tag{17.99}$$

$$+ \left[k_S^{(1)} C_S^{(1)} + k_P^{(0)} C_P^{(0)} \right] - \left[k_T^{(0)} C_T^{(0)} + k_T^{(1)} C_T^{(1)} \right]. \tag{17.100}$$

In all these contributions one can exhibit $(v/\Lambda)^2$ terms. Indeed we have first

$$\rho_A^e d_e = \beta_A^{e\gamma} \left(\frac{v}{\Lambda} \right)^2 \mathrm{Im} C_{e\gamma}, \tag{17.101}$$

$$d_N = \alpha_N \bar{\theta} + \left(\frac{v}{\Lambda} \right)^2 \sum_k \beta_N^{(k)} \mathrm{Im} C_k, \tag{17.102}$$

$$\bar{g}^{(i)} = \lambda_{(i)} \bar{\theta} + \left(\frac{v}{\Lambda} \right)^2 \sum_k \gamma_{(i)}^{(k)} \mathrm{Im} C_k, \tag{17.103}$$

and $\bar{g}^{(i)} \equiv \bar{g}_{\pi NN}^{(i)}$ inserted into (17.69) allows to express also Schiff moment in terms of $(v/\Lambda)^2$. The coefficients α_N, $\beta_N^{(k)}$, $\lambda_{(i)}$, and $\gamma_{(i)}^{(k)}$ are subject to hadronic uncertainties and result from the calculations of hadronic matrix elements. This is the place where hadronic scales enter the game.

Next we have

$$C_S^{(0)} = -g_S^{(0)} \left(\frac{v}{\Lambda} \right)^2 \mathrm{Im} C_{eq}^{(-)}, \qquad C_S^{(1)} = g_S^{(1)} \left(\frac{v}{\Lambda} \right)^2 \mathrm{Im} C_{eq}^{(+)}, \tag{17.104}$$

$$C_P^{(0)} = g_P^{(0)} \left(\frac{v}{\Lambda} \right)^2 \mathrm{Im} C_{eq}^{(+)}, \qquad C_P^{(1)} = -g_P^{(1)} \left(\frac{v}{\Lambda} \right)^2 \mathrm{Im} C_{eq}^{(-)}, \tag{17.105}$$

$$C_T^{(0)} = -g_T^{(0)} \left(\frac{v}{\Lambda} \right)^2 \mathrm{Im} C_{lequ}^{(3)}, \qquad C_S^{(1)} = -g_T^{(1)} \left(\frac{v}{\Lambda} \right)^2 \mathrm{Im} C_{lequ}^{(3)}, \tag{17.106}$$

where

$$C_{eq}^{(\pm)} = C_{ledq} \pm C_{lequ}^{(1)} \tag{17.107}$$

and $g^{(i)}_{S,P,T}$ are the isoscalar and isovector formfactors in the limit of isospin symmetry

$$\frac{1}{2}\langle N|[\bar{u}\Gamma u + \bar{d}\Gamma d]|N\rangle \equiv g^{(0)}_\Gamma \bar{\psi}_N \Gamma \psi_N, \qquad \frac{1}{2}\langle N|[\bar{u}\Gamma u - \bar{d}\Gamma d]|N\rangle \equiv g^{(0)}_\Gamma \bar{\psi}_N \Gamma \tau_3 \psi_N,$$
(17.108)

where $\Gamma = 1, \gamma_5, \sigma_{\mu\nu}$.

The coefficients ρ^N_Z, κ_S, $k^{(0)}_{S,P,T}$, $k^{(1)}_{S,P,T}$, $g^{(0)}_{S,P,T}$, and $g^{(1)}_{S,P,T}$ are subject to nonperturbative uncertainties. A compilation of their values can be found in [1120]. In that paper also applications of this formula and examples can be found. Here we just quote the result for *diamagnetic* atoms presented in [1188]:

$$d_A = \sum_{N=p,n} \rho^N_Z d_N + \kappa_S S - \left[k^{(0)}_T C^{(0)}_T + k^{(1)}_T C^{(1)}_T \right],$$
(17.109)

with the various entries given in eqs. (II.14)–(II.16) of that paper.

17.3.5 General Picture and New Physics

One can then ask which short-distance scales can be probed by EDMs. As discussed in [1120] by naive dimension analysis EDMs probe a NP scale of several TeV. This assumes order one CP-violating phases ϕ_{CP} for the electron EDM that arises at one loop order:

$$d_e \approx e\frac{m_e}{\Lambda^2}\frac{\alpha_e}{4\pi}\sin\phi_{CP} \approx \frac{1}{2}\left(\frac{1\,\text{TeV}}{\Lambda}\right)^2 \sin\phi_{CP} \cdot 10^{-26} e\,\text{cm}\,.$$
(17.110)

Recently, the upper bound on d_e has been improved by an order of magnitude with respect to the previous bound in [1201] and reads [1177]

$$|d_e| \leq 1.1 \cdot 10^{-29} e\,\text{e cm}.$$
(17.111)

This then implies for the CP-violating phase $|\sin\phi_{CP}| \lesssim \left(\frac{\Lambda}{15\,\text{TeV}}\right)^2$.

The scale of NP can be even higher for the neutron and ^{199}Hg EDMs as they are sensitive to the chromo-magnetic EDM, which enters with a factor of α_s rather than the fine structure constant α_e. As one can see from (17.110) the sensitivity to the NP scale goes as $1/\Lambda^2$, whereas in many other cases such as lepton flavor violation the sensitivity goes as $1/\Lambda^4$. Future EDM measurements aim to improve their sensitivity by approximately two orders of magnitude, which will then push the mass scale sensitivity into the (30–100) TeV range.

Flavor-diagonal CP violating phases as needed for electroweak baryogenesis can be strongly constrained by EDMs. In the MSSM, for example, this requires rather heavy first- and second-generation sfermions but at the same time light electroweak gauginos below one TeV as well as a subset of the third-generation sfermions (see [1202] for details). However, as can be deduced from the plots in [1203] the improved bound on d_e in (17.111) nearly excludes this possibility. While the bino-driven baryogenesis analyzed in [1204] is still allowed by this new measurements, it further constraints this scenario.

A new and largely unexplored direction for electroweak baryogenesis is flavor nondiagonal CPV that would enter the B or D meson systems [1205–1207]. Flavor nondiagonal CP violation is far less susceptible to EDM constraints than flavor-diagonal phases because it arises at multiloop order. In the SM, for example, it is a two-loop effect that involves the one-loop CP-violating penguin operator and a hadronic loop with two $\Delta S = 1$ weak interactions.

Until now our discussion was quite general, and it is time to stress which contributions are largest. We stated already that in the SM the EDMs are tiny. When new sources of CP violation are present in NP models, large contributions to EDMs can be generated, and unless some suppression mechanisms are invoked even one-loop contributions are too large to satisfy the present upper bounds. Once these suppressions are invoked the question arises of whether two-loop diagrams could be important. In fact, as realized first by Weinberg [1208], in a wide class of models two-loop diagrams generating Weinberg-operator can be a very important contribution. In fact, these contributions would be by far dominant if they were not strongly suppressed by RG effects. Another type of two-loop diagrams are Barr–Zee(-type) diagrams [1209–1211] that are often dominant in models with new heavy scalars.

It is not possible to review the studies of EDMs in various extensions of the SM. A brief review of EDMs in supersymmetric models, in models with extended gauge symmetry, in particular left-right symmetric models and in models with extra dimensions can be found in [1120]. Studies of EDMs in 2HDM models with flavor blind phases can be found in [908, 909] and supersymmetry [1212], where further references to the rich literature can be found. A general description [1213] of the search for NP with the EDMs can be found in [1213]. The implications of the improved upper bound on d_e [1177] on various NP models has been studied in [1214]. Correlations between tautonic B meson decays like $B \to \tau\nu$, $B \to D^{(*)}\tau\nu$, $B \to \pi\tau\nu$, and EDMs, in particular the one of the neutron have been investigated in the context of S_1 LQ model in [1215].

As EDMs are sensitive to CP-violating phases, they should provide bounds on new phases in the Yukawa couplings of quarks to the Higgs boson. Indeed, as shown in [1216–1219] these bounds could be rather stringent. The same is true for CP-phases in leptoquark models as demonstrated for R_2 and S_1 models in [1076, 1220]. In fact, generally the bounds on EDMs could have an impact on the explanations of various anomalies, like B physics anomalies, as stressed recently in [719, 1076, 1200, 1221]. The correlation between the CP violation in the Kaon sector, in particular ε'/ε, with the neutron EDM in the context of left-right symmetric models, has been analyzed in [1318]. Further references to older discussions of this important issue can be found in these papers.

17.3.6 EDMs of Heavy Fermions

As we have seen until now, the most stringent limits for EDMs stem presently from experiments with heavy atoms and molecules as well as neutrons, built from first-generation fermions. However, there is the distinct possibility that NP affects primarily heavier fermions. CP-violating moments of these heavier fermions have been investigated in the past, and improvements are possible at current and future colliders. However, these rather direct limits are typically many orders of magnitude weaker than those for the moments of first-generation fermions. Recently it has been demonstrated in several papers that much more stringent indirect limits on the EDMs, chromo-EDMs, and weak dipole moments of heavy particles like charm, beauty, and top quarks, as well as tau lepton and the muon, can be obtained from existing high-precision measurements with atoms, molecules, and neutrons. In particular in [781, 1192, 1222–1224] bounds on d_t from the bounds on neutron's EDM d_n have been derived. Even stronger constraints result from the bounds on d_e and when the bounds on d_n and d_e are simultaneously considered [1192]:

$$|d_t| \le 1.3 \times 10^{-19} \, \mathrm{e\,cm}. \tag{17.112}$$

With future improved bounds on d_e, d_n, and d_p this bound could be improved by two orders of magnitude.

Here I would like to describe the basic structure of a recent study by Martin Jung, Emilie Passemar, and myself that generalizes the analyses just mentioned with the aim to constrain the EDMs of all quarks and leptons beyond d, u, and e that is also the (s, c, b) and (μ, τ).

This is an arena for the application of the SMEFT and the related RG technology, which we presented at length in Chapter 14. To this end we postulate the existence of the operators at dimension-6 involving heavy fermions at a high-energy scale Λ that are responsible for EDMs of these fermions. We then investigate RG-induced effects of these new operators on observables at lower energies, like the EDMs of atoms, molecules from which d_e can be extracted and of the neutron and proton. The determination of the dominant mechanism for a given fermion is complicated by the presence of various hierarchies in couplings, masses, mixing angles, hadronic matrix elements, and loop factors. The possible contributions enter at one, two, and three loops. This is illustrated in Table 17.8, where q_i are electric fermion charges and $N_C = 3$ the number of colors. We will list the main patterns in this table after presenting formulas behind these results.

In addition to the evolution of the SMEFT coefficients, which can be performed perturbatively, the connection to measured observables, as we have seen earlier, requires the matching onto hadronic quantities, which involves matrix elements at the atomic, QCD, and nuclear levels, often involving sizable theoretical uncertainties. These steps finally result in expressions for observables in terms of NP coefficients involving heavy fermions.

More explicitly, the CP-violating dipole moments of a fermion f (quark or lepton), defined via

$$\mathcal{L}_f \supset -\frac{i}{2} \left(d_f^{\gamma} \bar{f} \sigma^{\mu\nu} \gamma_5 f F_{\mu\nu} + d_f^Z \bar{f} \sigma^{\mu\nu} \gamma_5 f Z_{\mu\nu} + g_s \tilde{d}_q \bar{q} \sigma^{\mu\nu} \gamma_5 q G_{\mu\nu} \right), \tag{17.113}$$

are related to the coefficients of the SMEFT (in the mass eigenbasis) as follows:

$$d_{f \in D,E}^{\gamma} = -\sqrt{2} e v \, \mathrm{Im}\left[\frac{C_{fB}}{g_1} - \frac{C_{fW}}{g_2} \right], \qquad d_{f \in U}^{\gamma} = -\sqrt{2} e v \, \mathrm{Im}\left[\frac{C_{fB}}{g_1} + \frac{C_{fW}}{g_2} \right], \tag{17.114}$$

$$d_{f \in D,E}^{Z} = -\sqrt{2} e v \, \mathrm{Im}\left[-\frac{C_{fB}}{g_2} - \frac{C_{fW}}{g_1} \right], \qquad d_{f \in U}^{Z} = -\sqrt{2} e v \, \mathrm{Im}\left[-\frac{C_{fB}}{g_2} + \frac{C_{fW}}{g_1} \right], \tag{17.115}$$

$$\tilde{d}_{q \in U,D} = -\frac{\sqrt{2} v}{g_s} \, \mathrm{Im}\, C_{qG}. \tag{17.116}$$

Here E, D, U refer to the corresponding families, i.e., $E = \{e, \mu, \tau\}$, $D = \{d, s, b\}$, and $U = \{u, c, t\}$. C_{fX} stand here for the corresponding entry on the diagonal of the coefficient matrix, e.g., $C_{cB} = (C_{uB})_{22}$. We do not consider sizable nondiagonal entries in the following, corresponding to large flavor-changing neutral currents at tree-level.

On the other hand, we recall that the relevant Lagrangian at low energies (~ 1 GeV) can be written as follows:

$$\mathcal{L}_{\text{low}} = \sum_{\substack{f,f' = u,d,s,e,\mu \\ q = u,d,s}} \left[-\frac{i}{2} d_f^\gamma \bar{f} \sigma^{\mu\nu} \gamma_5 f F_{\mu\nu} - \frac{i}{2} g_s \tilde{d}_q \bar{q} \sigma^{\mu\nu} \gamma_5 q G_{\mu\nu} + C_{ff'} O_{ff'} \right] + C_W O_W,$$

(17.117)

where $O_{ff'}$ denote relevant four-fermion operators without derivatives and O_W the Weinberg operator.

Now, starting with the WCs of a top quark C_{tB} and C_{tW} that determine d_t and using the RG equations of the SMEFT and subsequently the ones of LEFT, one can determine the contribution of d_t to d_d and d_u and consequently to neutron electric dipole moment. The bounds on the latter imply the bounds on d_t. Similarly, the nonvanishing coefficients $C_{\tau B}$ and $C_{\tau W}$ that determine d_τ can have impact not only on the electron EDM d_e but also on $d_{d,u}$. The bounds on $d_{e,d,u}$ imply then the bounds on d_τ.

As an example we quote one-loop contributions to electric and weak dipole moments of d-quark and u-quark at the electroweak scale μ_{EW} ($t_W = \tan\vartheta_W$)

$$d_d^\gamma \supset \frac{\alpha}{16\pi \sin^2\vartheta_W} \frac{m_d m_i}{M_W^2} |V_{id}|^2 \ln\left(\frac{\Lambda^2}{\mu_{EW}^2}\right) \left[\frac{1 - t_W^2}{1 + t_W^2} d_i^\gamma - \sin 2\vartheta_W d_i^Z\right], \quad i = c, t, \quad (17.118)$$

$$d_d^Z \supset \frac{-\alpha}{16\pi \sin^2\vartheta_W} \frac{m_d m_i}{M_W^2} |V_{id}|^2 \ln\left(\frac{\Lambda^2}{\mu_{EW}^2}\right) \left[\frac{1 - t_W^2}{1 + t_W^2} d_i^Z + \sin 2\vartheta_W d_i^\gamma\right], \quad i = c, t, \quad (17.119)$$

$$d_u^\gamma \supset \frac{\alpha}{16\pi \sin^2\vartheta_W} \frac{m_u m_i}{M_W^2} |V_{ui}|^2 \ln\left(\frac{\Lambda^2}{\mu_{EW}^2}\right) \left[\frac{1 - t_W^2}{1 + t_W^2} d_i^\gamma - \sin 2\vartheta_W d_d^Z\right], \quad i = b, \quad (17.120)$$

$$d_u^Z \supset \frac{-\alpha}{16\pi \sin^2\vartheta_W} \frac{m_u m_i}{M_W^2} |V_{ui}|^2 \ln\left(\frac{\Lambda^2}{\mu_{EW}^2}\right) \left[\frac{1 - t_W^2}{1 + t_W^2} d_i^Z + \sin 2\vartheta_W d_i^\gamma\right], \quad i = b. \quad (17.121)$$

Calculating C_{fB} and C_{fW} for a given heavy fermion f as functions of fundamental parameters of a given theory and inserting them into (17.114) and (17.115) allows to calculate d_f^γ and d_f^Z of this fermion. Inserting the results into formulas just listed allows to calculate $d_{d,u}^\gamma$ and $d_{d,u}^Z$ as functions of these fundamental parameters and consequently neutron and proton EDMs, although the last step is tough because of theoretical uncertainties. Comparing the results with existing experimental bounds on the EDMs in question allows to put constraints on the fundamental parameters and consequently on the EDMs of heavy fermions.

On the other hand, the presence of operators $O_{\tau B}$ and $O_{\tau W}$ responsible for the d_τ cannot generate d_e, d_d, and d_u at one-loop level, but can do it via two-step RG running, discussed in Section 14.4, resulting in two-loop contributions to d_e, d_d, and d_u. The results for the imaginary parts of the relevant coefficients C_{fB} and C_{fW} with $f = e, d, u$ can be obtained using the technology outlined in Appendix E together with the RGE of [758–760]. We just give the final results in the approximations made in Appendix E. More accurate numerical results are given in our paper, which should appear in hep-arxiv in 2020.

We find then

$$\text{Im}(C_{eB}) = F_e\left[\left(\frac{3}{2}g_2^2 + 4g_1^2\bar{y}_e^2\right)\text{Im}(C_{\tau B}) + 3g_1g_2\bar{y}_e\text{Im}(C_{\tau W})\right], \tag{17.122}$$

$$\text{Im}(C_{eW}) = F_e\left[(g_2^2 + 2g_1^2\bar{y}_e^2)\text{Im}(C_{\tau W}) + g_1g_2\bar{y}_e\text{Im}(C_{\tau B})\right], \tag{17.123}$$

$$\text{Im}(C_{dB}) = F_d\left[\left(\frac{3}{2}g_2^2 + 4g_1^2\bar{y}_e\bar{y}_d\right)\text{Im}(C_{\tau B}) + 3g_1g_2\bar{y}_e\text{Im}(C_{\tau W})\right], \tag{17.124}$$

$$\text{Im}(C_{dW}) = F_d\left[(g_2^2 + 2g_1^2\bar{y}_e\bar{y}_d)\text{Im}(C_{\tau W}) + g_1g_2\bar{y}_d\text{Im}(C_{\tau B})\right], \tag{17.125}$$

$$\text{Im}(C_{uB}) = F_u\left[\left(-\frac{3}{2}g_2^2 + 6g_1^2\bar{y}_e\bar{y}_u\right)\text{Im}(C_{\tau B}) - 6g_1g_2\bar{y}_e\text{Im}(C_{\tau W})\right], \tag{17.126}$$

$$\text{Im}(C_{uW}) = F_u\left[\left(\frac{5}{2}g_2^2 - 2g_1^2\bar{y}_e\bar{y}_u\right)\text{Im}(C_{\tau W}) - 2g_1g_2\bar{y}_u\text{Im}(C_{\tau B})\right], \tag{17.127}$$

where

$$F_i = 2\frac{m_im_\tau}{v^2}\left[\frac{1}{16\pi^2}\ln\left(\frac{\Lambda}{\mu}\right)\right]^2, \qquad \bar{y}_f = \begin{cases} (Y_q + Y_f) \text{ for } f = u,d \\ (Y_l + Y_f) \text{ for } f = e \end{cases}. \tag{17.128}$$

Here Y_k are the hypercharges, that is $Y_q = 1/6$, $Y_l = -1/2$, $Y_e = -1$, $Y_d = -1/3$, and $Y_u = 2/3$.

With these formulas at hand one can not only get bounds on d_τ^γ and d_τ^Z but also include two-loop effects in the calculations of $d_{d,u}^\gamma$ and $d_{d,u}^Z$ generated by τ's EDMs. Moreover, now d_e^γ and d_e^Z enter the analysis providing additional constraints on NP.

The only important three-loop contribution is a contribution to the light-fermion EDMs via diagrams analogous to the light-by-light contributions to $(g-2)_\mu$ [1225]. The key property of these diagrams is that the only relevant mass scales are the fermion masses. In the case of a mass hierarchy between the intermediate and external fermions F and f, dimensional analysis implies $d_f \sim m_f/m_F(\alpha/\pi)^3$. In the case of fermions with intermediate masses (like the charm quark) this mechanism can be competitive to the one- and two-loop contributions.

We can now return to Table 17.8. It illustrates the scaling of contributions of the CP-violating dipole moments of light fermions in terms of those of heavier ones, displaying explicitly loop factors, fermion and gauge boson masses, gauge couplings, electric charges, the number of colors, $N_C = 3$, and potentially large logarithms. The various hierarchies involved complicate the identification of the dominant contributions for a given fermion. Yet, some properties are evident:

- d_e does not receive any one-loop contributions,
- $d_{\mu,\tau}$ do not contribute to d_e, d_d, d_u at one-loop level,
- The only one-loop contributions are present in the generation of d_d from $d_{U=(c),t}$ and of d_u from $d_{D=(s),b}$
- All entries in the table that involve two- and three-loop contributions are nonvanishing.

Table 17.8 Scaling for contributions at one, two, and three loops (first, second, and third line of each entry) to the light fermion EDMs.

$d_{U=(c),t}$	$d_{D=(s),b}$	$d_{L=(\mu),\tau}$
—	—	—
d_e $\dfrac{m_e m_U G_F}{(4\pi)^2}\dfrac{\alpha}{\pi}\dfrac{N_C}{\tan^2\vartheta_W}\ln^2\left(\dfrac{\Lambda}{\mu_{EW}}\right)$	$\dfrac{m_e m_D G_F}{(4\pi)^2}\dfrac{\alpha}{\pi}\dfrac{N_C}{\tan^2\vartheta_W}\ln^2\left(\dfrac{\Lambda}{\mu_{EW}}\right)$	$\dfrac{m_e m_L G_F}{(4\pi)^2}\dfrac{\alpha}{\pi}\dfrac{1}{\tan^2\vartheta_W}\ln^2\left(\dfrac{\Lambda}{\mu_{EW}}\right)$
$\dfrac{m_e}{m_U}q_U^3\left(\dfrac{\alpha}{\pi}\right)^3 N_C$	$\dfrac{m_e}{m_D}q_D^3\left(\dfrac{\alpha}{\pi}\right)^3 N_C$	$\dfrac{m_e}{m_L}\left(\dfrac{\alpha}{\pi}\right)^3$
$\dfrac{m_d m_U G_F}{(4\pi)^2}|V_{Ud}|^2\ln\left(\dfrac{\Lambda}{\mu_{EW}}\right)$	—	—
d_d $\dfrac{m_d m_U G_F}{(4\pi)^2}\dfrac{\alpha}{\pi}\dfrac{N_C}{\tan^2\vartheta_W}\ln^2\left(\dfrac{\Lambda}{\mu_{EW}}\right)$	$\dfrac{m_d m_D G_F}{(4\pi)^2}\dfrac{\alpha}{\pi}\dfrac{N_C}{\tan^2\vartheta_W}\ln^2\left(\dfrac{\Lambda}{\mu_{EW}}\right)$	$\dfrac{m_d m_L G_F}{(4\pi)^2}\dfrac{\alpha}{\pi}\dfrac{1}{\tan^2\vartheta_W}\ln^2\left(\dfrac{\Lambda}{\mu_{EW}}\right)$
$\dfrac{m_d}{m_U}q_U^3 q_d^3\left(\dfrac{\alpha}{\pi}\right)^3 N_C$	$\dfrac{m_d}{m_D}q_D^3 q_d^3\left(\dfrac{\alpha}{\pi}\right)^3 N_C$	$\dfrac{m_d}{m_L}\left(\dfrac{\alpha}{\pi}\right)^3$
—	$\dfrac{m_u m_D G_F}{(4\pi)^2}|V_{uD}|^2\ln\left(\dfrac{\Lambda}{\mu_{EW}}\right)$	—
d_u $\dfrac{m_u m_U G_F}{(4\pi)^2}\dfrac{\alpha}{\pi}\dfrac{N_C}{\tan^2\vartheta_W}\ln^2\left(\dfrac{\Lambda}{\mu_{EW}}\right)$	$\dfrac{m_u m_D G_F}{(4\pi)^2}\dfrac{\alpha}{\pi}\dfrac{N_C}{\tan^2\vartheta_W}\ln^2\left(\dfrac{\Lambda}{\mu_{EW}}\right)$	$\dfrac{m_u m_L G_F}{(4\pi)^2}\dfrac{\alpha}{\pi}\dfrac{1}{\tan^2\vartheta_W}\ln^2\left(\dfrac{\Lambda}{\mu_{EW}}\right)$
$\dfrac{m_u}{m_U}q_U^3 q_u^3\left(\dfrac{\alpha}{\pi}\right)^3 N_C$	$\dfrac{m_u}{m_D}q_D^3 q_u^3\left(\dfrac{\alpha}{\pi}\right)^3 N_C$	$\dfrac{m_u}{m_L}q_u^3\left(\dfrac{\alpha}{\pi}\right)^3$

- The hierarchies implied by the scalings in Table 17.8 are drastically different for the different flavors. For the top quark, the two-loop contributions are clearly dominant because the one-loop ones are suppressed by $|V_{td}|^2$ and the three-loop ones are not enhanced compared to $m_t^2/v^2 \sim 1$. On the other hand, for the charm quark, all three loop levels are comparable, despite each class receiving different suppression factors.

Numerical analysis of these formulas with the bounds on EDMs of heavy fermions should appear in hep-arxiv in 2020.

17.4 Anomalous Magnetic Moments $(g-2)_{\mu,e}$

17.4.1 Preliminaries

Although $a_{e,\mu}$ are both flavor- and CP-conserving, they also offer powerful probes to test the SM and its extensions. In particular the anomalous magnetic moment of the muon

$$a_\mu = \frac{(g-2)_\mu}{2},$$

(17.129)

which we defined in Section 17.1.3, provides an excellent test for physics beyond the SM.

On the theory side a_μ receives within the SM four dominant contributions:

$$a_\mu^{\text{SM}} = a_\mu^{\text{QED}} + a_\mu^{\text{ew}} + a_\mu^{\gamma\gamma} + a_\mu^{\text{hvp}}. \tag{17.130}$$

While the QED [1226–1229] and EW contributions [803, 1230] to a_μ^{SM} are known very precisely and the light-by-light contribution $a_\mu^{\gamma\gamma}$ is currently known with a reasonable accuracy [1231, 1232], the theoretical uncertainty is dominated by the hadronic vacuum polarization. Review of the relevant calculations of all these contributions and references to related extensive analyses can be found in [803, 1119, 1233–1236].

The analyses by the master of this field, Fred Jegerlehner, just quoted, show that the very precise measurement of a_μ by the E821 experiment [1115] in Brookhaven,

$$a_\mu^{\text{exp}} = (11659209.1 \pm 5.4 \pm 3.3[6.3]) \times 10^{-10}, \tag{17.131}$$

differs from its SM prediction by roughly 4σ

$$\boxed{a_\mu^{\text{exp}} - a_\mu^{\text{SM}} = (31.3 \pm 7.7) \times 10^{-10}.} \tag{17.132}$$

Many NP models have been used by many theorists with the goal to explain this discrepancy. Especially supersymmetric models were very popular [1237–1243] ten years ago. In SUSY the discrepancy could easily be accommodated for relatively light smuon masses and large $\tan\beta$. However, so far no light SUSY particles have been discovered. Another approach was followed in [1244], where the interplay of $(g-2)_\mu$ and a soft muon Yukawa coupling that is generated radiatively in the MSSM was studied. With the increased SUSY mass scale the explanation of $(g-2)_\mu$ anomaly became difficult already in 2012 [1245]. We will return to other extensions of the SM in this context.

On the experimental side, the $g-2$ ring at BNL has been disassembled and brought to Fermilab for a run, which started in 2017. The data are expected in 2020. An experiment at J-PARC should provide an independent measurement in the coming years.

The anomalous magnetic moment of the muon a_μ is more sensitive to lepton flavor-conserving NP than a_e, and consequently the latter was not as popular as a_μ until recently. However, as emphasized in [1123] and in references quoted therein, because of the fact that a_e is very precisely measured and very precisely calculated within the SM, it can also be used to probe NP, even if the theory agrees very well with experiment. Indeed, a_e plays a central role in QED because its precise measurement provides the best source of α_{em} assuming the validity of QED [1246]. Conversely, one can use a value of α_{em} from a less precise measurement and insert it into the theory prediction for a_e to probe NP. The 2007 calculation implied [1247]

$$a_e = 1\,159\,652\,182.79\,(7.71) \times 10^{-12}, \tag{17.133}$$

where the largest uncertainty came from the second-best measurement of α_{em}, which is $\alpha_{\text{em}}^{-1} = 137.03599884(91)$ from a Rubidium atom experiment [1248]. The difference between experiment and data turned out to be 1.7σ.

However, recently a measurement using atomic Cesium [1249] resulted in the new most precise value

$$\alpha_{\text{em}}^{-1} = 137.035999046(27). \tag{17.134}$$

Comparison of the theoretical SM prediction for a_e^{SM} [1250] with the existing experimental measurement a_e^{exp} [1251] now leads to a discrepancy of 2.4 σ,

$$\boxed{\Delta a_e \equiv a_e^{\text{exp}} - a_e^{\text{SM}} = (-87 \pm 36) \times 10^{-14},} \tag{17.135}$$

with errors added in quadrature. I can recommend a very nice paper by Davoudiasl and Marciano [1252] that discusses the implications of this new anomaly. See also [1253], where a combined explanation of $(g-2)_{\mu,e}$ anomalies with implications for a large muon EDM has been presented.

In what follows we will collect one-loop formulas for a_μ, which allow to calculate the contribution to a_μ in any model with fermions, gauge bosons, and scalars. These formulas can easily be generalized to a_ℓ with $\ell = e, \tau$. Such formulas were first derived in [802] and subsequently by many authors. See, e.g., [803]. In particular systematic presentation of various cases has been done recently in [804]. Exact results in terms of integrals can be found both in [802] and [804].

Here we will confine our presentation to approximate formulas derived in [804] that have been verified by Robert Szafron. We should, however, stress that these formulas are listed here to primarily give the reader a rough idea on the structure of various contributions as functions of the involved masses and couplings. For phenomenology it is advisable to use exact expressions given in [802] and [804].

The contributing diagrams are given in Figure 17.4 with all particles being physical. Indeed, as there are no tree-level contributions, results of one-loop calculations must be finite so that even a nonrenormalizable gauge like unitary gauge can be used for calculations.

17.4.2 Gauge Boson Contributions

These are obtained from diagrams (a) and (b) in Figure 17.4, where F denotes any fermion and X a gauge boson with electric charges Q_F and Q_X, respectively. They will be given in units of $e > 0$ so that $Q_\mu = -1$. The interaction Lagrangian is given as

$$\mathcal{L}_{\text{int}} = \sum_{F,X} \bar{F}_i[C_V^{ij}(X)\gamma^\mu + C_A^{ij}(X)\gamma^\mu\gamma^5]l_j X_\mu, \tag{17.136}$$

where $l_j = e, \mu, \tau$. The sum is over all fermions F including the SM fermions and over all gauge bosons X that also include SM gauge bosons. Model dependence enters through the coefficients C_V and C_A. In a given model they can be read off the interaction Lagrangian in the *unitary gauge*.

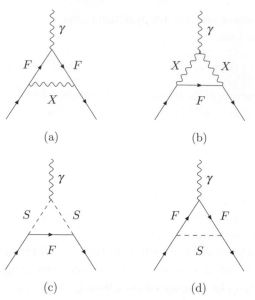

Figure 17.4 Diagrams contributing to $(g-2)_\mu$. F denotes a fermion, X a gauge boson, and S a scalar or pseudoscalar. The outgoing fermion is μ, but could also be e or τ.

The formulas listed following can easily be generalized to a_e and a_τ, but to avoid proliferation of the indices we will give all formulas for the case of the muon. Moreover, it is sufficient to expand in m_μ/M_X.

Charged Fermion and Neutral Gauge Boson

As a simple example we consider the case of a heavy neutral gauge boson Z' and a charged fermion $F_i = E_i$ with a mass m_{E_i}. In this case, with $Q_E = -1$ from charge conservation and $Q_X = Q_{Z'} = 0$, only the diagram (a) in Figure 17.4 contributes. For a Z' much heavier than μ and E_i one finds [804] ($C_V = \Delta_V/2$ and $C_A = \Delta_A/2$ in our notation)

$$\Delta a_\mu(E_i, Z') \approx -\frac{1}{4\pi^2} \frac{m_\mu^2}{M_{Z'}^2} \left[\left|C_V^{i\mu}(Z')\right|^2 \left(\frac{2}{3} - \frac{m_{E_i}}{m_\mu}\right) + \left|C_A^{i\mu}(Z')\right|^2 \left(\frac{2}{3} + \frac{m_{E_i}}{m_\mu}\right) \right].$$

(17.137)

Usually contributions from $E_i = e, \mu, \tau$ have to be added. If E_i is a new heavy lepton and $m_{E_i}/M_{Z'} \geq 0.2$, exact expressions in [804] have to be used.

In terms of Z' couplings used in this book

$$C_{V,A}^{ij}(Z') = \frac{1}{2}\Delta_{V,A}^{ij}(Z') = \frac{1}{2}\left[\Delta_R^{ij}(Z') \pm \Delta_L^{ij}(Z')\right],$$

(17.138)

see (15.54).

Neutral Fermion and Charged Gauge Boson

We consider next the case of a heavy charged gauge boson W' and a neutral fermion N_i with a mass m_{N_i}. In this case, with $Q_F = Q_N = 0$, only the diagram (b) in Figure 17.4 contributes. For W' much heavier than μ and N_i one finds [804]

$$\Delta a_\mu(N_i, W') \approx \frac{1}{4\pi^2} \frac{m_\mu^2}{M_{W'}^2} \left[\left| C_V^{i\mu}(W') \right|^2 \left(\frac{5}{6} - \frac{m_{N_i}}{m_\mu} \right) + \left| C_A^{i\mu}(W') \right|^2 \left(\frac{5}{6} + \frac{m_{N_i}}{m_\mu} \right) \right].$$

(17.139)

As this simple formula works very well only for $m_{N_i}/M_{W'} \leq 0.3$, for a heavy N_i exact expressions in [804] have to be used.

17.4.3 Scalar Contributions

These are obtained from diagrams (c) and (d) in Figure 17.4, where F denotes any fermion and S a scalar or pseudoscalar with electric charges Q_F and Q_S

The interaction Lagrangian for a charged scalar S^+ and a neutral fermion N_i is given as follows

$$\mathcal{L}_{\text{Yukawa}} = \sum_{N,S} S^+ \bar{N}_i [C_S^{ij}(S^+) + C_P^{ij}(S^+)\gamma^5] l_j, \tag{17.140}$$

where again $l_j = e, \mu, \tau$. In the case of a neutral scalar S^0 and a charged fermion E_i we have

$$\mathcal{L}_{\text{Yukawa}} = \sum_{E,S} S^0 \bar{E}_i [C_S^{ij}(S^0) + C_P^{ij}(S^0)\gamma^5] l_j, \tag{17.141}$$

where again $l_j = e, \mu, \tau$.

The sum is over all fermions $F = N, E$ that is also the SM fermions and over all scalars and pseudoscalars S that is also Higgs scalar. Model dependence enters through the coefficients C_S and C_P. In a given model they can be read off the interaction Lagrangian in the unitary gauge.

Neutral Fermion and Charged Scalar

As a simple application we consider the case of a heavy charged scalar S^+ and a neutral fermion N_i with a mass m_{N_i}. In this case, with $Q_F = Q_{N_i} = 0$, only diagram (c) in Figure 17.4 contributes. For S^+ much heavier than μ and N_i one finds [804]

$$\Delta a_\mu(N_i, S^+) \approx -\frac{1}{4\pi^2} \frac{m_\mu^2}{M_{S^+}^2} \left[\left| C_S(S^+) \right|^2 \left(\frac{1}{12} + \frac{m_{N_i}}{4m_\mu} \right) + \left| C_P(S^+) \right|^2 \left(\frac{1}{12} - \frac{m_{N_i}}{4m_\mu} \right) \right].$$

(17.142)

Exact result in terms of integrals can be found both in [802] and [804]. It turns out that the preceding simple formula works very well only for $m_{N_i}/M_{S^+} \leq 0.1$.

Charged Fermion and Neutral Scalar

Finally, we consider the case of a heavy neutral scalar S^0 and a charged fermion E_i with a mass m_{E_i}. In this case, with $Q_F = Q_E = -1$ and $Q_S = 0$, only diagram (d) in Figure 17.4 contributes. For S^0 much heavier than μ and E_i one finds [804]

$$\Delta a_\mu(E_i, S^0) \approx \frac{1}{4\pi^2} \frac{m_\mu^2}{M_{S^0}^2} \left[|C_S(S^0)|^2 \left(\frac{1}{6} - \frac{m_{E_i}}{m_\mu} P \right) + |C_P(S^0)|^2 \left(\frac{1}{6} + \frac{m_{E_i}}{m_\mu} P \right) \right],$$

(17.143)

with

$$P = \frac{3}{4} + \ln\left(\frac{m_E}{m_S} \right).$$

(17.144)

Again exact results in terms of integrals can be found both in [802] and [804]. It turns out that the preceding simple formula works very well only for $m_{E_i}/M_{S^0} \leq 0.1$.

17.4.4 Beyond the SM

With these formulas at hand one can study basically all extensions of the SM that contain new heavy fermions, new heavy gauge bosons, and new heavy scalars. A very nice review in this context is the one in [804] in which $(g - 2)_\mu$ in numerous simplified models and concrete models with UV completions has been analyzed in detail. In addition to examples presented here also examples with doubly charged scalars, fermions, and gauge bosons present in some extensions of the SM have been considered. In particular correlations between $(g - 2)_\mu$ and $\mu \rightarrow e\gamma$ and also $\mu \rightarrow 3e$ have been analyzed in these models in detail. I can strongly recommend this review as it is very systematically written and contains over 500 references, which certainly will motivate the readers to learn more about this subject. In particular, the desire to explain the $(g - 2)_\mu$ anomaly led various authors to invoke vector-like leptons. See, e.g., [943, 1034–1038] and references therein. Prospects for vector-like leptons at future proton-proton colliders are reviewed in [1254], where further references to theoretical papers can be found. This review indicates that vector-like leptons with masses above 10 TeV will be probed only by indirect searches even in the presence of a 100 TeV collider. The explanation of a_μ and a_e anomalies with the help of axion-like particles has been suggested in [1315].

For me the main messages from all these studies is the following one: Provided the $(g-2)_\mu$ anomaly will be confirmed by the Fermilab experiment, we will have fun, using the expressions and plots in [804] and other papers listed earlier, to find out which mediators are hiding behind this anomaly. But simultaneously one should look at correlations with other processes considered in this book.

Last, but certainly not least, the formfactors listed in Table 17.4 for $f = i$ can be directly used to calculate NP contributions to anomalous magnetic moments of charged leptons [795]

$$a_{l_i} = \frac{2m_{l_i}}{e\Lambda^2} \text{Re}[F_{TR}^{ii}]. \tag{17.145}$$

17.5 Neutrino Oscillations

The observation of neutrino oscillations is a clear signal of physics beyond the SM and so far together with Dark Matter and the matter-antimatter asymmetry observed in our universe the only clear sign of NP.

To accommodate neutrino masses one needs to extend the SM. The most straightforward way is to proceed in the same manner as for quark and charged lepton masses and just introduce three right-handed neutrinos that are singlets under the SM gauge group anyway. A Dirac mass term is then generated via the usual Higgs coupling $\bar{\nu}_L Y_\nu H \nu_R$. However, then there is also the possibility for a Majorana mass term for the right-handed neutrinos because it is gauge invariant. One would need to introduce or postulate a further symmetry to forbid this term, which is also already an extension of the SM. Furthermore, this Majorana mass term introduces an additional scale M_R and because it is not protected by any symmetry it could be rather high. Then with the help of the so-called seesaw mechanism one can generate light neutrino masses as observed in nature. Another possibility to get neutrino masses without right-handed neutrinos is the introduction of an additional Higgs-triplet field. Either way, the accommodation of neutrino masses requires an extension of the SM.

Here, in addition to CP-violation in the lepton sector, which has not been discovered yet, neutrinoless double-beta decay plays a very prominent role. It will tell us whether a neutrino is a Dirac or Majorana particle. All these topics are very important. Unfortunately the author of this book never explored this valley of NP and can only direct interested readers to the excellent textbook by Bilenky [23] on massive neutrino physics and most recent reviews and papers on the neutrinoless double-beta decay [797, 1255–1258]. Finally, I enjoyed very much the history of neutrino masses and mixing presented in [1259].

18 Grand Summary of New Physics Models

18.1 Preliminaries

In Chapter 12 we listed various anomalies that included in particular:

- Violation of the lepton flavor universality in $b \to c\tau\nu$ and $b \to s\ell\bar{\ell}$ transitions with LQs being the most favorite mediators behind these anomalies. We have discussed various LQ models in this context in Section 16.4.11. Z' models could also help in the case of $b \to s\ell\bar{\ell}$ transitions, but to explain also $b \to c\tau\bar{\nu}$ anomalies, additional heavy particles had to be added.

- As discussed in detail in Chapter 10, the experimental value of the ratio ε'/ε appears, according to DQCD and 2015 RBC-UKQCD results, to be significantly larger than the SM prediction. Here LQs are unlikely to be the origin of this anomaly as summarized in Section 16.4.13, and Z' models as well as models with VLQs generating FCNC mediated by the SM Z boson can easier address this anomaly. We have discussed this at various places in this book, in particular in Chapters 15 and 16.

- Possible tension between ΔM_s, ΔM_d, and the angle γ is intriguing as it involves observables that were with us for many years, and this tension only arose due to improved LQCD calculations of the relevant hadronic matrix elements and the improved measurements of the angle γ. We have discussed this tension in some detail in Section 15.1. It can be easier solved than the other two tensions, and its structure is simpler so that various NP scenarios can be considered here.

- Possible violation of CKM unitarity mentioned briefly in Chapter 12. If confirmed in the future, it would have implications on a number of observables.

- $(g - 2)_\mu$ anomaly, which we considered in Section 17.4. New experiment at Fermilab should be able to clarify the situation. The recent $(g - 2)_e$ anomaly is also very interesting.

Now, the ε'/ε anomaly and the tension between $\Delta M_{s,d}$ and γ were in the last years under the shadow of B physics anomalies and to a lesser extent under the shadow of the $(g-2)_\mu$ anomaly. Partly this is related to the fact that ε'/ε and $\Delta M_{s,d}$ were precisely measured a long time ago. On the other hand, B physics anomalies and the $(g-2)_\mu$ anomaly are still in the hands of experimentalists, and it is, of course, very exciting what the future will bring us here.

In particular the anomalies in B decays, that is in $b \to s\ell\bar{\ell}$ and $b \to c\ell\bar{\nu}$ transitions, led to many theoretical analyses, and their number exploded in April 2017 after the announcement by the LHCb collaboration that not only the ratio $R(K)$ but also the ratio $R(K^*)$ deviates significantly from the SM predictions. The messages from LHCb on $R(K)$ and from Belle on $R(K^*)$, $R(D)$, and $R(D^*)$ at the 2019 Moriond meeting indicate that the B physics anomalies are not as pronounced as expected in 2017. This had a significant impact on the flux of papers on anomalies in hep-arxiv in 2019 that decreased relative to the years 2017 and 2018 by much.

Still, when we look back in time, B physics anomalies governed the flavor physics in the 2010s. As by now several hundreds of papers have been published addressing all or part of anomalies in question, it was impossible to refer to all of them in this book, although many references can be found in previous chapters. Yet, we would like now to draw here a general picture of what these anomalies seem to tell us. While it is possible that in the Belle-II and LHCb upgrade era some of them could disappear, the ideas and the models studied with the goal to explain them could still turn out to be useful in the future. Let us begin with general comments, which express the view of the author of this book and are possibly not shared by everybody.

First of all, a number of analyses, even before the measurement of $R(K^*)$, have argued not only on the basis of $R(K)$ and P_5' anomalies, but also $R(D)$ and $R(D^*)$ anomalies, that they all can be eliminated through the modifications of Wilson coefficients of four-fermion effective operators $(\bar{s}\gamma_\mu P_L b)(\bar{\mu}\gamma^\mu P_L \mu)$ and $(\bar{c}_L \gamma_\mu b_L)(\bar{\tau}_L \gamma_\mu \nu_L)$. This picture has been confirmed by finding $R(K^*) \approx R(K)$ as emphasized in particular in [609, 1066]. A subset of other relevant papers in this context can be found in [486, 494, 565, 710, 721, 723, 725, 730, 776, 1260–1262]. Others have been listed in previous chapters. Indeed, in the case of $b \to s\ell\bar{\ell}$ transitions an effective operator $(\bar{s}\gamma_\mu P_L b)(\bar{\mu}\gamma^\mu P_L \mu)$ can also help in understanding the tensions in the measurements of angular observables and branching ratios involving the $b \to s$ transition, although these are subject to sizeable hadronic uncertainties [485, 499, 1263, 1264].

Critical analysis of $R(K)$ and $R(K^*)$ anomalies can be found in [723, 726]. In [1265] it has been found that tensor operators can explain both low and high q^2 of $R(K^*)$ anomaly, but to get simultaneous explanation of $R(K)$ anomaly one needs also vector or axial vector operators. See also [1266].

While 2017 global fits suggested that NP in $(\bar{s}\gamma_\mu P_L b)(\bar{\mu}\gamma^\mu P_L \mu)$ is preferred at between 4.2σ and 6.2σ [51, 494, 721, 723, 725, 730] over the SM, other solutions could not be excluded, in particular when several WCs were included in the fit. In these papers the values of the WCs of the relevant low-energy effective Hamiltonians can be found. They will be surely changing with time. This is confirmed by the modifications of various fits after Moriond 2019 as summarized in Section 12.3.

It should be again emphasized that a simultaneous explanation of $R(K)$ and $R(K^*)$ anomalies in neutral-current $b \to s\ell\bar{\ell}$ transitions and of $R(D)$ and $R(D^*)$ anomalies in charged-current $b \to c\ell\bar{\nu}$ transitions with the help of left-handed currents is very natural because a left-handed operator $(\bar{c}_L \gamma_\mu b_L)(\bar{\tau}_L \gamma_\mu \nu_L)$ is related to $(\bar{s}_L \gamma_\mu b_L)(\bar{\mu}_L \gamma_\mu \mu_L)$ operator by the $SU(2)_L$ gauge symmetry [777]. However, as summarized in [765], this picture might work only provided NP couples much more strongly to the third generation than to the first two. Such a requirement can be naturally accomplished in two ways:

- Assuming that NP is coupled, in the flavor basis, only to the third generation of quarks and leptons. Couplings to lighter generations are then generated by the misalignment between the mass and the flavor bases through small flavor mixing angles [776],
- If NP couples to different fermion generations proportionally to their mass squared [778].

In the first scenario LFU violation necessarily implies lepton flavor-violating (LFV) phenomena. But as emphasized in [778], the same is not true in the second scenario if the lepton family numbers are preserved. Most papers follow the first route, which as pointed out in [1067] and [765] may disagree with direct LHC searches and EW precision tests, unless as demonstrated in [719] particular flavor structure is chosen.

While discussing various models in previous chapters, we have already described how they face various flavor anomalies. In particular we discussed the two most popular classes of models, as far as B physics anomalies are concerned, the Z' models and leptoquark models in Sections 15.4 and 16.4.11, respectively. But here we would like still to make several comments and discuss other solutions to the anomalies in question. In this context the reviews of NP models in [305, 718, 719, 1160, 1267] are very useful. In particular section 7 in [305] discusses all B physics anomalies and lists many more references than we can list here. I can also recommend two recent papers [1321, 1322].

18.2　General Observations on B Physics Anomalies

Let us begin with several papers that have not been mentioned in previous chapters.

Interestingly the MSSM has problems with generating sufficient nonuniversality in lepton couplings. Also measurements of the dilepton invariant mass distribution disfavor many popular NP scenarios (e.g., type-II two Higgs doublet models [583]) as candidate explanations. The situation improves in a general 2HDM model (type-III) as shown in [1268, 1269], although the explanation of the $R(D)$ and $R(D^*)$ anomalies is under pressure because of the B_c-lifetime. The most recent analysis of $b \to s\ell^+\ell^-$ transitions in 2HDM models with many references can be found in [1270]. Moreover, $R(K)$ and $R(K^*)$ anomalies in a Aligned 2HDM with right-handed neutrinos have been addressed [1271]. Here a large Yukawa coupling allows for LFUV generated through box diagrams mediated by charged Higgs bosons and right-handed neutrinos.

Next the anomalies have been addressed in models with extra dimensions [1100, 1272–1274]. In [1275] the $b \to s\ell^+\ell^-$ anomalies have been addressed in Z' models without quark flavor violation. But these models must clearly fail if the ε'/ε anomaly will be fully established.

Connection of $R(K^{(*)})$ anomalies to the generation of neutrino masses in the context of Zee–Babu model that contains S_3 LQ and scalar diquark has been considered in [1276]. Another analysis of $(g-2)_\mu$, $R(D^{(*)})$ and $R(K)$ in conjunction with neutrino mass generation has been presented in [964].

There is an important and interesting question of what scale of NP is responsible for the B physics anomalies. Ideally one would like to discover this physics directly at the LHC, but this becomes harder and harder, although as described in [1277] the Run III could bring surprises.

The NP scale explored indirectly in B physics depends on the couplings involved and on whether the $b \rightarrow c\ell\bar{\nu}$ or $b \rightarrow s\ell\bar{\ell}$ anomalies are considered. For a fixed coupling, the scale of NP must clearly be lower in the first case than in the latter case, simply because the $R(D^{(*)})$ anomalies affect tree-level decays in the SM but $R(K^{(*)})$ anomalies the loop-induced ones. In an interesting paper the constraints of perturbative unitarity on NP interpretation of the anomalies in $b \rightarrow c\ell\bar{\nu}$ and $b \rightarrow s\ell\bar{\ell}$ transitions have been considered [1278]. They turn out to be stronger than naive perturbative bounds with couplings significantly larger than unity. Within an EFT approach the authors find that $2 \rightarrow 2$ fermion scattering amplitudes saturate the unitarity bound below 9 TeV and 80 TeV for $b \rightarrow c\ell\bar{\nu}$ and $b \rightarrow s\ell\bar{\ell}$, respectively. Even stronger bounds, up to few TeV, are obtained when leading effective operators couple dominantly to the third generation. These two numbers increased after Moriond 2019, but a detailed analysis has still to be done.

Inspecting the tables in [1278] it appears that in NP scenarios with couplings close to unity and no stringent flavor structure, the B physics anomalies can be addressed successfully by tree-level mediators without conflicts with perturbative unitarity and LHC lower bounds on the masses of these mediators. But in models in which the couplings are suppressed by some symmetries the necessary scales are so low that they violate the LHC lower bounds. In any case the main message of this paper is that when trying to explain various anomalies one has to be careful with unitarity bounds that played an important role in the 1970s in the context of the SM.

The question then arises of how various explanations of B-physics anomalies could be distinguished on the basis of indirect searches alone. As pointed out by several authors, $R(K^{(*)})$ anomalies alone constrain the chiralities of the mediators responsible for these anomalies. In particular in [730] Z' models, LQs, loop mediators, and composite dynamics have been discussed, and their distinctive features in terms of chiralities and flavor structure have been highlighted. Also in [724] comparison of Z' and LQs has been made, and in [1279] it has been pointed out that the CP-violating asymmetries in $B \rightarrow K^{(*)}\mu^+\mu^-$ could distinguish between LQ and Z' models.

In my view the main differences between Z' and LQ scenarios are as follows:

- LQs have bigger potential to explain both neutral and charged B physics anomalies than Z' models.
- The constraints from $\Delta F = 2$ transitions can be much easier satisfied in LQ models than in Z' models because NP contributions enter at one-loop and tree-level in these scenarios, respectively.
- But if ε'/ε anomaly will stay with us forever, Z' models will be able to explain it, while LQs will have a very hard time to do it as demonstrated in [667].

If both B physics anomalies and ε'/ε anomaly will be established, possibly LQs are responsible for the B physics ones while Z' for the ε'/ε anomaly. But in both cases one should not forget the role of VLQs that we analyzed in detail in Section 16.3.

Right-handed currents alone cannot explain these anomalies, and scalar and pseudoscalar currents have problems with B_c-lifetime [571]. This then excludes the 2HDM model. On the other hand, models with tensor and scalars operators seem to be viable [768].

Adding kinematic distributions in $\tau^- \rightarrow \ell^- \bar{\nu}_\ell \nu_\tau$ like angular distributions and longitudinal polarization in $\tau^- \rightarrow \pi^- \nu_\tau$ allows to distinguish between various Dirac structures [592].

A general analysis of the implications of lepton flavor universality violation for $b \rightarrow s\tau^+\tau^-$ can be found in [1280].

18.3 Tree-Level Mediators

Possible mediators, assuming they couple only to SM fields, were classified in [541]. Moreover, flavor models for $R(D^*)$ anomalies have been considered. Classification of operators that could be responsible for them, and corresponding completions have been discussed. A very useful paper in this context is [782], where the complete matching of any NP model onto the SMEFT at tree-level has been calculated. We described it already in Section 14.6.2.

The tree-level mediators that can explain the $R(D)$-$R(D^*)$ anomaly, as summarized in [305], are

- LQs, as discussed in Section 16.4.11, generating various couplings. See in particular the papers after Moriond 2019 [552–556].
- A W' [719, 933, 934, 944, 1272].
- A charged color-neutral scalar [574, 719, 997, 1268, 1269, 1281–1284].

For comparisons between these models, see, for instance, [535, 541, 549, 551, 573, 708, 719, 1285, 1286]. In particular in [1286] it has been demonstrated that CP-violating observables could distinguish between different models. Allowing for additional light particles opens up the possibility to address the anomalies with contributions involving right-handed neutrinos [558–560, 1047, 1104, 1287–1289] because the neutrino is not detected.

18.4 Kaon Physics

In Kaon physics the hints for the ε'/ε anomaly dominated the analyses found in the literature, and we refer to Section 14.7, Table 14.3, and various analyses in Chapter 16 in specific models. Here the correlations with $K^+ \rightarrow \pi^+ \nu \bar{\nu}$ and $K_L \rightarrow \pi^0 \nu \bar{\nu}$ will play an important role in the identification of NP, the point stressed at various places in our book. But beyond the ε'/ε anomaly and its relation to $K^+ \rightarrow \pi^+ \nu \bar{\nu}$ and $K_L \rightarrow \pi^0 \nu \bar{\nu}$ there are hopes that NP will also be found in other rare kaon decays.

In this context the hints for the breakdown of LFU in B decays motivated various authors to analyze their implications for K decays. In particular in [1166] the authors address LFUV in $K \to \pi\ell\ell'$ and $K \to \ell\ell'$ with $\ell = \mu, e$ and improve the analysis of long-distance contributions to $K_L \to \ell^+\ell^-$. They discuss the corresponding opportunities at the NA62 experiment and with the help of MFV make the first steps toward relating this LFUV with the one observed in B decays. These ideas have been extended in [1290]. On the other hand, in [1168] LFUV in $K^+ \to \pi^+\nu\bar\nu$ and $K_L \to \pi^0\nu\bar\nu$ has been analyzed using the fact that these decays are sensitive to the physics of the third generation, not only through loop effects, as usually discussed, but through ν_τ that can be produced in these decays.

A totally different proposal for the explanation of several anomalies has been made in [1291]: they are supposed to be generated by nonperturbative SM effects. Finally I find model constructions in [1321, 1322] very interesting, but there is no space to describe them here.

18.5 Generalities

It should be stressed that in many papers found in the literature RG effects present in the SMEFT have not been included under the assumption that NP will be found still at scales explored by the LHC. But if NP is beyond the reach of the LHC, the inclusion of such effects is mandatory. The codes generated already by various authors and listed in Chapter 14 allow the inclusion of such effects in an efficient manner.

One should also realize that if ε'/ε anomaly is fully established one day, the couplings of NP to first generation cannot be set to zero as done in most papers that study the B physics anomalies. Whether new CP-violating phases have to be included or not depends on the model considered. As an example, in $U(2)^3$ models, it is possible to explain this anomaly by just suppressing the EWP contribution without modifying its SM complex phase [1326, 1327].

From my point of view, which I am sure is shared by those who read the previous pages, the picture of NP above the 1 TeV scale is still very messy at present. While my intention was to close this chapter with a grand picture of most important favorite NP scenarios on the basis of present experimental data and theoretical studies, it appears now that it is much too early to draw clear-cut conclusions not only about NP involved but also about the scale at which it enters. The recent decrease of the $R(K)$, $R(D)$, and $R(D^*)$ anomalies convinced me that I should wait with such a grand picture at least for the second edition of this book, which hopefully will materialize one day.

The final two chapters of this book have the goal to convince the readers that at the end of the 2020s the painting of such a grand picture could turn out to be possible. The five recent reports [305, 428, 1277, 1292, 1293] give a strong support for these expectations.

19 Flavor Expedition to the Zeptouniverse

19.1 Preliminaries

Despite tremendous efforts of experimentalists and theorists to find NP beyond the SM, no clear indications for NP beyond dark matter, neutrino masses, and matter-antimatter asymmetry in the universe have been observed directly in high-energy experiments. Yet, the discovery of a Higgs-like particle and the overall agreement of the SM with the present data shows that our general approach of describing physics at very short-distance scales with the help of exact (QED and QCD) and spontaneously broken (for weak interactions) gauge theory is correct.

As the SM on the theoretical side is not fully satisfactory and the three NP signals mentioned earlier are already present, we know that some new particles and new forces have to exist. We have given a number of reasons for this at the beginning of this book. As far as the topics discussed by us are concerned, the upgrade of the LHCb, Belle II, and dedicated kaon physics experiments at CERN and J-PARC, as well as improved measurements of charged lepton flavor violation (CLFV), electric dipole moments (EDMs), and $(g-2)_{\mu,e}$ will definitely shed light on the question of whether NP is present not only below, say, $10\,\mathrm{TeV}$. In this decade these rare processes will allow us to probe much higher energy scales. These are in particular particle-antiparticle mixings ($\Delta F = 2$ processes), rare decays of mesons ($\Delta F = 1$ processes), CLFV, EDMs, and $(g-2)_{\mu,e}$. All of them have been discussed in detail in this book both within the SM and a number of its extensions.

As this is an indirect search for NP, one has to develop special strategies to reach the Zeptouniverse, that is, scales as short as $10^{-21}\mathrm{m}$ or equivalently energy scales as high as several hundreds of TeV. In what follows we will present some of such strategies developed in my group at the Technical University in Munich during the last 15 years. Some of them are summarized in [208, 1294], but since then a number of deviations from SM expectations have been found by various experimental groups. We encountered them already at various places in this book and identified in the last chapter a number of NP models that appear to be most successful in addressing them. Yet, in the coming years we should hope to find more departures from SM predictions so that it should be possible to reduce the number of possibilities. With this goal in mind, let us then recall the strategies presented in [208, 1294] and suggest improvements on them.

19.2 Basic Requirements for a Successful Zeptouniverse Expedition

The coming 10 years (2020–2030) of flavor precision era invites us to attempt an *expedition from the Attouniverse to the Zeptouniverse*, that is, from scales $O(10^{-18})$ m to scales $O(10^{-21})$ m. For such an expedition to have a chance to be successful at least the following requirements have to be fulfilled:

- Many precise measurements of many observables.
- Precise extraction of CKM parameters from those tree-level decays, which are expected to have at most tiny NP contributions.[1] Here, the main targets for the coming years are

$$|V_{ub}|, \qquad |V_{cb}|, \qquad \gamma, \tag{19.1}$$

where γ, one of the angles in the Unitarity Triangle, is up to the sign the complex phase of V_{ub}. The fourth parameter of the CKM matrix, $|V_{us}|$, is already precisely known.

- Precise LQCD calculations of weak decay constants F_{B_s} and F_{B_d}, of various hadronic matrix elements for $\Delta F = 2$ transitions in the SM and beyond it and of form factors for various semileptonic transitions, in particular for $B \rightarrow K(K^*)$ transitions with leptons in the final state and those relevant for the determinations of $|V_{ub}|$ and $|V_{cb}|$ that are also relevant for $\bar{B} \rightarrow D(D^*)\ell\bar{\nu}$ decays. In the case of nonleptonic decays of B-mesons LQCD appears to be less useful, and in this case approaches like QCDF approach, LCSR, and those based on flavor symmetries and their breakdown will continue to play important roles. We discussed them in Sections 7.4 and 8.6.
- Precise LQCD calculations of hadronic matrix elements in the SM and beyond it that are relevant for ε_K, the ratio ε'/ε, the $\Delta I = 1/2$ rule, the mass difference ΔM_K, and long-distance contributions to $K^+ \rightarrow \pi^+ \nu\bar{\nu}$, $K_L \rightarrow \pi^0 \ell^+ \ell^-$, and $K_{L,S} \rightarrow \mu^+ \mu^-$.
- NLO and NNLO QCD corrections and NLO electroweak corrections to various Wilson coefficients. Among the tasks listed here I would claim that at least within the SM this task has been completed after 30 years of efforts by several theorists (1988–2019). An updated review of these efforts can be found in [77]. I do not think we need more precision here within the SM, and these calculations are sufficiently demanding that there is no point in doing them in extensions of the SM before we know what these extensions are. An exception are tree-level flavor-changing neutral currents mediated by Z', Z, G', or heavy neutral scalars. Their structure is sufficiently simple so that NLO QCD corrections to some of these exchanges could be already calculated [749, 1296].

Let us then assume that the CKM parameters have been determined with high precision and nonperturbative parameters and hadronic matrix elements, relevant both for the SM and its extensions, have been calculated accurately. Having then precise SM predictions let us assume that future precise measurements of various observables have identified a number of deviations from SM predictions so that without any doubt we can conclude that some NP has been discovered. The question then arises of what kind of NP could be responsible for these deviations. In my view the most powerful approach is based on

---

[1] Analyses of the room left for NP in tree-level decays can be found in [393, 413, 1295].

- Intensive studies of correlations between many observables in a given extension of the SM with the goal to identify patterns of deviations from the SM expectations characteristic for this extension.
- Intensive studies of correlations between low-energy precision measurements, including electroweak precision tests and the measurements at the highest available energy, that is in the coming decades the measurements of a multitude of observables in proton-proton collisions at the LHC and hopefully even more energetic colliders.

But when studying these correlations numerically, it is crucial to take into account

- RG evolution from NP scale down to EW scale that includes not only QCD effects but also electroweak and Yukawa coupling effects that are described by the SMEFT.
- RG evolution from EW scale down to low-energy scales relevant for meson decays that includes QCD and QED effects in the context of the LEFT.
- Matching between the SMEFT and the LEFT and other possible matchings within these effective theories when some heavy particles are integrated out.

This is clearly an ambitious goal, but with the help of the technology described in Chapter 14, represented in particular by the results listed in Section 14.6.2 and by the numerical codes listed in Section 14.6.4, reaching this goal one day appears to me realistic.

Yet, I would like to emphasize that the strategy presented here differs from the usual one in which a very intensive numerical analysis of hundreds of observables is first made with the goal to identify pulls and sigmas and to draw allowed regions in the parameter space of a given model. Clearly, such an approach will provide a global view of a given extension of the SM. But I think that studying directly observables, in particular correlations between them, is more efficient as the observables and not the parameters are directly measured in experiments.

We will first classify various correlations. Subsequently we will summarize the role of RG evolutions within SMEFT and LEFT that have been already presented at length in Chapter 14 with applications in Chapters 15–17.

19.3 Classifying Correlations between Various Observables

In studying correlations between various observables it is important to distinguish between two first types of correlations [718] to which I would like to add third one

- Correlations between observables within a given meson system (B, K, D) or within a given lepton system (e, μ, τ) give information on the operator structure of the relevant operators. In this manner one can distinguish between vector and scalar mediators at the

tree-level and tensor operators generated at one-loop level. This would then allow the determination of the chirality of NP couplings to the SM quarks and leptons.
- Correlations between observables in different meson systems and separately different lepton systems that allow in principle the identification of the underlying flavor symmetry within meson systems and separately within lepton systems. As discussed in Chapter 15 very specific patterns of NP effects are predicted by MFV and the $U(2)^3$ symmetry but also correlations beyond these simplest frameworks can provide useful information.
- Correlations between meson and lepton observables characteristic for grand unification frameworks and the ones like Pati–Salam scenarios. Such correlations can also follow from the required cancellation of gauge anomalies as mentioned in Section 15.4.9.

We have seen in Section 15.1 that in the SM and MFV models based on $U(3)^3$ flavor symmetry the correlations between various observables are indeed very stringent and could be considered as standard candles of flavor physics. They can then be used as the starting point for the classification of correlations between various observables.

Let us recall that the stringent correlations between observables in the SM and in models with CMFV originate in the fact that the observables in question depend on a selected number of basic universal functions that are the same for K and $B_{s,d}$ decays. In particular $\Delta F = 2$ processes depend only on the function $S(\omega)$, while the most important rare K and $B_{s,d}$ decays depend on three universal functions, $X(\omega)$, $Y(\omega)$, and $Z(\omega)$, with ω denoting symbolically the parameters of a given model, $\omega = x_t$ in the SM. Consequently, a number of correlations exist between various observables not only within the K and B systems but also between K and B systems. In particular the latter correlations are very interesting as they are characteristic for this class of models. Moreover, while the various observables depend generally on ω, the correlations in question are often ω independent.

In [920] a classification of correlations following from CMFV has been presented. It has been extended in [208], and we will further extend it here. We distinguish the following classes of correlations in CMFV models. They all involve the main players introduced at the beginning of this book. Therefore, some repetitions cannot be avoided. But it is good to see these players in different roles. We have then:

Class 1: Correlations implied by the universality of the real function X. They involve rare K and B decays with $\nu\bar{\nu}$ in the final state. These are

$$K^+ \to \pi^+\nu\bar{\nu}, \quad K_L \to \pi^0\nu\bar{\nu}, \quad B \to X_{s,d}\nu\bar{\nu}, \quad B \to K^*(K)\nu\bar{\nu}. \tag{19.2}$$

We refer to Figure 15.6, related discussion, as well as to various correlations presented in the context of simplified models in [601].

Class 2: Correlations implied by the universality of the real function Y. They involve rare K and B decays with $\ell^+\ell^-$ in the final state. These are in particular

$$B_{s,d} \to \mu^+\mu^-, \quad K_{L,S} \to \mu^+\mu^-, \quad K_L \to \pi^0\mu^+\mu^-, \quad K_L \to \pi^0 e^+e^-. \tag{19.3}$$

In this context, the present bounds on $B_s \to \mu^+\mu^-$ give very strong constraints on the scalar and pseudoscalar operators [504, 505, 565, 731].

Class 3: In models with CMFV NP contributions enter the functions X and Y approximately in the same manner as at least in the Feynman gauge they come dominantly from Z-penguin diagrams. This implies correlations between rare decays with $\mu^+\mu^-$ and $\nu\bar{\nu}$ in the final state. It should be emphasized that this is a separate class as NP can generally have a different impact on decays with $\nu\bar{\nu}$ and $\mu^+\mu^-$ in the final state. This class involves simply the decays of Class 1 and Class 2. But it should be stressed that if left-handed couplings of charged leptons are involved there is automatically a correlation between these two classes following from $SU(2)_L$ symmetry, even beyond MFV framework. We have stressed it in Chapter 14 in the context of the SMEFT.

Class 4: Here we group correlations between $\Delta F = 2$ and $\Delta F = 1$ transitions in which the one-loop functions S and (X, Y), respectively, cancel out and the correlations follow from the fact that the CKM parameters extracted from tree-level decays are universal. One known correlation of this type involves [193, 216]

$$\boxed{K^+ \to \pi^+\nu\bar{\nu}, \quad K_L \to \pi^0\nu\bar{\nu} \quad \text{and} \quad S_{\psi K_S},} \tag{19.4}$$

as already discussed in Section 9.5 and another one [503] involving

$$\boxed{B_{s,d} \to \mu^+\mu^- \quad \text{and} \quad \Delta M_{s,d},} \tag{19.5}$$

discussed already in Section 15.1.

Now, in Chapter 15, we described briefly correlations in models with tree-level FCNC mediated by neutral gauge bosons and scalars that go beyond the CMFV framework. In these models multicorrelations between various observables in a given meson system are predicted, and it is useful to group these processes in the following three classes. These are

Class 5:

$$\boxed{\varepsilon_K, \quad K^+ \to \pi^+\nu\bar{\nu}, \quad K_L \to \pi^0\nu\bar{\nu}, \quad K_{L,S} \to \mu^+\mu^-, \quad K_L \to \pi^0\ell^+\ell^-, \quad \varepsilon'/\varepsilon.}$$
$$\tag{19.6}$$

The study of correlations of $K \to \pi\nu\bar{\nu}$ decays and ϵ'/ϵ in simplified NP models in [601, 851] demonstrates the usefulness of such studies.

Class 6:

$$\boxed{\Delta M_d, \quad S_{\psi K_S}, \quad B_d \to \mu^+\mu^-, \quad S_{\mu\mu}^d, \quad B \to \pi\nu\bar{\nu}, \quad B \to \varrho\nu\bar{\nu}, \quad B \to X_d\nu\bar{\nu},}$$
$$\tag{19.7}$$

where the CP-violating asymmetry $S_{\mu\mu}^d$ can only be obtained from time-dependent rate of $B_d \to \mu^+\mu^-$ and will remain in the realm of theory for the foreseeable future.

Class 7:

$$\Delta M_s, \quad S_{\psi\phi}, \quad B_s \to \mu^+\mu^-, \quad S^s_{\mu\mu}, \quad B \to K\nu\bar{\nu}, \quad B \to K^*\nu\bar{\nu}, \quad B \to X_s\nu\bar{\nu}, \tag{19.8}$$

where the measurement of $S^s_{\mu\mu}$ will require heroic efforts from experimentalists but apparently its measurement is not totally hopeless.

Class 8:

$$B \to X_s\gamma, \quad B \to X_d\gamma, \quad B \to K^*\gamma, \quad B \to \varrho\gamma, \tag{19.9}$$

in which new charged gauge bosons and heavy scalars can play a significant role. They differ from previous decays as they are governed by dipole operators.

Class 9:

$$B \to K\ell^+\ell^-, \quad B \to K^*\ell^+\ell^-, \quad B \to X_s\ell^+\ell^-, \tag{19.10}$$

to which several operators contribute and for which multitude of observables, as discussed in Section 9.1, can be defined. Moreover, in the case of FCNC mediated by tree-level neutral gauge boson exchanges, interesting correlations between these observables and the ones of Class 7 exist. Furthermore in the case of left-handed couplings of charged leptons the correlations with Class 7 are implied by the $SU(2)_L$ gauge invariance. Finally, hints for significant violation of lepton flavor universality have been observed in this class.

Class 10: As discussed in Section 9.4, of particular interest are correlations involving observables that can be measured in

$$\bar{B} \to D\ell\bar{\nu}_l, \quad \bar{B} \to D^*\ell\bar{\nu}_l, \quad B_c \to J/\psi\tau\bar{\nu}, \quad B_c^+ \to \tau^+\nu_\tau \quad B^+ \to \tau^+\nu_\tau. \tag{19.11}$$

In particular, longitudinal polarizations of D^* described by $F_L(D^*)$ and the one of τ, given by $P_\tau(D^{(*)})$, are powerful means to distinguish between various models. See in particular the analyses in [551–556]. Also correlations between this class and class 9 implied by $SU(2)_L$ gauge invariance are present.

Class 11: Correlations not only between K and D observables but also D and $B_{s,d}$ observables discussed in Section 11.4. The recent news from the LHCb on enhanced CP violation in D decays combined with ε'/ε anomaly and B physics anomalies makes these studies exciting.

Class 12: Correlations between quark flavor violation, charged lepton flavor violation, $\mu-e$ conversion, electric dipole moments, and $(g-2)_{e,\mu}$. We have stressed it in Chapter 17.

Clearly the full picture is only obtained by looking simultaneously at patterns of violations of the correlations in question in a given NP scenario. The numerous codes listed in Section 14.6.4 are very helpful in such studies.

This list will hopefully motivate the readers to look for such correlations in specific models. Many of them have been presented in [208, 851, 1294] and in a number of papers I was involved in, which can easily be found with the help of INSPIRE. Possibly the best visualization of these correlations can be made with the help of the colorful DNA charts that have been presented in [208]. Here we want to describe them in less-exciting colors.

19.4 DNA Charts

19.4.1 General Idea

As reviewed in [870, 1000], extensive studies of many models allowed to construct various classifications of NP contributions in the form of "DNA" tables [529] and *flavor codes* [1000]. The "DNA" tables in [529] had as a goal to indicate whether in a given theory a value of a given observable can differ by a large, moderate, or only tiny amount from the prediction of the SM. The *flavor codes* [1000] were more a description of a given model in terms of the presence or absence of left- or right-handed currents in it and the presence or absence of new CP phases, flavor violating, and/or flavor conserving.

Certainly in both cases there is room for improvements. In particular in the case of the "DNA" tables in [529], we know now that in most quark flavor observables considered there, NP effects can be at most by a factor of two larger than the SM contributions. Exceptions are the cases in which some branching ratios or asymmetries vanish in the SM or are very strongly suppressed. This is in particular the case of LFV and EDMs. But the particular weakness of this approach is the difficulty in depicting the correlations between various observables that could be characteristic for a given theory. Such correlations are much easier to show on a circle and in this spirit, following [208], we would like to formulate this idea and illustrate it with few examples.

Step 1

We construct a chart showing different observables, typically a branching ratio for a given decay or an asymmetry, like CP-asymmetries $S_{\psi K_S}$ and $S_{\psi\phi}$ and quantities ΔM_s, ΔM_d, ε_K, and ε'/ε. The important point is to select the optimal set of observables, which are simple enough so that definite predictions in a given theory can be made.

Step 2

In a given theory we calculate the selected observables and investigate whether a given observable is enhanced or suppressed relative to the SM prediction or is basically unchanged. What this means requires a measure, like one or two σ. In the case of asymmetries we will proceed in the same manner if its sign remains unchanged relative to the one in the SM, but otherwise we define the change of its sign from + to − as a suppression and the change from − to + as an enhancement. For these three situations we will use the following color coding:

$$\boxed{\text{enhancement}} = \text{light gray}, \quad \boxed{\text{no change}} = \text{white}, \quad \boxed{\text{suppression}} = \text{black}.$$

$$(19.12)$$

To this end the predictions within the SM have to be known precisely.

Step 3

It is only seldom that a given observable in a given theory is uniquely suppressed or enhanced, but frequently two observables are correlated or anticorrelated, that is the enhancement of one observable implies uniquely an enhancement (correlation) or suppression (anticorrelation) of another observable. It can also happen that no change in the value of a given observable implies no change in another observable. There are, of course, other possibilities. The idea then is to connect in our DNA chart a given pair of observables that are correlated with each other by a line. Absence of a line means that two given observables are uncorrelated. To distinguish the correlation from anticorrelation we will use the following color coding for the lines in question:

$$\boxed{\text{correlation}} = \text{dark gray} \Leftrightarrow, \qquad \boxed{\text{anticorrelation}} = \text{light gray} \Leftrightarrow. \qquad (19.13)$$

Let us illustrate this strategy on examples of a few simplified models discussed in Chapter 15. This was already done in [208], and in fact the colors in DNA charts presented there are much nicer than in this book.

19.4.2 Examples of DNA Charts

The first DNA chart that one should in principle construct is the one dictated by experiment. This chart will have no correlation lines but will show where the SM disagrees with the data, and comparing it with DNA chart specific to a given theory will indicate which theories survived and which have been excluded. In 2013 when this idea has been proposed such an *experimental* chart was rather boring, and being basically white it has not been shown. However, in the second half of the 2010s several anomalies emerged, theoretical precision improved through new lattice calculations and new higher-order QCD calculations. Therefore, it will be tempting to construct such an experimental DNA chart in the coming years when these anomalies will be established. For now on we will present four examples of DNA charts representing specific simplified models. In Figure 19.1 we show the DNA chart of CMFV and the corresponding chart for $U(2)^3$ models is shown in Figure 19.2. The DNA charts representing models in LH scenario (LHS) and RH scenario (RHS) for flavor-violating couplings of Z' or Z can be found in Figures 19.3 and 19.4, respectively.

These charts summarize compactly the correlations present in these four simplified models that were presented in detail in Chapter 15. In particular we observe the following features:

- When going from the DNA chart of CMFV in Figure 19.1 to the one for the $U(2)^3$ models in Figure 19.2, the correlations between K and $B_{s,d}$ systems are broken as the symmetry is reduced from $U(3)^3$ down to $U(2)^3$. The anticorrelation between $S_{\psi\phi}$ and $S_{\psi K_S}$ is just the one represented by (15.46).
- As the decays $K^+ \to \pi^+ \nu \bar{\nu}$, $K_L \to \pi^0 \nu \bar{\nu}$, and $B \to K \nu \bar{\nu}$ are only sensitive to the vector quark currents, they do not change when the couplings are changed from LH to RH ones. On the other hand, the remaining three decays in Figures 19.3 and 19.4 are sensitive to axial-vector couplings, implying interchange of enhancements and suppressions when

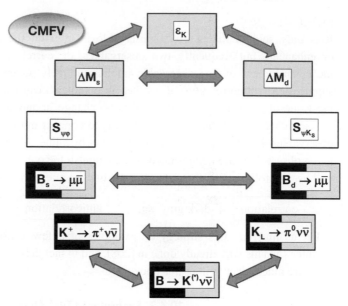

Figure 19.1 DNA chart of CMFV models. See text for explanation of various colors and arrows.

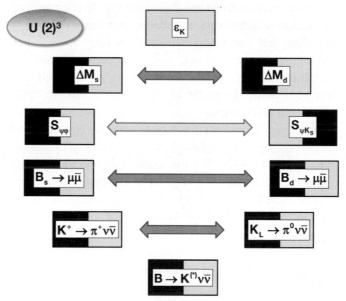

Figure 19.2 DNA chart of U(2)^3 models. See text for explanation of various colors and symbols.

going from L to R and also change of correlations to anticorrelations between the latter three and the former three decays. Note that the correlation between $B_s \rightarrow \mu^+\mu^-$ and $B \rightarrow K^*\mu^+\mu^-$ does not change as both decays are sensitive only to axial-vector coupling.

- However, it should be remarked that to obtain the correlations or anticorrelations in LHS and RHS scenarios, it was assumed that the signs of the LH couplings to neutrinos and

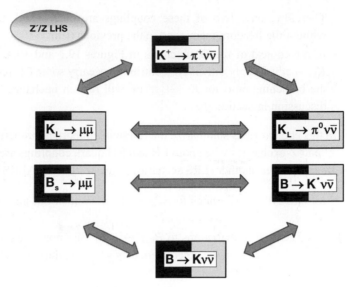

Figure 19.3   DNA charts of Z' and Z models with LH currents.

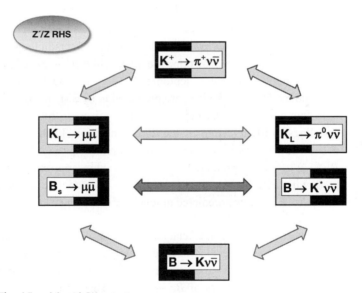

Figure 19.4   DNA charts of Z' and Z models with RH currents.

the axial-vector couplings to muons are the same, which does not have to be the case. If they are opposite, the correlations between the decays with neutrinos and muons in the final state change to anticorrelations and vice versa.

- On the other hand, due to $SU(2)_L$ symmetry, the LH Z' couplings to muons and neutrinos are equal, and this implies the relation

$$\Delta_L^{\nu\bar{\nu}}(Z') = \frac{\Delta_V^{\mu\bar{\mu}}(Z') - \Delta_A^{\mu\bar{\mu}}(Z')}{2}. \tag{19.14}$$

Therefore, once two of these couplings are determined, the third follows uniquely without the freedom mentioned in the previous item.

- In the context of the DNA charts in Figures 19.3 and 19.4, the correlations involving $K_L \to \pi^0 \nu \bar{\nu}$ apply only if NP contributions carry some CP-phases. If this is not the case, the branching ratio for $K_L \to \pi^0 \nu \bar{\nu}$ will remain unchanged. This is evident from our discussion in Section 9.5.

In this context let as summarize the following important properties of the case of tree-level Z' and Z exchanges when both LH and RH quark couplings are present, which in addition are equal to each other (LRS scenario) or differ by sign (ALRS scenario):

- In LRS NP contributions to $B_{s,d} \to \mu^+ \mu^-$ vanish but not to $K_L \to \pi^0 \nu \bar{\nu}$ and $K^+ \to \pi^+ \nu \bar{\nu}$.
- In ALRS NP contributions to $B_{s,d} \to \mu^+ \mu^-$ are nonvanishing, and this also applies to $B_d \to K^* \mu^+ \mu^-$. On the other hand, they vanish in the case of $K_L \to \pi^0 \nu \bar{\nu}$, $K^+ \to \pi^+ \nu \bar{\nu}$, and $B_d \to K \mu^+ \mu^-$

Let us still give a few examples how one could distinguish between different NP scenarios.

Example 1   As summarized in [305], the Z' models based on the gauged $L_\mu - L_\tau$ symmetry, used to explain $b \to s \mu^+ \mu^-$ anomalies, predict effects in the semileptonic $b \to s \mu^+ \mu^-$ and $b \to s \tau^+ \tau^-$ transitions of opposite sign, while $b \to s e^+ e^-$ transitions remain SM-like. In these models, the purely leptonic $B_s \to \mu^+ \mu^-$ and $B_s \to \tau^+ \tau^-$ decays, as well as the neutrino modes $B \to K^{(*)} \nu \bar{\nu}$, are predicted to be SM-like [925].

In this context, totally different pattern arises in the Z' scenarios based on dominant couplings to LH fermions of the third generation. In these models, the $b \to s \tau^+ \tau^-$ and $B \to K^{(*)} \nu \bar{\nu}$ rates are significantly enhanced over the SM predictions. Next, the $B_s \to \mu^+ \mu^-$ rate is predicted to be suppressed by approximately 25% compared to the SM prediction. Finally, in contrast to the $L_\mu - L_\tau$ models, rare lepton flavor-violating decays like $B \to K^{(*)} \tau \mu$ are predicted at levels of $O(10^{-8})$ [776].

Example 2   The extensive analyses in [504, 509] exhibit striking differences between implications of tree-level Z', pseudoscalar and scalar tree-level exchanges for correlations between various observables that can be extracted from flavor untagged and tagged time-dependent measurements of $B_s \to \mu^+ \mu^-$. These are $\mathcal{A}_{\Delta\Gamma}^{\mu\mu}$ and $\mathcal{S}_{\mu\mu}^s$ that we discussed in detail in Section 9.2.2. Moreover, the branching ratio can only be enhanced through tree-level scalar exchanges but also suppressed through Z' and pseudoscalar exchanges. There is no space to present here numerous colorful plots given in this context in [208, 504, 509].

Example 3   In this decade the decays $B \to K \nu \bar{\nu}$ and $B \to K^* \nu \bar{\nu}$ will play a significant role in the tests of NP models because of the efforts at Belle II to measure them. The study of the correlations between the branching ratios for these two decays has been performed in a number of models in [607]. These are Z' scenarios, MSSM, partial compositeness, LQs and tree-level FCNC of the Z boson. Figure 11 in that paper demonstrates that already the signs of NP effects will be able to distinguish between these scenarios.

Example 4   Future measurements of rare processes will allow us to distinguish between various vectorlike-quark models that we discussed in detail in Section 16.3. DNA tables 6 and 7 in [666] allow right away to see in which processes large NP effects are to be expected in a given model.

19.5 Can We Reach the Zeptouniverse with Rare *K* and $B_{s,d}$ Decays?

It is well known by now that $\Delta F = 2$ observables can, in the presence of left-right operators with chirally enhanced hadronic matrix elements, be sensitive to scales as high as 10^4 TeV, or even higher scales [686, 1297–1299]. Yet, as pointed out in [1294] the study of $\Delta F = 2$ observables alone, as done in these papers, will not really give us significant information about the particular nature of this NP. To this end also $\Delta F = 1$ processes, in particular rare *K* and $B_{s,d}$ decays, have to be considered. As left-right operators involving four quarks are not the driving force in these decays, which generally contain operators built out of one quark current and one lepton current, it is not evident that these decays can help us in reaching the Zeptouniverse even in the flavor precision era. In fact, as found in [1294], NP at scales well above 1000 TeV cannot be probed by rare meson decays. However, such scales can be in principle probed in the future with the help of lepton flavor-violating decays such as $\mu \rightarrow e\gamma$ and $\mu \rightarrow 3e$, $\mu \rightarrow e$ conversion in nuclei, and electric dipole moments [1118, 1120, 1121, 1212, 1300–1305]. A recent study of model-independent bounds within the SMEFT that includes both $\Delta F = 2$ and $\Delta F = 1$ processes has been performed in [770].

Here I would like to summarize the results of a detailed analysis in [1294], which used dominantly a heavy Z' gauge boson to probe very short distance scales, assuming that all its couplings to SM fermions take values of at most 3.0. Interested readers can have a look at numerous plots in this paper. The main lessons from this analysis are as follows:

1. $\Delta F = 2$ processes alone cannot give us any concrete information about the nature of NP at short-distance scales beyond the reach of the LHC. In particular if some deviations from SM expectations will be observed, it will not be possible to find out whether they come from LH currents, RH currents, or both.

2. On the other hand future precise measurements of several $\Delta F = 1$ observables and in particular correlations between them can distinguish between LH and RH currents, but the maximal resolution consistent with perturbativity strongly depends on whether only LH or only RH or both LH and RH flavor-changing Z' couplings to quarks are present in nature. If only LH or RH couplings are present in nature, we can in principle reach scales of 200 TeV and 15 TeV for *K* and $B_{s,d}$, respectively. These numbers depend on the room left for NP in $\Delta F = 2$ observables, which have an important impact on the resolution available in these NP scenarios.

3. Smaller distance scales can only be resolved if both RH and LH couplings are present to cancel the NP effects on the $\Delta F = 2$ observables. Moreover, to achieve the necessary tuning, the couplings should differ considerably from each other. This large hierarchy

of couplings is dictated, as seen in Table 13.2, primarily by the ratio of hadronic matrix elements of LR $\Delta F = 2$ operators and those for LL and RR operators and by the room left for NP in $\Delta F = 2$ processes. One finds that in this case the scales as high as 2000 TeV and 160 TeV for K and $B_{s,d}$ systems, respectively, could be in principle resolved.

4. A study of tree-level neutral (pseudo-)scalar exchanges showed that $B_{s,d} \rightarrow \mu^+\mu^-$ can probe scales close to 1000 TeV, both for scenarios with purely LH or RH scalar couplings to quarks and for scenarios allowing for both LH and RH couplings. In the limit of degenerate scalar and pseudoscalar masses, NP effects in $\Delta F = 2$ observables can cancel even without imposing a tuning on the couplings.

5. If several gauge bosons are considered, it is easier to break the stringent correlation between $\Delta F = 1$ and $\Delta F = 2$ processes and to suppress NP contributions to the latter without suppressing NP contributions to rare decays. The presence of a second heavy neutral gauge boson allows us to probe very high scales with only LH or RH currents by applying an appropriate tuning. While the highest achievable resolution in the presence of several gauge bosons is comparable to the case of a single gauge boson because of the perturbativity bound, the correlations between $\Delta F = 1$ observables could differ from the ones in the case of a single gauge boson. This would be in particular the case if LH and RH couplings of these bosons where of similar size. A detailed study of such scenarios would require the formulation of concrete models.

6. If FCNC only occur at one loop level the highest energy scales that can be resolved for maximal couplings are typically reduced relative to the case of tree-level FCNC by a factor of at least 3 and 6 for $\Delta F = 1$ and $\Delta F = 2$ processes, respectively.

In summary the analysis in [1294] demonstrated that NP with a particular pattern of dynamics could be investigated through rare K and $B_{s,d}$ decays even if the scale of this NP would be close to the Zeptouniverse. However, one should emphasize that although the main goal of that paper was to reach the highest energy scales with the help of rare decays, it will, of course, be exciting to explore any scale of NP above the LHC scales in this decade. Moreover, we still hope that high-energy proton-proton collisions at the LHC will exhibit at least some footprints of new particles and forces. This would greatly facilitate flavor analyses as the one presented in [1294].

Yet, after these rather enthusiastic statements, a warning is in order. The RG analysis in [1294] took only QCD effects into account. For a proper assessment of NP scales, that one can reach with the help of the processes listed earlier, a full SMEFT analysis is required. Such an analysis is missing in the literature but again similar to other comments made at other places of this book should be available in the hep-arxiv latest in the Summer of the year 2020.

20 Summary and Shopping List

It looks like our expedition is coming to the end. We have seen that weak decays of mesons and leptons provide an excellent arena for the tests of the SM and of its extensions. While the flavor-violating decays played already a crucial role in the construction of the SM and in testing its extensions, flavor-conserving transitions like EDMs and anomalous magnetic moments will also play in the coming years an important role in identifying NP beyond the LHC scales. In view of the unfortunate fact that during the Run II of the LHC so far no new particles have been directly identified in high-energy collisions, the role of the so-called high-intensity frontier to which the study of strongly suppressed rare decays of mesons and leptons, of EDMs and other high-precision studies belong got enhanced relative to what we thought just after the Higgs discovery. Clearly LHC will continue to play a very important, complementary role, in the search for NP, but it appears from various studies that some important questions, like the presence of new sources of CP violation, both in flavor-violating and flavor-conserving processes, will be more efficiently addressed at the high-intensity frontier rather than the high-energy frontier. Yet High-Luminosity and High-Energy upgrades of LHC, as summarized in [1277], still could hopefully allow us to see NP directly.

In the last decades tremendous progress has been made by experimentalists in measuring hundreds of observables related to the physics presented in this book. Not all could be discussed by us, and we have concentrated on those observables on which also impressive progress has been made on the theory side. In addition, those observables have been considered that should become important players in this decade when both experiment and theory further improve. The main players have been already introduced in the first chapter, but it is useful to return to them in the final chapter and discuss them briefly again in view of the knowledge gained in this book. This will then automatically constitute a *shopping list* for the period 2020–2030, at least as seen from my point of view. Here we go.

Step 1: Very important is the simultaneous study of most relevant observables in $\Delta F = 2$ processes, that is,

$$
\boxed{\Delta M_s, \quad \Delta M_d, \quad S_{\psi K_S}, \quad S_{\psi \phi}, \quad \varepsilon_K, \quad \Delta M_K} \tag{20.1}
$$

in conjunction with tree-level determinations of the parameters of the CKM matrix, that is,

$$
\boxed{|V_{us}|, \quad |V_{ub}|, \quad |V_{cb}|, \quad \gamma.} \tag{20.2}
$$

In this manner one will be able to determine both UT and the full CKM matrix independent of NP contributions, thereby testing the unitarity of the CKM matrix. Moreover, the test of various stringent relations between the observables in (20.1) in models with MFV and CMFV, listed in Section 15.1, will be possible. The clarification of the tensions between inclusive and exclusive determinations of $|V_{ub}|$ and $|V_{cb}|$ will be important, but I expect that with much improved lattice calculations of the relevant formfactors, exclusive determinations will be favored. Knowing these elements and the angle γ precisely one will be able to make accurate predictions for all observables in (20.1) provided the relevant nonperturbative hadronic matrix elements will be known from LQCD. I am rather optimistic that within the coming years the status of the observables and the parameters in (20.1) and (20.2) will allow us to make a clear-cut conclusion whether non-MFV sources in this sector are present or not. In fact, as discussed in Section 15.1, presently a tension between tree-level value for the angle γ and ΔM_d appears to be present, but in order to reach clear-cut conclusions the error on γ should be reduced down to $1°$, which should be possible in the Belle II era [428] and in the Run III of the LHC.

It would be important to reduce the error on the QCD factor $\eta_1 = \eta_{cc}$, which affects both ε_K and ΔM_K. This seems to have been achieved very recently [1317]. As we stressed at various places in this book, precise value of ΔM_K in the SM obtained one day by LQCD will provide an additional observable that could play an important role in the search for NP.

Step 2: The clarification of violation of lepton flavor universality presently hinted by angular observables (P_5' among others) and branching ratios in the decays ($\ell = e, \mu, \tau$)

$$\boxed{B \to K^* \ell^+ \ell^-, \qquad B \to K \ell^+ \ell^-, \qquad B_s \to \phi \ell^+ \ell^-} \tag{20.3}$$

and also in

$$\boxed{\mathcal{B}(\bar{B} \to D \tau \nu_\tau), \qquad \mathcal{B}(\bar{B} \to D^* \tau \nu_\tau),} \tag{20.4}$$

should hopefully be clarified soon. If the future improved data will confirm the values of

$$\boxed{R(K), \qquad R(K^*), \qquad R(D), \qquad R(D^*), \qquad F_L(D^*), \qquad P_\tau(D), \qquad P_\tau(D^*)}$$
$$\tag{20.5}$$

collected in Section 12.2, this will have many implications for many decays of $B_{s,d}$ and K mesons. Some of them have been discussed already by us while discussing various BSM scenarios. Yet, despite many efforts by theorists, I do not think that the picture of NP responsible for these anomalies is transparent. A heavy Z' and LQs, supported by vector-like quarks and heavy scalars, could in principle explain all these anomalies, but to make progress we need badly confirmation of $R(K)$ and $R(K^*)$ anomalies by LHCb itself and in particular by Belle II. Recent messages from LHCb imply $R(K)$ closer to its SM value, but the anomalies in P_5' and $R(K^*)$ could still remain. Let us hope that by 2021–2022 we will be able to make a clear-cut conclusion on this issue. Similar comments apply to $R(D)$ and $R(D^*)$ and the remaining quantities in (20.5). In this context improved data on $\mathcal{B}(B^+ \to \tau^+ \nu_\tau)$ accompanied by a precise value of $|V_{ub}|$ will help in distinguishing various models.

Step 3: The branching ratios

$$\mathcal{B}(B_s \to \mu^+\mu^-), \qquad \mathcal{B}(B_d \to \mu^+\mu^-) \qquad (20.6)$$

will continue to play an important role. They should be known very precisely in the SM once the accuracy on the CKM elements $|V_{ts}|$ and $|V_{td}|$ will be improved at Belle II in the coming years. Experimentally $\mathcal{B}(B_s \to \mu^+\mu^-)$ should be known rather precisely by 2025, while it may take a longer time for $\mathcal{B}(B_d \to \mu^+\mu^-)$. These decays, in particular taken together, offer very powerful tests of the SM and together with Step 1 tests of various MFV and CMFV relations. Whether $\mathcal{B}(B_d \to \tau^+\tau^-)$ will also be known in the next 10 years remains to be seen. One should also hope that the observables related to the time evolution in $B_s \to \mu^+\mu^-$ like effective lifetime and $S^s_{\mu\mu}$ will be known with sufficient precision to provide good tests. All these decays are very sensitive to heavy scalars. The collection of talks on the role of these decays in the search for NP can be found in [1267].

Step 4: Simultaneous study of the number of branching ratios in correlation with ε'/ε:

$$\mathcal{B}(B \to K\nu\bar{\nu}), \qquad \mathcal{B}(B \to K^*\nu\bar{\nu}), \qquad \mathcal{B}(B \to X_s\nu\bar{\nu}), \qquad (20.7)$$

$$\mathcal{B}(K^+ \to \pi^+\nu\bar{\nu}), \qquad \mathcal{B}(K_{\mathrm{L}} \to \pi^0\nu\bar{\nu}), \qquad \varepsilon'/\varepsilon, \qquad (20.8)$$

$$\mathcal{B}(K_L \to \pi^0\ell^+\ell^-), \qquad \mathcal{B}(K_L \to \mu^+\mu^-), \qquad \mathcal{B}(K_S \to \mu^+\mu^-) \qquad (20.9)$$

will offer excellent tests of right-handed currents and new sources of CP violation. With the possible exception of ε'/ε, SM predictions for all these observables should be known precisely by 2025 due to improved formfactors from lattice QCD relevant for (20.7) and reduced uncertainties on CKM parameters that are in particular very important for $K^+ \to \pi^+\nu\bar{\nu}$ and $K_{\mathrm{L}} \to \pi^0\nu\bar{\nu}$. Precise measurement of $\mathcal{B}(K^+ \to \pi^+\nu\bar{\nu})$ by NA62 collaboration combined with improved lattice calculations of hadronic matrix elements relevant for ε'/ε will tell us whether the ε'/ε anomaly is really there and whether this NP has implications for $K^+ \to \pi^+\nu\bar{\nu}$ and $K_{\mathrm{L}} \to \pi^0\nu\bar{\nu}$. The branching ratio for $K_{\mathrm{L}} \to \pi^0\nu\bar{\nu}$ should be measured by KOTO in the first half of this decade. Also KLEVER experiment at CERN will contribute here. It should also be emphasized that both $K_{\mathrm{L}} \to \pi^0\ell^+\ell^-$ and $K_S \to \mu^+\mu^-$ are CP-violating and consequently correlated in various NP scenarios with ε'/ε and $K_{\mathrm{L}} \to \pi^0\nu\bar{\nu}$. Invoking in addition the $b \to s\nu\bar{\nu}$ transitions will provide powerful tests of MFV and new sources of flavor violation beyond it.

As far as ε'/ε is concerned, we need badly another result from a second LQCD collaboration to not only be confident about the ε'/ε anomaly but also to know precisely its size. Whether this will be possible by 2025 is not certain from the present perspective.

Step 5: The branching ratios

$$\mathcal{B}(B \to X_s\gamma), \qquad \mathcal{B}(B \to K^*(\varrho)\gamma) \qquad (20.10)$$

and other radiative decays should not be forgotten. The very good agreement of the SM prediction with the data puts very strong constraints on the WCs of the relevant dipole operators.

Step 6: Simultaneous study of the branching ratios

$$\mathcal{B}(\mu \to e\gamma), \qquad \mathcal{B}(\tau \to e\gamma), \qquad \mathcal{B}(\tau \to \mu\gamma), \tag{20.11}$$

$$\mathcal{B}(\mu^- \to e^- e^+ e^-), \qquad \mathcal{B}(\tau^- \to \mu^- \mu^+ \mu^-), \qquad \mathcal{B}(\tau^- \to e^- e^+ e^-), \tag{20.12}$$

$$\mathcal{B}(\tau^- \to \mu^- e^+ e^-), \qquad \mathcal{B}(\tau^- \to e^- \mu^+ \mu^-) \tag{20.13}$$

$$\mathcal{B}(\tau^- \to e^- \mu^+ e^-), \qquad \mathcal{B}(\tau^- \to \mu^- e^+ \mu^-), \tag{20.14}$$

and of the $\mu - e$ conversion in nuclei will hopefully open for us a new world that is very different from the one described by the SM. The dedicated J-PARC experiment PRISM/PRIME should reach a sensitivity of $O(10^{-18})$. Also, semileptonic τ decays like $\tau \to \pi \mu e$ should not be forgotten.

It should be emphasized that in contrast to meson decays, these decays are free from hadronic uncertainties, but the fact that only upper bounds on their branching ratios are known makes the distinction between models that satisfy these bounds impossible. Fortunately within the SMEFT enriched by some dynamical assumptions and in specific NP models, correlations between quark and lepton processes are possible. They constitute an important tool for selecting successful models.

Step 7: Assuming that Step 2 confirms the violation of lepton flavor universality the following decays that proceed through both quark flavor and lepton flavor-violating transitions will belong to stars of flavor physics.

$$K_{L,S} \to \mu e, \qquad K_{L,S} \to \pi^0 \mu e, \tag{20.15}$$

$$B_{d,s} \to \mu e, \qquad B_{d,s} \to \tau e, \qquad B_{d,s} \to \tau \mu, \tag{20.16}$$

$$B_d \to K^{(*)} \tau^\pm \mu^\mp, \qquad B_d \to K^{(*)} \mu^\pm e^\mp. \tag{20.17}$$

Whether tree-level exchanges of leptoquarks, tree-level Z' exchanges, or one-loop contributions with vector-like fermions and new scalars will be behind possible NP effects should be possible to find out provided the previous steps will bring sufficient information.

Step 8: As we have seen in Sections 17.3 and 17.4,

$$\text{Electric Dipole Moments}, \qquad (g-2)_{e,\mu} \tag{20.18}$$

can offer in principle impressive information on NP. In particular EDMs, similar to lepton flavour violation, if measured, should open a new world for us. But also the new measurement of $(g-2)_\mu$ at Fermilab will clarify whether the anomaly found at Brookhaven in 2001 is true or not. Unfortunately, in particular in the case of the EDMs, there are still large hadronic uncertainties so that to be able to distinguish between various NP models significant progress from theorists is required.

Step 9: In addition to all these processes, charm physics will play a significant role in constraining NP models but I did not list any processes here to emphasize that from my point of view the observables listed in preceding steps will be more powerful because of smaller hadronic uncertainties. An exception here is CP violation in D decays, and the recent messages from the LHCb summarized at the end of Chapter 11 confirms it. Also some rare decays, like $D \to \mu^+\mu^-$, could offer useful information as their branching ratio are predicted to be tiny in the SM. Needless to say, charm plays an important role in B physics as already seen at various places in this book and recently reviewed in [1320].

Step 10: CP violation in neutrino physics, neutrinoless double β decay, and related observables will clearly give us additional information about NP. However, I never worked in this field and would prefer not to make unqualified comments here.

While we have made here a list of the most important separate steps, it is evident from our discussion of SMEFT that through renormalization group effects several of these steps are correlated with each other and also with collider physics, including Higgs and top quark physics and also with electroweak precision tests. Although powerful codes have been developed in recent years that allow studying the implications of SMEFT, including all these steps, efficiently, the large number of Wilson coefficients involved makes the identification of NP above LHC scales more difficult than we thought, at least I thought, 10 years ago.

In my view we need some new ideas for transparently presenting various different patterns of correlations between various observables in different NP scenarios. Some ideas have been presented in the previous chapter, but certainly we have to develop and improve such ideas by much. I still think that the construction of new models, that hopefully will have sufficiently low number of parameters, will continue to be important to be able to draw pictures of oases well above the LHC scales and if we are lucky to find out what happens in the Zeptouniverse. In any case, we can look forward to very exciting times ahead of us. The flavor clock in Figure 20.1 makes this evident.

In particular the coming years should be very exciting for flavor phenomenology, not only at Belle II [428] and LHCb [305, 1293], but also generally at CERN, where also ATLAS and CMS will contribute in an important manner [305, 1277] and generally through BSM searches beyond colliders at CERN [1292]. Moreover, with the advances in LQCD it will be possible to make clear-cut conclusions about the presence of NP in processes in which hadronic effects play an important role [305–308].

In this spirit I expect that the results on $K^+ \to \pi^+ \nu\bar{\nu}$ from NA62 collaboration at CERN SPS [204] and on $K_L \to \pi^0 \nu\bar{\nu}$ from KOTO [598] at J-PARC and KLEVER [599] at CERN SPS, when correlated with ε'/ε, will allow a deep insight into possible NP at short distances [851]. This insight will be enriched by $(g-2)_\mu$ experiments at Fermilab and J-PARC and in the context of various experiments probing charged lepton flavor violation at CERN, PSI, KEK, and J-PARC. Finally, improved LQCD calculations and other nonperturbative

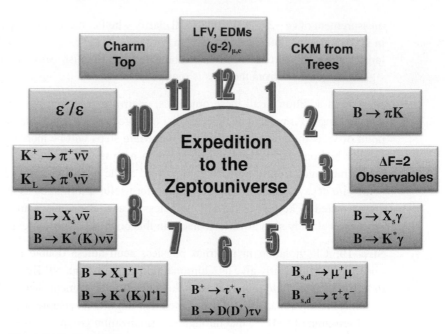

Figure 20.1   Expedition to the Zeptouniverse.

methods, relevant for many of these processes, will allow us clear-cut conclusions about the presence of NP [1306] and important insights into the anomalies listed in Section 12.2. I also hope that my prophetic statements about ε'/ε in [674], very appropriate for a 2018 Christmas story, will be confirmed by several lattice QCD groups before my 80th birthday in October 2026. Until then we will have a great time in making progress in gauge theories for weak decays and for the multitude of observables considered in this book. Therefore despite the first sentence of our summary, this is not the end of the story.

OUR EXPEDITION WILL CONTINUE FOR MANY YEARS!

Appendix A Dirac Algebra, Spinors, Pauli and Gell-Mann Matrices

A.1 Dirac Algebra and Spinors

A.1.1 Gamma Matrices

The Dirac matrices γ_μ with $\mu = (0, 1, 2, 3)$ satisfy anticommutation relation

$$\gamma_\mu \gamma_\nu + \gamma_\nu \gamma_\mu = 2g_{\mu\nu}, \tag{A.1}$$

with the metric tensor defined as $g_{\mu\mu} = g^{\mu\mu} \sim (+, -, -, -)$. This implies Dirac algebra with 16 elements

$$S = \mathbb{1}_{4\times4}, \quad P = \gamma_5, \quad V = \gamma_\mu, \quad A = \gamma_\mu \gamma_5, \quad T = \sigma_{\mu\nu}, \tag{A.2}$$

where[1]

$$\gamma_5 = i\gamma^0 \gamma^1 \gamma^2 \gamma^3 = \gamma^5, \qquad \sigma_{\mu\nu} \equiv \frac{i}{2}(\gamma_\mu \gamma_\nu - \gamma_\nu \gamma_\mu). \tag{A.3}$$

In (A.2) S stands for *scalar*, P for *pseudoscalar*, V for *vector*, A for *axial-vector*, and T for *tensor*.

The explicit expressions for γ_μ and γ_5 can be found in (1.83) and any textbook for relativistic quantum field theory like [13, 14], but we will not need them here. The relations given following are sufficient for our purposes.

The γ_μ can be chosen unitary, if Hermitian conjugation is defined as

$$(\gamma^\mu)^\dagger = \gamma_\mu \Rightarrow (\gamma^\mu)^\dagger = g_{\mu\nu}\gamma^\nu. \tag{A.4}$$

We have then a set of useful properties:

$$(\gamma^0)^\dagger = g_{0\nu}\gamma^\nu = +\gamma^0, \qquad (\gamma^i)^\dagger = g_{i\nu}\gamma^\nu = -\gamma^i, \tag{A.5}$$

where $i = (1, 2, 3)$. Moreover,

$$(\gamma^\mu)^\dagger = \gamma^0 \gamma^\mu \gamma^0, \quad (\gamma_0)^2 = 1, \quad (\gamma_i)^2 = -1, \tag{A.6}$$

and very important relations involving γ_5

$$\gamma_5 \gamma_\mu = -\gamma_\mu \gamma_5, \qquad \gamma_5^\dagger = \gamma^5 = \gamma_5. \tag{A.7}$$

---

[1] In the literature one finds sometimes $\sigma_{\mu\nu}$ defined without the i factor.

Next come chirality projectors ($A = L, R$)

$$P_{L,R} \equiv \frac{1}{2}(1 \mp \gamma_5), \quad (P_A)^\dagger = P_A, \quad P_A \gamma_\mu = \gamma_\mu P_{-A}. \tag{A.8}$$

Here the notation $-A$ means simply $-L = R$ and $-R = L$.

Of importance is also the charge conjugation matrix

$$C = i\gamma^2\gamma^0, \tag{A.9}$$

which has a number of properties

$$C^T = -C, \quad (C^{-1})^T = -C^{-1}, \quad C^\dagger = C^{-1}, \quad CC^\dagger = 1, \tag{A.10}$$

$$C(\gamma^\mu)^T C^{-1} = -\gamma^\mu, \quad C(\gamma_5)^T C^{-1} = +\gamma_5, \quad C(\gamma^0)^* C^\dagger = -\gamma^0, \quad C(P_A)^T = P_A C. \tag{A.11}$$

A.1.2 Dirac Spinors

Dirac adjoint spinor is given as usual by

$$\overline{\psi} = (\psi)^\dagger \gamma_0. \tag{A.12}$$

Spinors with chiral fields ($A = L, R$) are given by

$$\psi_A \equiv P_A \psi, \qquad\qquad \overline{\psi_A} = \overline{\psi} P_{-A} \tag{A.13}$$

and the charge-conjugated spinors by

$$\psi^c \equiv C\overline{\psi}^T, \qquad (\psi^c)^c = \psi, \tag{A.14}$$

with useful properties

$$\overline{\psi^c} = -\psi^T C^\dagger, \qquad \psi_A^c = P_{-A}\psi^c, \qquad \overline{\psi_A^c} = \overline{\psi^c} P_A. \tag{A.15}$$

Furthermore, Hermitian conjugation of spinors gives

$$\psi^\dagger = \overline{\psi}\gamma_0, \quad \overline{\psi}^\dagger = \gamma_0\psi, \quad (\psi^c)^\dagger = \overline{\psi^c}\gamma_0, \quad \overline{\psi^c}^\dagger = \gamma_0\psi^c. \tag{A.16}$$

A.2 Pauli and Gell-Mann Matrices

The Pauli matrices σ^a are given as follows

$$\sigma^1 = \begin{pmatrix} 0 & 1 \\ 1 & 0 \end{pmatrix}, \quad \sigma^2 = \begin{pmatrix} 0 & -i \\ i & 0 \end{pmatrix}, \quad \sigma^3 = \begin{pmatrix} 1 & 0 \\ 0 & -1 \end{pmatrix}. \tag{A.17}$$

The Gell-Mann matrices λ^a satisfying

$$\mathrm{Tr}(\lambda^a\lambda^b) = 2\delta^{ab} \tag{A.18}$$

are given as follows

$$
\lambda^1 = \begin{pmatrix} 0 & 1 & 0 \\ 1 & 0 & 0 \\ 0 & 0 & 0 \end{pmatrix}, \qquad
\lambda^2 = \begin{pmatrix} 0 & -i & 0 \\ i & 0 & 0 \\ 0 & 0 & 0 \end{pmatrix}, \qquad
\lambda^3 = \begin{pmatrix} 1 & 0 & 0 \\ 0 & -1 & 0 \\ 0 & 0 & 0 \end{pmatrix}, \tag{A.19}
$$

$$
\lambda^4 = \begin{pmatrix} 0 & 0 & 1 \\ 0 & 0 & 0 \\ 1 & 0 & 0 \end{pmatrix}, \qquad
\lambda^5 = \begin{pmatrix} 0 & 0 & -i \\ 0 & 0 & 0 \\ i & 0 & 0 \end{pmatrix}, \tag{A.20}
$$

$$
\lambda^6 = \begin{pmatrix} 0 & 0 & 0 \\ 0 & 0 & 1 \\ 0 & 1 & 0 \end{pmatrix}, \qquad
\lambda^7 = \begin{pmatrix} 0 & 0 & 0 \\ 0 & 0 & -i \\ 0 & i & 0 \end{pmatrix}, \qquad
\lambda^8 = \frac{1}{\sqrt{3}} \begin{pmatrix} 1 & 0 & 0 \\ 0 & 1 & 0 \\ 0 & 0 & -2 \end{pmatrix}. \tag{A.21}
$$

A.3 Fierz Identities

Fierz identities [1307], often called also Fierz relations or Fierz transformations, allow to transfer a given chain of spinors into a another one of which matrix elements between external states are easier to evaluate than of the original chain. It should be stressed that the usual Fierz identities are only valid in $D = 4$ dimensions. If the calculations are performed in $D \neq 4$ dimensions, these identities have to involve evanescent operators. We refer to [744] for details warning the reader that there $\sigma_{\mu\nu} \equiv [\gamma_\mu, \gamma_\nu]/2$, which implies additional $i^2 = -1$ in the terms involving contractions of two $\sigma_{\mu\nu}$. Here we stick to $D = 4$ and $\sigma_{\mu\nu}$ as defined in (A.3).

In what follows we will list three classes of Fierz identities that take into account the fact that we are dealing with Grassmann-valued operators. This means that they include additional minus signs whenever the fermion fields are anticommuted odd times. This is the case of classes A and B but not C, where the fields are anticommuted twice. In the literature such identities are usually given for spinors, but it is more useful to have them for fermion fields. The results presented here use the information presented in [744, 1308, 1309] and have also been derived by Christoph Bobeth. Useful information can also be found in [1310]. In all these formulas P_A and P_B stand for projectors in (A.8) but in a given relation $P_A \neq P_B$. This means that $P_B = P_{-A}$ or more explicitly if $P_A = P_L$, then $P_B = P_R$ and vice versa.

Class A
These are the original Fierz identities of type $(12)(34) \rightarrow (14)(32)$ in which the exchange of fermion fields $2 \leftrightarrow 4$ (or equivalently $1 \leftrightarrow 3$) takes place. We have then

$$
(\bar\psi_1 P_A \psi_2)(\bar\psi_3 P_A \psi_4) = -\frac{1}{2}(\bar\psi_1 P_A \psi_4)(\bar\psi_3 P_A \psi_2) - \frac{1}{8}(\bar\psi_1 \sigma_{\mu\nu} P_A \psi_4)(\bar\psi_3 \sigma^{\mu\nu} P_A \psi_2),
$$
$$
\tag{A.22}
$$

$$(\bar{\psi}_1 P_A \psi_2)(\bar{\psi}_3 P_B \psi_4) = -\frac{1}{2}(\bar{\psi}_1 \gamma_\mu P_B \psi_4)(\bar{\psi}_3 \gamma^\mu P_A \psi_2), \tag{A.23}$$

$$(\bar{\psi}_1 \gamma_\mu P_A \psi_2)(\bar{\psi}_3 \gamma^\mu P_A \psi_4) = (\bar{\psi}_1 \gamma_\mu P_A \psi_4)(\bar{\psi}_3 \gamma^\mu P_A \psi_2), \tag{A.24}$$

$$(\bar{\psi}_1 \gamma_\mu P_A \psi_2)(\bar{\psi}_3 \gamma^\mu P_B \psi_4) = -2(\bar{\psi}_1 P_B \psi_4)(\bar{\psi}_3 P_A \psi_2), \tag{A.25}$$

$$(\bar{\psi}_1 \sigma_{\mu\nu} P_A \psi_2)(\bar{\psi}_3 \sigma^{\mu\nu} P_A \psi_4) = -6(\bar{\psi}_1 P_A \psi_4)(\bar{\psi}_3 P_A \psi_2) + \frac{1}{2}(\bar{\psi}_1 \sigma_{\mu\nu} P_A \psi_4)(\bar{\psi}_3 \sigma^{\mu\nu} P_A \psi_2), \tag{A.26}$$

$$(\bar{\psi}_1 \sigma_{\mu\nu} P_A \psi_2)(\bar{\psi}_3 \sigma^{\mu\nu} P_B \psi_4) = 0. \tag{A.27}$$

Class B

These are Fierz transformations *with* charge-conjugated fermion fields in which after Fierz transformation charged-conjugated fields are absent, which simplifies the calculations. The transformation is given by $(12^c)(3^c 4) \to (13)(24)$ so that the fields ψ_1 and ψ_4 do not move, and ψ_2 and ψ_3 are not only interchanged but also charged conjugation on them is removed. We have then

$$(\bar{\psi}_1 P_A \psi_2^c)(\bar{\psi}_3^c P_A \psi_4) = -\frac{1}{2}(\bar{\psi}_1 P_A \psi_3)(\psi_2 P_A \psi_4) + \frac{1}{8}(\bar{\psi}_1 \sigma_{\mu\nu} P_A \psi_3)(\bar{\psi}_2 \sigma^{\mu\nu} P_A \psi_4), \tag{A.28}$$

$$(\bar{\psi}_1 P_A \psi_2^c)(\bar{\psi}_3^c P_B \psi_4) = \frac{1}{2}(\bar{\psi}_1 \gamma_\mu P_B \psi_3)(\bar{\psi}_2 \gamma^\mu P_B \psi_4), \tag{A.29}$$

$$(\bar{\psi}_1 \gamma_\mu P_A \psi_2^c)(\bar{\psi}_3^c \gamma^\mu P_A \psi_4) = 2(\bar{\psi}_1 P_B \psi_3)(\bar{\psi}_2 P_A \psi_4), \tag{A.30}$$

$$(\bar{\psi}_1 \gamma_\mu P_A \psi_2^c)(\bar{\psi}_3^c \gamma^\mu P_B \psi_4) = (\bar{\psi}_1 P_A \psi_3)(\bar{\psi}_2 P_B \psi_4), \tag{A.31}$$

$$(\bar{\psi}_1 \sigma_{\mu\nu} P_A \psi_2^c)(\bar{\psi}_3^c \sigma^{\mu\nu} P_A \psi_4) = 6(\bar{\psi}_1 P_A \psi_3)(\bar{\psi}_2 P_A \psi_4) + \frac{1}{2}(\bar{\psi}_1 \sigma_{\mu\nu} P_A \psi_3)(\bar{\psi}_2 \sigma^{\mu\nu} P_A \psi_4). \tag{A.32}$$

Class C

In some leptoquark models (for example, S_3) with charge-conjugated fields the transformation $(12^c)(3^c 4) \to (14)(23)$ is useful. We have then

$$(\bar{\psi}_1 P_A \psi_2^c)(\bar{\psi}_3^c P_A \psi_4) = -\frac{1}{2}(\bar{\psi}_1 P_A \psi_4)(\bar{\psi}_2 P_A \psi_3) + \frac{1}{8}(\bar{\psi}_1 \sigma_{\mu\nu} P_A \psi_4)(\bar{\psi}_2 \sigma^{\mu\nu} P_A \psi_3), \tag{A.33}$$

$$(\bar{\psi}_1 P_A \psi_2^c)(\bar{\psi}_3^c P_B \psi_4) = \frac{1}{2}(\bar{\psi}_1 \gamma_\mu P_B \psi_4)(\bar{\psi}_2 \gamma^\mu P_B \psi_3), \tag{A.34}$$

$$(\bar{\psi}_1 \gamma_\mu P_A \psi_2^c)(\bar{\psi}_3^c \gamma^\mu P_A \psi_4) = -(\bar{\psi}_1 P_A \psi_4)(\bar{\psi}_2 P_B \psi_3), \tag{A.35}$$

$$(\bar{\psi}_1 \gamma_\mu P_A \psi_2^c)(\bar{\psi}_3^c \gamma^\mu P_B \psi_4) = -2(\bar{\psi}_1 P_B \psi_4)(\bar{\psi}_2 P_A \psi_3), \tag{A.36}$$

$$(\bar{\psi}_1 \sigma_{\mu\nu} P_A \psi_2^c)(\bar{\psi}_3^c \sigma^{\mu\nu} P_A \psi_4) = 6(\bar{\psi}_1 P_A \psi_4)(\bar{\psi}_2 P_A \psi_3) + \frac{1}{2}(\bar{\psi}_1 \sigma_{\mu\nu} P_A \psi_4)(\bar{\psi}_2 \sigma^{\mu\nu} P_A \psi_3). \tag{A.37}$$

B.1 Preliminaries

In this appendix we list all SM Feynman rules relevant for this book. These rules are simply derived by multiplying SM Lagrangian by i and reading off the interactions. We note that given the Feynman rule for a vertex obtained from one particular term in the Lagrangian, the rule for the conjugate vertex is obtained from the Hermitian conjugate of this particular term, subsequently multiplied by i. In this manner the elements of the CKM matrix V_{ij} and of PMNS matrix U_{ij} are replaced by V_{ij}^* and U_{ij}^*, respectively. Moreover, while the vector interactions like $V \pm A$ are unchanged by Hermitian conjugation, there is a chirality flip from $P_{L,R}$ to $P_{R,L}$ in the case of scalar interactions.

For anybody doing Feynman diagram calculations and comparing intermediate results with those given in the literature, a nightmare are various possible signs in the Feynman rules. A nice collection of various possibilities and comparison of them can be found in [1311]. The Feynman rules for QCD and electroweak interactions are given there in terms of various $\eta_i = \pm$ factors. Our rules correspond to

$$\eta_s = \eta = \eta_e = \eta' = -1, \qquad \eta_Z = \eta_Y = \eta_\theta = +1 \tag{B.1}$$

and are the same as in the book by Peskin and Schroeder [14]. We give all these factors here but for Feynman rules really only η_s, η, η_e, and η_Z are relevant. The remaining Feynman rules can be found in [1311].

Some vertices depend on the four-momenta of particles. All momenta in our rules are, as in [1311], *incoming*. A list of papers with Feynman rules in a number of extensions of the SM has been given in Section 16.1.

B.2 Gauge Boson Propagators

For gauge bosons we give them setting all gauge parameters to unity. General expressions can be found in [1311]. We have then ($a, b = 1, ..8$):

$$G_{\mu\nu}^{ab}(k) : -i\delta_{ab}\frac{g_{\mu\nu}}{k^2 + i\epsilon}, \qquad A_{\mu\nu}(k) : -i\frac{g_{\mu\nu}}{k^2 + i\epsilon}, \tag{B.2}$$

$$(W_{\mu\nu}^{\pm})(k) : -i\frac{g_{\mu\nu}}{k^2 - M_W^2 + i\epsilon}, \qquad Z_{\mu\nu}(k) : -i\frac{g_{\mu\nu}}{k^2 - M_Z^2 + i\epsilon}. \tag{B.3}$$

B.3 Fermion and Scalar Propagators

For quarks q and leptons l the propagators read ($\alpha, \beta = 1, 2, 3$)

$$q(p)_{\alpha\beta} : i\delta_{\alpha\beta}\frac{\slashed{p} + m_q}{p^2 - m_q^2 + i\epsilon}, \qquad l(p) : i\frac{\slashed{p} + m_l}{p^2 - m_l^2 + i\epsilon}. \tag{B.4}$$

For pseudo-Goldstone bosons $G^{\pm,0}$ and Higgs H they are

$$G^\pm(p) : \frac{i}{p^2 - M_W^2 + i\epsilon}, \qquad G^0(p) : \frac{i}{p^2 - M_Z^2 + i\epsilon}, \tag{B.5}$$

$$H(p) : \frac{i}{p^2 - M_H^2 + i\epsilon}. \tag{B.6}$$

B.4 Fermion–Gauge Boson Couplings

$$\bar{q}_\alpha G_\mu^a q_\beta : ig_3\gamma_\mu t_{\alpha\beta}^a, \tag{B.7}$$

$$\bar{f}A_\mu f : ieQ_f\gamma_\mu, \tag{B.8}$$

$$\bar{u}_i Z_\mu u_i : \frac{ig_2}{\cos\vartheta_W}\gamma_\mu\left[\left(\frac{1}{2} - \frac{2}{3}\sin^2\vartheta_W\right)P_L - \frac{2}{3}\sin^2\vartheta_W P_R\right], \tag{B.9}$$

$$\bar{d}_i Z_\mu d_i : \frac{ig_2}{\cos\vartheta_W}\gamma_\mu\left[\left(-\frac{1}{2} + \frac{1}{3}\sin^2\vartheta_W\right)P_L + \frac{1}{3}\sin^2\vartheta_W P_R\right], \tag{B.10}$$

$$\bar{\nu}_i Z_\mu \nu_i : \frac{ig_2}{\cos\vartheta_W}\gamma_\mu\left[\frac{1}{2}P_L\right], \tag{B.11}$$

$$\bar{\ell}_i Z_\mu \ell_i : \frac{ig_2}{\cos\vartheta_W}\gamma_\mu\left[\left(-\frac{1}{2} + \sin^2\vartheta_W\right)P_L + \sin^2\vartheta_W P_R\right], \tag{B.12}$$

$$\bar{u}_i W_\mu^+ d_j : \frac{ig_2}{\sqrt{2}}V_{ij}\gamma_\mu P_L, \qquad \bar{d}_j W_\mu^- u_i : \frac{ig_2}{\sqrt{2}}V_{ij}^*\gamma_\mu P_L, \tag{B.13}$$

$$\bar{\nu}_i W_\mu^+ l_j : \frac{ig_2}{\sqrt{2}}U_{ij}\gamma_\mu P_L, \qquad \bar{l}_j W_\mu^- \nu_i : \frac{ig_2}{\sqrt{2}}U_{ij}^*\gamma_\mu P_L. \tag{B.14}$$

B.5 Fermion–Goldstone (Higgs) Boson Couplings

$$\bar{u}^i G^+ d^j : i\frac{g_2}{\sqrt{2}M_W}\left(m_u^i P_L - m_d^j P_R\right)V_{ij}, \tag{B.15}$$

$$\bar{d}^j G^- u^i \,:\, i\frac{g_2}{\sqrt{2}M_W}\left(m_u^i P_R - m_d^j P_L\right) V_{ij}^*, \tag{B.16}$$

$$\bar{\nu}^i G^+ \ell^j \,:\, -i\frac{g_2}{\sqrt{2}}\frac{m_\ell^j}{M_W} P_R U_{ij}, \qquad \bar{\ell}^j G^- \nu^i \,:\, -i\frac{g_2}{\sqrt{2}}\frac{m_\ell^j}{M_W} P_L U_{ij}^*, \tag{B.17}$$

$$\bar{u}^i G^0 u^j \,:\, -\frac{g_2 m_u^i}{2M_W}\gamma_5 \delta_{ij}, \qquad \bar{d}^i G^0 d^j \,:\, \frac{g_2 m_d^i}{2M_W}\gamma_5 \delta_{ij}, \tag{B.18}$$

$$\bar{\ell}^i G^0 \ell^j \,:\, \frac{g_2 m_\ell^i}{2M_W}\gamma_5 \delta_{ij}, \qquad \bar{f} H f = -i\frac{g}{2}\frac{m_f}{M_W}. \tag{B.19}$$

We have set $m_\nu^i = 0$ and $f = q, l$.

B.6 Gauge Boson Self-Interactions

$$G_\mu^a(p_1) G_\nu^b(p_2) G_\rho^c(p_3) \,:\, g_3 f^{abc}\left[g_{\mu\nu}(p_1 - p_2)_\rho + g_{\nu\rho}(p_2 - p_3)_\mu + g_{\rho\mu}(p_3 - p_1)_\nu\right], \tag{B.20}$$

$$W_\sigma^-(p_-) W_\rho^+(p_+) Z_\mu(q) \,:\, ig_2 \cos\vartheta_W\left[g_{\sigma\rho}(p_- - p_+)_\mu + g_{\rho\mu}(p_+ - q)_\sigma + g_{\mu\sigma}(q - p_-)_\rho\right], \tag{B.21}$$

$$W_\sigma^-(p_-) W_\rho^+(p_+) A_\mu(q) \,:\, ie\left[g_{\sigma\rho}(p_- - p_+)_\mu + g_{\rho\mu}(p_+ - q)_\sigma + g_{\mu\sigma}(q - p_-)_\rho\right]. \tag{B.22}$$

B.7 Gauge Boson–Goldstone (Higgs) Interactions

$$W_\mu^\pm Z_\nu G^\mp \,:\, -ig_2 M_W \frac{\sin^2\vartheta_W}{\cos\vartheta_W} g^{\mu\nu}, \qquad W_\mu^\pm A_\nu G^\mp \,:\, ie M_W g^{\mu\nu}, \tag{B.23}$$

$$G^+(p_+) G^-(p_-) Z_\mu \,:\, ig_2 \frac{1 - 2\sin^2\vartheta_W}{2\cos\vartheta_W}(p_+ - p_-)_\mu, \qquad G^+(p_+) G^-(p_-) A_\mu \,:\, ie(p_+ - p_-)_\mu, \tag{B.24}$$

$$W_\mu^\pm H W_\nu^\mp \,:\, ig_2 M_W g_{\mu\nu}, \qquad Z_\mu H Z_\nu \,:\, i\frac{g_2}{\cos\vartheta_W} M_Z g^{\mu\nu}. \tag{B.25}$$

Appendix C Massive Loop Integrals

The list of integrals given in this appendix is sufficient for the calculation of one-loop diagrams with vanishing external momenta. The latter are only kept in the calculation of self-energy diagrams on the external lines, like the ones in Figure 6.8. All formulas have been derived by the author and checked by Tobias Theil.

C.1 Integrals with Two Propagators

$$I_1(m, M) = \int \frac{d^D k}{(2\pi)^D} \frac{m}{[(k+p)^2 - m^2][k^2 - M^2]} = m \frac{i}{16\pi^2} \left[\frac{2}{\bar{\varepsilon}} + \frac{3}{2} + F_1(x) + F_2(x) \right],$$
(C.1)

$$I_2(m, M) = \int \frac{d^D k}{(2\pi)^D} \frac{(p+k)_\mu}{[(k+p)^2 - m^2][k^2 - M^2]} = p_\mu \frac{i}{16\pi^2} \left[\frac{1}{\bar{\varepsilon}} + \frac{3}{4} + F_2(x) \right],$$
(C.2)

where p_μ is an external momentum, and we set $p^2 = 0$. $F_{1,2}(x)$ are given by

$$F_1(x) = -\frac{1}{2(1-x)^2} \left[x^2 \log x - 2x \log x - x(1-x) \right],$$
(C.3)

$$F_2(x) = -\frac{1}{2(1-x)^2} \left[x^2 \log x + (1-x) \right],$$
(C.4)

and ($D = 4 - 2\varepsilon$)

$$x = \frac{m^2}{M^2}, \qquad \frac{1}{\bar{\varepsilon}} = \frac{1}{2\varepsilon} + \frac{1}{2} \left[\log 4\pi - \gamma_E + \log \left(\frac{\mu^2}{M^2} \right) \right].$$
(C.5)

C.2 Integrals with Three Propagators

$$I_3(m_1, m_2, M) = \int \frac{d^4 k}{(2\pi)^4} \frac{1}{[k^2 - m_1^2][k^2 - m_2^2][k^2 - M^2]} \tag{C.6}$$

$$= \frac{i}{16\pi^2} \frac{1}{M^2} \left[\frac{x_1 \log x_1}{(1 - x_1)(x_1 - x_2)} + \frac{x_2 \log x_2}{(1 - x_2)(x_2 - x_1)} \right], \tag{C.7}$$

where

$$x_i = \frac{m_i^2}{M^2}, \qquad x = \frac{m^2}{M^2}. \tag{C.8}$$

Special cases:

$$I_3(m, m, M) = \frac{i}{16\pi^2} \frac{1}{M^2} \left[\frac{\log x}{(1 - x)^2} + \frac{1}{(1 - x)} \right], \tag{C.9}$$

$$I_3(m, M, M) = -\frac{i}{16\pi^2} \frac{1}{M^2} \left[\frac{x \log x}{(1 - x)^2} + \frac{1}{(1 - x)} \right], \tag{C.10}$$

$$I_3(M, M, M) = -\frac{i}{32\pi^2} \frac{1}{M^2}. \tag{C.11}$$

$$I_4(m_1, m_2, M) = \int \frac{d^D k}{(2\pi)^D} \frac{k_\mu k_\nu}{[k^2 - m_1^2][k^2 - m_2^2][k^2 - M^2]} \tag{C.12}$$

$$= \frac{i g_{\mu\nu}}{32\pi^2} \left[\frac{1}{\varepsilon} + \frac{3}{4} + \frac{x_1^2 \log x_1}{2(1 - x_1)(x_1 - x_2)} + \frac{x_2^2 \log x_2}{2(1 - x_2)(x_2 - x_1)} \right]. \tag{C.13}$$

Special cases:

$$I_4(m, m, M) = \frac{i g_{\mu\nu}}{32\pi^2} \left[\frac{1}{\varepsilon} + \frac{3}{4} + F_1(x) \right], \tag{C.14}$$

$$I_4(m, M, M) = \frac{i g_{\mu\nu}}{32\pi^2} \left[\frac{1}{\varepsilon} + \frac{3}{4} + F_2(x) \right], \tag{C.15}$$

$$I_4(M, M, M) = \frac{i g_{\mu\nu}}{32\pi^2} \left[\frac{1}{\varepsilon} \right], \tag{C.16}$$

where $F_1(x)$ and $F_2(x)$ are given in (C.3) and (C.4), respectively.

C.3 Integrals with Four Propagators

$$I_5(m_1, m_2, M_1, M_2) = \int \frac{d^4k}{(2\pi)^4} \frac{1}{[k^2 - m_1^2][k^2 - m_2^2][k^2 - M_1^2][k^2 - M_2^2]} \quad \text{(C.17)}$$

$$= -\frac{i}{16\pi^2} \frac{1}{M_1^4} \left[F(x_1; x_2, \eta) + F(x_2; x_1, \eta) + F(\eta; x_1, x_2) \right] , \quad \text{(C.18)}$$

where

$$F(x; y, z) = \frac{x \log x}{(x-1)(x-y)(x-z)}, \qquad x_i = \frac{m_i^2}{M_1^2}, \qquad \eta = \frac{M_2^2}{M_1^2} . \quad \text{(C.19)}$$

Special cases:

$$I_5(m_1, m_2, M, M) = -\frac{i}{16\pi^2} \frac{1}{M^4} \left[H(x_1, x_2) + H(x_2, x_1) + \frac{1}{(1-x_1)(1-x_2)} \right] , \quad \text{(C.20)}$$

$$I_5(m, m, M, M) = -\frac{i}{16\pi^2} \frac{1}{M^4} \left[\frac{(1+x)\log x}{(1-x)^3} + \frac{2}{(1-x)^2} \right] , \quad \text{(C.21)}$$

where

$$H(x_1, x_2) = \frac{x_1 \log x_1}{(1-x_1)^2(x_1-x_2)} . \quad \text{(C.22)}$$

$$I_6(m_1, m_2, M_1, M_2) = \int \frac{d^4k}{(2\pi)^4} \frac{k_\mu k_\nu}{[k^2 - m_1^2][k^2 - m_2^2][k^2 - M_1^2][k^2 - M_2^2]} \quad \text{(C.23)}$$

$$= -\frac{ig_{\mu\nu}}{64\pi^2} \frac{1}{M_1^2} \left[\tilde{F}(x_1; x_2, \eta) + \tilde{F}(x_2; x_1, \eta) + \tilde{F}(\eta; x_1, x_2) \right] , \quad \text{(C.24)}$$

where

$$\tilde{F}(x; y, z) = \frac{x^2 \log x}{(x-1)(x-y)(x-z)}, \qquad x_i = \frac{m_i^2}{M_1^2}, \qquad \eta = \frac{M_2^2}{M_1^2} . \quad \text{(C.25)}$$

Special cases:

$$I_6(m_1, m_2, M, M) = -\frac{ig_{\mu\nu}}{64\pi^2} \frac{1}{M^2} \left[\tilde{H}(x_1, x_2) + \tilde{H}(x_2, x_1) + \frac{1}{(1-x_1)(1-x_2)} \right] , \quad \text{(C.26)}$$

$$I_6(m, m, M, M) = -\frac{ig_{\mu\nu}}{64\pi^2} \frac{1}{M^2} \left[\frac{2x \log x}{(1-x)^3} + \frac{1+x}{(1-x)^2} \right] , \quad \text{(C.27)}$$

where

$$\tilde{H}(x_1, x_2) = \frac{x_1^2 \log x_1}{(1-x_1)^2(x_1-x_2)} . \quad \text{(C.28)}$$

C.4 More Complicated Integrals

C.4.1 Preliminaries

The integrals in previous sections have the general structure

$$I = \int \frac{d^D k}{(2\pi)^D} f(k^2), \tag{C.29}$$

with the integrand depending only on k^2. This is the case of one-loop diagrams in which external momenta are set to zero. Such integrals are much easier to evaluate than those with external momenta like self-energy diagrams or vertex diagrams, which we calculated in the context of renormalization of QCD and QED in Chapter 4. In particular shifts in momenta and Feynman integrals can be avoided. Integrating over angular variables one finds first

$$I = \frac{i}{(4\pi)^{D/2}} \frac{1}{\Gamma(D/2)} \int_0^\infty ds\, s^{1-\varepsilon} f(k^2 = -s) \tag{C.30}$$

with $D = 4 - 2\varepsilon$. It is useful to perform subsequently a change of variables:

$$s = rM^2, \qquad ds\, s^{1-\varepsilon} = dr\, r^{1-\varepsilon} (M^2)^{2-\varepsilon} \tag{C.31}$$

so that the calculation is reduced to the evaluation of the following r integrals

$$R = \int_0^\infty dr\, F(r, x_i), \qquad x_i = \frac{m_i^2}{M^2} \tag{C.32}$$

with m_i denoting the masses of particles different than the one having the mass M. These particles can be scalars, fermions, and gauge bosons. Depending on the diagram considered, one obtains a function

$$F(r, x_i) = \frac{r^{n-\varepsilon}}{[r+a][r+b]...[r+e]}, \tag{C.33}$$

where in most cases the parameters $(a, b, c, ..)$ are just equal to different x_i or 1. See examples at the end of this appendix. The number of factors in the denominator could be larger than four considered in Appendix C.3. This is the case if calculations are done not in the Feynman gauge but in R_ξ gauges or unitary gauge. For these cases the list of integrals in Appendix C is insufficient. Therefore, following we give a list of r-integrals that should allow the evaluation of any box and Z-penguin diagram in any gauge and in any theory at one-loop level. We will also give integrals relevant for self-energy diagrams.

The basic technique to evaluate all integrals given here is to do first the decomposition

$$\frac{r^n}{[r+a][r+b]\cdots[r+e]} = \frac{A}{[r+a]} + \frac{B}{[r+b]} + \dots \frac{E}{[r+e]}, \tag{C.34}$$

where n is smaller than the number of factors in the denominator and

$$A = \frac{(-a)^n}{(b-a)(c-a)(d-a)(e-a)}, \tag{C.35}$$

$$B = \frac{(-b)^n}{(a-b)(c-b)(d-b)(e-b)}, \tag{C.36}$$

$$E = \frac{(-e)^n}{(a-e)(b-e)(c-e)(d-e)} \tag{C.37}$$

with analogous expressions for other terms. They have, for instance, $(-c)^n$ and $(-d)^n$ in the numerator.

Inserting (C.34) into (C.32) one obtains a sum of terms each multiplied by $r^{-\varepsilon}$. They are evaluated by using

$$\int_0^\infty dr \frac{r^{-\varepsilon}}{[r+a]} = \frac{1}{\varepsilon} - \gamma_E - \ln a. \tag{C.38}$$

To give compact expressions for the integrals listed here, we introduce the functions:

$$G_1(x) = \frac{x \ln x}{(y_1 - x)}, \qquad G_2(x) = \frac{x \ln x}{(y_1 - x)(y_2 - x)}, \tag{C.39}$$

$$G_3(x) = \frac{x \ln x}{(y_1 - x)(y_2 - x)(y_3 - x)}, \qquad G_4(x) = \frac{x^2 \ln x}{(y_1 - x)(y_2 - x)(y_3 - x)(y_4 - x)}, \tag{C.40}$$

$$G_5(x) = \frac{x^4 \ln x}{(y_1 - x)(y_2 - x)(y_3 - x)(y_4 - x)(y_5 - x)}. \tag{C.41}$$

Here x stands for one particular variable in the set $\{a, b, c, d, ..\}$, and y_i denote the remaining variables in this set. As an example, $G_5(c)$ given here reads

$$G_5(c) = \frac{c^4 \ln c}{(a-c)(b-c)(d-c)(e-c)(f-c)}. \tag{C.42}$$

C.4.2 Integrals for Box Diagrams

These integrals are defined as follows:

$$I_1^{(n)} \equiv \int_0^\infty dr \frac{r^{n-\varepsilon}}{[r+a][r+b][r+c][r+d]}, \qquad n = 1, 2, \tag{C.43}$$

$$I_2^{(n)} \equiv \int_0^\infty dr \frac{r^{n-\varepsilon}}{[r+a][r+b][r+c][r+d][r+e]}, \qquad n = 2, 3, \tag{C.44}$$

$$I_3^{(n)} \equiv \int_0^\infty dr \frac{r^{n-\varepsilon}}{[r+a][r+b][r+c][r+d][r+e][r+f]}, \qquad n = 4. \tag{C.45}$$

We find

$$I_1^{(1)} = G_3(a) + G_3(b) + G_3(c) + G_3(d) \,, \tag{C.46}$$

$$I_1^{(2)} = -aG_3(a) - bG_3(b) - cG_3(c) - dG_3(d) \,, \tag{C.47}$$

$$I_2^{(2)} = -G_4(a) - G_4(b) - G_4(c) - G_4(d) - G_4(e) \,, \tag{C.48}$$

$$I_2^{(3)} = aG_4(a) + bG_4(b) + cG_4(c) + dG_4(d) + eG_4(e) \,, \tag{C.49}$$

$$I_3^{(4)} = -G_5(a) - G_5(b) - G_5(c) - G_5(d) - G_5(e) - G_5(f) \,. \tag{C.50}$$

C.4.3 Integrals for Z-Penguins

These integrals are defined as follows:

$$\mathcal{J}_1^{(n)} \equiv \int_0^\infty dr \frac{r^{n-\varepsilon}}{[r+a][r+b][r+c]}, \qquad n = 1, 2, 3 \,, \tag{C.51}$$

$$\mathcal{J}_2^{(n)} \equiv \int_0^\infty dr \frac{r^{n-\varepsilon}}{[r+a][r+b][r+c][r+d]}, \qquad n = 1, 2, 3, 4 \,, \tag{C.52}$$

$$\mathcal{J}_3^{(n)} \equiv \int_0^\infty dr \frac{r^{n-\varepsilon}}{[r+a][r+b][r+c][r+d][r+e]}, \qquad n = 4 \,. \tag{C.53}$$

We find

$$\mathcal{J}_1^{(1)} = G_2(a) + G_2(b) + G_2(c) \,, \tag{C.54}$$

$$\mathcal{J}_1^{(2)} = -aG_2(a) - bG_2(b) - cG_2(c) + \frac{1}{\varepsilon} \,, \tag{C.55}$$

$$\mathcal{J}_1^{(3)} = a^2 G_2(a) + b^2 G_2(b) + c^2 G_2(c) - \frac{(a+b+c)}{\varepsilon} \,, \tag{C.56}$$

$$\mathcal{J}_2^{(1)} = I_1^{(1)} \,, \qquad \mathcal{J}_2^{(2)} = I_1^{(2)} \,, \qquad \mathcal{J}_3^{(3)} = I_2^{(3)} \,, \tag{C.57}$$

$$\mathcal{J}_2^{(3)} = a^2 G_3(a) + b^2 G_3(b) + c^2 G_3(c) + d^2 G_4(d) + \frac{1}{\varepsilon} \,, \tag{C.58}$$

$$\mathcal{J}_2^{(4)} = -a^3 G_3(a) - b^3 G_3(b) - c^3 G_3(c) - d^3 G_3(d) - \frac{(a+b+c+d)}{\varepsilon} \,, \tag{C.59}$$

$$\mathcal{J}_3^{(4)} = -a^2 G_4(a) - b^2 G_4(b) - c^2 G_4(c) - d^2 G_4(d) - e^2 G_4(e) + \frac{1}{\varepsilon} \,. \tag{C.60}$$

C.4.4 Integrals for Self-Energies

These integrals are defined as follows:

$$\mathcal{K}_1^{(n)} \equiv \int_0^\infty dr \frac{r^{n-\varepsilon}}{[r+a][r+b]}, \qquad n = 1, 2, 3 \,. \tag{C.61}$$

$$\mathcal{K}_2^{(n)} \equiv \int_0^\infty dr \frac{r^{n-\varepsilon}}{[r+a][r+b][r+c]}, \qquad n = 2 \,. \tag{C.62}$$

We find

$$\mathcal{K}_1^{(1)} = G_1(a) + G_1(b) + \frac{1}{\varepsilon}, \qquad \mathcal{K}_2^{(2)} = \mathcal{J}_1^{(2)}, \tag{C.63}$$

$$\mathcal{K}_1^{(2)} = -aG_1(a) - bG_1(b) - \frac{(a+b)}{\varepsilon}, \tag{C.64}$$

$$\mathcal{K}_1^{(3)} = a^2 G_1(a) + b^2 G_1(b) + \frac{(a^2 + b^2 + ab)}{\varepsilon}, \tag{C.65}$$

$$\mathcal{K}_2^{(2)} = \mathcal{J}_1^{(2)}. \tag{C.66}$$

Appendix D **Numerical Input**

In Table D.1 we collect the experimental values of most important parameters used in the numerical estimates in the book. If in certain cases other values are used, we warn the reader about it. In any case, these values change with time, and in doing phenomenology the readers should consult PDG, FLAG, and HFLAV.

Table D.1 Values of the experimental and theoretical quantities used as input parameters as of December 2018 based on PDG, FLAG, and HFLAV.

$G_F = 1.16637(1) \times 10^{-5}\,\mathrm{GeV}^{-2}$	$M_Z = 91.188(2)\,\mathrm{GeV}$	$M_W = 80.385(15)\,\mathrm{GeV}$
$\sin^2 \theta_W = 0.23116(13)$	$\alpha(M_Z) = 1/127.9$	$\alpha_s(M_Z) = 0.1184(7)$
$m_e = 0.511\,\mathrm{MeV}$	$m_\mu = 105.66\,\mathrm{MeV}$	$m_\tau = 1776.9(1)$
$m_u(2\,\mathrm{GeV}) = 2.16(11)\,\mathrm{MeV}$	$m_c(m_c) = 1.279(13)\,\mathrm{GeV}$	$m_t(m_t) = 163(1)\,\mathrm{GeV}$
$m_d(2\,\mathrm{GeV}) = 4.68(15)\,\mathrm{MeV}$	$m_s(2\,\mathrm{GeV}) = 93.8(24)\,\mathrm{MeV}$	$m_b(m_b) = 4.19^{+0.18}_{-0.06}\,\mathrm{GeV}$
$m_{\pi^\pm} = 139.57\,\mathrm{MeV}$	$m_{\pi^0} = 134.98\,\mathrm{MeV}$	$m_\eta = 547.86(2)\,\mathrm{MeV}$
$m_{K^\pm} = 493.68(2)\,\mathrm{MeV}$	$m_{K^0} = 497.61(1)\,\mathrm{MeV}$	$m_{K^*} = 895.81(19)\,\mathrm{MeV}$
$m_{D^\pm} = 1869.58(0.09)\,\mathrm{MeV}$	$m_{D^0} = 1864.83(05)\,\mathrm{MeV}$	$m_{D_s^\pm} = 1968.27(10)\,\mathrm{MeV}$
$m_{B_d} = 5279.62(15)\,\mathrm{MeV}$	$m_{B_s} = 5366.82(22)\,\mathrm{MeV}$	$m_{B^+} = 5279.31(15)\,\mathrm{MeV}$
$\tau_\mu = 2.197\,10^{-6}\,(s)$	$\tau_\tau = 2.903(5)\,10^{-13}\,(s)$	
$\tau_{\pi^\pm} = 2.603\,10^{-8}\,(s)$	$\tau_{\pi^0} = 8.52(18)\,10^{-17}\,(s)$	
$\tau_{K^\pm} = 1.238(2)\,10^{-8}\,(s)$	$\tau_{K_L} = 5.116(21)\,10^{-8}\,(s)$	$\tau_{K_S} = 0.895\,10^{-10}\,(s)$
$\tau_{D^\pm} = 10.4(1)\,10^{-13}\,(s)$	$\tau_{D^0} = 4.10(2)\,10^{-13}\,(s)$	$\tau_{D_s^\pm} = 5.0(1)\,10^{-13}\,(s)$
$\tau_{B_d} = 1.520(4)\,\mathrm{ps}$	$\tau_{B_s} = 1.510(5)\,\mathrm{ps}$	$\tau_{B^\pm} = 1.638(4)\,\mathrm{ps}$
$\Delta M_K = 0.5292(9) \times 10^{-2}\,\mathrm{ps}^{-1}$	$\Delta M_d = 0.5055(20)\,\mathrm{ps}^{-1}$	$\Delta M_s = 17.757(21)\,\mathrm{ps}^{-1}$
$|\varepsilon_K| = 2.228(11) \times 10^{-3}$	$S_{\psi K_S} = 0.699(17)$	$S_{\psi\phi} = 0.054(20)$

Table D.2 Values of selected experimental branching ratios as of April 2019 based on PDG.

$\mathcal{B}(\pi^+ \to \mu^+ \nu_\mu) = 0.9999$	$\mathcal{B}(K^+ \to \mu^+ \nu_\mu) = 0.6356(11)$	$\mathcal{B}(\mu^- \to e^- \bar{\nu}_e \nu_\mu) \approx 1.0$
$\mathcal{B}(\tau^- \to e^- \bar{\nu}_e \nu_\tau) = 0.1782(4)$	$\mathcal{B}(\tau^- \to \mu^- \bar{\nu}_\nu \nu_\tau) = 0.1739(4)$	$\mathcal{B}(K^+ \to \pi^0 e^+ \nu_e) = 0.0507(4)$
$\mathcal{B}(K^+ \to \pi^0 \mu^+ \nu_\nu) = 0.0335(3)$	$\mathcal{B}(K_S^0 \to \pi^0 \pi^0) = 0.3069(5)$	$\mathcal{B}(K_S^0 \to \pi^+ \pi^-) = 0.6920(5)$
$\mathcal{B}(K_L^0 \to \pi^\pm e^\mp \nu_e) = 0.4055(11)$	$\mathcal{B}(K_L^0 \to \pi^\pm \mu^\mp \nu_\mu) = 0.2704(7)$	$\mathcal{B}(K_L^0 \to \pi^0 \pi^0 \pi^0) = 0.1952(12)$
$\mathcal{B}(K_L^0 \to \pi^+ \pi^+ \pi^0) = 0.1254(5)$	$\mathcal{B}(B^0 \to D^- l^+ \nu_l) = 0.0225(8)$	$\mathcal{B}(B^+ \to X_c l^+ \nu_e) = 0.1079(37)$

Table D.3 Values of theoretical quantities used as input parameters as of April 2019 based on FLAG, HFLAV.

$F_\pi = 130.41(20)$ MeV	$F_K = 156.1(11)$ MeV	$\hat{B}_K = 0.766(10)$
$F_{B_d} = 190.5(1.3)$ MeV	$F_{B_s} = 230.7(1.2)$ MeV	$F_{B^+} = 189.4(1.3)$ MeV
$\hat{B}_{B_d} = 1.27(10)$	$\hat{B}_{B_s} = 1.33(6)$	$\hat{B}_{B_s}/\hat{B}_{B_d} = 1.01(2)$
$F_{B_d}\sqrt{\hat{B}_{B_d}} = 216(15)$ MeV	$F_{B_s}\sqrt{\hat{B}_{B_s}} = 266(18)$ MeV	$\xi = 1.21(2)$
$\eta_{cc} = 1.87(76)$	$\eta_{ct} = 0.496(47)$	$\eta_{tt} = 0.5765(65)$
$\eta_B = 0.55(1)$	$\Delta\Gamma_s/\Gamma_s = 0.124(9)$	$\kappa_\epsilon = 0.94(2)$
$|V_{us}| = 0.2254(4)$	$|V_{ub}^{\text{nom}}| = 3.7 \times 10^{-3}$	$|V_{cb}^{\text{nom}}| = 42.0 \times 10^{-3}$

For conversion from seconds to GeV, the following relations are useful

$$6.582 \, \text{ps} \, \text{GeV} \cdot 10^{-13} = 6.582 \, \text{sec} \, \text{GeV} \cdot 10^{-25} = 1 \,. \tag{D.1}$$

In Table D.2 we collect the values of selected experimental branching ratios used in book. Complete collection can be found in PDG. In Table D.3 we collect the values of most important parameters that we used in this book. Other, more specific, parameters are given while presenting numerical output for various observables.

Appendix E   Analytic Solutions to SMEFT RG Equations

We have given a number of RG equations and provided analytic solutions for them as in (14.39) and (14.43). In doing this we have assumed that the anomalous dimensions are scale independent. If the anomalous dimensions are just functions of gauge couplings g_1 and g_2 or lepton Yukawa couplings that all are only weakly scale dependent, this is a good approximation. Moreover, if the leading logarithms is not large, one does not have to perform the summation of leading logarithms like we did in other parts of our book. But if top-quark Yukawa couplings are involved and $\ln \mu_\Lambda / \mu_{ew}$ is large, one would like to do it better as the quark masses are renormalized through QCD and moreover the approximation of keeping single logarithm could turn out to be unsatisfactory. Clearly, one can just set all these equations into a computer code (see Section 14.6.4) and let the computer improve the accuracy of solutions, but with the technology developed in this book we can still make a few analytic improvements by summing the logarithms and including the scale dependence of quark masses.

For pedagogical reasons we will first do the derivations for a single coefficient C, and we will later generalize the analysis to arbitrary number of coefficients. Consider then the equation

$$\mu \frac{dC}{d\mu} = \frac{g_W^2}{16\pi^2} \gamma\, C\,, \tag{E.1}$$

with g_W denoting $g_{1,2}$.

Formally, the solution can be written as

$$C(\mu) = \exp\left[\int_{\ln \Lambda}^{\ln \mu} \frac{g_W^2}{16\pi^2} \gamma(\tilde{\mu})\, d\ln \tilde{\mu}\right] C(\Lambda)\,. \tag{E.2}$$

We next neglect scale dependence of g_W and keep it, say, at the value μ_{ew}. As far as γ is concerned, we can consider three cases that one can encounter in the SMEFT.

Case A: γ involves only lepton Yukawa couplings and is only very weakly scale dependent. Neglecting this scale dependence and performing the integral we find

$$C(\mu) = \exp\left[-\frac{g_W^2}{16\pi^2} \gamma \ln\frac{\Lambda}{\mu}\right] C(\Lambda)\,. \tag{E.3}$$

The improvement over the results, like the ones in (14.39) and (14.43), is the summation of logarithms. Expanding the exponential we obtain the leading terms as found previously and in addition we obtain higher-order terms.

Case B: We next assume that

$$\gamma = a\,m_q(\mu) = a\,m_q(\mu_0)\left[\frac{\alpha_s(\mu)}{\alpha_s(\mu_0)}\right]^{\frac{\gamma_m^{(0)}}{2\beta_0}} \equiv a\,\tilde{\gamma}_a\left[\alpha_s(\mu)\right]^{\frac{\gamma_m^{(0)}}{2\beta_0}},$$

$$\tilde{\gamma}_a = a\,m_q(\mu_0)\left[\alpha_s(\mu_0)\right]^{-\frac{\gamma_m^{(0)}}{2\beta_0}}, \tag{E.4}$$

where we used (4.153) with $\gamma_m^{(0)} = 8$ keeping only LO term. The coefficient a is some numerical factor and could involve leptonic Yukawa couplings. But

$$d\ln\tilde{\mu} = -16\pi^2\frac{dg_s}{\beta_0 g_s^3} = -2\pi\frac{d\alpha_s}{\beta_0\alpha_s^2}. \tag{E.5}$$

Making this change of variables in (E.2) and inserting there (E.4) a simple integration gives

$$C(\mu) = \exp\left[X(\mu,\Lambda)\right]C(\Lambda) \tag{E.6}$$

with

$$X(\mu,\Lambda) = \frac{1}{4\pi}\frac{\tilde{\gamma}_a g_W^2}{\gamma_m^{(0)} - 2\beta_0}\left[\left[\alpha_s(\Lambda)\right]^{\frac{\gamma_m^{(0)}}{2\beta_0}-1} - \left[\alpha_s(\mu)\right]^{\frac{\gamma_m^{(0)}}{2\beta_0}-1}\right]. \tag{E.7}$$

One can check that setting $\gamma_m^{(0)} = 0$, and using the LO result for $\alpha_s(\mu)$ one finds (E.3).

Case C: We next assume that

$$\gamma = b\,m_q^2(\mu) = b\,m_q^2(\mu_0)\left[\frac{\alpha_s(\mu)}{\alpha_s(\mu_0)}\right]^{\frac{\gamma_m^{(0)}}{\beta_0}} \equiv b\,\tilde{\gamma}_b\left[\alpha_s(\mu)\right]^{\frac{\gamma_m^{(0)}}{\beta_0}},$$

$$\tilde{\gamma}_b = bm_q^2(\mu_0)\left[\alpha_s(\mu_0)\right]^{-\frac{\gamma_m^{(0)}}{\beta_0}}, \tag{E.8}$$

with b being a μ independent factor analogous to a. Proceeding as in the previous case, we find this time

$$X(\mu,\Lambda) = \frac{1}{8\pi}\frac{\tilde{\gamma}_b g_W^2}{\gamma_m^{(0)} - \beta_0}\left[\left[\alpha_s(\Lambda)\right]^{\frac{\gamma_m^{(0)}}{\beta_0}-1} - \left[\alpha_s(\mu)\right]^{\frac{\gamma_m^{(0)}}{\beta_0}-1}\right]. \tag{E.9}$$

If there are several Wilson coefficients and the RG equations are given as in (14.28)

$$C_a(\mu) = \left(\exp\left[X(\mu,\Lambda)\right]\right)_{ab}C_b(\Lambda) \tag{E.10}$$

with different entries in the matrix calculated as in (E.3), (E.7), and (E.7) dependently on the structure of the elements of the matrix $\hat{\gamma}$.

If a single logarithm is sufficient we find

$$C_a(\mu) = (\delta_{ab} + X_{ab}(\mu, \Lambda)) C_b(\Lambda) \qquad \text{(E.11)}$$

with X_{ab} calculated using (E.3), (E.7), and (E.9).

There exist situations in which a given operator of interest does not mix directly with a second operator that we are interested in as it enters an experimental observable, but does so via a third "mediator operator" [781, 1224, 1312]. In this case, the single logarithmic contribution vanishes, and one has to expand the exponential to second order to obtain the nonvanishing result. To describe this situation in simple terms, we confine our discussion to the simplest case of three operators with anomalous dimension matrix $\hat{\gamma}$ that is μ-independent, as in case A. The generalization to cases B and C is straightforward using the preceding expressions and is discussed in detail in [1313], where even more complicated cases are considered.

We consider then the following situation: the coefficient of an operator O_1 is the only one with a nonzero value at some high scale Λ. At some much lower scale μ an observable is determined by the value of a coefficient of an operator O_3, which does not mix directly with O_1, but O_1 mixes into a third operator, O_2, and O_2 mixes into O_3.

The solution to RG is just the generalization of (E.3) to include three operators

$$\vec{C}(\mu) = \exp\left[-\frac{g_W^2}{16\pi^2}\hat{\gamma}\ln\frac{\Lambda}{\mu}\right]\vec{C}(\Lambda), \qquad \text{(E.12)}$$

where $\vec{C} = (C_1, C_2, C_3)^T$. The situation described corresponds to

$$\vec{C}(\Lambda) = (C_1(\Lambda), 0, 0)^T, \qquad \hat{\gamma}_{31} = 0 \qquad \hat{\gamma}_{21} \neq 0 \qquad \text{and} \qquad \hat{\gamma}_{32} \neq 0 \qquad \text{(E.13)}$$

with other entries of $\hat{\gamma}$ not playing any role at this level so that they can be set to zero.

Expanding the exponential in (E.12) in powers of g_W^2 we find that a nonvanishing result for $C_3(\mu)$ is obtained first at second-order ($O(g_W^4)$), which introduces a factor $1/2$. One can also check that as long as conditions in (E.13) are satisfied, the result for $C_3(\mu)$ is independent of other entries in $\hat{\gamma}$. The latter enter first at g_W^6 and can be neglected. We thus find

$$C_3(\mu) = \frac{1}{2}\hat{\gamma}_{32}\,\hat{\gamma}_{21}\left[\frac{g^2}{16\pi^2}\ln\left(\frac{\Lambda}{\mu}\right)\right]^2 C_1(\Lambda). \qquad \text{(E.14)}$$

In practice the following heuristic method can be used instead. One uses RG evolution to calculate $C_2(\mu)$ from $C_1(\Lambda)$. This brings in one logarithm $\ln(\Lambda/\mu)$ and the factor $\hat{\gamma}_{21}g^2/16\pi^2$. We then equate $C_2(\mu) = C_2(\Lambda)$ and run again down to μ generating $C_3(\mu)$ from $C_2(\Lambda)$. This brings in second logarithm $\ln(\Lambda/\mu)$ and the factor $\hat{\gamma}_{32}g^2/16\pi^2$. Now this heuristic procedure is not quite right as $C_2(\Lambda)$ is in the second step nonzero, but in (E.13) we took it to be vanishing. The result in (E.14) tells us how to correct for this. We just have to multiply the product of these two factors by $1/2$ to obtain the correct result. Personally I find this procedure easier than multiplying matrices by hand, and in the case of complicated ADMs one obtains the results using this simple trick much faster.

References

[1] S. L. Glashow, Partial symmetries of weak interactions, *Nucl. Phys.* **22** (1961) 579–588.

[2] A. Salam, Weak and electromagnetic interactions, *Conf. Proc.* **C680519** (1968) 367–377.

[3] S. Weinberg, A model of leptons, *Phys. Rev. Lett.* **19** (1967) 1264–1266.

[4] G. 't Hooft, *Naturalness, chiral symmetry, and spontaneous chiral symmetry breaking*, NATO Sci. Ser. B **59** (1980) 135–157.

[5] M. Gaillard and B. W. Lee, Rare decay modes of the K-mesons in gauge theories, *Phys. Rev.* **D10** (1974) 897.

[6] S. L. Glashow, J. Iliopoulos, and L. Maiani, Weak interactions with Lepton-Hadron symmetry, *Phys. Rev.* **D2** (1970) 1285–1292.

[7] G. C. Branco, L. Lavoura, and J. P. Silva, CP violation, *Int. Ser. Monogr. Phys.* **103** (1999) 1–536.

[8] J. F. Donoghue, E. Golowich, and B. R. Holstein, Dynamics of the standard model, *Camb. Monogr. Part. Phys. Nucl. Phys. Cosmol.* **2** (1992) 1–540. [*Camb. Monogr. Part. Phys. Nucl. Phys. Cosmol.* 35(2014)].

[9] I. I. Bigi and A. Sanda, CP violation, *Camb. Monogr. Part. Phys. Nucl. Phys. Cosmol.* **9** (2000) 1–382.

[10] P. Langacker, *The Standard Model and Beyond*. CRC Press, 2010.

[11] L. H. Ryder, *Quantum Field Theory*. Cambridge University Press, 1996.

[12] M. Gell-Mann, P. Ramond, and R. Slansky, Color embeddings, charge assignments, and proton stability in unified gauge theories, *Rev. Mod. Phys.* **50** (1978) 721.

[13] J. D. Bjorken and S. D. Drell, *Relativistic Quantum Fields*. McGraw-Hill Book Company, 1965.

[14] M. E. Peskin and D. V. Schroeder, *An Introduction to Quantum Field Theory*. Addison-Wesley Publishing Company, 1995.

[15] S. Weinberg, *The Quantum Theory of Fields. Vol. 2: Modern Applications*. Cambridge University Press, 1996.

[16] D. Bailin and A. Love, *Introduction to Gauge Field Theory*. Adam Hilger, Bristol and Boston, 1986.

[17] S. Weinberg, Implications of dynamical symmetry breaking, *Phys. Rev.* **D13** (1976) 974–996.

[18] S. Pokorski, *Gauge Field Theories*. Cambridge University Press, 2005.

[19] N. Cabibbo, Unitary symmetry and leptonic decays, *Phys. Rev. Lett.* **10** (1963) 531–533. [648(1963)].

[20] M. Kobayashi and T. Maskawa, CP violation in the renormalizable theory of weak interaction, *Prog. Theor. Phys.* **49** (1973) 652–657.

[21] B. Pontecorvo, Mesonium and antimesonium, *Sov. Phys. JETP* **6** (1957) 429.

[22] Z. Maki, M. Nakagawa, and S. Sakata, Remarks on the unified model of elementary particles, *Prog. Theor. Phys.* **28** (1962) 870.

[23] S. Bilenky, Introduction to the physics of massive and mixed neutrinos, *Lect. Notes Phys.* **817** (2010) 1–255.

[24] W. Altmannshofer, C. Frugiuele, and R. Harnik, Fermion hierarchy from sfermion anarchy, *JHEP* **12** (2014) 180, [arXiv:1409.2522].

[25] L.-L. Chau and W.-Y. Keung, Comments on the parametrization of the Kobayashi-Maskawa matrix, *Phys. Rev. Lett.* **53** (1984) 1802.

[26] L. Wolfenstein, Parametrization of the Kobayashi-Maskawa matrix, *Phys. Rev. Lett.* **51** (1983) 1945.

[27] A. J. Buras, M. E. Lautenbacher, and G. Ostermaier, Waiting for the top quark mass, $K^+ \to \pi^+ \nu \bar{\nu}$, $B_s^0 - \bar{B}_s^0$ mixing and CP asymmetries in B decays, *Phys. Rev.* **D50** (1994) 3433–3446, [hep-ph/9403384].

[28] G. C. Branco and L. Lavoura, Wolfenstein type parametrization of the quark mixing matrix, *Phys. Rev.* **D38** (1988) 2295.

[29] A. J. Buras, Weak Hamiltonian, CP violation and rare decays, in Probing the Standard Model of Particle Interactions. *Proceedings, Summer School in Theoretical Physics*, NATO Advanced Study Institute, 68th session, Les Houches, France, July 28–September 5, 1997. Pt. 1, 2, pp. 281–539, 1998. hep-ph/9806471.

[30] C. Jarlskog and R. Stora, Unitarity polygons and CP violation areas and phases in the standard electroweak model, *Phys. Lett.* **B208** (1988) 268–274.

[31] R. Aleksan, B. Kayser, and D. London, Determining the quark mixing matrix from CP violating asymmetries, *Phys. Rev. Lett.* **73** (1994) 18–20, [hep-ph/9403341].

[32] C. Jarlskog, Commutator of the quark mass matrices in the standard electroweak model and a measure of maximal CP violation, *Phys. Rev. Lett.* **55** (1985) 1039.

[33] C. Jarlskog, A basis independent formulation of the connection between quark mass matrices, *CP violation and experiment*, *Z. Phys.* **C29** (1985) 491–497.

[34] A. J. Buras, F. Parodi, and A. Stocchi, The CKM matrix and the unitarity triangle: Another look, *JHEP* **0301** (2003) 029, [hep-ph/0207101].

[35] Particle Data Group Collaboration, M. Tanabashi et al., Review of particle physics, *Phys. Rev.* **D98** (2018), no. 3 030001.

[36] H. D. Politzer, Reliable perturbative results for strong interactions? *Phys. Rev. Lett.* **30** (1973) 1346–1349.

[37] D. J. Gross and F. Wilczek, Ultraviolet behavior of nonabelian gauge theories, *Phys. Rev. Lett.* **30** (1973) 1343–1346.

[38] D. J. Gross and F. Wilczek, Asymptotically free gauge theories. 1, *Phys. Rev.* **D8** (1973) 3633–3652.

[39] D. J. Gross and F. Wilczek, Asymptotically Free gauge theories. 2., *Phys. Rev.* **D9** (1974) 980–993.

[40] N. Brambilla, A. Pineda, J. Soto, and A. Vairo, Effective field theories for heavy Quarkonium, *Rev. Mod. Phys.* **77** (2005) 1423, [hep-ph/0410047].

[41] N. Brambilla et al., Heavy quarkonium: Progress, puzzles, and opportunities, *Eur. Phys. J.* **C71** (2011) 1534, [arXiv:1010.5827].

[42] N. Brambilla et al., QCD and strongly coupled gauge theories: Challenges and perspectives, *Eur. Phys. J.* **C74** (2014), no. 10 2981, [arXiv:1404.3723].

[43] M. Blanke, Introduction to flavour physics and CP violation, *CERN Yellow Rep. School Proc.* **1705** (2017) 71–100, [arXiv:1704.03753].

[44] J. Zupan, Introduction to flavour physics, 2019. arXiv:1903.05062.

[45] Y. Nir, CP violation in and beyond the standard model, in *CP Violation: In and beyond the Standard Model: Proceedings, 27th SLAC Summer Institute on Particle Physics (SSI 99): Stanford, USA, Jul 7–16 1999*, 1999. hep ph/0011321.

[46] L. Silvestrini, Effective theories for quark flavour physics, 2019. arXiv:1905.00798.

[47] M. D. Schwartz, *Quantum Field Theory and the Standard Model*. Cambridge University Press, 2014.

[48] S. Aoki et al., Review of lattice results concerning low-energy particle physics, *Eur. Phys. J.* **C77** (2017), no. 2 112, [arXiv:1607.00299].

[49] TUMQCD, Fermilab Lattice, MILC Collaboration, A. Bazavov et al., B- and D-meson leptonic decay constants and quark masses from four-flavor lattice QCD, in *13th Conference on the Intersections of Particle and Nuclear Physics (CIPANP 2018) Palm Springs, California, USA, May 29–June 3, 2018*, 2018. arXiv:1810.00250.

[50] V. Cirigliano, G. Ecker, H. Neufeld, A. Pich, and J. Portoles, Kaon decays in the standard model, *Rev. Mod. Phys.* **84** (2012) 399, [arXiv:1107.6001].

[51] F. U. Bernlochner, Z. Ligeti, M. Papucci, and D. J. Robinson, Combined analysis of semileptonic B decays to D and D^*: $R(D^{(*)})$, $|V_{cb}|$, and new physics, *Phys. Rev.* **D95** (2017), no. 11 115008, [arXiv:1703.05330].

[52] P. Gambino, Inclusive semileptonic B decays and $|V_{cb}|$: In memoriam Kolya Uraltsev, *Int. J. Mod. Phys.* **A30** (2015), no. 10 1543002, [arXiv:1501.00314].

[53] P. Colangelo and F. De Fazio, Tension in the inclusive versus exclusive determinations of $|V_{cb}|$: A possible role of new physics, *Phys. Rev.* **D95** (2017), no. 1 011701, [arXiv:1611.07387].

[54] D. Bigi and P. Gambino, Revisiting $B \to D\ell\nu$, *Phys. Rev.* **D94** (2016), no. 9 094008, [arXiv:1606.08030].

[55] B. Grinstein and A. Kobach, Model-independent extraction of $|V_{cb}|$ from $\bar{B} \to D^*\ell\bar{\nu}$, *Phys. Lett.* **B771** (2017) 359–364, [arXiv:1703.08170].

[56] G. Ricciardi, Semileptonic decays and $|V_{xb}|$ determinations, *EPJ Web Conf.* **182** (2018) 02104, [arXiv:1712.06988].

[57] D. Bigi, P. Gambino, and S. Schacht, A fresh look at the determination of $|V_{cb}|$ from $B \to D^*\ell\nu$, *Phys. Lett.* **B769** (2017) 441–445, [arXiv:1703.06124].

[58] D. Bigi, P. Gambino, and S. Schacht, $R(D^*)$, $|V_{cb}|$, and the heavy quark symmetry relations between form factors, *JHEP* **11** (2017) 061, [arXiv:1707.09509].

[59] F. De Fazio, Theory overview of tree-level B decays, in *2017 European Physical Society Conference on High Energy Physics (EPS-HEP 2017) Venice, Italy, July 5–12, 2017*, 2017. arXiv:1710.10017.

[60] M. Jung and D. M. Straub, Constraining new physics in $b \to c\ell\nu$ transitions, *JHEP* **01** (2019) 009, [arXiv:1801.01112].

[61] Fermilab Lattice, MILC Collaboration, J. A. Bailey et al., Update of $|V_{cb}|$ from the $\bar{B} \to D^*\ell\bar{\nu}$ form factor at zero recoil with three-flavor lattice QCD, *Phys. Rev.* **D89** (2014) 114504, [arXiv:1403.0635].

[62] Fermilab Lattice, MILC Collaboration, D. Du et al., $B \to \pi\ell\nu$ semileptonic form factors from unquenched lattice QCD and determination of $|V_{ub}|$, *PoS* **LATTICE2014** (2014) 385, [arXiv:1411.6038].

[63] J. Bailey, A. Bazavov, C. Bernard, C. Bouchard, C. DeTar, et al., $B \to \pi\ell\nu$ semileptonic form factors from unquenched lattice QCD and determination of $|V_{ub}|$, arXiv:1411.6038.

[64] S. Aoki, Y. Aoki, C. Bernard, T. Blum, G. Colangelo, et al., Review of lattice results concerning low-energy particle physics, *Eur. Phys. J.* **C74** (2014), no. 9 2890, [arXiv:1310.8555].

[65] A. Alberti, P. Gambino, K. J. Healey, and S. Nandi, Precision determination of the Cabibbo-Kobayashi-Maskawa element V_{cb}, *Phys. Rev. Lett.* **114** (2015) 061802, [arXiv:1411.6560].

[66] I. Caprini, L. Lellouch, and M. Neubert, Dispersive bounds on the shape of $\bar{B} \to D^{(*)}\ell\bar{\nu}$ form-factors, *Nucl. Phys.* **B530** (1998) 153–181, [hep-ph/9712417].

[67] C. G. Boyd, B. Grinstein, and R. F. Lebed, Constraints on form-factors for exclusive semileptonic heavy to light meson decays, *Phys. Rev. Lett.* **74** (1995) 4603–4606, [hep-ph/9412324].

[68] C. Bouchard, L. Cao, and P. Owen, Summary of the 2018 CKM working group on semileptonic and leptonic b-hadron decays, 2019. arXiv:1902.09412.

[69] P. Gambino, M. Jung, and S. Schacht, The V_{cb} puzzle: An update, arXiv:1905.08209.

[70] T. Muta, *Foundations of Quantum Chromodynamics: An Introduction to Perturbative Methods in Gauge Theories (3rd ed.)*, vol. 78 of *World Scientific Lecture Notes in Physics*. World Scientific, Hackensack, NJ, 2010.

[71] A. J. Buras and P. H. Weisz, QCD nonleading corrections to weak decays in dimensional regularization and 't Hooft-Veltman schemes, *Nucl. Phys.* **B333** (1990) 66–99.

[72] P. Breitenlohner and D. Maison, Dimensionally renormalized Green's functions for theories with massless particles. 1, *Commun. Math. Phys.* **52** (1977) 39.

[73] P. Breitenlohner and D. Maison, Dimensionally renormalized Green's functions for theories with massless particles. 2, *Commun. Math. Phys.* **52** (1977) 55.

[74] P. Breitenlohner and D. Maison, Dimensional renormalization and the action principle, *Commun. Math. Phys.* **52** (1977) 11–38.

[75] G. Bonneau, Consistency in dimensional regularization with γ_5, *Phys. Lett.* **B96** (1980) 147–150.

[76] G. Bonneau, Preserving canonical ward identities in dimensional regularization with a nonanticommuting γ_5, *Nucl. Phys.* **B177** (1981) 523–527.

[77] A. J. Buras, Climbing NLO and NNLO summits of weak decays, arXiv:1102.5650.

[78] W. Siegel, Supersymmetric dimensional regularization via dimensional reduction, *Phys. Lett.* **B84** (1979) 193–196.

[79] G. Altarelli, G. Curci, G. Martinelli, and S. Petrarca, QCD nonleading corrections to weak decays as an application of regularization by dimensional reduction, *Nucl. Phys.* **B187** (1981) 461–513.

[80] H. Nicolai and P. K. Townsend, Anomalies and supersymmetric regularization by dimensional reduction, *Phys. Lett.* **B93** (1980) 111–115.

[81] P. Majumdar, E. C. Poggio, and H. J. Schnitzer, The supersymmetry ward identity for the supersymmetric nonabelian gauge theory, *Phys. Rev.* **D21** (1980) 2203.

[82] R. Grigjanis, P. J. O'Donnell, M. Sutherland, and H. Navelet, QCD corrected effective Lagrangian for $B \to s$ processes, *Phys. Lett.* **B213** (1988) 355. [Erratum: Phys. Lett.B286,413(1992)].

[83] M. Misiak, On the dimensional methods in rare b decays, *Phys. Lett.* **B321** (1994) 113–120, [hep-ph/9309236].

[84] G. 't Hooft and M. J. G. Veltman, Regularization and renormalization of gauge fields, *Nucl. Phys.* **B44** (1972) 189–213.

[85] D. A. Akyeampong and R. Delbourgo, Anomalies via dimensional regularization, *Nuovo Cim.* **A19** (1974) 219–224.

[86] D. A. Akyeampong and R. Delbourgo, Dimensional regularization and PCAC, *Nuovo Cim.* **A18** (1973) 94–104.

[87] D. A. Akyeampong and R. Delbourgo, Dimensional regularization, abnormal amplitudes and anomalies, *Nuovo Cim.* **A17** (1973) 578–586.

[88] J. G. Körner, N. Nasrallah, and K. Schilcher, Evaluation of the flavor changing vertex $b \to sH$ using the Breitenlohner-Maison-'t Hooft-Veltman γ_5 scheme, *Phys. Rev.* **D41** (1990) 888.

[89] M. Jamin and M. E. Lautenbacher, TRACER: Version 1.1: A Mathematica package for gamma algebra in arbitrary dimensions, *Comput. Phys. Commun.* **74** (1993) 265–288.

[90] J. C. Collins, *Renormalization*, vol. 26 of *Cambridge Monographs on Mathematical Physics*. Cambridge University Press, Cambridge, UK, 1986.

[91] S. Weinberg, Phenomenological Lagrangians, *Physica* **A96** (1979) 327–340.

[92] G. 't Hooft, Dimensional regularization and the renormalization group, *Nucl. Phys.* **B61** (1973) 455–468.

[93] S. Weinberg, New approach to the renormalization group, *Phys. Rev.* **D8** (1973) 3497–3509.

[94] W. A. Bardeen, A. J. Buras, D. W. Duke, and T. Muta, Deep inelastic scattering beyond the leading order in asymptotically free gauge theories, *Phys. Rev.* **D18** (1978) 3998.

[95] D. R. T. Jones, Two loop diagrams in Yang-Mills theory, *Nucl. Phys.* **B75** (1974) 531.

[96] R. Tarrach, The pole mass in perturbative QCD, *Nucl. Phys.* **B183** (1981) 384–396.

[97] O. V. Tarasov, A. A. Vladimirov, and A. Yu. Zharkov, The Gell-Mann-Low function of QCD in the three loop approximation, *Phys. Lett.* **B93** (1980) 429–432.

[98] S. A. Larin and J. A. M. Vermaseren, The three loop QCD Beta function and anomalous dimensions, *Phys. Lett.* **B303** (1993) 334–336, [hep-ph/9302208].

[99] T. van Ritbergen, J. A. M. Vermaseren, and S. A. Larin, The four loop beta function in quantum chromodynamics, *Phys. Lett.* **B400** (1997) 379–384, [hep-ph/9701390].

[100] M. Czakon, The four-loop QCD beta-function and anomalous dimensions, *Nucl. Phys.* **B710** (2005) 485–498, [hep-ph/0411261].

[101] P. A. Baikov, K. G. Chetyrkin, and J. H. Kühn, Five-loop running of the QCD coupling constant, *Phys. Rev. Lett.* **118** (2017), no. 8 082002, [arXiv:1606.08659].

[102] F. Herzog, B. Ruijl, T. Ueda, J. A. M. Vermaseren, and A. Vogt, The five-loop beta function of Yang-Mills theory with fermions, *JHEP* **02** (2017) 090, [arXiv:1701.01404].

[103] S. A. Larin, The renormalization of the axial anomaly in dimensional regularization, *Phys. Lett.* **B303** (1993) 113–118, [hep-ph/9302240].

[104] K. G. Chetyrkin, Quark mass anomalous dimension to $O(\alpha_s^4)$, *Phys. Lett.* **B404** (1997) 161–165, [hep-ph/9703278].

[105] J. A. M. Vermaseren, S. A. Larin, and T. van Ritbergen, The four loop quark mass anomalous dimension and the invariant quark mass, *Phys. Lett.* **B405** (1997) 327–333, [hep-ph/9703284].

[106] P. A. Baikov, K. G. Chetyrkin, and J. H. Kühn, Quark mass and field anomalous dimensions to $O(\alpha_s^5)$, *JHEP* **10** (2014) 076, [arXiv:1402.6611].

[107] K. G. Wilson, Nonlagrangian models of current algebra, *Phys. Rev.* **179** (1969) 1499–1512.

[108] K. G. Wilson and W. Zimmermann, Operator product expansions and composite field operators in the general framework of quantum field theory, *Commun. Math. Phys.* **24** (1972) 87–106.

[109] W. Zimmermann, Normal products and the short distance expansion in the perturbation theory of renormalizable interactions, *Annals Phys.* **77** (1973) 570–601. [Lect. Notes Phys. 558, 278(2000)].

[110] E. Witten, Short distance analysis of weak interactions, *Nucl. Phys.* **B122** (1977) 109–143.

[111] G. Buchalla, A. J. Buras, and M. E. Lautenbacher, Weak decays beyond leading logarithms, *Rev. Mod. Phys.* **68** (1996) 1125–1144, [hep-ph/9512380].

[112] A. A. Petrov and A. E. Blechman, *Effective Field Theories*. WSP, 2016.

[113] T. Appelquist and J. Carazzone, Infrared singularities and massive fields, *Phys. Rev.* **D11** (1975) 2856.

[114] K. G. Wilson, The renormalization group and critical phenomena, *Rev. Mod. Phys.* **55** (1983) 583–600.

[115] W. J. Marciano, Dimensional Regularization and Mass Singularities, *Phys. Rev.* **D12** (1975) 3861.

[116] G. Buchalla and A. J. Buras, QCD Corrections to the $\bar{s}dZ$ vertex for arbitrary top quark mass, *Nucl. Phys.* **B398** (1993) 285–300.

[117] C. Greub and T. Hurth, Two loop matching of the dipole operators for $b \to s\gamma$ and $b \to sg$, *Phys. Rev.* **D56** (1997) 2934–2949, [hep-ph/9703349].

[118] A. J. Buras, A. Kwiatkowski, and N. Pott, Next-to-leading order matching for the magnetic photon penguin operator in the $B \to X_s\gamma$ decay, *Nucl. Phys.* **B517** (1998) 353–373, [hep-ph/9710336].

[119] G. Altarelli and L. Maiani, Octet enhancement of nonleptonic weak interactions in asymptotically free gauge theories, *Phys. Lett.* **B52** (1974) 351–354.

[120] M. Gaillard and B. W. Lee, $\Delta I = 1/2$ rule for nonleptonic decays in asymptotically free field theories, *Phys. Rev. Lett.* **33** (1974) 108.

[121] A. J. Buras, Asymptotic freedom in deep inelastic processes in the leading order and beyond, *Rev. Mod. Phys.* **52** (1980) 199.

[122] A. J. Buras, M. Jamin, M. E. Lautenbacher, and P. H. Weisz, Effective Hamiltonians for $\Delta S = 1$ and $\Delta B = 1$ nonleptonic decays beyond the leading logarithmic approximation, *Nucl. Phys.* **B370** (1992) 69–104. [Addendum: Nucl. Phys. B375, 501(1992)].

[123] A. J. Buras, M. Jamin, and M. E. Lautenbacher, The anatomy of ε'/ε beyond leading logarithms with improved hadronic matrix elements, *Nucl. Phys.* **B408** (1993) 209–285, [hep-ph/9303284].

[124] M. Ciuchini, E. Franco, G. Martinelli, and L. Reina, ε'/ε at the next-to-leading order in QCD and QED, *Phys. Lett.* **B301** (1993) 263–271, [hep-ph/9212203].

[125] M. Ciuchini, E. Franco, G. Martinelli, and L. Reina, The $\Delta S = 1$ effective Hamiltonian including next-to-leading order QCD and QED corrections, *Nucl. Phys.* **B415** (1994) 403–462, [hep-ph/9304257].

[126] N. Tracas and N. Vlachos, Two loop calculations in QCD and the $\Delta I = 1/2$ rule in nonleptonic weak decays, *Phys. Lett.* **B115** (1982) 419.

[127] A. J. Buras and M. Münz, Effective Hamiltonian for $B \to X_s e^+ e^-$ beyond leading logarithms in the NDR and HV schemes, *Phys. Rev.* **D52** (1995) 186–195, [hep-ph/9501281].

[128] A. J. Buras, A. Kwiatkowski, and N. Pott, On the scale uncertainties in the $B \to X_s \gamma$ decay, *Phys. Lett.* **B414** (1997) 157–165, [hep-ph/9707482]. [Erratum: Phys. Lett.B434,459(1998)].

[129] M. J. Dugan and B. Grinstein, On the vanishing of evanescent operators, *Phys. Lett.* **B256** (1991) 239–244.

[130] S. Herrlich and U. Nierste, Evanescent operators, scheme dependences and double insertions, *Nucl. Phys.* **B455** (1995) 39–58, [hep-ph/9412375].

[131] A. J. Buras, M. E. Lautenbacher, M. Misiak, and M. Münz, Direct CP violation in $K_L \to \pi^0 e^+ e^-$ beyond leading logarithms, *Nucl. Phys.* **B423** (1994) 349–383, [hep-ph/9402347].

[132] T. Inami and C. Lim, Effects of superheavy quarks and leptons in low-energy weak processes $K_L \to \mu^+ \mu^-, K^+ \to \pi^+ \nu \bar{\nu}$ and $K^0 - \bar{K}^0$, *Prog. Theor. Phys.* **65** (1981) 297.

[133] G. Buchalla, A. J. Buras, and M. K. Harlander, Penguin box expansion: Flavor changing neutral current processes and a heavy top quark, *Nucl. Phys.* **B349** (1991) 1–47.

[134] G. Buchalla and A. J. Buras, Two-loop large-m_t electroweak corrections to $K \to \pi \nu \bar{\nu}$ for arbitrary Higgs boson mass, *Phys. Rev.* **D57** (1998) 216–223, [hep-ph/9707243].

[135] J. Brod, M. Gorbahn, and E. Stamou, Two-loop electroweak corrections for the $K \to \pi \nu \bar{\nu}$ decays, *Phys. Rev.* **D83** (2011) 034030, [arXiv:1009.0947].

[136] J. Brod and M. Gorbahn, The Z penguin in generic extensions of the standard model, arXiv:1903.05116.

[137] A. J. Buras, M. Jamin, and P. H. Weisz, Leading and next-to-leading QCD corrections to ε parameter and $B^0 - \bar{B}^0$ mixing in the presence of a heavy top quark, *Nucl. Phys.* **B347** (1990) 491–536.

[138] A. Lenz, U. Nierste, J. Charles, S. Descotes-Genon, A. Jantsch, et al., Anatomy of new physics in $B - \bar{B}$ mixing, *Phys. Rev.* **D83** (2011) 036004, [arXiv:1008.1593].

[139] S. Herrlich and U. Nierste, Enhancement of the $K_L - K_S$ mass difference by short distance QCD corrections beyond leading logarithms, *Nucl. Phys.* **B419** (1994) 292–322, [hep-ph/9310311].

[140] S. Herrlich and U. Nierste, Indirect CP violation in the neutral kaon system beyond leading logarithms, *Phys. Rev.* **D52** (1995) 6505–6518, [hep-ph/9507262].

[141] S. Herrlich and U. Nierste, The Complete $|\Delta S| = 2$ Hamiltonian in the next-to-leading order, *Nucl. Phys.* **B476** (1996) 27–88, [hep-ph/9604330].

[142] J. Urban, F. Krauss, U. Jentschura, and G. Soff, Next-to-leading order QCD corrections for the $B^0 - \bar{B}^0$ mixing with an extended Higgs sector, *Nucl. Phys.* **B523** (1998) 40–58, [hep-ph/9710245].

[143] J. Brod and M. Gorbahn, ϵ_K at Next-to-next-to-leading order: The charm-top-quark contribution, *Phys. Rev.* **D82** (2010) 094026, [arXiv:1007.0684].

[144] J. Brod and M. Gorbahn, Next-to-next-to-leading-order charm-quark contribution to the CP violation parameter ε_K and ΔM_K, *Phys. Rev. Lett.* **108** (2012) 121801, [arXiv:1108.2036].

[145] M. A. Shifman, A. I. Vainshtein, and V. I. Zakharov, Nonleptonic decays of K mesons and hyperons, *Sov. Phys. JETP* **45** (1977) 670. [Zh. Eksp. Teor. Fiz.72,1275(1977)].

[146] A. J. Buras, M. Jamin, M. E. Lautenbacher, and P. H. Weisz, Two loop anomalous dimension matrix for $\Delta S = 1$ weak nonleptonic decays. 1. $O(\alpha_s^2)$, *Nucl. Phys.* **B400** (1993) 37–74, [hep-ph/9211304].

[147] K. G. Chetyrkin, M. Misiak, and M. Münz, Weak radiative B meson decay beyond leading logarithms, *Phys. Lett.* **B400** (1997) 206–219, [hep-ph/9612313]. [Erratum: Phys. Lett.B425,414(1998)].

[148] R. Fleischer, CP violation and the role of electroweak penguins in nonleptonic *B* decays, *Int. J. Mod. Phys.* **A12** (1997) 2459–2522, [hep-ph/9612446].

[149] M. Misiak et al., Updated NNLO QCD predictions for the weak radiative B-meson decays, *Phys. Rev. Lett.* **114** (2015), no. 22 221801, [arXiv:1503.01789].

[150] HFLAV Collaboration, Y. Amhis et al., Averages of *b*-hadron, *c*-hadron, and τ-lepton properties as of summer 2016, *Eur. Phys. J.* **C77** (2017), no. 12 895, [arXiv:1612.07233].

[151] S. Bertolini, F. Borzumati, and A. Masiero, QCD enhancement of radiative b decays, *Phys. Rev. Lett.* **59** (1987) 180.

[152] N. G. Deshpande, P. Lo, J. Trampetic, G. Eilam, and P. Singer, $B \to K^*\gamma$ and the top quark mass, *Phys. Rev. Lett.* **59** (1987) 183–185.

[153] M. Ciuchini, E. Franco, G. Martinelli, L. Reina, and L. Silvestrini, Scheme independence of the effective Hamiltonian for $b \to s\gamma$ and $b \to sg$ decays, *Phys. Lett.* **B316** (1993) 127–136, [hep-ph/9307364].

[154] M. Ciuchini, E. Franco, L. Reina, and L. Silvestrini, Leading order QCD corrections to $b \to s\gamma$ and $b \to sg$ decays in three regularization schemes, *Nucl. Phys.* **B421** (1994) 41–64, [hep-ph/9311357].

[155] G. Cella, G. Curci, G. Ricciardi, and A. Vicere, The $b \to s\gamma$ decay revisited, *Phys. Lett.* **B325** (1994) 227–234, [hep-ph/9401254].

[156] G. Cella, G. Curci, G. Ricciardi, and A. Vicere, QCD corrections to electroweak processes in an unconventional scheme: Application to the $b \to s\gamma$ decay, *Nucl. Phys.* **B431** (1994) 417–452, [hep-ph/9406203].

[157] M. Misiak, The $b \to se^+e^-$ and $b \to s\gamma$ decays with next-to-leading logarithmic QCD corrections, *Nucl. Phys.* **B393** (1993) 23–45. [Erratum: Nucl. Phys.B439,461(1995)].

[158] A. Ali and C. Greub, A determination of the CKM matrix element ratio $|V_{ts}|/|V_{cb}|$ from the rare B decays $B \to K^*\gamma$ and $B \to X_s\gamma$, *Z. Phys.* **C60** (1993) 433–442.

[159] A. J. Buras, M. Misiak, M. Münz, and S. Pokorski, Theoretical uncertainties and phenomenological aspects of $B \to X_s\gamma$ decay, *Nucl. Phys.* **B424** (1994) 374–398, [hep-ph/9311345].

[160] A. J. Buras and M. Misiak, $\bar{B} \to X_s\gamma$ after completion of the NLO QCD calculations, *Acta Phys. Polon.* **B33** (2002) 2597–2612, [hep-ph/0207131].

[161] K. Adel and Y.-P. Yao, Exact α_s calculation of $b \to s\gamma$ and $b \to sg$, *Phys. Rev.* **D49** (1994) 4945–4948, [hep-ph/9308349].

[162] M. Ciuchini, G. Degrassi, P. Gambino, and G. F. Giudice, Next-to-leading QCD corrections to $B \to X_s\gamma$: Standard model and two Higgs doublet model, *Nucl. Phys.* **B527** (1998) 21–43, [hep-ph/9710335].

[163] M. Misiak and M. Münz, Two loop mixing of dimension five flavor changing operators, *Phys. Lett.* **B344** (1995) 308–318, [hep-ph/9409454].

[164] P. Gambino, M. Gorbahn, and U. Haisch, Anomalous dimension matrix for radiative and rare semileptonic B decays up to three loops, *Nucl. Phys.* **B673** (2003) 238–262, [hep-ph/0306079].

[165] K. G. Chetyrkin, M. Misiak, and M. Münz, Beta functions and anomalous dimensions up to three loops, *Nucl. Phys.* **B518** (1998) 473–494, [hep-ph/9711266].

[166] A. Ali and C. Greub, Photon energy spectrum in $B \to X_s\gamma$ and comparison with data, *Phys. Lett.* **B361** (1995) 146–154, [hep-ph/9506374].

[167] N. Pott, Bremsstrahlung corrections to the decay $b \to s\gamma$, *Phys. Rev.* **D54** (1996) 938–948, [hep-ph/9512252].

[168] C. Greub, T. Hurth, and D. Wyler, Virtual Corrections to the decay $b \to s\gamma$, *Phys. Lett.* **B380** (1996) 385–392, [hep-ph/9602281].

[169] C. Greub, T. Hurth, and D. Wyler, Virtual $O(\alpha_s)$ Corrections to the inclusive decay $b \to s\gamma$, *Phys. Rev.* **D54** (1996) 3350–3364, [hep-ph/9603404].

[170] A. J. Buras, A. Czarnecki, M. Misiak, and J. Urban, Two-loop matrix element of the current-current operator in the decay $\bar{B} \to X_s\gamma$, *Nucl. Phys.* **B611** (2001) 488–502, [hep-ph/0105160].

[171] A. J. Buras, A. Czarnecki, M. Misiak, and J. Urban, Completing the NLO QCD Calculation of $\bar{B} \to X_s\gamma$, *Nucl. Phys.* **B631** (2002) 219–238, [hep-ph/0203135].

[172] P. Gambino and M. Misiak, Quark mass effects in $\bar{B} \to X_s\gamma$, *Nucl. Phys.* **B611** (2001) 338–366, [hep-ph/0104034].

[173] M. Misiak and M. Steinhauser, NNLO QCD corrections to the $\bar{B} \to X_s\gamma$ matrix elements using interpolation in m_c, *Nucl. Phys.* **B764** (2007) 62–82, [hep-ph/0609241].

[174] B. Grinstein, R. P. Springer, and M. B. Wise, Strong interaction effects in weak radiative $\bar{B}$ meson decay, *Nucl. Phys.* **B339** (1990) 269–309.

[175] N. Cabibbo and L. Maiani, The lifetime of charmed particles, *Phys. Lett.* **B79** (1978) 109–111.

[176] C. S. Kim and A. D. Martin, On the determination of V_{ub} and V_{cb} from semileptonic B decays, *Phys. Lett.* **B225** (1989) 186–190.

[177] Y. Nir, The mass ratio m_c/m_b in semileptonic B decays, *Phys. Lett.* **B221** (1989) 184–190.

[178] A. L. Kagan and M. Neubert, QCD anatomy of $\bar{B} \to X_s\gamma$ decays, *Eur. Phys. J.* **C7** (1999) 5–27, [hep-ph/9805303].

[179] F. Borzumati and C. Greub, 2HDMs predictions for $\bar{B} \to X_s\gamma$ in NLO QCD, *Phys. Rev.* **D58** (1998) 074004, [hep-ph/9802391].

[180] A. Czarnecki and W. J. Marciano, Electroweak radiative corrections to $b \to s\gamma$, *Phys. Rev. Lett.* **81** (1998) 277–280, [hep-ph/9804252].

[181] G. Buchalla and A. J. Buras, The rare decays $K \to \pi\nu\bar{\nu}$, $B \to X\nu\bar{\nu}$ and $B \to \ell^+\ell^-$: An update, *Nucl. Phys.* **B548** (1999) 309–327, [hep-ph/9901288].

[182] M. Misiak and J. Urban, QCD corrections to FCNC decays mediated by Z penguins and W boxes, *Phys. Lett.* **B451** (1999) 161–169, [hep-ph/9901278].

[183] C. Bobeth, P. Gambino, M. Gorbahn, and U. Haisch, Complete NNLO QCD analysis of $\bar{B} \to X_s\ell^+\ell^-$ and higher order electroweak effects, *JHEP* **04** (2004) 071, [hep-ph/0312090].

[184] T. Huber, E. Lunghi, M. Misiak, and D. Wyler, Electromagnetic logarithms in $\bar{B} \to X_s l^+ l^-$, *Nucl. Phys.* **B740** (2006) 105–137, [hep-ph/0512066].

[185] M. Misiak, Rare B-meson decays, in *Proceedings, 15th Lomonosov Conference on Elementary Particle Physics (LomCon): Particle Physics at the Tercentenary of Mikhail Lomonosov*, pp. 301–305, 2013. arXiv:1112.5978.

[186] A. J. Buras, J. Girrbach, D. Guadagnoli, and G. Isidori, On the standard model prediction for $\mathcal{B}(B_{s,d} \to \mu^+\mu^-)$, *Eur. Phys. J.* **C72** (2012) 2172, [arXiv:1208.0934].

[187] C. Bobeth, M. Gorbahn, and E. Stamou, Electroweak corrections to $B_{s,d} \to \ell^+\ell^-$, *Phys. Rev.* **D89** (2014) 034023, [arXiv:1311.1348].

[188] T. Hermann, M. Misiak, and M. Steinhauser, Three-loop QCD corrections to $B_s \to \mu^+\mu^-$, *JHEP* **1312** (2013) 097, [arXiv:1311.1347].

[189] C. Bobeth, M. Gorbahn, T. Hermann, M. Misiak, E. Stamou, et al., $B_{s,d} \to \ell^+\ell^-$ in the standard model with reduced theoretical uncertainty, *Phys. Rev. Lett.* **112** (2014) 101801, [arXiv:1311.0903].

[190] A. J. Buras, F. De Fazio, and J. Girrbach, 331 models facing new $b \to s\mu^+\mu^-$ data, *JHEP* **1402** (2014) 112, [arXiv:1311.6729].

[191] C. Dib, I. Dunietz, and F. J. Gilman, CP violation in the $K_L \to \pi^0\ell^+\ell^-$ decay amplitude for large m_t, *Phys. Lett.* **B218** (1989) 487–492.

[192] J. Flynn and L. Randall, The CP violating contribution to the decay $K_L \to \pi^0 e^+ e^-$, *Nucl. Phys.* **B326** (1989) 31. [Erratum: Nucl. Phys.B334,580(1990)].

[193] G. Buchalla and A. J. Buras, $\sin 2\beta$ from $K \to \pi\nu\bar{\nu}$, *Phys. Lett.* **B333** (1994) 221–227, [hep-ph/9405259].

[194] J. R. Ellis and J. S. Hagelin, Constraints on light particles from kaon decays, *Nucl. Phys.* **B217** (1983) 189–214.

[195] C. Dib, I. Dunietz, and F. J. Gilman, Strong interaction corrections to the decay $K \to \pi$ neutrino antineutrino for large m_t, *Mod. Phys. Lett.* **A6** (1991) 3573–3582.

[196] G. Buchalla and A. J. Buras, QCD corrections to rare K and B decays for arbitrary top quark mass, *Nucl. Phys.* **B400** (1993) 225–239.

[197] G. Buchalla and A. J. Buras, The rare decays $K^+ \to \pi^+\nu\bar{\nu}$ and $K_L \to \mu^+\mu^-$ beyond leading logarithms, *Nucl. Phys.* **B412** (1994) 106–142, [hep-ph/9308272].

[198] A. J. Buras, M. Gorbahn, U. Haisch, and U. Nierste, The rare decay $K^+ \to \pi^+\nu\bar{\nu}$ at the next-to-next-to-leading order in QCD, *Phys. Rev. Lett.* **95** (2005) 261805, [hep-ph/0508165].

[199] A. J. Buras, M. Gorbahn, U. Haisch, and U. Nierste, Charm quark contribution to $K^+ \to \pi^+\nu\bar{\nu}$ at next-to-next-to-leading order, *JHEP* **11** (2006) 002, [hep-ph/0603079].

[200] M. Gorbahn and U. Haisch, Effective Hamiltonian for non-leptonic $|\Delta F| = 1$ decays at NNLO in QCD, *Nucl. Phys.* **B713** (2005) 291–332, [hep-ph/0411071].

[201] G. Isidori, F. Mescia, and C. Smith, Light-quark loops in $K \to \pi\nu\bar{\nu}$, *Nucl. Phys.* **B718** (2005) 319–338, [hep-ph/0503107].

[202] F. Mescia and C. Smith, Improved estimates of rare K decay matrix-elements from $K_{\ell 3}$ decays, *Phys. Rev.* **D76** (2007) 034017, [arXiv:0705.2025].

[203] J. Brod and M. Gorbahn, Electroweak corrections to the charm quark contribution to $K^+ \to \pi^+ \nu \bar{\nu}$, *Phys. Rev.* **D78** (2008) 034006, [arXiv:0805.4119].

[204] A. Ceccucci, Review and outlook on kaon physics, *Acta Phys. Polon.* **B49** (2018) 1079–1086.

[205] T. Komatsubara, Experiments with K-meson decays, *Prog. Part. Nucl. Phys.* **67** (2012) 995–1018, [arXiv:1203.6437].

[206] KOTO Collaboration, K. Shiomi, $K_L^0 \to \pi^0 \nu \bar{\nu}$ at KOTO, in *8th International Workshop on the CKM Unitarity Triangle (CKM 2014) Vienna, Austria, September 8–12, 2014*, 2014. arXiv:1411.4250.

[207] A. J. Buras, F. Schwab, and S. Uhlig, Waiting for precise measurements of $K^+ \to \pi^+ \nu \bar{\nu}$ and $K_L \to \pi^0 \nu \bar{\nu}$, *Rev. Mod. Phys.* **80** (2008) 965–1007, [hep-ph/0405132].

[208] A. J. Buras and J. Girrbach, Towards the identification of new physics through quark flavour violating processes, *Rept. Prog. Phys.* **77** (2014) 086201, [arXiv:1306.3775].

[209] M. Blanke, New physics signatures in kaon decays, *PoS* **KAON13** (2013) 010, [arXiv:1305.5671].

[210] C. Smith, Rare K decays: Challenges and perspectives, arXiv:1409.6162.

[211] A. J. Buras, D. Buttazzo, J. Girrbach-Noe, and R. Knegjens, Can we reach the Zeptouniverse with rare K and $B_{s,d}$ decays? *JHEP* **1411** (2014) 121, [arXiv:1408.0728].

[212] A. J. Buras, D. Buttazzo, J. Girrbach-Noe, and R. Knegjens, $K^+ \to \pi^+ \nu \bar{\nu}$ and $K_L \to \pi^0 \nu \bar{\nu}$ in the standard model: Status and perspectives, *JHEP* **11** (2015) 033, [arXiv:1503.02693].

[213] G. Isidori, G. Martinelli, and P. Turchetti, Rare kaon decays on the lattice, *Phys. Lett.* **B633** (2006) 75–83, [hep-lat/0506026].

[214] RBC, UKQCD Collaboration, N. H. Christ, X. Feng, A. Portelli, and C. T. Sachrajda, Prospects for a lattice computation of rare kaon decay amplitudes II $K \to \pi \nu \bar{\nu}$ decays, *Phys. Rev.* **D93** (2016), no. 11 114517, [arXiv:1605.04442].

[215] K. Chetyrkin, J. Kühn, A. Maier, P. Maierhofer, P. Marquard, et al., Charm and bottom quark masses: An update, *Phys. Rev.* **D80** (2009) 074010, [arXiv:0907.2110].

[216] A. J. Buras and R. Fleischer, Bounds on the unitarity triangle, $\sin 2\beta$ and $K \to \pi \nu \bar{\nu}$ decays in models with minimal flavor violation, *Phys. Rev.* **D64** (2001) 115010, [hep-ph/0104238].

[217] G. 't Hooft, A planar diagram theory for strong interactions, *Nucl. Phys.* **B72** (1974) 461.

[218] G. 't Hooft, A two-dimensional model for mesons, *Nucl. Phys.* **B75** (1974) 461.

[219] E. Witten, Baryons in the $1/N$ expansion, *Nucl. Phys.* **B160** (1979) 57.

[220] S. B. Treiman, E. Witten, R. Jackiw, and B. Zumino, *Current algebra and anomalies.* 1986.

[221] W. A. Bardeen, A. J. Buras, and J.-M. Gérard, A consistent analysis of the $\Delta I = 1/2$ rule for K decays, *Phys. Lett.* **B192** (1987) 138.

[222] M. Fukugita, T. Inami, N. Sakai, and S. Yazaki, Nonleptonic decays of kaons in the $1/N$ expansion, *Phys. Lett.* **B72** (1977) 237.

[223] H. P. Nilles and V. Visnjic-Triantafillou, Nonleptonic weak decays in QCD in two-dimensions, *Phys. Rev.* **D19** (1979) 969.

[224] D. Tadic and J. Trampetic, Weak meson decays and the 1/N expansion, *Phys. Lett.* **B114** (1982) 179.

[225] A. J. Buras, J.-M. Gérard, and R. Rückl, 1/N expansion for exclusive and inclusive charm decays, *Nucl. Phys.* **B268** (1986) 16.

[226] M. Wirbel, B. Stech, and M. Bauer, Exclusive semileptonic decays of heavy mesons, *Z. Phys.* **C29** (1985) 637.

[227] W. A. Bardeen, A. J. Buras, and J.-M. Gérard, The $\Delta I = 1/2$ rule in the large N limit, *Phys. Lett.* **B180** (1986) 133.

[228] W. A. Bardeen, A. J. Buras, and J.-M. Gérard, The $K \to \pi \pi$ decays in the large N limit: Quark evolution, *Nucl. Phys.* **B293** (1987) 787.

[229] W. A. Bardeen, A. J. Buras, and J.-M. Gérard, The B parameter beyond the leading order of 1/N expansion, *Phys. Lett.* **B211** (1988) 343.

[230] A. J. Buras and J.-M. Gérard, 1/N expansion for kaons, *Nucl. Phys.* **B264** (1986) 371.

[231] B. D. Gaiser, T. Tsao, and M. B. Wise, Parameters of the six quark model, *Annals Phys.* **132** (1981) 66.

[232] A. J. Buras, The 1/N approach to nonleptonic weak interactions, *Adv. Ser. Direct. High Energy Phys.* **3** (1989) 575–645.

[233] A. J. Buras, Phenomenological applications of the 1/N expansion, *Nucl. Phys. Proc. Suppl.* **10A** (1989) 199–267.

[234] W. A. Bardeen, Weak decay amplitudes in large N QCD, *Nucl. Phys. Proc. Suppl.* **7A** (1989) 149.

[235] A. J. Buras, Strangeness and the large N expansion, *Nucl. Phys.* **A479** (1988) 399C–421C.

[236] J.-M. Gérard, Electroweak interactions of hadrons, *Acta Phys. Polon.* **B21** (1990) 257–305.

[237] W. A. Bardeen, Weak matrix elements in the large N_c limit, *Proceedings* **KAON99** (1999) 171–176.

[238] W. A. Bardeen, On the large N_c expansion in quantum chromodynamics, *Fortsch. Phys.* **50** (2002) 483–488, [hep-ph/0112229].

[239] A. J. Buras, J.-M. Gérard, and W. A. Bardeen, Large N approach to kaon decays and mixing 28 years later: $\Delta I = 1/2$ rule, $\hat{B}_K$ and ΔM_K, *Eur. Phys. J.* **C74** (2014), no. 5 2871, [arXiv:1401.1385].

[240] A. J. Buras, $\Delta I = 1/2$ rule and $\hat{B}_K$: 2014, in *Proceedings, 7th International Workshop on Quantum Chromodynamics Theory and Experiment (QCD@Work 2014): Giovinazzo, Bari, Italy, June 16–19, 2014*, 2014. arXiv:1408.4820.

[241] M. A. Shifman, A. Vainshtein, and V. I. Zakharov, Light quarks and the origin of the $\Delta I = 1/2$ rule in the nonleptonic decays of strange particles, *Nucl. Phys.* **B120** (1977) 316.

[242] S. Fajfer and J.-M. Gérard, A simple chiral Lagrangian approach to $K \rightarrow \pi\pi\pi$ decays and $\varepsilon'+0$-, *Z. Phys.* **C42** (1989) 425.

[243] J. Bijnens, J.-M. Gérard, and G. Klein, The $K_L - K_S$ mass difference, *Phys. Lett.* **B257** (1991) 191–195.

[244] A. J. Buras and J.-M. Gérard, Upper bounds on ε'/ε parameters $B_6^{(1/2)}$ and $B_8^{(3/2)}$ from large N QCD and other news, *JHEP* **12** (2015) 008, [arXiv:1507.06326].

[245] A. J. Buras and J.-M. Gérard, Final state interactions in $K \rightarrow \pi\pi$ decays: $\Delta I = 1/2$ rule vs. ε'/ε, *Eur. Phys. J.* **C77** (2017), no. 1 10, [arXiv:1603.05686].

[246] J. Aebischer, A. J. Buras, and J.-M. Gérard, BSM Hadronic matrix elements for ε'/ε and $K \rightarrow \pi\pi$ decays in the dual QCD approach, *JHEP* **02** (2019) 021, [arXiv:1807.01709].

[247] H. Gisbert and A. Pich, Direct CP violation in $K^0 \rightarrow \pi\pi$: Standard model status, *Rept. Prog. Phys.* **81** (2018), no. 7 076201, [arXiv:1712.06147].

[248] J. Bijnens and B. Guberina, Chiral perturbation theory and the evaluation of $1/N_c$ corrections to non-leptonic decays, *Phys. Lett.* **B205** (1988) 103.

[249] A. Pich and E. de Rafael, Weak K amplitudes in the chiral and $1/N$ expansions, *Phys. Lett.* **B374** (1996) 186–192, [hep-ph/9511465].

[250] J. Bijnens and J. Prades, The B_K parameter in the $1/N$ expansion, *Nucl. Phys.* **B444** (1995) 523–562, [hep-ph/9502363].

[251] J. Bijnens and J. Prades, The $\Delta I = 1/2$ rule in the chiral limit, *JHEP* **9901** (1999) 023, [hep-ph/9811472].

[252] T. Hambye, G. Kohler, E. Paschos, P. Soldan, and W. A. Bardeen, $1/N$ corrections to the hadronic matrix elements of Q_6 and Q_8 in $K \rightarrow \pi\pi$ decays, *Phys. Rev.* **D58** (1998) 014017, [hep-ph/9802300].

[253] T. Hambye, G. Kohler, and P. Soldan, New analysis of the $\Delta I = 1/2$ rule in kaon decays and the $\hat{B}_K$ parameter, *Eur. Phys. J.* **C10** (1999) 271–292, [hep-ph/9902334].

[254] S. Peris and E. de Rafael, $K^0 - \bar{K}^0$ mixing in the $1/N$ expansion, *Phys. Lett.* **B490** (2000) 213–222, [hep-ph/0006146].

[255] V. Cirigliano, J. F. Donoghue, E. Golowich, and K. Maltman, Improved determination of the electroweak penguin contribution to ε'/ε in the chiral limit, *Phys. Lett.* **B555** (2003) 71–82, [hep-ph/0211420].

[256] T. Hambye, S. Peris, and E. de Rafael, $\Delta I = 1/2$ and ε'/ε in large N QCD, *JHEP* **05** (2003) 027, [hep-ph/0305104].

[257] B. Lucini and M. Panero, SU(N) gauge theories at large N, *Phys. Rept.* **526** (2013) 93–163, [arXiv:1210.4997].

[258] S. R. Coleman and E. Witten, Chiral symmetry breakdown in large N chromodynamics, *Phys. Rev. Lett.* **45** (1980) 100.

[259] E. Witten, Anti-de Sitter space, thermal phase transition, and confinement in gauge theories, *Adv. Theor. Math. Phys.* **2** (1998) 505–532, [hep-th/9803131].

[260] J. Polchinski and M. J. Strassler, Hard scattering and gauge / string duality, *Phys. Rev. Lett.* **88** (2002) 031601, [hep-th/0109174].

[261] T. Hambye, B. Hassanain, J. March-Russell, and M. Schvellinger, On the $\Delta I = 1/2$ rule in holographic QCD, *Phys. Rev.* **D74** (2006) 026003, [hep-ph/0512089].

[262] T. Hambye, B. Hassanain, J. March-Russell, and M. Schvellinger, Four-point functions and kaon decays in a minimal AdS/QCD model, *Phys. Rev.* **D76** (2007) 125017, [hep-ph/0612010].

[263] J. P. Fatelo and J. M. Gérard, Current current operator evolution in the chiral limit, *Phys. Lett.* **B347** (1995) 136–142.

[264] R. S. Chivukula, J. Flynn, and H. Georgi, Polychromatic penguins don't fly, *Phys. Lett.* **B171** (1986) 453–458.

[265] J. Gasser and H. Leutwyler, Chiral perturbation theory to one loop, *Annals Phys.* **158** (1984) 142.

[266] M. Gell-Mann and A. Pais, Behavior of neutral particles under charge conjugation, *Phys. Rev.* **97** (1955) 1387–1389.

[267] M. Gell-Mann and A. Rosenfeld, Hyperons and heavy mesons (systematics and decay), *Ann. Rev. Nucl. Part. Sci.* **7** (1957) 407–478.

[268] E. Fermi, An attempt of a theory of beta radiation. 1., *Z. Phys.* **88** (1934) 161–177.

[269] R. P. Feynman and M. Gell-Mann, Theory of Fermi interaction, *Phys. Rev.* **109** (1958) 193–198.

[270] E. C. G. Sudarshan and R. e. Marshak, Chirality invariance and the universal Fermi interaction, *Phys. Rev.* **109** (1958) 1860–1860.

[271] A. J. Buras, F. De Fazio, and J. Girrbach, $\Delta I = 1/2$ rule, ε'/ε and $K \to \pi\nu\bar{\nu}$ in $Z'(Z)$ and G' models with FCNC quark couplings, *Eur. Phys. J.* **C74** (2014) 2950, [arXiv:1404.3824].

[272] J.-M. Gérard, An upper bound on the kaon B-parameter and Re(ϵ_K), *JHEP* **1102** (2011) 075, [arXiv:1012.2026].

[273] A. J. Buras and J. M. Gérard, Isospin breaking contributions to ε'/ε, *Phys. Lett.* **B192** (1987) 156.

[274] T. Blum, P. Boyle, N. Christ, N. Garron, E. Goode, et al., Lattice determination of the $K \to (\pi\pi)_{I=2}$ decay amplitude A_2, *Phys. Rev.* **D86** (2012) 074513, [arXiv:1206.5142].

[275] T. Blum et al., $K \to \pi\pi \, \Delta I = 3/2$ decay amplitude in the continuum limit, *Phys. Rev.* **D91** (2015), no. 7 074502, [arXiv:1502.00263].

[276] RBC, UKQCD Collaboration, Z. Bai et al., Standard model prediction for direct CP violation in K decay, *Phys. Rev. Lett.* **115** (2015), no. 21 212001, [arXiv:1505.07863].

[277] J. Bijnens and J. Prades, ε'/ε in the chiral limit, *JHEP* **06** (2000) 035, [hep-ph/0005189].

[278] J. Bijnens, E. Gamiz, and J. Prades, Matching the electroweak penguins Q_7, Q_8 and spectral correlators, *JHEP* **10** (2001) 009, [hep-ph/0108240].

[279] V. Cirigliano, J. F. Donoghue, E. Golowich, and K. Maltman, Determination of $\langle(\pi\pi)I = 2|Q_{7,8}|K^0\rangle$ in the chiral limit, *Phys. Lett.* **B522** (2001) 245–256, [hep-ph/0109113].

[280] V. Antonelli, S. Bertolini, M. Fabbrichesi, and E. I. Lashin, The $\Delta I = 1/2$ selection rule, *Nucl. Phys.* **B469** (1996) 181–201, [hep-ph/9511341].

[281] S. Bertolini, J. O. Eeg, and M. Fabbrichesi, A new estimate of ε'/ε, *Nucl. Phys.* **B476** (1996) 225–254, [hep-ph/9512356].

[282] E. Pallante and A. Pich, Strong enhancement of ε'/ε through final state interactions, *Phys. Rev. Lett.* **84** (2000) 2568–2571, [hep-ph/9911233].

[283] E. Pallante and A. Pich, Final state interactions in kaon decays, *Nucl. Phys.* **B592** (2001) 294–320, [hep-ph/0007208].

[284] M. Buchler, G. Colangelo, J. Kambor, and F. Orellana, A note on the dispersive treatment of $K \to \pi\pi$ with the kaon off-shell, *Phys. Lett.* **B521** (2001) 29–32, [hep-ph/0102289].

[285] M. Buchler, G. Colangelo, J. Kambor, and F. Orellana, Dispersion relations and soft pion theorems for $K \to \pi\pi$, *Phys. Lett.* **B521** (2001) 22–28, [hep-ph/0102287].

[286] E. Pallante, A. Pich, and I. Scimemi, The standard model prediction for ε'/ε, *Nucl. Phys.* **B617** (2001) 441–474, [hep-ph/0105011].

[287] S. Aoki et al., FLAG review 2019, arXiv:1902.08191.

[288] Y. Aoki, R. Arthur, T. Blum, P. Boyle, D. Brommel, et al., Continuum limit of B_K from 2+1 flavor domain wall QCD, *Phys. Rev.* **D84** (2011) 014503, [arXiv:1012.4178].

[289] T. Bae, Y.-C. Jang, C. Jung, H.-J. Kim, J. Kim, et al., B_K using HYP-smeared staggered fermions in $N_f = 2 + 1$ unquenched QCD, *Phys. Rev.* **D82** (2010) 114509, [arXiv:1008.5179].

[290] ETM Collaboration, M. Constantinou, et al., B_K-parameter from $N_f = 2$ twisted mass lattice QCD, *Phys. Rev.* **D83** (2011) 014505, [arXiv:1009.5606].

[291] G. Colangelo, S. Durr, A. Juttner, L. Lellouch, H. Leutwyler, et al., Review of lattice results concerning low energy particle physics, *Eur. Phys. J.* **C71** (2011) 1695, [arXiv:1011.4408].

[292] J. A. Bailey, T. Bae, Y.-C. Jang, H. Jeong, C. Jung, et al., Beyond the standard model corrections to $K^0 - \bar{K}^0$ mixing, *PoS* **LATTICE2012** (2012) 107, [arXiv:1211.1101].

[293] S. Durr, Z. Fodor, C. Hoelbling, S. Katz, S. Krieg, et al., Precision computation of the kaon bag parameter, *Phys. Lett.* **B705** (2011) 477–481, [arXiv:1106.3230].

[294] SWME Collaboration, T. Bae, et al., Improved determination of $\hat{B}_K$ with staggered quarks, *Phys. Rev.* **D89** (2014), no. 7 074504, [arXiv:1402.0048].

[295] ETM Collaboration, N. Carrasco, P. Dimopoulos, R. Frezzotti, V. Lubicz, G. C. Rossi, S. Simula, and C. Tarantino, $\Delta S = 2$ and $\Delta C = 2$ bag parameters in the standard model and beyond from N_f=2+1+1 twisted-mass lattice QCD, *Phys. Rev.* **D92** (2015), no. 3 034516, [arXiv:1505.06639].

[296] A. J. Buras, M. Gorbahn, S. Jäger, and M. Jamin, Improved anatomy of ε'/ε in the standard model, *JHEP* **11** (2015) 202, [arXiv:1507.06345].

[297] Z. Bai, N. H. Christ, X. Feng, A. Lawson, A. Portelli, and C. T. Sachrajda, Exploratory lattice QCD study of the rare kaon decay $K^+ \to \pi^+ \nu \bar{\nu}$, *Phys. Rev. Lett.* **118** (2017), no. 25 252001, [arXiv:1701.02858].

[298] RBC, UKQCD Collaboration, N. H. Christ, X. Feng, A. Portelli, and C. T. Sachrajda, Prospects for a lattice computation of rare kaon decay amplitudes: $K \to \pi \ell^+ \ell^-$ decays, *Phys. Rev.* **D92** (2015), no. 9 094512, [arXiv:1507.03094].

[299] N. H. Christ, X. Feng, A. Juttner, A. Lawson, A. Portelli, and C. T. Sachrajda, First exploratory calculation of the long-distance contributions to the rare kaon decays $K \to \pi \ell^+ \ell^-$, *Phys. Rev.* **D94** (2016), no. 11 114516, [arXiv:1608.07585].

[300] RBC, UKQCD Collaboration, N. H. Christ, T. Izubuchi, C. T. Sachrajda, A. Soni, and J. Yu, Long distance contribution to the $K_L - K_S$ mass difference, *Phys. Rev.* **D88** (2013) 014508, [arXiv:1212.5931].

[301] Z. Bai, N. H. Christ, T. Izubuchi, C. T. Sachrajda, A. Soni, and J. Yu, $K_L - K_S$ Mass difference from lattice QCD, *Phys. Rev. Lett.* **113** (2014) 112003, [arXiv:1406.0916].

[302] N. Carrasco, V. Lubicz, G. Martinelli, C. T. Sachrajda, N. Tantalo, C. Tarantino, and M. Testa, QED Corrections to hadronic processes in lattice QCD, *Phys. Rev.* **D91** (2015), no. 7 074506, [arXiv:1502.00257].

[303] V. Lubicz, G. Martinelli, C. T. Sachrajda, F. Sanfilippo, S. Simula, and N. Tantalo, Finite-volume QED corrections to decay amplitudes in lattice QCD, *Phys. Rev.* **D95** (2017), no. 3 034504, [arXiv:1611.08497].

[304] D. Giusti, V. Lubicz, G. Martinelli, C. T. Sachrajda, F. Sanfilippo, S. Simula, N. Tantalo, and C. Tarantino, First lattice calculation of the QED corrections to leptonic decay rates, *Phys. Rev. Lett.* **120** (2018), no. 7 072001, [arXiv:1711.06537].

[305] A. Cerri, V. V. Gligorov, S. Malvezzi, J. Martin Camalich, and J. Zupan, Opportunities in flavour physics at the HL-LHC and HE-LHC, arXiv:1812.07638.

[306] C. Lehner et al., Opportunities for lattice QCD in quark and lepton flavor physics, arXiv:1904.09479.

[307] V. Cirigliano, Z. Davoudi, T. Bhattacharya, T. Izubuchi, P. E. Shanahan, S. Syritsyn, and M. L. Wagman, The role of lattice QCD in searches for violations of fundamental symmetries and signals for new physics, arXiv:1904.09704.

[308] B. Jo, C. Jung, N. H. Christ, W. Detmold, R. Edwards, M. Savage, and P. Shanahan, Status and future perspectives for lattice gauge theory calculations to the exascale and beyond, arXiv:1904.09725.

[309] M. Beneke, G. Buchalla, M. Neubert, and C. T. Sachrajda, QCD factorization for $B \to K\pi, \pi\pi$ Decays: Strong phases and CP violation in the heavy quark limit, *Phys. Rev. Lett.* **83** (1999) 1914–1917, [hep-ph/9905312].

[310] M. Beneke, G. Buchalla, M. Neubert, and C. T. Sachrajda, QCD factorization for exclusive, nonleptonic B meson decays: General arguments and the case of heavy light final states, *Nucl. Phys.* **B591** (2000) 313–418, [hep-ph/0006124].

[311] D. Fakirov and B. Stech, *F and D decays*, *Nucl. Phys.* **B133** (1978) 315–326.

[312] J. D. Bjorken, Topics in B physics, *Nucl. Phys. Proc. Suppl.* **11** (1989) 325–341.

[313] H. D. Politzer and M. B. Wise, Perturbative corrections to factorization in $\bar{B}$ decay, *Phys. Lett.* **B257** (1991) 399–402.

[314] C. W. Bauer, S. Fleming, D. Pirjol, and I. W. Stewart, An effective field theory for collinear and soft gluons: Heavy to light decays, *Phys. Rev.* **D63** (2001) 114020, [hep-ph/0011336].

[315] C. W. Bauer, D. Pirjol, and I. W. Stewart, Soft collinear factorization in effective field theory, *Phys. Rev.* **D65** (2002) 054022, [hep-ph/0109045].

[316] C. W. Bauer, D. Pirjol, and I. W. Stewart, A proof of factorization for $B \to D\pi$, *Phys. Rev. Lett.* **87** (2001) 201806, [hep-ph/0107002].

[317] M. Beneke, Soft-collinear factorization in B decays, *Nucl. Part. Phys. Proc.* **261–262** (2015) 311–337, [arXiv:1501.07374].

[318] R. Fleischer, Flavour physics and CP violation: Expecting the LHC, in *High-Energy Physics. Proceedings, 4th Latin American CERN-CLAF School, Vina del Mar, Chile, February 18–March 3, 2007*, pp. 105–157, 2008. arXiv:0802.2882.

[319] M. Bartsch, G. Buchalla, and C. Kraus, $B \to V_L V_L$ decays at next-to-leading order in QCD, arXiv:0810.0249.

[320] G. Bell, M. Beneke, T. Huber, and X.-Q. Li, Two-loop current–current operator contribution to the non-leptonic QCD penguin amplitude, *Phys. Lett.* **B750** (2015) 348–355, [arXiv:1507.03700].

[321] S. Alte, M. König, and M. Neubert, Effective field theory after a new-physics discovery, *JHEP* **08** (2018) 095, [arXiv:1806.01278].

[322] S. Alte, M. König, and M. Neubert, Effective theory for a heavy scalar coupled to the SM via vector-like quarks, *Eur. Phys. J.* **C79** (2019), no. 4 352, [arXiv:1902.04593].

[323] Y. Y. Keum, H.-N. Li, and A. I. Sanda, Penguin enhancement and $B \to K\pi$ decays in perturbative QCD, *Phys. Rev.* **D63** (2001) 054008, [hep-ph/0004173].

[324] C.-D. Lu, K. Ukai, and M.-Z. Yang, Branching ratio and CP violation of B —> pi pi decays in perturbative QCD approach, *Phys. Rev.* **D63** (2001) 074009, [hep-ph/0004213].

[325] H.-n. Li, QCD aspects of exclusive B meson decays, *Prog. Part. Nucl. Phys.* **51** (2003) 85–171, [hep-ph/0303116].

[326] D.-C. Yan, X. Liu, and Z.-J. Xiao, Anatomy of $B_s \to PP$ decays and effects of the next-to-leading contributions in the perturbative QCD approach, arXiv:1906.01442.

[327] N. Isgur and M. B. Wise, Weak decays of heavy mesons in the static quark approximation, *Phys. Lett.* **B232** (1989) 113–117.

[328] N. Isgur and M. B. Wise, Weak transition form-factors between heavy mesons, *Phys. Lett.* **B237** (1990) 527–530.

[329] B. Grinstein, The static quark effective theory, *Nucl. Phys.* **B339** (1990) 253–268.

[330] A. F. Falk, H. Georgi, B. Grinstein, and M. B. Wise, Heavy meson form-factors from QCD, *Nucl. Phys.* **B343** (1990) 1–13.

[331] H. Georgi, An effective field theory for heavy quarks at low-energies, *Phys. Lett.* **B240** (1990) 447–450.

[332] N. Isgur and M. B. Wise, Heavy quark symmetry, *Adv. Ser. Direct. High Energy Phys.* **10** (1992) 234–285.

[333] T. Mannel, W. Roberts, and Z. Ryzak, A derivation of the heavy quark effective Lagrangian from QCD, *Nucl. Phys.* **B368** (1992) 204–217.

[334] M. Neubert, Heavy quark symmetry, *Phys. Rept.* **245** (1994) 259–396, [hep-ph/9306320].

[335] A. V. Manohar and M. B. Wise, Heavy quark physics, *Camb. Monogr. Part. Phys. Nucl. Phys. Cosmol.* **10** (2000) 1–191.

[336] V. A. Khoze and M. A. Shifman, Heavy quarks, *Sov. Phys. Usp.* **26** (1983) 387.

[337] M. A. Shifman and M. B. Voloshin, Preasymptotic effects in inclusive weak decays of charmed particles, *Sov. J. Nucl. Phys.* **41** (1985) 120. [Yad. Fiz.41,187(1985)].

[338] I. I. Y. Bigi and N. G. Uraltsev, Gluonic enhancements in non-spectator beauty decays: An inclusive mirage though an exclusive possibility, *Phys. Lett.* **B280** (1992) 271–280.

[339] B. Blok and M. A. Shifman, The rule of discarding $1/N$ in inclusive weak decays. 1., *Nucl. Phys.* **B399** (1993) 441–458, [hep-ph/9207236].

[340] I. I. Y. Bigi, N. G. Uraltsev, and A. I. Vainshtein, Nonperturbative corrections to inclusive beauty and charm decays: QCD versus phenomenological models, *Phys. Lett.* **B293** (1992) 430–436, [hep-ph/9207214]. [Erratum: Phys. Lett.B297,477(1992)].

[341] B. Blok and M. A. Shifman, The rule of discarding $1/N$ in inclusive weak decays. 2., *Nucl. Phys.* **B399** (1993) 459–476, [hep-ph/9209289].

[342] I. I. Y. Bigi, B. Blok, M. A. Shifman, and A. I. Vainshtein, The baffling semileptonic branching ratio of B mesons, *Phys. Lett.* **B323** (1994) 408–416, [hep-ph/9311339].

[343] I. I. Y. Bigi, V. A. Khoze, N. G. Uraltsev, and A. I. Sanda, The question of CP noninvariance – as seen through the eyes of neutral beauty, *Adv. Ser. Direct. High Energy Phys.* **3** (1989) 175–248.

[344] I. I. Y. Bigi, M. A. Shifman, and N. Uraltsev, Aspects of heavy quark theory, *Ann. Rev. Nucl. Part. Sci.* **47** (1997) 591–661, [hep-ph/9703290].

[345] A. Lenz, Lifetimes and heavy quark expansion, *Int. J. Mod. Phys.* **A30** (2015), no. 10 1543005, [arXiv:1405.3601]. [,63(2014)].

[346] M. Artuso, G. Borissov, and A. Lenz, CP violation in the B_s^0 system, *Rev. Mod. Phys.* **88** (2016), no. 4 045002, [arXiv:1511.09466].

[511] LHCb Collaboration, R. Aaij et al., Measurement of the $B_s^0 \to \mu^+\mu^-$ branching fraction and effective lifetime and search for $B^0 \to \mu^+\mu^-$ decays, *Phys. Rev. Lett.* **118** (2017), no. 19 191801, [arXiv:1703.05747].

[512] S. Descotes-Genon, J. Matias, and J. Virto, An analysis of $B_{d,s}$ mixing angles in presence of new physics and an update of $B_s \to K^{0*}\bar{K}^{0*}$, *Phys. Rev.* **D85** (2012) 034010, [arXiv:1111.4882].

[513] R. Fleischer, D. G. Espinosa, R. Jaarsma, and G. Tetlalmatzi-Xolocotzi, CP violation in leptonic rare B_s^0 decays as a probe of new physics, *Eur. Phys. J.* **C78** (2018), no. 1 1, [arXiv:1709.04735].

[514] G. Tetlalmatzi-Xolocotzi, Rare leptonic B decays, *PoS* **BEAUTY2018** (2018) 043, [arXiv:1809.00637].

[515] M. Beneke, C. Bobeth, and R. Szafron, Enhanced electromagnetic correction to the rare B-meson decay $B_{s,d} \to \mu^+\mu^-$, *Phys. Rev. Lett.* **120** (2018), no. 1 011801, [arXiv:1708.09152].

[516] J. Aebischer, W. Altmannshofer, D. Guadagnoli, M. Reboud, P. Stangl, and D. M. Straub, B-decay discrepancies after Moriond 2019, arXiv:1903.10434.

[517] CMS Collaboration, S. Chatrchyan et al., Measurement of the $B_s \to \mu^+\mu^-$ branching fraction and search for $B_d \to \mu^+\mu^-$ with the CMS experiment, *Phys. Rev. Lett.* **111** (2013) 101804, [arXiv:1307.5025].

[518] LHCb, CMS Collaboration, V. Khachatryan et al., Observation of the rare $B_s^0 \to \mu^+\mu^-$ decay from the combined analysis of CMS and LHCb data, *Nature* **522** (2015) 68–72, [arXiv:1411.4413].

[519] ATLAS Collaboration, M. Aaboud et al., Study of the rare decays of B_s^0 and B^0 mesons into muon pairs using data collected during 2015 and 2016 with the ATLAS detector, *JHEP* **04** (2019) 098, [arXiv:1812.03017].

[520] A. J. Buras, P. Gambino, M. Gorbahn, S. Jäger, and L. Silvestrini, Universal unitarity triangle and physics beyond the standard model, *Phys. Lett.* **B500** (2001) 161–167, [hep-ph/0007085].

[521] A. J. Buras, Minimal flavor violation, *Acta Phys. Polon.* **B34** (2003) 5615–5668, [hep-ph/0310208].

[522] RBC/UKQCD Collaboration, P. A. Boyle, L. Del Debbio, N. Garron, A. Juttner, A. Soni, J. T. Tsang, and O. Witzel, SU(3)-breaking ratios for $D_{(s)}$ and $B_{(s)}$ mesons, arXiv:1812.08791.

[523] D. King, A. Lenz, and T. Rauh, B_s mixing observables and $-V_{td}/V_{ts}-$ from sum rules, *JHEP* **05** (2019) 034, [arXiv:1904.00940].

[524] LHCb Collaboration, R. Aaij et al., Search for the decays $B_s^0 \to \tau^+\tau^-$ and $B^0 \to \tau^+\tau^-$, *Phys. Rev. Lett.* **118** (2017), no. 25 251802, [arXiv:1703.02508].

[525] BaBar Collaboration, B. Aubert et al., A search for $B^+ \to \tau^+\nu$ with hadronic B tags, *Phys. Rev.* **D77** (2008) 011107, [arXiv:0708.2260].

[526] Belle Collaboration, K. Ikado et al., Evidence of the purely leptonic decay $B^- \to \tau^-\bar{\nu}_\tau$, *Phys. Rev. Lett.* **97** (2006) 251802, [hep-ex/0604018].

[527] Belle Collaboration, I. Adachi et al., Measurement of $B^- \to \tau^-\bar{\nu}_\tau$ with a hadronic tagging method using the full data sample of Belle, *Phys. Rev. Lett.* **110** (2013) 131801, [arXiv:1208.4678].

[528] UTfit Collaboration, M. Bona et al., An improved standard model prediction of $BR(B \to \tau\nu)$ and its implications for new physics, *Phys. Lett.* **B687** (2010) 61–69, [arXiv:0908.3470].

[529] W. Altmannshofer, A. J. Buras, S. Gori, P. Paradisi, and D. M. Straub, Anatomy and phenomenology of FCNC and CPV effects in SUSY theories, *Nucl. Phys.* **B830** (2010) 17–94, [arXiv:0909.1333].

[530] G. Banelli, R. Fleischer, R. Jaarsma, and G. Tetlalmatzi-Xolocotzi, Decoding (pseudo)-scalar operators in leptonic and semileptonic B decays, *Eur. Phys. J.* **C78** (2018), no. 11 911, [arXiv:1809.09051].

[531] Belle Collaboration, A. Sibidanov et al., Search for $B^- \to \mu^-\bar{\nu}_\mu$ decays at the Belle experiment, *Phys. Rev. Lett.* **121** (2018), no. 3 031801, [arXiv:1712.04123].

[532] W.-S. Hou, M. Kohda, T. Modak, and G.-G. Wong, Enhanced $B \to \mu\bar{\nu}$ decay at tree level, arXiv:1903.03016.

[533] J. G. Körner and G. A. Schuler, Exclusive semileptonic heavy meson decays including lepton mass effects, *Z. Phys.* **C46** (1990) 93.

[534] B. Dassinger, R. Feger, and T. Mannel, Complete Michel parameter analysis of inclusive semileptonic $b \to c$ transition, *Phys. Rev.* **D79** (2009) 075015, [arXiv:0803.3561].

[535] M. Tanaka and R. Watanabe, New physics in the weak Interaction of $\bar{B} \to D^{(*)}\tau\bar{\nu}$, *Phys. Rev.* **D87** (2013), no. 3 034028, [arXiv:1212.1878].

[536] S. Fajfer, J. F. Kamenik, and I. Nisandzic, On the $\bar{B} \to D^*\tau\bar{\nu}_\tau$ sensitivity to new physics, *Phys. Rev.* **D85** (2012) 094025, [arXiv:1203.2654].

[537] Y. Sakaki, M. Tanaka, A. Tayduganov, and R. Watanabe, Testing leptoquark models in $\bar{B} \to D^{(*)}\tau\bar{\nu}$, *Phys. Rev.* **D88** (2013), no. 9 094012, [arXiv:1309.0301].

[538] I. Dorsner, S. Fajfer, N. Kosnik, and I. Nisandzic, Minimally flavored colored scalar in $\bar{B} \to D^{(*)}\tau\bar{\nu}$ and the mass matrices constraints, *JHEP* **11** (2013) 084, [arXiv:1306.6493].

[539] M. Duraisamy and A. Datta, The Full $\bar{B} \to D^*\tau^-\bar{\nu}_\tau$ angular distribution and CP violating triple products, *JHEP* **09** (2013) 059, [arXiv:1302.7031].

[540] M. Duraisamy, P. Sharma, and A. Datta, Azimuthal $B \to D^*\tau^-\bar{\nu}_\tau$ angular distribution with tensor operators, *Phys. Rev.* **D90** (2014), no. 7 074013, [arXiv:1405.3719].

[541] M. Freytsis, Z. Ligeti, and J. T. Ruderman, Flavor models for $\bar{B} \to D^{(*)}\tau\bar{\nu}$, *Phys. Rev.* **D92** (2015), no. 5 054018, [arXiv:1506.08896].

[542] S. Bhattacharya, S. Nandi, and S. K. Patra, Optimal-observable analysis of possible new physics in $\bar{B} \to D^{(*)}\tau\nu_\tau$, *Phys. Rev.* **D93** (2016), no. 3 034011, [arXiv:1509.07259].

[543] D. Becirevic, S. Fajfer, I. Nisandzic, and A. Tayduganov, Angular distributions of $\bar{B} \to D^{(*)}\ell\bar{\nu}_\ell$ decays and search of New Physics, arXiv:1602.03030.

[544] D. Bardhan, P. Byakti, and D. Ghosh, A closer look at the R_D and R_{D^*} anomalies, *JHEP* **01** (2017) 125, [arXiv:1610.03038].

[545] Z. Ligeti, M. Papucci, and D. J. Robinson, New physics in the visible final states of $\bar{B} \to D^{(*)}\tau\nu$, *JHEP* **01** (2017) 083, [arXiv:1610.02045].

[546] R. Alonso, A. Kobach, and J. Martin Camalich, New physics in the kinematic distributions of $\bar{B} \to D^{(*)}\tau^-(\to \ell^-\bar{\nu}_\ell\nu_\tau)\bar{\nu}_\tau$, *Phys. Rev.* **D94** (2016), no. 9 094021, [arXiv:1602.07671].

[547] A. K. Alok, D. Kumar, S. Kumbhakar, and S. U. Sankar, D^* polarization as a probe to discriminate new physics in $\bar{B} \to D^*\tau\bar{\nu}$, *Phys. Rev.* **D95** (2017), no. 11 115038, [arXiv:1606.03164].

[548] R. Alonso, J. Martin Camalich, and S. Westhoff, Tau properties in $\bar{B} \to D\tau\nu$ from visible final-state kinematics, *Phys. Rev.* **D95** (2017), no. 9 093006, [arXiv:1702.02773].

[549] M. A. Ivanov, J. G. Körner, and C.-T. Tran, Probing new physics in $\bar{B}^0 \to D^{(*)}\tau^-\bar{\nu}_\tau$ using the longitudinal, transverse, and normal polarization components of the tau lepton, *Phys. Rev.* **D95** (2017), no. 3 036021, [arXiv:1701.02937].

[550] P. Colangelo and F. De Fazio, Scrutinizing $\overline{B} \to D^*(D\pi)\ell^-\overline{\nu}_\ell$ and $\overline{B} \to D^*(D\gamma)\ell^-\overline{\nu}_\ell$ in search of new physics footprints, *JHEP* **06** (2018) 082, [arXiv:1801.10468].

[551] M. Blanke, A. Crivellin, S. de Boer, M. Moscati, U. Nierste, I. Nisandzic, and T. Kitahara, Impact of polarization observables and $B_c \to \tau\nu$ on new physics explanations of the $b \to c\tau\nu$ anomaly, *Phys. Rev.* **D99** (2019), no. 7 075006, [arXiv:1811.09603].

[552] D. Bardhan and D. Ghosh, B-meson charged current anomalies: The post-Moriond status, arXiv:1904.10432.

[553] C. Murgui, A. Peuelas, M. Jung, and A. Pich, Global fit to $b \to c\tau\nu$ transitions, arXiv:1904.09311.

[554] M. Blanke, A. Crivellin, T. Kitahara, M. Moscati, U. Nierste, and I. Nisandzic, Addendum: "Impact of polarization observables and $B_c \to \tau\nu$ on new physics explanations of the $b \to c\tau\nu$ anomaly," arXiv:1905.08253.

[555] R.-X. Shi, L.-S. Geng, B. Grinstein, S. Jäger, and J. Martin Camalich, Revisiting the new-physics interpretation of the $b \to c\tau\nu$ data, arXiv:1905.08498.

[556] P. Asadi and D. Shih, Maximizing the impact of new physics in $b \to c\tau\nu$ anomalies, arXiv:1905.03311.

[557] S. Iguro and Y. Omura, Status of the semileptonic B decays and muon g-2 in general 2HDMs with right-handed neutrinos, *JHEP* **05** (2018) 173, [arXiv:1802.01732].

[558] A. Greljo, D. J. Robinson, B. Shakya, and J. Zupan, R(D^0) from W and right-handed neutrinos, *JHEP* **09** (2018) 169, [arXiv:1804.04642].

[559] D. J. Robinson, B. Shakya, and J. Zupan, Right-handed neutrinos and $R(D^{(*)})$, *JHEP* **02** (2019) 119, [arXiv:1807.04753].

[560] A. Azatov, D. Barducci, D. Ghosh, D. Marzocca, and L. Ubaldi, Combined explanations of B-physics anomalies: The sterile neutrino solution, *JHEP* **10** (2018) 092, [arXiv:1807.10745].

[561] W. Buchmuller and D. Wyler, Effective Lagrangian analysis of new interactions and flavor conservation, *Nucl. Phys.* **B268** (1986) 621–653.

[562] B. Grzadkowski, M. Iskrzynski, M. Misiak, and J. Rosiek, Dimension-six terms in the standard model lagrangian, *JHEP* **1010** (2010) 085, [arXiv:1008.4884].

[563] J. Aebischer, A. Crivellin, M. Fael, and C. Greub, Matching of gauge invariant dimension-six operators for $b \to s$ and $b \to c$ transitions, *JHEP* **05** (2016) 037, [arXiv:1512.02830].

[564] V. Cirigliano, J. Jenkins, and M. Gonzalez-Alonso, Semileptonic decays of light quarks beyond the standard model, *Nucl. Phys.* **B830** (2010) 95–115, [arXiv:0908.1754].

[565] R. Alonso, B. Grinstein, and J. Martin Camalich, $SU(2) \times U(1)$ gauge invariance and the shape of new physics in rare B decays, *Phys. Rev. Lett.* **113** (2014) 241802, [arXiv:1407.7044].

[566] O. Cata and M. Jung, Signatures of a nonstandard Higgs boson from flavor physics, *Phys. Rev.* **D92** (2015), no. 5 055018, [arXiv:1505.05804].

[567] R. Watanabe, New physics effect on $B_c \to J/\psi \tau \bar\nu$ in relation to the $R_{D^{(*)}}$ anomaly, *Phys. Lett.* **B776** (2018) 5–9, [arXiv:1709.08644].

[568] W. Detmold, C. Lehner, and S. Meinel, $\Lambda_b \to p\ell^-\bar\nu_\ell$ and $\Lambda_b \to \Lambda_c \ell^-\bar\nu_\ell$ form factors from lattice QCD with relativistic heavy quarks, *Phys. Rev.* **D92** (2015), no. 3 034503, [arXiv:1503.01421].

[569] A. Datta, S. Kamali, S. Meinel, and A. Rashed, Phenomenology of $\Lambda_b \to \Lambda_c \tau \bar\nu_\tau$ using lattice QCD calculations, *JHEP* **08** (2017) 131, [arXiv:1702.02243].

[570] X.-Q. Li, Y.-D. Yang, and X. Zhang, Revisiting the one leptoquark solution to the $R(D^{(*)})$ anomalies and its phenomenological implications, *JHEP* **08** (2016) 054, [arXiv:1605.09308].

[571] R. Alonso, B. Grinstein, and J. Martin Camalich, Lifetime of B_c^- constrains explanations for anomalies in $\bar B \to D^{(*)}\tau\bar\nu$, *Phys. Rev. Lett.* **118** (2017), no. 8 081802, [arXiv:1611.06676].

[572] Y. Sakaki, M. Tanaka, A. Tayduganov, and R. Watanabe, Probing new physics with q^2 distributions in $\bar B \to D^{(*)}\tau\bar\nu$, *Phys. Rev.* **D91** (2015), no. 11 114028, [arXiv:1412.3761].

[573] S. Bhattacharya, S. Nandi, and S. K. Patra, Looking for possible new physics in $\bar B \to D^{(*)}\tau\bar\nu_\tau$ in light of recent data, *Phys. Rev.* **D95** (2017), no. 7 075012, [arXiv:1611.04605].

[574] A. Celis, M. Jung, X.-Q. Li, and A. Pich, Scalar contributions to $b \to c(u)\tau\bar\nu$ transitions, *Phys. Lett.* **B771** (2017) 168–179, [arXiv:1612.07757].

[575] P. Colangelo, F. De Fazio, and F. Loparco, Probing new physics with $\bar B \to \rho(770)\ell^-\bar\nu_\ell$ and $\bar B \to a_1(1260)\ell^-\bar\nu_\ell$, arXiv:1906.07068.

[576] BaBar Collaboration, J. P. Lees et al., Measurement of an excess of $\bar B \to D^{(*)}\tau^-\bar\nu_\tau$ decays and implications for charged Higgs bosons, *Phys. Rev.* **D88** (2013), no. 7 072012, [arXiv:1303.0571].

[577] Belle Collaboration, M. Huschle et al., Measurement of the branching ratio of $\bar B \to D^{(*)}\tau^-\bar\nu_\tau$ relative to $\bar B \to D^{(*)}\ell^-\bar\nu_\ell$ decays with hadronic tagging at Belle, *Phys. Rev.* **D92** (2015), no. 7 072014, [arXiv:1507.03233].

[578] Belle Collaboration, S. Hirose et al., Measurement of the τ lepton polarization and $R(D^*)$ in the decay $\bar B \to D^* \tau^-\bar\nu_\tau$, *Phys. Rev. Lett.* **118** (2017), no. 21 211801, [arXiv:1612.00529].

[579] Belle Collaboration, Y. Sato et al., Measurement of the branching ratio of $\bar B^0 \to D^{*+}\tau^-\bar\nu_\tau$ relative to $\bar B^0 \to D^{*+}\ell^-\bar\nu_\ell$ decays with a semileptonic tagging method, *Phys. Rev.* **D94** (2016), no. 7 072007, [arXiv:1607.07923].

[580] LHCb Collaboration, R. Aaij et al., Measurement of the ratio of branching fractions $\mathcal{B}(\bar B^0 \to D^{*+}\tau^-\bar\nu_\tau)/\mathcal{B}(\bar B^0 \to D^{*+}\mu^-\bar\nu_\mu)$, *Phys. Rev. Lett.* **115** (2015), no. 11 111803, [arXiv:1506.08614]. [Erratum: Phys. Rev. Lett.115,no.15,159901(2015)].

[581] G. Ciezarek, M. Franco Sevilla, B. Hamilton, R. Kowalewski, T. Kuhr, V. Lüth, and Y. Sato, A challenge to lepton universality in B meson decays, *Nature* **546** (2017) 227–233, [arXiv:1703.01766].

[582] LHCb Collaboration, R. Aaij et al., Measurement of the ratio of branching fractions $\mathcal{B}(B_c^+ \to J/\psi\tau^+\nu_\tau)/\mathcal{B}(B_c^+ \to J/\psi\mu^+\nu_\mu)$, *Phys. Rev. Lett.* **120** (2018), no. 12 121801, [arXiv:1711.05623].

[583] BaBar Collaboration, J. Lees et al., Evidence for an excess of $\bar B \to D^{(*)}\tau^-\bar\nu_\tau$ decays, *Phys. Rev. Lett.* **109** (2012) 101802, [arXiv:1205.5442].

[584] LHCb Collaboration, R. Aaij et al., Measurement of the ratio of the $B^0 \to D^{*-}\tau^+\nu_\tau$ and $B^0 \to D^{*-}\mu^+\nu_\mu$ branching fractions using three-prong τ-lepton decays, *Phys. Rev. Lett.* **120** (2018), no. 17 171802, [arXiv:1708.08856].

[585] MILC Collaboration, J. A. Bailey et al., $B \to D\ell\nu$ form factors at nonzero recoil and $|V_{cb}|$ from 2+1-flavor lattice QCD, *Phys. Rev.* **D92** (2015), no. 3 034506, [arXiv:1503.07237].

[586] G. Hiller, Lepton nonuniversality anomalies and implications, in *53rd Rencontres de Moriond on QCD and high energy interactions (Moriond QCD 2018) La Thuile, Italy, March 17–24, 2018*, 2018. arXiv:1804.02011.

[587] Belle Collaboration, A. Abdesselam et al., Measurement of $\mathcal{R}(D)$ and $\mathcal{R}(D^*)$ with a semileptonic tagging method, arXiv:1904.08794.

[588] A. Crivellin and F. Saturnino, Explaining the flavor anomalies with a vector leptoquark (Moriond 2019 update), 2019. arXiv:1906.01222.

[589] Belle Collaboration, A. Abdesselam et al., Measurement of the D^{*-} polarization in the decay $B^0 \to D^{*-}\tau^+\nu_\tau$, in *10th International Workshop on the CKM Unitarity Triangle (CKM 2018) Heidelberg, Germany, September 17–21, 2018*, 2019. arXiv:1903.03102.

[590] S. Bhattacharya, S. Nandi, and S. Kumar Patra, $b \to c\tau\nu_\tau$ Decays: A catalogue to compare, constrain, and correlate new physics effects, *Eur. Phys. J.* **C79** (2019), no. 3 268, [arXiv:1805.08222].

[591] Z.-R. Huang, Y. Li, C.-D. Lu, M. A. Paracha, and C. Wang, Footprints of new physics in $b \to c\tau\nu$ transitions, *Phys. Rev.* **D98** (2018), no. 9 095018, [arXiv:1808.03565].

[592] R. Alonso, J. Martin Camalich, and S. Westhoff, Tau polarimetry in B meson decays, *Sci Post Phys. Proc.* **1** (2019) 012, [arXiv:1811.05664].

[593] H. Yan, Y.-D. Yang, and X.-B. Yuan, Phenomenology of $b \to c\tau\bar{\nu}$ decays in a scalar leptoquark model, arXiv:1905.01795.

[594] E949 Collaboration, A. V. Artamonov et al., New measurement of the $K^+ \to \pi^+ \nu\bar{\nu}$ branching ratio, *Phys. Rev. Lett.* **101** (2008) 191802, [arXiv:0808.2459].

[595] L. S. Littenberg, The CP violating decay $K_L \to \pi^0 \nu\bar{\nu}$, *Phys. Rev.* **D39** (1989) 3322–3324.

[596] G. Buchalla and G. Isidori, The CP conserving contribution to $K_L \to \pi^0 \nu\bar{\nu}$ in the standard model, *Phys. Lett.* **B440** (1998) 170–178, [hep-ph/9806501].

[597] G. Buchalla and A. J. Buras, $K \to \pi\nu\bar{\nu}$ and high precision determinations of the CKM matrix, *Phys. Rev.* **D54** (1996) 6782–6789, [hep-ph/9607447].

[598] KOTO Collaboration, J. K. Ahn et al., Search for the $K_L \to \pi^0 \nu\bar{\nu}$ and $K_L \to \pi^0 X^0$ decays at the J-PARC KOTO experiment, *Phys. Rev. Lett.* **122** (2019), no. 2 021802, [arXiv:1810.09655].

[599] KLEVER Project Collaboration, F. Ambrosino et al., KLEVER: An experiment to measure BR($K_L \to \pi^0 \nu\bar{\nu}$) at the CERN SPS, arXiv:1901.03099.

[600] X.-G. He, J. Tandean, and G. Valencia, Charged-lepton-flavor violation in $|\Delta S| = 1$ hyperon decays, arXiv:1903.01242.

[601] A. J. Buras, D. Buttazzo, and R. Knegjens, $K \to \pi\nu\bar{\nu}$ and ϵ'/ϵ in simplified new physics models, *JHEP* **11** (2015) 166, [arXiv:1507.08672].

[602] Y. Grossman and Y. Nir, $K_L \to \pi^0 \nu\bar{\nu}$ beyond the standard model, *Phys. Lett.* **B398** (1997) 163–168, [hep-ph/9701313].

[603] P. Colangelo, F. De Fazio, P. Santorelli, and E. Scrimieri, Rare $B \to K^{(*)}\nu\bar{\nu}$ decays at B factories, *Phys. Lett.* **B395** (1997) 339–344, [hep-ph/9610297].

[604] W. Altmannshofer, A. J. Buras, D. M. Straub, and M. Wick, New strategies for new physics search in $B \to K^* \nu\bar{\nu}$, $B \to K\nu\bar{\nu}$ and $B \to X_s \nu\bar{\nu}$ decays, *JHEP* **04** (2009) 022, [arXiv:0902.0160].

[605] C. Bouchard, G. P. Lepage, C. Monahan, H. Na, and J. Shigemitsu, Rare decay $B \to Kll$ form factors from lattice QCD, *Phys. Rev. D 88,* **054509** (2013) 054509, [arXiv:1306.2384].

[606] R. R. Horgan, Z. Liu, S. Meinel, and M. Wingate, Lattice QCD calculation of form factors describing the rare decays $B \to K^* \ell^+ \ell^-$ and $B_s \to \phi \ell^+ \ell^-$, *Phys. Rev.* **D89** (2014) 094501, [arXiv:1310.3722].

[607] A. J. Buras, J. Girrbach-Noe, C. Niehoff, and D. M. Straub, $B \to K^{(*)}\nu\bar{\nu}$ decays in the standard model and beyond, *JHEP* **1502** (2015) 184, [arXiv:1409.4557].

[608] A. Crivellin and S. Pokorski, Can the differences in the determinations of V_{ub} and V_{cb} be explained by new physics? *Phys. Rev. Lett.* **114** (2015), no. 1 011802, [arXiv:1407.1320].

[609] G. Hiller and M. Schmaltz, R_K and future $b \to s\ell\ell$ BSM opportunities, *Phys. Rev.* **D90** (2014) 054014, [arXiv:1408.1627].

[610] A. J. Buras, K. Gemmler, and G. Isidori, Quark flavour mixing with right-handed currents: An effective theory approach, *Nucl. Phys.* **B843** (2011) 107–142, [arXiv:1007.1993].

[611] A. J. Buras, F. De Fazio, and J. Girrbach, The anatomy of Z' and Z with flavour changing neutral currents in the flavour precision era, *JHEP* **1302** (2013) 116, [arXiv:1211.1896].

[612] A. J. Buras and J. Girrbach, Left-handed Z' and Z FCNC quark couplings facing new $b \to s\mu^+\mu^-$ data, *JHEP* **1312** (2013) 009, [arXiv:1309.2466].

[613] A. J. Buras, F. De Fazio, J. Girrbach, and M. V. Carlucci, The anatomy of quark flavour observables in 331 models in the flavour precision era, *JHEP* **1302** (2013) 023, [arXiv:1211.1237].

[614] D. M. Straub, Anatomy of flavour-changing Z couplings in models with partial compositeness, *JHEP* **1308** (2013) 108, [arXiv:1302.4651].

[615] A. J. Buras, F. De Fazio, and J. Girrbach-Noe, Z-Z' mixing and Z-mediated FCNCs in $SU(3)_C \times SU(3)_L \times U(1)_X$ models, *JHEP* **1408** (2014) 039, [arXiv:1405.3850].

[616] P. Biancofiore, P. Colangelo, F. De Fazio, and E. Scrimieri, Exclusive $b \to s\nu\bar{\nu}$ induced transitions in RS_c model, *Eur. Phys. J.* **C75** (2015) 134, [arXiv:1408.5614].

[617] Y. Grossman, Z. Ligeti, and E. Nardi, First limit on inclusive $b \to x_s \nu\bar{\nu}$ decay and constraints on new physics, *Nucl. Phys.* **B465** (1996) 369–398, [hep-ph/9510378].

[618] N. Gubernari, A. Kokulu, and D. van Dyk, $B \to P$ and $B \to V$ form factors from B-meson light-cone sum rules beyond leading twist, *JHEP* **01** (2019) 150, [arXiv:1811.00983].

[619] Belle Collaboration, J. Grygier et al., Search for $B \to h\nu\bar{\nu}$ decays with semileptonic tagging at Belle, *Phys. Rev.* **D96** (2017), no. 9 091101, [arXiv:1702.03224]. [Addendum: Phys. Rev. D97, no.9, 099902(2018)].

[620] G. Hiller and R. Zwicky, (A)symmetries of weak decays at and near the kinematic endpoint, *JHEP* **03** (2014) 042, [arXiv:1312.1923].

[621] D. Melikhov, N. Nikitin, and S. Simula, Right-handed currents in rare exclusive $B \to (K, K^*)$ neutrino anti-neutrino decays, *Phys. Lett.* **B428** (1998) 171–178, [hep-ph/9803269].

[622] K. Schmidt-Hoberg, F. Staub, and M. W. Winkler, Constraints on light mediators: Confronting dark matter searches with B physics, *Phys. Lett.* **B727** (2013) 506–510, [arXiv:1310.6752].

[623] T. Hurth, G. Isidori, J. F. Kamenik, and F. Mescia, Constraints on new physics in MFV models: A model-independent analysis of $\Delta F =1$ processes, *Nucl. Phys.* **B808** (2009) 326–346, [arXiv:0807.5039].

[624] G. D'Ambrosio, G. F. Giudice, G. Isidori, and A. Strumia, Minimal flavour violation: An effective field theory approach, *Nucl. Phys.* **B645** (2002) 155–187, [hep-ph/0207036].

[625] M. Gorbahn and U. Haisch, Charm quark contribution to $K_L \to \mu^+\mu^-$ at next-to-next-to-leading order, *Phys. Rev. Lett.* **97** (2006) 122002, [hep-ph/0605203].

[626] G. Isidori and R. Unterdorfer, On the short-distance constraints from $K_{L,S} \to \mu^+\mu^-$, *JHEP* **01** (2004) 009, [hep-ph/0311084].

[627] G. D'Ambrosio and J. Portoles, Vector meson exchange contributions to $K \to \pi\gamma\gamma$ and $K_L \to \gamma\ell^+\ell^-$, *Nucl. Phys.* **B492** (1997) 417–454, [hep-ph/9610244].

[628] J.-M. Gérard, C. Smith, and S. Trine, Radiative kaon decays and the penguin contribution to the $\Delta I = 1/2$ rule, *Nucl. Phys.* **B730** (2005) 1–36, [hep-ph/0508189].

[629] D. Gomez Dumm and A. Pich, Long distance contributions to the $K_L \to \mu^+\mu^-$ decay width, *Phys. Rev. Lett.* **80** (1998) 4633–4636, [hep-ph/9801298].

[630] G. Ecker and A. Pich, The longitudinal muon polarization in $K_L \to \mu^+\mu^-$, *Nucl. Phys.* **B366** (1991) 189–205.

[631] LHCb Collaboration, R. Aaij et al., Improved limit on the branching fraction of the rare decay $K_S^0 \to \mu^+\mu^-$, *Eur. Phys. J.* **C77** (2017), no. 10 678, [arXiv:1706.00758].

[632] G. D'Ambrosio and T. Kitahara, Direct CP violation in $K \to \mu^+\mu^-$, *Phys. Rev. Lett.* **119** (2017), no. 20 201802, [arXiv:1707.06999].

[633] G. D'Ambrosio, G. Ecker, G. Isidori, and J. Portoles, The decays $K \to \pi\ell^+\ell^-$ beyond leading order in the chiral expansion, *JHEP* **08** (1998) 004, [hep-ph/9808289].

[634] G. Buchalla, G. D'Ambrosio, and G. Isidori, Extracting short-distance physics from $K_{L,S} \to \pi^0 e^+ e^-$ decays, *Nucl. Phys.* **B672** (2003) 387–408, [hep-ph/0308008].

[635] G. Isidori, C. Smith, and R. Unterdorfer, The rare decay $K_L \to \pi^0\mu^+\mu^-$ within the SM, *Eur. Phys. J.* **C36** (2004) 57–66, [hep-ph/0404127].

[636] S. Friot, D. Greynat, and E. De Rafael, Rare kaon decays revisited, *Phys. Lett.* **B595** (2004) 301–308, [hep-ph/0404136].

[637] F. Mescia, C. Smith, and S. Trine, $K_L \to \pi^0 e^+ e^-$ and $K_L \to \pi^0\mu^+\mu^-$: A binary star on the stage of flavor physics, *JHEP* **08** (2006) 088, [hep-ph/0606081].

[638] KTeV Collaboration, A. Alavi-Harati et al., Search for the rare decay $K_L \to \pi^0 e^+ e^-$, *Phys. Rev. Lett.* **93** (2004) 021805, [hep-ex/0309072].

[639] KTeV Collaboration, A. Alavi-Harati et al., Search for the decay $K_L \to \pi^0\mu^+\mu^-$, *Phys. Rev. Lett.* **84** (2000) 5279–5282, [hep-ex/0001006].

[640] J. Prades, ChPT progress on non-leptonic and radiative kaon decays, *PoS* **KAON** (2008) 022, [arXiv:0707.1789].

[641] C. Bruno and J. Prades, Rare kaon decays in the $1/N_c$-expansion, *Z. Phys.* **C57** (1993) 585–594, [hep-ph/9209231].

[642] G. D'Ambrosio, D. Greynat, and M. Knecht, On the amplitudes for the CP-conserving $K^\pm(K_S) \to \pi^\pm(\pi^0)\ell^+\ell^-$ rare decay modes, *JHEP* **02** (2019) 049, [arXiv:1812.00735].

[643] G. D'Ambrosio, D. Greynat, and M. Knecht, Matching long and short distances at order $O(\alpha_s)$ in the form factors for $K \to \pi\ell^+\ell^-$, arXiv:1906.03046.

[644] NA48 Collaboration, J. Batley et al., A precision measurement of direct CP violation in the decay of neutral kaons into two pions, *Phys. Lett.* **B544** (2002) 97–112, [hep-ex/0208009].

[645] KTeV Collaboration, A. Alavi-Harati et al., Measurements of direct CP violation, CPT symmetry, and other parameters in the neutral kaon system, *Phys. Rev.* **D67** (2003) 012005, [hep-ex/0208007].

[646] KTeV Collaboration, E. Abouzaid et al., Precise measurements of direct CP violation, CPT symmetry, and other parameters in the neutral kaon system, *Phys. Rev.* **D83** (2011) 092001, [arXiv:1011.0127].

[647] A. J. Buras, P. Gambino, and U. A. Haisch, Electroweak penguin contributions to non-leptonic $\Delta F = 1$ decays at NNLO, *Nucl. Phys.* **B570** (2000) 117–154, [hep-ph/9911250].

[648] M. Cerdá-Sevilla, M. Gorbahn, S. Jäger, and A. Kokulu, Towards NNLO accuracy for ε'/ε, *J. Phys. Conf. Ser.* **800** (2017), no. 1 012008, [arXiv:1611.08276].

[649] M. Cerdá-Sevilla, NNLO QCD contributions to ε'/ε, *Acta Phys. Polon.* **B49** (2018) 1087–1096.

[650] J. R. Ellis, M. K. Gaillard, and D. V. Nanopoulos, Left-handed currents and CP violation, *Nucl. Phys.* **B109** (1976) 213–243.

[651] F. J. Gilman and M. B. Wise, The $\Delta I = 1/2$ rule and violation of CP in the six quark model, *Phys. Lett.* **B83** (1979) 83–86.

[652] B. Guberina and R. D. Peccei, Quantum chromodynamic effects and CP violation in the Kobayashi-Maskawa model, *Nucl. Phys.* **B163** (1980) 289–311.

[653] J. F. Donoghue, E. Golowich, B. R. Holstein, and J. Trampetic, Electromagnetic and isospin breaking effects decrease ε'/ε, *Phys. Lett.* **B179** (1986) 361. [Erratum: Phys. Lett.B188,511(1987)].

[654] J. Bijnens and M. B. Wise, Electromagnetic contribution to ε'/ε, *Phys. Lett.* **B137** (1984) 245–250.

[655] J. M. Flynn and L. Randall, The electromagnetic penguin contribution to ε'/ε for large top quark mass, *Phys. Lett.* **B224** (1989) 221.

[656] G. Buchalla, A. J. Buras, and M. K. Harlander, The anatomy of ε'/ε in the standard model, *Nucl. Phys.* **B337** (1990) 313–362.

[657] E. A. Paschos and Y. L. Wu, Correlations between ε'/ε and heavy top, *Mod. Phys. Lett.* **A6** (1991) 93–106.

[658] M. Lusignoli, L. Maiani, G. Martinelli, and L. Reina, Mixing and CP violation in K and B mesons: A lattice QCD point of view, *Nucl. Phys.* **B369** (1992) 139–170.

[659] A. J. Buras, M. Jamin, and M. E. Lautenbacher, Two loop anomalous dimension matrix for $\Delta S = 1$ weak nonleptonic decays. 2. $O(\alpha\alpha_s)$, *Nucl. Phys.* **B400** (1993) 75–102, [hep-ph/9211321].

[660] S. Bertolini, M. Fabbrichesi, and J. O. Eeg, Theory of the CP violating parameter ε'/ε, *Rev. Mod. Phys.* **72** (2000) 65–93, [hep-ph/9802405].

[661] A. J. Buras and M. Jamin, ε'/ε at the NLO: 10 years later, *JHEP* **01** (2004) 048, [hep-ph/0306217].

[662] S. Bertolini, J. O. Eeg, A. Maiezza, and F. Nesti, New physics in ε' from gluomagnetic contributions and limits on left-right symmetry, *Phys. Rev.* **D86** (2012) 095013, [arXiv:1206.0668].

[663] V. Cirigliano, A. Pich, G. Ecker, and H. Neufeld, Isospin violation in ε', *Phys. Rev. Lett.* **91** (2003) 162001, [hep-ph/0307030].

[664] V. Cirigliano, G. Ecker, H. Neufeld, and A. Pich, Isospin breaking in $K \to \pi\pi$ decays, *Eur. Phys. J.* **C33** (2004) 369–396, [hep-ph/0310351].

[665] J. Bijnens and F. Borg, Isospin breaking in $K \to 3\pi$ decays III: Bremsstrahlung and fit to experiment, *Eur. Phys. J.* **C40** (2005) 383–394, [hep-ph/0501163].

[666] C. Bobeth, A. J. Buras, A. Celis, and M. Jung, Patterns of flavour violation in models with vector-like quarks, *JHEP* **04** (2017) 079, [arXiv:1609.04783].

[667] C. Bobeth and A. J. Buras, Leptoquarks meet ε'/ε and rare kaon processes, *JHEP* **02** (2018) 101, [arXiv:1712.01295].

[668] M. Ciuchini, E. Franco, G. Martinelli, L. Reina, and L. Silvestrini, An upgraded analysis of ε'/ε at the next-to-leading order, *Z. Phys.* **C68** (1995) 239–256, [hep-ph/9501265].

[669] S. Bosch et al., Standard model confronting new results for ε'/ε, *Nucl. Phys.* **B565** (2000) 3–37, [hep-ph/9904408].

[670] T. Kitahara, U. Nierste, and P. Tremper, Singularity-free next-to-leading order $\Delta S = 1$ renormalization group evolution and $\varepsilon'_K/\varepsilon_K$ in the standard model and beyond, *JHEP* **12** (2016) 078, [arXiv:1607.06727].

[671] G. Colangelo, J. Gasser, and H. Leutwyler, $\pi\pi$ scattering, *Nucl. Phys.* **B603** (2001) 125–179, [hep-ph/0103088].

[672] T. Wang and C. Kelly, Studies of I= 0 and 2 $\pi\pi$ scattering with physical pion mass, *PoS* **LATTICE2018** (2019) 276.

[673] C. Kelly and T. Wang, Update on the improved lattice calculation of direct CP-violation in K decays, *PoS* **LATTICE2018** (2019) 277.

[674] A. J. Buras, ϵ'/ϵ-2018: A Christmas story, arXiv:1812.06102.

[675] M. Blanke, A. J. Buras, S. Recksiegel, C. Tarantino, and S. Uhlig, Littlest Higgs model with T-parity confronting the new data on $D^0 - \bar{D}^0$ mixing, *Phys. Lett.* **B657** (2007) 81–86, [hep-ph/0703254].

[676] I. I. Bigi, M. Blanke, A. J. Buras, and S. Recksiegel, CP violation in $D^0 - \bar{D}^0$ oscillations: General considerations and applications to the littlest Higgs model with T-parity, *JHEP* **07** (2009) 097, [arXiv:0904.1545].

[677] G. Blaylock, A. Seiden, and Y. Nir, The role of CP violation in $D^0 - \bar{D}^0$ mixing, *Phys. Lett.* **B355** (1995) 555–560, [hep-ph/9504306].

[678] S. Bianco, F. L. Fabbri, D. Benson, and I. Bigi, A Cicerone for the physics of charm, *Riv. Nuovo Cim.* **26N7** (2003) 1–200, [hep-ex/0309021].

[679] E. Golowich, S. Pakvasa, and A. A. Petrov, New physics contributions to the lifetime difference in $D^0 - \bar{D}^0$ mixing, *Phys. Rev. Lett.* **98** (2007) 181801, [hep-ph/0610039].

[680] M. Ciuchini, E. Franco, D. Guadagnoli, V. Lubicz, M. Pierini, V. Porretti, and L. Silvestrini, $D - \bar{D}$ mixing and new physics: General considerations and constraints on the MSSM, *Phys. Lett.* **B655** (2007) 162–166, [hep-ph/0703204].

[681] E. Golowich, J. Hewett, S. Pakvasa, and A. A. Petrov, Implications of $D^0 - \bar{D}^0$ mixing for new physics, *Phys. Rev.* **D76** (2007) 095009, [arXiv:0705.3650].

[682] M. Artuso, B. Meadows, and A. A. Petrov, Charm meson decays, *Ann. Rev. Nucl. Part. Sci.* **58** (2008) 249–291, [arXiv:0802.2934].

[683] K. Blum, Y. Grossman, Y. Nir, and G. Perez, Combining $K^0 - \bar{K}^0$ mixing and $D^0 - \bar{D}^0$ mixing to constrain the flavor structure of new physics, *Phys. Rev. Lett.* **102** (2009) 211802, [arXiv:0903.2118].

[684] O. Gedalia, Y. Grossman, Y. Nir, and G. Perez, Lessons from recent measurements of $D^0 - \bar{D}^0$ Mixing, *Phys. Rev.* **D80** (2009) 055024, [arXiv:0906.1879].

[685] A. L. Kagan and M. D. Sokoloff, On indirect CP violation and implications for $D^0 - \bar{D}^0$ and $B^0 - \bar{B}^0$ mixing, *Phys. Rev.* **D80** (2009) 076008, [arXiv:0907.3917].

[686] G. Isidori, Y. Nir, and G. Perez, Flavor physics constraints for physics beyond the standard model, *Ann. Rev. Nucl. Part. Sci.* **60** (2010) 355, [arXiv:1002.0900].

[687] UTfit Collaboration, A. J. Bevan et al., The UTfit collaboration average of D meson mixing data: Winter 2014, *JHEP* **03** (2014) 123, [arXiv:1402.1664].

[688] E. Golowich and A. A. Petrov, Short distance analysis of $D^0 - \bar{D}^0$ mixing, *Phys. Lett.* **B625** (2005) 53–62, [hep-ph/0506185].

[689] M. Bobrowski, A. Lenz, and T. Rauh, Short distance $D^0 - \bar{D}^0$ mixing, in *Proceedings, 5th International Workshop on Charm Physics (Charm 2012): Honolulu, Hawaii, USA, May 14–17, 2012*, 2012. arXiv:1208.6438.

[690] A. J. Lenz, Selected topics in heavy flavour physics, *J. Phys.* **G41** (2014) 103001, [arXiv:1404.6197].

[691] S. Fajfer and N. Kosnik, Prospects of discovering new physics in rare charm decays, *Eur. Phys. J.* **C75** (2015), no. 12 567, [arXiv:1510.00965].

[692] V. Bhardwaj, M. Dorigo, and F.-S. Yu, Summary of WG7 at CKM 2018: "Mixing and CP violation in the D system: $x_D, y_D, |q/p|_D, \phi_D$, and direct CP violation in D decays," 2019. arXiv:1901.08131.

[693] LHCb Collaboration, R. Aaij et al., Measurement of the mass difference between neutral charm-meson eigenstates, *Phys. Rev. Lett.* **122** (2019), no. 23 231802, [arXiv:1903.03074].

[694] I. Dunietz and J. L. Rosner, Time dependent CP violation effects in $B^0 - \bar{B}^0$ systems, *Phys. Rev.* **D34** (1986) 1404.

[695] Y. Grossman, Y. Nir, and G. Perez, Testing new indirect CP violation, arXiv:0904.0305.

[696] Z. Ligeti, M. Papucci, and G. Perez, Implications of the measurement of the $B_s^0 - \bar{B}_s^0$ mass difference, *Phys. Rev. Lett* **97** (2006) 101801, [hep-ph/0604112].

[697] M. Blanke, A. J. Buras, D. Guadagnoli, and C. Tarantino, Minimal flavour violation waiting for precise measurements of ΔM_s, $S_{\psi\phi}$, A_{SL}^s, $|V_{ub}|$, γ and $B_{s,d}^0 \rightarrow \mu^+\mu^-$, *JHEP* **10** (2006) 003, [hep-ph/0604057].

[698] M. Blanke and A. J. Buras, A guide to flavour changing neutral currents in the littlest Higgs model with T-parity, *Acta Phys. Polon.* **B38** (2007) 2923, [hep-ph/0703117].

[699] J. Hubisz, S. J. Lee, and G. Paz, The flavor of a little Higgs with T-parity, *JHEP* **06** (2006) 041, [hep-ph/0512169].

[700] M. Blanke et al., Rare and CP-violating K and B decays in the littlest Higgs model with T-parity, *JHEP* **01** (2007) 066, [hep-ph/0610298].

[701] J. Aebischer, C. Bobeth, A. J. Buras, and D. M. Straub, Anatomy of ε'/ε beyond the standard model, *Eur. Phys. J.* **C79** (2019), no. 3 219, [arXiv:1808.00466].

[702] LHCb Collaboration, R. Aaij et al., Observation of *CP* violation in charm decays, arXiv:1903.08726.

[703] M. Chala, A. Lenz, A. V. Rusov, and J. Scholtz, ΔA_{CP} within the standard model and beyond, arXiv:1903.10490.

[704] H.-N. Li, C.-D. L, and F.-S. Yu, Implications on the first observation of charm CPV at LHCb, arXiv:1903.10638.

[705] Y. Grossman and S. Schacht, The Emergence of the $\Delta U = 0$ rule in charm physics, arXiv:1903.10952.

[706] J. Brod, Y. Grossman, A. L. Kagan, and J. Zupan, A consistent picture for large penguins in $D \rightarrow \pi^+\pi^-, K^+K^-$, *JHEP* **10** (2012) 161, [arXiv:1203.6659].

[707] LHCb Collaboration, F. Ferrari, Charm mixing and CPV, 2019. arXiv:1906.10952.

[708] S. Bifani, S. Descotes-Genon, A. Romero Vidal, and M.-H. Schune, Review of lepton universality tests in *B* decays, *J. Phys.* **G46** (2019), no. 2 023001, [arXiv:1809.06229].

[709] M. De Cian, S. Descotes-Genon, and K. Massri, Rare B, D and K decays, radiative and electroweak penguin decays, including constraints on V_{td}/V_{ts} and ϵ'/ϵ: Summary of CKM 2018 Working Group 3, 2019. arXiv:1901.04541.

[710] W. Altmannshofer and D. M. Straub, New physics in $b \rightarrow s$ transitions after LHC run 1, *Eur. Phys. J.* **C75** (2015), no. 8 382, [arXiv:1411.3161].

[711] Fermilab Lattice, MILC Collaboration, A. Bazavov et al., $|V_{us}|$ from $K_{\ell 3}$ decay and four-flavor lattice QCD, arXiv:1809.02827.

[712] C.-Y. Seng, M. Gorchtein, H. H. Patel, and M. J. Ramsey-Musolf, Reduced hadronic uncertainty in the determination of V_{ud}, *Phys. Rev. Lett.* **121** (2018), no. 24 241804, [arXiv:1807.10197].

[713] B. Belfatto, R. Beradze, and Z. Berezhiani, The CKM unitarity problem: A trace of new physics at the TeV scale? arXiv:1906.02714.

[714] LHCb Collaboration, A. C. S. Davis, Experimental prospects for V_{ud}, V_{us}, V_{cd}, V_{cs} and (semi-)leptonic decays at LHCb, in *10th International Workshop on the CKM Unitarity Triangle (CKM 2018) Heidelberg, Germany, September 17–21, 2018*, 2019. arXiv:1901.04785.

[715] A. Filipuzzi, J. Portoles, and M. Gonzalez-Alonso, U(2)^5 flavor symmetry and lepton universality violation in $W \rightarrow \tau\nu_\tau$, *Phys. Rev.* **D85** (2012) 116010, [arXiv:1203.2092].

[716] A. Pich, Precision tau physics, *Prog. Part. Nucl. Phys.* **75** (2014) 41–85, [arXiv:1310.7922].

[717] PiENu Collaboration, A. Aguilar-Arevalo et al., Improved measurement of the $\pi \rightarrow e\nu$ branching ratio, *Phys. Rev. Lett.* **115** (2015), no. 7 071801, [arXiv:1506.05845].

[718] M. Blanke, Quo vadis flavour physics? – FPCP2017 theory summary and outlook, *PoS* **FPCP2017** (2017) 042, [arXiv:1708.06326].

[719] D. Buttazzo, A. Greljo, G. Isidori, and D. Marzocca, B-physics anomalies: A guide to combined explanations, *JHEP* **11** (2017) 044, [arXiv:1706.07808].

[720] W. Altmannshofer, C. Niehoff, P. Stangl, and D. M. Straub, Status of the $B \rightarrow K^*\mu^+\mu^-$ anomaly after Moriond 2017, *Eur. Phys. J.* **C77** (2017), no. 6 377, [arXiv:1703.09189].

[721] W. Altmannshofer, P. Stangl, and D. M. Straub, Interpreting hints for lepton flavor universality violation, *Phys. Rev.* **D96** (2017), no. 5 055008, [arXiv:1704.05435].

[722] F. Beaujean, C. Bobeth, and D. van Dyk, Comprehensive Bayesian analysis of rare (semi)leptonic and radiative *B* decays, *Eur. Phys. J.* **C74** (2014) 2897, [arXiv:1310.2478]. [Erratum: Eur. Phys. J.C74,3179(2014)].

[723] M. Ciuchini, A. M. Coutinho, M. Fedele, E. Franco, A. Paul, L. Silvestrini, and M. Valli, On flavourful Easter eggs for new physics hunger and lepton flavour universality violation, *Eur. Phys. J.* **C77** (2017), no. 10 688, [arXiv:1704.05447].

[724] A. K. Alok, B. Bhattacharya, A. Datta, D. Kumar, J. Kumar, and D. London, New Physics in $b \rightarrow s\mu^+\mu^-$ after the measurement of R_{K^*}, *Phys. Rev.* **D96** (2017), no. 9 095009, [arXiv:1704.07397].

[725] L.-S. Geng, B. Grinstein, S. Jäger, J. Martin Camalich, X.-L. Ren, and R.-X. Shi, Towards the discovery of new physics with lepton-universality ratios of $b \rightarrow s\ell\ell$ decays, *Phys. Rev.* **D96** (2017), no. 9 093006, [arXiv:1704.05446].

[726] T. Hurth, F. Mahmoudi, D. Martinez Santos, and S. Neshatpour, Lepton nonuniversality in exclusive $b \rightarrow s\ell\ell$ decays, *Phys. Rev.* **D96** (2017), no. 9 095034, [arXiv:1705.06274].

[727] A. Celis, J. Fuentes-Martin, A. Vicente, and J. Virto, Gauge-invariant implications of the LHCb measurements on lepton-flavor nonuniversality, *Phys. Rev.* **D96** (2017), no. 3 035026, [arXiv:1704.05672].

[728] D. Choudhury, A. Kundu, R. Mandal, and R. Sinha, Minimal unified resolution to $R(K^{(*)})$ and $R(D^{(*)})$ anomalies with lepton mixing, *Phys. Rev. Lett.* **119** (2017), no. 15 151801, [arXiv:1706.08437].

[729] G. Hiller and I. Nisandzic, R_K and R_{K^*} beyond the standard model, *Phys. Rev.* **D96** (2017), no. 3 035003, [arXiv:1704.05444].

[730] G. D'Amico, M. Nardecchia, P. Panci, F. Sannino, A. Strumia, R. Torre, and A. Urbano, Flavour anomalies after the R_{K^*} measurement, *JHEP* **09** (2017) 010, [arXiv:1704.05438].

[731] A. Arbey, T. Hurth, F. Mahmoudi, and S. Neshatpour, Hadronic and new physics contributions to $b \to s$ transitions, *Phys. Rev.* **D98** (2018), no. 9 095027, [arXiv:1806.02791].

[732] B. Capdevila, U. Laa, and G. Valencia, Anatomy of a six-parameter fit to the $b \to s\ell^+\ell^-$ anomalies, arXiv:1811.10793.

[733] J. Kumar and D. London, New physics in $b \to se^+e^-$? *Phys. Rev.* **D99** (2019), no. 7 073008, [arXiv:1901.04516].

[734] M. Algueró, B. Capdevila, S. Descotes-Genon, P. Masjuan, and J. Matias, Are we overlooking lepton flavour universal new physics in $b \to s\ell\ell$? *Phys. Rev.* **D99** (2019), no. 7 075017, [arXiv:1809.08447].

[735] M. Alguer, B. Capdevila, S. Descotes-Genon, P. Masjuan, and J. Matias, What R_K and Q_5 can tell us about new physics in $b \to s\ell\ell$ transitions? arXiv:1902.04900.

[736] A. Datta, J. Kumar, and D. London, The B anomalies and new physics in $b \to se^+e^-$, arXiv:1903.10086.

[737] M. Ciuchini, A. M. Coutinho, M. Fedele, E. Franco, A. Paul, L. Silvestrini, and M. Valli, New Physics in $b \to s\ell^+\ell^-$ confronts new data on Lepton Universality, arXiv:1903.09632.

[738] M. Alguer, B. Capdevila, A. Crivellin, S. Descotes-Genon, P. Masjuan, J. Matias, and J. Virto, Addendum: "Patterns of New Physics in $b \to s\ell^+\ell^-$ transitions in the light of recent data," arXiv:1903.09578.

[739] K. Kowalska, D. Kumar, and E. M. Sessolo, Implications for New Physics in $b \to s\mu\mu$ transitions after recent measurements by Belle and LHCb, arXiv:1903.10932.

[740] A. K. Alok, A. Dighe, S. Gangal, and D. Kumar, Continuing search for new physics in $b \to s\mu\mu$ decays: two operators at a time, *JHEP* **06** (2019) 089, [arXiv:1903.09617].

[741] A. Arbey, T. Hurth, F. Mahmoudi, D. Martinez Santos, and S. Neshatpour, Update on the $b \to s$ anomalies, arXiv:1904.08399.

[742] C. Bobeth and U. Haisch, New Physics in Γ_{12}^s: $(\bar{s}b)(\bar{\tau}\tau)$ Operators, *Acta Phys. Polon.* **B44** (2013) 127–176, [arXiv:1109.1826].

[743] A. Crivellin, C. Greub, D. Müller, and F. Saturnino, Importance of Loop Effects in Explaining the Accumulated Evidence for New Physics in B Decays with a Vector Leptoquark, *Phys. Rev. Lett.* **122** (2019), no. 1 011805, [arXiv:1807.02068].

[744] A. J. Buras, M. Misiak, and J. Urban, Two loop QCD anomalous dimensions of flavor changing four quark operators within and beyond the standard model, *Nucl. Phys.* **B586** (2000) 397–426, [hep-ph/0005183].

[745] M. Ciuchini, E. Franco, V. Lubicz, G. Martinelli, I. Scimemi, et al., Next-to-leading order QCD corrections to $\Delta F = 2$ effective Hamiltonians, *Nucl. Phys.* **B523** (1998) 501–525, [hep-ph/9711402].

[746] A. J. Buras, S. Jäger, and J. Urban, Master formulae for $\Delta F = 2$ NLO QCD factors in the standard model and beyond, *Nucl. Phys.* **B605** (2001) 600–624, [hep-ph/0102316].

[747] P. Boyle, N. Garron, J. Kettle, A. Khamseh, and J. T. Tsang, BSM kaon mixing at the physical point, *EPJ Web Conf.* **175** (2018) 13010, [arXiv:1710.09176].

[748] A. Bazavov et al., B- and D-meson leptonic decay constants from four-flavor lattice QCD, *Phys. Rev.* **D98** (2018), no. 7 074512, [arXiv:1712.09262].

[749] A. J. Buras and J. Girrbach, Complete NLO QCD corrections for tree level $\Delta F = 2$ FCNC processes, *JHEP* **1203** (2012) 052, [arXiv:1201.1302].

[750] G. Beall, M. Bander, and A. Soni, Constraint on the mass scale of a left-right symmetric electroweak theory from the $K_L - K_S$ mass difference, *Phys. Rev. Lett.* **48** (1982) 848.

[751] J. A. Bagger, K. T. Matchev, and R.-J. Zhang, QCD corrections to flavor changing neutral currents in the supersymmetric standard model, *Phys. Lett.* **B412** (1997) 77–85, [hep-ph/9707225].

[752] SWME Collaboration, B. J. Choi et al., Kaon BSM B-parameters using improved staggered fermions from $N_f = 2 + 1$ unquenched QCD, *Phys. Rev.* **D93** (2016), no. 1 014511, [arXiv:1509.00592].

[753] RBC/UKQCD Collaboration, N. Garron, R. J. Hudspith, and A. T. Lytle, Neutral kaon mixing beyond the standard model with $n_f = 2 + 1$ chiral fermions Part 1: Bare matrix elements and physical results, *JHEP* **11** (2016) 001, [arXiv:1609.03334].

[754] RBC, UKQCD Collaboration, P. A. Boyle, N. Garron, R. J. Hudspith, C. Lehner, and A. T. Lytle, Neutral kaon mixing beyond the standard model with $n_f = 2 + 1$ chiral fermions. Part 2: Non perturbative renormalisation of the $\Delta F = 2$ four-quark operators, *JHEP* **10** (2017) 054, [arXiv:1708.03552].

[755] A. J. Buras and J.-M. Gérard, Dual QCD insight into BSM hadronic matrix elements for $K^0 - \bar{K}^0$ mixing from lattice QCD, *Acta Phys. Polon.* **B50** (2019) 121, [arXiv:1804.02401].

[756] J. Aebischer, M. Fael, C. Greub, and J. Virto, B physics beyond the standard model at one loop: Complete renormalization group evolution below the electroweak scale, *JHEP* **09** (2017) 158, [arXiv:1704.06639].

[757] I. Brivio and M. Trott, The standard model as an effective field theory, *Phys. Rept.* **793** (2019) 1–98, [arXiv:1706.08945].

[758] E. E. Jenkins, A. V. Manohar, and M. Trott, Renormalization group evolution of the standard model dimension six operators I: Formalism and lambda dependence, *JHEP* **10** (2013) 087, [arXiv:1308.2627].

[759] E. E. Jenkins, A. V. Manohar, and M. Trott, Renormalization group evolution of the standard model dimension six operators II: Yukawa dependence, *JHEP* **01** (2014) 035, [arXiv:1310.4838].

[760] R. Alonso, E. E. Jenkins, A. V. Manohar, and M. Trott, Renormalization group evolution of the standard model dimension six operators III: Gauge coupling dependence and phenomenology, *JHEP* **04** (2014) 159, [arXiv:1312.2014].

[761] R. Alonso, H.-M. Chang, E. E. Jenkins, A. V. Manohar, and B. Shotwell, Renormalization group evolution of dimension-six baryon number violating operators, *Phys. Lett.* **B734** (2014) 302–307, [arXiv:1405.0486].

[762] S. Weinberg, Baryon and lepton nonconserving processes, *Phys. Rev. Lett.* **43** (1979) 1566–1570.

[763] F. Wilczek and A. Zee, Operator analysis of nucleon decay, *Phys. Rev. Lett.* **43** (1979) 1571–1573.

[764] L. F. Abbott and M. B. Wise, The effective Hamiltonian for nucleon decay, *Phys. Rev.* **D22** (1980) 2208.

[765] F. Feruglio, P. Paradisi, and A. Pattori, Revisiting lepton flavor universality in B decays, *Phys. Rev. Lett.* **118** (2017), no. 1 011801, [arXiv:1606.00524].

[766] C. Bobeth, A. J. Buras, A. Celis, and M. Jung, Yukawa enhancement of Z-mediated new physics in $\Delta S = 2$ and $\Delta B = 2$ processes, *JHEP* **07** (2017) 124, [arXiv:1703.04753].

[767] F. Feruglio, P. Paradisi, and A. Pattori, On the Importance of electroweak corrections for B anomalies, *JHEP* **09** (2017) 061, [arXiv:1705.00929].

[768] M. Gonzlez-Alonso, J. Martin Camalich, and K. Mimouni, Renormalization-group evolution of new physics contributions to (semi)leptonic meson decays, *Phys. Lett.* **B772** (2017) 777–785, [arXiv:1706.00410].

[769] J. Kumar, D. London, and R. Watanabe, Combined explanations of the $b \to s\mu^+\mu^-$ and $b \to c\tau^-\bar{\nu}$ anomalies: A general model analysis, *Phys. Rev.* **D99** (2019), no. 1 015007, [arXiv:1806.07403].

[770] L. Silvestrini and M. Valli, Model-independent bounds on the standard model effective theory from flavour physics, arXiv:1812.10913.

[771] P. Langacker and M. Plumacher, Flavor changing effects in theories with a heavy Z' boson with family nonuniversal couplings, *Phys. Rev.* **D62** (2000) 013006, [hep-ph/0001204].

[772] M. Blanke, A. J. Buras, K. Gemmler, and T. Heidsieck, $\Delta F = 2$ observables and $B \to X_q \gamma$ in the left-right asymmetric model: Higgs particles striking back, *JHEP* **1203** (2012) 024, [arXiv:1111.5014].

[773] J. Aebischer et al., WCxf: An exchange format for Wilson coefficients beyond the standard model, *Comput. Phys. Commun.* **232** (2018) 71–83, [arXiv:1712.05298].

[774] V. Gherardi, D. Marzocca, M. Nardecchia, and A. Romanino, Rank-one flavor violation and B-meson anomalies, arXiv:1903.10954.

[775] S. Fajfer, J. F. Kamenik, I. Nisandzic, and J. Zupan, Implications of lepton flavour universality violations in B decays, *Phys. Rev. Lett.* **109** (2012) 161801, [arXiv:1206.1872].

[776] S. L. Glashow, D. Guadagnoli, and K. Lane, Lepton flavor violation in B decays? *Phys. Rev. Lett.* **114** (2015) 091801, [arXiv:1411.0565].

[777] B. Bhattacharya, A. Datta, D. London, and S. Shivashankara, Simultaneous explanation of the R_K and $R(D^{(*)})$ puzzles, *Phys. Lett.* **B742** (2015) 370–374, [arXiv:1412.7164].

[778] R. Alonso, B. Grinstein, and J. Martin Camalich, Lepton universality violation and lepton flavor conservation in B-meson decays, *JHEP* **10** (2015) 184, [arXiv:1505.05164].

[779] L. Calibbi, A. Crivellin, and T. Ota, Effective field theory approach to $b \to s\ell\ell'$, $B \to K^{(*)}\nu\bar{\nu}$ and $B \to D^{(*)}\tau\nu$ with third generation couplings, *Phys. Rev. Lett.* **115** (2015) 181801, [arXiv:1506.02661].

[780] E. E. Jenkins, A. V. Manohar, and P. Stoffer, Low-energy effective field theory below the electroweak scale: Anomalous dimensions, *JHEP* **01** (2018) 084, [arXiv:1711.05270].

[781] V. Cirigliano, W. Dekens, J. de Vries, and E. Mereghetti, Constraining the top-Higgs sector of the standard model effective field theory, *Phys. Rev.* **D94** (2016), no. 3 034031, [arXiv:1605.04311].

[782] J. de Blas, J. C. Criado, M. Perez-Victoria, and J. Santiago, Effective description of general extensions of the standard model: The complete tree-level dictionary, *JHEP* **03** (2018) 109, [arXiv:1711.10391].

[783] F. del Aguila, M. Perez-Victoria, and J. Santiago, Observable contributions of new exotic quarks to quark mixing, *JHEP* **0009** (2000) 011, [hep-ph/0007316].

[784] F. del Aguila, J. de Blas, and M. Perez-Victoria, Effects of new leptons in electroweak precision data, *Phys. Rev.* **D78** (2008) 013010, [arXiv:0803.4008].

[785] F. del Aguila, J. de Blas, and M. Perez-Victoria, Electroweak limits on general new vector bosons, *JHEP* **09** (2010) 033, [arXiv:1005.3998].

[786] J. de Blas, M. Chala, M. Perez-Victoria, and J. Santiago, Observable effects of general new scalar particles, *JHEP* **04** (2015) 078, [arXiv:1412.8480].

[787] A. Dedes, W. Materkowska, M. Paraskevas, J. Rosiek, and K. Suxho, Feynman rules for the standard model effective field theory in R_ξ-gauges, *JHEP* **06** (2017) 143, [arXiv:1704.03888].

[788] A. Dedes, M. Paraskevas, J. Rosiek, K. Suxho, and L. Trifyllis, SmeftFR – Feynman rules generator for the standard model effective field theory, arXiv:1904.03204.

[789] I. Brivio, Y. Jiang, and M. Trott, The SMEFTsim package, theory and tools, *JHEP* **12** (2017) 070, [arXiv:1709.06492].

[790] M. Misiak, M. Paraskevas, J. Rosiek, K. Suxho, and B. Zglinicki, Effective field theories in R_ξ gauges, *JHEP* **02** (2019) 051, [arXiv:1812.11513].

[791] M. Jiang, N. Craig, Y.-Y. Li, and D. Sutherland, Complete one-loop matching for a singlet scalar in the standard model EFT, *JHEP* **02** (2019) 031, [arXiv:1811.08878].

[792] P. Arnan, L. Hofer, F. Mescia, and A. Crivellin, Loop effects of heavy new scalars and fermions in $b \to s\mu^+\mu^-$, *JHEP* **04** (2017) 043, [arXiv:1608.07832].

[793] P. Arnan, A. Crivellin, M. Fedele, and F. Mescia, Generic loop effects of new scalars and fermions in $b \to s\ell^+\ell^-$ and a vector-like 4th generation, arXiv:1904.05890.

[794] J. Aebischer, A. Crivellin, and C. Greub, QCD improved matching for semi-leptonic B decays with leptoquarks, arXiv:1811.08907.

[795] A. Crivellin, S. Najjari, and J. Rosiek, Lepton flavor violation in the standard model with general dimension-six operators, *JHEP* **04** (2014) 167, [arXiv:1312.0634].

[796] A. Crivellin, S. Davidson, G. M. Pruna, and A. Signer, Renormalisation-group improved analysis of $\mu \to e$ processes in a systematic effective-field-theory approach, *JHEP* **05** (2017) 117, [arXiv:1702.03020].

[797] V. Cirigliano, W. Dekens, J. de Vries, M. L. Graesser, and E. Mereghetti, A neutrinoless double beta decay master formula from effective field theory, *JHEP* **12** (2018) 097, [arXiv:1806.02780].

[798] J. Aebischer, C. Bobeth, A. J. Buras, J.-M. Gérard, and D. M. Straub, Master formula for ε'/ε beyond the standard model, *Phys. Lett.* **B792** (2019) 465–469, [arXiv:1807.02520].

[799] M. Endo, T. Kitahara, S. Mishima, and K. Yamamoto, Revisiting kaon physics in general Z scenario, *Phys. Lett.* **B771** (2017) 37–44, [arXiv:1612.08839].

[800] M. Endo, T. Kitahara, and D. Ueda, SMEFT top-quark effects on $\Delta F = 2$ observables, arXiv:1811.04961.

[801] X.-G. He, J. Tandean, and G. Valencia, Lepton-flavor-violating semileptonic τ decay and $K \to \pi\nu\bar{\nu}$, arXiv:1904.04043.

[802] J. P. Leveille, The second order weak correction to $g - 2$ of the muon in arbitrary gauge models, *Nucl. Phys.* **B137** (1978) 63–76.

[803] F. Jegerlehner and A. Nyffeler, The Muon $g - 2$, *Phys. Rept.* **477** (2009) 1–110, [arXiv:0902.3360].

[804] M. Lindner, M. Platscher, and F. S. Queiroz, A call for new physics : The muon anomalous magnetic moment and Lepton flavor violation, *Phys. Rept.* **731** (2018) 1–82, [arXiv:1610.06587].

[805] V. Cirigliano, M. Gonzalez-Alonso, and M. L. Graesser, Non-standard charged current interactions: Beta decays versus the LHC, *JHEP* **02** (2013) 046, [arXiv:1210.4553].

[806] W. Dekens and J. de Vries, Renormalization group running of dimension-six sources of parity and time-reversal violation, *JHEP* **05** (2013) 149, [arXiv:1303.3156].

[807] T. Bhattacharya, V. Cirigliano, R. Gupta, E. Mereghetti, and B. Yoon, Dimension-5 CP-odd operators: QCD mixing and renormalization, *Phys. Rev.* **D92** (2015), no. 11 114026, [arXiv:1502.07325].

[808] C. Grojean, E. E. Jenkins, A. V. Manohar, and M. Trott, Renormalization group scaling of Higgs operators and $\Gamma(h \to \gamma\gamma)$, *JHEP* **04** (2013) 016, [arXiv:1301.2588].

[809] V. Cirigliano, S. Davidson, and Y. Kuno, Spin-dependent $\mu \to e$ conversion, *Phys. Lett.* **B771** (2017) 242–246, [arXiv:1703.02057].

[810] Q.-Y. Hu, X.-Q. Li, and Y.-D. Yang, $b \to c\tau\nu$ transitions in the standard model effective field theory, *Eur. Phys. J.* **C79** (2019), no. 3 264, [arXiv:1810.04939].

[811] E. E. Jenkins, A. V. Manohar, and P. Stoffer, Low-energy effective field theory below the electroweak scale: Operators and matching, *JHEP* **03** (2018) 016, [arXiv:1709.04486].

[812] T. Hurth, S. Renner, and W. Shepherd, Matching for FCNC effects in the flavour-symmetric SMEFT, *JHEP* **06** (2019) 029, [arXiv:1903.00500].

[813] D. M. Straub, Flavio: A Python package for flavour and precision phenomenology in the standard model and beyond, arXiv:1810.08132.

[814] D. van Dyk et al., EOS – A HEP Programm for Flavour Observables.

[815] W. Porod, F. Staub, and A. Vicente, A Flavor Kit for BSM models, *Eur. Phys. J.* **C74** (2014), no. 8 2992, [arXiv:1405.1434].

[816] W. Porod and F. Staub, SPheno 3.1: Extensions including flavour, CP-phases and models beyond the MSSM, *Comput. Phys. Commun.* **183** (2012) 2458–2469, [arXiv:1104.1573].

[817] J. A. Evans and D. Shih, FormFlavor manual, arXiv:1606.00003.

[818] F. Mahmoudi, SuperIso: A program for calculating the isospin asymmetry of $B \to K^*\gamma$ in the MSSM, *Comput. Phys. Commun.* **178** (2008) 745–754, [arXiv:0710.2067].

[819] J. de Blas et al., HEPfit: A code for the combination of indirect and direct constraints on high energy physics models, arXiv:1910.14012.

[820] The GAMBIT Flavour Workgroup Collaboration, F. U. Bernlochner et al., FlavBit: A GAMBIT module for computing flavour observables and likelihoods, *Eur. Phys. J.* **C77** (2017), no. 11 786, [arXiv:1705.07933].

[821] A. Celis, J. Fuentes-Martin, A. Vicente, and J. Virto, DsixTools: The standard model effective field theory toolkit, *Eur. Phys. J.* **C77** (2017), no. 6 405, [arXiv:1704.04504].

[822] J. Aebischer, J. Kumar, and D. M. Straub, Wilson: A Python package for the running and matching of Wilson coefficients above and below the electroweak scale, *Eur. Phys. J.* **C78** (2018), no. 12 1026, [arXiv:1804.05033].

[823] J. C. Criado, MatchingTools: A Python library for symbolic effective field theory calculations, *Comput. Phys. Commun.* **227** (2018) 42–50, [arXiv:1710.06445].

[824] F. Bishara, J. Brod, B. Grinstein, and J. Zupan, DirectDM: A tool for dark matter direct detection, arXiv:1708.02678.

[825] F. Staub, Exploring new models in all detail with SARAH, *Adv. High Energy Phys.* **2015** (2015) 840780, [arXiv:1503.04200].

[826] S. Das Bakshi, J. Chakrabortty, and S. K. Patra, CoDEx: Wilson coefficient calculator connecting SMEFT to UV theory, *Eur. Phys. J.* **C79** (2019), no. 1 21, [arXiv:1808.04403].

[827] J. C. Criado, BasisGen: Automatic generation of operator bases, *Eur. Phys. J.* **C79** (2019), no. 3 256, [arXiv:1901.03501].

[828] S. Descotes-Genon, A. Falkowski, M. Fedele, M. Gonzlez-Alonso, and J. Virto, The CKM parameters in the SMEFT, *JHEP* **05** (2019) 172, [arXiv:1812.08163].

[829] J. de Blas, O. Eberhardt, and C. Krause, Current and future constraints on Higgs couplings in the nonlinear effective theory, *JHEP* **07** (2018) 048, [arXiv:1803.00939].

[830] F. Feruglio, The chiral approach to the electroweak interactions, *Int. J. Mod. Phys.* **A8** (1993) 4937–4972, [hep-ph/9301281].

[831] J. Bagger, V. D. Barger, K.-M. Cheung, J. F. Gunion, T. Han, G. A. Ladinsky, R. Rosenfeld, and C. P. Yuan, The strongly interacting W-W system: Gold plated modes, *Phys. Rev.* **D49** (1994) 1246–1264, [hep-ph/9306256].

[832] V. Koulovassilopoulos and R. S. Chivukula, The phenomenology of a nonstandard Higgs boson in W(L) W(L) scattering, *Phys. Rev.* **D50** (1994) 3218–3234, [hep-ph/9312317].

[833] L.-M. Wang and Q. Wang, Electroweak chiral Lagrangian for neutral Higgs boson, *Chin. Phys. Lett.* **25** (2008) 1984, [hep-ph/0605104].

[834] B. Grinstein and M. Trott, A Higgs-Higgs bound state due to new physics at a TeV, *Phys. Rev.* **D76** (2007) 073002, [arXiv:0704.1505].

[835] R. Alonso, M. B. Gavela, L. Merlo, S. Rigolin, and J. Yepes, The effective chiral Lagrangian for a light dynamical "Higgs particle," *Phys. Lett.* **B722** (2013) 330–335, [arXiv:1212.3305]. [Erratum: Phys. Lett.B726,926(2013)].

[836] R. Contino, The Higgs as a composite Nambu-Goldstone boson, in *Physics of the Large and the Small, TASI 09, proceedings of the Theoretical Advanced Study Institute in Elementary Particle Physics, Boulder, Colorado, USA, June 1–26, 2009*, pp. 235–306, 2011. arXiv:1005.4269.

[837] R. L. Delgado, A. Dobado, and F. J. Llanes-Estrada, One-loop $W_L W_L$ and $Z_L Z_L$ scattering from the electroweak chiral Lagrangian with a light Higgs-like scalar, *JHEP* **02** (2014) 121, [arXiv:1311.5993].

[838] G. Buchalla, O. Cata, and C. Krause, Complete electroweak chiral Lagrangian with a light Higgs at NLO, *Nucl. Phys.* **B880** (2014) 552–573, [arXiv:1307.5017]. [Erratum: Nucl. Phys.B913,475(2016)].

[839] G. Buchalla, O. Cata, and C. Krause, A systematic approach to the SILH Lagrangian, *Nucl. Phys.* **B894** (2015) 602–620, [arXiv:1412.6356].

[840] G. Buchalla, O. Cata, A. Celis, M. Knecht, and C. Krause, Complete one-loop renormalization of the Higgs-electroweak chiral Lagrangian, *Nucl. Phys.* **B928** (2018) 93–106, [arXiv:1710.06412].

[841] R. Alonso, K. Kanshin, and S. Saa, Renormalization group evolution of Higgs effective field theory, *Phys. Rev.* **D97** (2018), no. 3 035010, [arXiv:1710.06848].

[842] G. C. Branco, J. M. Frere, and J. M. Gérard, The value of ϵ'/ϵ in models based on $SU(2)_L \times SU(2)_R \times U(1)$, *Nucl. Phys.* **B221** (1983) 317–330.

[843] ETM Collaboration, M. Constantinou, M. Costa, R. Frezzotti, V. Lubicz, G. Martinelli, D. Meloni, H. Panagopoulos, and S. Simula, $K \to \pi$ matrix elements of the chromomagnetic operator on the lattice, *Phys. Rev.* **D97** (2018), no. 7 074501, [arXiv:1712.09824].

[844] A. J. Buras and J.-M. Gérard, $K \to \pi\pi$ and $K - \pi$ matrix elements of the chromomagnetic operators from dual QCD, *JHEP* **07** (2018) 126, [arXiv:1803.08052].

[845] S. Bertolini, J. O. Eeg, and M. Fabbrichesi, Studying ϵ'/ϵ in the chiral quark model: γ_5 scheme independence and NLO hadronic matrix elements, *Nucl. Phys.* **B449** (1995) 197–228, [hep-ph/9409437].

[846] C.-H. Chen and T. Nomura, ϵ'/ϵ from charged-Higgs-induced gluonic dipole operators, *Phys. Lett.* **B787** (2018) 182–187, [arXiv:1805.07522].

[847] A. J. Buras, The return of kaon flavour physics, *Acta Phys. Polon.* **B49** (2018) 1043, [arXiv:1805.11096].

[848] F. Feruglio, P. Paradisi, and O. Sumensari, Implications of scalar and tensor explanations of $R_{D^{(*)}}$, *JHEP* **11** (2018) 191, [arXiv:1806.10155].

[849] D. Becirevic, I. Dorsner, S. Fajfer, N. Kosnik, D. A. Faroughy, and O. Sumensari, Scalar leptoquarks from grand unified theories to accommodate the B-physics anomalies, *Phys. Rev.* **D98** (2018), no. 5 055003, [arXiv:1806.05689].

[850] M. Blanke, A. J. Buras, and S. Recksiegel, Quark flavour observables in the littlest Higgs model with T-parity after LHC Run 1, *Eur. Phys. J.* **C76** (2016), no. 4 182, [arXiv:1507.06316].

[851] A. J. Buras, New physics patterns in ϵ'/ϵ and ϵ_K with implications for rare kaon decays and ΔM_K, *JHEP* **04** (2016) 071, [arXiv:1601.00005].

[852] A. J. Buras and F. De Fazio, ϵ'/ϵ in 331 models, *JHEP* **03** (2016) 010, [arXiv:1512.02869].

[853] A. J. Buras and F. De Fazio, 331 models facing the tensions in $\Delta F = 2$ processes with the impact on ϵ'/ϵ, $B_s \to \mu^+\mu^-$ and $B \to K^*\mu^+\mu^-$, *JHEP* **08** (2016) 115, [arXiv:1604.02344].

[854] M. Tanimoto and K. Yamamoto, Probing SUSY with 10 TeV stop mass in rare decays and CP violation of kaon, *PTEP* **2016** (2016), no. 12 123B02, [arXiv:1603.07960].

[855] T. Kitahara, U. Nierste, and P. Tremper, Supersymmetric explanation of CP violation in $K \to \pi\pi$ decays, *Phys. Rev. Lett.* **117** (2016), no. 9 091802, [arXiv:1604.07400].

[856] M. Endo, S. Mishima, D. Ueda, and K. Yamamoto, Chargino contributions in light of recent ϵ'/ϵ, *Phys. Lett.* **B762** (2016) 493–497, [arXiv:1608.01444].

[857] A. Crivellin, G. D'Ambrosio, T. Kitahara, and U. Nierste, $K \to \pi\nu\bar{\nu}$ in the MSSM in light of the ϵ'_K/ϵ_K anomaly, *Phys. Rev.* **D96** (2017), no. 1 015023, [arXiv:1703.05786].

[858] M. Endo, T. Goto, T. Kitahara, S. Mishima, D. Ueda, and K. Yamamoto, Gluino-mediated electroweak penguin with flavor-violating trilinear couplings, *JHEP* **04** (2018) 019, [arXiv:1712.04959].

[859] C.-H. Chen and T. Nomura, $Re(\epsilon'_K/\epsilon_K)$ and $K \to \pi\nu\bar{\nu}$ in a two-Higgs doublet model, *JHEP* **08** (2018) 145, [arXiv:1804.06017].

[860] S. Iguro and Y. Omura, The direct CP violation in a general two Higgs doublet model, arXiv:1905.11778.

[861] V. Cirigliano, W. Dekens, J. de Vries, and E. Mereghetti, An ε' improvement from right-handed currents, *Phys. Lett.* **B767** (2017) 1–9, [arXiv:1612.03914].

[862] S. Alioli, V. Cirigliano, W. Dekens, J. de Vries, and E. Mereghetti, Right-handed charged currents in the era of the large hadron collider, *JHEP* **05** (2017) 086, [arXiv:1703.04751].

[863] N. Haba, H. Umeeda, and T. Yamada, ϵ'/ϵ anomaly and neutron EDM in $SU(2)_L \times SU(2)_R \times U(1)_{B-L}$ model with charge symmetry, *JHEP* **05** (2018) 052, [arXiv:1802.09903].

[864] N. Haba, H. Umeeda, and T. Yamada, Direct CP violation in Cabibbo-favored charmed meson decays and ϵ'/ϵ in $SU(2)_L \times SU(2)_R \times U(1)_{B-L}$ Model, *JHEP* **10** (2018) 006, [arXiv:1806.03424].

[865] S. Matsuzaki, K. Nishiwaki, and K. Yamamoto, Simultaneous interpretation of K and B anomalies in terms of chiral-flavorful vectors, *JHEP* **11** (2018) 164, [arXiv:1806.02312].

[866] C.-H. Chen and T. Nomura, ϵ_K and ϵ'/ϵ in a diquark model, *JHEP* **03** (2019) 009, [arXiv:1808.04097].

[867] C.-H. Chen and T. Nomura, Left-handed color-sextet diquark in the kaon system, *Phys. Rev.* **D99** (2019), no. 11 115006, [arXiv:1811.02315].

[868] C. Marzo, L. Marzola, and M. Raidal, Common explanation to the $R_{K^{(*)}}$, $R_{D^{(*)}}$ and ϵ'/ϵ anomalies in a 3HDM+ν_R and connections to neutrino physics, arXiv:1901.08290.

[869] S. Matsuzaki, K. Nishiwaki, and K. Yamamoto, Simultaneous explanation of K and B anomalies in vectorlike compositeness, in *18th Hellenic School and Workshops on Elementary Particle Physics and Gravity (CORFU2018) Corfu, Corfu, Greece, August 31–September 28, 2018*, 2019. arXiv:1903.10823.

[870] A. J. Buras and J. Girrbach, BSM models facing the recent LHCb data: A first look, *Acta Phys. Polon.* **B43** (2012) 1427, [arXiv:1204.5064].

[871] A. J. Buras and R. Buras, A lower bound on sin 2β from minimal flavor violation, *Phys. Lett.* **B501** (2001) 223–230, [hep-ph/0008273].

[872] M. Blanke and A. J. Buras, Lower bounds on $\Delta M_{s,d}$ from constrained minimal flavour violation, *JHEP* **0705** (2007) 061, [hep-ph/0610037].

[873] Fermilab Lattice, MILC Collaboration, A. Bazavov et al., $B^0_{(s)}$-mixing matrix elements from lattice QCD for the standard model and beyond, *Phys. Rev.* **D93** (2016), no. 11 113016, [arXiv:1602.03560].

[874] M. Blanke and A. J. Buras, Emerging ΔM_d-anomaly from tree-level determinations of $|V_{cb}|$ and the angle γ, *Eur. Phys. J.* **C79** (2019), no. 2 159, [arXiv:1812.06963].

[875] T. Gershon, $\Delta\Gamma_d$: A forgotten null test of the standard model, *J. Phys.* **G38** (2011) 015007, [arXiv:1007.5135].

[876] T. Feldmann and T. Mannel, Minimal flavour violation and beyond, *JHEP* **0702** (2007) 067, [hep-ph/0611095].

[877] G. Colangelo, E. Nikolidakis, and C. Smith, Supersymmetric models with minimal flavour violation and their running, *Eur. Phys. J.* **C59** (2009) 75–98, [arXiv:0807.0801].

[878] P. Paradisi, M. Ratz, R. Schieren, and C. Simonetto, Running minimal flavor violation, *Phys. Lett.* **B668** (2008) 202–209, [arXiv:0805.3989].

[879] L. Mercolli and C. Smith, EDM constraints on flavored CP-violating phases, *Nucl. Phys.* **B817** (2009) 1–24, [arXiv:0902.1949].

[880] T. Feldmann, M. Jung, and T. Mannel, Sequential flavour symmetry breaking, *Phys. Rev.* **D80** (2009) 033003, [arXiv:0906.1523].

[881] A. L. Kagan, G. Perez, T. Volansky, and J. Zupan, General minimal flavor violation, *Phys. Rev.* **D80** (2009) 076002, [arXiv:0903.1794].

[882] P. Paradisi and D. M. Straub, The SUSY CP problem and the MFV principle, *Phys. Lett.* **B684** (2010) 147–153, [arXiv:0906.4551].

[883] G. Isidori, B physics in the LHC era, arXiv:1001.3431.

[884] Y. Nir, Probing new physics with flavor physics (and probing flavor physics with new physics), in *Prospects in Theoretical Physics (PiTP) Summer Program on The Standard Model and Beyond IAS, Princeton, NJ, June 16–27, 2007*, 2007. arXiv:0708.1872.

[885] G. Isidori and D. M. Straub, Minimal flavour violation and beyond, *Eur. Phys. J.* **C72** (2012) 2103, [arXiv:1202.0464].

[886] S. Baek and P. Ko, Probing SUSY induced CP violations at B factories, *Phys. Rev. Lett.* **83** (1999) 488–491, [hep-ph/9812229].

[887] S. Baek and P. Ko, Effects of supersymmetric CP violating phases on $B \to X_s \ell^+ \ell^-$ and ϵ_K, *Phys. Lett.* **B462** (1999) 95–102, [hep-ph/9904283].

[888] A. Bartl, T. Gajdosik, E. Lunghi, A. Masiero, W. Porod, et al., General flavor blind MSSM and CP violation, *Phys. Rev.* **D64** (2001) 076009, [hep-ph/0103324].

[889] J. Ellis, J. S. Lee, and A. Pilaftsis, B-meson observables in the maximally CP-violating MSSM with minimal flavour violation, *Phys. Rev.* **D76** (2007) 115011, [arXiv:0708.2079].

[890] W. Altmannshofer, A. Buras, and P. Paradisi, Low energy probes of CP violation in a flavor blind MSSM, *Phys. Lett.* **B669** (2008) 239–245, [arXiv:0808.0707].

[891] A. Pich and P. Tuzon, Yukawa alignment in the two-Higgs-doublet model, *Phys. Rev.* **D80** (2009) 091702, [arXiv:0908.1554].

[892] K. Blum, Y. Hochberg, and Y. Nir, Implications of large dimuon CP asymmetry in $B_{d,s}$ decays on minimal flavor violation with low $\tan\beta$, *JHEP* **1009** (2010) 035, [arXiv:1007.1872].

[893] B. A. Dobrescu, P. J. Fox, and A. Martin, CP violation in B_s mixing from heavy Higgs exchange, *Phys. Rev. Lett.* **105** (2010) 041801, [arXiv:1005.4238].

[894] W. Altmannshofer and M. Carena, B meson mixing in effective theories of supersymmetric Higgs bosons, *Phys. Rev.* **D85** (2012) 075006, [arXiv:1110.0843].

[895] W. Altmannshofer, M. Carena, S. Gori, and A. de la Puente, Signals of CP violation beyond the MSSM in Higgs and flavor physics, *Phys. Rev.* **D84** (2011) 095027, [arXiv:1107.3814].

[896] A. J. Buras, M. V. Carlucci, S. Gori, and G. Isidori, Higgs-mediated FCNCs: Natural flavour conservation vs. minimal flavour violation, *JHEP* **1010** (2010) 009, [arXiv:1005.5310].

[897] G. Branco, P. Ferreira, L. Lavoura, M. Rebelo, M. Sher, et al., Theory and phenomenology of two-Higgs-doublet models, *Phys. Rept.* **516** (2012) 1–102, [arXiv:1106.0034].

[898] T. Feldmann and T. Mannel, Large top mass and non-linear representation of flavour symmetry, *Phys. Rev. Lett.* **100** (2008) 171601, [arXiv:0801.1802].

[899] S. L. Glashow and S. Weinberg, Natural Conservation Laws for Neutral Currents, *Phys. Rev.* **D15** (1977) 1958.

[900] E. Paschos, Diagonal neutral currents, *Phys. Rev.* **D15** (1977) 1966.

[901] G. C. Branco, W. Grimus, and L. Lavoura, Relating the scalar flavor changing neutral couplings to the CKM matrix, *Phys. Lett.* **B380** (1996) 119–126, [hep-ph/9601383].

[902] A. S. Joshipura and B. P. Kodrani, Minimal flavour violations and tree level FCNC, *Phys. Rev.* **D77** (2008) 096003, [arXiv:0710.3020].

[903] F. J. Botella, G. C. Branco, and M. N. Rebelo, Minimal flavour violation and multi-Higgs models, *Phys. Lett.* **B687** (2010) 194–200, [arXiv:0911.1753].

[904] A. Celis, J. Fuentes-Martin, M. Jung, and H. Serodio, Family nonuniversal Z models with protected flavor-changing interactions, *Phys. Rev.* **D92** (2015), no. 1 015007, [arXiv:1505.03079].

[905] J. M. Alves, F. J. Botella, G. C. Branco, F. Cornet-Gomez, and M. Nebot, Controlled flavour changing neutral couplings in two Higgs doublet models, *Eur. Phys. J.* **C77** (2017), no. 9 585, [arXiv:1703.03796].

[906] M. Nebot, F. J. Botella, and G. C. Branco, Vacuum induced CP violation generating a complex CKM matrix with controlled scalar FCNC, arXiv:1808.00493.

[907] M. Jung, A. Pich, and P. Tuzon, Charged-Higgs phenomenology in the aligned two-Higgs-doublet model, *JHEP* **1011** (2010) 003, [arXiv:1006.0470].

[908] M. Jung and A. Pich, Electric dipole moments in two-Higgs-doublet models, *JHEP* **04** (2014) 076, [arXiv:1308.6283].

[909] A. J. Buras, G. Isidori, and P. Paradisi, EDMs versus CPV in $B_{s,d}$ mixing in two Higgs doublet models with MFV, *Phys. Lett.* **B694** (2011) 402–409, [arXiv:1007.5291].

[910] R. Barbieri, G. Isidori, J. Jones-Perez, P. Lodone, and D. M. Straub, U(2) and minimal flavour violation in supersymmetry, *Eur. Phys. J.* **C71** (2011) 1725, [arXiv:1105.2296].

[911] R. Barbieri, P. Campli, G. Isidori, F. Sala, and D. M. Straub, B-decay CP-asymmetries in SUSY with a $U(2)^3$ flavour symmetry, *Eur. Phys. J.* **C71** (2011) 1812, [arXiv:1108.5125].

[912] R. Barbieri, D. Buttazzo, F. Sala, and D. M. Straub, Flavour physics from an approximate $U(2)^3$ symmetry, *JHEP* **1207** (2012) 181, [arXiv:1203.4218].

[913] R. Barbieri, D. Buttazzo, F. Sala, and D. M. Straub, Less minimal flavour violation, *JHEP* **1210** (2012) 040, [arXiv:1206.1327].

[914] A. Crivellin, L. Hofer, and U. Nierste, The MSSM with a softly broken $U(2)^3$ flavor symmetry, *PoS* **EPS-HEP2011** (2011) 145, [arXiv:1111.0246].

[915] A. Crivellin, L. Hofer, U. Nierste, and D. Scherer, Phenomenological consequences of radiative flavor violation in the MSSM, *Phys. Rev.* **D84** (2011) 035030, [arXiv:1105.2818].

[916] A. Crivellin and U. Nierste, Supersymmetric renormalisation of the CKM matrix and new constraints on the squark mass matrices, *Phys. Rev.* **D79** (2009) 035018, [arXiv:0810.1613].

[917] A. J. Buras and J. Girrbach, On the correlations between flavour observables in minimal $U(2)^3$ models, *JHEP* **1301** (2013) 007, [arXiv:1206.3878].

[918] P. Langacker, The physics of heavy Z' gauge bosons, *Rev. Mod. Phys.* **81** (2009) 1199–1228, [arXiv:0801.1345].

[919] J. Erler, P. Langacker, S. Munir, and E. Rojas, Improved constraints on Z' bosons from electroweak precision data, *JHEP* **0908** (2009) 017, [arXiv:0906.2435].

[920] M. Blanke, A. J. Buras, B. Duling, K. Gemmler, and S. Gori, Rare K and B decays in a warped extra dimension with custodial protection, *JHEP* **03** (2009) 108, [arXiv:0812.3803].

[921] M. Blanke, A. J. Buras, B. Duling, S. Recksiegel, and C. Tarantino, FCNC processes in the littlest Higgs model with T-parity: A 2009 look, *Acta Phys. Polon.* **B41** (2010) 657–683, [arXiv:0906.5454].

[922] M. Bauer, S. Casagrande, U. Haisch, and M. Neubert, Flavor physics in the Randall-Sundrum model: II. Tree-level weak-interaction processes, *JHEP* **1009** (2010) 017, [arXiv:0912.1625].

[923] M. Blanke, A. J. Buras, S. Recksiegel, and C. Tarantino, The littlest Higgs model with T-parity facing CP-violation in $B_s - \bar{B}_s$ mixing, arXiv:0805.4393.

[924] M. Blanke, Insights from the interplay of $K \to \pi \nu \bar{\nu}$ and ϵ_K on the new physics flavour structure, *Acta Phys. Polon.* **B41** (2010) 127, [arXiv:0904.2528].

[925] W. Altmannshofer, S. Gori, M. Pospelov, and I. Yavin, Quark flavor transitions in $L_\mu - L_\tau$ models, *Phys. Rev.* **D89** (2014) 095033, [arXiv:1403.1269].

[926] A. Crivellin, G. D'Ambrosio, and J. Heeck, Explaining $h \to \mu^\pm \tau^\mp$, $B \to K^* \mu^+ \mu^-$ and $B \to K \mu^+ \mu^- / B \to K e^+ e^-$ in a two-Higgs-doublet model with gauged $L_\mu - L_\tau$, *Phys. Rev. Lett.* **114** (2015) 151801, [arXiv:1501.00993].

[927] A. Crivellin, G. D'Ambrosio, and J. Heeck, Addressing the LHC flavor anomalies with horizontal gauge symmetries, *Phys. Rev.* **D91** (2015), no. 7 075006, [arXiv:1503.03477].

[928] A. Crivellin, J. Fuentes-Martin, A. Greljo, and G. Isidori, Lepton flavor non-universality in B decays from dynamical Yukawas, *Phys. Lett.* **B766** (2017) 77–85, [arXiv:1611.02703].

[929] W. Altmannshofer and I. Yavin, Predictions for lepton flavor universality violation in rare B decays in models with gauged $L_\mu - L_\tau$, *Phys. Rev.* **D92** (2015), no. 7 075022, [arXiv:1508.07009].

[930] K. Fuyuto, W.-S. Hou, and M. Kohda, Z-induced FCNC decays of top, beauty, and strange quarks, *Phys. Rev.* **D93** (2016), no. 5 054021, [arXiv:1512.09026].

[931] C.-H. Chen and T. Nomura, Penguin $b \to s\ell'^+ \ell'^-$ and B-meson anomalies in a gauged $L_\mu - L_\tau$, *Phys. Lett.* **B777** (2018) 420–427, [arXiv:1707.03249].

[932] A. Falkowski, M. Nardecchia, and R. Ziegler, Lepton flavor non-universality in B-meson decays from a U(2) Flavor Model, *JHEP* **11** (2015) 173, [arXiv:1509.01249].

[933] S. M. Boucenna, A. Celis, J. Fuentes-Martin, A. Vicente, and J. Virto, Non-abelian gauge extensions for B-decay anomalies, *Phys. Lett.* **B760** (2016) 214–219, [arXiv:1604.03088].

[934] S. M. Boucenna, A. Celis, J. Fuentes-Martin, A. Vicente, and J. Virto, Phenomenology of an $SU(2) \times SU(2) \times U(1)$ model with lepton-flavour non-universality, *JHEP* **12** (2016) 059, [arXiv:1608.01349].

[935] R. Alonso, P. Cox, C. Han, and T. T. Yanagida, Anomaly-free local horizontal symmetry and anomaly-full rare B-decays, *Phys. Rev.* **D96** (2017), no. 7 071701, [arXiv:1704.08158].

[936] J. Ellis, M. Fairbairn, and P. Tunney, Anomaly-free models for flavour anomalies, *Eur. Phys. J.* **C78** (2018), no. 3 238, [arXiv:1705.03447].

[937] R. Alonso, P. Cox, C. Han, and T. T. Yanagida, Flavoured B-L local symmetry and anomalous rare B decays, *Phys. Lett.* **B774** (2017) 643–648, [arXiv:1705.03858].

[938] C. Bonilla, T. Modak, R. Srivastava, and J. W. F. Valle, $U(1)_{B_3-3L_\mu}$ gauge symmetry as a simple description of $b \to s$ anomalies, *Phys. Rev.* **D98** (2018), no. 9 095002, [arXiv:1705.00915].

[939] K. S. Babu, A. Friedland, P. A. N. Machado, and I. Mocioiu, Flavor gauge models below the Fermi scale, *JHEP* **12** (2017) 096, [arXiv:1705.01822].

[940] L. Bian, S.-M. Choi, Y.-J. Kang, and H. M. Lee, A minimal flavored $U(1)'$ for B-meson anomalies, *Phys. Rev.* **D96** (2017), no. 7 075038, [arXiv:1707.04811].

[941] Y. Tang and Y.-L. Wu, Flavor non-universal gauge interactions and anomalies in B-meson decays, *Chin. Phys.* **C42** (2018), no. 3 033104, [arXiv:1705.05643].

[942] J. M. Cline and J. Martin Camalich, B decay anomalies from nonabelian local horizontal symmetry, *Phys. Rev.* **D96** (2017), no. 5 055036, [arXiv:1706.08510].

[943] G. Blanger, C. Delaunay, and S. Westhoff, A dark matter relic from muon anomalies, *Phys. Rev.* **D92** (2015) 055021, [arXiv:1507.06660].

[944] A. Greljo, G. Isidori, and D. Marzocca, On the breaking of lepton flavor universality in B decays, *JHEP* **07** (2015) 142, [arXiv:1506.01705].

[945] B. Bhattacharya, A. Datta, J.-P. Guvin, D. London, and R. Watanabe, Simultaneous explanation of the R_K and $R_{D^{(*)}}$ puzzles: A model analysis, *JHEP* **01** (2017) 015, [arXiv:1609.09078].

[946] C.-W. Chiang, X.-G. He, J. Tandean, and X.-B. Yuan, $R_{K^{(*)}}$ and related $b \to s\ell\bar{\ell}$ anomalies in minimal flavor violation framework with Z' boson, *Phys. Rev.* **D96** (2017), no. 11 115022, [arXiv:1706.02696].

[947] A. Datta, J. Kumar, J. Liao, and D. Marfatia, New light mediators for the R_K and R_{K^*} puzzles, *Phys. Rev.* **D97** (2018), no. 11 115038, [arXiv:1705.08423].

[948] S. Di Chiara, A. Fowlie, S. Fraser, C. Marzo, L. Marzola, M. Raidal, and C. Spethmann, Minimal flavor-changing Z' models and muon $g - 2$ after the R_{K^*} measurement, *Nucl. Phys.* **B923** (2017) 245–257, [arXiv:1704.06200].

[949] W. Altmannshofer, C.-Y. Chen, P. S. Bhupal Dev, and A. Soni, Lepton flavor violating Z' explanation of the muon anomalous magnetic moment, *Phys. Lett.* **B762** (2016) 389–398, [arXiv:1607.06832].

[950] A. Datta, J. Liao, and D. Marfatia, A light Z' for the R_K puzzle and nonstandard neutrino interactions, *Phys. Lett.* **B768** (2017) 265–269, [arXiv:1702.01099].

[951] F. Sala and D. M. Straub, A new light particle in B decays? *Phys. Lett.* **B774** (2017) 205–209, [arXiv:1704.06188].

[952] W. Altmannshofer, M. J. Baker, S. Gori, R. Harnik, M. Pospelov, E. Stamou, and A. Thamm, Light resonances and the low-q^2 bin of R_{K^*}, *JHEP* **03** (2018) 188, [arXiv:1711.07494].

[953] L. Di Luzio, M. Kirk, and A. Lenz, Updated B_s-mixing constraints on new physics models for $b \to s\ell^+\ell^-$ anomalies, *Phys. Rev.* **D97** (2018), no. 9 095035, [arXiv:1712.06572].

[954] L. Di Luzio, M. Kirk, and A. Lenz, B_s-$\bar{B}_s$ mixing interplay with B anomalies, in *10th International Workshop on the CKM Unitarity Triangle (CKM 2018) Heidelberg, Germany, September 17–21, 2018*, 2018. arXiv:1811.12884.

[955] B. C. Allanach, J. M. Butterworth, and T. Corbett, Collider constraints on Z' models for neutral current B-anomalies, arXiv:1904.10954.

[956] R. Gauld, F. Goertz, and U. Haisch, On minimal Z' explanations of the $B \to K^*\mu^+\mu^-$ anomaly, *Phys. Rev.* **D89** (2014) 015005, [arXiv:1308.1959].

[957] W. Altmannshofer, S. Gori, J. Martin-Albo, A. Sousa, and M. Wallbank, Neutrino tridents at DUNE, arXiv:1902.06765.

[958] M. Carena, A. Daleo, B. A. Dobrescu, and T. M. P. Tait, Z' gauge bosons at the Tevatron, *Phys. Rev.* **D70** (2004) 093009, [hep-ph/0408098].

[959] J. Ellis, M. Fairbairn, and P. Tunney, Anomaly-free dark matter models are not so simple, *JHEP* **08** (2017) 053, [arXiv:1704.03850].

[960] B. Allanach, F. S. Queiroz, A. Strumia, and S. Sun, Z models for the LHCb and $g - 2$ muon anomalies, *Phys. Rev.* **D93** (2016), no. 5 055045, [arXiv:1511.07447]. [Erratum: Phys. Rev.D95,no.11,119902(2017)].

[961] F. Kahlhoefer, K. Schmidt-Hoberg, T. Schwetz, and S. Vogl, Implications of unitarity and gauge invariance for simplified dark matter models, *JHEP* **02** (2016) 016, [arXiv:1510.02110].

[962] A. Ekstedt, R. Enberg, G. Ingelman, J. Löfgren, and T. Mandal, Constraining minimal anomaly free U(1) extensions of the standard model, *JHEP* **11** (2016) 071, [arXiv:1605.04855].

[963] A. Ismail, W.-Y. Keung, K.-H. Tsao, and J. Unwin, Axial vector Z and anomaly cancellation, *Nucl. Phys.* **B918** (2017) 220–244, [arXiv:1609.02188].

[964] O. Popov and G. A. White, One leptoquark to unify them? Neutrino masses and unification in the light of $(g - 2)_\mu$, $R_{D^{(\star)}}$ and R_K anomalies, *Nucl. Phys.* **B923** (2017) 324–338, [arXiv:1611.04566].

[965] B. C. Allanach, J. Davighi, and S. Melville, An anomaly-free atlas: Charting the space of flavour-dependent gauged $U(1)$ extensions of the standard model, *JHEP* **02** (2019) 082, [arXiv:1812.04602].

[966] B. C. Allanach and J. Davighi, Third family hypercharge model for $R_{K^{(*)}}$ and aspects of the fermion mass problem, *JHEP* **12** (2018) 075, [arXiv:1809.01158].

[967] K. Ishiwata, Z. Ligeti, and M. B. Wise, New vector-like fermions and flavor physics, *JHEP* **10** (2015) 027, [arXiv:1506.03484].

[968] J. C. Pati and A. Salam, Lepton number as the fourth color, *Phys. Rev.* **D10** (1974) 275–289. [Erratum: Phys. Rev.D11,703(1975)].

[969] R. N. Mohapatra and J. C. Pati, A natural left-right symmetry, *Phys. Rev.* **D11** (1975) 2558.

[970] R. N. Mohapatra and J. C. Pati, Left-right gauge symmetry and an isoconjugate model of CP violation, *Phys. Rev.* **D11** (1975) 566–571.

[971] G. Senjanovic and R. N. Mohapatra, Exact left-right symmetry and spontaneous violation of parity, *Phys. Rev.* **D12** (1975) 1502.

[972] G. Senjanovic, Spontaneous breakdown of parity in a class of gauge theories, *Nucl. Phys.* **B153** (1979) 334.

[973] R. N. Mohapatra, F. E. Paige, and D. P. Sidhu, Symmetry breaking and naturalness of parity conservation in weak neutral currents in left-right symmetric gauge theories, *Phys. Rev.* **D17** (1978) 2462.

[974] D. Chang, A minimal model of spontaneous CP violation with the gauge group $SU(2)_L \times SU(2)_R \times U(1)_{B-L}$, *Nucl. Phys.* **B214** (1983) 435.

[975] H. Harari and M. Leurer, Left-right symmetry and the mass scale of a possible right-handed weak boson, *Nucl. Phys.* **B233** (1984) 221.

[976] K. Kiers, J. Kolb, J. Lee, A. Soni, and G.-H. Wu, Ubiquitous CP violation in a top inspired left-right model, *Phys. Rev.* **D66** (2002) 095002, [hep-ph/0205082].

[977] G. Ecker and W. Grimus, ϵ, ϵ' in a model with spontaneous P and CP violation, *Phys. Lett.* **B153** (1985) 279–285.

[978] J. M. Frere et al., $K^0 - \bar{K}^0$ in the $SU(2)_L \times SU(2)_R \times U(1)$ model of CP violation, *Phys. Rev.* **D46** (1992) 337–353.

[979] G. Barenboim, J. Bernabeu, and M. Raidal, Spontaneous CP-violation in the left-right model and the kaon system, *Nucl. Phys.* **B478** (1996) 527–543, [hep-ph/9608450].

[980] R. N. Mohapatra, G. Senjanovic, and M. D. Tran, Strangeness changing processes and the limit on the right-handed gauge boson mass, *Phys. Rev.* **D28** (1983) 546.

[981] Y. Zhang, H. An, X. Ji, and R. N. Mohapatra, General CP violation in minimal left-right symmetric model and constraints on the right-handed scale, *Nucl. Phys.* **B802** (2008) 247–279, [arXiv:0712.4218].

[982] P. Ball, J. M. Frere, and J. Matias, Anatomy of mixing-induced CP asymmetries in left-right-symmetric models with spontaneous CP violation, *Nucl. Phys.* **B572** (2000) 3–35, [hep-ph/9910211].

[983] P. Langacker and S. Uma Sankar, Bounds on the mass of W_R and the $W_L - W_R$ mixing angle ξ in general $SU(2)_L \times SU(2)_R \times U(1)$ models, *Phys. Rev.* **D40** (1989) 1569–1585.

[984] G. Barenboim, J. Bernabeu, J. Prades, and M. Raidal, Constraints on the W_R mass and CP violation in left-right models, *Phys. Rev.* **D55** (1997) 4213–4221, [hep-ph/9611347].

[985] Y. Zhang, H. An, X. Ji, and R. N. Mohapatra, Right-handed quark mixings in minimal left-right symmetric model with general CP violation, *Phys. Rev.* **D76** (2007) 091301, [arXiv:0704.1662].

[986] A. Maiezza, M. Nemevsek, F. Nesti, and G. Senjanovic, Left-right symmetry at LHC, *Phys. Rev.* **D82** (2010) 055022, [arXiv:1005.5160].

[987] K. Hsieh, K. Schmitz, J.-H. Yu, and C. P. Yuan, Global analysis of general $SU(2) \times SU(2) \times U(1)$ models with precision data, *Phys. Rev.* **D82** (2010) 035011, [arXiv:1003.3482].

[988] A. Crivellin and L. Mercolli, $B \to X_d \gamma$ and constraints on new physics, *Phys. Rev.* **D84** (2011) 114005, [arXiv:1106.5499].

[989] R. N. Mohapatra, G. Yan, and Y. Zhang, Ameliorating Higgs induced flavor constraints on TeV scale W_R, arXiv:1902.08601.

[990] A. Crivellin, Effects of right-handed charged currents on the determinations of $|V_{ub}|$ and $|V_{cb}|$, *Phys. Rev.* **D81** (2010) 031301, [arXiv:0907.2461].

[991] C.-H. Chen and S.-H. Nam, Left-right mixing on leptonic and semileptonic $b \to u$ decays, *Phys. Lett.* **B666** (2008) 462–466, [arXiv:0807.0896].

[992] R. Feger, T. Mannel, V. Klose, H. Lacker, and T. Luck, Limit on a right-handed admixture to the weak $b \to c$ current from semileptonic decays, *Phys. Rev.* **D82** (2010) 073002, [arXiv:1003.4022].

[993] G. Senjanovi and V. Tello, Right handed quark mixing in left-right symmetric theory, *Phys. Rev. Lett.* **114** (2015), no. 7 071801, [arXiv:1408.3835].

[994] K. S. Babu, R. N. Mohapatra, and B. Dutta, A theory of $R(D^*, D)$ anomaly with right-handed currents, *JHEP* **01** (2019) 168, [arXiv:1811.04496].

[995] A. Crivellin, A. Kokulu, and C. Greub, Flavor-phenomenology of two-Higgs-doublet models with generic Yukawa structure, *Phys. Rev.* **D87** (2013) 094031, [arXiv:1303.5877].

[996] U. Nierste, S. Trine, and S. Westhoff, Charged-Higgs effects in a new $B \to D\tau\nu$ differential decay distribution, *Phys. Rev.* **D78** (2008) 015006, [arXiv:0801.4938].

[997] A. Crivellin, C. Greub, and A. Kokulu, Explaining $B \to D\tau\nu$, $B \to D^*\tau\nu$ and $B \to \tau\nu$ in a 2HDM of type III, *Phys. Rev.* **D86** (2012) 054014, [arXiv:1206.2634].

[998] P. Ko, Y. Omura, and C. Yu, $B \to D^{(*)}\tau\nu$ and $B \to \tau\nu$ in chiral U(1)' models with flavored multi Higgs doublets, *JHEP* **1303** (2013) 151, [arXiv:1212.4607].

[999] A. Crivellin, C. Greub, and A. Kokulu, Flavour-violation in two-Higgs-doublet models, *PoS* **EPS-HEP2013** (2013) 338, [arXiv:1309.4806].

[1000] A. J. Buras, Minimal flavour violation and beyond: Towards a flavour code for short distance dynamics, *Acta Phys. Polon.* **B41** (2010) 2487–2561, [arXiv:1012.1447].

[1001] A. J. Buras, A. Poschenrieder, S. Uhlig, and W. A. Bardeen, Rare K and B decays in the littlest Higgs model without T-parity, *JHEP* **11** (2006) 062, [hep-ph/0607189].

[1002] A. J. Buras, B. Duling, T. Feldmann, T. Heidsieck, C. Promberger, et al., Patterns of flavour violation in the presence of a fourth generation of quarks and leptons, *JHEP* **1009** (2010) 106, [arXiv:1002.2126].

[1003] A. J. Buras, M. Nagai, and P. Paradisi, Footprints of SUSY GUTs in flavour physics, *JHEP* **1105** (2011) 005, [arXiv:1011.4853].

[1004] M. Albrecht, W. Altmannshofer, A. J. Buras, D. Guadagnoli, and D. M. Straub, Challenging $SO(10)$ SUSY GUTs with family symmetries through FCNC processes, *JHEP* **10** (2007) 055, [arXiv:0707.3954].

[1005] A. J. Buras, M. Spranger, and A. Weiler, The impact of universal extra dimensions on the unitarity triangle and rare K and B decays, *Nucl. Phys.* **B660** (2003) 225–268, [hep-ph/0212143].

[1006] A. J. Buras, A. Poschenrieder, M. Spranger, and A. Weiler, The impact of universal extra dimensions on $B \to X_s\gamma$, $B \to X_s$gluon, $B \to X_s\mu^+\mu^-$, $K_L \to \pi^0 e^+e^-$, and ε'/ε, *Nucl. Phys.* **B678** (2004) 455–490, [hep-ph/0306158].

[1007] M. Blanke, A. J. Buras, B. Duling, S. Gori, and A. Weiler, $\Delta F = 2$ observables and fine-tuning in a warped extra dimension with custodial protection, *JHEP* **03** (2009) 001, [arXiv:0809.1073].

[1008] M. E. Albrecht, M. Blanke, A. J. Buras, B. Duling, and K. Gemmler, Electroweak and flavour structure of a warped extra dimension with custodial protection, *JHEP* **09** (2009) 064, [arXiv:0903.2415].

[1009] G. Cacciapaglia et al., A GIM mechanism from extra dimensions, *JHEP* **04** (2008) 006, [arXiv:0709.1714].

[1010] C. Csaki, A. Falkowski, and A. Weiler, A simple flavor protection for RS, *Phys. Rev.* **D80** (2009) 016001, [arXiv:0806.3757].

[1011] S. Casagrande, F. Goertz, U. Haisch, M. Neubert, and T. Pfoh, Flavor physics in the Randall-Sundrum model: I. Theoretical setup and electroweak precision tests, *JHEP* **10** (2008) 094, [arXiv:0807.4937].

[1012] A. J. Buras, M. V. Carlucci, L. Merlo, and E. Stamou, Phenomenology of a gauged $SU(3)^3$ flavour model, *JHEP* **1203** (2012) 088, [arXiv:1112.4477].

[1013] A. J. Buras, C. Grojean, S. Pokorski, and R. Ziegler, FCNC effects in a minimal theory of fermion masses, *JHEP* **1108** (2011) 028, [arXiv:1105.3725].

[1014] C. Niehoff, P. Stangl, and D. M. Straub, Violation of lepton flavour universality in composite Higgs models, *Phys. Lett.* **B747** (2015) 182–186, [arXiv:1503.03865].

[1015] C. Niehoff, P. Stangl, and D. M. Straub, Direct and indirect signals of natural composite Higgs models, *JHEP* **01** (2016) 119, [arXiv:1508.00569].

[1016] F. Sannino, P. Stangl, D. M. Straub, and A. E. Thomsen, Flavor physics and flavor anomalies in minimal fundamental partial compositeness, *Phys. Rev.* **D97** (2018), no. 11 115046, [arXiv:1712.07646].

[1017] P. P. Stangl, *Direct constraints, flavor physics, and flavor anomalies in composite Higgs models*. PhD thesis, Munich, Tech. U., 2018. arXiv:1811.11750.

[1018] F. Pisano and V. Pleitez, An SU(3) x U(1) model for electroweak interactions, *Phys. Rev.* **D46** (1992) 410–417, [hep-ph/9206242].

[1019] P. H. Frampton, Chiral dilepton model and the flavor question, *Phys. Rev. Lett.* **69** (1992) 2889–2891.

[1020] R. A. Diaz, R. Martinez, and F. Ochoa, $SU(3)_c \times SU(3)_L \times U(1)_X$ models for beta arbitrary and families with mirror fermions, *Phys. Rev.* **D72** (2005) 035018, [hep-ph/0411263].

[1021] R. Gauld, F. Goertz, and U. Haisch, An explicit Z'-boson explanation of the $B \to K^*\mu^+\mu^-$ anomaly, *JHEP* **1401** (2014) 069, [arXiv:1310.1082].

[1022] L. T. Hue and L. D. Ninh, The simplest 3-3-1 model, *Mod. Phys. Lett.* **A31** (2016), no. 10 1650062, [arXiv:1510.00302].

[1023] A. Carcamo Hernandez, R. Martinez, and F. Ochoa, Z and Z' decays with and without FCNC in 331 models, *Phys. Rev.* **D73** (2006) 035007, [hep-ph/0510421].

[1024] D. T. Huong, D. N. Dinh, L. D. Thien, and P. Van Dong, Dark matter and flavor changing in the flipped 3-3-1 model, arXiv:1906.05240.

[1025] Y. Nir and D. J. Silverman, *Z* mediated flavor changing neutral currents and their implications for CP asymmetries in B^0 decays, *Phys. Rev.* **D42** (1990) 1477–1484.

[1026] G. C. Branco, T. Morozumi, P. A. Parada, and M. N. Rebelo, CP asymmetries in B^0 decays in the presence of flavor changing neutral currents, *Phys. Rev.* **D48** (1993) 1167–1175.

[1027] G. Barenboim, F. J. Botella, and O. Vives, Constraining models with vector-like fermions from FCNC in *K* and *B* physics, *Nucl. Phys.* **B613** (2001) 285–305, [hep-ph/0105306].

[1028] A. J. Buras, B. Duling, and S. Gori, The impact of Kaluza-Klein fermions on standard model fermion couplings in a RS model with custodial protection, *JHEP* **0909** (2009) 076, [arXiv:0905.2318].

[1029] F. Botella, G. Branco, and M. Nebot, The hunt for new physics in the flavour sector with up vector-like quarks, *JHEP* **1212** (2012) 040, [arXiv:1207.4440].

[1030] S. Fajfer, A. Greljo, J. F. Kamenik, and I. Mustac, Light Higgs and vector-like quarks without prejudice, *JHEP* **07** (2013) 155, [arXiv:1304.4219].

[1031] A. J. Buras, J. Girrbach, and R. Ziegler, Particle-antiparticle mixing, CP violation and rare K and B decays in a minimal theory of fermion masses, *JHEP* **1304** (2013) 168, [arXiv:1301.5498].

[1032] A. K. Alok, S. Banerjee, D. Kumar, S. U. Sankar, and D. London, New-physics signals of a model with a vector-singlet up-type quark, *Phys. Rev.* **D92** (2015) 013002, [arXiv:1504.00517].

[1033] D. Barducci, M. Fabbrichesi, C. M. Nieto, R. Percacci, and V. Skrinjar, In search of a UV completion of the standard model 378,000 models that don't work, *JHEP* **11** (2018) 057, [arXiv:1807.05584].

[1034] R. Dermisek and A. Raval, Explanation of the muon g-2 anomaly with vectorlike leptons and its implications for Higgs decays, *Phys. Rev.* **D88** (2013) 013017, [arXiv:1305.3522].

[1035] D. Aristizabal Sierra, F. Staub, and A. Vicente, Shedding light on the $b \to s$ anomalies with a dark sector, *Phys. Rev.* **D92** (2015), no. 1 015001, [arXiv:1503.06077].

[1036] W. Altmannshofer, M. Carena, and A. Crivellin, $L_\mu - L_\tau$ theory of Higgs flavor violation and $(g-2)_\mu$, *Phys. Rev.* **D94** (2016), no. 9 095026, [arXiv:1604.08221].

[1037] K. Kowalska and E. M. Sessolo, Expectations for the muon g-2 in simplified models with dark matter, *JHEP* **09** (2017) 112, [arXiv:1707.00753].

[1038] L. Darm, K. Kowalska, L. Roszkowski, and E. M. Sessolo, Flavor anomalies and dark matter in SUSY with an extra U(1), *JHEP* **10** (2018) 052, [arXiv:1806.06036].

[1039] A. Falkowski, D. M. Straub, and A. Vicente, Vector-like leptons: Higgs decays and collider phenomenology, *JHEP* **05** (2014) 092, [arXiv:1312.5329].

[1040] J. Kawamura, S. Raby, and A. Trautner, Complete vector-like fourth family and new U(1)' for muon anomalies, arXiv:1906.11297.

[1041] W. Buchmüller, R. Rückl, and D. Wyler, Leptoquarks in lepton-quark collisions, *Phys. Lett.* **B191** (1987) 442–448. [Erratum: Phys. Lett.B448,320(1999)].

[1042] A. J. Davies and X.-G. He, Tree level scalar fermion interactions consistent with the symmetries of the standard model, *Phys. Rev.* **D43** (1991) 225–235.

[1043] S. Davidson, D. C. Bailey, and B. A. Campbell, Model independent constraints on leptoquarks from rare processes, *Z. Phys.* **C61** (1994) 613–644, [hep-ph/9309310].

[1044] I. Dorsner, S. Fajfer, A. Greljo, J. F. Kamenik, and N. Kosnik, Physics of leptoquarks in precision experiments and at particle colliders, *Phys. Rept.* **641** (2016) 1–68, [arXiv:1603.04993].

[1045] N. Košnik, Model independent constraints on leptoquarks from $b \to s\ell^+\ell^-$ processes, *Phys. Rev.* **D86** (2012) 055004, [arXiv:1206.2970].

[1046] I. Dorsner, S. Fajfer, and A. Greljo, Cornering scalar leptoquarks at LHC, *JHEP* **10** (2014) 154, [arXiv:1406.4831].

[1047] D. Becirevic, S. Fajfer, N. Kosnik, and O. Sumensari, Leptoquark model to explain the *B*-physics anomalies, R_K and R_D, *Phys. Rev.* **D94** (2016), no. 11 115021, [arXiv:1608.08501].

[1048] D. Becirevic, N. Kosnik, O. Sumensari, and R. Zukanovich Funchal, Palatable leptoquark scenarios for lepton flavor violation in exclusive $b \to s\ell_1\ell_2$ modes, *JHEP* **11** (2016) 035, [arXiv:1608.07583].

[1049] A. Angelescu, D. Becirevic, D. A. Faroughy, and O. Sumensari, Closing the window on single leptoquark solutions to the *B*-physics anomalies, *JHEP* **10** (2018) 183, [arXiv:1808.08179].

[1050] J. M. Arnold, B. Fornal, and M. B. Wise, Phenomenology of scalar leptoquarks, *Phys. Rev.* **D88** (2013) 035009, [arXiv:1304.6119].

[1051] M. Hirsch, H. V. Klapdor-Kleingrothaus, and S. G. Kovalenko, New low-energy leptoquark interactions, *Phys. Lett.* **B378** (1996) 17–22, [hep-ph/9602305].

[1052] A. Crivellin, D. Müller, and T. Ota, Simultaneous explanation of $R(D^{(*)})$ and $b \to s\mu^+\mu^-$: The last scalar leptoquarks standing, *JHEP* **09** (2017) 040, [arXiv:1703.09226].

[1053] I. Dorsner, S. Fajfer, D. A. Faroughy, and N. Kosnik, The role of the S_3 GUT leptoquark in flavor universality and collider searches, arXiv:1706.07779. [JHEP10,188(2017)].

[1054] A. Denner, H. Eck, O. Hahn, and J. Kublbeck, Feynman rules for fermion number violating interactions, *Nucl. Phys.* **B387** (1992) 467–481.

[1055] M. J. Baker, J. Fuentes-Martin, G. Isidori, and M. König, High-p_T signatures in vector-leptoquark models, arXiv:1901.10480.

[1056] D. Becirevic and O. Sumensari, A leptoquark model to accommodate $R_K^{\mathrm{exp}} < R_K^{\mathrm{SM}}$ and $R_{K^*}^{\mathrm{exp}} < R_{K^*}^{\mathrm{SM}}$, *JHEP* **08** (2017) 104, [arXiv:1704.05835].

[1057] M. Bauer and M. Neubert, Minimal Leptoquark Explanation for the $R_{D^{(*)}}$, R_K, and $(g-2)_g$ Anomalies, *Phys. Rev. Lett.* **116** (2016), no. 14 141802, [arXiv:1511.01900].

[1058] B. Chauhan, B. Kindra, and A. Narang, Discrepancies in simultaneous explanation of flavor anomalies and IceCube PeV events using leptoquarks, *Phys. Rev.* **D97** (2018), no. 9 095007, [arXiv:1706.04598].

[1059] IceCube Collaboration, M. G. Aartsen et al., Search for sterile neutrino mixing using three years of IceCube DeepCore data, *Phys. Rev.* **D95** (2017), no. 11 112002, [arXiv:1702.05160].

[1060] I. Dorsner, S. Fajfer, and M. Patra, A comparative study of the S_1 and U_1 leptoquark effects at IceCube, arXiv:1906.05660.

[1061] B. Gripaios, M. Nardecchia, and S. A. Renner, Composite leptoquarks and anomalies in B-meson decays, *JHEP* **05** (2015) 006, [arXiv:1412.1791].

[1062] I. de Medeiros Varzielas and G. Hiller, Clues for flavor from rare lepton and quark decays, *JHEP* **06** (2015) 072, [arXiv:1503.01084].

[1063] S. Fajfer and N. Kosnik, Vector leptoquark resolution of R_K and $R_{D^{(*)}}$ puzzles, *Phys. Lett.* **B755** (2016) 270–274, [arXiv:1511.06024].

[1064] R. Barbieri, G. Isidori, A. Pattori, and F. Senia, Anomalies in B-decays and U(2) flavour symmetry, *Eur. Phys. J.* **C76** (2016), no. 2 67, [arXiv:1512.01560].

[1065] P. Cox, A. Kusenko, O. Sumensari, and T. T. Yanagida, SU(5) Unification with TeV-scale leptoquarks, *JHEP* **03** (2017) 035, [arXiv:1612.03923].

[1066] G. Hiller and M. Schmaltz, Diagnosing lepton-nonuniversality in $b \to s\ell\ell$, *JHEP* **02** (2015) 055, [arXiv:1411.4773].

[1067] D. A. Faroughy, A. Greljo, and J. F. Kamenik, Confronting lepton flavor universality violation in B decays with high-p_T tau lepton searches at LHC, *Phys. Lett.* **B764** (2017) 126–134, [arXiv:1609.07138].

[1068] E. Coluccio Leskow, G. D'Ambrosio, A. Crivellin, and D. Muller, $(g-2)_\mu$, lepton flavor violation, and Z decays with leptoquarks: Correlations and future prospects, *Phys. Rev.* **D95** (2017), no. 5 055018, [arXiv:1612.06858].

[1069] Y. Cai, J. Gargalionis, M. A. Schmidt, and R. R. Volkas, Reconsidering the one leptoquark solution: Flavor anomalies and neutrino mass, *JHEP* **10** (2017) 047, [arXiv:1704.05849].

[1070] T. Mandal, S. Mitra, and S. Raz, $R_{D^{(*)}}$ motivated S_1 leptoquark scenarios: Impact of interference on the exclusion limits from LHC data, *Phys. Rev.* **D99** (2019), no. 5 055028, [arXiv:1811.03561].

[1071] O. Catá and T. Mannel, Linking lepton number violation with B anomalies, arXiv:1903.01799.

[1072] O. Popov, M. A. Schmidt, and G. White, R_2 as a single leptoquark solution to $R_{D^{(*)}}$ and $R_{K^{(*)}}$, arXiv:1905.06339.

[1073] D. Becirevic, S. Fajfer, and N. Kosnik, Lepton flavor nonuniversality in $b \to s\ell^+\ell^-$ processes, *Phys. Rev.* **D92** (2015), no. 1 014016, [arXiv:1503.09024].

[1074] S. Sahoo and R. Mohanta, Leptoquark effects on $b \to s\nu\bar{\nu}$ and $B \to Kl^+l^-$ decay processes, *New J. Phys.* **18** (2016), no. 1 013032, [arXiv:1509.06248].

[1075] S. Sahoo and R. Mohanta, Study of the rare semileptonic decays $B_d^0 \to K^*l^+l^-$ in scalar leptoquark model, *Phys. Rev.* **D93** (2016), no. 3 034018, [arXiv:1507.02070].

[1076] W. Dekens, J. de Vries, M. Jung, and K. K. Vos, The phenomenology of electric dipole moments in models of scalar leptoquarks, *JHEP* **01** (2019) 069, [arXiv:1809.09114].

[1077] B. Chauhan and B. Kindra, Invoking chiral vector leptoquark to explain LFU violation in B Decays, arXiv:1709.09989.

[1078] A. K. Alok, D. Kumar, J. Kumar, and R. Sharma, Lepton flavor non-universality in the B-sector: A global analyses of various new physics models, arXiv:1704.07347.

[1079] S. Sahoo and R. Mohanta, Impact of vector leptoquark on $\bar{B} \to \bar{K}^* l^+ l^-$ anomalies, *J. Phys.* **G45** (2018), no. 8 085003, [arXiv:1806.01048].

[1080] L. Calibbi, A. Crivellin, and T. Li, Model of vector leptoquarks in view of the *B*-physics anomalies, *Phys. Rev.* **D98** (2018), no. 11 115002, [arXiv:1709.00692].

[1081] L. Di Luzio, A. Greljo, and M. Nardecchia, Gauge leptoquark as the origin of B-physics anomalies, *Phys. Rev.* **D96** (2017), no. 11 115011, [arXiv:1708.08450].

[1082] M. Bordone, C. Cornella, J. Fuentes-Martin, and G. Isidori, A three-site gauge model for flavor hierarchies and flavor anomalies, *Phys. Lett.* **B779** (2018) 317–323, [arXiv:1712.01368].

[1083] R. Barbieri, C. W. Murphy, and F. Senia, B-decay anomalies in a composite leptoquark model, *Eur. Phys. J.* **C77** (2017), no. 1 8, [arXiv:1611.04930].

[1084] A. Biswas, A. Shaw, and A. K. Swain, Collider signature of V_2 Leptoquark with $b \to s$ flavour observables, arXiv:1811.08887.

[1085] S. Sahoo, R. Mohanta, and A. K. Giri, Explaining the R_K and $R_{D^{(*)}}$ anomalies with vector leptoquarks, *Phys. Rev.* **D95** (2017), no. 3 035027, [arXiv:1609.04367].

[1086] J. E. Camargo-Molina, A. Celis, and D. A. Faroughy, Anomalies in bottom from new physics in top, *Phys. Lett.* **B784** (2018) 284–293, [arXiv:1805.04917].

[1087] A. Crivellin, D. Müller, A. Signer, and Y. Ulrich, Correlating lepton flavor universality violation in *B* decays with $\mu \to e\gamma$ using leptoquarks, *Phys. Rev.* **D97** (2018), no. 1 015019, [arXiv:1706.08511].

[1088] A. Greljo and D. Marzocca, High-p_T dilepton tails and flavor physics, *Eur. Phys. J.* **C77** (2017), no. 8 548, [arXiv:1704.09015].

[1089] B. Diaz, M. Schmaltz, and Y.-M. Zhong, The leptoquark hunters guide: Pair production, *JHEP* **10** (2017) 097, [arXiv:1706.05033].

[1090] G. Hiller, D. Loose, and I. Nisandzic, Flavorful leptoquarks at hadron colliders, *Phys. Rev.* **D97** (2018), no. 7 075004, [arXiv:1801.09399].

[1091] S. Bansal, R. M. Capdevilla, A. Delgado, C. Kolda, A. Martin, and N. Raj, Hunting leptoquarks in monolepton searches, *Phys. Rev.* **D98** (2018), no. 1 015037, [arXiv:1806.02370].

[1092] M. Schmaltz and Y.-M. Zhong, The leptoquark hunters guide: Large coupling, *JHEP* **01** (2019) 132, [arXiv:1810.10017].

[1093] A. Greljo, J. Martin Camalich, and J. D. Ruiz-Ivarez, Mono-τ signatures at the LHC constrain explanations of *B*-decay anomalies, *Phys. Rev. Lett.* **122** (2019), no. 13 131803, [arXiv:1811.07920].

[1094] N. Assad, B. Fornal, and B. Grinstein, Baryon number and lepton universality violation in leptoquark and diquark models, *Phys. Lett.* **B777** (2018) 324–331, [arXiv:1708.06350].

[1095] M. Bordone, C. Cornella, J. Fuentes-Martn, and G. Isidori, Low-energy signatures of the PS^3 model: From *B*-physics anomalies to LFV, *JHEP* **10** (2018) 148, [arXiv:1805.09328].

[1096] R. Barbieri and A. Tesi, *B*-decay anomalies in Pati-Salam SU(4), *Eur. Phys. J.* **C78** (2018), no. 3 193, [arXiv:1712.06844].

[1097] L. Di Luzio, J. Fuentes-Martin, A. Greljo, M. Nardecchia, and S. Renner, Maximal flavour violation: A Cabibbo mechanism for leptoquarks, *JHEP* **11** (2018) 081, [arXiv:1808.00942].

[1098] D. Marzocca, Addressing the B-physics anomalies in a fundamental composite Higgs model, *JHEP* **07** (2018) 121, [arXiv:1803.10972].

[1099] A. Greljo and B. A. Stefanek, Third family quarklepton unification at the TeV scale, *Phys. Lett.* **B782** (2018) 131–138, [arXiv:1802.04274].

[1100] M. Blanke and A. Crivellin, *B* meson anomalies in a Pati-Salam model within the Randall-Sundrum background, *Phys. Rev. Lett.* **121** (2018), no. 1 011801, [arXiv:1801.07256].

[1101] B. Fornal, S. A. Gadam, and B. Grinstein, Left-right SU(4) vector leptoquark model for flavor anomalies, *Phys. Rev.* **D99** (2019), no. 5 055025, [arXiv:1812.01603].

[1102] S. Trifinopoulos, Revisiting R-parity violating interactions as an explanation of the B-physics anomalies, *Eur. Phys. J.* **C78** (2018), no. 10 803, [arXiv:1807.01638].

[1103] T. Faber, M. Hudec, M. Malinsk, P. Meinzinger, W. Porod, and F. Staub, A unified leptoquark model confronted with lepton non-universality in *B*-meson decays, *Phys. Lett.* **B787** (2018) 159–166, [arXiv:1808.05511].

[1104] J. Heeck and D. Teresi, Pati-Salam explanations of the B-meson anomalies, *JHEP* **12** (2018) 103, [arXiv:1808.07492].

[1105] H. Georgi and Y. Nakai, Diphoton resonance from a new strong force, *Phys. Rev.* **D94** (2016), no. 7 075005, [arXiv:1606.05865].

[1106] M. Bordone, G. Isidori, and S. Trifinopoulos, Semileptonic *B*-physics anomalies: A general EFT analysis within U(2)^n flavor symmetry, *Phys. Rev.* **D96** (2017), no. 1 015038, [arXiv:1702.07238].

[1107] C. Cornella, J. Fuentes-Martin, and G. Isidori, Revisiting the vector leptoquark explanation of the B-physics anomalies, arXiv:1903.11517.

[1108] J. Bernigaud, I. de Medeiros Varzielas, and J. Talbert, Finite family groups for fermionic and leptoquark mixing patterns, arXiv:1906.11270.

[1109] L. Da Rold and F. Lamagna, A vector leptoquark for the B-physics anomalies from a composite GUT, arXiv:1906.11666.

[1110] D. Das, C. Hati, G. Kumar, and N. Mahajan, Towards a unified explanation of $R_{D^{(*)}}$, R_K and $(g - 2)_\mu$ anomalies in a left-right model with leptoquarks, *Phys. Rev.* **D94** (2016) 055034, [arXiv:1605.06313].

[1111] C.-H. Chen, T. Nomura, and H. Okada, Excesses of muon $g - 2$, $R_{D^{(*)}}$, and R_K in a leptoquark model, *Phys. Lett.* **B774** (2017) 456–464, [arXiv:1703.03251].

[1112] C. Cornella, F. Feruglio, and P. Paradisi, Low-energy effects of lepton flavour universality violation, *JHEP* **11** (2018) 012, [arXiv:1803.00945].

[1113] D. Aloni, A. Dery, C. Frugiuele, and Y. Nir, Testing minimal flavor violation in leptoquark models of the $R_{K^{(*)}}$ anomaly, *JHEP* **11** (2017) 109, [arXiv:1708.06161].

[1114] S. Bansal, R. M. Capdevilla, and C. Kolda, Constraining the minimal flavor violating leptoquark explanation of the $R_{D^{(*)}}$ anomaly, *Phys. Rev.* **D99** (2019), no. 3 035047, [arXiv:1810.11588].

[1115] Muon G-2 Collaboration, G. W. Bennett et al., Final report of the muon E821 anomalous magnetic moment measurement at BNL, *Phys. Rev.* **D73** (2006) 072003, [hep-ex/0602035].

[1116] Y. Kuno and Y. Okada, Muon decay and physics beyond the standard model, *Rev. Mod. Phys.* **73** (2001) 151–202, [hep-ph/9909265].

[1117] M. Raidal et al., Flavour physics of leptons and dipole moments, *Eur. Phys. J.* **C57** (2008) 13–182, [arXiv:0801.1826].

[1118] J. Hewett, H. Weerts, R. Brock, J. Butler, B. Casey, et al., Fundamental physics at the intensity frontier, arXiv:1205.2671.

[1119] F. Jegerlehner, The anomalous magnetic moment of the muon, *Springer Tracts Mod. Phys.* **274** (2017) pp.1–693.

[1120] J. Engel, M. J. Ramsey-Musolf, and U. van Kolck, Electric dipole moments of nucleons, nuclei, and atoms: The Standard Model and Beyond, *Prog. Part. Nucl. Phys.* **71** (2013) 21–74, [arXiv:1303.2371].

[1121] R. H. Bernstein and P. S. Cooper, Charged lepton flavor violation: An experimenter's guide, *Phys. Rept.* **532** (2013) 27–64, [arXiv:1307.5787].

[1122] J. Hisano, M. Nagai, P. Paradisi, and Y. Shimizu, Waiting for $\mu \to e\gamma$ from the MEG experiment, *JHEP* **0912** (2009) 030, [arXiv:0904.2080].

[1123] J. Girrbach, S. Mertens, U. Nierste, and S. Wiesenfeldt, Lepton flavour violation in the MSSM, *JHEP* **05** (2010) 026, [arXiv:0910.2663].

[1124] A. Czarnecki and W. J. Marciano, Electromagnetic dipole moments and new physics, *Adv. Ser. Direct. High Energy Phys.* **20** (2009) 11–67.

[1125] M. Blanke, A. J. Buras, B. Duling, A. Poschenrieder, and C. Tarantino, Charged lepton flavour violation and $(g - 2)_\mu$ in the littlest Higgs model with T-parity: A clear distinction from supersymmetry, *JHEP* **05** (2007) 013, [hep-ph/0702136].

[1126] J. R. Ellis, J. Hisano, M. Raidal, and Y. Shimizu, A new parametrization of the seesaw mechanism and applications in supersymmetric models, *Phys. Rev.* **D66** (2002) 115013, [hep-ph/0206110].

[1127] E. Arganda and M. J. Herrero, Testing supersymmetry with lepton flavor violating tau and mu decays, *Phys. Rev.* **D73** (2006) 055003, [hep-ph/0510405].

[1128] A. Brignole and A. Rossi, Anatomy and phenomenology of $\mu\tau$ lepton flavour violation in the MSSM, *Nucl. Phys.* **B701** (2004) 3–53, [hep-ph/0404211].

[1129] P. Paradisi, Higgs-mediated $\tau \to \mu$ and $\tau \to e$ transitions in II Higgs doublet model and supersymmetry, *JHEP* **02** (2006) 050, [hep-ph/0508054].

[1130] P. Paradisi, Higgs-mediated $e \to \mu$ transitions in II Higgs doublet model and supersymmetry, *JHEP* **08** (2006) 047, [hep-ph/0601100].

[1131] P. Paradisi, Constraints on SUSY lepton flavour violation by rare processes, *JHEP* **10** (2005) 006, [hep-ph/0505046].

[1132] F. del Aguila, J. I. Illana, and M. D. Jenkins, Precise limits from lepton flavour violating processes on the littlest Higgs model with T-parity, *JHEP* **01** (2009) 080, [arXiv:0811.2891].

[1133] T. Goto, Y. Okada, and Y. Yamamoto, Tau and muon lepton flavor violations in the littlest Higgs model with T-parity, *Phys. Rev.* **D83** (2011) 053011, [arXiv:1012.4385].

[1134] L. Calibbi and G. Signorelli, Charged lepton flavour violation: An experimental and theoretical introduction, *Riv. Nuovo Cim.* **41** (2018), no. 2 1, [arXiv:1709.00294].

[1135] A. J. Buras, B. Duling, T. Feldmann, T. Heidsieck, and C. Promberger, Lepton flavour violation in the presence of a fourth generation of quarks and leptons, *JHEP* **1009** (2010) 104, [arXiv:1006.5356].

[1136] V. Cirigliano, A. Falkowski, M. González-Alonso, and A. Rodríguez-Sánchez, Hadronic tau decays as new physics probes in the LHC era, *Phys. Rev. Lett.* **122** (2019), no. 22 221801, [arXiv:1809.01161].

[1137] A. Celis, V. Cirigliano, and E. Passemar, Model-discriminating power of lepton flavor violating τ decays, *Phys. Rev.* **D89** (2014), no. 9 095014, [arXiv:1403.5781].

[1138] B. M. Dassinger, T. Feldmann, T. Mannel, and S. Turczyk, Model-independent analysis of lepton flavour violating tau decays, *JHEP* **10** (2007) 039, [arXiv:0707.0988].

[1139] A. Baldini, F. Cei, C. Cerri, S. Dussoni, L. Galli, et al., MEG Upgrade proposal, arXiv:1301.7225.

[1140] A. Blondel, A. Bravar, M. Pohl, S. Bachmann, N. Berger, et al., Research proposal for an experiment to search for the decay $\mu \to eee$, arXiv:1301.6113.

[1141] R. Barlow, The PRISM/PRIME project, *Nucl. Phys. Proc. Suppl.* **218** (2011) 44–49.

[1142] SINDRUM II Collaboration, J. Kaulard et al., Improved limit on the branching ratio of $\mu \to e$ conversion on titanium, *Phys. Lett.* **B422** (1998) 334–338.

[1143] COMET Collaboration, G. Adamov et al., COMET Phase-I technical design report, arXiv:1812.09018.

[1144] COMET Collaboration, J. C. Anglique et al., COMET – A submission to the 2020 update of the European Strategy for Particle Physics on behalf of the COMET collaboration, arXiv:1812.07824.

[1145] Mu2e Collaboration, R. Abrams et al., Mu2e conceptual design report, arXiv:1211.7019.

[1146] V. Cirigliano, R. Kitano, Y. Okada, and P. Tuzon, On the model discriminating power of $\mu \to e$ conversion in nuclei, *Phys. Rev.* **D80** (2009) 013002, [arXiv:0904.0957].

[1147] A. Baldini et al., A submission to the 2020 update of the European Strategy for Particle Physics on behalf of the COMET, MEG, Mu2e and Mu3e collaborations, arXiv:1812.06540.

[1148] T. Feldmann, Lepton flavour violation theory, *PoS* **BEAUTY2011** (2011) 017, [arXiv:1105.2139].

[1149] A. Ibarra, Neutrino physics and lepton flavour violation: A theoretical overview, *Nuovo Cim.* **C033N5** (2010) 67–75.

[1150] BaBar Collaboration, B. Aubert et al., Searches for lepton flavor violation in the decays $\tau^\pm \to e^\pm \gamma$ and $\tau^\pm \to \mu^\pm \gamma$, *Phys. Rev. Lett.* **104** (2010) 021802, [arXiv:0908.2381].

[1151] Belle Collaboration, K. Hayasaka, Recent LFV results on tau lepton from Belle, *Nucl. Phys. Proc. Suppl.* **225-227** (2012) 169–172.

[1152] MEG Collaboration, A. M. Baldini et al., Search for the lepton flavour violating decay $\mu^+ \to e^+ \gamma$ with the full dataset of the MEG experiment, *Eur. Phys. J.* **C76** (2016), no. 8 434, [arXiv:1605.05081].

[1153] SINDRUM II Collaboration, W. H. Bertl et al., A search for muon to electron conversion in muonic gold, *Eur. Phys. J.* **C47** (2006) 337–346.

[1154] COMET Collaboration, Y. Kuno, A search for muon-to-electron conversion at J-PARC: The COMET experiment, *PTEP* **2013** (2013) 022C01.

[1155] Mu2e Collaboration, L. Bartoszek et al., Mu2e technical design report, arXiv:1501.05241.

[1156] SINDRUM Collaboration, U. Bellgardt et al., Search for the decay $\mu^+ \to e^+e^+e^-$, *Nucl. Phys.* **B299** (1988) 1–6.

[1157] T. Aushev, W. Bartel, A. Bondar, J. Brodzicka, T. Browder, et al., Physics at super B factory, arXiv:1002.5012.

[1158] J. Hisano, T. Moroi, K. Tobe, and M. Yamaguchi, Lepton-flavor violation via right-handed neutrino Yukawa couplings in supersymmetric standard model, *Phys. Rev.* **D53** (1996) 2442–2459, [hep-ph/9510309].

[1159] M. B. Einhorn and J. Wudka, The bases of effective field theories, *Nucl. Phys.* **B876** (2013) 556–574, [arXiv:1307.0478].

[1160] A. Crivellin, L. Hofer, J. Matias, U. Nierste, S. Pokorski, and J. Rosiek, Lepton-flavour violating B decays in generic Z' models, *Phys. Rev.* **D92** (2015), no. 5 054013, [arXiv:1504.07928].

[1161] J. Bernabeu, E. Nardi, and D. Tommasini, μ - e conversion in nuclei and Z' physics, *Nucl. Phys.* **B409** (1993) 69–86, [hep-ph/9306251].

[1162] J. C. Pati and A. Salam, Unified lepton-hadron symmetry and a gauge theory of the basic interactions, *Phys. Rev.* **D8** (1973) 1240–1251.

[1163] BNL Collaboration, D. Ambrose et al., New limit on muon and electron lepton number violation from $K_L^0 \to \mu^\pm e^\mp$ decay, *Phys. Rev. Lett.* **81** (1998) 5734–5737, [hep-ex/9811038].

[1164] P. Q. Hung, A. J. Buras, and J. D. Bjorken, Petite unification of quarks and leptons, *Phys. Rev.* **D25** (1982) 805.

[1165] A. J. Buras and P. Q. Hung, Petite unification of quarks and leptons: Twenty two years after, *Phys. Rev.* **D68** (2003) 035015, [hep-ph/0305238].

[1166] A. Crivellin, G. D'Ambrosio, M. Hoferichter, and L. C. Tunstall, Violation of lepton flavor and lepton flavor universality in rare kaon decays, *Phys. Rev.* **D93** (2016), no. 7 074038, [arXiv:1601.00970].

[1167] M. Borsato, V. V. Gligorov, D. Guadagnoli, D. Martinez Santos, and O. Sumensari, The strange side of LHCb, arXiv:1808.02006.

[1168] M. Bordone, D. Buttazzo, G. Isidori, and J. Monnard, Probing lepton flavour universality with $K \to \pi \nu \bar{\nu}$ decays, *Eur. Phys. J.* **C77** (2017), no. 9 618, [arXiv:1705.10729].

[1169] A. Sher et al., An improved upper limit on the decay $K^+ \to \pi^+ \mu^+ e^-$, *Phys. Rev.* **D72** (2005) 012005, [hep-ex/0502020].

[1170] R. Appel et al., Search for lepton flavor violation in K^+ decays, *Phys. Rev. Lett.* **85** (2000) 2877–2880, [hep-ex/0006003].

[1171] KTeV Collaboration, E. Abouzaid et al., Search for lepton flavor violating decays of the neutral kaon, *Phys. Rev. Lett.* **100** (2008) 131803, [arXiv:0711.3472].

[1172] LHCb Collaboration, R. Aaij et al., Search for the lepton-flavour violating decays $B^0_{(s)} \to e^\pm \mu^\mp$, *JHEP* **03** (2018) 078, [arXiv:1710.04111].

[1173] LHCb Collaboration, R. Aaij et al., Search for the lepton-flavour-violating decays $B^0_s \to \tau^\pm \mu^\mp$ and $B^0 \to \tau^\pm \mu^\mp$, arXiv:1905.06614.

[1174] E. P. Shabalin, Electric dipole moment of quark in a gauge theory with left-handed currents, *Sov. J. Nucl. Phys.* **28** (1978) 75. [Yad. Fiz.28,151(1978)].

[1175] E. P. Shabalin, The electric dipole moment of the neutron in a gauge theory, *Sov. Phys. Usp.* **26** (1983) 297. [Usp. Fiz. Nauk139,561(1983)].

[1176] W. Bernreuther and M. Suzuki, The electric dipole moment of the electron, *Rev. Mod. Phys.* **63** (1991) 313–340. [Erratum: Rev. Mod. Phys.64,633(1992)].

[1177] ACME Collaboration, V. Andreev et al., Improved limit on the electric dipole moment of the electron, *Nature* **562** (2018), no. 7727 355–360.

[1178] Muon (g-2) Collaboration, G. W. Bennett et al., An improved limit on the muon electric dipole moment, *Phys. Rev.* **D80** (2009) 052008, [arXiv:0811.1207].

[1179] Belle Collaboration, K. Inami et al., Search for the electric dipole moment of the tau lepton, *Phys. Lett.* **B551** (2003) 16–26, [hep-ex/0210066].

[1180] J. M. Pendlebury et al., Revised experimental upper limit on the electric dipole moment of the neutron, *Phys. Rev.* **D92** (2015), no. 9 092003, [arXiv:1509.04411].

[1181] B. Graner, Y. Chen, E. G. Lindahl, and B. R. Heckel, Reduced limit on the permanent electric dipole moment of Hg199, *Phys. Rev. Lett.* **116** (2016), no. 16 161601, [arXiv:1601.04339]. [Erratum: Phys. Rev. Lett.119,no.11,119901(2017)].

[1182] B. C. Regan, E. D. Commins, C. J. Schmidt, and D. DeMille, New limit on the electron electric dipole moment, *Phys. Rev. Lett.* **88** (2002) 071805.

[1183] N. Sachdeva et al., A new measurement of the permanent electric dipole moment of ^{129}Xe using ^{3}He comagnetometry and SQUID detection, arXiv:1902.02864.

[1184] R. H. Parker et al., First measurement of the atomic electric dipole moment of ^{225}Ra, *Phys. Rev. Lett.* **114** (2015), no. 23 233002, [arXiv:1504.07477].

[1185] M. Bishof et al., Improved limit on the ^{225}Ra electric dipole moment, *Phys. Rev.* **C94** (2016), no. 2 025501, [arXiv:1606.04931].

[1186] M. Pospelov and A. Ritz, Electric dipole moments as probes of new physics, *Annals Phys.* **318** (2005) 119–169, [hep-ph/0504231].

[1187] B. Batell, Flavor-diagonal CP violation, *Eur. Phys. J.* **C72** (2012) 2127.

[1188] T. Chupp and M. Ramsey-Musolf, Electric dipole moments: A global analysis, *Phys. Rev.* **C91** (2015), no. 3 035502, [arXiv:1407.1064].

[1189] N. Yamanaka, B. K. Sahoo, N. Yoshinaga, T. Sato, K. Asahi, and B. P. Das, Probing exotic phenomena at the interface of nuclear and particle physics with the electric dipole moments of diamagnetic atoms: A unique window to hadronic and semi-leptonic CP violation, *Eur. Phys. J.* **A53** (2017) 54, [arXiv:1703.01570].

[1190] J. S. M. Ginges and V. V. Flambaum, Violations of fundamental symmetries in atoms and tests of unification theories of elementary particles, *Phys. Rept.* **397** (2004) 63–154, [physics/0309054].

[1191] V. F. Dmitriev and R. A. Sen'kov, P violating and T violating Schiff moment of the mercury nucleus, *Phys. Atom. Nucl.* **66** (2003) 1940–1945, [nucl-th/0304048]. [Yad. Fiz.66,1988(2003)].

[1192] K. Fuyuto and M. Ramsey-Musolf, Top down electroweak dipole operators, *Phys. Lett.* **B781** (2018) 492–498, [arXiv:1706.08548].

[1193] R. D. Peccei and H. R. Quinn, Constraints imposed by CP conservation in the presence of instantons, *Phys. Rev.* **D16** (1977) 1791–1797.

[1194] R. D. Peccei and H. R. Quinn, CP Conservation in the presence of instantons, *Phys. Rev. Lett.* **38** (1977) 1440–1443.

[1195] S. Weinberg, A new light boson? *Phys. Rev. Lett.* **40** (1978) 223–226.

[1196] F. Wilczek, Problem of strong *P* and *T* invariance in the presence of instantons, *Phys. Rev. Lett.* **40** (1978) 279–282.

[1197] A. E. Nelson, Naturally weak CP violation, *Phys. Lett.* **136B** (1984) 387–391.

[1198] S. M. Barr, Solving the strong CP problem without the Peccei-Quinn symmetry, *Phys. Rev. Lett.* **53** (1984) 329.

[1199] D. J. E. Marsh, Axion cosmology, *Phys. Rept.* **643** (2016) 1–79, [arXiv:1510.07633].

[1200] W. Dekens, J. de Vries, J. Bsaisou, W. Bernreuther, C. Hanhart, U.-G. Meiner, A. Nogga, and A. Wirzba, Unraveling models of CP violation through electric dipole moments of light nuclei, *JHEP* **07** (2014) 069, [arXiv:1404.6082].

[1201] ACME Collaboration, J. Baron et al., Order of magnitude smaller limit on the electric dipole moment of the electron, *Science* **343** (2014) 269–272, [arXiv:1310.7534].

[1202] D. E. Morrissey and M. J. Ramsey-Musolf, Electroweak baryogenesis, *New J.Phys.* **14** (2012) 125003, [arXiv:1206.2942].

[1203] J. Kozaczuk, S. Profumo, M. J. Ramsey-Musolf, and C. L. Wainwright, Supersymmetric electroweak baryogenesis via resonant sfermion sources, *Phys. Rev.* **D86** (2012) 096001, [arXiv:1206.4100].

[1204] Y. Li, S. Profumo, and M. Ramsey-Musolf, Bino-driven electroweak baryogenesis with highly suppressed electric dipole moments, *Phys. Lett.* **B673** (2009) 95–100, [arXiv:0811.1987].

[1205] T. Liu, M. J. Ramsey-Musolf, and J. Shu, Electroweak beautygenesis: From $b \to s$ CP-violation to the cosmic baryon asymmetry, *Phys. Rev. Lett.* **108** (2012) 221301, [arXiv:1109.4145].

[1206] S. Tulin and P. Winslow, Anomalous *B* meson mixing and baryogenesis, *Phys. Rev.* **D84** (2011) 034013, [arXiv:1105.2848].

[1207] J. M. Cline, K. Kainulainen, and M. Trott, Electroweak baryogenesis in two Higgs doublet models and *B* meson anomalies, *JHEP* **1111** (2011) 089, [arXiv:1107.3559].

[1208] S. Weinberg, Larger Higgs exchange terms in the neutron electric dipole moment, *Phys. Rev. Lett.* **63** (1989) 2333.

[1209] S. M. Barr and A. Zee, Electric dipole moment of the electron and of the neutron, *Phys. Rev. Lett.* **65** (1990) 21–24. [Erratum: Phys. Rev. Lett.65,2920(1990)].

[1210] J. F. Gunion and D. Wyler, Inducing a large neutron electric dipole moment via a quark chromoelectric dipole moment, *Phys. Lett.* **B248** (1990) 170–176.

[1211] D. Chang, W.-Y. Keung, and T. C. Yuan, Chromoelectric dipole moment of light quarks through two loop mechanism, *Phys. Lett.* **B251** (1990) 608–612.

[1212] W. Altmannshofer, R. Harnik, and J. Zupan, Low energy probes of PeV scale sfermions, *JHEP* **1311** (2013) 202, [arXiv:1308.3653].

[1213] T. Fukuyama, Searching for new physics beyond the standard model in electric dipole moment, *Int. J. Mod. Phys.* **A27** (2012) 1230015, [arXiv:1201.4252].

[1214] G. Panico, A. Pomarol, and M. Riembau, EFT approach to the electron electric dipole moment at the two-loop level, *JHEP* **04** (2019) 090, [arXiv:1810.09413].

[1215] A. Crivellin and F. Saturnino, Correlating tauonic B decays to the neutron EDM via a scalar leptoquark, arXiv:1905.08257.

[1216] J. Brod, U. Haisch, and J. Zupan, Constraints on CP-violating Higgs couplings to the third generation, *JHEP* **11** (2013) 180, [arXiv:1310.1385].

[1217] J. Brod and E. Stamou, Electric dipole moment constraints on CP-violating heavy-quark Yukawas at next-to-leading order, arXiv:1810.12303.

[1218] J. Brod and D. Skodras, Electric dipole moment constraints on CP-violating light-quark Yukawas, *JHEP* **01** (2019) 233, [arXiv:1811.05480].

[1219] V. Cirigliano, A. Crivellin, W. Dekens, J. de Vries, M. Hoferichter, and E. Mereghetti, CP violation in Higgs–gauge interactions: from tabletop experiments to the LHC, arXiv:1903.03625.

[1220] K. Fuyuto, M. Ramsey-Musolf, and T. Shen, Electric dipole moments from CP-violating scalar leptoquark interactions, *Phys. Lett.* **B788** (2019) 52–57, [arXiv:1804.01137].

[1221] H. Gisbert and J. Ruiz Vidal, Improved bounds on heavy quark electric dipole moments, arXiv:1905.02513.

[1222] A. Cordero-Cid, J. M. Hernandez, G. Tavares-Velasco, and J. J. Toscano, Bounding the top and bottom electric dipole moments from neutron experimental data, *J. Phys.* **G35** (2008) 025004, [arXiv:0712.0154].

[1223] J. F. Kamenik, M. Papucci, and A. Weiler, Constraining the dipole moments of the top quark, *Phys. Rev.* **D85** (2012) 071501, [arXiv:1107.3143]. [Erratum: Phys. Rev.D88,no.3,039903(2013)].

[1224] V. Cirigliano, W. Dekens, J. de Vries, and E. Mereghetti, Is there room for CP violation in the top-Higgs sector? *Phys. Rev.* **D94** (2016), no. 1 016002, [arXiv:1603.03049].

[1225] A. G. Grozin, I. B. Khriplovich, and A. S. Rudenko, Upper limits on electric dipole moments of tau-lepton, heavy quarks, and W-boson, *Nucl. Phys.* **B821** (2009) 285–290, [arXiv:0902.3059].

[1226] T. Kinoshita and M. Nio, Improved α^4 term of the muon anomalous magnetic moment, *Phys. Rev.* **D70** (2004) 113001, [hep-ph/0402206].

[1227] M. Passera, Precise mass-dependent QED contributions to leptonic g-2 at order α^2 and α^3, *Phys. Rev.* **D75** (2007) 013002, [hep-ph/0606174].

[1228] T. Aoyama, M. Hayakawa, T. Kinoshita, and M. Nio, Tenth-order QED contribution to the electron g-2 and an improved value of the fine structure constant, *Phys. Rev. Lett.* **109** (2012) 111807, [arXiv:1205.5368].

[1229] T. Aoyama, M. Hayakawa, T. Kinoshita, and M. Nio, Complete tenth-order QED contribution to the muon g-2, *Phys. Rev. Lett.* **109** (2012) 111808, [arXiv:1205.5370].

[1230] A. Czarnecki, W. J. Marciano, and A. Vainshtein, Refinements in electroweak contributions to the muon anomalous magnetic moment, *Phys. Rev.* **D67** (2003) 073006, [hep-ph/0212229].

[1231] J. Prades, E. de Rafael, and A. Vainshtein, The hadronic light-by-light scattering contribution to the muon and electron anomalous magnetic moments, *Adv. Ser. Direct. High Energy Phys.* **20** (2009) 303–317, [arXiv:0901.0306].

[1232] J. Prades, Standard model prediction of the muon anomalous magnetic moment, *Acta Phys. Polon.Supp.* **3** (2010) 75–86, [arXiv:0909.2546].

[1233] M. Benayoun, P. David, L. DelBuono, and F. Jegerlehner, An update of the HLS estimate of the muon g-2, *Eur. Phys. J.* **C73** (2013) 2453, [arXiv:1210.7184].

[1234] F. Jegerlehner, Muon $g - 2$ theory: The hadronic part, *EPJ Web Conf.* **166** (2018) 00022, [arXiv:1705.00263].

[1235] F. Jegerlehner, The muon g-2 in progress, *Acta Phys. Polon.* **B49** (2018) 1157, [arXiv:1804.07409].

[1236] F. Jegerlehner, The role of mesons in muon $g - 2$, *EPJ Web Conf.* **199** (2019) 01010, [arXiv:1809.07413].

[1237] D. Stockinger, $(g - 2)_\mu$ and supersymmetry: Status and prospects, in *SUSY 2007 Proceedings, 15th International Conference on Supersymmetry and Unification of Fundamental Interactions, July 26– August 1, 2007, Karlsruhe, Germany*, pp. 720–723, 2007. arXiv:0710.2429.

[1238] S. Marchetti, S. Mertens, U. Nierste, and D. Stockinger, tan β-enhanced supersymmetric corrections to the anomalous magnetic moment of the muon, *Phys. Rev.* **D79** (2009) 013010, [arXiv:0808.1530].

[1239] F. Feroz, B. C. Allanach, M. Hobson, S. S. AbdusSalam, R. Trotta, et al., Bayesian selection of $sign(\mu)$ within mSUGRA in global fits including WMAP5 results, *JHEP* **0810** (2008) 064, [arXiv:0807.4512].

[1240] M. M. Nojiri et al., Physics beyond the standard model: Supersymmetry, in *Physics at TeV Colliders, La physique du TeV aux collisionneurs, Les Houches 2007: June 11–29, 2007*, pp. 291–361, 2008. arXiv:0802.3672.

[1241] G. Degrassi and G. Giudice, QED logarithms in the electroweak corrections to the muon anomalous magnetic moment, *Phys. Rev.* **D58** (1998) 053007, [hep-ph/9803384].

[1242] S. Heinemeyer, D. Stockinger, and G. Weiglein, Two loop SUSY corrections to the anomalous magnetic moment of the muon, *Nucl. Phys.* **B690** (2004) 62–80, [hep-ph/0312264].

[1243] S. Heinemeyer, D. Stockinger, and G. Weiglein, Electroweak and supersymmetric two-loop corrections to $(g - 2)_\mu$, *Nucl. Phys.* **B699** (2004) 103–123, [hep-ph/0405255].

[1244] A. Crivellin, J. Girrbach, and U. Nierste, Yukawa coupling and anomalous magnetic moment of the muon: An update for the LHC era, *Phys. Rev.* **D83** (2011) 055009, [arXiv:1010.4485].

[1245] F. Jegerlehner, Implications of low and high energy measurements on SUSY models, *Frascati Phys. Ser.* **54** (2012) 42–51, [arXiv:1203.0806].

[1246] D. Hanneke, S. Fogwell, and G. Gabrielse, New measurement of the electron magnetic moment and the fine structure constant, *Phys. Rev. Lett.* **100** (2008) 120801, [arXiv:0801.1134].

[1247] T. Aoyama, M. Hayakawa, T. Kinoshita, and M. Nio, Revised value of the eighth-order QED contribution to the anomalous magnetic moment of the electron, *Phys. Rev.* **D77** (2008) 053012, [arXiv:0712.2607].

[1248] P. Clade, E. de Mirandes, M. Cadoret, S. Guellati-Khelifa, C. Schwob, et al., Determination of the fine structure constant based on Bloch oscillations of ultracold atoms in a vertical optical lattice, *Phys. Rev. Lett.* **96** (2006) 033001.

[1249] R. H. Parker, C. Yu, W. Zhong, B. Estey, and H. Mller, Measurement of the fine-structure constant as a test of the standard model, *Science* **360** (2018) 191, [arXiv:1812.04130].

[1250] T. Aoyama, T. Kinoshita, and M. Nio, Revised and improved value of the QED tenth-order electron anomalous magnetic moment, *Phys. Rev.* **D97** (2018), no. 3 036001, [arXiv:1712.06060].

[1251] D. Hanneke, S. F. Hoogerheide, and G. Gabrielse, Cavity control of a single-electron quantum cyclotron: Measuring the electron magnetic moment, *Phys. Rev.* **A83** (2011) 052122, [arXiv:1009.4831].

[1252] H. Davoudiasl and W. J. Marciano, Tale of two anomalies, *Phys. Rev.* **D98** (2018), no. 7 075011, [arXiv:1806.10252].

[1253] A. Crivellin, M. Hoferichter, and P. Schmidt-Wellenburg, Combined explanations of $(g - 2)_{\mu,e}$ and implications for a large muon EDM, *Phys. Rev.* **D98** (2018), no. 11 113002, [arXiv:1807.11484].

[1254] P. N. Bhattiprolu and S. P. Martin, Prospects for vectorlike leptons at future proton–proton colliders, arXiv:1905.00498.

[1255] J. Engel and J. Menández, Status and future of nuclear matrix elements for neutrinoless double-beta decay: A review, *Rept. Prog. Phys.* **80** (2017), no. 4 046301, [arXiv:1610.06548].

[1256] L. Cardani, Neutrinoless double beta decay overview, *SciPost Phys. Proc.* **1** (2019) 024, [arXiv:1810.12828].

[1257] V. Cirigliano, W. Dekens, J. De Vries, M. L. Graesser, E. Mereghetti, S. Pastore, and U. Van Kolck, New leading contribution to neutrinoless double-beta Decay, *Phys. Rev. Lett.* **120** (2018), no. 20 202001, [arXiv:1802.10097].

[1258] M. J. Dolinski, A. W. P. Poon, and W. Rodejohann, Neutrinoless Double-beta decay: Status and prospects, *Submitted to: Ann. Rev. Nucl. Part. Phys.* (2019) [arXiv:1902.04097].

[1259] M. C. Gonzalez-Garcia, Neutrino masses and mixing: A little history for a lot of fun, 2019. arXiv:1902.04583.

[1260] S. Descotes-Genon, J. Matias, and J. Virto, Understanding the $B \to K^* \mu^+ \mu^-$ anomaly, *Phys. Rev. D 88*, **074002** (2013) [arXiv:1307.5683].

[1261] D. Ghosh, M. Nardecchia, and S. A. Renner, Hint of lepton flavour non-universality in B meson decays, *JHEP* **12** (2014) 131, [arXiv:1408.4097].

[1262] T. Hurth, F. Mahmoudi, and S. Neshatpour, Global fits to $b \to s\ell\ell$ data and signs for lepton non-universality, *JHEP* **12** (2014) 053, [arXiv:1410.4545].

[1263] J. Lyon and R. Zwicky, Resonances gone topsy turvy – the charm of QCD or new physics in $b \to s\ell^+\ell^-$? arXiv:1406.0566.

[1264] S. Jäger and J. Martin Camalich, Reassessing the discovery potential of the $B \to K^*\ell^+\ell^-$ decays in the large-recoil region: SM challenges and BSM opportunities, *Phys. Rev.* **D93** (2016), no. 1 014028, [arXiv:1412.3183].

[1265] D. Bardhan, P. Byakti, and D. Ghosh, Role of tensor operators in R_K and R_{K^*}, *Phys. Lett.* **B773** (2017) 505–512, [arXiv:1705.09305].

[1266] D. Ghosh, Explaining the R_K and R_{K^*} anomalies, *Eur. Phys. J.* **C77** (2017), no. 10 694, [arXiv:1704.06240].

[1267] A. Crivellin et al., PSI/UZH Workshop: Impact of $B \to \mu^+\mu^-$ on new physics searches, arXiv:1803.10097.

[1268] A. Crivellin, J. Heeck, and P. Stoffer, A perturbed lepton-specific two-Higgs-doublet model facing experimental hints for physics beyond the standard model, *Phys. Rev. Lett.* **116** (2016), no. 8 081801, [arXiv:1507.07567].

[1269] S. Iguro and K. Tobe, $R(D^{(*)})$ in a general two Higgs doublet model, *Nucl. Phys.* **B925** (2017) 560–606, [arXiv:1708.06176].

[1270] A. Crivellin, D. Mller, and C. Wiegand, $b \to s\ell^+\ell^-$ transitions in two-Higgs-doublet models, arXiv:1903.10440.

[1271] L. Delle Rose, S. Khalil, S. J. D. King, and S. Moretti, R_K and R_{K^*} in an aligned 2HDM with right-handed neutrinos, arXiv:1903.11146.

[1272] E. Megias, M. Quiros, and L. Salas, Lepton-flavor universality violation in R_K and $R_{D^{(*)}}$ from warped space, *JHEP* **07** (2017) 102, [arXiv:1703.06019].

[1273] E. Megias, G. Panico, O. Pujolas, and M. Quiros, A natural origin for the LHCb anomalies, *JHEP* **09** (2016) 118, [arXiv:1608.02362].

[1274] A. Biswas, A. Shaw, and S. K. Patra, $\mathcal{R}(D^{(*)})$ anomalies in light of a nonminimal universal extra dimension, *Phys. Rev.* **D97** (2018), no. 3 035019, [arXiv:1708.08938].

[1275] J. F. Kamenik, Y. Soreq, and J. Zupan, Lepton flavor universality violation without new sources of quark flavor violation, *Phys. Rev.* **D97** (2018), no. 3 035002, [arXiv:1704.06005].

[1276] S.-Y. Guo, Z.-L. Han, B. Li, Y. Liao, and X.-D. Ma, Interpreting the $R_{K^{(*)}}$ anomaly in the colored ZeeBabu model, *Nucl. Phys.* **B928** (2018) 435–447, [arXiv:1707.00522].

[1277] X. Cid Vidal et al., Beyond the standard model physics at the HL-LHC and HE-LHC, arXiv:1812.07831.

[1278] L. Di Luzio and M. Nardecchia, What is the scale of new physics behind the B-flavour anomalies? *Eur. Phys. J.* **C77** (2017), no. 8 536, [arXiv:1706.01868].

[1279] A. K. Alok, B. Bhattacharya, D. Kumar, J. Kumar, D. London, and S. U. Sankar, New physics in $b \to s\mu^+\mu^-$: Distinguishing models through CP-violating effects, *Phys. Rev.* **D96** (2017), no. 1 015034, [arXiv:1703.09247].

[1280] B. Capdevila, A. Crivellin, S. Descotes-Genon, L. Hofer, and J. Matias, Searching for new physics with $b \to s\tau^+\tau^-$ processes, *Phys. Rev. Lett.* **120** (2018), no. 18 181802, [arXiv:1712.01919].

[1281] A. Celis, M. Jung, X.-Q. Li, and A. Pich, Sensitivity to charged scalars in $B \to D^{(*)}\tau\nu_\tau$ and $B \to \tau\nu_\tau$ decays, *JHEP* **01** (2013) 054, [arXiv:1210.8443].

[1282] C.-H. Chen and T. Nomura, Charged-Higgs on $R_{D^{(*)}}$, τ polarization, and FBA, *Eur. Phys. J.* **C77** (2017), no. 9 631, [arXiv:1703.03646].

[1283] C.-H. Chen and T. Nomura, Charged Higgs boson contribution to $B_q^- \to \ell\bar{\nu}$ and $\bar{B} \to (P,V)\ell\bar{\nu}$ in a generic two-Higgs doublet model, *Phys. Rev.* **D98** (2018), no. 9 095007, [arXiv:1803.00171].

[1284] S.-P. Li, X.-Q. Li, Y.-D. Yang, and X. Zhang, $R_{D^{(*)}}$, $R_{K^{(*)}}$ and neutrino mass in the 2HDM-III with right-handed neutrinos, *JHEP* **09** (2018) 149, [arXiv:1807.08530].

[1285] A. K. Alok, D. Kumar, J. Kumar, S. Kumbhakar, and S. U. Sankar, New physics solutions for R_D and R_{D^*}, *JHEP* **09** (2018) 152, [arXiv:1710.04127].

[1286] B. Bhattacharya, A. Datta, S. Kamali, and D. London, CP violation in $\bar{B}^0 \to D^{*+}\mu^-\bar{\nu}_\mu$, *JHEP* **05** (2019) 191, [arXiv:1903.02567].

[1287] X.-G. He and G. Valencia, B decays with τ leptons in nonuniversal left-right models, *Phys. Rev.* **D87** (2013), no. 1 014014, [arXiv:1211.0348].

[1288] X.-G. He and G. Valencia, Lepton universality violation and right-handed currents in $b \to c\tau\nu$, *Phys. Lett.* **B779** (2018) 52–57, [arXiv:1711.09525].

[1289] P. Asadi, M. R. Buckley, and D. Shih, It's all right(-handed neutrinos): A new W model for the $R_{D^{(*)}}$ anomaly, *JHEP* **09** (2018) 010, [arXiv:1804.04135].

[1290] L. C. Tunstall, A. Crivellin, G. D'Ambrosio, and M. Hoferichter, Probing lepton flavour (universality) violation at NA62 and future kaon experiments, *J. Phys. Conf. Ser.* **800** (2017), no. 1 012014, [arXiv:1611.00495].

[1291] H. B. Nielsen and C. D. Froggatt, Anomalies from non-perturbative standard model effects, in *18th Hellenic School and Workshops on Elementary Particle Physics and Gravity (CORFU2018) Corfu, Corfu, Greece, August 31–September 28, 2018*, 2019. arXiv:1905.00070.

[1292] J. Beacham et al., Physics beyond colliders at CERN: Beyond the Standard Model Working Group Report, arXiv:1901.09966.

[1293] LHCb Collaboration, R. Aaij et al., Physics case for an LHCb Upgrade II – Opportunities in flavour physics, and beyond, in the HL-LHC era, arXiv:1808.08865.

[1294] A. Buras, Flavour expedition to the Zeptouniverse, *PoS* **FWNP** (2015) 003, [arXiv:1505.00618].

[1295] J. Brod, A. Lenz, G. Tetlalmatzi-Xolocotzi, and M. Wiebusch, New physics effects in tree-level decays and the precision in the determination of the quark mixing angle γ, *Phys. Rev.* **D92** (2015) 033002, [arXiv:1412.1446].

[1296] A. J. Buras and J. Girrbach, Completing NLO QCD Corrections for tree level non-leptonic $\Delta F = 1$ decays beyond the standard model, *JHEP* **02** (2012) 143, [arXiv:1201.2563].

[1297] F. Gabbiani, E. Gabrielli, A. Masiero, and L. Silvestrini, A complete analysis of FCNC and CP constraints in general SUSY extensions of the standard model, *Nucl. Phys.* **B477** (1996) 321–352, [hep-ph/9604387].

[1298] UTfit Collaboration, M. Bona et al., Model-independent constraints on $\Delta F = 2$ operators and the scale of new physics, *JHEP* **0803** (2008) 049, [arXiv:0707.0636]. Updates available on http://www.utfit.org.

[1299] J. Charles, S. Descotes-Genon, Z. Ligeti, S. Monteil, M. Papucci, et al., Future sensitivity to new physics in B_d, B_s and K mixings, *Phys. Rev.* **D89** (2014) 033016, [arXiv:1309.2293].

[1300] D. McKeen, M. Pospelov, and A. Ritz, Electric dipole moment signatures of PeV-scale superpartners, *Phys. Rev.* **D87** (2013), no. 11 113002, [arXiv:1303.1172].

[1301] T. Moroi and M. Nagai, Probing supersymmetric model with heavy sfermions using leptonic flavor and CP violations, *Phys. Lett.* **B723** (2013) 107–112, [arXiv:1303.0668].

[1302] T. Moroi, M. Nagai, and T. T. Yanagida, Lepton flavor violations in high-scale SUSY with right-handed neutrinos, *Phys. Lett.* **B728** (2014) 342–346, [arXiv:1305.7357].

[1303] L. Eliaz, A. Giveon, S. B. Gudnason, and E. Tsuk, Mild-split SUSY with flavor, *JHEP* **1310** (2013) 136, [arXiv:1306.2956].

[1304] A. S. Kronfeld, R. S. Tschirhart, U. Al-Binni, W. Altmannshofer, C. Ankenbrandt, et al., Project X: Physics opportunities, arXiv:1306.5009.

[1305] A. de Gouvea and P. Vogel, Lepton flavor and number conservation, and physics beyond the standard model, *Prog. Part. Nucl. Phys.* **71** (2013) 75–92, [arXiv:1303.4097].

[1306] S. Hashimoto, Hints and challenges in heavy flavor physics, *PoS* **LATTICE2018** (2018) 008, [arXiv:1902.09119].

[1307] M. Fierz, Force-free particles with any spin, *Helv. Phys. Acta* **12** (1939) 3–37.

[1308] J. F. Nieves and P. B. Pal, Generalized Fierz identities, *Am. J. Phys.* **72** (2004) 1100–1108, [hep-ph/0306087].

[1309] Y. Liao and J.-Y. Liu, Generalized Fierz identities and applications to spin-3/2 particles, *Eur. Phys. J. Plus* **127** (2012) 121, [arXiv:1206.5141].

[1310] C. C. Nishi, Simple derivation of general Fierz-like identities, *Am. J. Phys.* **73** (2005) 1160–1163, [hep-ph/0412245].

[1311] J. C. Romao and J. P. Silva, A resource for signs and Feynman diagrams of the standard model, *Int. J. Mod. Phys.* **A27** (2012) 1230025, [arXiv:1209.6213].

[1312] J. Hisano, K. Tsumura, and M. J. S. Yang, QCD corrections to neutron electric dipole moment from dimension-six four-quark operators, *Phys. Lett.* **B713** (2012) 473–480, [arXiv:1205.2212].

[1313] A. J. Buras and M. Jung, Analytic inclusion of the scale dependence of the anomalous dimension matrix in standard model effective theory, *JHEP* **06** (2018) 067, [arXiv:1804.05852].

[1314] V. Cirigliano, W. Dekens, J. De Vries, M. L. Graesser, E. Mereghetti, S. Pastore, M. Piarulli, U. Van Kolck, and R. B. Wiringa, A renormalized approach to neutrinoless double-beta decay, arXiv:1907.11254.

[1315] M. Bauer, M. Neubert, S. Renner, M. Schnubel, and A. Thamm, Axion-like particles, lepton-flavor violation and a new explanation of $a\mu$ and ae, arXiv:1908.00008.

[1316] J. Fuentes-Martín, G. Isidori, J. Pagès, and K. Yamamoto, With or without U(2)? Probing non-standard flavor and helicity structures in semileptonic B decays, *Phys. Lett.* **B800** (2020) 135080, [arXiv:1909.02519].

[1317] J. Brod, M. Gorbahn, and E. Stamou, Standard-model prediction of ε_K with manifest CKM unitarity, arXiv:1911.06822.

[1318] S. Bertolini, A. Maiezza, and F. Nesti, Kaon CP violation and neutron EDM in the minimal left-right symmetric model, arXiv:1911.09472.

[1319] J. Fuentes-Martín, G. Isidori, M. König, and N. Selimovi, Vector Leptoquarks Beyond Tree Level, arXiv:1910.13474.

[1320] S. Jäger, M. Kirk, A. Lenz, and K. Leslie, Charming New B-Physics, arXiv:1910.12924.

[1321] W. Altmannshofer, J. Davighi, and M. Nardecchia, Gauging the accidental symmetries of the Standard Model, and implications for the flavour anomalies, arXiv:1909.02021.

[1322] R. Coy, M. Frigerio, F. Mescia, and O. Sumensari, New physics in $b \to s\ell\ell$ transitions at one loop, arXiv:1909.08567.

[1323] J. Rathsman and F. Tellander, Anomaly-free Model Building with Algebraic Geometry, *Phys. Rev.* **D100** (2019), no. 5 055032, [arXiv:1902.08529].

[1324] A. Smolkovi, M. Tammaro, and J. Zupan, Anomaly free Froggatt-Nielsen models of flavor, *JHEP* **10** (2019) 188, [arXiv:1907.10063].

[1325] A. Ordell, R. Pasechnik, H. Serdio, and F. Teichmann, Classification of anomaly-free 2HDMs with a gauged U(1)' symmetry, arXiv:1909.05548.

[1326] A. Crivellin, C. Gross, S. Pokorski, and L. Vernazza, Correlating ε'/ε to hadronic B decays via U(2)^3 flavour symmetry, 2019. arXiv:1909.02101.

[1327] L. Calibbi, A. Crivellin, F. Kirk, C. A. Manzari, and L. Vernazza, Z' models with less-minimal flavour violation, arXiv:1910.00014.

[1328] A. Czarnecki, W. J. Marciano, and A. Sirlin, Radiative corrections to neutron and nuclear beta decays revisited, *Phys. Rev.* **D100** (2019), no. 7 073008, [arXiv:1907.06737].

[1329] J. C. Hardy and I. S. Towner, Nuclear beta decays and CKM unitarity, in *13th Conference on the Intersections of Particle and Nuclear Physics (CIPANP 2018) Palm Springs, California, USA, May 29– June 3, 2018*, 2018. arXiv:1807.01146.

[1330] B. Wang, Results for the mass difference between the long- and short-lived K mesons for physical quark masses, *PoS* LATTICE2018 (2019) 286, [arXiv:1812.05302].

[1331] A. Dery and Y. Nir, Implications of the LHCb discovery of CP violation in charm decays, arXiv:1909.11242.

[1332] I. Brivio et al., *Computing Tools for the SMEFT*, in *Computing Tools for the SMEFT* (J. Aebischer, M. Fael, A. Lenz, M. Spannowsky, and J. Virto, eds.), 2019. `arXiv:1910.11003`.

[1333] J. Aebischer, C. Bobeth, and A. J. Buras, On the Importance of NNLO QCD and Isospin-breaking Corrections in ε'/ε, arXiv:1909.05610.

[1334] N. Sachdeva et al., New Limit on the Permanent Electric Dipole Moment of ^{129}Xe using ^{3}He Comagnetometry and SQUID Detection, arXiv:1909.12800.

[1335] I. Esteban, M. C. Gonzalez-Garcia, A. Hernandez-Cabezudo, M. Maltoni, and T. Schwetz, Global analysis of three-flavour neutrino oscillations: Synergies and tensions in the determination of θ_{23}, δ_{CP}, and the mass ordering, *JHEP* **01** (2019) 106, [arXiv:1811.05487].

[1336] O. V. Tarasov, Anomalous dimensions of quark masses in the three-loop approximation, arXiv:1910.12231.

[1337] N. H. Christ, X. Feng, and C. T. Sachrajda, Lattice QCD study of the rare kaon decay $K^+ \to \pi^+ \nu\bar{\nu}$ at a near-physical pion mass, arXiv:1910.10644.

[1338] T. Kitahara, T. Okui, G. Perez, Y. Soreq, and K. Tobioka, New physics implications of recent search for $K_L \to \pi^0 \nu\bar{\nu}$ at KOTO, arXiv:1909.11111.

[1339] V. Cirigliano, H. Gisbert, A. Pich, and A. Rodríguez-Sánchez, Isospin-violating contributions to ε'/ε, arXiv:1911.01359.

[1340] L. Di Luzio, M. Kirk, A. Lenz, and T. Rauh, ΔM_s theory precision confronts flavour anomalies, arXiv:1909.11087.

[1341] M. Beneke, C. Bobeth, and R. Szafron, Power-enhanced leading-logarithmic QED corrections to $B_q \to \mu^+\mu^-$, *JHEP* **10** (2019) 232, [arXiv:1908.07011].

[1342] CMS Collaboration, A. M. Sirunyan et al., Measurement of properties of $B_s^0 \to \mu^+\mu^-$ decays and search for $B_d^0 \to \mu^+\mu^-$ with the CMS experiment, arXiv:1910.12127.

[1343] D. Bečirević, M. Fedele, I. Nišandžić, and A. Tayduganov, Lepton flavor universality tests through angular observables of $B \to D(*)\ell\nu$ decay modes, arXiv:1907.02257.

[1344] Belle Collaboration, G. Caria et al., Measurement of $R(D)$ and $R(D^*)$ with a semileptonic tagging method, arXiv:1910.05864.

[1345] W.-S. Hou, T. Modak, and G.-G. Wong, Scalar leptoquark effects on $B \to \mu\bar{\nu}$ decay, arXiv:1909.00403.

[1346] R. Mandal and A. Pich, Constraints on scalar leptoquarks from lepton and kaon physics, arXiv:1908.11155.

[1347] T. Chupp, P. Fierlinger, M. Ramsey-Musolf, and J. Singh, Electric dipole moments of atoms, molecules, nuclei, and particles, *Rev. Mod. Phys.* **91** (2019), no. 1 015001, [arXiv: 1710.02504].

Index